Great Mathematics Books of the Twentieth Century

                                    ] . ης̣ Ἡρα[
                                    ].ιλ̣η.[
          ].χ̣ο̣μ̣ε̣ν̣.[   ] . ρα ̣ . ἐπιταγ[
          ].λ̣λ̣α.[   ]υ̣σαι μοχθ[
          ]τ̣η̣ς . μ̣ε̣.[
                                    ] ἀποσχοίμεθα̣ . .[
                                    ] . ανθ᾽ ἡμερη̣σ̣[
                                    ]η̣ τοῖος ἀν̣[
          ] . θανητα̣   ] . τῶ στρατη̣γο̣ς̣
          ] . οντες τ̣[
          ]ε̣σθη̣ε̣ρ̣γ̣[   ] . ω̣ς̣ . [
                                    ]ν̣ε̣ . [
                                    ]τυ̣[

                                    ].τ̣α̣[
                                    ]γ̣α̣ . [

                                    ]π̣ι̣ . [

These two pages are from a papyrus manuscript dated in early second century A.D. They contain tables of fractions with prime denominators, followed by arithmetical problems concerning conversions between silver and copper money and between wheat standards, as well as calculations of carriage charges and of interests. They are in the special collection of University of Michigan.

Books also by Lizhen Ji

Compactifications of Symmetric Spaces
    by Yves Guivarc'h, Lizhen Ji, and John C. Taylor, Birkhäuser, 1998

Compactifications of Symmetric and Locally Symmetric Spaces
    by Armand Borel and Lizhen Ji, Birkhäuser, 2005

Lie Groups and Automorphic Forms
    by Lizhen Ji, Jian-Shu Li, H.W. Xu, and Shing-Tung Yau (Ed.), AMS & IP, 2006

Proceedings of The 4th International Congress of Chinese Mathematicians
    by Lizhen Ji, Kefeng Liu, Lo Yang, and Shing-Tung Yau (Ed.), HEP & IP, 2007

Arithmetic Groups and Their Generalizations: what, why, how?
    by Lizhen Ji, AMS & IP, 2008

Geometry, Analysis and Topology of Discrete Groups
    by Lizhen Ji, Kefeng Liu, Lo Yang, and Shing-Tung Yau (Ed.), HEP & IP, 2008

Handbook of Geometric Analysis Vol. I
    by Lizhen Ji, Peter Li, Richard Schoen, and Leon Simon (Ed.), HEP & IP, 2008

Automorphic Forms and the Langlands Program
    by Lizhen Ji, Kefeng Liu, and Shing-Tung Yau (Ed.), HEP & IP, 2009

Cohomology of Groups and Algebraic K-theory
    by Lizhen Ji, Kefeng Liu, and Shing-Tung Yau (Ed.), HEP & IP, 2009

Handbook of Geometric Analysis Vol. II, III
    by Lizhen Ji, Peter Li, Richard Schoen, and Leon Simon (Ed.), HEP & IP, 2010

Transformation Groups and Moduli Spaces of Curves
    by Lizhen Ji and Shing-Tung Yau (Ed.), HEP & IP, 2010

Geometry and Analysis Vol. I, II
    by Lizhen Ji (Ed.), HEP & IP, 2010

Fourth International Congress of Chinese Mathematicians
    by Lizhen Ji, Kefeng Liu, Lo Yang, and Shing-Tung Yau (Ed.), AMS & IP, 2010

Geometry of Riemann Surfaces and Their Moduli Spaces
    by Lizhen Ji, Scott A. Wolpert, and Shing-Tung Yau (Ed.), AMS & IP, 2010

Frontiers of Mathematical Sciences
    by Huai-Dong Cao, Shiu-Yuen Cheng, Binglin Gu, Lizhen Ji, and Shing-Tung Yau (Ed.), IP, 2011

Fifth International Congress of Chinese Mathematicians
    by Lizhen Ji, Yat Sun Poon, Lo Yang, and Shing-Tung Yau (Ed.), AMS & IP, 2012

Open Problems and Surveys of Contemporary Math
    by Lizhen Ji, Yat Sun Poon, and Shing-Tung Yau (Ed.), HEP & IP, 2013

Great Mathematics Books of the Twentieth Century

A Personal Journey

Lizhen Ji
Department of Mathematics
University of Michigan

Great Mathematics Books of the Twentieth Century: A Personal Journal
by Lizhen Ji (Department of Mathematics, University of Michigan)

Copyright © 2014 by International Press, Somerville, Massachusetts, U.S.A., and by Higher Education Press, Beijing, China.

This work is published and sold in China exclusively by Higher Education Press of China.

All rights reserved. Individual readers of this publication, and non-profit libraries acting for them, are permitted to make fair use of the material, such as to copy a chapter for use in teaching or research. Permission is granted to quote brief passages from this publication in reviews, provided the customary acknowledgement of the source is given. Republication, systematic copying, or mass reproduction of any material in this publication is permitted only under license from International Press. Excluded from these provisions is material in articles to which the author holds the copyright. (If the author holds copyright, notice of this will be given with the article.) In such cases, requests for permission to use or reprint should be addressed directly to the author.

ISBN: 978-1-57146-283-1

Printed in the United States of America.

18 17 16 15 14   1 2 3 4 5 6 7 8 9

The reading of all good books is like a conversation with the finest men of past centuries.
—— René Descartes

There is no friend as loyal as a book.
—— Ernest Hemingway

If we encounter a man of rare intellect, we should ask him what books he reads.
—— Ralph Waldo Emerson

A room without books is like a body without a soul.
—— Marcus Tullius Cicero

A house without books is like a room without windows.
—— Horace Mann

I guess there are never enough books.
—— John Steinbeck

The odd thing about people who had many books was how they always wanted more.
—— Patricia A. McKillip

I have always imagined that Paradise will be a kind of library.
—— Jorge Luis Borges

My best friend is a person who will give me a book I have not read.
—— Abraham Lincoln

I cannot live without books.
—— Thomas Jefferson

There are two motives for reading a book: one, that you enjoy it: the other, that you can boast about it.
—— Bertrand Russell

書富如入海，百貨皆有。人之精力，不能兼收盡取，
但得春所欲求者爾。故願學者每次作一意求之。
— 蘇軾

發奮識遍天下字，立志讀盡人間書。
— 蘇軾

路漫漫其修道遠，吾將上下而求索。
— 屈原

飯可以一日不吃，覺可以一日不睡，書不可以一日不讀。
— 毛澤東

好讀書，不求甚解；每有會意，便欣然忘食。
— 陶淵明

This is a very impressive job for both educators and researchers in mathematics. It collects basically all the important nontechnical books written by great mathematicians. The author also gave insightful comments on these books. This is especially important for those who want to get a global view about mathematics. The author writes with humors and so the book is not dry to read. I am amazed by the author's energy in preparing this book.

Shing-Tung Yau

Fields Medalist, Wolf Prize Winner, Harvard University

海量的數學書，哪些值得我們認真讀，哪些讀後讓我們對數學有更好的認識，這些對門外漢、學生、年輕的學者和專家都是非常需要解決的問題。季理真教授的新作《二十世紀偉大的數學書——個人之旅》(Great Mathematics Books of the Twentieth Century: A Personal Journey) 在這個問題上為我們帶來了極大的便利。本書比較全面收列了二十世紀以來最有影響的數學書並恰當地加以簡評和引述其他評論。本書收列的書目範圍之廣，數量之大令人吃驚，這需要作者廣闊的視野、艱辛的工作，並花大量的時間請教很多不同方向的專家。季理真教授完成了一項很有意義的工作。我相信大家都會歡迎這本書並能從中獲益。

席南華

中國科學院院士，中國科學院數學與系統科學研究院

This book provides an excellent and comprehensive map for your mathematical journey. It will guide you to the right direction and path for advanced studies in almost all mathematical fields.

Lo Yang

Member of Chinese Academy of Sciences (CAS)

阿貝爾有句名言："向大師學習！"本書正是通往大師作品的極佳引導，相信會使廣大數學工作者，無論是初始的學生還是成熟的學者，都受益匪淺。

張偉平

中國科學院院士，南開大學陳省身數學研究所

Unlike other sciences, old mathematical literature has a life of its own because it records not only the background, but also so much of the detailed technical knowledge, for the mathematics of today. This remarkable book is the first one to attempt to analyse the vast literature arising from 20th century mathematics. It lists and comments on a selection of the most influential books written during this time in all of the major fields of mathematics.

John Henry Coates
Fellow of the Royal Society, Senior Whitehead Prize Winner, Cambridge University

This book gives a survey of the influential books in mathematics, touching upon almost all fields of mathematics. This is a very impressive piece of work. It reminds me of the book: A Panorama of Pure Mathematics, written by Dieudonne, except that it is of more practical value since it tells readers where to look further.

Unlike other scientific disciplines that give introductory courses such as "General Physics", "General Chemistry" or "Introduction to Molecular Biology", mathematics does not offer such a course. Consequently, it is very difficult for beginning mathematics students to get an overview of mathematics, what each subject is about, and how different subjects are put together. Books like this one is of great value for filling in that gap.

Weinan E
Member of Chinese Academy of Sciences (CAS), Peking University

An unusual idea: to write a catalog of the math books you love and/or respect. Anyone's personal list of "great" books is bound to be idiosyncratic but can also be a useful complement for students to other entry points into the world of scholarly books.

David Mumford
Fields Medalist, Shaw Prize Winner, Brown University

About the book

Mathematics has a long history and is also forever new. A lot of work had been done in the ancient times, many deep results were obtained in the twentieth century, and exciting theories are being developed. What are important concepts, theories, theorems and conjectures in mathematics? One effective way is to read great books in mathematics. Which are great books, especially books which are read by the working mathematicians now? Like everything else under the sun, it takes time to gain perspective and judge the importance of mathematical results. This book on books provides a guide to the great mathematics books in the twentieth century. It also provides concise summaries of all major subjects in contemporary mathematics. Unlike other disciplines in science, mathematics enjoys strong continuity: whatever was proved always stays true and can serve as the solid foundation for future theories, and old mathematics books are often of more than only historical interests. In view of this, this book also contains many pictures of ancient mathematics books by the old masters such as Euler, Galileo, Gauss, Kepler, Leibniz, Newton, Poincare et al, which can be both informative and enjoyable.

About the author

Lizhen Ji is a professor of mathematics at University of Michigan and studies subjects related to Lie groups, discrete subgroups of Lie groups, transformation groups and related spaces. He loves books and is a chief-editor of four book series: Advanced Lectures in Mathematics, Mathematics and Humanities, Panorama of Mathematics, Surveys of Modern Mathematics, and of the journal Pure and Applied Mathematics Quarterly. He is also an editor of journals Asian Journal of Mathematics and Science in China: Mathematics.

He was a Sloan Fellow and received the NSF postdoctoral Fellowship and the Morningside Silver Medal of Mathematics. He enjoys listening to good mathematics talks on diverse topics and has organized over 30 summer schools, conferences or workshops. He is also an active organizer of seminars and colloquiums. For example, he is the organizer of one of the first seminars called "What is ..." in the world.

The picture on the right was drawn by my daughter Lena Ji for this book on mathematics books.

The characters in the picture are based on the following famous books: Cat in the Hat, One Fish Two Fish Red Fish Blue Fish, Life of Pi, Peter Pan, Alice in Wonderland, and Harry Potter.

Pythagoras might be the most famous and oldest mathematician and philosopher, well before Archimedes and Euclid. He is best known for the Pythagorean Theorem. But more importantly, he was a philosopher. His ideas had an enormous influence on Plato, and through him, on all of the western philosophy and civilization.

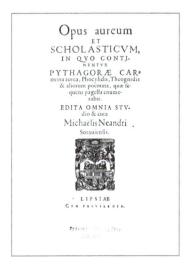

■ Pythagoras (about 570 –495 BC)

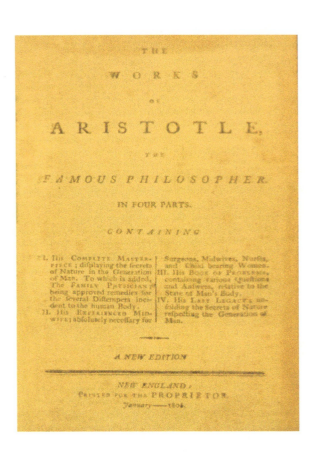

Aristotle is one of the few people in the history who had a long-lasting on the world civilization, especially the western civilization. He wrote extensively and his writings cover many subjects such as physics, logic, philosophy, poetry, theater, music, rhetoric, linguistics, biology, zoology etc. His theories and writings were often cited as authorities to such an extent that they could be harmful. Besides logic, he did not contribute too much to mathematics. But he used mathematics in several important ways: mathematics serves as a model for his philosophy of science and provides him with some important techniques and arguments for his theories in subjects such as physics, biology and ethics.

■ Aristotle (384 BC – 322 BC)

Which book is the most influential textbook since the beginning of civilization? The answer is clearly Euclid's *Elements*. Up to the 20th century, every educated person was supposed to be familiar with the Euclidean geometry, and this book was used until 1950. As the famous Bacon said famously: *mathematics makes one subtle*.

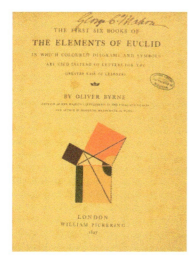 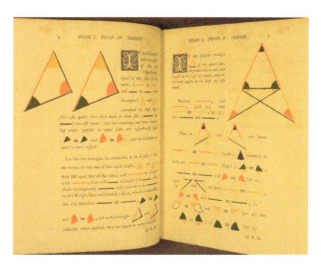

Euclid's classical textbook *Elements* was printed in color in 1847.

Remember that only recently mathematics textbooks are printed in color.

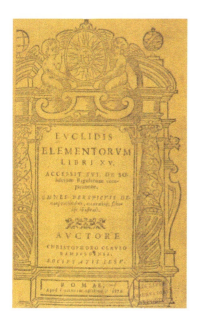

A version of Euclid's book *Elements* published in 1574. This edition is the earliest one I have seen in the original book form. The first printed edition appeared in 1482, based on an Arabic edition from the nineth century.

■ Euclid(325 BC – 265 BC)

People often like to pick the top five greatest mathematicians, and Archimedes is usually on the short list. His interests were broad and people still discover "new" ideas in his works.

A book by Archimedes.

Archimedes' collected works.

■ Archimedes (287 BC – 212 BC)

Though algebraic geometry is abstract, conic sections are something most students learn in high school and college and probably the most basic examples in analytic geometry. The terminologies such as ellipse, parabola, and hyperbola came from books of Apollonius, which have been in print for over 2000 years.

A book by Apollonius on conics.

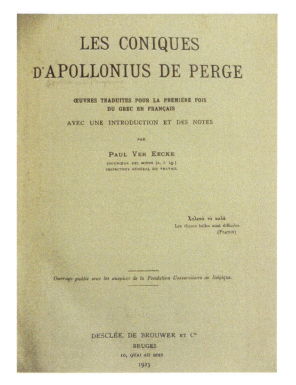

■ Apollonius (262 BC –190 BC)

This is the cover page of the famous Diophantus's *Arithmetica*. Fermat wrote down his famous Fermat Last Theorem in the margin of his copy of this book.

The first pages of Diophantus' book on Arithmetica. This book has been so original that the subject on integral solutions of indeterminate polynomial equations is called Diophantine equations, and the study of such equations is called Diophantine geometry. Probably this is the second mathematics book next to Euclid's *Elements* which brought its author so much credit.

■ Diophantus (200~214 – 284~298)

Fibonacci is probably most famous for the Fibonacci number due to many popular books. His most important contribution is the spreading of the Hindu–Arabic numeral system in Europe through his book *Liber Abaci* published in 1202. Here is a picture of Fibonacci.

■ Fibonacci (1170 –1250)

Da Vinci is called a universal man. To most people, his fame rests on his paintings such as Mona Lisa and the Last Supper. His contribution to mathematics and sciences is probably not clear to most people, except for an illustration of the Golden Ratio by a naked man standing on a circle.

  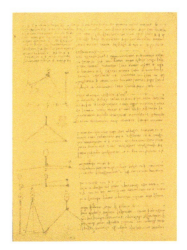

■ Da Vinci (1452 –1519)

When one thinks of algebra and how to solve equations, x and y come to one's mind. But some people may not be aware that they were first introduced by Viete, in particular through his book on new algebra. The power and convenience of these symbols in describing and solving problems cannot be overestimated.

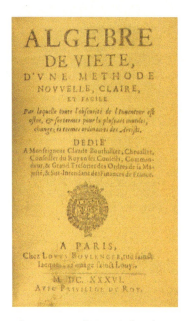

The cover of Viete's book on new algebra. It was published in 1636.

The first page of Viete's book.

Another book cover of Viete's book on new algebra.

■ Viete (1540 –1603)

Galileo is a great theorist and experimentist. His experiment in the leaning tower in Pisa has made the tower one of the most famous tourist attractions in Italy. He was also the first one who used telescope to explore the Moon.

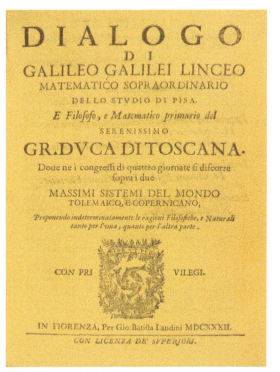

The book by Galileo *Dialogue Concerning the Two Chief World Systems*. This book costed Galileo his freedom: he was under house arrest, and he was not allowed to publish anything more in the future. This book was banned and placed on Index of Forbidden Books, from which it was not removed until 1835.

■ Galileo (1564–1642)

Most people have heard of Kepler's three laws on the motion of planets around the Sun. But it is less known how geometry, in particular, Platonic (or regular) solids, were used in his theory of the universe. His theory did not succeed or survive in spite of being beautiful. Recently, the Platonic solids resurfaced in the string theory, the newest theory of the universe.

In this picture, Platonic solids are nested.

Illustration of Kepler's book on orbits in the solar system.

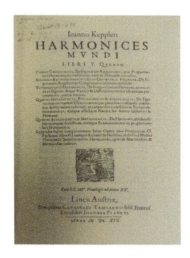 

Illustrations in Kepler's book *The Harmony of the World*.

■ Kepler (1571–1630)

Pascal is probably most famous for two things: Pascal triangles and his mathematical reasoning, Pascal's Wager, for believing in the God. He made many substantial contributions to mathematics, for example, probability.

A book by Pascal on arithmetic.

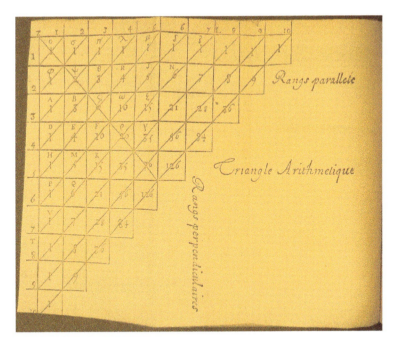

The Pascal triangle is shown in the original book.

■ Pasal (1623 –1662)

If we order mathematicians according to the ratio of the output (or impact) over the number of years devoted to mathematics research, Newton would come out the absolute first. He did all his scientific work in the usual sense in three years in his early twenty. Geometry and Euclid's book impressed Newton so much that he wrote his masterpiece in the language of geometry and in a similar format. If we compare the first edition with a later edition, we will see that the title was printed in red later. Newton's book was one of the books in the Great Book Program advocated by some leading universities in USA for the college student. It is not clear that how many of them can and have read it.

This is a portrait of the great Newton, probably in his old age.

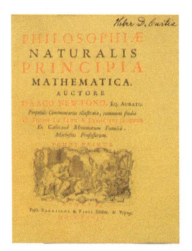

Mathematical Principles of Natural Philosophy published in 1739.

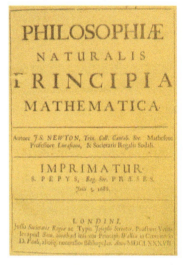

This Newton's classical book Mathematical Principles of Natural Philosophy published in 1686. This is the first edition of the book.

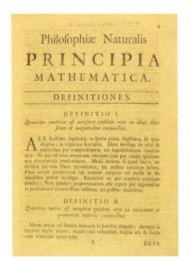

The first definition in Newton's book.

■ Newton (1643 –1727)

Most people have heard of the Bernoulli family and some results due to them. But it is often difficult to tell which is which. It is simply amazing that so many outstanding mathematicians and scientists came from a single family in Basel, Switzerland. Jacob, Johann, and Daniel are probably the most famous among them. The family was rich. One can see portraits of some of them in a church in Basel now, and their graves are in the backyard.

Daniel Bernoulli.

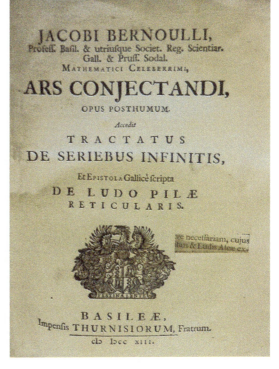

A book by Jacobi Bernoulli, where he proved the law of large numbers.

- Daniel Bernoulli (1700 –1782)
- Jacobi Bernoulli (1654 –1705)

This is the complete works of Johann Bernoulli. He is known for his contributions to infinitesimal calculus and for being a teacher of Leonhard Euler.

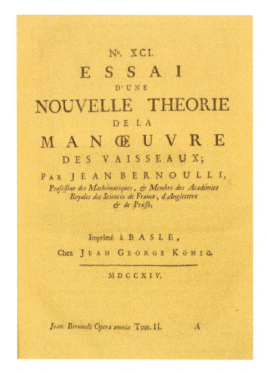

Another volume of the complete works by Johann Bernoulli, who was also called Jean Bernoulli.

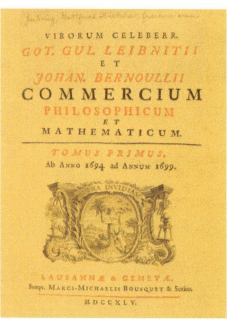

A book by Johann Bernoulli published in 1745.

■ Johann Bernoulli (1667 –1748)

It is well-known that Euler is the most productive mathematician in the history of mathematics in spite of his blindness. But it is probably less known that some of his books have been successful textbooks and are still in print today.

Euler's book on infinitesimal analysis.

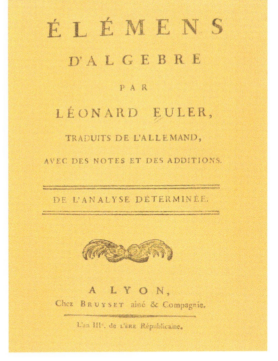

Euler's textbook on algebra.

■ Euler (1707–1783)

People all agree that mathematics is useful, especially in the Internet age. How about 200 years ago?
Monge showed concretely how mathematics could be used effectively in military.

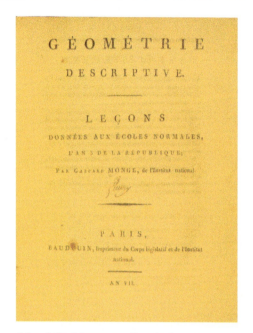 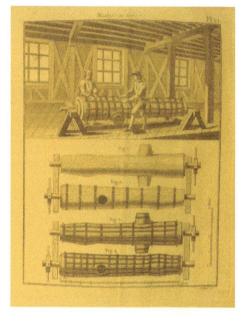

A book by Monge on descriptive geometry.   Illustrations in a book by Monge.

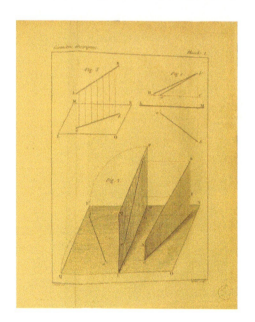 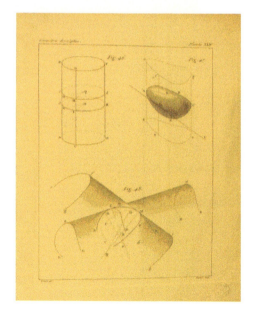

■ Monge (1746 –1818)

Laplace is a great mathematician, maybe the greatest French scientist in the history, sometime called the French Newton. Many things are named after him such as the Laplace operator and the Laplace transform. He contributed greatly to both mathematics and physics. Though he was very interested in and ambitious about politics and was favored by Napoleon, he was not successful as a politician. Here is Laplace's famous book on celestial mechanics.

■ Laplace (1749–1827)

Gauss is called the prince of mathematics, and is often credited with the quote "Mathematics is the queen of sciences". People often forget the second part "Mathematics is the servant of sciences". According to the legend, his talent shone when he was only a few years old. But his real talent shone when he published the classic book *Disquisitiones Arithmeticae* at the age of 24. This book brought in the modern age of number theory. It is still a very valuable reference and a source of inspiration.

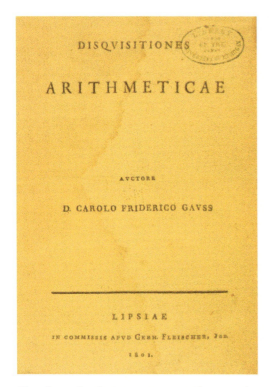

The classic book *Disquisitiones Arithmeticae* by Gauss in 1801.

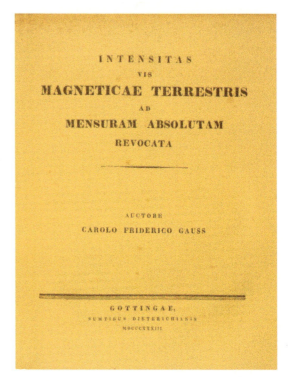

A book by the great Gauss in Latin with a title *The Intensity of The Earth's Magnetic Force Reduced to Absolute Measurement*.

Gauss' talent is not only limited to mathematics. He also contributed to astronomy and physics. For example, he was the director of the observatory, and together with Weber, designed the first telegraph.

■ Gauss (1777 –1855)

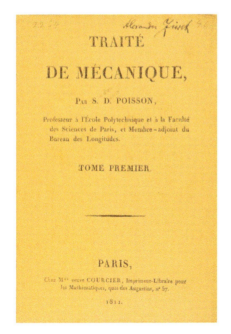
Poisson is a great French mathematician. There are many results named after him, such as the Poisson summation, Poisson formula, Poisson processes, Poisson bracket etc. His book on mechanics has also been a standard book in the subject and had a huge impact.

■ Poisson (1781–1840)

Cauchy is a familiar name to every mathematics student right from the beginning of calculus. Besides analysis, he contributed to many subjects in mathematics and physics. It is probably less known that Cauchy was one of the first who formally introduced the definition of group.

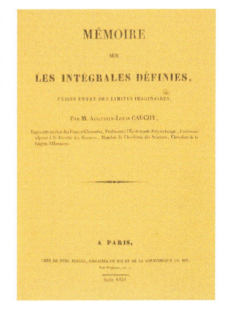

A book by Cauchy on definite integral.

■ Cauchy (1789– 1857)

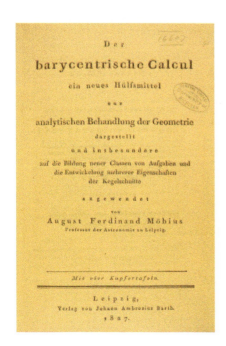

Mobius is probably best known for the Mobius band, which has been depicted in pictures and sculptures and can be explained to children. Besides its importance in mathematics, the fact that the Mobius band has only one side has been used effectively in some practical situations such as conveyor belts, fabric computers printer and typewriter ribbons in order to use the surfaces evenly. To students in complex analysis, Mobius is also well-known for Mobius transformations. It is probably less known that he was also an astronomer and studied mathematics under Gauss, and later studied mathematics under Gauss' teacher.

■ Mobius (1790– 1868)

Dirichlet is probably most famous for the Dirichlet principle (which he did not do and Riemann coined the name) and the Dirichlet L-function in analytic number theory. He was rich and this can be seen from his elaborate tomb in Gottengen.

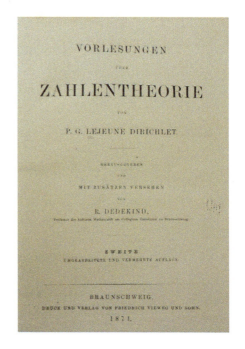

■ Dirichlet (1805– 1859)

If we make a list of top five greatest mathematicians, Galois probably could not make it. But if we make a list of the top five most original mathematicians, Galois should be there (may even at the top?). The notion of group has played such a basic and fundamental role in mathematics, and one cannot imagine how mathematics would look like now without it. But Galois had little publication before he died, and even his manuscripts were buried for a long time after he died.

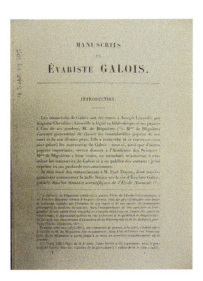

A book consisting of manuscripts of Galois.

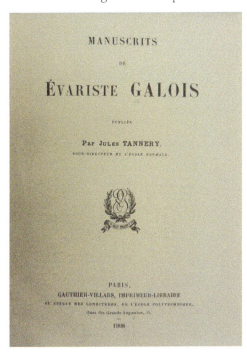

A letter by Galois.

■ Galois (1811–1832)

Hilbert is a global mathematician with a broad vision. Besides his famous list of 23 open problems and his failed attempt, i.e. Hilbert's program to put mathematics on a solid and complete logical foundation, he also wrote many influential books and lecture notes, which are still in print. The best known book is *Geometry and Imagination*. He first became famous for solving the major open problem in invariant theory by a completely new method (or from a new perspective), but he also killed the subject.

Hilbert had many good students. Probably the two most famous are Herman Weyl for his mathematical achievements and Richard Courant for the Courant Institute and his books. He had another special student Max Dehn. Dehn was the first person who solved one of Hilbert's 23 open problems, and in 1907 he wrote with Heegaard the first book on algebraic topology, called analysis situs at that time. He made deep contributions to low dimensional topology and geometric group theory. But at his last job, he was paid only $40 per month and was the only mathematician at his college. He taught both mathematics, philosophy and other subjects. Andre Weil praised Dehn highly and compared him with Socrates.

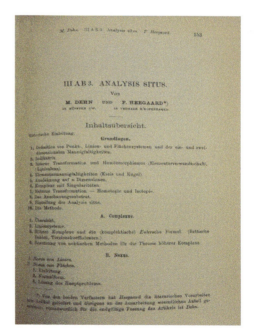

■ Hilbert (1862–1943)

To my great and beautiful wife Lan Wang

Preface

> *Some books are to be tasted, others to be swallowed,*
> *and some few to be chewed and digested.*
>
> —Francis Bacon

In this book, we list and introduce some interesting, important or useful mathematics books. Most selected books were published during the twentieth century. For the convenience of the reader, we have arranged books according to topics. Besides some introductions and comments, we also quote from informative reviews of these books from sources including MathSciNet, Zentralblatt Math and the Bulletin of the American Mathematica Society. A common way for people to pick out books to read is to follow recommendations of either book reviews or experts. *The list of books is probably the most interesting part of this book.*[1] Once the titles or authors' names are known, it is relatively easy to find valuable information and reviews about the book from many different sources (but it might take some efforts to find good books on subjects outside one's expertise.) In spite of this, we hope that additional information provided here about these books might be helpful and convenient.

We hope that such a list of books in contemporary mathematics might be helpful to other people and students who are interested in finding out what kind of mathematics books exist and have been read or enjoyed by others, or who are simply interested in mathematics books.

---

[1] Many people enjoy browsing books in bookstores and new book shelves in libraries without having any particular books in mind. New and interesting books are often found this way. It is also probably true that when people go to libraries to pick up books from bookshelves, they often find other interesting books on related topics on the same bookshelf. Having related books all in one place is a crucial point for this to happen. The number of books in mathematics has been increasing rapidly in the past few decades. How to organize them and to introduce them to a broader readership is becoming more important.

This book could not exist without the kind help from many experts in various fields. I would like to thank the following people for their opinions, suggestions, comments, criticism, corrections, interests and encouragement: Maxim Arap, Tim Austin, Jinho Baik, Oliver Baues, Vitaly Bergelson, Jean-Michel Bismut, Andreas Blass, Patrick Boland, Martin Bridson, Dan Burns, Peter Buser, Richard Canary, John Coates, Joseph Conlon, S.G. Dani, Igor Dolgachev, Peter Duren, Weinan E, Alexandre Eremenko, Tom Farrell, Sergey Fomin, Jacques Franchi, Kenji Fukaya, Bill Fulton, Francis Fung, Sergei Gelfand, David Harbater, William Harvey, Elton Hsu, Mattias Jonsson, Manfred Karbe, Linda Keen, Hans Koelsch, Kai Köhler, Igor Kriz, Jeff Lagarias, Robert Lazarsfeld, Enrico Leuzinger, Tien-Yien Li, Wenbo Li, Eduard Looijenga, Hugh Montgomery, Kumar Murty, Louis Nirenberg, Peter Olver, Athanase Papadopoulos, Hugo Parlier, Katrik Prasana, Stratos Prassidis, Mikael Ragstedt, Andrei Rapinchuk, Frank Raymond, John Schotland, Leonard Scott, Mei-Chi Shaw, Reyer Sjamaar, Peter Smereka, Ralf Spatzier, Christopher Stark, Alejandro Uribe, Roman Vershynin, Divakar Viswanath, Charles Weibel, Trevor Wooley, Scott Wolpert, Ping Zhang, Weiping Zhang, Michael Zieve, and Steve Zucker.

I would also like to thank Dr. Graeme Fairweather, the director of MathSciNet, for his permission to quote from reviews in MathSciNet and Zentralblatt Math for providing me the full access to its data base during the preparation of this book.

Especially I would like to thank my wife, Lan Wang, for suggesting the key word "journey" in the subtitle and for her encouragement, my oldest daughter Lena for drawing the picture for this book which is based on many famous books related to mathematics, and my youngest daughter Karen for proof-reading the preface and introduction.

Finally I would like to thank Liping Wang of the Higher Education Press for her interests in this unusual project, and her time and efforts in carefully editing this book. She made suggestions for many new features and improvements to this book. For example, the idea of inclusion of pictures of libraries and other buildings of University of Michigan came from her, and the idea of inclusion of many pictures of old and rare mathematics books arose from discussion with her. I would also like to thank Shannian Lu of the Higher Education Press for preparing the index of books which have been translated into Chinese or reprinted in China, and Yushan Deng of International Press of Boston for carefully proofreading this book and adjusting the pictures.

One reason for including these pictures of mathematics books is that though ancient and classical mathematics books are not read by many people now, they had greatly influenced the development of mathematics and are still interesting to many mathematicians. According to one Chinese proverb, "One picture is worth ten thousand words". We hope that these pictures and the accompanying comments might be both interesting and informative to the reader. For example, by putting them together, we can see from these pictures of old books how book printing has changed over the past centuries, and how the authors of these books have left their permanent marks on the history of mathematics. At the beginning of each chapter and section, we have tried to select pictures which fit the topics under discussion, but it is not clear whether we have succeeded due to many obvious constraints and the lack of knowledge of the author.

All pictures used in this book come from the Special Collections Library and the other libraries of the University of Michigan. Except for the picture of a collection of books on the cover and for the picture of the Galileo manuscript at the beginning of Chapter 10, almost all other pictures were taken by the author. I would like to thank the staff members of the Special Collections Library at the University of Michigan, in particular the curator Peggy Daub, for providing these two special pictures and for their help which made it possible for me to view and take pictures of over one hundred rare mathematics books at the Special Collections Library. It was the first time I came in close contact with these great mathematics books by old masters, and flipping through these books was both a humbling and inspiring experience for me.

Lizhen Ji
May 1, 2013

Contents

1 Introduction 1
2 Expository books on mathematics and mathematicians 5
  2.1 Popular and expository books on mathematics 6
    2.1.1 R. Courant, H. Robbins, What Is Mathematics? Oxford University Press, New York, 1941. xix+521 pp. 6
    2.1.2 A.D. Aleksandrov, A.N. Kolmogorov, M.A. Lavrent'ev, Mathematics: Its Content, Methods, and Meaning. Vol. I, Vol. II, Vol. III, The M.I.T. Press, Cambridge, Mass., 1963, xi+359 pp.; xi+377 pp.; xi+356 pp.. Translated by S.H. Gould and T. Bartha; S.H. Gould; K. Hirsch. 12
    2.1.3 G. Pólya, How to Solve it. A New Aspect of Mathematical Method. Expanded version of the 1988 edition, with a new foreword by John H. Conway, Princeton Science Library, Princeton University Press, 2004. xxviii+253 pp. 15
    2.1.4 G.H. Hardy, A Mathematician's Apology, With a foreword by C.P. Snow, Reprint of the 1967 edition, Canto, Cambridge University Press, Cambridge, 1992. 19
    2.1.5 J.E. Littlewood, Littlewood's Miscellany, Edited and with a foreword by Béla Bollobás, Cambridge University Press, Cambridge, 1986. vi+200 pp. 21
    2.1.6 Autobiographies of mathematicians 22
    2.1.7 H. Weyl, Symmetry. Reprint of the 1952 original. Princeton Science Library. Princeton University Press, Princeton, N.J., 1989. 27
    2.1.8 D. Hilbert, S. Cohn-Vossen, Geometry and the Imagination, American Mathematical Society, 1, 1999. 357 pages 37
  2.2 Biographies of mathematicians and history of mathematics 42
    2.2.1 E.T. Bell, Men of Mathematics, Touchstone, 1986. 608 pages. 42

2.2.2	C. Reid, Hilbert, Springer-Verlag, New York-Berlin, 1970. xi+290 pp.	44
2.2.3	More biographies of mathematicians	46
2.2.4	A. Weil, Number Theory. An Approach through History. From Hammurapi to Legendre. Birkhäuser Boston, MA, 1984. xxi+375 pp.	55
2.2.5	W. Scharlau, H. Opolka, From Fermat to Minkowski. Lectures on the Theory of Numbers and its Historical Development, Undergraduate Texts in Mathematics. Springer-Verlag, New York, 1985. xi+184 pp.	59
2.2.6	History of mathematics	60
2.2.7	More mathematical history books	63
2.2.8	J. Dieudonné, A History of Algebraic and Differential Topology 1900–1960, Reprint of the 1989 edition, Modern Birkhäuser Classics. Birkhäuser Boston, Inc., Boston, MA, 2009. xxii+648 pp.	72
2.2.9	History of Lie groups and related spaces	73
2.2.10	F. Klein, Development of Mathematics in the 19th Century, With a preface and appendices by Robert Hermann, Translated from the German by M. Ackerman. Math Sci Press, Brookline, Mass., 1979. ix+630 pp.	76

3 Analysis — 79

3.1 Calculus and real analysis — 79

3.1.1	G.H. Hardy, A Course of Pure Mathematics, Reprint of the (1952) tenth edition, Cambridge Mathematical Library, Cambridge University Press, Cambridge, 1992. xii+509 pp.	80
3.1.2	Calculus	82
3.1.3	E.T. Whittaker and G.N. Watson, A Course of Modern Analysis, An Introduction to the General Theory of Infinite Processes and of Analytic Functions: with an Account of the Principal Transcendental Functions, Fourth edition. Reprinted, Cambridge University Press, New York, 1962, vii+608 pp.	85
3.1.4	Special functions	85
3.1.5	W. Rudin, Real and Complex Analysis, Third edition, McGraw-Hill Book Co., New York, 1987. xiv+416 pp.	90
3.1.6	More books on real analysis	90
3.1.7	Measure theory	97
3.1.8	Analysis on general spaces	100

	3.1.9	G. Pólya, G. Szegö, Problems and Theorems in Analysis. Vol. I. Series, Integral calculus, Theory of Functions. Vol. II. Theory of Functions, Zeros, Polynomials, Determinants, Number Theory, Geometry, Grundlehren der mathematischen Wissenschaften, Classics in Mathematics, Springer-Verlag, Berlin, 1998. xii+392 pp. xx+389 pp.	104
	3.1.10	Orthogonal polynomials	106
	3.1.11	Problem books	107
	3.1.12	G.H. Hardy, J.E. Littlewood, G. Pólya, Inequalities, Reprint of the 1952 edition, Cambridge Mathematical Library, Cambridge University Press, Cambridge, 1988. xii+324 pp.	110
3.2	Complex analysis		114
	3.2.1	H. Cartan, Elementary Theory of Analytic Functions of One or Several Complex Variables. Éditions Scientifiques Hermann, Paris; Addison-Wesley Publishing Co., Inc., Reading, Mass.-Palo Alto, Calif.-London, 1963. 228 pp.	115
	3.2.2	L. Ahlfors, Complex Analysis, Third edition, International Series in Pure and Applied Mathematics, McGraw-Hill Book Co., New York, 1978. xi+331 pp.	115
	3.2.3	R. Remmert, Theory of Complex Functions, Translated from the second German edition by Robert B. Burckel, Graduate Texts in Mathematics, 122, Readings in Mathematics, Springer-Verlag, New York, 1991. xx+453 pp.	116
	3.2.4	More books on complex analysis	117
	3.2.5	Analytic and meromorphic functions, and conformal maps	121
	3.2.6	Several complex variables	123
	3.2.7	Sheaf theories	126
	3.2.8	H. Weyl, The Concept of a Riemann Surface, Translated from the third German edition by Gerald R. MacLane, ADIWES International Series in Mathematics, Addison-Wesley Publishing Co., Inc., Reading, Mass.-London, 1964. xi+191 pp.	128
	3.2.9	More books on Riemann surfaces	130
	3.2.10	Quasiconformal mappings	132
	3.2.11	Teichmüller theory	135
3.3	Harmonic analysis		140
	3.3.1	Y. Katznelson, An Introduction to Harmonic Analysis, 3rd Edition, Cambridge Mathematical Library, Cambridge University Press, 2004. xviii+314 pp.	141

	3.3.2	More books on harmonic analysis and Fourier series	142
	3.3.3	A. Zygmund, Trigonometric Series. 2nd ed. Vols. I, II. Cambridge University Press, New York, 1959, Vol. I., xii+383 pp.; Vol. II. vii+354 pp.	147
	3.3.4	Potential theory	149
	3.3.5	J. Garnett, Bounded Analytic Functions, Pure and Applied Mathematics, 96, Academic Press, Inc., New York-London, 1981. xvi+467 pp.	152
	3.3.6	More books on Hardy spaces and function spaces	153
	3.3.7	E. Stein, Singular Integrals and Differentiability Properties of Functions, Princeton Mathematical Series, No. 30, Princeton University Press, 1970, xiv+290 pp.	157
	3.3.8	G. Folland, Harmonic Analysis in Phase Space, Annals of Mathematics Studies, 122, Princeton University Press, Princeton, N.J., 1989. x+277 pp.	160
	3.3.9	I. Daubechies, Ten Lectures on Wavelets, CBMS-NSF Regional Conference Series in Applied Mathematics, 61, Society for Industrial and Applied Mathematics (SIAM), Philadelphia, PA, 1992. xx+357 pp.	160
3.4	Functional analysis and operator theory		162
	3.4.1	S. Banach, Theory of Linear Operations, Translated from the French by F. Jellett, With comments by A. Peczyski and Cz. Bessaga, North-Holland Mathematical Library, 38, North-Holland Publishing Co., Amsterdam, 1987. x+237 pp.	163
	3.4.2	More basic books on functional analysis	164
	3.4.3	More books on Banach spaces	167
	3.4.4	More books on Hilbert spaces	169
	3.4.5	L. Schwartz, Théorie des Distributions, Hermann, Paris, 1966, xiii+420 pp.	171
	3.4.6	I.M. Gelfand, G.E. Shilov, Generalized Functions. Vol. 1. Properties and Operations, Vol. 2. Spaces of Fundamental and Generalized Functions, Vol. 3. Theory of Differential Equations, I.M. Gelfand, N.Ya. Vilenkin, Vol. 4. Applications of Harmonic Analysis, I.M. Gelfand, M.I. Graev, N.Ya. Vilenkin, Vol. 5. Integral Geometry and Representation Theory, Academic Press, New York-London, 1964, xviii+423 pp., x+261 pp., x+222 pp., xiv+384 pp., xvii+449 pp.	172

	3.4.7	K. Yosida, Functional analysis, reprint of the sixth (1980) edition. Classics in Mathematics, Springer-Verlag, Berlin, 1995. xii+501 pp.	174
	3.4.8	N. Dunford, J. Schwartz, Linear Operators, Part I, General Theory; Part II, Spectral theory. Selfadjoint operators in Hilbert space. Part III, Spectral operators, Wiley Classics Library, A Wiley-Interscience Publication, John Wiley & Sons, Inc., New York, 1988. xiv+858 pp., x+859–1923, xx+1925–2592.	175
	3.4.9	Semigroups and functional analysis	176
	3.4.10	A. Connes, Noncommutative Geometry, Academic Press, Inc., San Diego, CA, 1994. xiv+661 pp.	178
	3.4.11	Operator algebras	179
	3.4.12	H. Brezis, Analyse Fonctionnelle, Théorie et Applications, Collection Mathématiques Appliquées pour la Matrise. Masson, Paris, 1983. xiv+234 pp.	184
	3.4.13	K. Deimling, Nonlinear Functional Analysis, Springer-Verlag, Berlin, 1985. xiv+450 pp.	185

4 Algebra   187
4.1 Abstract algebras and finite groups . . . 188
4.1.1 Linear algebra . . . 188
4.1.2 Advanced linear algebra and matrix algebra . . . 191
4.1.3 van der Waerden, Algebra. Vol. I., Algebra. Vol. II., Based in part on lectures by E. Artin and E. Noether, Translated from the fifth German edition by John R. Schulenberge, Springer-Verlag, New York, 1991. xiv+265 pp., xii+284 pp. . . . 194
4.1.4 More books on abstract algebra . . . 196
4.1.5 I.R. Shafarevich, Basic Notions of Algebra, Springer, Berlin, 1990; Springer-Verlag, Berlin, 1997. iv+258 pp. . . . 204
4.1.6 R. Carter, Simple Groups of Lie Type, Pure and Applied Mathematics, Vol. 28, Reprint of the 1972 original, Wiley Classics Library, A Wiley-Interscience Publication, John Wiley & Sons, Inc., New York, 1989. x+335 pp . . . 205
4.1.7 Finite groups . . . 206
4.1.8 Finite groups and their representation theories . . . 213
4.1.9 Representation theories and associate algebras . . . 214
4.1.10 Rings and modules . . . 216
4.2 Commutative algebras . . . 218
4.3 Homological algebra . . . 225

	4.3.1	H. Cartan, S. Eilenberg, Homological Algebra, With an appendix by David A. Buchsbaum, Reprint of the 1956 original, Princeton Landmarks in Mathematics, Princeton University Press, Princeton, NJ, 1999. xvi+390 pp.	226
	4.3.2	More books on homological algebra and related subjects	228

5 Geometry 233
5.1 Differential geometry . . . 233
- 5.1.1 H. Hopf, Differential Geometry in the Large, Notes taken by Peter Lax and John Gray, With a preface by S. S. Chern, Lecture Notes in Mathematics, 1000, Springer-Verlag, Berlin, 1983. vii+184 pp. . . . 234
- 5.1.2 Introduction to differential geometry and Riemannian geometry . . 235
- 5.1.3 More advanced books on Riemannian geometry . . . 240
- 5.1.4 Special topics in differential geometry . . . 245
- 5.1.5 M. Gromov, Metric Structures for Riemannian and Non-Riemannian Spaces, Based on the 1981 French original, With appendices by M. Katz, P. Pansu and S. Semmes, Translated from the French by Sean Michael Bates, Progress in Mathematics, 152, Birkhäuser Boston, Inc., Boston, MA, 1999. xx+585 pp. . . . 253
- 5.1.6 M. Bridson, A. Haefliger, Metric Spaces of Non-positive Curvature, Grundlehren der Mathematischen Wissenschaften, 319, Springer-Verlag, Berlin, 1999. xxii+643 pp. . . . 255
- 5.1.7 S. Helgason, Differential Geometry, Lie Groups, and Symmetric Spaces, Corrected reprint of the 1978 original, Graduate Studies in Mathematics, 34, American Mathematical Society, Providence, RI, 2001. xxvi+641 pp. . . . 256

5.2 Geometric analysis . . . 258
- 5.2.1 R. Schoen, S.T. Yau, Lectures on Differential Geometry, International Press, 1994. v+235 pp. . . . 259
- 5.2.2 T. Aubin, Nonlinear Analysis on Manifolds, Monge Ampére equations, Grundlehren der Mathematischen Wissenschaften, 252, Springer-Verlag, New York, 1982. xii+204 pp. . . . 260
- 5.2.3 More books on geometric analysis . . . 260
- 5.2.4 M. Gromov, Partial Differential Relations, Ergebnisse der Mathematik und ihrer Grenzgebiete (3), 9, Springer-Verlag, Berlin, 1986. x+363 pp. . . . 264
- 5.2.5 H. Federer, Geometric Measure Theory, Die Grundlehren der mathematischen Wissenschaften, Band 153, Springer-Verlag New York Inc., New York, 1969, xiv+676 pp. . . . 265

	5.2.6	More books on geometric measure theory	266
	5.2.7	Calculus of variation	267
	5.2.8	Fractal geometry	270
5.3	Complex geometry and complex analysis	272
	5.3.1	L. Hörmander, An Introduction to Complex Analysis in Several Variables, Third edition, North-Holland Publishing Co., Amsterdam, 1990. xii+254 pp.	272
	5.3.2	More books on complex geometry	273
	5.3.3	C.L. Siegel, Topics in Complex Function Theory. Vol. I. Elliptic Functions and Uniformization Theory. Vol. II. Automorphic Functions and Abelian Integrals. Vol. III. Abelian Functions and Modular Functions of Several Variables, Interscience Tracts in Pure and Applied Mathematics, No. 25, Wiley-Interscience, 1969. ix+186 pp., 1988. xii+193 pp., 1989. x+244 pp.	276
5.4	Algebraic geometry	277
	5.4.1	A. Grothendieck, The Éléments de géométrie algébrique, Inst. Hautes Études Sci. Publ. Math., 1960–1967.	279
	5.4.2	I. Shafarevich, Basic Algebraic Geometry. 1. Varieties in Projective Space. 2. Schemes and Complex Manifolds. Second edition, Springer-Verlag, Berlin, 1994. xx+303 pp., xiv+269 pp.	281
	5.4.3	D. Mumford, The Red book of Varieties and Schemes, Second, expanded edition. Includes the Michigan lectures (1974) on curves and their Jacobians, With contributions by Enrico Arbarello. Lecture Notes in Mathematics, 1358. Springer-Verlag, Berlin, 1999. x+306 pp.	283
	5.4.4	R. Hartshorne, Algebraic Geometry, Graduate Texts in Mathematics, No. 52, Springer-Verlag, New York-Heidelberg, 1977. xvi+496 pp.	285
	5.4.5	P. Griffiths, J. Harris, Principles of Algebraic Geometry, Pure and Applied Mathematics, Wiley-Interscience, New York, 1978. xii+813 pp.	286
	5.4.6	Introduction to algebraic geometry and algebraic curves	288
	5.4.7	Topology of algebraic varieties	289
	5.4.8	Symplectic geometry and symplectic topology	290
	5.4.9	D. Mumford, J. Fogarty, F. Kirwan, Geometric invariant theory, Third edition, Ergebnisse der Mathematik und ihrer Grenzgebiete (2), 34, Springer-Verlag, Berlin, 1994. xiv+292 pp.	293
	5.4.10	Classification of varieties and moduli spaces	294

- 5.4.11 Algebraic curves 297
- 5.4.12 Algebraic surfaces 299
- 5.4.13 W. Fulton, Intersection Theory, Second edition, Ergebnisse der Mathematik und ihrer Grenzgebiete. 3. Folge, Springer-Verlag, Berlin, 1998. xiv+470 pp. 302
- 5.4.14 R. Lazarsfeld, Positivity in Algebraic Geometry. I. Classical Setting: Line Bundles and Linear Series; II. Positivity for Vector Bundles, and Multiplier Ideals, Ergebnisse der Mathematik und ihrer Grenzgebiete. 3. Folge, Springer-Verlag, Berlin, 2004. xviii+387 pp., xviii+385 pp. 305
- 5.4.15 Toric varieties 306
- 5.5 Convex geometry and discrete geometry 309
  - 5.5.1 R.T. Rockafellar, Convex Analysis, Princeton Mathematical Series, No. 28, Princeton University Press, Princeton, N.J., 1970. xviii+451 pp. 309
  - 5.5.2 Convex functions and convex geometry 311

6 Topology 315
- 6.1 More classical topology 316
  - 6.1.1 P. Alexandroff, H. Hopf, Topologie I. Berlin, Springer, 1935. xiii+636 pp. 316
  - 6.1.2 K. Kuratowski, Topology. Vol. I., Vol. II., New edition, Revised and augmented, Translated from the French by J. Jaworowski Academic Press, New York-London; Państwowe Wydawnictwo Naukowe, Warsaw, 1966. xx+560 pp., 1968. xiv+608 pp. 318
  - 6.1.3 More books on topology 318
  - 6.1.4 Fixed point theory 321
  - 6.1.5 Dimension theory 322
- 6.2 Algebraic topology 323
  - 6.2.1 J. Milnor, J. Stasheff, Characteristic Classes, Annals of Mathematics Studies, No. 76, Princeton University Press, 1974. vii+331 pp. 324
  - 6.2.2 S. Eilenberg, N. Steenrod, Foundations of Algebraic Topology, Princeton University Press, 1952. xv+328 pp. 325
  - 6.2.3 More books on algebraic topology 326
  - 6.2.4 D. Rolfsen, Knots and Links, Corrected reprint of the 1976 original, Mathematics Lecture Series, 7, Publish or Perish, Inc., Houston, TX, 1990. xiv+439 pp. 328
  - 6.2.5 More knots and their invariants books 329

	6.3	Generalized cohomology theory and homotopy theories	332
	6.3.1	J. Adams, Stable Homotopy and Generalised Homology, Chicago Lectures in Mathematics, University of Chicago Press, Chicago, Ill.-London, 1974. x+373 pp.	332
	6.3.2	K-theory	333
	6.3.3	K. Brown, Cohomology of Groups, Graduate Texts in Mathematics, 87, Corrected reprint of the 1982 original, Graduate Texts in Mathematics, 87, Springer-Verlag, New York, 1994. x+306 pp.	335
	6.3.4	G.W. Whitehead, Elements of Homotopy Theory, Graduate Texts in Mathematics, 61, Springer-Verlag, New York-Berlin, 1978. xxi+744 pp.	337
	6.4	Differential topology	338
	6.4.1	J. Milnor, Morse Theory, Based on lecture notes by M. Spivak and R. Wells, Annals of Mathematics Studies, No. 51, Princeton University Press, Princeton, N.J., 1963. vi+153 pp.	339
	6.4.2	Differential topology	340
	6.4.3	R. Bott, L. Tu, Differential Forms in Algebraic Topology, Graduate Texts in Mathematics, 82, Springer-Verlag, New York-Berlin, 1982. xiv+331 pp.	341
	6.4.4	W.V.D. Hodge, The Theory and Applications of Harmonic Integrals, Reprint of the 1941 original, With a foreword by Michael Atiyah, Cambridge Mathematical Library, Cambridge University Press, Cambridge, 1989. xiv+284 pp.	342
	6.4.5	M. Goresky, R. MacPherson, Stratified Morse theory, Ergebnisse der Mathematik und ihrer Grenzgebiete (3), 14, Springer-Verlag, Berlin, 1988. xiv+272 pp.	343
	6.5	Geometric topology	345
	6.5.1	W. Thurston, The Geometry and Topology of Three-Manifolds, Lecture notes at Princeton University, 1978–1980.	345
	6.5.2	Four dimensional manifolds	350
	6.5.3	Geometric topology and surgery theory	351

7 Number theory   355

7.1		Number theory	356
	7.1.1	G.H. Hardy, E. Wright, An Introduction to the Theory of Numbers, Sixth edition, Revised by D. R. Heath-Brown and J. H. Silverman, With a foreword by Andrew Wiles, Oxford University Press, Oxford, 2008. xxii+621 pp.	356
	7.1.2	Books on basic number theory	357

- 7.1.3 H. Davenport, The Higher Arithmetic. An Introduction to the Theory of Numbers, Eighth edition, With editing and additional material by James H. Davenport, Cambridge University Press, Cambridge, 2008. x+239 pp. 361
- 7.1.4 H. Hasse, Number Theory, Translated from the third German edition and with a preface by Horst Günter Zimmer. Grundlehren der Mathematischen Wissenschaften, 229. Springer-Verlag, Berlin-New York, 1980. xvii+638 pp. 362
- 7.1.5 A. Borevich, I. Shafarevich, Number Theory, Translated from the Russian by Newcomb Greenleaf, Pure and Applied Mathematics, Vol. 20, Academic Press, New York-London, 1966, x+435 pp. ... 362
- 7.1.6 A.Y. Khinchin, Three Pearls of Number Theory, Translated from the Russian by F. Bagemihl, H. Komm, and W. Seidel, Reprint of the 1952 translation, Dover Publications, Inc., Mineola, NY, 1998. 64 pp. 363
- 7.1.7 E. Artin, Galois Theory, Notre Dame Mathematical Lectures, No. 2, Edited and with a supplemental chapter by Arthur N. Milgram, Reprint of the 1944 second edition, Dover Publications, Inc., Mineola, N.Y., 1998. iv+82 pp. 364
- 7.2 Algebraic number theory 365
  - 7.2.1 D. Hilbert, The Theory of Algebraic Number Fields, Translated from the German and with a preface by Iain T. Adamson, With an introduction by Franz Lemmermeyer and Norbert Schappacher, Springer-Verlag, Berlin, 1998. xxxvi+350 pp. 365
  - 7.2.2 E. Hecke, Lectures on the Theory of Algebraic Numbers, Translated from the German by George U. Brauer, Jay R. Goldman and R. Kotzen, Graduate Texts in Mathematics, 77, Springer-Verlag, New York-Berlin, 1981. xii+239 pp. 366
  - 7.2.3 J.W.S. Cassels, A. Fröhlich, Algebraic Number Theory, Proceedings of the instructional conference held at the University of Sussex, Brighton, September 1–17, 1965, Academic Press, London; Thompson Book Co., Inc., Washington, D.C., 1967, xviii+366 pp. 367
  - 7.2.4 S. Lang, Algebraic Number Theory, Second edition, Graduate Texts in Mathematics, 110, Springer-Verlag, New York, 1994. xiv+357 pp. 368
  - 7.2.5 More books on algebraic number theory 368
  - 7.2.6 Computational algebraic number theory 370

		7.2.7	J.P. Serre, Local Fields, Translated from the French by Marvin Jay Greenberg, Graduate Texts in Mathematics, 67, Springer-Verlag, New York-Berlin, 1979. viii+241 pp.	371
		7.2.8	Galois cohomology	372
		7.2.9	Geometry of numbers	373
	7.3	Analytic number theory		374
		7.3.1	E.C. Titchmarsh, The Theory of the Riemann Zeta-Function, Second edition. Edited and with a preface by D. R. Heath-Brown, The Clarendon Press, Oxford University Press, New York, 1986. x+412 pp.	375
		7.3.2	More books on the Riemann zeta function	375
		7.3.3	Analytic number theory	377
		7.3.4	Additive number theory	380
		7.3.5	Multiplicative number theory	381
	7.4	Transcendental number theory		382
	7.5	Arithmetic algebraic geometry		384
		7.5.1	G. Shimura, Introduction to the Arithmetic Theory of Automorphic Functions, Iwanami Shoten, Publishers, Tokyo; Princeton University Press, Princeton, N.J., 1971. xiv+267 pp.	384
		7.5.2	J.P. Serre, A Course in Arithmetic, Translated from the French, Graduate Texts in Mathematics, No. 7, Springer-Verlag, New York-Heidelberg, 1973. viii+115 pp.	385
		7.5.3	J. Silverman, The Arithmetic of Elliptic Curves, Graduate Texts in Mathematics, 106, Second edition, Graduate Texts in Mathematics, 106, Springer, Dordrecht, 2009. xx+513 pp.	385
		7.5.4	D. Mumford, Abelian Varieties. With appendices by C. P. Ramanujam and Yuri Manin, Corrected reprint of the second (1974) edition, Tata Institute of Fundamental Research Studies in Mathematics, 5, Hindustan Book Agency, New Delhi, 2008. xii+263 pp.	387
		7.5.5	Abelian varieties and theta functions	387
		7.5.6	Diophantine geometry	389
		7.5.7	Étale cohomology	390
		7.5.8	Quadratic forms	391
		7.5.9	More books on number theory	393
	7.6	Modular forms and automorphic representations		395

- 7.6.1 R. Fricke, F. Klein, Vorlesungen uber die Theorie der automorphen Funktionen. Band 1: Die gruppentheoretischen Grundlagen. Band II: Die funktionentheoretischen Ausfhrungen und die Andwendungen. Bibliotheca Mathematica Teubneriana, Bande 3, 4, Johnson Reprint Corp., New York; B. G. Teubner Verlagsgesellschaft, Stuttg art 1965. Band I: xiv+634 pp.; Band II: xiv+668 pp. . . . 396
- 7.6.2 I.M. Gelfand, M. Graev, I.I. Pyatetskii-Shapiro, Representation Theory and Automorphic Functions, W. B. Saunders Co., Philadelphia, Pa.-London-Toronto, Ont. 1969. xvi+426 pp. 399
- 7.6.3 H. Jacquet, R. Langlands, Automorphic Forms on $GL_(2)$, Lecture Notes in Mathematics, Vol. 114, Springer-Verlag, Berlin-New York, 1970. vii+548 pp. 400
- 7.6.4 More books on modular forms and automorphic forms 401
- 7.6.5 R. Langlands, On the Functional Equations Satisfied by Eisenstein Series, Lecture Notes in Mathematics, Vol. 544, Springer-Verlag, Berlin-New York, 1976. v+337 pp. 402
- 7.6.6 A. Borel, W. Casselman, Automorphic Forms, Representations and L-functions. Part 1, Part 2, Proceedings of Symposia in Pure Mathematics, XXXIII. American Mathematical Society, Providence, R.I., 1979. x+322 pp., vii+382 pp. 403
- 7.6.7 More books on modular forms, automorphic representations and cohomology of arithmetic groups 404
- 7.6.8 Hypergeometric series and theory of partitions 405

8 Differential equations 407
8.1 Ordinary differential equations 407
- 8.1.1 E. Coddington, N. Levinson, Theory of Ordinary Differential Equations, McGraw-Hill Book Company, Inc., New York-Toronto-London, 1955. xii+429 pp. 408
- 8.1.2 More books on ordinary differential equations 409
- 8.1.3 V.I. Arnold, Ordinary Differential Equations, Translated from the third Russian edition by Roger Cooke, Springer Textbook, Springer-Verlag, Berlin, 1992. 334 pp. 410
- 8.1.4 More books on ordinary differential equations and dynamical systems 411
8.2 Linear differential operators 413
- 8.2.1 L. Evans, Partial Differential Equations, Graduate Studies in Mathematics, 19, Second edition, Graduate Studies in Mathematics, 19, American Mathematical Society, Providence, RI, 2010. xxii+749 pp. 413

	8.2.2	L. Hörmander, Linear Partial Differential Operators, Springer Verlag, Berlin-New York, 1976. vii+285 pp.	414
	8.2.3	More books on linear differential equations	416
	8.2.4	Inverse problems	418
	8.2.5	Critical point theory and minimax methods	419
	8.2.6	D. Gilbarg, N. Trudinger, Elliptic Partial Differential Equations of Second Order, Reprint of the 1998 edition, Classics in Mathematics, Springer-Verlag, Berlin, 2001. xiv+517 pp.	421
	8.2.7	More books on elliptic differential equations	421
	8.2.8	Pseudodifferential operators	423
	8.2.9	Parabolic equations	425
	8.2.10	R. Adams, John J.H. Fournier, Sobolev Spaces, Second edition, Pure and Applied Mathematics (Amsterdam), 140, Elsevier/Academic Press, Amsterdam, 2003. xiv+305 pp.	427
	8.2.11	More books on Sobolev spaces	427
	8.2.12	T. Kato, Perturbation Theory for Linear Operators, Reprint of the 1980 edition, Classics in Mathematics, Springer-Verlag, Berlin, 1995.	428
8.3	Nonlinear differential equations		429
	8.3.1	More geometric nonlinear differential equations	429
	8.3.2	Nonlinear differential equations and fluid mechanics	432

9 Lie theories   439

9.1	Lie groups and Lie algebras		440
	9.1.1	C. Chevalley, Theory of Lie Groups. I, Fifteenth printing, Princeton Mathematical Series, 8, Princeton Landmarks in Mathematics, Princeton University Press, Princeton, NJ, 1999. xii+217 pp.	440
	9.1.2	J.P. Serre, Complex Semisimple Lie Algebras, Translated from the French by G.A. Jones, Reprint of the 1987 edition, Springer Monographs in Mathematics, Springer-Verlag, Berlin, 2001. x+74 pp.	441
	9.1.3	N. Bourbaki, Lie Groups and Lie Algebras. Chapters 4–6, Translated from the 1968 French original by Andrew Pressley. Elements of Mathematics (Berlin), Springer-Verlag, Berlin, 2002. xii+300 pp.	442
	9.1.4	More books on Lie algebras and Lie groups	443
	9.1.5	A. Borel, Linear Algebraic Groups, Second edition, Graduate Texts in Mathematics, 126, Springer-Verlag, New York, 1991. xii+288 pp.	445
	9.1.6	More books on algebraic groups, algebraic geometry and number theory	446

	9.1.7	Algebraic invariant theories and representations of algebraic groups	450
	9.1.8	E. Artin, Geometric Algebra, Reprint of the 1957 original, Wiley Classics Library, A Wiley-Interscience Publication, John Wiley & Sons, Inc., New York, 1988. x+214 pp.	452
	9.1.9	J. Tits, Buildings of Spherical Type and Finite BN-pairs, Lecture Notes in Mathematics, Vol. 386, Springer-Verlag, Berlin-New York, 1974. x+299 pp.	453
	9.1.10	More books on buildings and finite geometries	454
	9.1.11	Applications of Lie theories to differential equations	457
	9.1.12	Discrete subgroups of Lie groups and algebraic groups	463
	9.1.13	J.P. Serre, Trees, Translated from the French by John Stillwell, Springer-Verlag, Corrected 2nd printing of the 1980 English translation, Springer Monographs in Mathematics, Springer-Verlag, Berlin, 2003. x+142 pp.	466
	9.1.14	Discrete subgroups of low rank Lie groups and algebraic groups	467
	9.1.15	Combinatorial groups and geometric group theory	469
	9.1.16	Coxeter groups	474
	9.1.17	Transformation groups	475
	9.1.18	V. Kac, Infinite-dimensional Lie Algebras, Third edition, Cambridge University Press, Cambridge, 1990. xxii+400 pp.	479
	9.1.19	Loop groups, quantum groups, Hopf algebras and vertex operator algebras	479
	9.1.20	Applications of Lie groups in sciences	482
9.2	Representation theory		483
	9.2.1	J.P. Serre, Linear Representations of Finite Groups, Translated from the second French edition by Leonard L. Scott, Graduate Texts in Mathematics, Vol. 42, Springer-Verlag, New York-Heidelberg, 1977. x+170 pp.	484
	9.2.2	I.G. Macdonald, Symmetric Functions and Hall Polynomials, Oxford Mathematical Monographs, The Clarendon Press, Oxford University Press, New York, 1979. viii+180 pp.	484
	9.2.3	Representation theory of the symmetric group	485
	9.2.4	H. Weyl, The Classical Groups. Their Invariants and Representations, Fifteenth printing, Princeton Landmarks in Mathematics, Princeton University Press, Princeton, NJ, 1997. xiv+320 pp.	487
	9.2.5	Representation theories of Lie groups	489

10 Mathematical physics, dynamical systems and ergodic theory   **495**
    10.1 Classical mathematical physics 495

		10.1.1	R. Courant, D. Hilbert, Methods of Mathematical Physics. Vol. I., Vol. II., Interscience Publishers, Inc., New York, 1953. xv+561 pp., 1962. xxii+830 pp. 496

       10.1.1 R. Courant, D. Hilbert, Methods of Mathematical Physics. Vol. I., Vol. II., Interscience Publishers, Inc., New York, 1953. xv+561 pp., 1962. xxii+830 pp. 496

       10.1.2 H. Weyl, The Theory of Groups and Quantum Mechanics, from the 2d rev., German ed., by H. P. Robertson, Dover Publications, 1949. 448 pp. 497

       10.1.3 L.D. Landau, E.M. Lifshitz, Course of Theoretical Physics. Vol. 1. Mechanics, Third edition, Pergamon Press, Oxford-New York-Toronto, Ont., 1976. xxvii+169 pp. 499

10.2 More modern mathematical physics 503

       10.2.1 General relativity and gravitation 503

       10.2.2 S. Hawking, G. Ellis, The Large Scale Structure of Space-time, Cambridge Monographs on Mathematical Physics, No. 1, Cambridge University Press, London-New York, 1973. xi+391 pp. . . . 506

       10.2.3 Statistical mechanics 509

       10.2.4 Quantum field theory 511

       10.2.5 V.I. Arnold, Mathematical Methods of Classical Mechanics, Second edition, Graduate Texts in Mathematics, 60, Springer-Verlag, New York, 1989. xvi+508 pp. 512

       10.2.6 M. Reed, B. Simon, Methods of Modern Mathematical Physics. I. Functional Analysis. Second edition; II. Fourier Analysis, Self-adjointness; III. Methods of Modern Mathematical Physics; IV. Analysis of Operators, Academic Press, Inc., New York, 1980. xv+400 pp., 1975. xv+361 pp., 1979. xv+463 pp., 1978. xv+396 pp. 514

       10.2.7 Scattering theory 515

10.3 Dynamical systems 518

       10.3.1 Dynamics and celestial mechanics 518

       10.3.2 Dynamical systems 519

       10.3.3 Infinite-dimensional dynamical systems 526

       10.3.4 R. Thom, Structural Stability and Morphogenesis. An Outline of a General Theory of Models, Translated from the French by D. H. Fowler, With a foreword by C. H. Waddington, Advanced Book Classics, Addison-Wesley Publishing Company, Advanced Book Program, Redwood City, CA, 1989. xxxvi+348 pp. 529

       10.3.5 Functional-differential equations 530

       10.3.6 Complex dynamics 530

10.4 Ergodic theory 533

11 Discrete mathematics and combinatorics   537
11.1 Combinatorics 537
11.1.1 R. Stanley, Enumerative Combinatorics, Vol. I., Vol. 2., With a foreword by Gian-Carlo Rota, The Wadsworth & Brooks/Cole Mathematics Series, 1986. xiv+306 pp., 1999. xii+581 pp. 538
11.1.2 Polytopes, convex polytopes and geometric arrangements 539
11.1.3 Gröbner bases 541
11.1.4 Matroid theory 542
11.1.5 L. Lovász, Combinatorial Problems and Exercises, Corrected reprint of the 1993 second edition, AMS Chelsea Publishing, Providence, RI, 2007. 642 pp. 544
11.2 Discrete mathematics 545
11.2.1 J. Conway, N. Sloane, Sphere Packings, Lattices and Groups, Third edition, With additional contributions by E. Bannai, R. E. Borcherds, J. Leech, S. P. Norton, A. M. Odlyzko, R. A. Parker, L. Queen and B. B. Venkov, Grundlehren der Mathematischen Wissenschaften, 290. Springer-Verlag, New York, 1999. lxxiv+703 pp. 545
11.2.2 Graph theory 546
11.2.3 Graphs and their spectra 551
11.2.4 Random graphs 552

12 Probability and applications   553
12.1 Probability 553
12.1.1 A.N. Kolmogorov, Foundations of the Theory of Probability, Translation edited by Nathan Morrison, with an added bibliography by A. T. Bharucha-Reid, Chelsea Publishing Co., New York, 1956. viii+84 pp. Translation of Grundbegriffe der Wahrscheinlichkeits-rechnung, Springer, Berlin, 1933. 554
12.1.2 W. Feller, An Introduction to Probability Theory and its Applications, Vol. I, Vol. II. Third edition, John Wiley & Sons, Inc., New York-London-Sydney, 1968. xviii+509 pp., 1971, xxiv+669 pp. . . 555
12.1.3 More classical books on probability 556
12.1.4 More modern books on probability 558
12.1.5 Probability and analysis 559
12.1.6 More specialized books in probability 560
12.1.7 Random walks 562
12.2 Stochastic analysis 564

CONTENTS

- 12.2.1 J.L. Doob, Stochastic Processes, Reprint of the 1953 original, Wiley Classics Library, A Wiley-Interscience Publication, John Wiley & Sons, Inc., New York, 1990. viii+654 pp. 565
- 12.2.2 Brownian motions and stochastic processes 567
- 12.2.3 Stochastic calculus and equations 572
- 12.2.4 Large deviations 574
- 12.2.5 Malliavin calculus 575
- 12.3 Applications of probability 576
  - 12.3.1 Probabilistic methods and applications 576
  - 12.3.2 Random matrices 580

13 Foundations of math, computer science, numerical math — 583

- 13.1 Mathematical logic 583
  - 13.1.1 Mathematical logic 584
  - 13.1.2 D. Hofstadter, Gödel, Escher, Bach: an Eternal Golden Braid, Basic Books, Inc., Publishers, New York, 1979. xxi+777 pp. 585
  - 13.1.3 B. Russell, Introduction to Mathematical Philosophy, Reprint of the 1920 second edition, Dover Publications, Inc., New York, 1993. viii+208 pp. 586
  - 13.1.4 Set theory 586
  - 13.1.5 Model theory, non-standard analysis, recursive functions 591
- 13.2 Computer science 594
  - 13.2.1 D. Knuth, The Art of Computer Programming. Vol. 1-IV, Second printing, Addison-Wesley Publishing Co., 1969. xxi+634 pp. . . . 595
  - 13.2.2 R. Graham, D. Knuth, O. Patashnik, Concrete Mathematics. A Foundation for Computer Science, Second edition, Addison-Wesley Publishing Company, Reading, MA, 1994. xiv+657 pp. 595
  - 13.2.3 T. Cover, J. Thomas, Elements of Information Theory, Second edition, Wiley-Interscience [John Wiley & Sons], Hoboken, NJ, 2006. xxiv+748 pp. 596
  - 13.2.4 N. Wiener, Cybernetics, or Control and Communication in the Animal and the Machine, Actualités Sci. Ind., no. 1053, Hermann et Cie., Paris; The Technology Press, Cambridge, Mass.; John Wiley & Sons, Inc., New York, 1948. 194 pp. 597
  - 13.2.5 M. Petkovsek, H. Wilf, D. Zeilberger, $A = B$, With a foreword by Donald E. Knuth, With a separately available computer disk, A K Peters, Ltd., Wellesley, MA, 1996. xii+212 pp. 599

13.2.6 I. MacWilliams, N. Sloane, The Theory of Error-correcting Codes, I, II, North-Holland Mathematical Library, Vol. 16. North-Holland Publishing Co., Amsterdam-New York-Oxford, 1977. pp. i–xv and 1–369, pp. i–ix and 370–762. 600

13.2.7 More books on coding theory 601

13.2.8 Algorithm and automata 602

13.3 Game theory and optimization 604

13.3.1 J. von Neumann, O. Morgenstern, Theory of Games and Economic Behavior, Fourth printing of the 2004 sixtieth-anniversary edition, With an introduction by Harold W. Kuhn and an afterword by Ariel Rubinstein, Princeton University Press, Princeton, NJ, 2007. xxxii+739 pp. 605

13.3.2 More books on game theory and optimization 606

13.4 Numerical analysis and matrix computation 610

13.4.1 Numerical analysis 611

13.4.2 Finite element methods and finite difference methods 617

13.4.3 Approximation theory 620

Appendix A: Books in Chinese version — 623

Appendix B: Books reprinted in the mainland of China — 631

Index of books — 643

1. The Old Law Library of the University of Michigan

Chapter 1

Introduction

I have been interested in books for as long as I can remember, and during childhood, I used all my allowance on children's books. Before I went to college, I did not have access to any library and had no idea what a library really was or looked like. It was difficult for me to imagine a room full of books!

I have been studying mathematics for a long time and will probably spend the rest of my life doing so. In this lifetime adventure, reading books in mathematics is certainly important.

A natural question that occurred to me recently is what interesting or important mathematics books I could or I should have read.[1] To put it another way, if one is going

---

[1] Mathematics has developed to a very advanced degree, and there are many specialized subjects and topics in mathematics. But it is still fun and possible to learn and enjoy things from different areas in mathematics. Reading books is probably one of the most effective ways to learn from and communicate with people from the past and far away. It is perhaps helpful to point out that among all subjects, mathematics books are special and have played a more important role than in other fields due to the nature of mathematics. As Asimov wrote in the Foreword to the second edition of the book titled *A History of Mathematics* by Carl B. Boyer:

"Mathematics is a unique aspect of human thought, and its history differs in essence from all other histories.

As time goes on, nearly every field of human endeavor is marked by changes which can be considered as correction and/or extension. Thus, the changes in the evolving history of political and military events are always chaotic; ... But only among the sciences is there true progress; only there is the record one of continuous advance toward ever greater heights. And yet, among most branches of science, the process of progress is one of both correction and extension...

Now we can see what makes mathematics unique. Only in mathematics is there no significant correction—only extension. Once the Greeks had developed the deductive method, they were correct in what they did, correct for all time. Euclid was incomplete and his work has been extended enormously, but it has not had to be corrected. His theorems are, every one of them, valid to this day.

Ptolemy may have developed an erroneous picture of the planetary system, but the system of trigonom-

to live in a place for life, it is perhaps reasonable to explore neighboring places. As in real life, people often do not ask the obvious questions or do the obvious things.

Therefore, I have set out on a personal journey to find and make a list of great mathematics books which are read and used by working mathematicians.[2] The key word *great* is not well defined and could mean different things. In this book, by a *great* book, we mean a book that satisfies one or more of the following conditions:

1. It has introduced many people to some important topics or fields and has contributed to establishing the subject discussed in the book, i.e., it is a well-written expository book of some mainstream topics in mathematics. It may not have always been in use since publication, but at least, it has been influential and popular in the formative years of the subject under discussion. (Examples include some books by J.P. Serre, J. Milnor and B.L. van der Waerden.)

2. It is an important book that has introduced some new fundamental concepts, ideas and methods systematically for the first time and has continued to be used (or still remains as a reliable reference). Meanwhile it has made a huge impact, or it is the only book available about some important concepts, ideas or methods, and widely used or cited (though it might be difficult to read). (Examples include some books by A. Grothendieck and R. Langlands.)

3. It is an enjoyable book about some interesting themes, or is an accessible and beautiful exposition of important and well-known fields or topics. (Examples include some popular books by H. Weyl, I. Shafarevich, R. Courant and H. Robbins, and C. Reid.)

It should be stressed that these conditions are often not simultaneously satisfied by the great books selected here. Checking whether a book satisfies such conditions or not is also a very **subjective** process. Indeed, a difficult book for one person might be easy for another person, and an exciting book to some people might be completely dry and boring to others. Furthermore, it is also difficult to check how many people have actually read a book or parts of it.

Given the above criterions, the next problem is how to find great books. Since there are so many books on so many different topics, there is no easy solution. One natural approach is based on the personal experiences of studying and learning mathematics. Another is to browse through prestigious book series of major mathematics publishers, or

---

etry he worked out to help him with his calculations remains correct forever. Each great mathematician adds to what came previously, but nothing needs to be uprooted..."

[2]Maybe one can ask a simple and practical question: what essential books should any mathematics library have?

books on display at conferences or in libraries. The third way is to search on MathSciNet and Zentralblatt Math under subjects and titles, or search on Amazon under titles. Probably the best way is to ask experts. All these approaches have been followed in this book. Given the large number of published mathematics books, no list can really be exhaustive no matter how hard one tries.

Since many classical mathematics books are interesting mainly for historical reasons, we have tried to concentrate on more recent ones. (As mentioned in the Preface, we have inserted many pictures of ancient and classical mathematics books at the beginning of the book, and at the beginnings of chapters and sections). All the books selected here were published after 1900,[3] and many were published or reprinted after 1980, and most are published before 2000. Though the title is *Great Mathematics Books in the Twentieth Century*, some books published after 2000 are also included.

As a general guiding principle, books on more general and foundational topics are emphasized. Besides the highlighted books which appear in the headings of subsections, there are many other books. For example, in some cases, there are several important books on the same topic, and it is not clear which one stands out as a great book. Whenever this happens, we use the subject as the heading of a section or a subsection, and put these books together under that heading.[4]

Even in subsections with highlighted books, there are other books included for completeness. These books might also complement each other, since some great books might not treat some topics or aspects as well as some not-so-great books. For example, in the Section 2.2.2 on the biography *Hilbert* by C. Reid, we also discuss other biographies written by C. Reid. Another example is the Section 2.1.7 on Weyl's classic book on symmetry and related books.

Many great books might not be selected due to the lack of knowledge of the author. In other words, they might be blocked or not visible to the author since they are far away from the paths taken by the author or due to the short-sightedness of the author even though they should be clear and noticed. (As it happens in life, people sometimes cannot see the most obvious or visible thing staring at them.) We would like to apologize to those people whose great books are not selected here.

Besides the table of contents, one easy way to find books discussed in this book is to look in the index. Books are listed according to last name(s) of their author(s). There is also an index of books which have been translated into Chinese or reprinted in China.

---

[3]Some important books whose first edition appeared before 1900 are not listed. For example, the famous book by D. Hilbert, *Grundlagen der Geometrie (Foundations of Geometry)*, was first published in 1899, and it is not listed here. Another famous book *Lehrbuch der Algebra (Textbook on algebra)* by Heinrich Weber published in 1895 is not listed here either.

[4]One reason why some books are listed is that the subjects or topics they study seem interesting or important. Of course, there is no simple or fair criterion for this either.

For each book, or a group of selected books, we would like to describe some motivations, summarize some features or the contents of the book in order to give a taste of the book or the subject of the book, or some useful and interesting information about their authors and the level of the book. Since we cannot read all these books or make sensible comments on them, we often use some book reviews and comments from various sources. In other words, we often quote from experts.

It is reasonable to expect that reviews in MathSciNet, Zentralblatt Math, and the Bulletin of the American Mathematical Society or newsletters of other mathematical societies are more reliable and authoritative. After reading many reviews and descriptions of books, it seems that descriptions of books by publishers and summaries on back-covers are often quite accurate and informative. Therefore, all these reviews and descriptions are considered and those that contain valuable and concrete information are quoted.

II. The reading room of the Harlan Hatcher Graduate Library of the University of Michigan

Archimedes is famous for many things, and his life and death provide good topics for biographies. For school children, he is famous for the story of the golden crown, i.e., to decide if the crown was made of pure gold. He jumped out of the bathtub and run to the street, calling "Eureka!", when he found out Archimedes' principle. People usually interpret "Eureka!" as "I have found it!" On the other hand, according to a recent autobiographic article of Michael Gromov titled "A few reflection", "Eureka!" means "Give me a towel!" The story of his death is really sad, and according the famous Roman orator Cicero, the tomb of Archimedes was surmounted by a sphere inscribed within a cylinder, which is regarded as the greatest of his mathematical achievements.

Adolf Hurwitz was a student of Felix Klein and a close friend of David Hilbert. Probably his best known result concerns the automorphism group of Riemann surfaces. The biography *Hilbert* by Constance Reid described many mathematical walks in Gottingen he took with Hilbert and Minkowski, which played an important role in the mathematical careers in all three of them and are still inspiring to people now. Hurwitz was older than the other two and could be considered as their mentor. It might be fair to say that if there were no Hurwitz, the biography *Hilbert* might not be so inspiring or interesting.

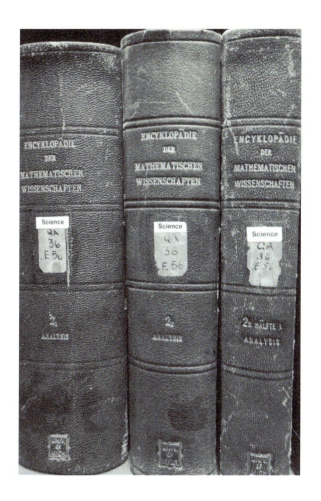

Felix Klein is probably most famous for two things: the Erlangen program and Kleinian groups. For the former, he did not do anything to substantiate it after proposing it, and for the second, he did not really discover them, and they should be called Poincare groups. The story of his academic life is inspiring. After he realized that he could not do research anymore after his competition with Poincare, he devoted his time to writing and lecturing. His last project was an all inclusive encyclopedia on mathematical sciences, often called Klein's encyclopedia.

Euler has been very famous since his life time. Recently, there is a surge of interests in his life and work. For example, this can be seen in recent popular books: *Euler: The Master of Us All*, *Eulers' Gem: The Polyhedron Formula and the Birth of Topology*, *Dr. Eulers' Fabulous Formula: Cures Many Mathematical Ills*, *The Genius of Euler: Reflections on his Life and Work*, *Gamma: Exploring Eulers' Constant*, etc. Maybe one reason is the celebration of the 300 birthday of Euler in 2007. He was not only productive in mathematics but also produced 13 children.

Chapter 2

Expository books on mathematics and mathematicians

Though everyone agrees that mathematics is important and mathematics should be a part of the education of every person, it is often believed that mathematics is difficult and is the subject for only selected few. It is often the case that people say proudly that mathematics is their worst subject in school and they are really bad at mathematics.

Given the rich structures of mathematics and logical or deductive nature of mathematics, it is not easy to write mathematics that is accessible and interesting to the general reader, or even to mathematicians who are not experts on the topic(s) under discussion. But it does not mean that it is impossible to do so. Mathematics and lives of mathematicians can provide excellent material and topics for great mathematics books and great stories.

Many people may not realize that mathematics is becoming more and more relevant to our life. Indeed, it is difficult to live in this world without using mathematics: think of GPS for driving directions, security for online shopping, the convenience of cell phones and readability of CAT scan, and improvement of weather forecasting, etc. Therefore, there are more reasons for people to understand mathematics and mathematicians now.

It is true that mathematics is an acquired or cultivated taste, and it takes time to learn all the proper terminology in order to understand mathematics and hence to appreciate its power and beauty. In this chapter, we take a look at some of the great books that try to explain mathematics and mathematicians to the general reader. They show that one can both learn and enjoy some essential mathematics by starting from the standard basics in mathematics.

Though there are many mathematicians, there are not too many biographies of them, and even fewer autobiographies. Each autobiography of mathematician is unique and interesting. We have selected some to show different kinds of stories about mathematicians.

The origin of mathematics came from counting and measuring, but mathematics has developed to a very high stage with many specialized subjects. It is often difficult for mathematicians to understand each other. In this chapter, we have also included several books that explain mathematics to the general mathematician through a historic perspective (or with more emphasis on history than standard mathematics books).

2.1 Popular and expository books on mathematics

2.1.1 R. Courant, H. Robbins, What Is Mathematics? Oxford University Press, New York, 1941. xix+521 pp.

The book *What Is Mathematics?* was first published in 1941 by Oxford University Press, revised in 1943, 1945, and 1947. A new edition with a new Chapter on Recent Developments by Ian Stewart was published in 1996.

Since its publication, it has been a best seller among mathematics and science books. Though it was written for the public, mathematicians can also read it with benefit. One strength is its broad coverage of topics (the new Chapter on recent developments is very valuable in this sense), and another is its accessibility. It can be recommended to people at all levels.

According to Herman Weyl, "It is a work of high perfection, whether judged by aesthetic, pedagogical or scientific standards. It is astonishing to what extent *What is Mathematics?* has succeeded in making clear by means of the simplest examples all the fundamental ideas and methods which we mathematicians consider the life blood of our science."

It is different from many popular books on mathematics by not avoiding discussion of mathematics. To explain the motivation and the purpose of this book, it is perhaps best to quote from the preface to the first edition:

"For more than two thousand years some familiarity with mathematics has been regarded as an indispensable part of the intellectual equipment of every cultured person. Today the traditional place of mathematics in education is in grave danger. Unfortunately, professional representatives of mathematics share in the responsibility. The teaching of mathematics has sometimes degenerated into empty drill in problem solving, which may develop formal ability but does not lead to real understanding or to greater intellectual independence. Mathematical research has shown a tendency toward overspecialization and overemphasis on abstraction. Applications and connections with other fields haven been neglected. On the other hand, the opposite reaction must and does arise from those who are aware of the value of intellectual discipline. Teachers, students, and the educated public demand constructive reform, not resignation along the line of

2.1. POPULAR AND EXPOSITORY BOOKS ON MATHEMATICS

least resistance. The goal is genuine comprehension of mathematics as an organic whole and as a basis for scientific thinking and acting.

Some splendid books on biography and history and some provocative popular writings have stimulated the latent general interest. But knowledge cannot be attained by indirect means alone. Understanding of mathematics cannot be transmitted by painless entertainment any more than education in music can be brought by the most brilliant journalism to those who have never listened intensively. Actual contact with the content of living mathematics is necessary. Nevertheless technicalities and detours should be avoided, and the presentation of mathematics should be just as free from emphasis on routine as from forbidding dogmatism which refuses to disclose motive or goal and which is an unfair obstacle to honest effort. It is possible to proceed on a straight road from the very elements to vantage points from which the substance and driving forces of modern mathematics can be surveyed.

The present book is an attempt in this direction..."

The problems raised and approaches proposed to solve them in the preface are still valid.

The simple question of "what is mathematics" is not easy to answer. It is true that mathematics deals with numbers, geometric shapes and relations between them. But it is much more than that. According to Courant and Robbins, "Mathematics as an expression of the human mind reflects the active will, the contemplative reason, and the desire for aesthetic perfection. Its basic elements are logic and intuition, analysis and construction, generality and individuality. Though different traditions may emphasize different aspects, it is only the interplay of these antithetic forces and the struggle for their synthesis that constitute the life, usefulness, and the supreme value of mathematical science."

Though the answer above is a bit abstract, after a four and half pages of introduction titled "What is mathematics?", this book gets into concrete and serious mathematics. The first two chapters on numbers say much about the style and structure of this book.

After introducing natural numbers and laws of arithmetic for them, it discusses the decimal system and other bases such as 7 by an explicit computation. Then the principle of mathematical induction is introduced and applied to compute the sum of arithmetic progression and geometric progression. The basic binomial theorem is also introduced. The supplement to Chapter 1 discusses factorization of integers into primes, and the problems of distribution of the primes, primes in arithmetic progressions are raised, and the prime number theorem is described. To show the reader that many basic problems in number theory are not solved, it raises two open problems: the Goldbach conjecture and the conjecture on existence of infinitely many twin primes. It also discusses a question of Fermat on finding simple arithmetic formulas that yield only primes.

A formula for primes is given in §1 of the new chapter *Recent developments*, and updates on the Goldbach conjecture and the twin prime conjecture are given in §2.

The book continues with the concept of congruence for integers, Fermat's theorem, quadratic residues, Pythagorean numbers, and the Euclidean algorithm and its applications to the fundamental theorem of arithmetic, continued fractions, and Diophantine equations.

The brief summary above shows how the book starts from basics and gets to the crucial results along the shortest path. The discussion is not technical yet quite accurate. The results on number theory in this book are more than enough for the education of a general mathematician who does not study number theory. It is enjoyable and informative to both the general public and mathematicians.

The second Chapter studies number systems. It starts with rational numbers. Geometric interpretation of rational numbers and the concept of commensurability leads to the notion of irrational numbers. Then the real and complex numbers are introduced. After algebraic numbers and transcendental numbers are defined, Liouville's theorem on existence and construction of transcendental numbers are discussed.

The contents of this book is summarized as follows: Chapter 3 deals with geometric constructions using ruler and compass, relations with number fields, and other geometric constructions such as mechanical curves and linkages. Chapter 4 studies projective geometry and non-Euclidean geometries, Chapter 5 is on topology, Chapter 6 is on limits and functions, Chapter 7 is on maxima and minima, and Chapter 8 is on calculus. The new Chapter 9 also includes updates on Fermat's last theorem, the continuum hypothesis, the Four color problem, Hausdorff dimension and fractals, Steiner's problem, soap films and minimal surfaces, and nonstandard analysis.

Mathematics described in this book gives a very concrete answer to the question "what is mathematics?"

H. Rademacher, O. Toeplitz, *The Enjoyment of Mathematics; Selections from Mathematics for the Amateur,* Princeton University Press, Princeton, 1957. 204 pp.

This is a probably less known popular book on mathematics without avoiding discussion of mathematics. The authors share a similar point of view with Courant and Robbins that real mathematics should be presented and can be made accessible, but the coverage seems to be less systematic.

As they explained in the preface, "It is the aim of these pages to show that the aversion towards mathematics banishes if only truly mathematical, essential ideas are presented. This book is intended to give samples of the diversified phenomena which comprise mathematics, of mathematics for its own sake, and of the intrinsic values which it possesses.

2.1. POPULAR AND EXPOSITORY BOOKS ON MATHEMATICS

The attempt to present mathematics to non-mathematicians has often been made, but this has usually been done by emphasizing the usefulness of mathematics in other fields of human endeavor in an effort to secure the comprehension and interest of the reader. Frequently the advantages which it offers in technological and other applications have been described and these advantages have been illustrated by numerous examples. On the other hand, many books have been written on mathematical games and pastimes. Although these books contain much interesting material, they give at best a very distorted picture of what mathematics really is. Finally, other books have discussed the foundations of mathematics with regard to their general philosophical validity. A reader of the following pages who is primarily interested in the pure, the absolute mathematics will naturally direct his attention toward just such an epistemological evaluation of mathematics. But this seems to us to be attaching an extraneous value to mathematics, to be judging its value according to measures outside itself... In other words, our presentation will emphasize not the facts as other sciences can disclose them to the outsider but the types of phenomena, the method of proposing problems, and the method of solving problems... We plan to select such 'smaller pieces' from the huge realm of mathematics: a sequence of subjects each one complete in itself, none requiring more than an hour to read and understand. The subjects are independent, so that one need not remember what has gone before when reading any chapter. Also, the reader is not required to remember what he may have been compelled to learn in his younger years..."

The "small pieces of mathematics" in this book are some of the essential parts of mathematics. Besides more standard topics such as existence of infinite primes, perfect numbers, the four-color problem, regular polyhedra, Pythagorean triples, periodic decimal fractions, Waring's problem, maximum and minimum problems, it also contains a discussion of a property of the integer 30 that all the numbers less than it and relatively prime to it are prime numbers. There is also a piece called "transverse nets of curves".

This book is enjoyable and instructive to both the public and the professional mathematicians.

According to an Amazon review, "This is a superb book, and the best way to first sample the delights of math for its own sake. Don't let the subtitle 'Selections from Mathematics for the Amateur' turn you off. True, the math involved is elementary, not going beyond high school algebra and geometry. But the material won't be found in most textbooks, having been chosen with elegance and beauty, rather than utility, in mind (which isn't to say that it doesn't have its important uses). Even the professional mathematician who has already seen most of the contents will profit from studying the book's exemplary treatment of its topics. The authors are some of the best math expositors who have ever lived. Each idea in the book is developed admirably, and strikes the perfect balance between conciseness and lucidity."

The next book on popular mathematics written by two experts contains real mathematics. It can be enjoyed by both mathematicians and non-mathematicians who are willing to think mathematically.

M. Kac, S. Ulam, *Mathematics and Logic: Retrospect and Prospects,* Dover Publications, Inc., New York, 1992. x+170 pp.

The motivation for writing this book is explained in the preface: "What is mathematics? How was it created and who were and are the people creating and practicing it? Can one describe its development and its role in the history of scientific thinking and can one predict its future? This book is an attempt to provide a few glimpses into the nature of such questions and the scope and depth of the subject."

The authors of this book are famous for many things. Among them, Kac is famous for writing an expository paper *Can you hear the shape of a drum?*, which initiated an important branch of modern differential geometry—spectral geometry, and Ulam is famous for being a co-inventor of the H-bomb. This is an inside guide to mathematics and doing mathematics. They show how simple questions and solutions can lead to more sophisticated problems and deep theories. They did not define what mathematics is but show the reader what is essential with many interesting examples. For example, starting from the existence of infinite primes, they moved to many questions about distributions of primes and formulas that gives prime numbers, approximation of irrational numbers by rational numbers, and transcendental numbers etc. The notion of groups is treated in a similar way. This is a fun and stimulating read.

According to MathSciNet, "This book is one of the 'Perspectives' made to commemorate the 200th anniversary of the Encyclopedia Britannica. It aims to give for intelligent nonspecialists some idea of the nature of mathematics and its role in human thought. It does this by surveying a sample of the more spectacular accomplishments of mathematics, some old and some quite new, and draws a general conclusion from discussion of the examples. It does not have as much to say about logic as its title would indicate..."

T. Körner, *The Pleasures of Counting*, Cambridge University Press, Cambridge, 1996. x+534 pp.

This more recent popular book on mathematics contains real mathematics. Although it was meant for good school students of age above 14, professional mathematicians can read it with benefit and maybe with surprise too. There are many not so common applications of mathematics. It is one strength of the book that very different kinds of applications are put together. From the book description, "The topics range from the design of anchors and the Battle of the Atlantic to the outbreak of cholera in Victorian Soho. The author uses relatively simple terms and ideas, yet explains difficulties and avoids condescension. If you are a mathematician who wants to explain to others how

2.1. POPULAR AND EXPOSITORY BOOKS ON MATHEMATICS

you spend your working days, then seek inspiration here. This book will appeal to everyone interested in the uses of mathematics."

According to MathSciNet, "This extremely enjoyable book aims to introduce the casual or 'mathematics for poets' reader to some interesting ways in which mathematics arises in the real world. Most of the book requires little beyond secondary school mathematics (and 'mathematical maturity'). There are five principal sections. In the first the author discusses Dr. John Snow's analysis showing the transmission of cholera, and some applications of operations research (not called that) in marine warfare during World Wars I and II... The second section contains discussions of measurement in biology (metabolic rate vs. size) and physics (the nature of dimensions), special relativity, and the work of L. F. Richardson (1881–1953) on numerical weather forecasting, war (and the beginnings of chaos theory). Section Three is concerned with algorithms and what makes for a good approximation. Section Four contains a discussion of coding, beginning with simple substitution codes and progressing through Enigma to Shannon's theorem. The final section is devoted to questions of genetics and some Greek mathematics (Euclid, infinity, etc.)... The book contains a large number of entertaining anecdotes and quotations, especially about military history..."

The appendix "Further reading" contains a list of "Some interesting books" and another of "Some hard but interesting books". The brief comments there are both entertaining and valuable.

T. Körner, *Fourier Analysis*, Second edition. Cambridge University Press, Cambridge, 1989. xii+591 pp.

This is not a standard text book on Fourier analysis. It can serve as a good supplement to a textbook. It consists of many short pieces which can more or less be read independently and is particularly enjoyable for browsing.

According to the introduction, "This book is meant neither as a drill book for the successful nor as a lifebelt for the unsuccessful student. Rather, it is intended as shop window for some of the ideas, techniques and elegant results of Fourier analysis. I have tried to write a series of interlinked essays accessible to a student with a good general background in mathematics such as an undergraduate at a British university is supposed to have after two years of study. If the reader has not covered the relevant topic, say contour integration or probability, then she can usually omit, or better, skim, through any chapters which involve this topic without impairing her ability to cope with subsequent chapters."

From the book description, the topics "... range from number theory, numerical analysis, control theory and statistics, to earth science, astronomy, and electrical engineering. Each application is placed in perspective with a short essay."

According to an Amazon review, "This book makes great reading. There is a fair amount of (well written) high level mathematics, but also a number of sections of a more historical or narrative nature, and a wonderful sense of humor pervades the work. The account of the laying of the transatlantic cable in the nineteenth century and the technical problems associated with it is priceless. Several sections are devoted to the life of Fourier."

J. von Neumann, *The Computer and the Brain*, Silliman Memorial Lectures, Vol. 36, Yale University Press, New Haven, Conn, 1958, xiv+82 pp.

In the early days, computer science was a part of mathematics. von Neumann made enormous contributions to several parts of mathematics, and also made crucial contribution to the design of the modern computer. He is often called the father of modern computer, because he proposed now the basic principle of computer design known as the "von Neumann architecture." Computers defined by this principle are the ancestors of the widely used desktop computers and laptop PCs. This book was the last book by von Neumann written for a public lecture series at Yale, which he could not deliver due to his untimely death. von Neumann has written many influential books. Though his last book was not properly finished, it still shines with brilliance.

From the book description, "This book represents the views of one of the greatest mathematicians of the twentieth century on the analogies between computing machines and the living human brain. John von Neumann concludes that the brain operates in part digitally, in part analogically, but uses a peculiar statistical language unlike that employed in the operation of man-made computers..."

According to MathSciNet, "In a language that any intelligent layman can understand, he speaks in the first part, concisely and precisely, of analog and digital computers; in the second, of neuro-anatomical and physiological data concerning nervous systems. Similarities and dissimilarities are brought forth vividly on the basis of quantitative estimates of speed, complexity and energy consumption in both; particularly stimulating is the analysis of the different logical and arithmetical depth of operations in the machine and in the brain, as related to their different structures..."

2.1.2 A.D. Aleksandrov, A.N. Kolmogorov, M.A. Lavrent'ev, Mathematics: Its Content, Methods, and Meaning. Vol. I, Vol. II, Vol. III, The M.I.T. Press, Cambridge, Mass., 1963, xi+359 pp.; xi+377 pp.; xi+356 pp.. Translated by S.H. Gould and T. Bartha; S.H. Gould; K. Hirsch.

Though this book was written for students and the public, it is also a good read for mathematicians, since it presents a global and essential picture of mathematics at an

2.1. POPULAR AND EXPOSITORY BOOKS ON MATHEMATICS

elementary level. As it is written in the introduction, "An adequate presentation of any science cannot consist of detailed information alone, however extensive. It must also provide a proper view of the essential nature of science as a whole... elementary mathematics and the history of the science already provide a sufficient foundation for general conclusions." The title of the book "Mathematics: Its content, methods, and meaning" gives a very good summary of its contents.

This unique book was created by a team of top experts at the height of the Soviet Union. It is difficult to imagine that another book of the same kind and quality can be produced again.

According to MathSciNet, it is "An example of popularization of the best kind. There are twenty chapters written by about the same number of outstanding Russian mathematicians. Algebra, geometry, analysis and topology are well represented and there are chapters on number theory, numerical analysis and computing machines."

Three authors A. D. Aleksandrov, A. N. Kolmogorov and M. A. Lavrent'ev are all very famous. Certainly, the great Kolmogorov needs no introduction. Aleksandrov is famous for his work in geometry. His approach to Riemannian geometry through length function can be applied to more general metric spaces, and he is one of the people whose names appear in the important class of CAT(0)-spaces. Among his Ph. D. students, the last one is probably the most famous: Grigori Perelman, who in 2002 made a breakthrough and completed the proof of Thurston's geometrization conjecture, which contains the Poincaré conjecture as a special case.

Lavrent'ev is one of the creators of the theory of incorrect problems of mathematical physics. He also produced a number of important results on multidimensional inverse problems for differential equations that arise in the study of the structure of the earth on the basis of observed geophysical fields.

This book consists of a superb series of essays on major areas of mathematics by top experts. As explained in the preface, "the present book will give an idea of the present state of mathematics, its origins, and its probable future development."

These essays provide a quick but serious introduction to each area. They provide motivations and point out connections between different areas of mathematics. It is not so formal as research monographs, and ties various concepts to geometry, provides the reader with an intuitive view of the methods and the meaning of basic results of modern mathematical. For example, the essay on probability by Kolmogorov started out by comparing two pieces of information and asking which is more valuable: (1) the life of a lamp is between 500 hours and 10,000 hours, (2) in approximately 80% of the cases, the life of a lamp is at least 2,000 hours.

Therefore, it is valuable and helpful to people who are reading it to learn a subject for the first time or to review some concepts. Though it is 50 over years old, most of the topics discussed is still central to the modern mathematics. There are recent books of related nature. One important example is the next book.

Timothy Gowers, June Barrow-Green and Imre Leader (Ed.), *The Princeton Companion to Mathematics,* Princeton University Press, Princeton, N.J., 2008. xxii+1034 pp.

It covers more advanced mathematics than the previous books. As the title suggests, it is a companion and one can pick up this book and open it and read some selected pages for pleasure or for information. Though it is comprehensive both in terms of contents and format/style, it does not cover all important subjects in mathematics or with depth. But one can gain a good picture of many important topics in modern mathematics by reading this unique book.

From the book description, "it presents nearly two hundred entries, written especially for this book by some of the world's leading mathematicians, that introduce basic mathematical tools and vocabulary; trace the development of modern mathematics; explain essential terms and concepts; examine core ideas in major areas of mathematics; describe the achievements of scores of famous mathematicians; explore the impact of mathematics on other disciplines such as biology, finance, and music–and much, much more."

The purpose of this book is explained well by two passages from its preface: "[The companion] simply aims to present for the reader a large and representative sample of the ideas that mathematicians are grappling with at the beginning of the twenty-first century." "The companion is not an encyclopedia: ... the book is like a human companion, complete with gaps in its knowledge and views on some topics that may not be universally shared."

According to MathSciNet, "The Princeton companion to mathematics is a unique text, which does not fall neatly into any of the usual categories of mathematical writing. It is not quite a mathematical encyclopedia, it is not quite a collection of mathematical surveys, it is not quite a popular introduction to mathematics, and it is certainly not a mathematics textbook; and yet it is still an immensely rich and valuable reference work that covers almost all aspects of modern mathematics today (although there is certainly an emphasis on pure mathematics at the research level). An encyclopedia might focus primarily on definitions, a survey article might focus on history or on the latest research, and a popular introduction might focus on analogies, personalities or entertaining narrative; in contrast, this book is intended to answer (or at least address) basic questions about mathematics, such as 'What is arithmetic geometry?', 'Why do we care about function spaces?', 'How is mathematics used today in biology?', 'What is the significance of the Riemann hypothesis?', 'Why are there so many number systems?', or 'Is mathematical research all about proving theorems rigorously?' Even at over 1000 pages, there is certainly not enough space in the book to answer these sorts of questions

2.1. POPULAR AND EXPOSITORY BOOKS ON MATHEMATICS 15

in full detail, but there is usually enough to get a glimpse of the 'big picture' that one usually cannot obtain otherwise, except by long experience and by conversation with other mathematicians."

According to the book review of Bulletin of AMS, "Timothy Gowers and his associate editors have aimed to give an account of as much of mathematics as can reasonably be made accessible; in particular, students at school should be helped to understand what mathematics is about, intending graduate students should be helped to decide what topics to research in, and established mathematicians should be helped to understand what their colleagues are doing."

Given the diversity and broadness of modern mathematics, it is almost impossible to cover all major topics and subjects in mathematics in a single volume. According to the review in Bulletin of AMS, "It is easy to complain about what is not covered in the book, although such criticism is largely deflected by the not-an-encyclopedia quote above. There is very little on differential geometry. I was hoping to find a broad discussion of the influence of cohomology in various guises (surely one of the main developments of the twentieth century), but was disappointed. It would have been interesting and topical to see more on quantum field theory, as a notable idea 'that mathematicians are grappling with at the beginning of the twenty-first century' (although there is some coverage of this under the headings 'Mirror symmetry' and 'Vertex operator algebras')... Overall this book is an enormous achievement for which the authors deserve to be thanked. It contains a wealth of material, much of a kind one would not find elsewhere, and can be enjoyed by readers with many different backgrounds."

2.1.3 G. Pólya, How to Solve it. A New Aspect of Mathematical Method. Expanded version of the 1988 edition, with a new foreword by John H. Conway, Princeton Science Library, Princeton University Press, 2004. xxviii+253 pp.

This book was first published in 1945. It was expanded in 1948 and has been reprinted many times. It has never been out of print and is a charming book. Though many suggestions seem like common sense and people are subconsciously following them, a high percentage of mathematicians and mathematics students probably have not thought of them so systematically. Reading it is both fun and enlightening, especially for the first time. It helps one to learn how to solve problems and to do research in a systematic way and might serve as a guide, but actually solving problems and doing it is the best and probably the only way to really learn it.

According to Pólya, "Solving problems is a practical art, like swimming, or skiing, or playing the piano: You can learn it only by imitation and practice. This book cannot offer you a magic key that opens all the doors and solves all the problems, but it offers you good examples for imitation and many opportunities for practice: If you wish to learn swimming you have to go into the water and if you wish to become a problem solver you have to solve problems."

The review of the 1948 edition in MathSciNet by H. Weyl describes the book well: "Searching for answers to problems that arise in the conduct of our lives or arouse our curiosity is a constant occupation of our minds. Philosophers and mathematicians have tried to devise rules for the direction of mind in this its effort, as witnessed by Pappus's report on the ancient art of heuristics, by Descartes's and Leibniz's attempts, and those of many minor lights. Indeed, if there are rules for truth (logic), one may expect that there are also rules for finding the truth. How much thought the author has given to the ways and means by which the problem-solving capacity of a mathematician in statu nascendi [in the state of being born] may be trained is evidenced by the preface and the two volumes of Pólya and Szegő [Aufgaben und Lehrsätze aus der Analysis, Springer, Berlin, 1925]. The present book, a sort of elementary textbook on heuristic reasoning, shows anew how keen its author is on questions of method and the formulation of methodological principles. His standpoint here is chiefly pedagogical; heuristics is conceived as a half-logical, half-psychological affair. Exposition and illustrative material are of a disarmingly elementary character, but very carefully thought out and selected.

A one-page, concise list of questions and suggestions about the four phases of Understanding the problem, Devising a plan, Carrying out the plan, Looking back, opens and ends the book. Its main part is in the form of a dictionary. Analysis, auxiliary elements, generalization, induction and mathematical induction, are among the items discussed, but there are many less obvious ones. Here are a few of the author's characteristic aphorisms. 'It is foolish to answer a question that you do not understand. It is sad to work for an end that you do not desire.' 'Good ideas are based on past experience and formerly acquired knowledge.' 'Rules of style: The first rule of style is to have something to say. The second rule of style is to control yourself when, by chance, you have two things to say; say first one, then the other, not both at the same time.' 'The first rule of teaching is to know what you are supposed to teach. The second rule of teaching is to know a little more than you are supposed to teach.' "

An expanded version is enhanced with a foreword by John H. Conway, where he writes: "How to solve it is a wonderful book! This I realized when I first read right through it as a student many years ago, but it has taken me a long time to appreciate just how wonderful it is. Why is that? One part of the answer is that the book is unique. In all my years as a student and teacher, I have never seen another that lives up to George Pólya's title by teaching you how to go about solving problems."

G. Pólya, *Mathematics and Plausible Reasoning, Vol. I. Induction and Analogy in Mathematics, Vol. II. Patterns of Plausible Inference,* Reprint of the 1954 original, Princeton University Press, Princeton, NJ, 1990. xvi+280 pp., x+225 pp.

This is a sequel to "How to solve it". It is known that rigorous proof and demonstrating reasoning is an essential part or a unique feature of mathematics. One point of this book is to show that nonrigorous plausible reasoning is at least equally important in creating mathematics. Questions are raised and conjectures and solutions are suggested by plausible reasoning first. Probably most research mathematicians will agree with these points.

According to MathSciNet, "A great number of interesting examples from various parts of mathematics are discussed in order to point out the way which leads from a problem to its solution and, more generally, the manner in which mathematical research should be conducted; in connection with these examples, a number of 'loci', which more or less vaguely shows the right direction (such as: generalisation, specialisation, analogy), and of devices (mathematical tools which can be applied in a straightforward fashion: mathematical induction, generating functions) are explained."

According to an Amazon review, "Analogies are frequently the key to a discovery, but it is rare that this essential step receives credit. Here there is a collection of them: some of the most beautiful. Perhaps the most famous is Bernoulli's solution of the brachistochrone problem, based on an analogy with the path of light in the atmosphere. But there are many others, with comments and analysis by Pólya, who spent a life thinking at these things..."

These two books by Pólya were followed by two less known books. They are also of similar nature. If you like the earlier ones, you will like them too.

G. Pólya, *Mathematical Discovery. On Understanding, Learning, and Teaching Problem Solving. Vol. I, Vol. II*, John Wiley & Sons, Inc., New York-London, 1962, xv+216 pp., 1965 xxii+191 pp.

According to MathSciNet, "These volumes continue the author's study of heuristic. The first part (a subset of Vol. I) discusses four important patterns of problem solving and reinforces the discussion with a wealth of problems for the reader (solutions are given at the back of the book). Part II looks at the principles of discovering solutions in a more general and philosophical way (still with many problems for the reader), and ends with chapters 'On learning, teaching, and learning teaching' and 'Guessing and scientific method'... Suffice it to say that the whole is presented with the author's usual charm, cogency, and distaste for pedantry, and can be read with profit by any mathematician; at the same time, it should be taken to heart by everyone concerned with the training of high school mathematics teachers (the improvement of this training being one of the author's avowed aims)."

There is a related book of Pólya with G. Siegö, *Problems and theorems in analysis* discussed in §3.1.9, where many of the ideas on how to solve problems are demonstrated in many well-arranged problems and theorems. The earlier books are more for general mathematics students and teachers, this book is challenging and is for serious students and mathematicians.

Solving good problems is an important part of understanding mathematics, and unexpected solutions also shows the beauty of mathematics. Several books with more challenging mathematics problems will be mentioned later in §3.1.9 and §3.1.11. Here is a more elementary and accessible one.

V. Klee, S. Wagon, *Old and New Unsolved Problems in Plane Geometry and Number Theory*, The Dolciani Mathematical Expositions, 11, Mathematical Association of America, Washington, DC, 1991. xvi+333 pp.

The vitality of a subject can be measured by abundance of open problems. To people outside of mathematics, it might seem most parts of mathematics are understood and those which are still open should be very abstract and far away from our intuition. This book tries to show that the opposite is true. Most people are attracted by the beauty and rigor of plane geometry at a young age, and arithmetic of integers (or basic number theory) is also familiar to most people. This book presents many open problems related to plane geometry and basic number theory, which are easy to state and understand, but difficult to solve. For example, the Fermat last theorem is one of them.

The book consists of 24 principal problems and is arranged into three chapters. The first chapter contains 12 problems from plane geometry; the second contains eight problems from number theory; and the third chapter titled "Interesting real numbers" contains four additional problems.

It is not a standard book of problems. This can be seen from formulation of some of the problems. For example, the first one asks *whether each reflecting polygonal region illuminable?* The fourth asks *if a convex body C contains a translate of each plane set of unit diameter, how small can C's area be?*

According to MathSciNet, "Each of the principal problems within a given chapter is discussed in two parts. Each of the problems is presented initially in its historical and mathematical context, together with a discussion of the problem itself and, in most cases, other variants of the problem. The presentation in the first part is from a more intuitive or elementary point of view. The second part contains details at a deeper level about the problems and their variants. In some cases proofs of some of the variants are included. Each of the chapters includes an excellent set of references. Finally, there are several short sections of hints and solutions for each of the three chapters, a glossary, index of names, and a subject index..."

2.1. POPULAR AND EXPOSITORY BOOKS ON MATHEMATICS

2.1.4 G.H. Hardy, A Mathematician's Apology, With a foreword by C.P. Snow, Reprint of the 1967 edition, Canto, Cambridge University Press, Cambridge, 1992.

It was first published in 1940 and reissued with a new edition in 1967. It has been reprinted many times. This is a famous and unique book. Since it is famous to the public and is often quoted in discussion of usefulness and beauty of mathematics, it is worthwhile to read it. In any case, it is a fun read.

As the title of the book suggests, this is an apology (in the sense of a formal justification or defense) of an old established mathematician for his mathematical life, after a glorious and productive life. Instead of a happy and proud book, it is a book of "haunting sadness". Some quotes from the book will show this well. The book starts with:

"It is a melancholy experience for a professional mathematician to find himself writing about mathematics. The function of a mathematician is to do something, to prove new theorems, to add to mathematics, and not to talk about what he or other mathematicians have done. Statesmen despise publicists, painters despise art-critics, and physiologists, physicists, or mathematicians have usually similar feelings: there is no scorn more profound, or on the whole more justifiable, than that of the men who make for the men who explain. Exposition, criticism, appreciation, is work for second-rate minds."

"What is the proper justification of a mathematician's life? My answers will be, for the most part, such as are expected from a mathematician: I think that it is worthwhile, that there is ample justification. But I should say at once that my defense of mathematics will be a defense of myself, and that my apology is bound to be to some extent egotistical. I should not think it worth while to apologize for my subject if I regarded myself as one of its failures. Some egotism of this sort is inevitable, and I do not feel that it really needs justification. Good work is no done by 'humble' men. It is one of the first duties of a professor, for example, in any subject, to exaggerate a little both the importance of his subject and his own importance in it. A man who is always asking 'Is what I do worth while?' and 'Am I the right person to do it?' will always be ineffective himself and a discouragement to others. He must shut his eyes a little and think a little more of his subject and himself than they deserve. This is not too difficult: it is harder not to make his subject and himself ridiculous by shutting his eyes too tightly."

Here is another quote from the book. "I had better say something here about this question of age, since it is particularly important for mathematicians. No mathematician should ever allow himself to forget that mathematics, more than any other art or science, is a young man's game. To take a simple illustration at a comparatively humble level, the average age of election to the Royal Society is lowest in mathematics. We can naturally find much more striking illustrations. We may consider, for example, the career of a man who was certainly one of the world's three greatest mathematicians. Newton gave up mathematics at fifty, and had lost his enthusiasm long before; he had recognized no doubt by the time he was forty that his greatest creative days were over. His greatest idea of all, fluxions and the law of gravitation, came to him about 1666, when he was twenty four – 'in those days I was in the prime of my age for invention, and minded mathematics and philosophy more than at any time since'. He made big discoveries until he was nearly forty (the 'elliptic orbit' at thirty-seven), but after that he did little but polish and perfect.

Galois died at twenty-one, Abel at twenty-seven, Ramanujan at thirty-three, Riemann at forty. There have been men who have done great work a good deal later; Gauss's great memoir on differential geometry was published when he was fifty (though he had had the fundamental ideas ten years before). I do not know an instance of a major mathematical advance initiated by a man past fifty. If a man of mature age loses interest in and abandons mathematics, the loss is not likely to be very serious either for mathematics or for himself."

Another famous quote from the book is "Beauty is the first test: there is no permanent place in this world for ugly mathematics."

There have been many reviews of this book, and it might be interesting to take a look at some of them.

The MathSciNet review by Pólya of the first edition consists one short quote from the book. A review of the book in National Mathematics Magazine 16 (1942), no. 6, 311, is more detailed, and was reprinted in Mathematics Magazine 41 (1968), no. 3, 155–156. A much longer and serious review by Mondell appeared in Amer. Math. Monthly 77 (1970), no. 8, 831–836, and shed interesting light on this book and some opinions of Hardy.

A review from Amazon summaries this book well: "Every discipline has a list of items that must be read if one is to be considered educated in that field. There is no doubt that this book should be required reading for any degree in mathematics. Most of the soul of mathematics is contained in the 91 pages of the 'Apology' (the first 58 pages consists of the foreword by Snow). Written in his later years when Hardy knew his mathematical powers were failing, this is a superb exposition by a brilliant, eccentric personality. He not only captures the grandeur of mathematical discovery, but also clearly articulates the feelings of a man who knows that his time has passed. First published in 1940, the twin messages are timeless. Clearly distinguishing between the real mathematician and the puzzle solver, Hardy is exceptional in declaring what the real beauty of mathematics is. Among all the beautiful things that exist, the percentage of individuals that can truly appreciate an elegant theorem is among the smallest. However, anyone who can read this work and not see at least some of the poetic qualities of mathematics has a blind

2.1. POPULAR AND EXPOSITORY BOOKS ON MATHEMATICS

spot in their soul. One of the masterpieces of literature, this book can be understood and appreciated by anyone with an eye for the beautiful things that life has to offer."

2.1.5 J.E. Littlewood, Littlewood's Miscellany, Edited and with a foreword by Béla Bollobás, Cambridge University Press, Cambridge, 1986. vi+200 pp.

It was originally published as "A mathematician's miscellany" in 1953. It has been popular with both mathematicians and the public. "Every line will delight lovers of mathematics, but even readers who have never had a mathematical thought will – if they love evidence of humane intelligence and wise wit – find here much to savor. A very special classic."

Some sentences from this book have become standard quotes. One such example is "A good mathematical joke is better, and better mathematics, than a dozen mediocre papers." Since this book also contains a lot of genuine mathematics explained in a non-standard way, this book is a "must-read" for mathematicians.

According to the original introduction, it "is a collection without a natural ordering relation... Anyone open to the idea of looking through a popular book on mathematics should be able to get on with this one... The book contains pieces of technical mathematics, on occasion pieces that only a professional mathematician can follow; these have been included as contributing to the full picture of the moment as viewed by the professional, but they can all be skipped without prejudice to the rest, and a coherent story will remain... The qualities I have aimed at in selecting material are two. First relative unfamiliarity, even to some mathematicians. This is why some things receive only bare mention... The other quality is lightness, notwithstanding the highbrow pieces: my aim is entertainment and there will be no uplifting..."

For the new expanded edition, "curiosities, howlers, strange anecdotes and various recollections of life at Trinity College" and other materials were added.

In some sense, this is the wisdom of an old, established man which he would like to pass to the new generation.

Besides his own work, Littlewood is also famous for his legendary collaboration with Hardy. There was a saying that at one point in time, there were only three great English mathematicians: Hardy, Littlewood, and Hardy-Littlewood.

The book "A mathematician's apology" can also be looked upon as the collected wisdom of Hardy which he liked to pass to the future generation. Both books have been successful and are quite different in contents and style. It is interesting to compare these two books. An interesting question is what kind of book will the mathematician Hardy-Littlewood write.

2.1.6 Autobiographies of mathematicians

The two books by Hardy and Littlewood in the previous subsections contain autobiographic information of them in some sense. There are not too many autobiographies by mathematicians. Unlike scientists, mathematicians do not do enough public outreach and self-promotion. We mention several autobiographies here for completeness.

A. Weil, *The Apprenticeship of a Mathematician*, Translated from the 1991 French original by Jennifer Gage, Birkhäuser Verlag, Basel, 1992. 197 pp.

Weil is probably a mathematician who needs no introduction. He is both a deep and broad mathematician. According to EMS Newsletter, this book is "Extremely readable... rare testimony of a period of the history of 20th century mathematics. Includes very interesting recollections on the author's participation in the formation of the Bourbaki Group, tells of his meetings and conversations with leading mathematicians, reflects his views on mathematics. The book describes an extraordinary career of an exceptional man and mathematicians. Strongly recommended to specialists as well as to the general public." His description of his month long walk through the Alps and of his participation in the Hadamard seminar are some of the many unforgettable descriptions in this book.

P. Halmos, *I Want to Be a Mathematician, An Automathography in Three Parts*, MAA Spectrum, Mathematical Association of America, Washington, DC, 1985. xvi+421 pp.

Halmos is probably better known for his books and advice on writing than his original contribution to mathematics. His booklet on how to write mathematics is a classic.

Students and young mathematicians can learn many things about a career in mathematics from this book. Indeed, as the book description says, it "is an account of the author's life as a mathematician. It tells us what it is like to be a mathematician and to do mathematics. It will be read with interest and enjoyment by those in mathematics and by those who might want to know what mathematicians and mathematical careers are like."

This is a very interesting book to read for mathematicians both young and old. As Halmos said in his autobiography that he might not be a first-rated great mathematician, but certainly he is a great and engaging writer and a good story teller.

According to an Amazon review, "I do not think my words of praise would do justice to this wonderful book. Halmos has strong opinions almost about everything and the way he talks about his examples are very wise. You do not need to be a would-be mathematician to enjoy the book. If you have ever wondered or invested some time in the world of mathematics, science and academia, Halmos provides you a very good account. If you are more than interested in math or maybe thinking about pursuing a Ph.D. this book will be much more valuable for you... All along the book I had a feeling:

2.1. POPULAR AND EXPOSITORY BOOKS ON MATHEMATICS

it was more like a frank and witty dialogue between me and the great mathematician (and lecturer) who had been there and done that. I kept on asking questions and Halmos kept on giving answers..."

S.M. Ulam, *Adventures of a Mathematician,* Charles Scribner's Sons, New York, 1976. xi+317 pp.

Though Ulam made important contributions to mathematics, he is much more famous for his work on the H-bomb. He is certainly not an ordinary mathematician or scientist, and his life is also an unusual one. For many years after reading this book, his description of his life as a young mathematician at Harvard had remained fresh in my mind.

From the book description, "The autobiography of mathematician Stanislaw Ulam, one of the great scientific minds of the twentieth century, tells a story rich with amazingly prophetic speculations and peppered with lively anecdotes. As a member of the Los Alamos National Laboratory from 1944 on, Ulam helped to precipitate some of the most dramatic changes of the postwar world. He was among the first to use and advocate computers for scientific research, originated ideas for the nuclear propulsion of space vehicles, and made fundamental contributions to many of today's most challenging mathematical projects.

With his wide-ranging interests, Ulam never emphasized the importance of his contributions to the research that resulted in the hydrogen bomb. Now Daniel Hirsch and William Mathews reveal the true story of Ulam's pivotal role in the making of the "Super," in their historical introduction to this behind-the-scenes look at the minds and ideas that ushered in the nuclear age. An epilogue by Fransoise Ulam and Jan Mycielski sheds new light on Ulam's character and mathematical originality."

This book is mainly about Ulam's encounter with other scientists with great details about Ulam's three important friends: Banach, Von Neumann and Fermi. It is a must-have for all historians of science.

Ulam was a pure mathematician and had the ability to escape from formal abstract considerations to think about how other sciences could show him a path to new mathematical considerations.

For that matter, this book is of the greatest interest for he who wish to deepen his understanding of links between mathematics and physics, that are usually discussed by physicists who often have very poor idea of what mathematics is really about. The chapter "random reflections" is a jewel, explaining for instance how practical problems can lead to new mathematical concepts, how mathematic theories link altogether, or advocating the use of computers to help mathematicians view new spaces of new objects.

N. Wiener, *Ex-prodigy. My Childhood and Youth,* Simon and Schuster, New York, 1953. xii+309 pp. *I am a Mathematician. The later Life of a Prodigy,* Doubleday and Co., Garden City, N. Y., 1956. 380 pp.

Many mathematicians do their best work and become famous at a young age, but some do so at a younger age than others. It is also known or rather conceived by many that mathematicians are sort of special and have an unusual life, but some might have a more than unusual life. Both are probably true with Wiener. Wiener is a well-known name in mathematics and engineering, but the story behind his education, academic success and academic life may not be so well known. If the principle that knowing one's life helps one understand his work better is true, then it is valuable to read these two books by a genius.

According to an Amazon review, "This is the story of the childhood and youth of a genius. Norbert Wiener who would go on to become an important mathematician and one of the principal developers of communication theory, and his own specific discipline 'cybernetics' tells here the story of his most unusual childhood and youth. At the center is his relation to his father Leo Wiener who was a Professor of Slavic Languages, and an extraordinarily ambitious person. He pushed his son from an early age in much the same way that John Mill pushed John Stuart Mill. In the process he was often cruel. 'He would begin the discussion in an easy, conversational tone. This lasted exactly until I made the first mathematical mistake. Then the gentle and loving father was replaced by the avenger of the blood... Father was raging, I was weeping, and my mother did her best to defend me, although hers was a losing battle.' The father pushed Wiener so well that he enrolled in Tufts University at the age of eleven finished four years later with a degree in Mathematics. The father was also a publicity - hound who publicized his son the genius, and claimed it had nothing to do with any genetic quality or special gift of his son, but rather was solely attributable to his own educational methods. The father too saw to it that the son had a family life, and selected one of his students to be his son's wife, and practical daily life manager. Wiener despite all this went on to become a distinguished MIT professor of Mathematics, an original genius and a highly respected teacher..."

G. Shimura, *The Map of My Life,* Springer, New York, 2008. vi+211 pp.

Shimura is probably best known for Shimura varieties, which are fundamental in the arithmetic algebraic geometry and the Langlands program. He has strong opinions and unique points of views about mathematics and mathematicians, and these are reflected in this book. Some people might be turned away by some strong comments, but it is an interesting and historical read due to the fame and importance of the Taniyama-Shimura-Weil conjecture on elliptic curves.

From the book description, "In this book, the author writes freely and often humorously about his life, beginning with his earliest childhood days. He describes his survival of American bombing raids when he was a teenager in Japan, his emergence as a researcher in a post-war university system that was seriously deficient, and his life as a mature mathematician in Princeton and in the international academic community. Every page of this memoir contains personal observations and striking stories. Such luminaries as Chevalley, Oppenheimer, Siegel, and Weil figure prominently in its anecdotes."

According to MathSciNet, "This volume is an exciting autobiography of Goro Shimura (born 1930). The author relates not only his life and his significant mathematical achievements but also many aspects outside mathematics... The volume contains many personal feelings of the author towards the reaction of the mathematical community with respect to his work. On page 137, for example, he states, 'I was often the target of jealousy of other mathematicians, which I found strange.' He also relates some of his personal stories and feelings. This book is certainly exciting for any reader interested in mathematics or Japanese civilization."

It might be interesting to quote from the book about his motivation for writing this book: "... the purpose of this book is manifold, and I include many things that may not strictly be called my recollections. Therefore I need to explain why I write them. First of all, I want to write not only about myself but also about my time, or rather, the atmosphere of the time. Thus I include many things that are completely ordinary and known to almost everyone of my generation. I do so because they are so ordinary, they will never be written, and as a consequence, will be forgotten.

Next, I have included my opinions on various historical events and thoughts about human nature, most of which are what I wanted to say at least once, and for which most likely I would not find a platform elsewhere. Such opinions can be found in many autobiographies, and more of them appear in this book."

M. Kac, *Enigmas of Chance. An Autobiography,* Alfred P. Sloan Foundation, Harper & Row, Publishers, New York, 1985. xxviii+163 pp.

Though the Feymann-Kac formula has played an important role in the study of parabolic equations, Kac is probably better known for writing a famous paper titled "Can one hear the shape of a drum?" This paper opened up the subject of spectral geometry, which studies relations between the geometry and the spectral theory of the Laplacian of a Riemannian manifold.

This is an unusual autobiography and contains some genuine mathematics which are well explained. This book was clearly well-thought out and executed. The long introduction about the nature of autobiographies and his analysis of three autobiographies is fascinating. The fact that he is one of the founders of probability theory makes this interesting book also informative and instructive.

According to MathSciNet, "This book is part of the Alfred P. Sloan Foundation series of memoirs by eminent scientists; one of the series' purposes is to increase understanding of science by intelligent non-scientists. Mark Kac (1914–1984) succeeds very well in giving a good idea of what mathematicians do and who they are. In particular, he has written remarkably clear and attractive descriptions of each of his major discoveries... For the professional mathematician this autobiography is very much worth reading – it gives insight into the characters of many remarkable mathematicians, and it contains an orderly description of the events of Kac's life, including clear descriptions of his most important mathematical works. This book also includes an unusual and valuable section: To prepare for writing the book, Kac reviewed the autobiographies of Erwin Chargaff, Stanislaw Ulam, and Freeman J. Dyson; these reviews constitute the introduction."

According to a review in Bulletin of AMS, "The charm of Enigmas of chance cannot be even hinted at by surveying its contents. There is in it the spirit of a warm human being possessed by driving curiosity, by an urge to understand and clarify. There is an account of going through a stormy period in history, with personal tragedies and times of happiness. And there is the picture of a mathematician who, instead of clinging to mathematics as an abstract game, treated it as a bridge to reality; a mathematician who, as quoted by Gian-Carlo Rota, warned that 'axioms will change with the whims of time, but an application is forever'. To Kac the problem often was the reason for the theory; he admitted that 'almost everything new in mathematics I learned after getting my doctoral degree has been by being forced to learn it in trying to solve a problem.'

In the introduction to his book, Kac expressed the hope to be able to impart to the reader some feeling for the thrill that comes with getting a new idea, as well as for the frustrations and disappointments in the life of a scientist. He did it with charm and grace. He also succeeded in carrying out his other wish: the book gives a moving account of a rich life, and the way it was shaped by family, teachers, collaborators, history, and last but not least, by 'that powerful but capricious lady Chance.' "

W. Rudin, *The Way I Remember It,* History of Mathematics, 12, American Mathematical Society, Providence, RI; London Mathematical Society, London, 1997. x+191 pp.

Rudin is probably best known for his textbooks on analysis, and the first book was written when he was a new postdoctor at MIT. Probably every one of the recent generations of mathematicians has read or used one of his many excellent books in one way or other. Reading this book gives one an opportunity to meet the man behind them.

As in the case of the famous autobiography of Benjanmin Franklin, Rudin started out to write this book for children. Later more mathematics was added for publication. From the book description, "Characterized by his personal style of elegance, clarity, and brevity, Rudin presents in the first part of the book his early memories about his family history, his boyhood in Vienna throughout the 1920s and 1930s, and his experiences during World War II.

Part II offers samples of his work, in which he relates where problems came from, what their solutions led to, and who else was involved. As those who are familiar with Rudin's

2.1. POPULAR AND EXPOSITORY BOOKS ON MATHEMATICS

writing will recognize, he brings to this book the same care, depth, and originality that is the hallmark of his work."

According to MathSciNet, "With this memoir Rudin gives the entire mathematical community a chance to make his acquaintance both mathematically and personally, and a very worthwhile acquaintance it is. The biographical section covering the period from 19th-century ancestors down to the present is fascinating. I found it difficult to tear myself away from it when professorial and family duties called. It's what the literary critics call 'a good read', and it is accompanied by a family photo album... In summary, this book is a delight to read and will also help to inspire and guide young analysts in the path of wisdom. You will not want to miss a single page of it."

2.1.7 H. Weyl, Symmetry. Reprint of the 1952 original. Princeton Science Library. Princeton University Press, Princeton, N.J., 1989.

This book was first published in 1952 and has been reprinted many times and also translated into many languages.

It is a masterpiece of popular mathematics with real contents written by a master. This book contains many things and can be enjoyed at different levels and from different perspectives. It is not a very easy book as a typical current popular book, but it is an enjoyable and valuable book.

Symmetry is an integral part of our civilization and sciences since the beginning of the civilization. Many things exist due to symmetry, and we can understand and solve problems by symmetry. But the mathematics behind symmetry, i.e., group theory, is complicated and relatively recent. So it is not surprising that there are no older books on symmetry and its role in sciences and humanities.

In this wonderful book, according to Scientific American, "Dr. Weyl presents a masterful and fascinating survey of the applications of the principle of symmetry in sculpture, painting, architecture, ornament, and design; its manifestations in organic and inorganic nature; and its philosophical and mathematical significance."

The review by an expert on geometry and group theory, Coxeter, in MathSciNet, gives a very good summary: "The first lecture begins by showing how the idea of bilateral symmetry has influenced painting and sculpture, especially in ancient times. This leads naturally to a discussion of 'the philosophy of left and right', including such questions as the following. Is the occurrence in nature of one of the two enantiomorphous forms of an optically active substance characteristic of living matter? At what stage in the development of an embryo is the plane of symmetry determined?

The second lecture contains a neat exposition of the theory of groups of transformations, with special emphasis on the group of similarities and its subgroups...

The third lecture gives the essential steps in the enumeration of the seventeen space-groups of two-dimensional crystallography... The author remarks that examples of all these groups occur in the decorative patterns of the ancient Egyptians, as well as in those of the Arabs... The fourth lecture begins with the complete list of finite groups of congruent transformations in Euclidean 3-space, and the attenuated list produced by the crystallographic restriction... Turning from physics to mathematics, he gives an extraordinarily concise epitome of Galois theory, leading up to the statement of his guiding principle: 'Whenever you have to do with a structure-endowed entity, try to determine its group of automorphisms'..."

According to an Amazon review, "This delightful booklet motivates the study of symmetry by showing its presence in art and nature. This is a work of love, frequently bordering poetry. Yet, it is a scientific book of high class. Hermann Weyl, one of the very great mathematicians of this century, then explains the mathematics behind symmetry, mostly group theory, and obtains all forms that, by repetition, completely fill the plane and the space (the crystallographic groups). This is wonderful reading."

After this book, many expository books and thousands articles on symmetry have appeared. Different and new topics have been covered. But it is not clear if any has surpassed Weyl's book. We listed some of the books below for the convenience of the reader.

M. Du Sautoy, *Symmetry: A Journey into the Patterns of Nature*, Reprint edition (March 3, 2009), Harper Perennial. 384 pages.

This book explains many important theories and concepts in mathematics which are related to group theory and symmetry in an accessible way to the public. It is broad and the story is told in a more personal way. The chapter headings are unusual and follow the twelve months of a year.

From Publishers Weekly, "The author takes readers gently by the hand and leads them elegantly through some steep and rocky terrain as he explains the various kinds of symmetry and the objects they swirl around. Du Sautoy explains how this twirling world of geometric figures has strange but marvelous connections to number theory, and how the ultimate symmetrical object, nicknamed the Monster, is related to string theory. This book is also a memoir in which du Sautoy describes a mathematician's life and how one makes a discovery in these strange lands. He also blends in minibiographies of famous figures like Galois, who played significant roles in this field. This is mainly for science buffs, but fans of scientific biographies will also find it appealing."

I. Stewart, *Why Beauty is Truth. A History of Symmetry,* Basic Books, New York, 2007. xii+291 pp.

Stewart is a famous writer of popular mathematics, and this book is probably one of the best books he has written. Though it covers many standard topics related to group theory and symmetry, his broad knowledge about history and mathematics makes this a unique book among many books on symmetry.

According to MathSciNet, "Ian Stewart has written a story about symmetry and its role in mathematics and physics, beginning with the Babylonians and ending with modern physics. It's a book for the non-mathematician who would like to learn something about the nature of mathematics, and so perhaps there is no better subject for such a book than symmetry, an enticing property that is well known to most readers through art and music. But Stewart's book is not a picture book, though it is well illustrated. Neither does it contain many formulas, though the author treats mathematics in a serious way. The book's intention is to describe what symmetry means to mathematicians and why it has played such an important role throughout the ages in both mathematics and physics. The reader begins to see why and how abstract mathematical thought is intimately connected with truth, beauty, and the nature of the physical universe. The author also entertains the reader with stories about the 'muddle of mathematicians' we meet as the story unfolds."

According to Publishers Weekly, "while the math behind symmetry is important, the heart of this history lies in its characters, from a hypothetical Babylonian scribe with a serious case of math anxiety, through Évariste Galois (inventor of 'group theory'), killed at 21 in a duel, and William Hamilton, whose eureka moment came in 'a flash of intuition that caused him to vandalize a bridge,' to Albert Einstein and the quantum physicists who used group theory and symmetry to describe the universe."

J. Rosen, *Symmetry Discovered: Concepts and Applications in Nature and Science*, Revised reprint of the 1975 original, Dover Publications, Inc., Mineola, NY, 1998. xiv+152 pp.

This is a good introduction to group theory and applications of symmetry. It is elementary, and concepts and definitions are carefully explained. According to MathSciNet, "This is an entertaining introduction to geometrical representations of discrete groups, enlivened by abundant illustrations and delightfully relevant quotations from books by A. A. Milne. The concept of symmetry is extended from isometries to similarities, and from geometry to physics, music and biology. Thus the book might be regarded as a modern counterpart for H. Weyl's book *Symmetry* ..., though it lacks Weyl's polished style."

Rosen is a physicist and has also written other books on group theory and symmetry. The basic point of these books is that science does not only make use of symmetry, but is essentially symmetry. Indeed, science builds on the foundation of reproducibility, predictability, and reduction, all of which are symmetries.

J. Rosen, *Symmetry in Science. An Introduction to the General Theory,* Springer-Verlag, New York, 1995. xvi+213 pp.

According to MathSciNet, "This book is very exceptional in its conceptual approach to the use of symmetry in science as a whole, with emphasis on fundamental problems of quantum physics, relativity and cosmology. The concept of symmetry is rather elusive to define in a strict mathematical manner; its description as an immunity to a possible change, given in a brief introductory chapter, is probably the best ever invented. Group theory, the natural mathematical language for applications of symmetry considerations, is not the main concern of the author. Just about as much of it is given in three chapters as a willing nonspecialist without a mathematical background can digest. The rest of the book is a nice and deep piece of the philosophy of natural science, a firework of ideas which concern the fundamentals of a scientific view of nature wherein symmetry plays an indispensable role..."

J. Rosen, *Symmetry Rules. How Science and Nature Are Founded on Symmetry,* Frontiers Collection, Springer-Verlag, Berlin, 2008. xiv+304 pp.

From the book description, "Emphasizing the concepts, this book leads the reader coherently and comprehensively into the fertile field of symmetry and its applications. Among the most important applications considered are the fundamental forces of nature and the Universe."

R. McWeeny, *Symmetry: An Introduction to Group Theory and Its Applications,* The International Encyclopedia of Physical Chemistry and Chemical Physics, Mathematical techniques, Vol. 3, A Pergamon Press Book The Macmillan Co., New York, 1963. xiv+248 pp.

This is a mathematics book on group theory. It was written for users of group theory, i.e., chemistry and physics students, but it is also good for mathematics students and mathematicians. It is between standard textbooks on group theory and representation theory and popular books on symmetry.

From the book description, "This well-organized volume develops the elementary ideas of both group theory and representation theory in a progressive and thorough fashion. Designed to allow students to focus on any of the main fields of application, it is geared toward advanced undergraduate and graduate physics and chemistry students."

J.H. Conway, H. Burgiel, C. Goodman-Strauss, *The Symmetries of Things,* A K Peters, Ltd., Wellesley, MA, 2008. xviii+426 pp.

This is an unusual book and "has been germinating for a long time". It started with Conway's lectures at Cambridge University and his later talks in many places including the Princeton Rug Society and the International Congress of Mathematicians. Besides the usual topics on transformation groups and periodic patterns, it contains many new results, for example, hyperbolic groups and orbifolds. Both the style and contents are unique in comparison with other books on symmetry. A very interesting book!

From the book description, "Inspired by the geometric intuition of Bill Thurston and empowered by his own analytical skills, John Conway, with his coauthors, has developed a comprehensive mathematical theory of symmetry that allows the description and classification of symmetries in numerous geometric environments. This richly and compellingly illustrated book addresses the phenomenological, analytical, and mathematical aspects of symmetry on three levels that build on one another and will speak to interested lay people, artists, working mathematicians, and researchers."

M. Field, M. Golubitsky, *Symmetry in Chaos. A Search for Pattern in Mathematics, Art, and Nature,* Second edition. Society for Industrial and Applied Mathematics (SIAM), Philadelphia, PA, 2009. xiv+199 pp.

There are many books, both technical and popular, on chaos and fractals. A unique feature of this expository book is its emphasis on the role of symmetry in studying chaos and fractals. Indeed, the purposes of this book are to present pictures that show symmetry and dynamics coexist and to explain mathematical ideas behind these beautiful pictures. From the book description, "Mathematical symmetry and chaos come together to form striking, beautiful color images throughout this impressive work, which addresses how the dynamics of complexity can produce familiar universal patterns..."

According to MathSciNet, "The goal of the authors is twofold. First they simply want to illustrate the beauty of symmetric chaos, and secondly they use this fact as a motivation for the reader to understand the related mathematical background. Both of these goals are indeed smoothly accomplished with this book. It contains a careful introduction to the mathematical background of symmetric chaos, and, in particular, the crucial notions of symmetry—being a motion which leaves an object invariant—and chaos—representing complicated dynamical behavior—are explained in a very nice fashion. Moreover, spectacular pictures illustrating the different aspects of the interplay of symmetry and chaos accompany the entire text. In addition to the mathematical and the artistic aspects the authors indicate applications of symmetric chaos to real physical systems such as the Taylor-Couette experiment. Finally they present all the information necessary for the reader to reproduce the pictures on a personal computer.

Field and Golubitsky have written an excellent book which is entertaining and of interest not just to mathematicians working on dynamical systems but to everyone having some kind of scientific background or interest in complicated dynamics."

B. Grünbaum, G. Shephard, *Tilings and Patterns,* W. H. Freeman and Company, New York, 1987. xii+700 pp.

Since the beginning of the civilization, people have been interested in tilings and patterns. But the mathematical theory behind them is both elementary and profound. For example, the complete classification of crystallographic groups in higher dimension is still open, and the recent discovery of quasi-crystals indicates a wide open frontier. This is a comprehensive book covering all aspects of tilings and patterns at various levels. A shorter version, but not a short book, is given in the next book.

B. Grünbaum, G. Shephard, *Tilings and Patterns. An Introduction*, A Series of Books in the Mathematical Sciences, W. H. Freeman and Company, New York, 1989. xii+446 pp.

From the book description of the original book, "The definitive book on tiling and geometric patterns, this volume features 520 figures and over 100 tables. Accessible to anyone with a grasp of geometry, it offers illustrated examples of two-dimensional spaces covered with interlocking figures, plus related problems and references. Equally suitable for geometry courses and independent study."

According to MathSciNet, "Tilings and patterns have been made and enjoyed for thousands of years. Their mathematical treatment was begun by J. Kepler but was then forgotten until the nineteenth-century development of crystallography."

According to a review in Bulletin of AMS, "From time immemorial artisans and artists have constructed ingenious tilings and ornaments using repeated motives. This is demonstrated in the introduction of the beautiful volume under review by numerous examples from widely separated cultures. However, the importance of tilings and patterns in crystallography and some related branches of science was recognized only towards the end of the last century. From this time on many crystallographers, chemists, physicists, architects, engineers, and mathematicians have been working in this field. Although they accumulated a vast literature in books and periodicals, much effort has been wasted duplicating previously known results. When the authors started collecting material for this book, they were surprised to find 'how little about tilings and patterns is known,' and how many errors were made because of 'badly formulated definitions and lack of rigor.'

For more than a decade the authors were busy critically revising the earlier results and making significant contributions to the theory of tilings and ornaments in a series of papers of their own. Their effort is crowned by the unique comprehensive monograph *Tilings and Patterns*, which lays a solid foundation for one of the most attractive fields in geometry."

M. Ronan, *Symmetry and the Monster. One of the Greatest Quests of Mathematics*, Oxford University Press, Oxford, 2006. vi+251 pp.

One of the major achievements of the 20th century in mathematics is the classification of finite simple groups. Besides those associated with Lie groups, there are 26 sporadic finite simple groups. The largest of them is called the "monster" in the title. This book tells the story of finite simple groups starting from regular solids (or Platonic solids) to the monstrous moonshine. An interesting book about the biggest story in modern mathematics.

From the book description, "Ronan describes how the quest to understand symmetry really began with the tragic young genius Evariste Galois, who died at the age of 20 in a duel. Galois, who spent the night before he died frantically scribbling his unpublished discoveries, used symmetry to understand algebraic equations, and he discovered that there were building blocks or 'atoms of symmetry.' Most of these building blocks fit into a table, rather like the periodic table of elements, but mathematicians have found 26 exceptions. The biggest of these was dubbed 'the Monster'–a giant snowflake in $196,884$ dimensions. Ronan, who personally knows the individuals now working on this problem, reveals how the Monster was only dimly seen at first. As more and more mathematicians became involved, the Monster became clearer, and it was found to be not monstrous but a beautiful form that pointed out deep connections between symmetry, string theory, and the very fabric and form of the universe..."

A. Ash, R. Gross, *Fearless Symmetry. Exposing the Hidden Patterns of Numbers. With a foreword by Barry Mazur,* Princeton University Press, Princeton, NJ, 2006. xx+272 pp.

This is not an ordinary popular book on symmetry, but rather an accessible book to outline a proof of the Last Fermat Theorem by Andrew Wiles. Reciprocity laws, Modular forms, étale cohomology and Galois representations etc are natural parts of this book. It is an expository book on some important aspects of modern algebraic number theory for mathematics students and mathematicians. It deals with beautiful topics in mathematics, and the authors are "fearless" to take up this challenge to explain them well.

From the book description, "All too often, abstract mathematics, one of the most beautiful of human intellectual creations, is ground into the dry dust of drills and proofs. Useful, yes; exciting, no. Avner Ash and Robert Gross have done something different–by focusing *on the ideas that modern mathematicians actually care about.* Fearless Symmetry is a book about detecting hidden patterns, about finding definitions that clarify, about the study of numbers that has entranced some of our great thinkers for thousands of years. It is a book that takes on number theory in a way that a non-mathematician can follow systematically but without a barrage of technicalities. Ash and Gross are two terrific guides who take the reader, scientist or layman, on a wonderful hike through concepts that matter, culminating in the extraordinary peaks that surround the irresistible, beckoning claim of Fermat's Last Theorem."

A. Zee, *Fearful Symmetry. The Search for Beauty in Modern Physics*, Reprinted from the 1999 edition and with a new foreword by Roger Penrose, Princeton Science Library, Princeton University Press, 2007. xx+356 pp.

This is an expository book more about modern physics than symmetry. It tells a story of physicists following Einstein in their search for the beauty and simplicity of the nature and physics theories understanding it, and finite groups and Lie groups have played a fundamental role in this process. It is different from other more mathematical books on symmetry and could an eye opener for mathematicians about the importance and beauty of applications of groups.

From Library Journal, "The story of modern physics is the story of the triumph of reductionism. It follows that a study of symmetry (the observation that physical laws remain unchanged when action is viewed from different perspectives or transformations) is a useful way to demonstrate the development of contemporary theories like superstrings and grand unified out of the seeming chaos of the quantum world. Zee delineates classical mechanics, relativity, the quantum, and the strange new world of subatomic particles with style and wit. Non-mathematicians especially will appreciate his entirely comprehensible, virtually non-mathematical explanation of group theory. He acknowledges the philosophical implications of what has come to be called the 'new' physics, but does not stray too far from his purpose. The race for unity contains an important insight: that the nature of the physical world is abstract, rather than complicated notion with possibly vast significance for our understanding of first causes."

L.M. Lederman, C.T. Hill, *Symmetry and the Beautiful Universe,* Prometheus Books, Amherst, NY, 2004. 363 pp.

Emmy Noether is certainly well-known to most mathematicians for her fundamental work on modern algebra, but she rarely appears many books on symmetry and group theories, in particular those written by mathematicians. (In those books, Galois and Abel are the natural heros). This book is really an introduction to modern physics, and Emmy Noether figured prominently due to her important result on conservation laws in connection with symmetry. There is a single chapter devoted to her. This fact alone makes it a worthy book for mathematicians. Of course, this is a very good book on other aspects as well.

From Publishers Weekly, "The concept of symmetry has seen increasing service in science popularizations as a metaphor to convey the intuitive appeal of physics, a vogue that continues in this dense treatise. Nobel Laureate Lederman (The God Particle) and theoretical physicist Hill deploy mathematical symmetry as a unifying theme in a tour of physics from Newton's laws to quarks and superstrings. Sometimes, as in a demonstration that the invariance of physical laws through time implies the law of conservation of energy, this approach yields insights. But usually, as in their confusing exposition of special

2.1. POPULAR AND EXPOSITORY BOOKS ON MATHEMATICS

relativity, symmetry considerations get in the way. The authors keep things readable with lots of physics-for-poets bits, including some tie-ins to environmentalism, comparisons of modern cosmology with ancient Greek myths, and a fictional dialogue—partly in Italian—between two newlywed physicists and Galileo's ghost. Unfortunately, symmetry is a forbiddingly abstract branch of mathematics that was peripheral to the development of much of physics and gives little tangible feel for its substance, and the point where it becomes indispensable to discussions of modern physics is also the tipping point where the book, like many others, topples into total incomprehensibility to laypeople. Readers who think symmetry implies clarity and grace will be disappointed."

It is a pity that **Emmy Noether's deep contribution** to conservation laws and mechanics is not so well-known to mathematicians. There is a recent book devoted to this.

Y. Kosmann-Schwarzbach, *The Noether Theorems. Invariance and Conservation Laws in the Twentieth Century,* Translated, revised and augmented from the 2006 French edition by Bertram E. Schwarzbach, Sources and Studies in the History of Mathematics and Physical Sciences, Springer, New York, 2011. xiv+205 pp.

Besides discussing the history and development of Noether's ideas and her conservation theorems and their impacts in physics and mathematics, this book also contains an English translation of the original paper of Noether.

I. Hargittai, M. Hargittai, *Symmetry Through the Eyes of a Chemist*, Second edition, Plenum Press, New York, 1995. xiv+469 pp.

Probably it is better known that groups have important applications to physics. But they also have fundamental applications to chemistry. This book is a good introduction to chemistry and this can be seen from some chapter headings: *Molecules: shape and geometry, Molecular vibrations, Electronic structures of atoms and molecules, Chemical reactions, Symmetries in crystals etc.* This book complements other books on groups and symmetry and is fun and instructive to read. Indeed, according to MathSciNet, "This is a beautiful and unusual book on symmetry in chemistry, written by two chemists with an audience of chemists in mind. There are several books on symmetry for chemists, and many books on symmetry for the general reader, but this is the only book the reviewer knows in which the role of symmetry in a scientific subject is described in detail and in depth. The contents include an introduction to symmetry, the shape and geometry of molecules, matrices and group representations (sufficient for the purposes at hand), molecular vibrations, electronic structure of atoms and molecules, chemical reactions, space group symmetries, and symmetries in crystals. The mathematics is both elementary and standard; the mathematician will find no surprises here. But the illustrations are lavish and nonstandard, and above all the discussions of chemistry are, for once, finely tuned to mathematical sensibilities. A subtitle for this book could well be 'chemistry for mathematicians'."

Besides the above book, I. Hargittai, M. Hargittai have also written another book on symmetry.

I. Hargittai, M. Hargittai, *Symmetry: A Unifying Concept*, Shelter Publications, Inc.; First Printing edition, 1994. 224 pages.

This book can be described as a picture book on symmetry for adults, containing few words and many pictures. It covers most topics in popular books on symmetry one can imagine and more. Besides many stuning pictures, the chapter headings are also unusual: *Pinwheels & Windmills, Ballons, Walnuts & Molecules, Helix & Spiral, Bees & Engineering etc.*

According to the backcover summary, "A fascinating, highly-visual journey through the worlds of symmetry, with 550 photos and some 300 drawings.

From snowflakes to starfish, from dandelions to BMWS, from fences to Fibonacci, from crystals to Coke machines, this is a thought-provoking and stimulating display of visual imaginary with simple, clear text, describing 15 aspects of symmetry.

Here is a unique view of the subject that has fascinated humans for millennia. It will help you train your eye and mind to see new patterns and make new connections. It will provide a powerful unifying factor between seemingly disparate fields of human endeavor."

The next book does not deal exclusively with symmetry, but symmetry certainly plays an important role in it.

D. Mumford, C. Series, D. Wright, *Indra's Pearls. The Vision of Felix Klein*, Cambridge University Press, New York, 2002. xx+396 pp.

Felix Klein is probably best known for the Erlangen program which says roughly the geometry is to study invariants of groups. He is also famous for his work on Fuchsian groups, which should be called *Kleinian groups*, and the Kleinian groups should be called *Poincaré groups*. This unique book is related to works of Kleinian on Fuchsian groups and covers many other topics as well, and can be read by different people at different levels.

From the book description, "In this extraordinary book they explore the path from some basic mathematical ideas to the simple algorithms that create delicate fractal filigrees, most appearing in print for the first time. Step-by-step instructions for writing computer programs allow beginners to generate the images."

According to MathSciNet, "Felix Klein, one of the great nineteenth-century geometers, discovered in mathematics an idea prefigured in Buddhist mythology: the heaven of Indra contained a net of pearls, each of which was reflected in its neighbour, so that the whole Universe was mirrored in each pearl. Klein studied infinitely repeated reflections

2.1. POPULAR AND EXPOSITORY BOOKS ON MATHEMATICS

and was led to forms with multiple co-existing symmetries, each simple in itself, but whose interactions produce fractals on the edge of chaos. For a century these images, which were practically impossible to draw by hand, barely existed outside the imagination of mathematicians. However in the 1980s the authors embarked on the first computer exploration of Klein's vision, and in so doing found further extraordinary images of their own. The book is written as a guide to actually coding the algorithms which are used to generate the delicate fractal filigrees, most of which have never appeared in print before. ... Beginners can learn to understand what the images mean and follow the step-by-step instructions for writing computer programs that generate them. Experts in the geometry of discrete groups can see how the images relate to ideas that take them to the forefront of research."

2.1.8 D. Hilbert, S. Cohn-Vossen, Geometry and the Imagination, American Mathematical Society, 1, 1999. 357 pages

This book was first published in German in 1932 and the English translation was published in 1952. It was reissued in 1999. It has also been translated into several other languages.

The geometry in the title is interpreted very broadly and includes geometry of numbers, projective geometry, differential geometry, transformation group (geometry), and topology. It covers many topics in an accessible and yet profound way. It contains many gems. For example, the proof of the fact that the hexagon lattice packing is the densest lattice packing in dimension 2 is geometric and shorter than the usual proof using reduction of lattices (or binary quadratic forms). It is rare that a mathematician of status like Hilbert is willing to take time to write such an expository book. It is probably also helpful to note that when Hilbert gave a series of lectures around 1920–1921 on which this book was based, Hilbert was moving towards the most abstract and formal conception of mathematics. Perhaps he needed intuition to balance it.

According to the preface, "the tendency toward intuitive understanding fosters a more immediate grasp of the objects one studies, a live *rapport* with them, so to speak, which stresses the concrete meaning of their relations." This explains the word intuition in the title.

From the book description, "This remarkable book has endured as a true masterpiece of mathematical exposition. There are few mathematics books that are still so widely read and continue to have so much to offer– after more than half a century! The book is overflowing with mathematical ideas, which are always explained clearly and elegantly, and above all, with penetrating insight. It is a joy to read, both for beginners and experienced mathematicians. It is full of interesting facts, many of which you wish you had known before, or had wondered where they could be found... It would be hard to overestimate the continuing influence Hilbert-Cohn-Vossen's book has had on mathematicians of this century. It surely belongs in the 'pantheon' of great mathematics books."

According to an Amazon review, "David Hilbert liked to quote 'an old French mathematician' saying 'A mathematical theory should not be considered complete until you have made it so clear that you can explain it to the first man you meet on the street'. By that standard, this book by Hilbert was the first to complete several branches of geometry... In line with Hilbert's thinking, the results and the descriptions are beautiful because they are so clear.

More than that, this book is an accessible look at how Hilbert saw mathematics. In the preface he denounces 'the superstition that mathematics is but a continuation ... of juggling with numbers'. Ironically, some people today will tell you Hilbert thought math was precisely juggling with formal symbols. That is a misunderstanding of Hilbert's logical strategy of 'formalism' which he created to avoid various criticisms of set theory. This book is the only written work where Hilbert actually applied that strategy by dividing proofs up into intuitive and infinitary/set-theoretic parts. Alongside many thoroughly intuitive proofs, Hilbert gives several extensively intuitive proofs which also require detailed calculation with the infinite sets of real of complex numbers. In those cases Hilbert says 'we would use analysis to show ...' and then he wraps up the proof without actually giving the analytic part."

The following book was originally planned as an appendix of the above book *Geometry and the Imagination*. The preface was written by Hilbert and explains why it was not included as a part of the book as planned.

P. Alexandroff, *Elementary concepts of topology*. Translated by Alan E. Farley, Dover Publications, Inc., New York, 1961. 73 pp.

P. Alexandroff has made foundational contributions to general topology and algebraic topology. For example, the one point compactification is also called the Alexandroff compactification. He has written extensively both at the technical and expository levels. Together with Urysohn, he introduced the now standard notion of a compact space and a locally compact space. His life long relation with Kolmogorov is legendary but probably not so well-known.

From the book description, "Concise work presents topological concepts in clear, elementary fashion without sacrificing their profundity or exactness. Author proceeds from basics of set-theoretic topology, through topological theorems and questions based on concept of the algebraic complex, to the concept of Betti groups."

According to an Amazon review, "Alexandroff was a favorite student of Emmy Noether, and L.E.J. Brouwer, and followed Hilbert's lectures. The greatest algebraist, the greatest topologist, and the greatest mathematician of the early 20th century all had

2.1. POPULAR AND EXPOSITORY BOOKS ON MATHEMATICS 39

direct input into this book. All believed the most important, deepest mathematics can be made the clearest. They were right."

According to a review in Bulletin of AMS in 1933, "Alexandroff has written particularly for those who do not care to undertake a systematic study of topology, and his enthusiasm in exhibiting the beauties of the subject will not escape the reader. There are many simple examples and the accompanying drawings are so skillfully made that one can not fail to see how the homology group operates, or to appreciate its intuitive meaning. The author does not allow the proofs of important matters to depend on the reader's intuition; on the contrary, it is remarkable how much he has treated in complete scientific detail."

There are several more advanced books on topology by Alexandroff: *Topologie I* (with H. Hopf), and *Combinatorial topology*.

Another great book by a distinguished geometer is the following

H.S.M. Coxeter, *Introduction to Geometry*, Reprint of the 1969 edition, Wiley Classics Library, John Wiley & Sons, Inc., New York, 1989. xxii+469 pp.

Coxeter is probably best known to the modern reader in mathematics for Coxeter groups. Many other things are also named after him. Besides mathematics, he is also known for a very long productive life, being active till the end.

Though every mathematics student has learned plane geometry, a standard course does not cover too many topics. For example, it is probably not so well-known to many mathematicians that the last chapter, Chapter 13, of the book *Elements* by Euclid, deals with classification of regular solids. This book covers many important topics in both classical and more modern geometry. One can read it to make up one's education in geometry and also for pleasure, since geometry is one of the most beautiful and accessible subjects in mathematics.

The review in MathSciNet gives a good summary: "The solution, by Eudoxus about 370 B.C., of the difficulties arising from the discovery of the irrational, was purely geometric, with the result that the Greeks expressed and solved their algebraic problems in geometric guise. Thus there came into existence what may be described as a long-standing scandal in mathematics, arising from the dominance of geometry; for example, Newton felt compelled, at considerable cost, to write his *Principia* in ancient geometric form. This scandal may be said to have been brought to an end by the great analysts, Gauss and others, of the nineteenth century, since which time the pendulum has swung the other way... we have somehow lost interest in geometry, and the purpose of the present book, as the author states, is to revitalize this sadly neglected subject.

The reader will find not only that geometry is a subject of great interest in its own right, but that it is important for other branches of mathematics and science and has many active and interesting avenues of research open at the present time. Everywhere the author enlivens and clarifies his text with references to literature, history, architecture and the like. His diagrams and illustrations are numerous and in many cases beautiful... A single unifying thread runs throughout the book, from the simplest to the most advanced topics; namely, the idea of symmetry, or of a group of transformations."

According to an Amazon review, "This is the best book I've seen covering geometry at this level. Coxeter was known as an apostle of visualization in geometry; many other books that cover this material just give you page after page of symbols with no diagrams. He motivates all the topics well, and lays out the big picture for the reader rather than just presenting a compendium of facts. This is a survey of a huge field, but he does a great job of focusing on the most important results. As other reviewers have noted, this book is not 'introductory' in the sense of high school geometry; it's introductory in the sense of being the kind of book a college math major would use in his/her first upper-division geometry course. It doesn't presuppose a great deal of mathematical knowledge, but it probably isn't a book that one could appreciate without having already developed quite a high level of mathematical maturity."

Two other expository books by Coxeter on geometry are the following.

H.S.M. Coxeter, S.L. Greitzer, *Geometry Revisited*, The Mathematical Association of America; 1967. 208 pages.

The preface of this book starts with the quote "He who despises Euclidean Geometry is like a man who, returning from foreign parts, disparages his home."

It is probably "the book" on elementary geometry in high school (in the old days). It has been translated into several languages and has been popular. Some topics are also interested to more advanced readers, such as cross ratio and projective geometry. It discusses many beautiful and nontrivial theorems in geometry such as the theorems of Ceva, Menelaus, Pappus, Desargues, Pascal, Brianchon, and Morley's remarkable theorem on angle trisectors. The transformational point of view is emphasized: reflections, rotations, translations, similarities, inversions, and affine and projective transformations. Many fascinating properties of circles, triangles, quadrilaterals, and conics are developed.

Many mathematicians now probably do not learn these topics in standard courses, and this book can also be read by mathematicians for further education in geometry and pleasure.

H.S.M. Coxeter, *Projective Geometry*, Revised reprint of the second (1974) edition. Springer-Verlag, New York, 1994. xii+162 pp.

Projective geometry has a long history and its early study was closely related to perspective in drawing. It has experienced ups and downs. Besides its influence on projective algebraic geometry and projective differential geometry, it has also motivated

2.1. POPULAR AND EXPOSITORY BOOKS ON MATHEMATICS 41

the more recent theory of buildings for algebraic groups. This book is a classic and gives a systematic introduction to projective geometry.

According to MathSciNet, "The book is written with all the grace and lucidity that characterize the author's other writings. Regrettably, there is probably no time in either a high school or a college course for a unit which would cover the content of this book, but the book could be read with pleasure and profit by the student towards the end of his high school career or in his early years at college. To do so would add significantly to his appreciation of the ethos of mathematics."

H.S.M. Coxeter, *Regular Polytopes,* Third edition, Dover Publications, Inc., New York, 1973. xiv+321 pp.

This is a classical and still one of the best books on polytopes. As mentioned before, Coxeter is probably best known for Coxeter groups, and Coxeter groups are discussed in this book in the chapter under the heading *The Generalized Kaleidoscope.* We note that fundamental domains of Coxeter groups is closely related to regular polytopes. The study of the regular polyhedra, which was the culmination of Euclid's immortal book *Elements,* is continued in this book and extended to include the analogous shapes in four and high dimensions. Beginning with polygons and polyhedrons, the book moves on to multi-dimensional polytopes in a way that anyone with a basic knowledge of geometry and trigonometry can understand.

According to a review in Bulletin of AMS in 1949, "The study of polytopes (that is, polygons and polyhedra of three or higher dimensions) appears to interest more different kinds of people than any other branch of mathematics with the possible exception of number theory. Its beauties inspired the rug merchant, P. S. Donchian, patiently to construct a remarkable set of models representing all varieties of polytopes. It enabled the housewife, Alicia Boole Stott, a daughter of George Boole, to capitalize on her unusual powers of geometrical visualization in spite of her meager mathematical education. It provided the struggling young lawyer, Thorold Gosset, with an amusing and constructive pastime during his long waits between clients. And equally well it has attracted the attention of many famous mathematicians such as Klein, Poincaré, Poinsot, Schlafli, Cayley, Euler, and Goursat, to mention only a few. Nor is it solely a pure discipline devoted to beauty but not utility, for it has been cultivated by a number of crystallographers such as Fedorov.

Coxeter has spent a major portion of his mathematical career digging out the obscure references in early works, in making personal contact with contemporary gifted amateurs, and in developing his own outstanding contributions to the field. In this book he has poured forth all his devotion and scholarship and has produced a work which will be the standard treatise in this field for many years. It is beautifully illustrated with photographs of Donchian's models and with numerous drawings. Its value as a reference book is greatly enhanced by historical material at the end of each chapter, by tables giving the essential combinatorial and metric properties of polytopes of many varieties, by an exhaustive bibliography, and by a carefully constructed index. It is a particular pleasure to record this last feature; for its omission in so many mathematical books published in England greatly detracts from their value."

2.2 Biographies of mathematicians and history of mathematics

Many people will probably agree that knowledge of lives of mathematicians helps people to understand their mathematics better. Besides being inspiring, the lives of mathematicians are also interesting to read, as the following selection shows.

2.2.1 E.T. Bell, Men of Mathematics, Touchstone, 1986. 608 pages.

This book was first published in 1937 and has always been popular and in print. It has been read by many generations of mathematicians. Though this is not a history book on mathematics, it is probably the best known and most interesting book on mathematicians to many people. To young people, it can be very inspiring, and many people have indeed being inspired to pursue mathematics. For others, they are fantastic stories which are partially based on facts and can be used to tell stories about mathematicians.

Probably E.T. Bell is best (or only) known to many people now through this book, but he had a colorful life and was an established mathematician in his time. A biography about him was written by C. Reid and is titled *The search for E.T. Bell*, which will be discussed later in this book.

A review of this book in Natl. Math. Mag. 11 (1937), no. 8, 406–407, makes interesting reading many years later: "Footnotes and biographies are in general excluded and sources or authority for statements are usually not indicated. The author has purposely done this, since the aim of the book is popularization, or appeal to the general reader. He has specifically states that it is not intended to be a history of mathematics, or any section of it... Dr. Bell is a seasoned, skillful writer with a fluent style; he writes with a realistic, curt, potent wit and stark, frank humor which does not stop short of vigorous, rollicking slang... Doubtless this will amuse many casual, rapid readers, but it also leads him to a certain type of exaggeration... Dr. Bell frequently recounts the charming legends, interesting traditions, and melodramatic fiction concerning the various mathematicians, ... Dr. Bell allows his imagination to play; conjecture, personal opinion, and speculation are abundant throughout these chapters as to what the historical development would have been if conditions had been different, or some mathematicians had lived longer, behaved differently, or if mortals were constituted otherwise.

2.2. BIOGRAPHIES AND HISTORY

To write such a work requires a huge amount of labor, but the author will probably be rewarded by finding a large circle of readers... Its laudable metic of popularizing, if this succeeds, will justify its publication as well as the toil required to produce it..."

In spite of all its faults and mistakes, this book has succeeded beyond expectation. It is perhaps best to quote from the most popular review on Amazon on this book:

"This book has entertained, educated and intrigued two generations of young aspiring mathematicians, as well as people who would never grow up to do research mathematics, but who could see the beauty of number. Bell's style is addictive; he makes every personality come to life–from Galois, brilliant, unlucky and doomed, to Gauss, the 'Prince of Mathematicians', to Pascal, mystical and tormented. No one who reads this book can forget, for example, the section entitled 'Galois' last night', where, the night before Galois knows he will die, he spends "the fleeting hours feverishly dashing off his last will and testament, writing against time to glean a few of the great things in his teeming mind before the death which he foresaw could overtake him. Time after time he broke off to scribble in the margin 'I have not time; I have not time,' and passed on to the next frantically scrawled outline.

Which is sad, in a way, because it is, according to modern accounts of Galois' life, not accurate. The work of Galois Bell is describing was written before his last night, in no such hurry. This has been known for some time, and yet few who know, and who perhaps should know better, will relinquish their affection for this marvelous book. It so captures the enthusiasm one can feel for the beauty and poetry that mathematics brings to the mind that errors of fact are minor flaws.

And the errors are few enough that they really do not matter. In Galois' case, for example, one takes away a deeply etched portrait of an astonishing mind that descended on revolutionary France like a meteorite, and which had about as much chance of being understood. This is accurate, and Bell tells his stories so powerfully that they stay in the mind for decades.

Bell includes many wonderful quotes and stories. The whole first section of the book is just a series of quotes–my favourite is perhaps Weierstrass, 'A mathematician who is not also something of a poet will never be a complete mathematician.' But he lards the book with quotes, and since this book can profitably be read by an enthusiastic 12-year-old, and often has been, for many people this book is the first time they will meet with such famous quotes as Newton's line about being merely a child, playing with pretty pebbles on the seashore.

Bell claims that the book is not a history of mathematics, and he's right. It's a series of chapters that provide biographical–and mathematical–sketches of thirty-odd great mathematicians, from Archimedes to Cantor. You'll learn a lot about the history of mathematics from this book, but mostly you'll be infected by the passionate enthusiasm of someone who knows and loves his subject. Buy it; read it; if you love mathematics you won't regret it."

2.2.2 C. Reid, Hilbert, Springer-Verlag, New York-Berlin, 1970. xi+290 pp.

This is simply a fantastic book. It is more than a biography of an important mathematician, but rather a moving description of the center of mathematics and its decline. Since it was first published in 1970, it has been reprinted and retranslated into several languages. For some strange reason, it seems that this book is much better known in USA and other countries than in Germany. If one wants to read only one biography of mathematicians, this is it!

According to Science, "It presents a sensitive portrait of a great human being. It describes accurately and intelligibly on a nontechnical level the world of mathematical ideas in which Hilbert created his masterpieces. And it illuminates the background of German social history against which the drama of Hilbert's life was played. Beyond this, it is a poem in praise of mathematics."

It might be helpful to quote from the book. On page 47, it describes the entrance to Göttingen: "The red-tiled roofs of Göttingen are ringed by gentle hills which are broken here and there by the rugged silhouette of an ancient watch tower. Much of the old city wall still surrounds the inner town, and on Sunday afternoons the townspeople 'walk the wall' – it is an hour's walk. Outside the wall lie the yellow-brick buildings of the Georg August Universität, founded by the Elector of Hannover who was also George II of England. Inside, handsome half-timbered houses line crooked, narrow streets. Two thoroughfares, Prinzestrasse and Weender Strasse, intersect at a point which the mathematicians call the origin of the coordinates in Göttingen. The center of town, however, is the Rathaus, or town hall. On the wall of its Ratskeller there is a motto which states unequivocally: *Away from Göttingen there is no life*."

On page 102, it says "At the beginning of the twentieth century mathematics students all over the world were receiving the same advice: 'Pack your suitcase and take yourself to Göttingen!' Sometimes it seemed that the little city was entirely populated with mathematics... the mathematicians preferred to recount how Minkowski, walking along Weender Strasse, saw a young man pondering an obviously grave problem and patted him on the shoulder, saying, 'It is sure to converge'– and the young man smiled gratefully.

The days were long past when Hilbert had delivered his lectures on analytic functions for the sole benefit of Franklin. Frequently now several hundred people jammed the hall to hear him, some perching even on the window sills. He remained unaffected by the size or the importance of the audience. 'If the Emperor himself had come into the hall', says Hugo Steinhaus, who came to Göttingen at that time, 'Hilbert would not have changed'.

2.2. BIOGRAPHIES AND HISTORY

Was Hilbert so because of his position as the leading mathematician in Germany? 'No, Hilbert would have been the same if he had had only one piece of bread.' "

What has been achieved in mathematics by people in Göttingen including Gauss, Riemann, Hilbert and Klein, together with the above descriptions really inspired people's imagination, and it has also drawn many people to visit Göttingen to see the real place.

When I finally went to Göttingen, I carried a copy of this book "Hilbert" and walked along the paths described in the book and also visited the houses of Klein and Hilbert.

After reading this biography on Hilbert, I also tried to find and read other biographies written by C. Reid. Though they are not all up to the high standard set by *Hilbert*, they are all interesting. They include the following books:

C. Reid, *Courant in Göttingen and New York. The Story of an Improbable Mathematician*, Springer-Verlag, New York-Heidelberg, 1976. ii+314 pp.

This is closely related to the biography *Hilbert* and can be considered as a sequel. It also provides some helpful information about the life in Göttingen during the Hilbert time. But the style is different and one does not feel the excitement as one feels in reading *Hilbert*.

C. Reid, *Neyman–From Life*, Springer-Verlag, New York, 1982. ii+298 PP.

Neyman is apparently an important figure in statistics, and UC Berkeley has been a center of statistics. But it does not have the charm and appeal of *Hilbert*, maybe because Berkeley is not really equivalent to Göttingen in its golden days. But it is still an interesting read. Maybe there is no equivalent of *Hilbert* in statistics, but a good approximation to an equivalent in statistics of *Men of Mathematics* is the following book: **D. Salsburg**, *The Lady Tasting Tea. How Statistics Revolutionized Science in the Twentieth Century*. W.H. Freeman and Company, New York, 2001. xii+340 pp.

C. Reid, *Julia. A Life in Mathematics*, With Contributions by Lisl Gaal, Martin Davis and Yuri Matijasevich, MAA Spectrum, Mathematical Association of America, Washington, DC, 1996. xii+124 pp.

This is not a standard biography and but gives an overview of the life and work of Julia Robinson. The biographic part consists of two thirds of the book and the rest consists of commentaries by others on her work and life. This book is suitable for both mathematicians and also for mathematics students and the public.

From the book description, "Julia is the story of Julia Bowman Robinson, the gifted and highly original mathematician who during her lifetime was recognized in ways that no other woman mathematician had ever been recognized."

C. Reid, *The Search for E.T. Bell*, Also Known as John Taine, MAA Spectrum. Mathematical Association of America, Washington, DC, 1993. xii+372 pp.

Since Bell is called "the father of mathematical biography" according to C. Reid, and this book is naturally attractive to mathematicians. On top of that, the life of Bell is interesting and provides the right material to his successor in mathematical biography.

According to MathSciNet, "E.T. Bell is familiar to many as the author of the popular history *Men of Mathematics*, an entertaining (albeit not altogether factually correct) account of a number of famous mathematicians, as well as of a few other volumes of popular mathematics and mathematical history. However, it is less well known that Bell was also a highly regarded number theorist in his day; he even shared, in 1924, the second Bocher prize with Solomon Lefschetz! Some of his results are still important and frequently cited in the field of combinatorics. However, beyond writing these popular books and having a prolific mathematical career (he published over 200 papers!), he also wrote under the pseudonym John Taine a number of science fiction novels in the very early years of this genre, some of which were most favourably received, and composed several volumes of poetry. This last endeavour was perhaps nearest of all to his heart, but he never received any measure of recognition for it... E. T. Bell was a colorful man, and his life as documented here is unexpectedly fascinating. Also very interesting is the picture painted here of the mathematical profession, and the position of mathematics departments within universities, in the 1920s and 1930s."

From the book description, "Constance Reid has given us a compelling account of this complicated, difficult man who never divulged to anyone, not even to his wife and son, the story of his early life and family background. Her book is thus more of a mystery than a traditional biography. It begins with the discovery of an unexpected inscription in an English churchyard and a series of cryptic notations in a boy's schoolbook. Then comes an inadvertent revelation, by Bell himself, in a respected mathematical journal. You will have to read the book to learn the rest."

2.2.3 More biographies of mathematicians

There are many other excellent biographies of mathematicians. We have selected several on some special mathematicians and hope to convey that mathematicians are people too and have all the virtues and faults, happiness and sadness of the ordinary mortals. As in every walk of life, there are different kinds of mathematicians. It is their common love for mathematics (or pursuit of mathematics) that put them into a special group.

A. Stubhaug, *Niels Henrik Abel and His Times. Called Too Soon by Flames Afar*, Translated from the second Norwegian (1996) edition by Richard H. Daly, Springer-Verlag, Berlin, 2000. x+580 pp.

Niels Henrik Abel was born in Nedstrand, Norway, in 1802 and died at the age of 27. He is certainly well-known to many people for his proof of nonsolvability of the

2.2. BIOGRAPHIES AND HISTORY

general quintic equation in radicals, but he made major contributions to other subjects in mathematics, for example, his deep work on elliptic functions. Several basic concepts and theorems are named after him. The great mathematician Hermite said "Abel has left mathematicians enough to keep them busy for 500 years." This is indeed true up to now. When asked how he developed his mathematical abilities so rapidly, he replied "by studying the masters, not their pupils." Abel said famously of Carl Friedrich Gauss's writing style, "He is like the fox, who effaces his tracks in the sand with his tail." It is tempting to learn more about the life of such a genius.

According to MathSciNet, "This is a truly remarkable book that takes you inside the life and passion of a mathematician. Anybody with or without a mathematical background can enjoy it."

According to a review in American Monthly, "Arild Stubhaug has given us a work of lasting value. It will serve as an indispensable source book for any one wanting to learn about Abel, ..."

According to an Amazon review, "This is an excellent book about the Norwegian mathematical genius Niels Henrik Abel. This book is also very painful to read, since it is about a man who dies at only 26 years of age, just before he is to reach to glory. It's a tragedy. But none the less what he accomplished in his life is just tremendous. This is a real book. And it will not leave you unaffected."

J. Spicci, *Beyond the Limit: The Dream of Sofya Kovalevskaya,* Forge Books; 1st edition, 2002. 496 pp.

Sofya Kovalevskaya was born in 1850 in Moscow and died in 1891. She was the first major Russian female mathematician, and made important original contributions to analysis, differential equations and mechanics. She is the first woman appointed to a full professorship in Northern Europe. In many ways, she is not a typical mathematician, or even a typical female mathematician. This makes her biography special too.

From the book description, "Based on Kovalevskaya's own writings, and many other primary sources, the story of her life plays out against a panorama of the turbulent, intellectually challenging 1860s and 1870s, as it follows a brilliant, complex woman on a quest that seems almost impossible to imagine, more than a century later... This fascinating, intimate portrait of Sofya Kovalevskaya's life confronts issues of women's rights and feminism that continue to face women who pursue careers in the sciences in the twenty-first century."

A. Hodges, *Alan Turing: the Enigma*, A Touchstone Book, Simon & Schuster, New York, 1983. x+588 pp.

With ever increasing use and impact of computer on every aspect of life, the name Turning is becoming a household name in the world scientific community. Many people know that he was famous for cracking the Enigma during World War II. But his life is still an enigma. What this book reveals is certainly a highly nontrivial life already.

According to MathSciNet, "This biography of Turing, among other things, vividly recounts his pioneering work in the 1930s on computability, his war-time contributions to the deciphering of the Enigma code and his part in the postwar development of the modern computer. There is also a compassionate account of his private life leading to his untimely death at the age of 42."

From the book description, "Alan Turing (1912-54) was a British mathematician who made history. His breaking of the German U-boat Enigma cipher in World War II ensured Allied-American control of the Atlantic. But Turing's vision went far beyond the desperate wartime struggle. Already in the 1930s he had defined the concept of the universal machine, which underpins the computer revolution. In 1945 he was a pioneer of electronic computer design. But Turing's true goal was the scientific understanding of the mind, brought out in the drama and wit of the famous 'Turing test' for machine intelligence and in his prophecy for the twenty-first century.

Drawn in to the cockpit of world events and the forefront of technological innovation, Alan Turing was also an innocent and unpretentious gay man trying to live in a society that criminalized him. In 1952 he revealed his homosexuality and was forced to participate in a humiliating treatment program, and was ever after regarded as a security risk. His suicide in 1954 remains one of the many enigmas in an astonishing life story."

According to an Amazon review, "Without this book, the real Alan Turing might fade into obscurity or at least the easy caricature of an eccentric British mathematician. And to the relief of many, because Turing was a difficult person: an unapologetic homosexual in post-victorian England; ground-breaking mathematician; utterly indifferent to social conventions; arrogantly original (working from first principles, ignoring precedents); with no respect for professional boundaries (a 'pure' mathematician who taught himself engineering and electronics)... Andrew Hodges's remarkable insight weaves Turing's mathematical and computer work with his personal life to produce one of the best biographies of our time, and the basis of the Derek Jacobi movie Breaking the Code. Hodges has the mathematical knowledge to explain the intellectual significance of Turing's work, while never losing sight of the human and social picture."

B. Schechter, *My Brain is Open. The Mathematical Journeys of Paul Erdős*, Simon & Schuster, New York, 1998. 224 pp.

Paul Erdős was born in Budapest, Hungary in 1913 and died in 1996. He is one of the most prolific writers of papers in mathematical history, probably comparable with Leonhard Euler in terms of quantity. He published more papers than any other mathematician in history and worked with hundreds of collaborators. He made important contributions to combinatorics, graph theory, number theory, classical analysis, approximation theory,

2.2. BIOGRAPHIES AND HISTORY

set theory, and probability theory. He was also famous for asking questions, for example, the famous Erdős problems with monetary value attached.

To most people, mathematicians live and think in a different life from the rest of world. They probably do not understand why someone is willing to spend so much time to study something so abstract. Paul Erdős is certainly a special member of this special group, and his life is one of the best testimonies for the beauty and attraction of mathematics. This book is one of the best record of this unique colorful life of a person whose whole life is devoted to mathematics. To him, "the meaning is life was to prove and conjecture."

According to MathSciNet, "The great mathematical achievements of Paul Erdős were matched by a unique personality. Bruce Schechter's biography is aimed at a general audience. The theorems given—Königsberg bridges, infinity of primes, irrationality of square root of two, etc.—are basic stuff. Still, there is much here to intrigue and delight the professional mathematician.

There are stories galore. Ulam, Gödel, von Neumann, Cantor, Hilbert, Ramsey and many others fill these pages. Mathematicians from Erdős' homeland naturally get special attention and the 'mystery of Hungarian talent' (as Laura Fermi famously phrased it) is described and explored. What effect, to give one important example, did the magazine Komal for budding young Hungarian mathletes have on their remarkable development?

The core of the book is Erdős himself. His ceaseless traveling from place to place; his total dedication to 'proving and conjecturing'; his lifelong interest in prodigies; his numerous joint authors inspiring the Erdős number; his view of 'the Book' and the Book proof—Schechter attempts to weave these and other threads together to give a full picture of this special man. His success or failure must be decided by the individual reader."

P. Hoffman, *The Man Who Loved Only Numbers. The Story of Paul Erdős and the Search for Mathematical Truth,* Hyperion Books, New York, 1998. viii+302 pp.

The title of this book might be a bit misleading. Though Paul Erdős was single, he was always surrounded by people and working with people. He has the most collaborators among all mathematicians in the history, which caused the introduction of the Erdős number. This biography complements the earlier one (the earlier seems to cover more grounds and paints a bigger picture), and different people will prefer a different one.

It seems that the title of the book above *The man who loved only numbers* was motivated by the following book.

R. Kanigel, *The Man Who Knew Infinity. A Life of the Genius Ramanujan,* Charles Scribner's Sons, New York, 1991. x+438 pp.

Srinivasa Iyengar Ramanujan was born in India in 1887 and died in 1920. With almost no formal training in pure mathematics, he made highly original contributions to mathematical analysis, number theory, infinite series and continued fractions. According to the famous English mathematician G.H. Hardy, Ramanujan belonged to the same league as legendary mathematicians such as Gauss, Euler, Cauchy, Newton and Archimedes. Probably the best known statement named after him is the Ramanujan conjecture, which gives a sharp bound on the Fourier coefficients of a modular function of weight 12. Since the giants of mathematics such as Gauss, Euler, and Newton lived in ages quite far away from us and Ramanujan lived in the recent history, his story has a more contemporary appeal. The fact that he is not a mathematician trained in the standard way also makes this biography special.

According to MathSciNet, "The public life of Srinivas Ramanujan is recorded in his 21 major research papers. The author has thrown much light on the actual life of Ramanujan. He collected his data chiefly from Trinity College in Cambridge, Crownwall Road and Chambertown in London where Ramanujan spent a good number of days. He visited extensively the places associated with Ramanujan's memory, including Erode, the birthplace of Ramanujan. He laboured hard in writing the book with great care. The book depicts the life of a mathematician whose life span is anomalous and enigmatic. An assessment of such a life of a celebrity is an almost impossible task.

The author presents Ramanujan's happy and unhappy times from different viewpoints. The events of his personal life are depicted vividly in this book. Most of the space in this fascinating biography is taken up with an account of Ramanujan's search for patrons, his coming to England at the invitation of G. H. Hardy, and his life and work at Cambridge during the years 1913–1918."

From Publishers Weekly, "In 1913 Ramanujan, a 25-year-old clerk who had flunked out of two colleges, wrote a letter filled with startlingly original theorems to eminent English mathematician G. H. Hardy. Struck by the Indian's genius, Hardy, member of the Cambridge Apostles and an obsessive cricket aficionado, brought Ramanujan to England. Over the next five years, the vegetarian Brahmin who claimed his discoveries were revealed to him by a Hindu goddess turned out influential mathematical propositions. Cut off from his young Indian wife left at home and emotionally neglected by fatherly yet aloof Hardy, Ramanujan returned to India in 1919, depressed, sullen and quarrelsome; he died one year later of tuberculosis."

There are also **several other good books** that are partly mathematical and biographical. We mention some of them here.

J.W. Dauben, *Georg Cantor: His Mathematics and Philosophy of the Infinite*, Princeton University Press, Princeton, NJ, 1990. xiv+404 pp.

Georg Cantor was born in 1845 in the western merchant colony in Saint Petersburg, Russia, and died in 1918. His family moved to Germany in 1856, and he spent most of his academic life in Germany. He is best known as the inventor of set theory, which has become a fundamental theory in mathematics. When people think of Cantor, people think of theories of infinite sets. His earlier work on trigonometric series and his motivation might not be well-known to many people. This biography gives a careful comprehensive overview of his work and its development. For example, it contains 40 pages of discussion on trigonometric series with chapter headings: Preludes in Analysis, The origins of Cantorian Set Theory: Trigonometric Series, Real Numbers, and Derived Sets. Besides learning the usual life of a great mathematician (Cantor is sometimes compared with the famous painter Van Gogh), reading this book will help people to understand better a global picture of Cantor's theory of infinite sets.

According to MathSciNet, "It has been said that E. T. Bell's *Men of Mathematics* inspired a generation of historians of mathematics – who then spent their time cleaning up his errors. The evaluation of Cantor's mathematical life and work has required a second generation, but has now been brilliantly accomplished in the present book, which grew out of the author's dissertation, but includes twice the material.

The first half of the book treats the early development of Cantor's set theory and includes a particularly careful account of how it grew out of his study of trigonometric series. The advances and frustrations in his work are recounted, sometimes on a day-to-day basis, showing how he was led to the printed results. This part of the book concludes with an analysis of Cantor's philosophy of the infinite.

The story then continues to the famous Beiträge of 1895 and 1897. This later material is more widely known than the earlier, but here, too, the author manages, through a careful reading of published and unpublished material, to convey a sense of the motives guiding Cantor's research. The author then presents an analysis of the foundations of Cantor's set theory and of its philosophical and theological implications. The effect of the discovery of the antinomies in set theory is told in some detail. A description of Cantor's personality, including a sane and sober account of his nervous breakdowns – as opposed to Bell's wild sensationalism – round out this book, which will remain for many years the essential scholarly study of its subject."

From the book description, "One of the greatest revolutions in mathematics occurred when Georg Cantor (1845–1918) promulgated his theory of transfinite sets. This revolution is the subject of Joseph Dauben's important study the most thorough yet written of the philosopher and mathematician who was once called a 'corrupter of youth' for an innovation that is now a vital component of elementary school curricula.

Set theory has been widely adopted in mathematics and philosophy, but the controversy surrounding it at the turn of the century remains of great interest. Cantor's own faith in his theory was partly theological. His religious beliefs led him to expect paradoxes in any concept of the infinite, and he always retained his belief in the utter veracity of transfinite set theory. Later in his life, he was troubled by recurring attacks of severe depression. Dauben shows that these played an integral part in his understanding and defense of set theory."

I.M. Yaglom, *Felix Klein and Sophus Lie. Evolution of the Idea of Symmetry in the Nineteenth Century,* Translated from the Russian by Sergei Sossinsky, Translation edited by Hardy Grant and Abe Shenitzer, Birkhäuser Boston, Inc., Boston, MA, 1988. xii+237 pp.

As discussed before, there have been many popular books on Lie groups and symmetry. This wonderful book is different and covers the works and life of many major mathematicians in the 19th century who have contributed to the development of geometry and group theory. It is good for both pleasure reading and more serious studying, since more mathematical notes are added near the second half of the book and can be skipped, if one chooses, without causing interruption. On the other hand, these notes contain a lot of valuable information.

According to MathSciNet, "This is a lucid exposition of the interrelationship of geometry and the emerging group theory in the 19th century. It starts with a short exposition of Galois' work in modern terminology, connecting both the mathematical introduction of finite groups and historical information on their origin in early Galois theory and the investigation of permutation groups. The middle part of the book deals with three aspects of the development of geometry in the 19th century, the rise of projective geometry, of non-Euclidean geometries, and of higher-dimensional vector spaces (and hypercomplex number systems).

That sets the stage for an exposition of the central ideas of Lie's theory of continuous transformation groups and Felix Klein's 'Erlanger Programm' (Erlangen Program). Here again the author's style is that of a lucid blend of exposition of mathematical ideas in modern notation and terminology with historical and biographical information. In Lie's case he includes a short discussion of the latter's intended use of the new concept in analysis (differential equations) and geometry (contact transformations).

The author attains a very lucid narrative of his story directed to a mathematically literate, broad audience. This is partly due to the export of the bulk of the more detailed information, be it historical or mathematical, into an extensive appendix of notes, which in total is of the same length as the core text. These notes fill in many of the historical details and contain interesting side remarks referring to later mathematical developments and recent mathematical textbooks. The cited literature is of a remarkable international range and is thus as useful for the Western reader without any restriction."

M. Monastyrsky, *Riemann, Topology, and Physics. With a Foreword by Freeman J. Dyson,* Translated from the Russian by Roger Cooke, Second edition, Birkhäuser Boston, Inc., Boston, MA, 1999. xiv+215 pp.

2.2. BIOGRAPHIES AND HISTORY

Riemann is certainly a well-known name to mathematicians. Many objects named after him, such as Riemann zeta function, Riemannian manifolds, Riemann surfaces etc, come immediately to mathematician's mind. But his impact on physics is probably less known to them. This book is also a good introduction to some important topics in modern physics.

This book consists of two separate but related booklets. The first gives a concise account of the life and work of Riemann, the second an account of several different topics in contemporary physics which are illuminated by the introduction of topological ideas originated from Riemann's work.

According to MathSciNet, "It is particularly satisfactory to see how topological ideas, which have their origins in Riemann's work, explain substantial items of current work in physics. The result is a fine piece of exposition in applied mathematics and in the uses of history... The book does three things very well: it reminds us of the range and depth of Riemann's interests, which are emblematic of what the author values in mathematical physics; describes some of the many successes of Russian mathematicians and physicists; and it provides a lucid account of some modern work in which topology is genuinely applied. Books like this are vital for the health of mathematics and it is to be hoped that more will be written."

W.K. Bühler, *Gauss. A Biographical Study,* Springer-Verlag, Berlin-New York, 1981. viii+208 pp.

There are many biographies about Gauss, one of the most famous mathematicians in the history of mathematics, from some which are really simplified and elementary ones for children to others which are more sophisticated and specialized. This biography on Gauss by Bühler is for mathematicians, or rather for serious mathematicians. By reading this book, one can get a global picture of what the prince of mathematics has done.

According to MathSciNet, "This biography is addressed to the contemporary professional mathematician who is assumed to be a specialist with limited historical interests and knowledge. Although the fifteen chapters, supplemented by several shorter interchapters, are basically chronological, particular aspects of Gauss's work are stressed in particular chapters (e.g., potential theory in Chapter 11) so that for most of the book the reader will be outside his narrow specialty. Inevitably the life of the "Prince of Mathematicians" raises many important historical questions, such as the relationship between pure and applied mathematics and of both to the political and economic background, communication and cooperation, conservatism and innovation, and personal and social life, so no reader can avoid considering them. As Gauss spent most of his life directing the observatory at Göttingen, we can maintain that he established his reputation by predicting the position of the new planet Ceres in 1801. In astronomy and other applied fields Gauss was able to influence and cooperate with others. But the story is very different in his most important pure topic, non-Euclidean geometry. There he behaved as an aristocrat; although sufficiently interested to learn Russian to read Lobachevsky in the original, he always argued that he had discovered new results before these but gave no clear lead as to the revolutionary importance of the new work. The biography abounds with material on these and other important topics."

According to a review in Bulletin of AMS, it "is definitely worthwhile and interesting reading for any serious mathematician. The 25 chapters vary considerably in ease of reading. Many chapters read very well, but some of the chapters which go into mathematical detail are rather tough going..."

W. Dunham, *Euler: the Master of Us All*, The Dolciani Mathematical Expositions, 22, Mathematical Association of America, Washington, DC, 1999. xxviii+185 pp.

Euler is probably the most prolific mathematician in the history and his work has influenced almost all branches of mathematics. This can be seen by many things named after him. If we follow the advice of Abel "learn from the masters", Euler is certainly one of the masters we should learn from. How to get a glimpse of what he has done? This book provides a good snap view of the huge amount of output of Euler.

From the book description, "This book examines the huge scope of mathematical areas explored and developed by Euler, which includes number theory, combinatorics, geometry, complex variables and many more. The information known to Euler over 300 years ago is discussed, and many of his advances are reconstructed. Readers will be left in no doubt about the brilliance and pervasive influence of Euler's work."

According to MathSciNet, "Those who explore the prodigious output of Leonard Euler (1707–1783) are captivated by its richness, inventiveness and elegance. Dunham's fine book captures the spirit of Euler's achievements and enables the English reader with a solid background in calculus and school algebra to share the enjoyment. The author has sampled Euler's work in real and complex analysis, algebra, number theory, geometry and combinatorics. Each of the eight chapters opens with a prologue setting the context, paraphrases some contribution of Euler and concludes with a description of later developments. The first six chapters are a carefully orchestrated examination of Euler's use of algebra and analysis to study perfect numbers, primes, series expansions of transcendental functions, sums of certain series of integer reciprocals and the theory of equations. In the final two chapters, geometry is represented by Heron's area formula and the Euler line, and combinatorics by derangements and partitions. The volume is augmented by a brief biography and an overview of Euler's *Opera Omnia*."

T. Crilly, *Arthur Cayley*, Mathematician laureate of the Victorian age, Johns Hopkins University Press, Baltimore, MD, 2006. xxiv+610 pp.

2.2. BIOGRAPHIES AND HISTORY

As mentioned before, Euler is probably the most prolific and broad mathematician in the history. Another more recent productive and broad mathematician is Arthur Caley. His impact on the modern mathematics is also huge. For example, he was the first person who formally defined the notion of groups, and was one of the founders of invariant theory, and the Cayley graph of a finitely generated group is the bread and butter of geometric group theory. Outside mathematics, he lived a full and interesting life. Cayley was born in 1821. He is probably not at the level of Euler, but he can be considered as a modern master. This book gives a thorough nontechnical summary of the life and work of Cayley. To mathematicians, this book might not be mathematical or precise enough in the discussion of his work.

From the book description, "Arthur Cayley (1821–1895) was one of the most prolific and important mathematicians of the Victorian era. His influence still pervades modern mathematics, in group theory (Cayley's theorem), matrix algebra (the Cayley-Hamilton theorem), and invariant theory, where he made his most significant contributions."

According to MathSciNet, "The author presents a very thorough biography of Arthur Cayley. He appreciates Cayley as the central figure of British mathematics in the 19th century, as the man who led his compatriots out of the intellectual hibernation that had existed since the time of Newton, and as a typical representative of the Victorian Age in Britain... The author describes Cayley's life in a broad context. For example, he draws a clear picture of life in Russia in the 1820s, of the systems of competitions at Cambridge University, of the striving for reforms of mathematical studies at that university, which took place in several stages during the 19th century, and of the spirit of the Victorian Age. He also portrays many factors which influenced Cayley's investigations, such as the English algebraical school. The reader learns many facts about Cayley's achievements and obtains an impression of how Cayley could manage to produce nearly simultaneously so many mathematical results important to many different parts of mathematics. But all the mathematical parts are presented without extensive use of mathematical formalism. Therefore, the book can be recommended to those who are not specialists in mathematics and are interested in historico-cultural or social science aspects. However, due to the bibliographical references and Cayley's 'Collected Works', which are available in many libraries, mathematicians can easily study the details."

2.2.4 A. Weil, Number Theory. An Approach through History. From Hammurapi to Legendre. Birkhäuser Boston, MA, 1984. xxi+375 pp.

Who are the best writers of the history? Probably the history makers. Certainly they will have different perspectives from the professional historians. The same thing is probably true about the history of mathematics. It takes deep understanding of mathematics to see clearly evolution and development of ideas and theories in mathematics, and connections between them. This is a history book on an important field of mathematics by a master.

From the book cover, "Andre Weil, one of the outstanding contributors to number theory, has written an historical exposition of this subject; his study examines texts that span roughly thirty-six centuries of arithmetical work – from an Old Babylonian tablet, datable to the time of Hammurapi to Legendre's Essai sur la Théorie des Nombres (1798). Motivated by a desire to present the substance of his field to the educated reader, Weil employs an historical approach in the analysis of problems and evolving methods of number theory and their significance within mathematics. In the course of his study Weil accompanies the reader into the workshops of four major authors of modern number theory (Fermat, Euler, Lagrange and Legendre) and there he conducts a detailed and critical examination of their work. Enriched by a broad coverage of intellectual history, Number Theory represents a major contribution to the understanding of our cultural heritage."

According to MathSciNet, "As the author says, this is a historical treatment of that oldest and purest field of mathematics, the theory of numbers; his presentation is meticulous and scholarly... it is a discursive, expository, leisurely peek over the shoulders of several great authors in number theory, a subject 'conspicuous for the quality rather than for the number of its devotees; at the same time it is perhaps unique in the enthusiasm it has inspired', as Professor Weil says in his preface."

According to Zentralblatt Math, "The book makes a fascinating reading, permitting to perceive the birth of new ideas, and to understand why they should have been born... There are four chapters: Protohistory, Fermat and his correspondents, Euler and An age of transition: Lagrange and Legendre, and also several appendices, which introduce a modern point of view and provide proofs for many mentioned results. The book is strongly recommended to anybody interested in the history of mathematics and should be on the shelf of every number-theorist."

According to Periodica Mathematica Hungaria, "A very unusual book combining thorough philological exactness, keen observation, apt comments of the essential points, picturesque fantasy, enthusiastic love of the subject, and brilliant literary style: a romantic novel of documents. It is both number theory and its history in an inseparable oneness, helping us understand the very roots and the first big stage of progress of this discipline. The author, one of the most prominent number theorists, chose to give us a broad perspective of the birth of modern number theory."

According to an Amazon review, "When someone like Weil sets out to write a history of number theory it is destined to be the standard reference for decades to come. But this is not only an authoritative reference everyone loves to cite–it is also delightfully

2.2. BIOGRAPHIES AND HISTORY

readable. It is not a substitute for a textbook (although Weil hints at this possibility is the preface), but even for readers with only a modest background in number theory this book will be a source of insight and joy..."

According to a review in Bulletin of AMS, "It is written in a prose which is precise, with a pleasant rhythm, very agreeable to read. To state that the subject matter has been very well researched and the author has found the relevant documents—is obvious, but insufficient to express the lifelong familiarity of Weil with the historical development of number theory. Nourished in the mathematics of the past, Weil propelled the future. In number theory and algebraic geometry his well-known discoveries and conjectures have their roots in genuinely classical work.

Weil has chosen to develop his book around four mathematicians among past giants, Fermat, Euler, Lagrange and Legendre—the period to be covered excluded a priori their successors Gauss, Dirichlet, Kummer, Riemann, and others."

Weil also wrote another famous historic book on number theory.

A. Weil, *Elliptic Functions According to Eisenstein and Kronecker,* Ergebnisse der Mathematik und ihrer Grenzgebiete, Band 88, Springer-Verlag, Berlin-New York, 1976. ii+93 pp.

Kronecker is probably a household name in the mathematics community, but Eisenstein is not. One reason might be that he died young at the age of 29. Eisenstein is a major mathematician and his ideas and work have had a huge impact on the modern mathematics, and Eisenstein series is one such example. It is believed by some people that Gauss had claimed, "There have been only three epoch-making mathematicians: Archimedes, Newton, and Eisenstein".

This is a great historic (or mathematics) book by a great mathematician, and the work of Eisenstein is emphasized. According to Bulletin of the London Mathematical Society, "As a contribution to the history of mathematics, this is a model of its kind. While adhering to the basic outlook of Eisenstein and Kronecker, it provides new insight into their work in the light of subsequent developments, right up to the present day. As one would expect from this author, it also contains some pertinent comments looking into the future. It is not however just a chapter in the history of our subject, but a wide-ranging survey of one of the most active branches of mathematics at the present time. The book has its own very individual flavour, reflecting a sort of combined Eisenstein-Kronecker-Weil personality. Based essentially on Eisenstein's approach to elliptic functions via infinite series over lattices in the complex plane, it stretches back to the very beginnings on the one hand and reaches forward to some of the most recent research work on the other... The persistent reader will be richly rewarded."

Some quotes from the book might be interesting. "'When kings are building', says the German poet, 'carters have work to do'. Kronecker quoted this in his letter to Cantor of September 1891, only to add, thinking of himself no doubt, that each mathematician has to be king and carter at the same time.

But carters need roads. Not seldom, in the history of our science, has it happened that a king opened up a new road into the promised land and that his successors, intent upon their own paths, allowed it to be overrun by brambles and become unfit for transit.

To help clean up such a road is the purpose of this little book, arising out of lectures given at the Institute for Advanced Study in the Fall of 1974 ... Where the road will lead remains, in large part, to be seen, but indications are not lacking that fertile country lies ahead."

A conjecture stated in the book above by Weil inspired the following interesting book.

D. Cox, *Primes of the Form $x^2 + ny^2$. Fermat, class field theory and complex multiplication*, A Wiley-Interscience Publication, John Wiley & Sons, Inc., New York, 1989. xiv+351 pp.

The author of this book on number theory is more an algebraic geometer than number theorist, and his love for the subject is clearly visible. It is accessible and enjoyable to read, and a valuable introduction for students and non-experts.

From the book description, it "begins with Fermat and explains how his work ultimately gave birth to quadratic reciprocity and the genus theory of quadratic forms. Further, the book shows how the results of Euler and Gauss can be fully understood only in the context of class field theory. Finally, in order to bring class field theory down to earth, the book explores some of the magnificent formulas of complex multiplication..."

According to MathSciNet, "The principal theme is the following basic question of classical number theory: given a positive integer n, which primes p can be expressed in the form $p = x^2 + ny^2$, for integers x, y? The complete solution demands application of diverse material from rich areas of number theory. Quadratic reciprocity and elementary theory of binary quadratic forms over $\mathbb{Z}$ are sufficient to deal with the special cases of $n = 1, 2, 3$ considered by Fermat. Genus theory and cubic and biquadratic reciprocity extend the results, but more powerful machinery is needed to advance further. Class field theory provides an abstract solution to the problem, but only by use of modular functions and complex multiplication is a final effective algorithm shown to exist. The book accordingly covers material from the quite elementary to the rather advanced, but the author's style is totally lucid and very easy to read. One of the goals accomplished is to provide a well-motivated introduction to the classical formulation of class field theory (and accordingly adéle and idéle treatment is ignored). Throughout, great relevance is placed upon the explicit numerical example, illustrating the power of basic theorems in various concrete situations."

2.2.5 W. Scharlau, H. Opolka, From Fermat to Minkowski. Lectures on the Theory of Numbers and its Historical Development, Undergraduate Texts in Mathematics. Springer-Verlag, New York, 1985. xi+184 pp.

Even though this book is a volume in the book series *Undergraduate Texts in Mathematics*, research mathematicians can learn things from it and enjoy reading it. One of the important aspects of arithmetic subgroups of linear algebraic groups concerns the reduction theory, i.e., finding good fundamental domains for the arithmetic groups acting on associated symmetric spaces. This book explains clearly the origin and explicit results on the reduction theory in special cases. The historical and biographic material enhances the reading experience.

From the preface, "This book arose from a course of lectures given by the first author during the winter term 1977/1978 at University of Münster. The course was primarily addressed to future high school teachers of mathematics; it was not meant as a systematic introduction to number theory but rather as a historically motivated invitation to the subject, designed to interest the audience in number-theoretical questions and developments. This is also the objective of this book, which is certainly not meant to replace any of the existing excellent texts in number theory. Our selection of topics and examples tries to show how, in the historical development, the investigation of obvious or natural questions has led to more and more comprehensive and profound theories, how again and again, surprising connections between seemingly unrelated problems were discovered, and how the introduction of new methods and concepts led to the solution of hitherto unassailable questions. All these means that we do not present the student with polished proofs (which in turn are the fruit of a long historical development); rather, we try to show how these theorems are the necessary consequences of natural questions..."

According to MathSciNet, "It is an excellent, inspiring and original text on number theory and its development centering on the theme of quadratic forms. Following the historical evolution from the representation theory of Fermat up to the reduction theory of Minkowski, it points out how certain number-theoretic problems have given rise to steadily growing theories in number theory. It stresses the motivating ideas giving credit to the historical context and it shows, by means of many examples, how genuine arithmetical problems are very often related to other fields in mathematics such as algebra, geometry, analysis and even physics. The lectures require no prerequisites other than very basic knowledge in linear algebra, group theory and analysis. Some elementary Galois theory of the cyclotomic field is used at one place. The book is grouped into chapters each allotted to a mathematician containing also a summary of the latter's biography. It is enriched by portraits and by a facsimile of a letter by Gauss."

The short biographies of the major mathematicians inserted in the book put their works into a historical perspective and make the book interesting. They also motivate people to read on.

2.2.6 History of mathematics

The history of mathematics is long and broad, and there are many mathematics history books. It is difficult to pick one or two as the great book(s) on the history of mathematics. We list several of them which are popular with various people and have withstood the test of time.

D.J. Struik, *A Concise History of Mathematics*, Fourth edition, Dover Publications, Inc., New York, 1987. xiv+228 pp.

It was first published in 1948 and has been revised and reprinted many times. It has also been translated into several languages. Unlike many other books on the history of mathematics, this is a very short book and many comments are concise and to the point. It is amazing how much material can be covered in such a short book. It is a classic.

According to Nature, "This compact, well-written history covers major mathematical ideas and techniques from the ancient Near East to 20th-century computer theory, surveying the works of Archimedes, Pascal, Gauss, Hilbert, and many others. The author's ability as a first-class historian as well as an able mathematician has enabled him to produce a work which is unquestionably one of the best."

According to MathSciNet, it is "A very condensed survey of the history of mathematics covering the whole development from the earliest times up to 1900. The conciseness of the work imposed the necessity of a strict limitation to the main currents of thought. This, however, did not prevent the author from paying due attention to what always has been his favourite topic, viz., the relations of mathematics to the general cultural and sociological background of the period. In this way the two trends which can be distinguished in Renaissance mathematics, the arithmetical-algebraic and what may be called the fluctional one, are related, respectively, to the commercial and to the engineering interests of these centuries. In treating of Greek mathematics, that most important, but very imperfectly known, phase of the development of mathematics, the author takes care to distinguish between established facts, plausible theories, wild hypotheses and traditional ideas. The exposition of 19th century mathematics is based more on persons and schools than on subjects. The work is richly illustrated."

V. Katz, *A History of Mathematics. An Introduction*, Harper Collins College Publishers, New York, 1993. xiv+786 pp.

This book is a comprehensive and also contains some discussion of mathematics, even fairly recent results, which makes it different from other history books and attractive.

2.2. BIOGRAPHIES AND HISTORY

There are many books on the history of mathematics. In comparison with the previous book in terms of length and style, it lies at the other end of the spectrum of history books on mathematics. It is a standard textbook on history of mathematics.

From the book description: "This book's global perspective covers how contributions from Chinese, Indian, and Islamic mathematicians shaped our modern understanding of mathematics. This book also includes discussions of important historical textbooks and primary sources to help readers further understand the development of modern math."

The author of the previous book, Struik, reviewed this book in MathSciNet: "The author, who teaches at the University of the District of Columbia, has written this attractive, but heavy, tome in the 'conviction that not only prospective school teachers of mathematics, but also prospective college teachers of mathematics need a background in history'. It is therefore written with emphasis on the needs of secondary and undergraduate education, and also of students who are of varied ethnic backgrounds. This emphasis on the multi-cultural nature of the mathematics that is to be taught has led to extensive discussion of Chinese, Mesopotamian, Indian and Islamic mathematics, together with that of the European Renaissance up to Newton and Euler... This is a well-written and well-illustrated book with fair accounts of the important texts of the various periods, of Euclid, Archimedes and Apollonius, of Brahmagupta, al-Khwarizmi, al-Karaji, Fibonacci and the leading books of later years up to Euler, Gauss, Cantor and others... Though of the almost 800 pages almost 600 are devoted to the mathematics before the nineteenth century, the remaining 200 pages illustrate quite well the principal features of the last and even the present century, from Gauss to Turing and von Neumann. The mathematics of the British colonies in America and the USA also finds its place. There is an 'Interchapter' with information on the Mayas, the Incas, and ethnomathematics in general. Short biographies appear in special boxes, and there is a very interesting set of problems, mostly taken from the ancient texts themselves, from the Rhind papyrus to the Turing machine... The book is richly illustrated and has an excellent reference account, especially of recent literature (mostly in English)... We recommend this highly readable book to teachers, and indeed to all interested in the history of mathematics."

Another classical book on history of mathematics is the following.

C. Boyer, *A History of Mathematics. Second edition. Edited and with a preface by Uta C. Merzbach,* John Wiley & Sons, Inc., New York, 1989. xviii+762 pp. Updated new edition, 3 edition (January 11, 2011).

It is probably one of the most famous and popular books on history of mathematics. Many editions of it have been published. It does not cover anything about modern mathematics and is typical of most books on history of mathematics. Maybe it takes a century to really evaluate and put things into perspective.

From the book description, "Distills thousands of years of mathematics into a single, approachable volume Covers mathematical discoveries, concepts, and thinkers, from Ancient Egypt to the present Includes up-to-date references and an extensive chronological table of mathematical and general historical developments. Whether you're interested in the age of Plato and Aristotle or Poincaré and Hilbert, whether you want to know more about the Pythagorean theorem or the golden mean, A History of Mathematics is an essential reference that will help you explore the incredible history of mathematics and the men and women who created it."

According to MathSciNet, "This is a revised edition of the classic text by Boyer which first appeared over twenty years ago. As is noted in the preface, these years have seen a renewed interest in the history of mathematics, stimulated partly, no doubt, by Boyer's book itself. A steady stream of books and papers has revealed new insights and interpretations. The subject is alive and vibrant. Was the time ripe for a new edition of Boyer's text? To answer "Yes" to the question would be easy; to determine the form of the revision would require more thought... Finally, Boyer's bizarre chronological table has been retained and expanded... The reviser is to be complimented for being willing to take on the task of bringing up to date such a venerable tome, and for effecting the changes so successfully. The new edition deserves to prosper and no doubt will do so. Lecturers in the history of mathematics will have to get used to the new reference, 'Boyer and Merzbach'.

Its level is substantially higher than both D. E. Smith's *History of mathematics* and Howard Eves' *An introduction to the history of mathematics*."

H. Eves, *An Introduction to the History of Mathematics,* Sixth edition, With cultural connections by Jamie H. Eves. Saunders Series. Saunders College Publishing, Philadelphia, PA, 1990. xx+775 pp.

This is not a typical book on history of mathematics. It was written as a textbook and contains problems related to historical contexts. But others will enjoy reading it too.

According to MathSciNet, "This textbook on the history of mathematics is suitable for students with at least one college course in mathematics... The author might consider including material on applicable contemporary mathematics such as game theory, information theory or evolution of systems. He might also include a discussion of mathematics in various countries, such as Poland in the interwar period, or the U.S., particularly since 1930, or deal with recreational mathematics and the increase in publications directed towards a lay or general audience. In this way it would be possible to give a fuller sense of the variety of contemporary mathematics."

2.2. BIOGRAPHIES AND HISTORY

2.2.7 More mathematical history books

Mathematics and mathematicians are both important in the history of mathematics. Evolution of mathematical ideas and concepts, lives of mathematicians and mathematics can be discussed together to enhance all each other. Hence, there are different perspectives of history books on mathematics. The next several books in this subsection are different from the earlier ones by emphasizing some selected key ideas and concepts in mathematics, and hence are more *mathematical*.

N. Bourbaki, *Elements of the History of Mathematics,* Translated from the 1984 French original by John Meldrum. Springer-Verlag, Berlin, 1994. viii+301 pp.

Most books on history of mathematics do not discuss current topics in mathematics. To give a global historical picture of the current mathematics, it is difficult or impossible for a single person or a few people. But for a large group of expert mathematicians, there is a chance. This book is a result of such joint efforts. This is a very valuable and reliable reference. Small articles in this book are independent, and it is fun to read them.

According to MathSciNet, "This fascinating volume assembles the historical notes from Bourbaki's various Éléments... In virtue of its origin as appendices to separate texts on topics of current interest, these elements of history are just that: Former mathematics as it seems now to Bourbaki, and not as it seemed to its practitioners then. In the terminology of historiography, it is "Whig history". But for mathematicians, the various chapters are full of interesting connections and insights—Riemann's decisive influence on the origins of topology (analysis situs), the idea that various sets naturally have more than one topology, the fact that Leibniz had the notion of isomorphism, but that it was Emmy Noether who straightened up the terminology. On the way to these insights, there may be some minor annoyances. Thus the Gibbs-Wilson vector analysis (1900), which has dominated notation in Anglo-Saxon physics for 85 years, is dismissed as a vulgarization of the ideas of Hamilton and Grassmann—despite the fact that Gibbs understood the tensor product of vector spaces well before Bourbaki."

J. Stillwell, *Mathematics and Its History,* Third edition, Undergraduate Texts in Mathematics, Springer, New York, 2010. xxii+660 pp.

As the title indicates, this book unifies serious mathematics with historical information. It covers many important topics of elementary mathematics, though not in depth (which is impossible in a book at such a level). The biographic notes of the main figures at the end of each chapter really enhances the book. It is not a history book, but one can probably learn some highlights more easily this way. It is a fun book to read. Stillwell has written multiple books, and this is the first of his books.

According to MathSciNet, "The present book is a very interesting and useful attempt to give a unified view of undergraduate mathematics by approaching the subject through its history. The author motivates this project as follows: 'One of the disappointments experienced by most mathematics students is that they never get a course in mathematics. They get courses in calculus, algebra, topology, and so on, but the division of labor in teaching seems to prevent three different topics from being combined into a whole.' Indeed, there is a gap just of the kind described in most mathematics curricula, and a historical approach seems to be the right way to fill this gap... Each chapter provides a historical survey explaining the origins of the basic ideas, hints at the connections between the subject treated and other fields of mathematics and science, remarks on recent discoveries and results, nice exercises and some biographical notes.

The author presents the mathematics of the past in modern notation and explains it by using modern concepts and results. This has the risk that the students obtain a biased impression of the content and form of classical sources. On the other hand this presentation makes it much easier to grasp the main ideas. This book is highly recommended as the basis for courses, especially for students who want to become teachers at secondary schools. It should be used together with a good source book (those by Struik, Fauvel and Gray, for instance) in order to complete the historical approach to important mathematical fields and their mutual relations by studying selected sources."

According to an Amazon review, "This is a brilliant book that conveys a beautiful, unified picture of mathematics. It is not an encyclopedic history, it is history for the sake of understanding mathematics. There is an idea behind every topic, every section makes a mathematical point, showing how the mathematical theories of today has grown inevitably from the natural problems studied by the masters of the past.

Math history textbooks of today are often enslaved by the modern curriculum, which means that they spend lots of time on the question of rigor in analysis and they feel obliged to deal with boring technicalities of the history of matrix theory and so on. This is of course the wrong way to study history. Instead, one of the great virtues of a history such as Stillwell's is that it studies mathematics the way mathematics wants to be studied, which gives a very healthy perspective on the modern customs. Again and again topics which are treated unnaturally in the usual courses are seen here in their proper setting. This makes this book a very valuable companion over the years.

Another flaw of many standard history textbooks is that they spend too much time on trivial things like elementary arithmetic, because they think it is good for aspiring teachers and, I think, because it is fashionable to deal with non-western civilizations. It gives an unsound picture of mathematics if Gauss receives as much attention as abacuses, and it makes these books useless for understanding any of the really interesting mathematics, say after 1800. Here Stillwell saves us again. The chapter on calculus is

2.2. BIOGRAPHIES AND HISTORY

done by page 170, which is about a third of the book. A comparable point in the more mainstream book of Katz, for instance, is page 596 of my edition, which is more than two thirds into that book.

Petty details aside, the main point is the following: This is the single best book I have ever seen for truly understanding mathematics as a whole."

The following history book also emphasizes mathematical ideas rather than mathematicians, but it is more comprehensive and systematic. It is very popular but has also attracted criticism regarding its solid understanding of mathematics and accuracy of historical facts.

M. Kline, *Mathematical Thought from Ancient to Modern Times. Vol. 1., Vol. 2, Vol. 3,* Second edition. The Clarendon Press, Oxford University Press, New York, 1990. pp. i-xviii, 1-390 and i-xxii (Vol. 1); pp. i-xx, 391-812 and i-xxii (Vol. 2); pp. i-xvi, 813-1212 and i-xxii (Vol. 3).

By modern times, this book meant before 1900. It would be valuable to have a corresponding comprehensive book that gives a review of what has been achieved in mathematics after 1900. This would probably be very difficult, since mathematics has expanded and there are many more topics. Though recent mathematics might be more interesting to active mathematicians, it takes time and distance to properly judge impact and value of any theory.

Though this is not a mathematics book, it contains a lot of discussion of important results. By reading this book, one can learn a little bit of many topics in mathematics, some of which are usually not taught in standard courses. Reading it is an enjoyable experience.

From the book description, "This comprehensive history traces the development of mathematical ideas and the careers of the mathematicians responsible for them. Volume 1 looks at the discipline's origins in Babylon and Egypt, the creation of geometry and trigonometry by the Greeks, and the role of mathematics in the medieval and early modern periods. Volume 2 focuses on calculus, the rise of analysis in the 19th century, and the number theories of Dedekind and Dirichlet. The concluding volume covers the revival of projective geometry, the emergence of abstract algebra, the beginnings of topology, and the influence of Gödel on recent mathematical study."

According to MathSciNet, "The first, hardcover, edition of this book of more than 1200 pages came out in 1972. Now it is republished, unchanged, in three handsome paperback volumes, which makes this impressive work, result of a lifetime of critical study and research, by the now emeritus professor of New York University, again available. The book contains a wealth of information concerning the mathematics of more than two millennia, from the Babylonians to the present century. The well-known and appreciated attempts by the author in previous books to humanize the subject of mathematics are also characteristic for this book.

In this survey of the history of mathematics in 51 chapters, leading ideas are stressed rather than persons, biography is only subordinate. Some chapters, as those on differential equations in the eighteenth century, on integral equations and functional analysis in modern times, are monographs in their own right. Other chapters, as those on the Renaissance, are more descriptive. The chapters have their own selected bibliography, which, of course, does not reach beyond 1972.

Not only students of the history of mathematics can profit by Professor Kline's extensive and critical knowledge, but also students of more modern mathematics in specialized fields. Where the treatment of older mathematics is sometimes (but not always) a bit sketchy, it is in the more modern sections that we discover the particular strength of the book. Here and there we find passages that are mildly controversial, but this adds to the fun of reading the book. At various points the author pauses for summaries and conclusions, adding to the readability of the work. We recommend it to all who wish to understand the way our present mathematics has come into being..."

According to a review in Bulletin of AMS by Rota, "It is easy to find something to criticize in a treatise 1200 pages long and packed with information. But whatever we say for or against it, we had better treasure this book on our shelf, for as far as mathematical history goes, it is the best we have."

The following anthology of mathematics was edited a lawyer and amateur-mathematician who really loves mathematics.

J. Newman, *The World of Mathematics. Vols. I-IV. A small library of the literature of mathematics from A'h-mosé the Scribe to Albert Einstein, presented with commentaries and notes.* Simon & Schuster, New York, 1956. xviii+2535 pp.

We all know that every generation of mathematicians (or scientists) builds on the work of earlier generations. But mathematics in the ancient time is as valid today as in the past when it was first produced, and this feature makes mathematics unique. On the other hand, most mathematicians rarely read old writings by masters. We also believe that mathematics is important in sciences and has broad applications, but many applications of mathematics are not discussed in standard mathematics books. This book addresses the issues above and provides a convenient and comprehensive collection. It does not cover applications of more modern topics such as Lie groups. In fact, in spite of many applications of Lie groups in modern sciences and mathematics, most books on Lie groups do not discuss their applications.

Reading this book for the first time is an eye opener to many people. Some topics selected in this book might not occur to many mathematicians.

2.2. BIOGRAPHIES AND HISTORY

It was a work of love. According to the *Introduction*, "It is more than fifteen years since I began gathering the material for an anthology which I hoped would convey something of the diversity, the utility and the beauty of mathematics... It presents mathematics as a tool, a language and a map; as a work of art and as an end in itself; as a fillment of the passion for perfection. It is seen as an object of satire, a subject for humor and a source of controversy; as a spur to wit and a haven to the storyteller's imagination; as an activity which has driven men to frenzy and provided them with delight. It appears in broad view as a body of knowledge made by men, yet standing apart and independent of them. In this collection, I hope, will be found material to suit every taste and capacity..."

According to MathSciNet, "This anthology of mathematical essays by mathematicians and other writers has already been reviewed in many journals. Here it is only necessary to say that these essays will be as interesting to mathematical readers as to others. Many of them are examples of the art (more difficult in mathematics than in any other science?) of 'haute vulgarisation', and the editor's commentaries are a fitting accompaniment. Some of the subjects dealt with are: biography of mathematicians, counting, space and motion, the physical world, social science, chance, statistics, design of experiments, group theory, logic, mathematical machines, warfare, art, ethics, literature and music. The writers include Archimedes, Boole, Descartes, Eddington, Galileo, Halley, Hardy, Jeans, Keynes, Laplace, Mendel, Newton, Poincaré, Russell, Shaw, Whitehead and many others."

According to a review in Bulletin of AMS, "While it is not customary to review popular books on mathematics in this Bulletin, this one so far exceeds the norm both in range and in sales that it demands notice. (It is undoubtedly the all-time best-seller among mathematics books other than textbooks.) A non-mathematician with an amateur's interest in the subject might well wonder at first why he should buy these volumes rather than one of the more compact (and less expensive) popular books, of which there are many excellent ones that have enjoyed a far smaller sale. However, most short popular books on mathematics cover only a limited selection of topics that are not too technical to discuss superficially and are conceded to possess universal appeal. Most of these topics are included here too, but so is much more, and the reader can make his own choice. The subtitle is in a sense misleading, since the contents are much more literature about mathematics than mathematics as such. This is of course inevitable in any popular book. A non-mathematician will not learn much mathematics from these volumes, although he is told a great deal about mathematics and about cognate subjects, such as mathematicians, physics, logic, and foreign politics; whether this will help him understand what mathematics is about and what mathematicians do is not for a professional mathematician to say. However, there is also a great deal here of value for the professional mathematician, collected from sources that are not on everyone's bookshelf.

Some at least of this material will be helpful to teachers, and it would be hard to find any mathematician who will not be entertained by some of it, or who will not find something that is new to him... All in all, this is an anthology with the faults of its genre and more virtues than most specimens of its kind, especially in the set of mathematical anthologies of which it is almost the only example. It has as legitimate a place in any mathematician's library as the Oxford Book of English Verse has in that of a specialist in English literature."

Most mathematics books concentrate on mathematics created by mathematicians. But understanding evolution of mathematics from the perspective of life and work of mathematicians, in particular, the founders, can be instructive and inspiring. The following book is the history of a special topic, representation theories of finite groups, with biographies of the founders.

C. Curtis, *Pioneers of Representation Theory: Frobenius, Burnside, Schur, and Brauer,* History of Mathematics, 15, American Mathematical Society, Providence, RI; London Mathematical Society, London, 1999. xvi+287 pp.

Representation theories of groups are relatively recent theories in mathematics, and they are not covered in usual mathematics history books. They are becoming more important with the passing of time, but have a reputation of being not so easy. This book will make it easier by putting it into a historical perspective. One can learn both mathematics and lives of the founders by reading this book.

According to The LMS Newsletter, "This book is likely to be of interest to any mathematician who has had occasion in any of his/her own work to use group representation theory in any of its many contemporary guises. This is a beautiful and carefully written book, which succeeds at many levels. The mathematics discussed is powerful, and influences many areas of modern mathematics (and other sciences). The story of its evolution and its various diversifications has its own fascination, and serves to remind us how a single mathematical question can lead to the creation of vital new areas. The mathematical contribution of the main characters inspires admiration, while we also gain some insight into their lives at a human level. In short, the book fascinates both as mathematics and as history."

According to a review in Bulletin of AMS, "it may be worthwhile to pose a general question: Does one need to know anything about the history of mathematics (or the lives of individual mathematicians) in order to appreciate the subject matter? Most of us are complacent about quoting the usual sloppy misattributions of famous theorems, even if we are finicky about the details of proofs.

There seems to be a recent trend in undergraduate textbooks (especially in subjects like abstract algebra and number theory) to include snippets of history and biography. This is certainly a harmless way to add human interest to what might otherwise seem dry

2.2. BIOGRAPHIES AND HISTORY

axiomatics, but may not by itself make the subject matter more understandable. It is much easier to convey the facts of Emmy Noether's life than to explain to undergraduates what she accomplished mathematically.

Aside from the human interest involved in biographical studies, there may be some intellectual value in retracing the way mathematical ideas have developed. This development is often messy, however. Occasionally good ideas emerge prematurely in obscure places and are forgotten for a time, only to be rediscovered independently. Sometimes the original motivation for an investigation looks a bit eccentric to later generations, as in the case of Hamilton's approach to quaternions. But, in the end, one is often just curious to know where the currently accepted ideas came from.

Whatever one's view may be on the role of the history of mathematics in teaching or research, probably most people will agree that it is more challenging to deal with the twentieth century than with the immediately preceding centuries. Mathematics tends to be hierarchical, making it difficult to appreciate later work without a substantial foundation in earlier work... What makes the book by Curtis *especially attractive is the way it blends biography and the history of ideas with an explanation of the mathematics itself.* The author writes in a careful but readable scholarly style, with judicious footnotes and full references to the primary literature. He goes to considerable pains to explain the sometimes opaque-looking early literature in modern language and notation. While it is quite possible to learn the basic facts about finite group representations from a wide variety of modern textbooks (including those written by Curtis and the late Irving Reiner), those who are at all attracted to the subject will certainly enjoy spending time with Curtis' account. The only prerequisite is a standard mathematical education... All those with an interest in the representation theory of finite groups owe a debt of gratitude to Curtis for having written a thoughtful and informative account of this important chapter in twentieth-century mathematics."

From the book description, "The year 1897 was marked by two important mathematical events: the publication of the first paper on representations of finite groups by Ferdinand Georg Frobenius (1849–1917) and the appearance of the first treatise in English on the theory of finite groups by William Burnside (1852–1927). Burnside soon developed his own approach to representations of finite groups. In the next few years, working independently, Frobenius and Burnside explored the new subject and its applications to finite group theory. They were soon joined in this enterprise by Issai Schur (1875–1941) and some years later, by Richard Brauer (1901–1977). These mathematicians' pioneering research is the subject of this book. It presents an account of the early history of representation theory through an analysis of the published work of the principals and others with whom the principals' work was interwoven. Also included are biographical sketches and enough mathematics to enable readers to follow the development of the subject...

The volume would be a suitable text for a course on representations of finite groups, particularly one emphasizing an historical point of view."

The next book is a history of an important subject with substantial amount of mathematical content written by an expert.

L.E. Dickson, *History of the Theory of Numbers. Vol. I: Divisibility and Primality,* Chelsea Publishing Co., New York, 1966, xii+486 pp. *History of the Theory of Numbers. Vol. II: Diophantine analysis,* Chelsea Publishing Co., New York, 1966, xxv+803 pp. *History of the Theory of Numbers. Vol. III: Quadratic and Higher Forms, With a chapter on the class number by G. H. Cresse,* Chelsea Publishing Co., New York, 1966. v+313 pp.

These three books were first published in 1919, 1920 and 1923, and have been read and enjoyed by many generations of mathematicians (number theorists or not) since then. They are very broad and complete in regarding results in number theory up to 1923. No history books on mathematics can be compared with its depth and width as far as number theory is concerned.

Dickson is remembered for various things including his work on finite fields and finite classical groups, but these three books will be enough to secure his fame. As A. A. Albert remarked, this three volume work "would be a life's work by itself for a more ordinary man."

From the book description, "Dickson's History is truly a monumental account of the development of one of the oldest and most important areas of mathematics. It is remarkable today to think that such a complete history could even be conceived. That Dickson was able to accomplish such a feat is attested to by the fact that his History has become the standard reference for number theory up to that time. One need only look at later classics, such as Hardy and Wright, where Dickson's History is frequently cited, to see its importance... It is interesting to see the topics being resuscitated today that are treated in detail in Dickson... As usual with Dickson, the account is encyclopedic and the references are numerous."

O. Neugebauer, *A History of Ancient Mathematical Astronomy. Part One.* pp. i–xxii and 1–555; *Part Two.* pp. i–iv and 560–1058; *Part Three.* pp. i–iii and 1061–1456. Studies in the History of Mathematics and Physical Sciences, No. 1, Springer-Verlag, New York-Heidelberg, 1975.

In the ancient time, mathematics was often intertwined with sciences and practical applications, in particular with astronomy. Such interplay will never disappear and will probably intensify with time. The subject of mathematical astronomy is the science of using mathematical methods to describe the universe or to predict various aspects of the universe, and mathematics has been, is and will be of central importance to astronomy.

2.2. BIOGRAPHIES AND HISTORY

Astronomy started when the civilization started. Perhaps a natural question to many mathematicians is how mathematics was used in astronomy and how astronomy motivated the development of mathematics. This authoritative book on the history of mathematical astronomy probably provides the best answer one can get.

Neugebauer is a famous historian on astronomy and the other exact sciences in antiquity and into the Middle Ages. His main contribution to mathematics is the review journal *Zentralblatt für Mathematik und ihre Grenzgebiete (Zbl)* in 1931, and *Mathematical Reviews* the American Mathematical Society in 1939. He is also the only person who was honored with a plaque on the outside wall of the building of Mathematisches Institut, University of Göttingen.

According to Zentralblatt MATH, "The book is a classical work. It is unique amongst books which go as far as it does in presenting a history of ancient exact sciences, more exactly of mathematical astronomy. With great care the author, who is familiar with ancient sources, describes the methods employed by oldest astronomers to overcome fundamental problems and to create a rational system of astronomy. He deals here with the numerical, geometrical and graphical methods devised to control the mechanism of the planetary system. In this way the author has profoundly and thoroughly explored the arithmetical methods and cinematic models used from the latest period of Mesopotamian civilization (4 c. B.C.) to the period after Ptolemy (7 c. A.D.), to predict lune-solar phenomena and planetary phases and positions. Many problems are examined which have not been treated by other investigators. There are numerous examples to make the exposition as far as possible intelligible and interesting for readers."

According to Bibliotheca Orientalis, "This monumental work will henceforth be the standard interpretation of ancient mathematical astronomy. It is easy to point out its many virtues: comprehensiveness and common sense are two of the most important. Neugebauer has studied profoundly every relevant text in Akkadian, Egyptian, Greek, and Latin, no matter how fragmentary;... With the combination of mathematical rigor and a sober sense of the true nature of the evidence, he has penetrated the astronomical and the historical significance of his material... His work has been and will remain the most admired model for those working with mathematical and astronomical texts."

K. Parshall, D. Rowe, *The Emergence of the American Mathematical Research Community, 1876–1900: J. J. Sylvester, Felix Klein, and E. H. Moore*, History of Mathematics, 8, American Mathematical Society, Providence, RI; London Mathematical Society, London, 1994. xxiv+500 pp.

The mathematics community of USA has had a great influence on the development of mathematics in the twentieth century and will continue its leading role in the coming years. There were few mathematicians in USA in the early 1900s. To many mathematicians, it is interesting to figure out how the American mathematical research community could emerge in such a short time. This book gives a description of its emergence.

According to MathSciNet, "This fascinating book is a contribution to the history of American science, but is also written for a general mathematical audience. For those of us who have made our careers in American mathematics and are interested in understanding our intellectual heritage, it is essential reading.

The book is organized around three groups of American mathematicians in the period 1875–1910: the school founded by J. J. Sylvester at Johns Hopkins, the group of American students at Leipzig and Göttingen who studied under Felix Klein and brought home his influence, and the Chicago school founded by E. H. Moore. There is an informative discussion of the pre-history—the period in which isolated researchers like Benjamin Pierce and Josiah Willard Gibbs made their marks but failed to found schools of research.

The mathematics is treated fairly thoroughly, but in the main this is a work of educational history. It is a mixed history. We are shown impressive founding presidents cooperating with visionary mathematicians to establish intellectual centers that could hold their own with the best departments in Europe; but we are also told stories of failure whose contemporary relevance will bemuse modern readers.

By 1906, the U.S. had seven principal graduate programs in mathematics: in alphabetical order these were Chicago, Clark, Columbia, Cornell, Harvard, Johns Hopkins and Yale. Later, others joined the list. Why some institutions were more successful and enduring than others is discussed with sympathetic candor. Necessary conditions for success of an institution seem to have included: a large endowment, a president who was a good judge of men (the small place of women is very well discussed, but was not, alas, an important factor), and founding professors who believed firmly in research and publication as objectively good and worthy goals. Beyond the stories of individual institutions, one is struck by the interest which the leading mathematicians of those days took in the founding of the early journals and of the AMS.

Between the period described in this book and modern times lies another important chapter in American mathematical history, the arrival of the refugees from Hitler. The book under review describes, and explains in depth, what those refugees found when they arrived."

2.2.8 J. Dieudonné, A History of Algebraic and Differential Topology 1900–1960, Reprint of the 1989 edition, Modern Birkhäuser Classics. Birkhäuser Boston, Inc., Boston, MA, 2009. xxii+648 pp.

The author Dieudonné read widely, and wrote extensively and quickly. This is probably one of the most accessible and interesting books written by him. It is amazing that within a short time, algebraic topology has achieved such an important status and played such

2.2. BIOGRAPHIES AND HISTORY

an important role in mathematics. Besides history, one can also learn mathematics from this book, since the author of this book clearly loves and understands mathematics.

According to Zentralblatt MATH, "[The author] traces the development of algebraic and differential topology from the innovative work by Poincaré at the turn of the century to the period around 1960. [He] has given a superb account of the growth of these fields. The details are interwoven with the narrative in a very pleasant fashion. [The author] has previous written histories of functional analysis and of algebraic geometry, but neither book was on such a grand scale as this one. He has made it possible to trace the important steps in the growth of algebraic and differential topology, and to admire the hard work and major advances made by the founders."

According to MathSciNet, "This book is a well-informed and detailed analysis of the problems and development of algebraic topology, from Poincaré and Brouwer to Serre, Adams, and Thom. The author has examined each significant paper along this route and describes the steps and strategy of its proofs and its relation to other work. Previously, the history of the many technical developments of twentieth century mathematics had seemed to present insuperable obstacles to scholarship. This book demonstrates in the case of topology how these obstacles can be overcome, with enlightening results... Within its chosen boundaries the coverage of this book is superb. Read it!"

2.2.9 History of Lie groups and related spaces

The theory of Lie groups has played a fundamental role in the modern mathematics and sciences. Besides being useful, they are also beautiful due to their rich structures. The following two books examine the history of Lie groups from different perspectives:

T. Hawkins, *Emergence of the Theory of Lie Groups. An Essay in the History of Mathematics 1869–1926,* Sources and Studies in the History of Mathematics and Physical Sciences, Springer-Verlag, New York, 2000. xiv+564 pp.

Though there are many books on Lie groups and Lie algebras, most of them do not discuss their history. Though there are a lot of original materials on Lie groups, Lie algebras and related topics, for example, the collected works of Lie consist of 7 huge volumes plus several substantial books, and E. Cartan had also written extensively, it is difficult for people to read these original papers and books. This long essay is helpful to those who wants to understand the history of this essential theory of modern mathematics.

From the Back Cover, "Written by the recipient of the 1997 MAA Chauvenet Prize for mathematical exposition, this book tells how the theory of Lie groups emerged from a fascinating cross fertilization of many strains of nineteenth and early twentieth century geometry, analysis, mathematical physics, algebra and topology. The reader will meet a host of mathematicians from the period and become acquainted with the major mathematical schools... The book is written with the conviction that mathematical understanding is deepened by familiarity with underlying motivations and the less formal, more intuitive manner of original conception. The human side of the story is evoked through extensive use of correspondence between mathematicians. The book should prove enlightening to a broad range of readers, including prospective students of Lie theory, mathematicians, physicists and historians and philosophers of science."

According to MathSciNet, "The book under review is a very nice essay on the history of the theory of Lie groups during the period 1869–1926. It is focused upon the origins of the theory and on the subsequent developments of its structural aspects, particularly the structure and representations of semisimple groups. The book is divided into four parts, each bearing the name of a mathematician, who stands out as the central figure there. The first part is devoted to the geometrical and analytical origins of the theory of continuous transformation groups of Sophus Lie—the precursor of the modern theory of Lie groups. In the second part the central figure is Wilhelm Killing, who discovered almost all central concepts and theorems on the structure and classification of semisimple Lie algebras. The third part is named after Élie Cartan and is primarily concerned with developments that would now be interpreted as representations of Lie algebras, particularly simple and semisimple algebras. In the last part the main role is played by Hermann Weyl and this part itself is mainly focused on the development of representation theory of Lie groups and algebras."

A. Borel, *Essays in the History of Lie Groups and Algebraic Groups,* History of Mathematics, 21, American Mathematical Society, Providence, RI; London Mathematical Society, Cambridge, 2001. xiv+184 pp.

One of the most effective methods to construct and study Lie groups is to use linear algebraic groups. Probably more importantly, linear algebraic groups and their arithmetic subgroups are essentially the only way to construct interesting discrete subgroups of Lie groups in many cases. Borel is one of the founders of linear algebraic groups. For example, the Borel subgroups, named after him, are essential in the study of algebraic groups and associated spaces. In some sense, this book can be viewed as an eyewitness account of the modern development of Lie groups and algebraic groups. It is a valuable reference written by an authority.

From the book cover, "Lie groups and algebraic groups are important in many major areas of mathematics and mathematical physics. We find them in diverse roles, notably as groups of automorphisms of geometric structures, as symmetries of differential systems, or as basic tools in the theory of automorphic forms. The author looks at their development, highlighting the evolution from the almost purely local theory at the start to the global theory that we know today. Starting from Lie's theory of local analytic transformation groups and early work on Lie algebras, he follows the process of globalization in its two main frameworks: differential geometry and topology on one hand, algebraic geometry on the other..."

According to MathSciNet,"Borel shows, with great skill, sense of history, originality and enthusiasm, how a simple concept (the Galois group of an equation) can grow and play a central role in a number of contexts. The book contains different proofs of some important theorems of the theory, how the proofs arose and very interesting historical remarks on theorems (or proofs of theorems) that have being rediscovered after being hidden in the literature for many years. Moreover, Borel beautifully describes the development of some pieces of this mathematics before they went to print... The book possesses a unity of vision that gives it an intellectual coherence. It should be on the bookshelf of anyone who is curious about one of the last two centuries' wonderful contributions to mathematics. Without hesitation, we may say that this book is an important source for historians as well as for mathematicians."

M.A. Akivis, B.A. Rosenfeld, *Élie Cartan (1869–1951), Translated from the Russian manuscript by V. V. Goldberg,* Translations of Mathematical Monographs, 123, American Mathematical Society, Providence, RI, 1993. xii+317 pp.

Though E. Cartan is often called the father of modern geometry, his main contribution lies in Lie groups and their geometric applications. For example, he single-handedly created the theory of symmetric spaces, which are special and rigid Riemannian manifolds and also serve as a canonical space for Lie groups to act in order to understand properties of Lie groups. It is well-known that it is difficult to read Cartan's papers, and this book will be very helpful to gain a global picture of the scope of the work of Cartan. Due to the importance of Lie groups in Cartan's work, this book will go well with the above two books on Lie groups.

From the book description, "Here readers will find detailed descriptions of Cartan's discoveries in Lie groups and algebras, associative algebras, differential equations, and differential geometry, as well of later developments stemming from his ideas. There is also a biographical sketch of Cartan's life. A monumental tribute to a towering figure in the history of mathematics, this book will appeal to mathematicians and historians alike."

According to a review by S.S. Chern in Bulletin of AMS, "This is a carefully written and rather complete biography of one of the greatest mathematicians of the twentieth century."

2.2.10 F. Klein, Development of Mathematics in the 19th Century, With a preface and appendices by Robert Hermann, Translated from the German by M. Ackerman. Math Sci Press, Brookline, Mass., 1979. ix+630 pp.

Before translating into English, this book has long been the only one on the history of mathematics in the nineteenth century and is still one of the best books on this subject. Klein is one of the most capable expositors in the history of mathematics. (It is probably true that without writings by Klein, some highly original ideas and works of Riemann might not be understood and well-appreciated by the mathematical community and hence developed further). Many of his more elementary books are still translated into various languages and widely used around the world. (Books on more advanced topics can become outdated quicker due to development of subjects. The two huge volumes by Fricke and Klein mentioned later are good examples).

In the preface of the English translation, Herman wrote "This is one of the few classics in the history of science which is lively, interesting and readable; as a bonus, it was written by one of the greatest mathematical expositors and generalists, and contains sketches of mathematical ideas that were models of their kind and still very useful. As if this were not enough motivation for translation, it is also very valuable for the light it throws on two aspects of the 19th century science which are now almost completely lost to us: geometry (both differentiable and algebraic) and mathematical physics. We are so used to thinking of the "progress" of science that it is hard for us to understand that certain matters were understood better one hundred years ago! ... Notice that many of Klein's complaints about the intellectual atmosphere of his day have a contemporary sound!"

Indeed, Klein knew many of the main figures in the 19th century mathematics either personally or through their work and had also contributed substantially to mathematics. In this sense, this is an eye-witness account of the history, and is also an overview by an expert participant.

The origin of this book was explained in a review of this book in Bulletin of AMS in 1928: "Twenty years ago, during a series of walks in the forest about Hahnenklee, in the Hartz Mountains, the conversation between Klein and a companion covered, as would naturally be the case, a wide range. Three statements, however, impressed his listener very strongly. One was political: 'There was a time when we looked up to England socially, politically, and as a naval power,—but that is a thing of the past.' The second was political and of military significance: 'America has no standing army today; twenty-five years hence she will have a large one.' It is not strange that his auditor wondered at the real significance of these two statements by a man of Klein's vision and prominence. The third remark was in response to a statement to the effect that he of all men was the

2.2. BIOGRAPHIES AND HISTORY
77

one to write a history of mathematics in the 19th century. 'I am too old,' was the reply, 'It needs a young man who could devote years to its preparation.' When it was urged that he had seen the development and had taken part in it as few if any others living had done, he remarked, 'No, all that I could do would be to give a few lectures on the great events, but I am too much occupied to prepare even these.' Ten years later, when the war was on, and his family had been sorely stricken, he gave these very lectures in his home in Göttingen, before a small group of listeners anxious to receive from a master that which only a master could give."

It is known that Klein's creativity as a research mathematician was destroyed in his competition with Poincaré. It might be interesting to note that when he described Poincaré as a mathematician, he attributed Poincaré's success to one superiority of the French system: well-rounded comprehensive education, and lamented the narrow German tradition of following one's advisor.

In an article titled "Felix Klein and the history of modern mathematics", G.A.Miller wrote in 1927 that "[Klein's] His own work illustrates the fundamental principle of the history of mathematics that the great forward movements are due more and more to collective efforts, and that one of the most effective services which an individual can render is to point out resultants of such collective forces and to work in harmony therewith. This is the work of the scientific statesman, and in the mathematical world Klein was preeminently such a statesman. He took a leading part in an effort to collect and coordinate the extensive body of mathematical advances in the form of large encyclopedias, and during the last few years of his life he gave courses of lectures on the development of mathematics during the nineteenth century. While his work along this line was left in an incomplete form it was in a sufficiently advanced state to exhibit his views on the general methods which should be pursued in such a history. He aimed to popularize mathematics even at the risk of a kind of piafraus, due to making the subject appear easier than it really is. Just as in the case of his treatment of elementary mathematics, he proceeded in his history mainly from the higher point of view instead of toward the higher point of view in the sense of first laying an elaborate basis on which a superstructure leading to this point of view might be erected."

F. Klein, *Lectures on Mathematics*, (AMS Chelsea Publishing), American Mathematical Society, 2000. 109 pages

This is another book by Klein on mathematics and its history. Klein was a mathematician with a global perspective of mathematics. He was well educated and could detect essential connections between subjects. The best example is the famous Erlangen program proposed by him. This book conveys well his understanding of mathematics. To many young people, it is not easy to realize the impact of Klein on the development of mathematics in USA. Reading this book is one way to appreciate his impact. This book will also allow people to understand the status of mathematics at the end of the 19th century.

From the book description, "In the late summer of 1893, following the Congress of Mathematicians held in Chicago, Felix Klein gave two weeks of lectures on the current state of mathematics. Rather than offering a universal perspective, Klein presented his personal view of the most important topics of the time. It is remarkable how most of the topics continue to be important today. Originally published in 1893 and reissued by the AMS in 1911. Klein begins by highlighting the works of Clebsch and of Lie. In particular, he discusses Clebsch's work on Abelian functions and compares his approach to the theory with Riemann's more geometrical point of view. Klein devotes two lectures to Sophus Lie, focussing on his contributions to geometry, including sphere geometry and contact geometry. Klein's ability to connect different mathematical disciplines clearly comes through in his lectures on mathematical developments. For instance, he discusses recent progress in non-Euclidean geometry by emphasizing the connections to projective geometry and the role of transformation groups... Klein's look at mathematics at the end of the 19th Century remains compelling today, both as history and as mathematics. It is delightful and fascinating to observe from a one-hundred year retrospect, the musings of one of the masters of an earlier era."

III. The Ross School of Business of the University of Michigan

Leibniz made deep contributions to and now occupies a prominent place in the history of both mathematics and philosophy. Leibniz developed calculus independently of Newton. This caused a very harmful priority dispute in the history of mathematics. Leibniz's notation for derivatives is probably superior to the notation of Newton but the English mathematics community refused to adopt it for a long time because of the dispute. Consequently, the English mathematics was delayed for at least half century. This is probably an important lesson for people who want to get into such a priority dispute. Leibniz's contribution was not restricted to mathematics. For example, he was one of the innovators of mechanical calculators. In philosophy, Leibniz is best known for his belief that our universe is the best possible one that God could have created. Leibniz, Descartes and Spinoza were the three great advocates of rationalism in seventeeth century.

Though the notions of area and volume are obvious to people, making them rigorous is nontrivial. It is closely related to integration. Many great mathematicians have contributed to this subject such as Archimedes, Newton, Leibniz, and Riemann. The Lebesgue measure is now the standard way to measure subsets of the Euclidean space or manifolds, and it is Lebesgue's claim to fame. His contributions to other subjects are not so known. One story says that Grothendieck discovered the theory of Lebesgue measure by himself in college.

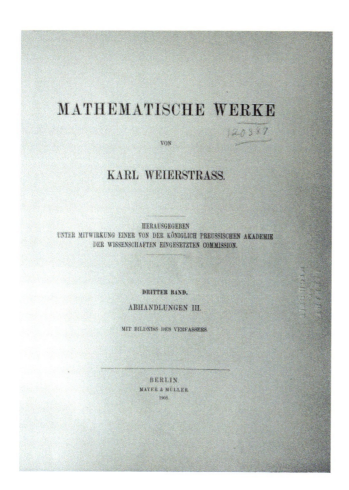

Weierstrass was the person who made analysis really rigorous and is often called "father of modern analysis". He also made significant contribution to the theory of calculus of variations. He studied law, economics, and finance in college but did not do well, and left it without a degree. After that he studied mathematics and became a high school teacher in mathematics later. After suffering a long period of illness around 1850, Weierstrass published original papers that brought him fame and distinction, resulting in a chair at the Technical University of Berlin. He had many famous students, and the most famous is probably the beautiful Sofia Kovalevskaya. It is said that he liked Sofia so much that he burned all correspondences with her after he learnt that Sofia was actually married.

Cavalieri is probably less known to most people in spite of his important contributions to mathematics. For example, building on the classic method of exhaustion, he developed a geometrical approach to calculus and published a treatise in 1635. In this work, an area is considered to consist of an indefinite number of parallel segments and a volume to consist of an indefinite number of parallel areas. As an application, he computed the areas under the curves, an important example of integral. Cavalieri is also known for Cavalieri's principle, which states that the volumes of two objects are equal if their corresponding crosssections have the same area. These represented a significant step towards to the modern calculus.

Chapter 3

Analysis

Analysis is one of the most basic branches of mathematics, though probably not as ancient as number theory and geometry. It includes the theories of differentiation, integration and measure, limits, infinite series, and analytic functions. It has origins in the early days of ancient Greek mathematics. For instance, an infinite geometric sum is implicit in Zeno's paradox of the dichotomy. Later, Greek mathematicians such as Eudoxus and Archimedes made more explicit, but still informal, use of the concepts of limits and convergence when they used the method of exhaustion to compute the area and volume of regions and solids. During the later half of the seventeen century, motivated by practical applications, Newton and Leibniz independently developed infinitesimal calculus. In the early nineteen century, Cauchy began to put calculus on a firm logical foundation. With the introduction of integration by Riemann in the middle of the nineteen century, the basic theory of analysis is complete.

Analysis includes (1) real analysis, which is the rigorous study of derivatives and integrals of functions of real variables, (2) complex analysis, (3) harmonic analysis and potential theory, (4) functional analysis, and (5) many other subjects such as differential equations, calculus of variations, and geometric analysis. In this chapter, we will be concerned with books that belong to the first four broad areas.

3.1 Calculus and real analysis

Calculus is a very basic course of mathematics and there have been many textbooks on calculus and real analysis at different levels. The more recent ones are becoming bigger, heavier and much more expensive. It is not easy to pick out good ones among so many of them. We have selected several that has withstood the test of time. Like everything else under the sun, the time is still the best test.

It is well-known that Newton and Leibinz independently discovered calculus. But they did not build the subject out of nothing. Eudoxus in the ancient Greece used the method of exhaustion to calculate areas and volumes, and Archimedes developed this idea into heuristics which resemble the methods of integral calculus. The method of exhaustion was later developed independently by Liu Hui to find the area of a circle, and Zu Chongzhi in the fifth century established a method to find the volume of a sphere, which was rediscovered by Bonaventura Cavalieri in the seventeenth century and is often called Cavalieri's principle. Newton's book Principia Mathematica *contained some basic concepts and methods of calculus, though not in the language as we know today. The purpose of this book was to state Newton's laws of motion, Newton's law of universal gravitation, and a derivation of Kepler's laws of planetary motion, obtained by Kepler empirically.*

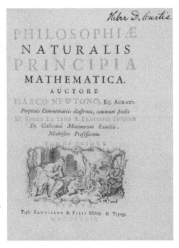

Figure 3.1: Newton's book.

3.1.1 G.H. Hardy, A Course of Pure Mathematics, Reprint of the (1952) tenth edition, Cambridge Mathematical Library, Cambridge University Press, Cambridge, 1992. xii+509 pp.

It was first published in 1908, and went through ten editions (up to 1952) and many more reprints. (Not many mathematics books can last so long and stay in print all the time.) It remains one of the most popular books on pure mathematics.

It was intended to help reform mathematics teaching in the UK, and more specifically in the University of Cambridge, and in schools preparing pupils to study mathematics at Cambridge. Therefore, it was aimed directly at serious mathematics students and contains a large number of difficult problems. It covers introductory calculus and the theory of infinite series. It starts out at a really elementary level, for example, with the definitions of rational numbers and irrational numbers, and their correspondence with points on a line. The exposition is quite leisurely, but rigorous. For example, on page 4, it says "Let us look at the matter for a moment with the eye of common sense, and consider some of the properties which we may reasonably expect a straight line to possess...", then it proves the existence of irrational numbers by geometric construction. Indeed, according to the book description, "There are few textbooks of mathematics as well-known as Hardy's Pure Mathematics. Since its publication in 1908, this classic book has inspired successive generations of budding mathematicians at the beginning of their undergraduate courses. In its pages, Hardy combines the enthusiasm of the missionary with the rigour of the purist in his exposition of the fundamental ideas of the differential

3.1. CALCULUS AND REAL ANALYSIS

and integral calculus, of the properties of infinite series and of other topics involving the notion of limit."

When he wrote it, Hardy had successfully implemented reforms of the Mathematical Tripos at Cambridge, making it less a test of sheer problem-solving technique. In writing his *Pure Mathematics* he was proposing a course of study preliminary to a French-style *Cours d'Analyse* by Camille Jordan, which was at the time a benchmark for a mathematical education leading to research in the field. (Note that *Cours d'Analyse* was first published in three volumes between 1882 and 1887).

The following book is an old and classic in mathematical analysis.

E. Goursat, *A Course in Mathematical Analysis: Vol 1: Derivatives and Differentials, Definite Integrals, Expansion in Series, Applications to Geometry. Vol. 2, Part 1: Functions of a Complex Variable. Vol. 2, Part 2: Differential Equations, Translated by E. R. Hedrick (Vol. 1), and E. R. Hedrick and O. Dunkel (Vol. 2)*, Dover Publications, Inc., New York, 1959, ix+548, x+259, viii+300 pp. *Vol. 3, Part I: Variation of Solutions. Partial Differential Equations of the Second Order, Vol. 3, Part Two: Integral Equations. Calculus of Variations*, Translated by Howard G. Bergmann, Dover Publications, Inc., New York, 1964, x+329 pp, Xi+389 pp.

The book of Goursat above was published in the beginning of 1900 and has not been out of print since then. It has made a huge impact on teaching mathematical analysis, by setting a standard for the high-level teaching of mathematical analysis, especially complex analysis, and Goursat is now remembered principally as the author of this book.

According to a review in Bulletin of AMS 1903 by Osgood, "The extraordinarily high standard of simplicity and attractiveness in style, combined with modern rigor, which Picard set in his Traité d'Analyse is fully maintained by Professor Goursat. The objects of the two works are quite different. Picard's purpose was to write a treatise on differential equations, and he developed only such parts of analysis and geometry as bear on this subject. Goursat, on the other hand, has set himself the task of writing a systematic treatment of the calculus, and thus the whole field of the calculus is included here.

A treatise on advanced calculus which should present the whole subject rigorously and attractively, and at the same time in the spirit of modern analysis, has been sorely needed by students of mathematics who intend to proceed to the study of mathematical physics or of some of the various branches of analysis — theory of functions, differential equations, calculus of variations, etc. Professor Goursat's work meets the needs of such students in a thoroughly satisfactory manner, and we recommend it to them most heartily. The teacher of calculus will find many suggestions in the book which will enable him to improve his course, and he may often with advantage refer even an elementary class to the more elementary parts of the book for collateral reading. The range of the book is wide.

While beginning with the elements of the calculus, it carries the reader to the point where he is prepared to use original sources and extracts from ε-proofs the underlying thought. When the future historian inquires how the calculus appeared to the mathematicians of the close of the nineteenth century, he may safely take Professor Goursat's book as an exponent of that which is central in the calculus conceptions and methods of this age."

According to an Amazon review, "... The proof are rigorous, and the development of proofs are much more make sense than today's delta-epsilon proofs. You could see the theorems in the book are proved in a much more natural and intellectual way. Of course delta-epsilon could bring you a 'rigorous' proof too, but sometimes the development of the proof is just so awkward."

3.1.2 Calculus

Every mathematician has learned calculus and every mathematics professor has probably taught calculus. Many mathematicians have also tried to write textbooks on calculus. Though the essential topics they cover are the same, there are still many different styles, choices of topics, and arrangements of materials. Even for such basic topics, it still takes real understanding and appreciation of the subject to write good books. In this subsection, we mention several more recent ones which seem to contain serious mathematics and have been used by many people.

M. Spivak, *Calculus*, 4th edition, Publish or Perish, 2008. 680 pages.

Besides the standard topics on differentiation, integration, and infinite series, it also contains more advanced topics: π is irrational, planetary motion, e is transcendental.

From the book description, "Spivak's celebrated Calculus combines leisurely explanations, a profusion of examples, a wide range of exercises and plenty of illustrations in an easy-going approach that enlightens difficult concepts and rewards effort. Ideal for honors students and mathematics majors seeking an alternative to doorstop textbooks and more formidable introductions to real analysis."

According to Zentralblatt MATH, "The really strong point of the book is a real effort to confront students with the necessity of rigorous treatment in order to broaden intuition. If students accept this challenge, they will find much help in this direction in the work itself."

A closely related but more advanced book on analysis by Spivak is

M. Spivak, *Calculus on Manifolds. A Modern Approach to Classical Theorems of Advanced Calculus*, W. A. Benjamin, Inc., New York-Amsterdam, 1965. xii+144 pp.

Indeed, from the preface, "This little book is especially concerned with those portions of 'advanced calculus' in which the subtlety of the concepts and methods makes rigor

3.1. CALCULUS AND REAL ANALYSIS

difficult to attain at an elementary level." This book culminates in Stokes' Theorem. The second page of the preface gives a precise and clear explanation of the fascinating history of this famous theorem.

According to MathSciNet, "It has been evident for at least ten years that the teaching of advanced calculus is badly out of step with the modern treatment of much of this subject as used by active workers in differential geometry and algebraic topology. Of particular importance is the divergence between the classical and modern approaches to the theorems of Gauss, Green, and Stokes and the corresponding replacement of the Gibbs notation for vector analysis with that of differential forms. The proposed revision of this material assumes that the students are already acquainted with linear algebra, and also involves the modernization of some of the fundamental ideas in terms of current set terminology. This brief text aims at such a major revision. It is not the first such attempt, nor should it be the last. The general sweep of ideas is excellent, and so the book is to be recommended to the mature mathematician who is not aware of advances along these lines. For sophomores and juniors, however, the exposition is excessively compact and is lacking in sufficient illustrative examples. Its use in a normal undergraduate class would lead to frustration if not chaos..."

Another classical book on calculus and analysis is

R. Courant, F. John, *Introduction to Calculus and Analysis. Vol. I., Vol. II*, Reprint of the 1965 edition, Springer-Verlag, New York, 1989. xxiv+661 pp. Reprint of the 1974 edition. Springer-Verlag, New York, 1989. xxvi+954 pp.

Courant is certainly a very well-known name and in particular well-known for his writing skill. Though Fritz John is not so well-known, he is a very powerful mathematician. In the Foreword to the Collected papers of Fritz John, Lars Garding wrote: "Every beginning mathematician eager to prove himself should from time to time study the serious thinking of some of the seminal papers of the past. This will give both a perspective on and a relief from his daily labour with the very latest developments in his field. For anyone interested in the analysis of partial differential equations, the work of Fritz John is especially rewarding. He wrote by now classical papers in convexity, ill-posed problems, the numerical treatment of partial differential equations, quasi-isometry and blow-up in nonlinear wave propagation."

According to the preface, "It is animated by the same intention: To lead the student directly to the heart of the subject and to prepare him for active application of his knowledge. It avoids the dogmatic style which conceals the motivation and the roots of the calculus in intuitive reality.... The book is addressed to students on various levels, to mathematicians, scientists, engineers. It does not pretend to make the subject easy by glossing over difficulties, but rather tries to help the genuinely interested reader by throwing light on the interconnections and purposes of the whole. Instead of obstructing the access to the wealth of facts by lengthy discussions of a fundamental nature we have sometimes postponed such discussions to appendices in the various chapters. Numerous examples and problems are given at the end of various chapters. Some are challenging, some are even difficult; most of them supplement the material in the text."

According to The Mathematical Gazette, "... It is well-illustrated, well-motivated and very well-provided with a multitude of unusually useful and accessible exercises.... There are three aspects of Courant and John [Vol 1] in which it outshines (some) contemporaries: (i) the extensive historical references, (ii) the chapter on numerical methods, and (iii) the two chapters on physics and geometry. The exercises in Courant and John are put together purposefully, and either look numerically interesting, or are intuitively significant, or lead to applications. It is the best text known to the reviewer for anyone trying to make an analysis course less abstract... The mathematics are rigorous but the many examples that are given and the applications that are treated make the books extremely readable and the arguments easy to understand."

Yet another classical book on calculus and mathematical analysis is the next book.

T. Apostol, *Calculus. Vol. I: One-variable Calculus, with an Introduction to Linear Algebra. Calculus. Vol. II: Multi-variable Calculus and Linear Algebra, with Applications to Differential Equations and Probability.* Second edition, Blaisdell Publishing Co. Ginn and Co., Waltham, Mass.-Toronto, Ont.-London, 1967, xx+666 pp.; 1969, xxi+673 pp.

This is not a standard textbook on calculus. As the author describes in the preface, there are two approaches to calculus: either start off with a thorough treatment of real numbers and develop the subject in a logical and rigorous way, or stress applications by appealing to intuition and by extensive drill on computational problems. This book strikes a balance between these approaches. The wide acceptance of this book shows that it has been successful in reaching this goal.

According to MathSciNet, "The book is written for those who wish to have some knowledge of calculus and analytic geometry. The presentation differs considerably from what is usually found in standard books of the same nature. Each important new concept is preceded by a historical introduction and, whenever possible, is explained geometrically. Each chapter begins with an introduction and is followed by unsolved examples, some of which are illuminating while others are routine. Also, there are several worked examples to illustrate new concepts."

3.1. CALCULUS AND REAL ANALYSIS 85

3.1.3 E.T. Whittaker and G.N. Watson, A Course of Modern Analysis, An Introduction to the General Theory of Infinite Processes and of Analytic Functions: with an Account of the Principal Transcendental Functions, Fourth edition. Reprinted, Cambridge University Press, New York, 1962, vii+608 pp.

This classical book was first published by Cambridge University Press in 1902. The fourth edition was released in 1927 and reissued many times. The subtitle of this book describes the content of this book well. The first part, about 200 pages, consists of standard topics in analysis, and the second part, about 350 pages, deal with special functions. The second is the really valuable part. According to Zentralblatt MATH, "The larger part of the book is devoted to part II dealing with the classical transcendental functions. Many generations of students (like this reviewer) were introduced to the special functions of mathematical physics by reading the pertinent chapters of this part II, and it is a pleasure to say that the work under review ever was and still is one of the best texts on this subject."

According to a review in Nature, this book "has entered and held the field as the standard book of reference in English on the applications of analysis to the transcendental functions. This end has been successfully achieved by following the sensible course of explaining the methods of modern analysis in the first part of the book and then proceeding to a detailed discussion of the transcendental function, unhampered by the necessity of continually proving new theorems for special applications. In this way the authors have succeeded in being rigorous without imposing on the reader the mass of detail which so often tends to make a rigorous demonstration tedious."

The respect which people have paid to this book can be seen from the following most popular review from Amazon: "If you read its page of contents, you'll call it prophetic! Every kind of function he studied became important in theoretical physics some time. String theory was started with an amplitude containing only Gamma functions. Renormalization, reborn from the ashes, discovered the Zeta-function (in Whittaker-Watson, for sure), Legendre's less familiar functions were prominent in Regge pole theory (again, the source was Whittaker), and even the Theta functions became important for some field theory skirmishes."

3.1.4 Special functions

Special functions are particular functions which have established names and notations due to their importance in mathematical analysis, functional analysis, physics, or other applications. They often arise from solutions of special differential equations and integrals. For example, sine and cosine functions and the exponential function are special functions too, though they are probably easier to understand than others. The book of

Whittaker-Watson is still one of the best books on special functions. Several more recent ones are listed below.

G.N. Watson, *A Treatise on the Theory of Bessel Functions*, Reprint of the second (1944) edition, Cambridge Mathematical Library, Cambridge University Press, Cambridge, 1995. viii+804 pp.

This book can be considered as a sequel to the famous book by Whittaker and Watson. For some people, this is the Bible on Bessel functions. According to Zentralblatt MATH, "Indeed, this work has been universally acknowledged as giving an excellent and systematic treatment, clearly developed, on the theory of Bessel functions."

According to MathSciNet, "Watson's Bessel functions was first published in 1922, and the second edition was published in 1944. If you think about this a minute, you are surprised. There was a war going on, paper was scarce, and the sales were not likely to be large. However, copies were needed. I have been told that when the work on the first successful atomic pile was being done at the University of Chicago, a copy of Watson's book was chained to a table and always open. Bessel functions also arose in work on radar and probably many other applied problems which needed to be solved rapidly. This book has been reprinted frequently, and now, almost 75 years after the initial publication, it has been reprinted again.

The first thing you notice is the scholarship. There are references to Bessel functions which Watson missed, but few of the important ones. Watson was 36 when this book appeared. The next thing you notice is the care with which Watson treats many topics. Sometimes his arguments are more perceptive than he realizes... One role this book has played is to highlight the important formulas known for Bessel functions, and so tell us what to look for other classes of functions... It is unlikely anyone will write a similar book about other functions, which is a shame. While Watson tried to tell all about Bessel functions, he had enough taste to highlight what he thought was of long-term importance. His judgment was usually right."

A. Erdélyi, W. Magnus, F. Oberhettinger, F. Tricomi, *Higher Transcendental Functions. Vols. I, II.* Based, in part, on notes left by Harry Bateman, McGraw-Hill Book Company, Inc., New York-Toronto-London, 1953. xxvi+302, xvii+396 pp. *Higher Transcendental Functions. Vol. III.* Based on notes left by Harry Bateman, Reprint of the 1955 original, Robert E. Krieger Publishing Co., Inc., Melbourne, Fla., 1981. xvii+292 pp.

This is a valuable handbook and has served an important purpose. From the preface, "It is true that much of this material was already in existence. However, anyone who has been faced with the task of handling and discussing and understanding in detail the solution to an applied problem which is described by a differential equation is painfully fa-

3.1. CALCULUS AND REAL ANALYSIS

miliar with the disproportionately large amount of scattered research on special functions one must wade through in the hope of extracting the desired information."

According to MathSciNet, "The late Harry Bateman, during his last years, planned an extensive compilation of the ' Special Functions'. He intended to investigate and tabulate their properties, the inter-relations between them, their representations in various forms, their macro- and microscopic behavior, and to construct tables of the definite integrals involving them. The whole project was to have been on a gigantic scale; it would have been an authoritative and definitive account of its vast subject.

While much of the material is available, it is not readily accessible, being scattered in books and journals on many fields. The 'Guide to the Functions' which Bateman planned would have been invaluable. The project was never completed, and, after his death, the California Institute of Technology and the U. S. Office of Naval Research pooled their resources to continue Bateman's task.

It turned out that no single section of Bateman's work was in a state suitable for immediate publication, and the field was so wide that it appeared essential to narrow it down if anything useful was to be accomplished. It was decided to concentrate on a three-volume work on the Higher transcendental functions (of which the first two are now under review), to be followed by two volumes of tables of integrals. The whole work has been carried out by the staff of the Bateman Manuscript Project, under the directorship of Arthur Erdélyi.

The volumes on the Higher transcendental functions might be described as an up-to-date version of Part II of Whittaker and Watson's *Modern Analysis*... To attempt to review such a work in detail is quite impossible. Suffice it to say that it is a valuable piece of work well done.

As Erdélyi says in his Introduction, the chapters are of widely varying characters... Each chapter presented its special problems, which had to be treated on its merits. But these differences of treatment in no way detract from the value of the work."

A. George, R. Askey, R. Roy, *Special Functions*, Encyclopedia of Mathematics and its Applications, 71, Cambridge University Press, Cambridge, 1999. xvi+664 pp.

Since the authors of this book have made fundamental contributions to special functions, this is an authoritative book. According to Zentralblatt MATH, "This book covers a wealth of material on special functions, notably knowledge which was developed by Richard Askey and his co-authors during the several decades of his contributions to this subject, but also material which connects special functions with combinatorial questions collected by George Andrews."

From the book description, "This treatise presents an overview of the area of special functions, focusing primarily on the hypergeometric functions and the associated hypergeometric series. It includes both important historical results and recent developments and shows how these arise from several areas of mathematics and mathematical physics. Particular emphasis is placed on formulas that can be used in computation... This clear, authoritative work will be a lasting reference for students and researchers in number theory, algebra, combinatorics, differential equations, applied mathematics, mathematical computing, and mathematical physics."

According to MathSciNet, "Special functions, which include the trigonometric functions, might be called 'useful' functions. For some years, the field was somewhat dormant, but in the past three decades, special functions have returned to the forefront of mathematical research. The many reasons for this resurgence cannot be sufficiently delineated in a brief review. But fresh new ideas, in large part due to the authors; the many new uses of special functions; and their connections with other branches of mathematics, physics, and other sciences have played leading roles in this revival. The field is so huge that it cannot be adequately covered under one book cover, but the authors do an excellent job in helping readers find inroads into developments which space limitations prevent them from treating. Naturally, the selection of topics reflects the authors' own interests.

The book genuinely reflects the authors' vast accumulated insights... All of the chapters are beautifully written, but if one were to choose the book's zenith, it would be the two chapters on hypergeometric functions, which provide the connecting thread for most of the book.

Special functions will certainly emerge as the chief textbook and reference on special functions for the next several years. Indeed, the historical insights and copious problems make it an excellent textbook for a beginning graduate course. This book joins F. W. J. Olver's *Asymptotics and special functions*, 1974, as the only general books on special functions during the past three decades that belong 'in the Hobbs class', to quote G. H. Hardy."

According to a review in Bulletin of AMS, "One measure of the success of a mathematical book is: does it give the reader ideas, ideas as lush and provocative as those one gets from a stimulating conference? By that criterion alone, this book is way over the top. This is a splendid work, and I predict that it will be a bestseller as well."

F.W. Olver, *Asymptotics and Special Functions*, Computer Science and Applied Mathematics, Academic Press, New York-London, 1974. xvi+572 pp.

This book was written for graduate students, physicists and engineers, i.e., users of mathematics. Because of this purpose, it was intended both as a textbook and also as a reference book, and contains good features of both. The historical notes and additional references at the end of each chapter make the book more interesting and useful.

According to MathSciNet, "This is an extremely well written book. The author's style is straight-forward and lucid. He often elucidates a method to be presented by a heuristic discussion which makes the following rigorous discussion easy to understand.

3.1. CALCULUS AND REAL ANALYSIS

The analysis of special functions becomes the thread which ties together asymptotic methods for integrals, differential equations and sequences and series. The last of these is not usually presented in a book of this type, but appears quite naturally here. An important contribution of this book is the author's concern with error bounds rather than the 'O' estimates found in most books on the subject... This book should most definitely be on the shelf of anyone working in asymptotic analysis and/or special functions."

According to an Amazon review, "What I enjoy about this book is its clear and complete exposition of the asymptotic theory of (mostly linear second-order) differential equations and special functions. The author has not just taken a chunk of material out of his research and redressed it as a monograph. There is a great deal of painstaking care evident in his definition of useful tools for error estimates and in the statement of highly useful and accurate asymptotic theorems. Although the result is necessarily much more detailed, even finicky, than a reader dipping in and out might want, I cannot think of a better reference for asymptotic theorems for linear ODEs, and special functions that solve them, than this one. One of the reasons for this is that the author gives very precise error estimates. Another reason is the concise and pithy exposition which is pitched just right for a mathematical reader (who is at least at the graduate student level). I recommend this book highly as a reference for very accurate, rigorous results in asymptotics, or to browse through as an example of masterly development of the subject of asymptotics."

N.N. Lebedev, *Special Functions and Their Applications*, Revised edition, Translated from the Russian and edited by Richard A. Silverman, Unabridged and corrected republication, Dover Publications, Inc., New York, 1972. xii+308 pp.

The purpose of this book is to give a sufficiently detailed exposition of those problems which are of the greatest practical interest. It is a classic and a particularly useful book to learn the subject.

According to Zentralblatt MATH, "This book is one of several now available which treat the subject of special functions in a manner suitable for use as a teaching text. It presupposes only a familiarity with basic complex function theory, ... the book provides an excellent introduction to important problems in several areas of mathematical physics and could be useful to the mathematician who wants to acquaint himself with the true font of all mathematical inspiration, the real world. The reviewer has the highest praises for this book. He has used it as a basis for two-term course in special function theory and found it almost entirely self-contained. The book is a model of lucidity and organization, in the best tradition of Russian textbook writing."

3.1.5 W. Rudin, Real and Complex Analysis, Third edition, McGraw-Hill Book Co., New York, 1987. xiv+416 pp.

This classical book was first published in 1966 and several editions have been published. It has also been translated over a dozen languages. Besides being well-written, the novel idea of combining real and complex analysis into a single book with extensive coverage of many important topics also makes it unique. Probably every serious mathematics student in the past few decades has read or used this book in one way or another.

From the book description, "This is an advanced text for the one- or two-semester course in analysis taught primarily to math, science, computer science, and electrical engineering majors at the junior, senior or graduate level. The basic techniques and theorems of analysis are presented in such a way that the intimate connections between its various branches are strongly emphasized. The traditionally separate subjects of 'real analysis' and 'complex analysis' are thus united in one volume. Some of the basic ideas from functional analysis are also included. This is the only book to take this unique approach. Proofs of theorems presented in the book are concise and complete and many challenging exercises appear at the end of each chapter. The book is arranged so that each chapter builds upon the other, giving students a gradual understanding of the subject."

According to MathSciNet, "The exposition is overall at once very clear, precise, stimulating and readable... This book deserves a warm welcome on the grounds of both design and execution."

W. Rudin, *Function Theory in the Hnit Ball of* $\mathbb{C}^n$, Grundlehren der Mathematischen Wissenschaften, 241, Springer-Verlag, New York-Berlin, 1980. xiii+436 pp.

The unit ball in $\mathbb{C}^n$ is an important and special example of two important classes of domains: bounded symmetric domains and strongly pseudoconvex domains. These two classes of domains have been studied by two camps of people who usually do not interact with each other. This book will be useful to both of them.

According to Zentralblatt Math, it is "an excellent introduction to one of the most active research fields of complex analysis." As the author emphasizes, the principal ideas can be presented clearly and explicitly in the ball, specific theorems can be quickly proved. Mathematics lives in the book: main ideas of theorems and proofs, essential features of the subjects, lines of further developments, problems and conjectures are continually underlined. Numerous examples throw light on the results as well as on the difficulties.

3.1.6 More books on real analysis

Analysis is a big subject and can be studied from different points of view. We listed several other good books on analysis.

3.1. CALCULUS AND REAL ANALYSIS

J. Dieudonné, *Foundations of Modern Analysis*, Pure and Applied Mathematics, Vol. X, Academic Press, New York-London, 1960, xiv+361 pp.

This is the first of a series of books by Dieudonné on analysis titled "Treatise on analysis. Vol. II – Vol. VIII", which covers many different aspects of analysis such as harmonic analysis and differential equations. The first book serves as a foundation and can also be read independently of other volumes.

According to the author, "... I became convinced that there was a place for a survey of modern analysis, somewhere between the minimum tool kit of an elementary nature which I had intended to write, and specialist monographs leading to the frontiers of research. My experience of teaching has also persuaded me that the mathematical apprentice, after taking the first step of Foundations, needs further guidance and a kind of general birds eye view of his subject before he is launched onto the ocean of mathematical literature or set on the narrow path of his own topic of research.Thus I have finally been led to attempt to write an equivalent, for the mathematicians of 1970, of what the *Cours d'Analyse* of Jordan, Picard, and Goursat were for mathematical students between 1880 and 1920. It is manifestly out of the question to attempt encyclopedic coverage, and certainly superfluous to rewrite the works of N. Bourbaki. I have therefore been obliged to cut ruthlessly in order to keep within limits comparable to those of the classical treatises. I have opted for breadth rather than depth, in the opinion that it is better to show the reader rudiments of many branches of modern analysis rather than to provide him with a complete and detailed exposition of a small number of topics. Experience seems to show that the student usually finds a new theory difficult to grasp at a first reading. He needs to return to it several times before he becomes really familiar with it and can distinguish for himself which are the essential ideas and which results are of minor importance, and only then will he be able to apply it intelligently. The chapters of this treatise are therefore samples rather than complete theories: indeed, I have systematically tried not to be exhaustive. The works quoted in the bibliography will always enable the reader to go deeper into any particular theory. However, I have refused to distort the main ideas of analysis by presenting them in too specialized a form, and thereby obscuring their power and generality. It gives a false impression, for example, if differential geometry is restricted to two or three dimensions, or if integration is restricted to Lebesgue measure, on the pretext of making these subjects more accessible or intuitive. On the other hand I do not believe that the essential content of the ideas involved is lost, in a first study, by restricting attention to separable metrizable topological spaces. The mathematicians of my own generation were certainly right to banish, hypotheses of countability wherever they were not needed: this was the only way to get a clear understanding."

According to MathSciNet, "The most remarkable feature of the text is the consistently geometrical formulation of the results. For example, the differential calculus is developed in terms of linear approximation to functions on an open subset of a Banach space to a Banach space. Yet it would be completely false to assert that the book contained a study of Banach spaces—no non-trivial proposition on such spaces is proved. The subject of study is indeed elementary analysis, and the theorems are theorems of analysis stated in geometrical terms. This geometrization is rather like the geometrization of linear algebra which occurred some years ago, and, as in the linear algebra case, there are enormous conceptual and technical advantages. A good deal is accomplished in the 350 pages of the text. The mathematical organization is superb, the presentation lucid, there are a large number of very good problems, and there are excellent expository introductions to each chapter (couched in the author's customary diffident style). In brief, it is a beautiful text... Finally, it is emphasized again that this is a text on elementary analysis. Thus, for example, there is no mention of the Lebesgue integral whose continuity and completeness properties make feasible the standard methods of Banach-space duality. The book is perhaps best characterized as a modern synthesis of basic classical analysis."

G. Folland, *Real Analysis. Modern Techniques and Their Applications*, Second edition, Pure and Applied Mathematics: A Wiley-Interscience Series of Texts, Monographs and Tracts, John Wiley & Sons, Inc., New York, 1999. xiv+386 pp.

Though this is a textbook on real analysis, it covers more advanced topics than other texts. This might be related to the author's other writings. He has written many books at different levels, which are mainly about harmonic analysis and differential equations. In his own words, "I enjoy writing and assembling mathematical material into a coherent package".

From the book description, "This new edition of the widely used analysis book continues to cover real analysis in greater detail and at a more advanced level than most books on the subject. Encompassing several subjects that underlie much of modern analysis, the book focuses on measure and integration theory, point set topology, and the basics of functional analysis. It illustrates the use of the general theories and introduces readers to other branches of analysis such as Fourier analysis, distribution theory, and probability theory."

According to MathSciNet, "To start with we single out the adequacy of the title. The book under review is not an ordinary textbook in 'real analysis', but a fascinating and useful journey through various fields of analysis. It provides the necessary rigorous basis for the correct attack on Fourier analysis, partial differential equations and probability theory. Thus, the student who learns 'real analysis' from this book also learns how to apply it in his 'mathematical life', becoming convinced once more of the splendid unity of mathematics... This long review is intended to show the characteristics of a remarkable analysis book. Mathematical objects are not only introduced in formal definitions and described in theorems, as in most mathematics books, but their motivation and provenance are also explained and, above all, their importance and interconnections are emphasized. Proofs are very clear and short, most of them requiring active cooperation on the part of the reader. As for the exercises, they are not extremely numerous, but carefully selected..."

According to Zentralblatt MATH, "An excellent book, particularly for bright and motivated students. The organization of the material and many of the proofs are very efficient, and so the reader is able to cover a lot of ground in a short time. A full description of the most important aspects of measure and integration is accompanied by the elements of point-set topology and functional analysis. These are then applied to topics in Fourier analysis and probability theory. The applications are very interesting and introduce the students to exciting mathematics. Moreover they serve to illustrate the need for the full power of the Lebesgue theory. The writing is fluid and, with few exceptions, the notation is natural and standard. This enables the reader interested in some particular topic to read the appropriate section without too much chasing after definitions in previous chapters. Each chapter is followed by a very valuable notes and references section. These sections contain interesting historical notes, discussions of alternative points of view and some additional results."

E. Lieb, M. Loss, *Analysis*, Second edition, Graduate Studies in Mathematics, 14, American Mathematical Society, Providence, RI, 2001. xxii+346 pp.

This is not a standard textbook on real analysis. According to the preface, it was written for students who "prefer to do something with the material, as it is learned, rather than wait for a full-fledged development of all basic principles". For example, in the first chapter, many technical details are left as exercises or referred to one of the many existing books.

According to Zentralblatt MATH, "This book is an unconventional introductory text on analysis starting with the most elementary facts on measure theory, on the Fourier transform, and on the commonly used function spaces, and ending with a short introduction to potential theory and to the calculus of variations. The essentials of modern analysis, needed to understand some modern developments, for instance, of quantum mechanics, are presented in a rigorous and pedagogical way. The readers (graduate students, mathematicians, physicists and other natural scientists) are guided to a level where they can read the current literature with understanding. After glancing or reading the material contained in the book, one can agree with the authors who say: '...relative beginners can get some flavour of research mathematics and the feeling that the subject is open-ended'. The treatment of the subject is as direct as possible, the most general and abstract formulation of a result and a theory is often avoided. Contrary to many texts, the authors did not make a big distinction between real analysis and functional analysis, ..."

G.H. Hardy, *Divergent Series*, With a Preface by J. E. Littlewood and a Note by L. S. Bosanquet, Reprint of the revised (1963) edition, Éditions Jacques Gabay, Sceaux, 1992. xvi+396 pp.

Hardy has written many good books, and this is the last one. From the Preface by J. E. Littlewood: "Hardy in his thirties held the view that the late years of a mathematician's life were spent most profitably in writing books:... All [Hardy's] books gave him some degree of pleasure, but this one, his last, was his favourite. When embarking on it he told me that he believed in its value (as well he might), and also that he looked forward to the task with enthusiasm. He had actually given lectures on the subject at intervals ever since his return to Cambridge in 1931, and he had at one time or another lectured on everything in the book except Chapter XIII [The Euler-MacLaurin sum formula] ... [I]n the early years of the century the subject [Divergent Series], while in no way mystical or unrigorous, was regarded as sensational, and about the present title, now colourless, there hung an aroma of paradox and audacity."

For most people who have studied infinite series in standard courses in calculus and analysis, there are only a few standard topics on criterions of convergence. A glance at the table of contents of this book will show that there are rich theories about infinite series, such as many ways to sum up divergent series, and the important Wiener's Tauberian theorems occupy one chapter.

According to MathSciNet, "This is an inspiring textbook for students who know the theory of functions of real and complex variables and wish further knowledge of mathematical analysis. There are no problems displayed and labelled 'problems,' but one who follows all of the arguments and calculations of the text will find use for his ingenuity and pencil. The book deals with interesting and important problems and topics in many fields of mathematical analysis to an extent very much greater than that indicated by the titles of the chapters. It is of course an indispensable handbook for those interested in divergent series. It assembles a considerable part of the theory of divergent series which has previously existed only in periodical literature. Hardy has greatly simplified and improved many theories, theorems and proofs. In addition, numerous acknowledgments show that the book incorporates many previously unpublished results, and improvements of old results, communicated to Hardy by his colleagues and by others interested in the book."

N.H. Bingham, C. Goldie, J. Teugels, *Regular Variation*, Encyclopedia of Mathematics and its Applications, 27, Cambridge University Press, Cambridge, 1987. xx+494 pp.

This book studies the theory of regular variation of positive functions of a real variable and its various applications, and is an excellent introduction to students and an indispensable reference for experts.

3.1. CALCULUS AND REAL ANALYSIS

According to MathSciNet, "In its basic form, the theory of regular variation is the study of the relation (1) $f(\lambda x)/f(x) \to \lambda \rho, x \to \infty, \lambda > 0$, for some $\rho \in (\infty, \infty)$, where f is a positive measurable real-valued function. Perhaps its most immediately useful application is Karamata's theorem... That is, the class of regularly varying functions is the class for which limiting behaviour of a certain type may be interchanged with integration. The theory in this weighty book (over 400 pages) has grown from this basic property, much generalized and elaborated... It is probably fair to say that the authors look at the subject from a probabilistic perspective, since they are well known in this area, and the book reflects this slant... But there is much here for the analyst too.. The book is beautifully written with an attractive style of presenting main results, then discussing variants immediately in a smaller typeface. The exposition is precise and succinct, yet enough detail is provided for main proofs to be verified. Thus the book will appeal to the student as much as to the specialist. With the importance of the subject to classical analysis as well as to the various fields of application, it seems destined to become a classic. Students in need of inspiration for problems will find plenty here as well."

S. Samko, A. Kilbas, O. Marichev, *Fractional Integrals and Derivatives. Theory and Applications*, Edited and with a foreword by S. M. Nikoliskii, Translated from the 1987 Russian original, Revised by the authors, Gordon and Breach Science Publishers, Yverdon, 1993. xxxvi+976 pp.

This comprehensive book presents for the first time systematically both the classical and modern theory of integrals, derivatives of fractional order, applications to differential and integral equations. This is an important and valuable reference for people at all levels.

According to MathSciNet, "The very manner in which it is written seems to be exactly adequate to an extremely difficult task of exposition of the theory of fractional calculus and its applications throughout its more than one-and-a-half century history, starting from the works of its founders—N. H. Abel, J. Liouville, B. Riemann, A. V. Letnikov, H. Weyl, J. Hadamard —and finishing with all the modern results on the subject, to which the authors of the monograph are known contributors... Finally, the reviewer dares to say that this book may be a major event in the modern mathematical literature and hopes that, like himself, many people will be able to get a lot of pleasure and profit out of reading it."

I. Podlubny, *Fractional Differential Equations. An introduction to Fractional Derivatives, Fractional Differential Equations, to Methods of Their Solution and some of Their Applications*, Mathematics in Science and Engineering, 198. Academic Press, Inc., San Diego, CA, 1999. xxiv+340 pp.

The subject of fractional calculus has becoming important for its successful applications in diverse fields of science and engineering such as fluid flow, dynamical processes in self-similar and porous structures, diffusive transport akin to diffusion, electrical networks, probability and statistics, control theory of dynamical systems, viscoelasticity, electrochemistry of corrosion, and chemical physics. One reason is that it provides useful tools for solving differential and integral equations and problems which involves special functions of mathematical physics as well as their extensions and generalizations. This book is a good supplement to the previous book.

J.A. Shohat, J.D. Tamarkin, *The Problem of Moments*, American Mathematical Society Mathematical Surveys, vol. II. American Mathematical Society, New York, 1943. xiv+140 pp.

The moment problem is as follows: Given a sequence of real numbers μ_n, $n = 0, 1, \cdots$, determine a nondecreasing function $\psi(t)$ such that $\mu_n = \int_a^b t^n d\psi(t)$, $n = 0, 1, 2, \cdots$. If $a = 0, b = \infty$, the problem is precisely as Stieltjes put it and is known as the Stieltjes problem. There are variants due to various hypotheses on the range of integration and the function $\psi(t)$.

According to MathSciNet, "The authors are principally concerned with the Hamburger moment problem, where $\psi(t)$ is non-decreasing and the range of integration is $(-\infty, \infty)$. While a few results on this problem, such as the determinant condition for the existence of a solution and Carleman's sufficient condition for its uniqueness, have become widely known, no unified and detailed presentation has hitherto been available of the deeper theory of the moment problem, which furnishes quite precise information about its solutions. This state of affairs is perhaps due to the large amount of analytic machinery, belonging to various fields, which has been used in investigations of the problem. The authors have succeeded to a large extent, in the first two chapters of this book, in presenting a unified and self-contained account of the principal results. The third chapter gives an idea of the wealth of generalizations and special problems to which the moment problem has given rise; the theory of many of these is still far from complete. The last chapter is devoted to a discussion of the approximate integration formulae arising from the moment problem..."

Another interesting, unique book on real analysis is the following.

B. Gelbaum, J. Olmsted, *Counterexamples in Analysis*, The Mathesis Series Holden-Day, Inc., San Francisco, Calif.-London-Amsterdam, 1964, xxiv+194 pp.

In calculus and real analysis, finding derivatives is much easier than integration, and differentiation is certainly a very basic concept of the subject. The first time encounter with a nowhere differential function might be shocking to many people. But these counterexamples to our intuition really deepen our understanding.

This is not a standard book. It consists of counterexamples to various statements in analysis which seem to be reasonable and true, and of examples to show that some

3.1. CALCULUS AND REAL ANALYSIS

unreasonable statements are actually true (for example, a space filling curve that is almost everywhere differentiable). The first part concerns functions of one variable, and the second concerns functions of two variables, plane sets, function spaces etc. It is not easy to solve problems, but it is also not easy either to make up problems. Hence, this is a good book for both students and teachers. The detailed table of contents is a good summary of important results of related analysis topics.

3.1.7 Measure theory

Measure theory generalizes the intuitive notions of length of intervals, area of regions on plane, and volume of solids in space. The best known and most important example is the Lebesgue measure, but other examples include Borel measure, Jordan measure, probability measure, Haar measure, and Hausdorff measure. They are foundations of many subjects in mathematics such as real analysis, harmonic analysis, probability and ergodic theory. We have selected several books dealing with different aspects of the measure theory at various levels.

P. Halmos, *Measure Theory*, D. Van Nostrand Company, Inc., New York, N. Y., 1950. xi+304 pp.

Halmos is famous for his writing, and this is the second of his many books. Besides the usual books on mathematics, he also wrote *How to read mathematics*, *How to write mathematics*, and *How to speak mathematics*. The measure theory plays an important role in subjects he studied: probability theory, statistics, operator theory, ergodic theory, and functional analysis (in particular, Hilbert spaces).

Though this book is an old book, it is still a good introduction and a useful reference. Besides the Lebesgue measure theory for subsets of $\mathbb{R}^n$, this book also discusses the Haar measure of locally compact topological groups. It is general and comprehensive.

According to MathSciNet, "This book presents a unified theory of the general theory of measure and is intended to serve both as a text book for students and as a reference book for advanced mathematicians. This book is written in a very clear style and will make an excellent text book for those graduate students who are already familiar with the theory of Lebesgue integration in a Euclidean space. But the treatment of the subject is rather abstract so that the book is perhaps not to be recommended for beginners... Those people who are familiar with the classical theory of measure and integration will find many improvements and simplifications in various parts of this book..."

According to Amazon reviews, "This book is an overview of measure theory that is somewhat dated in terms of the presentation, but could still be read profitably by someone interested in studying the subject with greater generality than more modern texts."

D. Cohn, *Measure Theory*, Reprint of the 1980 original, Birkhäuser Boston, Inc., Boston, MA, 1993. x+373 pp.

This is a clearly written, comprehensive introduction to measure and integration for beginners.

According to MathSciNet, "The author aims to present 'a straightforward treatment of the part of measure theory necessary for analysis and probability' assuming only basic knowledge of analysis and topology... It is the reviewer's opinion that the author has succeeded in his aim. In spite of its lack of new results, the selection and presentation of materials makes this a useful book for an introduction to measure and integration theory."

According to an Amazon review, "Cohn's book struck for me the right balance between expository clarity, mathematical rigor, and intuition. Though it makes no mention of probability theory, it [is] clearer than other more probability-oriented treatments of measure theory. The appendices do an excellent job of summarizing the required mathematical background."

R. Wheeden, A. Zygmund, *Measure and Integral. An Introduction to Real Analysis*, Pure and Applied Mathematics, Vol. 43, Marcel Dekker, Inc., New York-Basel, 1977. x+274 pp.

The second author Zygmund is one of the modern pioneers of Fourier series, Fourier analysis and harmonic analysis. This book was written as an introduction to measure and integral, or the classical theory of the Lebesgue integral together with some applications, for students with some emphasis on preparation for further study in Fourier analysis and harmonic analysis.

According to MathSciNet, "This textbook has been conceived with the explicit intention of providing an easy and quick access to the most useful techniques of measure and integration in the modern analysis of real variables. This goal has been achieved with very remarkable success... The textbook is written throughout from the viewpoint of real analysis. Throughout the exposition one can find in use, at the most opportune moment, basic tools of modern real analysis which cannot be easily found in other textbooks, such as the distribution function, the Hardy-Littlewood maximal operator, various covering theorems, convex functions, Jensen's inequality, Young's inequality for the norm of a convolution, and the Marcinkiewicz integral.

The style is characterized by its clarity and concreteness. The collection of more than two hundred exercises, with important complements to the theory, is excellent. For the difficult ones there are helpful hints... In summary, one can say that this work constitutes an excellent introductory textbook for those who wish to get acquainted with the modern methods of real variables."

3.1. CALCULUS AND REAL ANALYSIS

L. Evans, R. Gariepy, *Measure Theory and Fine Properties of Functions*, Studies in Advanced Mathematics, CRC Press, Boca Raton, FL, 1992. viii+268 pp.

The measure theory is basic to many subjects in mathematics, and almost every mathematics student has learned the Lebesgue measure on the real line, which is discussed in many books. There are also many books on general measure theories for abstract spaces, for example, locally compact topological groups. This book fills in a gap by studying some refined properties of the measure theory for subsets of $\mathbb{R}^n$, such as Hausdorff measures, capacities, Rademacher's theorem on almost everywhere differentiability of Lipschitz functions, the area and co-area formulas etc. This book can be titled *Measure Theories for Working Analysts*.

According to MathSciNet, "As the authors say, 'this book is definitely not for beginners'. While it does not contain a lot of motivational material, the exposition—though spare—is very well organized and quite complete, with few technicalities left untreated... The authors have done a splendid job of making this profound and interesting material accessible to an industrious reader."

C.A. Rogers, *Hausdorff Measures*, Reprint of the 1970 original, With a foreword by K. J. Falconer, Cambridge Mathematical Library, Cambridge University Press, Cambridge, 1998. xxx+195 pp.

A Hausdorff measure is a measure, named after Felix Hausdorff, that assigns a nonnegative number to each set in $\mathbb{R}^n$ or, more generally, in any metric space. For example, the zero dimensional Hausdorff measure is the number of points in the set (if the set is finite) or $+\infty$ if the set is infinite. The one and two dimensional Hausdorff measures are also the standard length and areas. But the important point is that for every $d \geq 0$, there are d-dimensional measure. For any subset A of a metric space, say $\mathbb{R}^n$, if the d-dimensional Hausdorff measure of A is finite and positive, then for any $d' > d$, the d'-dimensional measure of A is zero, and for any $d'' < d$, the d''-dimensional measure of A is ∞. Therefore, for every subset A, there is a unique dimension d to talk about its d-dimensional measure. This d is the Hausdorff dimension of the subset A. Hausdorff measures are fundamental in geometric measure theory, ergodic theory and dynamical systems.

The reprint of this book in 1998 contains a 22 page foreword by K. J. Falconer which gives a nice summary of the history of Hausdorff measures before the publication of this book (for example, the name came from a result of Hausdorff that the Cantor set has a finite positive $\log 2/\log 3$-dimensional measure), and developments after that. It also includes suggestions for further reading. The book of Rogers is the first one which discusses Hausdorff measures systematically. Together with the book, this foreword gives a comprehensive and updated view of Hausdorff measure theories. This is useful both as a textbook for students and a reference for experts.

According to MathSciNet, "This beautifully written and beautifully printed little book contains an astonishing amount of information. The author tacitly assumes that his reader knows elementary set theory, cardinal numbers and the elementary theory of functions of a real variable. Starting with only these tools, he constructs all of elementary measure theory in his first chapter, 49 pages. This is by all odds the most lucid and succinct presentation of abstract measure theory that the reviewer has seen. The proof that the family of measurable sets is closed under Suslin's operation is particularly elegant. Of course, in so brief a treatment it is impossible to cover all of measure theory, and the author plainly has his eye on Hausdorff measures, ..."

J. Diestel, J. Uhl, *Vector Measures, With a Foreword by B. J. Pettis*, Mathematical Surveys, No. 15, American Mathematical Society, Providence, R.I., 1977. xiii+322 pp.

The title *Vector Measures* means that it studies measurable functions with values in Banach spaces. It can be considered as a sequel to results on vector measures in Part I of the famous book *Linear Operators* by Dunford and Schwartz. Following Dunford and Schwartz, at the end of each chapter, there is a section called "Notes and remarks". It is a concise and reliable reference.

According to MathSciNet, "This text is a must for anyone interested in vector-valued measures or in Banach spaces, or both. It is devoted to the study of properties of measures with values in Banach spaces and the relationships between those properties and properties of the Banach spaces. It attempts to bring up to date the material on vector measures in Part I of the classic monograph by N. Dunford and J. T. Schwartz [Linear Operators, I. General Theory, Interscience, New York, 1958]. The authors successfully imitate Dunford and Schwartz's style of ending each chapter with a comprehensive and informative section of notes and remarks. For the researcher, these sections are the most important part of the book."

3.1.8 Analysis on general spaces

In calculus and real analysis, we study differentiation and integration of functions on $\mathbb{R}^n$ and smooth manifolds. But non-smooth spaces naturally arise from limits of smooth manifolds and subsets of smooth manifolds. A natural problem is to generalize the standard concepts and methods from analysis to these more general spaces. The following books deal with such theories.

J. Heinonen, *Lectures on Analysis on Metric Spaces*. Universitext. Springer-Verlag, New York, 2001. x+140 pp.

Analysis on metric spaces is a recent subject. Heinonen contributed substantially to differentiation of functions on metric spaces. This book is a good introduction to this new subject.

According to MathSciNet, "This very interesting book deals with how various concepts of Euclidean analysis (e.g., gradients, Sobolev spaces, Poincaré inequalities, quasi-conformal mappings, etc.) can be extended to more general metric spaces. The book is based on a 1996 course given by the author at the University of Michigan and it retains a comfortable, conversational style with numerous remarks, comments, references and conjectures... The author summarizes and describes an area which has been very active recently and does an excellent job of giving the flavor and main ideas in the field. It is a well-written and enjoyable book containing something for both beginners and experts and deserves to be on the shelf of anyone interested in the interplay of analysis, geometry and topology."

G. David, S. Semmes, *Analysis of and on Uniformly Rectifiable Sets*, Mathematical Surveys and Monographs, 38, American Mathematical Society, Providence, RI, 1993. xii+356 pp.

The two authors of this book have studied systematically the analysis of functions defined on a d-dimensional subset in $\mathbb{R}^n$, the behavior of certain linear operators on these functions such as variants of the Cauchy integral operators and on the related geometric properties of the subsets of $\mathbb{R}^n$. This book is a summary of their works and can serve as a valuable reference.

According to MathSciNet, "This is a research monograph devoted to the interaction between harmonic analysis and geometric measure theory. More precisely, it is concerned with quantitative notions of rectifiability of d-dimensional subsets of $\mathbb{R}^n$ which are related to the boundedness of certain singular integral operators, such as the Cauchy integral. It shows how a wide variety of powerful ideas used to analyze functions (e.g., Littlewood-Paley theory, Carleson measures, the corona construction, ...) have analogues for sets which are important for understanding certain linear operators, as well as understanding the geometry of rectifiable sets for their own sake... The content of this monograph is technical, but very interesting, and draws together many important ideas from classical and geometric analysis. The book is well written and the authors have made it more readable by starting with a careful summary of the background and main results and finishing with tables listing the acronyms and theorems. The book presents the state of the art for this kind of geometric analysis and should be valuable to researchers working in harmonic analysis, geometric measure theory and related topics, as well as students wishing to get an overview of recent work in this area."

R.T. Rockafellar, R. Wets, *Variational Analysis*, Grundlehren der Mathematischen Wissenschaften, 317, Springer-Verlag, Berlin, 1998. xiv+733 pp.

By "variational analysis", the authors mean an extension of the classic calculus of variations and convex analysis to more general problems of optimization theory, including topics in set-valued analysis such as generalized derivatives. It is an essential part of current researches in optimization, equilibrium, and stability of nonlinear systems. This comprehensive book is a standard reference.

According to MathSciNet,"Nonsmooth and set-valued analysis, for which the authors propose the all-encompassing term variational analysis, constitute the subject matter of this book. Nonsmooth analysis is that body of theory that is concerned with the calculus of functions and sets that do not admit linear approximations, in the sense of the customary derivative or the usual tangent space. This area of research has been a very active one in its own right in the last quarter-century, and its applications to other fields have greatly multiplied as well. This fact reflects the growing appreciation of the essential need to grapple with nonsmoothness (together with absence of convexity) in such topics as optimization, control, nonlinear partial differential equations, optimal design, elasticity, variational methods, and, in fact, in nonlinear analysis generally. In nonsmooth analysis there intervene certain set-valued functions which play the role of (for example) generalized derivatives or tangents. The study of multifunctions such as these and others, together with related and more general issues of convergence (e.g., graph or gamma-convergence), has given rise to that subject known as set-valued analysis."

J.P. Aubin, H. Frankowska, *Set-valued Analysis*, Systems & Control: Foundations & Applications, 2, Birkhäuser Boston, Inc., Boston, MA, 1990. xx+461 pp.

Analysis is certainly basic and important in mathematics. How about set-valued analysis? Indeed, the authors asked "Who needs set-valued analysis?" Their answer is "Everyone, we are tempted to say, and we shall state our case.... One can no longer afford the luxury of studying only well posed problems in Hadamard's sense: Ill prosed problems, inverse problems and many other unorthodox problems under other names are popping up in every domain of activity, ... Requiring that maps should be always single-valued, and even bijective, is too costly an attitude, above all in many applied fields, where we are not free to make such assumptions..."

This book is a comprehensive account of set-valued analysis and is a valuable introduction and reference on this topic.

According to MathSciNet, "It includes both old and new results with many historical comments giving the reader a sound perspective to look at the subject."

According to Zentralblatt Math, "The style is lively and vigorous, the relevant historical comments and suggestive overviews increase the interest for this work... Graduate students and mathematicians of every persuasion will welcome this unparalleled guide to set-valued analysis."

J.P. Aubin, I. Ekeland, *Applied Nonlinear Analysis*, Pure and Applied Mathematics (New York), A Wiley-Interscience Publication, John Wiley & Sons, Inc., New York, 1984. xi+518 pp.

This is an advanced introduction to applied nonlinear analysis. It includes both smooth and nonsmooth functions, convex variational (or optimization) problems to nonconvex ones, and applications to economics and Hamiltonian dynamical systems. It is a very useful reference written by two experts of the subject.

According to MathSciNet, "Multivalued nonsmooth analysis is much more recent and may yet undergo many important developments. The book under review presents a very broad range of topics and points of view of nonlinear analysis. Efforts are made throughout the book to exhibit the common aspects of a priori different problems. The presence of this book should contribute to amplifying interactions between various branches of nonlinear analysis and to a further growth of the theory."

F. Clarke, *Optimization and Nonsmooth Analysis*, Canadian Mathematical Society Series of Monographs and Advanced Texts, A Wiley-Interscience Publication, John Wiley & Sons, Inc., New York, 1983. xiii+308 pp.

As we are taught in elementary calculus, taking derivatives to find critical points is an essential step in finding extremum values of functions and hence solving optimization problems. One purpose of this book is to show that "that much of optimization and analysis which have evolved under traditional smoothness assumptions can be developed in a general nonsmooth setting". (By "nonsmooth", the author means "not necessarily smooth".) The second purpose is "to point out the benefits of doing so." This is a valuable and popular reference.

From the book description, "The author first develops a general theory of nonsmooth analysis and geometry. Clarke then applies these methods to obtain a powerful, unified approach to the analysis of problems in optimal control and mathematical programming. Examples are drawn from economics, engineering, mathematical physics, and various branches of analysis in this reprint volume."

According to MathSciNet, "This work of some 300 pages is devoted to the study of optimization problems (in the broad sense of the term) in which the data are not necessarily differentiable. One of the fundamental tools of the approach taken here is the notion of a generalized gradient of a function, a concept introduced by the author in 1973 that has led to numerous works, developments and applications... A real effort has been made to make the definitions and results accessible to those who are not experts in the field; the potential user of these concepts will have no problems using this book. Researchers wishing to invest time in nondifferentiable analysis and optimization will no doubt be better guided by the source articles; this is often the case when books present an elaborate finished product."

3.1.9 G. Pólya, G. Szegö, Problems and Theorems in Analysis. Vol. I. Series, Integral calculus, Theory of Functions. Vol. II. Theory of Functions, Zeros, Polynomials, Determinants, Number Theory, Geometry, Grundlehren der mathematischen Wissenschaften, Classics in Mathematics, Springer-Verlag, Berlin, 1998. xii+392 pp. xx+389 pp.

The first edition of this classical book was published in 1925 and it has been reprinted several times after that. When it first appeared, it was new and unique in its approach. After many years and many books on problems in analysis, there is probably still no other book which is comparable to it. The problems in this book are not routine ones. It takes efforts to solve them, and one can benefit a lot from doing that. The problems and accompanying materials are carefully thought out and arranged. This is a very valuable book for everyone who wants to learn and use analysis.

According to Bulletin of AMS, "The work is one of the real classics of this century; it has had much influence on teaching, on research in several branches of hard analysis, particularly complex function theory, and it has been an essential indispensable source book for those seriously interested in mathematical problems. These volumes contain many extraordinary problems and sequences of problems, mostly from some time past, well worth attention today and tomorrow. Written in the early twenties by two young mathematicians of outstanding talent, taste, breadth, perception, perseverance, and pedagogical skill, this work broke new ground in the teaching of mathematics and how to do mathematical research."

The purpose of this book is well-explained by the preface in the first German edition: "In the mathematical literature there exist a number of excellent and comprehensive collections of problems, books of exercises, review texts etc. The present book, in our view, differs from all of these, both in its aim and in the scope and arrangement of the material covered, as well as in the manner in which we envisage its use... The chief aim of this book, which we trust is not unrealistic, is to accustom advanced students of mathematics, through systematically arranged problems in some important fields of analysis, to the ways and means of independent thought and research. It is intended to serve the need for individual active study on the part of both the student and the teacher. The book may be used by the student to extend his own reading or lecture material, or he may work quite independently through selected portions of the book in detail. The instructor may use it as an aid in organizing tutorials or seminars.

This book is no mere collection of problems. Its most important feature is the systematic arrangement of the material which aims to stimulate the reader to independent work and to suggest to him useful lines of thoughts... Above all we aim to promote in the reader a correct attitude, a certain discipline of thought, which would appear to

3.1. CALCULUS AND REAL ANALYSIS

be of even more essential importance than in other scientific disciplines... Rather than knowing the correct rules of thought theoretically, one must have them assimilated into one's flesh and blood ready for instant and instinctive use. Therefore, for the schooling of one's powers of thought only the practice of thinking is useful. The independent solving of challenging problems will aid the reader far more than the aphorisms which follow, although as a start these can do him no harm.

One should try to understand everything: isolated facts by collating them with related facts, the newly discovered through its connection with the already assimilated, the unfamiliar by analogy with the accustomed, special results through generalization, general results by means of suitable specialization, complex solutions by dissecting them into their constituent parts, and details by comprehending them within a total picture..."

A companion or sequel to the book above is the following.

P. Borwein, T. Erdélyi, *Polynomials and Polynomial Inequalities*, Graduate Texts in Mathematics, 161, Springer-Verlag, New York, 1995. x+480 pp.

This is a useful introductory book for students and nonexperts. According to MathSciNet, "its focus is narrower than that of Pólya and Szegö, concentrating on polynomials, rational functions, trigonometric polynomials, and Muntz polynomials/rational functions, but its lively tutorial style makes it a delight to read. It highlights not only the authors' recent exciting results, but also many recent works that have not been surveyed anywhere in book form, while not neglecting the classics. Chebyshev, Remez, Bernstein, Cartan, Markov, Erdős, Turán, among others, are all represented, all with new twists, as are sharp inequalities for derivatives of polynomials, trigonometric polynomials, Muntz polynomials, rationals with pre-assigned poles, polynomials with zeros in some region, etc. There are delightful applications such as the length of a lemniscate, density of weighted polynomials, incomplete polynomials, irrationality and transcendence, etc."

According to Zentralblatt MATH, "This attractive book deals with the analytic theory of polynomials and rational functions of one variable. An interesting feature of the book is the presentation of many of the results as series of exercises. This enables the book to cover much more ground than otherwise and enhances its value as a text for an advanced undergraduate or graduate course in approximation theory. The more difficult exercises are provided with liberal hints. Although some of the material in the early chapters would be found in many texts on approximation theory, the later chapters contain many results that have not previously appeared in textbooks. Most of these are results to which the authors have made major contributions. Thus the book would be of interest to professional mathematicians as well as to students."

3.1.10 Orthogonal polynomials

An orthogonal polynomial sequence is a sequence of polynomials such that any two different polynomials in the sequence are orthogonal to each other under some inner product, which could be integration with respect to the Lebesgue measure or a weighted measure. They are fundamental tools in real analysis, harmonic analysis and related topics. The most widely used orthogonal polynomials are the classical orthogonal polynomials, such as the Hermite polynomials, the Jacobi polynomials together with their special cases the ultraspherical polynomials, the Chebyshev polynomials, and the Legendre polynomials. The field of orthogonal polynomials developed in the late 19th century from a study of continued fractions by P. L. Chebyshev and was pursued by A.A. Markov and T.J. Stieltjes. Szegö also made fundamental contribution to it, and an important book by Szegö on this topic is the next one.

G. Szegö, *Orthogonal Polynomials*, Fourth edition, American Mathematical Society, Colloquium Publications, Vol. XXIII, American Mathematical Society, Providence, R.I., 1975. xiii+432 pp.

The origin of orthogonal polynomials lies in the study of a certain type of continued fractions in the work of Stieltjes and others. This classic is the first book on orthogonal polynomials which starts out with the orthogonal property as the basic point and studies orthogonal polynomials systematically. Its importance can be seen by the many editions since its publication in 1939. It is still a very useful introduction and reference.

According to MathSciNet, "This is the first detailed systematic treatment of orthogonal polynomials... As stated in the Preface, no claim is made for completeness. Thus, very little space is devoted to the problem of moments and to the relation of orthogonal polynomials to continued fractions. However, these omissions are compensated by several interesting features: (a) an elaborate treatment of the asymptotic behaviour of orthogonal polynomials, by various methods, with applications, in particular, to the 'classical' polynomials of Legendre, Jacobi, Laguerre and Hermite; (b) a detailed study of expansions in series of orthogonal polynomials, regarding convergence and summability; (c) a detailed study of orthogonal polynomials in the complex domain; (d) a study of the zeros of orthogonal polynomials, particularly of the classical ones, based upon an extension of Sturm's theorem for differential equations. The book presents many new results; many results already known are presented in generalized or more precise form, with new simplified proofs."

T.S. Chihara, *An Introduction to Orthogonal Polynomials*, Mathematics and its Applications, Vol. 13, Gordon and Breach Science Publishers, New York-London-Paris, 1978. xii+249 pp.

3.1. CALCULUS AND REAL ANALYSIS

Due to active development in approximation theory and numerical analysis, interest in orthogonal polynomials has been revived. This book is a very good introduction to this classical and current subject.

According to MathSciNet, "The book presents, in a very readable way, background and development for an introduction to the general theory of orthogonal polynomials. A first undergraduate course in real analysis should be quite adequate preparation and, with the author's clear style and good selection of exercises for developing many of the established properties, the book gives a solid basis for (real) orthogonal polynomial theory."

3.1.11 Problem books

The importance of problems has been explained in the preface of the book *Problems and theorems in analysis* by Pólya and G. Szegö, which was quoted in §3.1.9. Another more recent, famous book based on mathematical problems (or rather known and important theorems) and their elegant proofs is the next one.

M. Aigner, G. Ziegler, *Proofs from The Book*, Fourth edition, Springer-Verlag, Berlin, 2010. viii+274 pp.

If the beauty of mathematics theorems and formulas is important, the beauty of their proofs probably matters too. It is perhaps helpful to point out that Gauss gave eight different proofs of the quadratic reciprocity law in his life. This unusual book contains elegant proofs of many beautiful results in elementary mathematics. It is more than proofs. Indeed, there is treasure buried in each page, one gem after another. Some of the proofs are classics, but many are new and brilliant. They are all beautiful. This is a fun book to read.

The review of the first edition in MathSciNet, "Paul Erdős maintained that God kept a Book with only the most elegant mathematical arguments. This volume, conceived in consultation with Erdős and published in his memory, suggests some of the Book's contents. Thirty sections treat results drawn from number theory, geometry (mainly combinatorial), analysis, combinatorics and graph theory; these can be followed by one versed in undergraduate mathematics including discrete topics. The proofs date mainly from the entire span of the twentieth century; many are due to Erdős himself... The presentation is clear and attractive with wide margins for portraits, diagrams and sketches."

Here is another review on this book. "This is one of those books that a serious mathematician will probably enjoy picking up and reading from time to time. It is neither a reference nor a textbook, but more a source of mathematical inspiration, a collection of particularly elegant mathematical results. My favourite aspect of this book is the way it focuses on proofs and results that draw connections between different areas of mathematics. The interconnectedness of mathematics is too often ignored by researchers nowadays, who have become specialized to the point that their work is often inaccessible. This book is a healthy step in the opposite direction.

My only complaint about this book (and it's not really about this book at all) is that there are not more books like this. The book is heavily slanted towards number theory, combinatorics, and graph theory. The chapter on analysis is beautiful, but atypical of analysis as a whole. I think that it would be wonderful for people to write similar books in other topics, in particular, analysis, algebra, or probability or topology like the other reviewer mentioned..."

The following book on problems in analysis is famous among students of several countries. For example, almost every mathematics student in China around 1980 owned a copy of it. Its impact on the mathematical education of these students is huge.

B.P. Demidovich, *Problems in Mathematical Analysis*, Mir Publishers, 1989. 496 pp.

It seems to be out of print in English, but a new Chinese translation based on a new Russian edition just appeared in 2010. The author is famous for this book much more than for his work in mathematics. It consists of many problems at various levels from all subjects of calculus and real analysis. In comparison with the book by Pólya and Szegö, problems are much easier. On the other hand, since it builds up from basic and easier ones, it is helpful to most students and encouraging. The comprehensive coverage of many basic topics makes it attractive to both diligent students and teachers. This is very useful for students who really want to horn their skills in calculus and real analysis.

According to Amazon reviews, "If you want to learn analysis with this book, forget it, but if you have a good text book, this is one of the best tools you'll need to master problem solving in calculus. Well explained and well organized problem solving tips and technics, step by step from the very beginning until more advanced topics, together with a large numbers of exercises, everyone with the proper result in the end of the book, make it a must have in the library of anyone who seriously needs calculus problem solving skills."

S.M. Ulam, *A Collection of Mathematical Problems*, Interscience Tracts in Pure and Applied Mathematics, no. 8, Interscience Publishers, New York-London, 1960, xiii+150 pp.

A better title for this book is perhaps *A Collection of* **Open** *Mathematical Problems*. They are not books of problems which have solutions. It is a valuable source for both ambitious students and mathematicians.

According to a review in Bulletin of AMS, "The problems considered are in the spirit of the so-called Scottish Book. During the 1930's a notably gregarious group of mathematicians in Lwow, Poland—including Banach, Steinhaus, Mazur, Orlicz, Schauder, Schreier, Ulam, and others—were accustomed to meet for long mathematical discussions in 'The Scottish Coffee Shop'. From time to time, problems which they posed to each other were written down in a notebook which was kept there for the purpose. (Sometimes the proposer would indicate his estimate of the difficulty of a problem by offering a prize for its solution; perhaps a bottle of wine, or two small beers !) Visiting mathematicians too were invited to add their problems to the collection. After the war this book was carried to Wroclaw, where the tradition was revived by some of the surviving members of the group. Many problems from the 'New Scottish Book' have appeared in the problem section of Colloquium Mathematicum. A few years ago, Ulam circulated privately a translation made from a copy of the original Scottish Book. The interest aroused by this encouraged him to write the present book. The problems include many which he first inscribed in the Scottish Book, but a greater number stem from later years. In fact, the book constitutes a kind of mathematical autobiography. Each of the various fields in which the author has worked has contributed its share of problems. But despite their diversity there is an underlying unity. As the author puts it, 'the motif of the collection is a set-theoretical point of view and a combinatorial approach to problems.' "

According to MathSciNet, "This fascinating little book is not exactly a collection of problems. It is of very uneven character ranging from very specific problems to minor discourses suggesting methods of attack on vast new topics. It is written in a kind of 'off the cuff' manner and very relevant references are occasionally missing, ... This, however, does not detract from the value of the book, in which every mathematician would like to browse and which should certainly stimulate many younger mathematicians. The topics covered are those encountered by the author throughout his work and are dominated by the set-theoretic and combinatorial point of view."

As mentioned before, **problems** are important in mathematics for both education and research. There are more problems in number theory and elementary geometry which are known to more people. Some of them can be easily formulated and understood, but proofs are difficult or not known yet. For example, the last Fermat's theorem and the Goldbach conjecture are such examples. The next two books are books on problems in number theory and geometry.

R. Guy, *Unsolved Problems in Number Theory*, Third edition, Problem Books in Mathematics, Springer-Verlag, New York, 2004. xviii+437 pp.

This unique book is a valuable source for beginning researchers and aspiring students. The motivation for writing this book is well explained in the preface, "Number theory has fascinated both the amateur and the professional for a longer time than any other branch of mathematics, so that much of it is now of considerable technical difficulty. However, there are more unsolved problems than ever before, and though many of these are unlikely to be solved in the next generation, this probably won't deter people from trying." "One purpose of this book is to provide beginning researchers, and others who are more mature, but isolated from adequate mathematical stimulus, with a supply of easily understood, if not easily solved, problems which they can consider in varying depth, and by making occasional partial progress, gradually acquire the interest, confidence and persistence that are essential to successful research."

According to MathSciNet, "The book serves this goal well. Perhaps a few of the problems are of secondary importance, and a few of them are perhaps hopelessly difficult. However, most of the problems are interesting and there is a nice chance to prove at least partial results so that the reviewer is sure that many talented young mathematicians will write their first papers starting out from problems found in this book."

According to MathSciNet, "Anecdotal evidence suggests that the earlier editions of this book are among the most-opened books on the shelves of many practicing number theorists. The descriptions of state-of-the-art results on every topic and the extensive bibliographies in each section provide valuable ports of entry to the vast literature... The prospects for the stimulation of the reader are great, whether that reader is an active researcher or an aspiring mathematician; the fact that most of the problems have elementary statements makes it unusually beneficial to the latter category."

H. Croft, K. Falconer, R. Guy, *Unsolved Problems in Geometry*, Corrected reprint of the 1991 original, Problem Books in Mathematics, *Unsolved Problems in Intuitive Mathematics, II*, Springer-Verlag, New York, 1994. xvi+198 pp.

This is a valuable source for people who are looking for problems in geometry of elementary and combinatorial nature. It can be read at different levels too.

From the book description, "Mathematicians and non-mathematicians alike have long been fascinated by geometrical problems, particularly those that are intuitive in the sense of being easy to state, perhaps with the aid of a simple diagram. Each section in the book describes a problem or a group of related problems."

The chapter headings can give a taste of the rich collection of problems of this book: Convexity; Polygons, Polyhedra and Polytopes; Tiling and Dissection; Packing and Covering; Combinatorial Geometry; Finite Sets of Points; General Geometric Problems.

3.1.12 G.H. Hardy, J.E. Littlewood, G. Pólya, Inequalities, Reprint of the 1952 edition, Cambridge Mathematical Library, Cambridge University Press, Cambridge, 1988. xii+324 pp.

An earlier edition of this famous book were written by Hardy, Littlewood, published in 1948. Starting with the edition in 1952, Pólya became a co-author. Simply put, it is a book written by master expositors and is considered the Bible on inequalities by people.

3.1. CALCULUS AND REAL ANALYSIS

Inequalities are clearly of fundamental importance in mathematics. Most people have learnt Cauchy-Schwarz inequality and Hölder's inequality. More advanced inequalities include Jensen's inequality and the Poincaré inequality. But there are many and many more inequalities, which are important for many applications. Picking and applying the right inequalities is an art, and having a well-organized and explained tool box is certainly helpful.

According to Bulletin of AMS by Zygmund, "In retrospect one sees that 'Hardy, Littlewood and Pólya' has been one of the most important books in analysis in the last few decades. It had an impact on the trend of research and is still influencing it. In looking through the book now, one realizes how little one would like to change the existing text."

E. Beckenbach, R. Bellman, *Inequalities*, Ergebnisse der Mathematik und ihrer Grenzgebiete, N. F., Bd. 30, Springer-Verlag, Berlin-Göttingen-Heidelberg, 1961, xii+198 pp.

This book can be considered as continuation of the book above by Hardy, Littlewood and Pólya. It is a very good reference on inequalities, in particular in combination with the previous book.

According to the preface, "Since the classical work on inequalities by Hardy, Littlewood and Pólya in 1934, an enormous amount of effort has been devoted to the sharpening and extension of the classical inequalities, to the discovery of new types of inequalities, and to the application of inequalities in many parts of analysis. ... The results presented in the following pages reflect to some extent these ramifications of inequalities into contiguous regions of analysis, but to a greater extent our concern is with inequalities in their native habitat..."

D.S. Mitrinović, *Analytic Inequalities*, In cooperation with P. M. Vasić, Die Grundlehren der mathematischen Wissenschaften, Band 165, Springer-Verlag, New York-Berlin, 1970, xii+400 pp.

This can be considered as a supplement of the two books above on inequalities.

According to MathSciNet, "There are now three major books on inequalities: this one, the classical work by G.H. Hardy, J.E. Littlewood and G. Pólya *Inequalities*, second edition, 1952, and the book by E. F. Beckenbach and R. Bellman *Inequalities*, 1965. Anyone interested in the subject will have to have all three: Hardy-Littlewood-Pólya for its exhaustive treatment of the classical inequalities and for its thorough discussion of advanced topics that do not appear in other books (even though some of this material could now be presented in more concise or more general form); Beckenbach-Bellman for its wide range both of methods and of topics; and Mitrinović for topics that are in neither of the other books; for its thorough bibliographies; and for an extensive collection of special inequalities, many of which are not otherwise easily accessible, and some of which appear here for the first time. By searching the literature the author has recovered many interesting inequalities that would otherwise have been forgotten. Although what appeals to one analyst need not appeal to another, almost anybody is sure to find interesting things in this book."

A. Marshall, I. Olkin, *Inequalities: Theory of Majorization and Its Applications*, Mathematics in Science and Engineering, 143, Academic Press, Inc. [Harcourt Brace Jovanovich, Publishers], New York-London, 1979. xx+569 pp.

This is another good, more modern book on inequalities.

From the preface, "Although they play a fundamental role in nearly all branches of mathematics, inequalities are usually obtained by ad hoc methods rather than as consequences of some underlying 'theory of inequalities'. For certain kinds of inequalities, the notion of majorization leads to such a theory that is sometimes extremely useful and powerful for deriving inequalities. Moreover, the derivation of an inequality by methods of majorization is often very helpful both for providing a deeper understanding and for suggesting natural generalizations."

According to MathSciNet, "It is innovative, coherent, well written and, most importantly, a pleasure to read. The authors are concerned with the history of their subject. They are generous with their insights. Definitions and conventions are recalled at strategic places and there is extensive cross-referencing so that the reader who is interested in a particular result can begin with it. This work is a valuable resource!"

Another more recent book on different types of inequalities is

Y. Burago, V. Zalgaller, *Geometric Inequalities*, Translated from the Russian by A. B. Sosinsky. Grundlehren der Mathematischen Wissenschaften, 285, Springer Series in Soviet Mathematics, Springer-Verlag, Berlin, 1988. xiv+331 pp.

Typical inequalities in the book by Hardy-Little-Pólya involve finite sums, infinite sums or integrals. This book studies inequalities between fundamental geometric invariants such as lengths, areas, volumes etc, which are called geometric inequalities. Probably the most famous geometric inequality is the isometric inequality $L^2 - 4\pi F \geq 0$, where L is the length of a simple closed curve on a plane, and F is the area of the region enclosed by the curve. The characterization of the equality (i.e., the equality is achieved only when the curve is a circle) is clean and beautiful. This book starts out with the isometric inequalities and many generalizations. It is a useful introduction and a systematic reference for people in geometric analysis and geometry.

According to a review in Bulletin of AMS, "This volume presents us with a masterful treatment of a subject that is not so easily treated. The basic difficulty is that 'geometric inequalities' is not so much a subject as a collection of topics drawing from diverse fields and using a wide variety of methods. One can therefore not expect the kind of cohesiveness or of structural development that is possible in a single-topic book. At most one hopes for a broadly representative selection of theorems organized by approach or content, with a good accounting of each and ample references for following up in any given direction; and that is just what we get. All the classical topics are found here: the isoperimetric inequality in its many guises, the Brunn-Minkowski inequality with its various consequences, area and volume bounds of different kinds. There are also many inequalities involving curvatures: Gauss, mean, Ricci, etc. The methods include those of differential geometry, geometric measure theory, and convex sets. In each of these areas, the book is right up to date, including the latest results to the time of writing. In addition to these classical topics, there are some more modern ones..."

S. Boyd, L. El Ghaoui, E. Feron, V. Balakrishnan, *Linear Matrix Inequalities in System and Control Theory*, SIAM Studies in Applied Mathematics, 15, Society for Industrial and Applied Mathematics (SIAM), Philadelphia, PA, 1994. xii+193 pp.

This is a book on inequalities for matrices. Its main topic is how to solve problems from system and control theory using convex optimization. It is a valuable reference for users in the subject.

According to MathSciNet, "The aim of the book is to show that a wide variety of problems in system and control theory can be reduced to a few standard convex or quasiconvex optimization problems involving linear matrix inequalities (LMI). Since the resulting optimization problems can be solved efficiently by interior-point methods, the above reduction gives a solution of the original problem... This is a very useful book for both researchers and beginners in the field. The style, however, is informal and in many places the details (that in fact make the results rigorous) are omitted..."

J. Garcia-Cuerva, J. Rubio de Francia, *Weighted norm inequalities and related topics*. North-Holland Mathematics Studies, 116. Notas de Matematica [Mathematical Notes], 104, North-Holland Publishing Co., Amsterdam, 1985. x+604 pp.

This is the first systematic account of weighted norm inequalities for singular integral operators and Hardy spaces. This accessible book shows interplays between different topics related to these inequalities. Since it organizes around weighted inequalities, it makes it easier to see the role of these inequalities, and it is a valuable reference for people in harmonic analysis.

According to MathSciNet, "The study of weighted norm inequalities for integral operators of interest in the harmonic analysis of $\mathbb{R}^n$ has been a particularly active field of research for the past twenty years. Some results of these endeavours have found their place in recently published books on harmonic analysis; however, the book under review is the first to have weighted inequalities as its main theme."

G. Duvaut, J.L. Lions, *Inequalities in Mechanics and Physics*, Translated from the French by C. W. John. Grundlehren der Mathematischen Wissenschaften, 219, Springer-Verlag, Berlin-New York, 1976, xvi+397 pp.

This book on inequalities is very different from the earlier ones and deals with problems in mechanics and physics that can be expressed in terms of inequalities where the constants, the equations of state, the physical laws change when certain thresholds are crossed or attained. It is a very specialized book and can serve as a good reference for these special inequalities.

3.2 Complex analysis

Cauchy was one of the first people who made analysis rigorous. For example, he introduced the definition of continuity using epsilon and delta, rejecting the heuristic principle exploited by earlier authors. He proved several important theorems in complex analysis such as the Cauchy residue formula, and also initiated the study of permutation groups in abstract algebra. Cauchy is a deep mathematician and had a great influence on the development of mathematics. For example, his writings cover the entire range of mathematics and mathematical physics in approximately 800 research articles and 5 complete textbooks. More concepts and theorems have been named after Cauchy than any other mathematician.

Figure 3.2: Cauchy's book.

Complex analysis started in the nineteenth century. Some important contributors include Euler, Gauss, Riemann, Cauchy and Weierstrass. There are some striking differences between real and complex analysis. For example, there are differential real functions of various degrees. But a complex function of one complex variable admitting a continuous complex derivative is automatically analytic and hence is differential in all degrees. The theory of conformal mappings is an important geometric aspect of complex analysis, and has many applications in mathematical physics. Complex analysis has also played a fundamental role in analytic number theory due to the pioneering of Riemann on the Riemann zeta function.

3.2. COMPLEX ANALYSIS

3.2.1 H. Cartan, Elementary Theory of Analytic Functions of One or Several Complex Variables. Éditions Scientifiques Hermann, Paris; Addison-Wesley Publishing Co., Inc., Reading, Mass.-Palo Alto, Calif.-London, 1963. 228 pp.

This book was based on the lectures by the author given during three academic years 1957–1960. The focus is on analytic functions of one variable. In order to "to give an insight into the harmonic functions of two real variables as analytic functions and to permit the treatment in Chapter VII of the existence theorem for the solutions of differential systems in cases where the data is analytic", analytic functions of several real or complex variables are also discussed briefly. This explains the title of this book. This is a classic written a master and is still a very good, efficient introduction to complex analysis.

According to an Amazon review, "This elegant little book by Henri Cartan covers both complex functions on one and several variables, and in that way (by the inclusion of several variables) it differs and stands out from most other books on complex variables at the beginning US-graduate level... It is readable, and the exercises are plenty and excellent... It is suitable as a text for a course or as a supplement in a standard beginning graduate course in complex function theory."

According to MathSciNet, "This is an excellent compact treatment of the basic ideas and techniques of elementary complex analysis. Economy is achieved by appropriate use of algebraic and topological concepts. The approach is that of Weierstrass, analytic functions being those that are locally power series."

3.2.2 L. Ahlfors, Complex Analysis, Third edition, International Series in Pure and Applied Mathematics, McGraw-Hill Book Co., New York, 1978. xi+331 pp.

Ahlfors made many fundamental contributions to geometric function theory, conformal geometry, quasi-conformal maps, Teichmüller theory, Kleinian groups etc. Complex analysis is the common underlying thread of these different subjects. This textbook on complex analysis emphasizes the geometric approach of Riemann and was written by a master. It has been the standard textbook for several decades and used by generations of mathematicians.

From the book description, "... this text has retained its wide popularity in this field by being consistently rigorous without becoming needlessly concerned with advanced or overspecialized material..."

According to a review in Bulletin of AMS, "In American universities the course in complex analysis is often used to review advanced calculus with modern standards of rigor. This procedure, which clearly has certain disadvantages, is to some extent forced on our graduate schools, and it is for such a course that this text is written. In complex analysis the temptation to base proofs on intuitive arguments is especially strong. It is a challenge to make the proofs simple and intuitively clear, and at the same time such that the reader can easily fill in the steps of a formal argument. The author has met this challenge in a masterful way... The book is an important contribution to mathematical literature. At every turn one sees the care and ingenuity which the author has used to make his proofs rigorous and readable. The book is intended for the conscientious student, and it will repay him well for the hours that he may spend with it."

According to an Amazon review, "This classic is a brilliant exposition of the Riemann (geometrical) method of complex analysis as opposed to the Weierstrassian (power series) method. The latter approach is done well by Whittaker & Watson or Henrici. Ahlfors' book is the best I know of for the geometrical approach. It is written for senior undergraduates or graduate students majoring in mathematics."

3.2.3 R. Remmert, Theory of Complex Functions, Translated from the second German edition by Robert B. Burckel, Graduate Texts in Mathematics, 122, Readings in Mathematics, Springer-Verlag, New York, 1991. xx+453 pp.

The fact that this book appeared as a volume of the series *Readings in Mathematics* is appropriate. The first seven pages give a historical introduction to complex functions. It starts with a letter of Gauss to Bessel in 1811, which marks the beginning of complex function theory. Near the end, it includes a picture of the tomb of Riemann. It contains a wealth of information and is a fun book to read too.

According to MathSciNet, "This is an elegant and thoroughly modern presentation of the basic theory of holomorphic and meromorphic functions. It is a masterpiece of concision, but at the same time includes much that is unusual in an introductory text, for example, Eisenstein's approach to the trigonometric functions, the evaluation of Gaussian sums, asymptotic series, theta functions, Bernoulli numbers. It is characteristic that the author presents four different proofs of the fundamental theorem of algebra in a page of text. He also gives extensive notes on the history of the subject... I would hesitate to recommend the book to a beginner, but anyone with some background should find it illuminating and informative."

"Its accessibility makes it very useful for a first graduate course on complex function theory, especially where there is an opportunity for developing an interest on the part of motivated students in the history of the subject. Historical remarks abound throughout the text. Short biographies of Abel, Cauchy, Eisenstein, Euler, Riemann and Weierstrass are given. There is an extensive bibliography of classical works on complex

3.2. COMPLEX ANALYSIS

function theory with comments on some of them. In addition, a list of modern complex function theory texts and books on the history of the subject and of mathematics is given. Throughout the book there are numerous interesting quotations. In brief, the book affords splendid opportunities for a rich treatment of the subject."

Another related book by Remmert studies functions of one and several complex variables.

R. Remmert, *Classical Topics in Complex Function Theory*, Translated from the German by Leslie Kay, Graduate Texts in Mathematics, 172, Springer-Verlag, New York, 1998. xx+349 pp.

The rich collections of topics, some of which are unusual for books at such a level, together with historical notes make this a very informative and enjoyable book for mathematicians and students. It is a good textbook for an advanced course.

According to MathSciNet, "This is an elegant and thoroughly modern presentation of the basic theory of holomorphic and meromorphic functions. It is a masterpiece of concision, but at the same time includes much that is unusual in an introductory text, for example, Eisenstein's approach to the trigonometric functions, the evaluation of Gaussian sums, asymptotic series, theta functions, Bernoulli numbers. It is characteristic that the author presents four different proofs of the fundamental theorem of algebra in a page of text. He also gives extensive notes on the history of the subject... The author says that he covers the material in a semester's lectures. I would hesitate to recommend the book to a beginner, but anyone with some background should find it illuminating and informative."

From the book description, "An ideal text for an advanced course in the theory of complex functions, this book leads readers to experience function theory personally and to participate in the work of the creative mathematician. The author includes numerous glimpses of the function theory of several complex variables, which illustrate how autonomous this discipline has become."

3.2.4 More books on complex analysis

There are many books on complex analysis of one variable from different perspectives. For example, the book *Complex Analysis* by Ahlfors has been a popular textbook for many years, while the book *Theory of Complex Functions* by Remmert is more suitable for self-studying and for leisurely reading. Several other excellent books on complex analysis are listed below.

E. Titchmarsh, *The Theory of Functions*, Second edition, Oxford University Press, USA, 1976, 464 pages.

This book consists of introductions to various branches of theory of functions, both real and complex. For example, it contains one chapter on Dirichlet series, and another chapter on residues, contour integration, and zeros. They are of real and current interests since Dirichlet series are fundamental objects in analytic number theory, and contour integration is one of the most effective computation tools in complex analysis. It is different from a usual book on real or complex analysis and is a very valuable book, especially the part on complex function theory.

According to an Amazon review, "It is the complex analysis part which is superb. Titchmarsh is one of those rare authors that manage to motivate the results, get them with rigour and clarity and, especially, select theorems so well that you always find what you need for applications... It is particularly good in asymptotic theorems like Phragmen-Lindelof's, or in the treatment of analytic continuation. But the whole book is a joy!"

According to a review in Nature, "The success which this book achieved proves that Professor Titchmarsh was right in anticipating that the collection which he describes too modestly as consisting of 'rather disconnected introductions to various branches of the theory of functions' would be of service to students to whom the mass of existing material appeared 'rather formidable'. It proves, too, that early reviewers were right in recognizing a quality of readability in his writing; in effect, the advice given to students was: 'This is a book from which you will be able to learn how to do mathematics', and we cannot do better than repeat this advice with all the confidence that comes from knowing that it has been endorsed by those best situated to discover that it was sound, namely, the students who have followed it."

K. Knopp, *Theory of Functions. I. Elements of the General Theory of Analytic Functions. II. Applications and Continuation of the General Theory*, Dover Publications, New York, 1945. vii+146 pp.; 1947. x+150 pp.

As the author explained, "Theory of functions" stands for theory of functions of a complex variable. This is a classic introduction and presents the essential material of basic complex analysis efficiently. This is a good first introduction to complex numbers and functions of one complex variable. Dover has also published his accompanying problem books.

According to an Amazon review, "This elegant little book covers the elements of a senior or 1st year graduate course on complex analysis, although a really good mathematics program like at Berkeley may look upon it as providing some material for a junior course in advanced calculus. It is not a new book, i.e. it predates the space age and computers, but the material is timeless and fundamental. Highly recommended for those who want some exposure to a first-class style in mathematics."

3.2. COMPLEX ANALYSIS

P. Henrici, *Applied and Computational Complex Analysis. Volume 1: Power Series–Integration–Conformal Mapping–Location of Zeros. Vol. 2. Special Functions–Integral Transforms–Asymptotics–Continued Fractions. Vol. 3. Discrete Fourier Analysis–Cauchy Integrals–Construction of Conformal Maps–Univalent Functions*, Pure and Applied Mathematics, Wiley-Interscience, New York-London-Sydney, 1974. xv+682 pp.; 1977. ix+662 pp.; 1986. xvi+637 pp.

Complex analysis is a basic subject for both pure and applied mathematics. There are many books on complex analysis, and most emphasize its beautiful theory. This book is different from others by emphasizing applications, numerical aspects and some additional topics besides those in standard books. For example, in Chapter 6 of Vol. 1 on polynomials, it starts with the Horner algorithm for the efficient evaluation of polynomials, and follows it with several sections on geometry and number of zeros of polynomials in different regions, and methods for determining zeros such as the general method of iteration and Newton's method for polynomials, and methods of descents.

This will supplement a standard book on complex analysis well. For those who are interested in the more computational aspect, it is probably the unique book.

According to a review in Bulletin of AMS, "General textbooks (I am not speaking of specialized monographs) on complex analysis fall into two broad classes of systematic treatises that develop the subject from the beginning and go more or less far, and (usually) briefer and more intuitive presentations that are oriented toward physical applications. Complex analysis is so vast a subject, and has such close connections to other parts of mathematics, that it is possible to write an interesting book, or teach an interesting course, on complex analysis without ever mentioning any of its applications, whether to other parts of mathematics or to the real world. It is equally possible to write a book or teach a course that deals primarily with applications of complex analysis to physics, engineering, and so on, without presenting more than a smattering of rigor, certainly with far less than will satisfy a student who has been trained to expect conclusions to follow logically from care- fully stated hypotheses. There is, in principle, nothing morally wrong about either approach. The subject possesses many deep and beautiful results, and an author or a teacher may want to concentrate on these... On the other hand, it is reasonable to ask that even an introductory course intended for mathematicians give the students some idea of how the subject is used... Henrici's book is that rare phenomenon, a really original textbook, and one that achieves originality honestly, by choice of topics and attitude, not merely by introducing neologisms. It is not a theoretical presentation in depth; your or my favorite profundity may well be missing; yet it does not at all slight the theoretical subtleties of the topics that it covers.

It is not a recipe book, but it has more details on many technical applications than we will find anywhere else outside the specialized journals (and more are promised for later volumes). Finally, as the word 'computational' should suggest, it differs in attitude from even the most 'applied' of its predecessors in adhering whenever possible to the principle 'not to consider a problem solved unless an algorithm for constructing the solution has been found.' By 'algorithm' we are to understand, not the word in its broad sense, but an algorithm that produces an answer in a reasonable time. An ordinary classical analyst would (I suppose) be happy to find a solution of a physical problem in the form of a convergent infinite series, without worrying about how fast it converges; Henrici, however, would clearly be unhappy with a soi-disant algorithm that involved so many steps (terms, in this case) that no computer now in existence could produce an acceptably accurate result in a year of computing time. Indeed, throughout the book he points out computational difficulties and gives error estimates. It is this algorithmic principle that gives this book its distinctive character, which I find quite intriguing..."

According to MathSciNet, "This volume may be used as a reference book by applied mathematicians, physicists or engineers, or as a textbook for sophisticated students of courses given over several semesters. An attractive feature is that, in addition to numerous exercises and problems, a list of seminar topics is given at the end of each chapter."

A.I. Markushevich, *Theory of Functions of a Complex Variable. Vol. I, II, III*, Translated and edited by Richard A. Silverman, Prentice-Hall, Inc., Englewood Cliffs, N.J., 1967, xi+360 pp.; 1965, xii+333 pp.;1965, xiv+459 pp.

This book covers all the major topics in classical function theory of a complex variable, for example, elliptic functions (Weierstrass theory and Jacobi theory), value distribution etc. It might contain too much material as the first textbook, but it certainly complements other books well. It can also serve as a comprehensive reference book.

From the book description of the second English edition printed in 2005, "The author was a leading Soviet function-theorist: It is seldom that an expert of his stature puts himself so wholly at the service of the student. This book includes over 150 illustrations and 700 exercises."

According to MathSciNet, "It covers the elementary theory (through Cauchy's theorem and power series) in great detail... The second volume covers isolated singular points and the theory of residues, inverse and implicit functions, univalent functions and the Schwarz-Christoffel transformation, harmonic and subharmonic functions, applications to fluid dynamics, the Poisson-Jensen formula and functions of bounded characteristic, and entire functions of finite order. This final volume covers conformal mapping, including boundary behavior and prime ends; approximation by rational functions and polynomials; elliptic functions; Riemann surfaces, analytic continuation, the symmetry principle (culminating with Picard's theorem and Julia directions)."

According to an Amazon review, "You cannot find modern things like covering spaces, cohomology, complex manifolds etc in this book. What you find is a page after page covering of classical function theory at its best. Nothing is left out. Lots of examples and exercises are included. A joy to read."

3.2.5 Analytic and meromorphic functions, and conformal maps

The theory of complex analysis is a basic subject and has applications to and connections with many other subjects. In this subsection, we list several books on more specialized and advanced topics.

Z. Nehari, *Conformal Mapping*, McGraw-Hill Book Co., Inc., New York, Toronto, London, 1952. viii+396 pp.

A conformal map is a map which preserves angles. The most basic case is when the map is between domains in the complex plane. If it is orientation preserving, then the map (or function) is holomorphic. Conformal maps provide a geometric method to study and understand holomorphic functions and are important for understanding domains. For example, the conformal homeomorphism between the unit disc and the upper half plane has been very useful to understand functions defined on them (think of constructing explicit bounded holomorphic functions on the upper half plane). By the Riemann mapping theorem, every two simple connected proper domains of the complex plane are homeomorphic by conformal maps. But actual construction of such conformal maps is not easy.

Conformal maps are important to both pure and applied mathematics, and this book tries to "bridge the gulf that many times divide these two disciplines" by combining both theoretical approach and practical approach (actual construction of conformal maps). This book is a classic.

From the preface, "The present book tries to bridge the gulf between the theoretical approach of the pure mathematician and the more practical interest of the engineer, physicist, and applied mathematician, who are concerned with the actual construction of conformal maps. Both the theoretical and the practical aspects of the subject are covered, . . . "

According to MathSciNet, "This book deals with an essential aspect of the theory of analytic functions. Designed for students with a good working knowledge of advanced calculus, the author discusses the subject theoretically as well as practically, ranging from its introductory part to modern developments. . . Altogether, the book is written in an elementary style and offers a presentation available to anyone possessing an elementary knowledge; its latter part will give a good introduction to the more recent advances of the theory of conformal mapping."

G.M. Goluzin, *Geometric Theory of Functions of a Complex Variable*, Translations of Mathematical Monographs, Vol. 26, American Mathematical Society, Providence, R.I., 1969. vi+676 pp.

Though this book is based on lectures in a course titled "the geometric theory of functions of a complex variable", and in part, in another course titled "supplementary topics in the theory of functions of a complex variable" given by the author at Leningrad State University, the first edition was published after his death. The second edition includes a survey of development by the works of his students after the first edition. This is a classic in geometric function theory.

From the book description, "It studies univalent conformal mapping of simply and multiply connected domains, conformal mapping of multiply connected domains onto a disk, applications of conformal mapping to the study of interior and boundary properties of analytic functions, and general questions of a geometric nature dealing with analytic functions... The book is intended for readers who are already familiar with the basics of the theory of functions of one complex variable."

According to MathSciNet, "This text is devoted to advanced topics in the theory of functions of a complex variable, and as the title indicates the emphasis is on the geometric parts of the theory.

Most of the material however has already appeared in the literature, and the author's chief contribution is in organizing and simplifying the material and making it readily accessible. The book is written in a clear and attractive style, due attention is given to motivation, and the selections from the literature were well made."

W.K. Hayman, *Meromorphic Functions*, Oxford Mathematical Monographs Clarendon Press, Oxford, 1964. xiv+191 pp.

A natural class of functions that generalize complex analytic functions consists of **meromorphic functions**. More conventional methods and techniques such as the maximum modulus principle cannot be applied to them due to presence of poles of the functions.

This is the first systematic account of the Nevanlinna theory of meromorphic functions of one complex variables, which deals roughly with relations between growth and value distribution of the functions. The Nevanlinna theory is a vast generalization of Picard's theorem that every transcendental entire function takes every complex value infinitely often, with at most two exceptions. It is a standard reference.

According to MathSciNet, "This attractively written monograph gives a rather complete account of the Nevanlinna theory of meromorphic functions. It also treats the Ahlfors theory of covering surfaces and the theory of invariant normal families. The book can be understood by a reader familiar with basic function theory and with the elements of Lebesgue theory. A very nice feature of the book is the way in which routine discussions of standard results are interspersed with non-standard applications."

C. Pommerenke, *Boundary Behaviour of Conformal Maps*, Grundlehren der Mathematischen Wissenschaften, 299, Springer-Verlag, Berlin, 1992. x+300 pp.

3.2. COMPLEX ANALYSIS

The theory of the boundary behavior of conformal maps studies relations between the analytic properties of mapping functions and the geometric properties of the image domains. This is a classical subject since the Riemann mapping theorem, and this book gives a modern presentation of this theory in geometric function theory.

According to MathSciNet, "Among the book's many excellent features is the way it brings out the kinship of boundary behavior of conformal maps to so many other subjects, such as harmonic analysis, probability, and plane topology. In fact, the book could well serve as a model for writers of advanced mathematics. The exposition is very clear and pleasant to read: relaxed but economical, informative but not loquacious. Each section within a chapter contains exercises designed to make sure the reader understands the immediately preceding text... Readers ranging from graduate students just beginning their research to experienced mathematicians from within and without complex analysis will find this book to be a most educational and agreeable companion."

3.2.6 Several complex variables

Holomorphic functions of several complex variables have very distinct properties from holomorphic function of one complex variable. After major differences were discovered by Friedrich Hartogs and Kiyoshi Oka in the 1930s, a general theory of several complex variables began to emerge. It has been one of the most active subjects in mathematics. On the other hand, in comparison with complex analysis of one variable, there are many fewer books on complex analysis of several variables.

R. Gunning, H. Rossi, *Analytic Functions of Several Complex Variables,* Prentice-Hall, Inc., Englewood Cliffs, N.J. 1965. xiv+317 pp.

The works of Oka and Cartan have shaped the relatively new subject of theory of analytic functions of several variables. This book is the first book which provides an extensive introduction to "the Oka-Cartan theory and some of its applications, and to the general theory of analytic spaces." It played a fundamental role in the development of the subject. It is still a very good introduction and a useful reference.

From the book description, "The theory of analytic functions of several complex variables enjoyed a period of remarkable development in the middle part of the twentieth century. After initial successes by Poincaré and others in the late 19th and early 20th centuries, the theory encountered obstacles that prevented it from growing quickly into an analogue of the theory for functions of one complex variable. Beginning in the 1930s, initially through the work of Oka, then H. Cartan, and continuing with the work of Grauert, Remmert, and others, new tools were introduced into the theory of several complex variables that resolved many of the open problems and fundamentally changed the landscape of the subject. These tools included a central role for sheaf theory and increased uses of topology and algebra. The book by Gunning and Rossi was the first of the modern era of the theory of several complex variables, which is distinguished by the use of these methods. The intention of Gunning and Rossi's book is to provide an extensive introduction to the Oka-Cartan theory and some of its applications, and to the general theory of analytic spaces. Fundamental concepts and techniques are discussed as early as possible. The first chapter covers material suitable for a one-semester graduate course, presenting many of the central problems and techniques, often in special cases. The later chapters give more detailed expositions of sheaf theory for analytic functions and the theory of complex analytic spaces. Since its original publication, this book has become a classic resource for the modern approach to functions of several complex variables and the theory of analytic spaces."

R. Gunning , *Introduction to Holomorphic Functions of Several Variables. Vol. I. Function Theory, Vol. II. Local Theory, Vol. III. Homological Theory*, The Wadsworth & Brooks/Cole Mathematics Series, Wadsworth & Brooks/Cole Advanced Books & Software, Pacific Grove, CA, 1990. xx+203 pp.; 1990. xx+218 pp.; 1990. xx+194 pp.

This three volume set is an update and expansion of the book by Gunning and Rossi. It is more modern and also covers more materials. It is an advanced introduction to several complex variables.

According to a review in Bulletin of AMS, "This is a book that every worker in several complex variables, indeed every analyst, should have on his shelf. It describes a central, influential, and important area of mathematical analysis and will be the reference of choice for some time. The ten years of attention that Gunning lavished on this project shows on each page. The book will serve as a model of elegance and clarity for future writers in several complex variables."

According to MathSciNet, "The book *Analytic Functions of Several Complex Variables* by the author and H. Rossi was one of the first presentations of this material [work of Oda, Cartan, Serre etc] in textbook form and was very influential in the further development of the subject. This material is given a new treatment in this book; this subject matter is divided naturally into three volumes, and the topics are covered in greater depth... This book is a pleasure to read for those acquainted with several complex variables and is a first choice for students seeking an introduction to the subject."

S. Krantz, *Function Theory of Several Complex Variables,* Pure and Applied Mathematics, A Wiley-Interscience Publication, John Wiley & Sons, Inc., New York, 1982. xiii+437 pp.

Krantz was motivated by "the strong interplay between harmonic analysis and the theory of several complex variables" and wanted to write a book which "introduces the classically oriented analysts to holomorphic functions on $\mathbb{C}^n$", in particular those "with an interest in partial differential equations, Fourier analysis, and integral operators." This makes the book more accessible to the otherwise difficult subject. It is a valuable introduction and a useful reference.

According to MathSciNet, "The author has written what is sure to become one of the standard graduate textbooks on several complex variables. His conversational style of writing makes this book a pleasure to read. He makes no attempt to hide mathematical scaffolding, and his pages are often bristling with supportive comments to help motivate proofs and to ease students into technical terrain. He has also compiled an excellent list of examples and exercises. For these reasons, his book will be valuable to graduate students seeking to learn the rudiments of the $\bar{\partial}$-approach to complex analysis."

According to a review of Bulletin of AMS, "One of the remarkable features of this book is that it contains a large number of Exercises and Problems. In fact this seems to be the first treatment of the subject to have a significant number of problems at all. This is particularly important since the subject abounds with abstract theorems and has relatively few concrete (and nontrivial) worked examples. Perhaps this is related to the fact that there are rather few connections with applied mathematics. Several of Krantz's problems are difficult; the student will probably have to look for hints in the references cited.

To be sure, this text is no replacement for the classic book of Hörmander, which is unsurpassed for its power and elegance. But Hörmander demands considerable sophistication from his reader, and the student needs a book like Krantz's, which is written as a text with explanations and exercises... As a text, this book should be excellent for a second course on complex analysis. It covers many of the basic results and connects them up with harmonic analysis and P.D.E.; and the final three chapters provide an introduction to more specialized material."

S.C. Chen, M.C. Shaw, *Partial Differential Equations in Several Complex Variables*, AMS/IP Studies in Advanced Mathematics, 19, American Mathematical Society, Providence, RI; International Press, Boston, MA, 2001. xii+380 pp.

This book studies two differential operators: Cauchy-Riemann and tangential Cauchy-Riemann operators, which are important in the context of functions of several complex variables. As it is well-known from complex analysis, the Cauchy-Riemann characterizes holomorphic functions, and its restriction to the boundary of pseudo-convex domains gives the tangential Cauchy-Riemann operators. This is a good introduction to several complex variables with an emphasis on partial differential equations.

According to a review in Bulletin of AMS, "For those who want to learn more about the interaction of partial differential equations and complex analysis, for those who want to learn the partial differential equations approach to solving the Levi problem, or for those hard analysts who want a sympathetic introduction to several complex variables, I can hardly think of a better source.

This book is deep and substantial and difficult. It does not conform to Sammy Eilenberg's dictum that doing mathematics should be 'like floating on your back downstream.' But its contribution will be a lasting one. The book elucidates and clarifies an important but heretofore obscure subject area. It has enriched the world of several complex variables."

S. Lojasiewicz, *Introduction to Complex Analytic Geometry*, Translated from the Polish by Maciej Klimek. Birkhäuser Verlag, Basel, 1991. xiv+523 pp.

Stanislaw Lojasiewicz is known for his work on semianalytic sets, for example, the Lojasiewicz inequality. This book also reflects his work such as results on constructible sets, removable singularities etc. This book studies the geometry of analytic spaces. It was written as an introduction to familiarize the reader with a basic range of problems of the subject at a level as elementary as possible. The book is almost self-contained and is a valuable introduction written by a specialist.

According to MathSciNet, "The book is an excellent introduction to the theory of complex analytic sets and spaces. The author presents a large amount of material, stressing the geometric aspects of the subject. The book assumes only a basic knowledge and includes the necessary background in three preliminary chapters: Algebra, Topology, and Complex analysis... The book is written tersely but clearly. All the statements have elementary and thorough proofs. The author intentionally omits the use of sheaves. The book includes no exercises."

3.2.7 Sheaf theories

Theory of several complex variables has played a crucial role in the emergence of sheaf theory, and sheaf theory has also clarified and put the theory of several complex variables in the proper framework. Sheaf theory is an important tool in studying complex spaces and allows one to pass from local to global properties. Due to their general nature and versatility, sheaves have many applications in topology and especially in algebraic and differential geometry. Here are some books on sheaf theory.

H. Grauert, R. Remmert, *Coherent Analytic Sheaves*, Grundlehren der Mathematischen Wissenschaften, 265, Springer-Verlag, Berlin, 1984. xviii+249 pp.

Coherent sheaves are sheaves which have played a foundational role in several complex variables, and complex geometry and algebraic geometry. Many results and properties in

3.2. COMPLEX ANALYSIS

complex analytic geometry and algebraic geometry are formulated in terms of coherent sheaves and their cohomology. In a sense, they are the most important customers and also developers of sheaf theories. Coherent sheaves can be seen as a generalization of vector bundles, since sections of an analytic bundle form a natural coherent sheaf. This book contains the general theory of complex analytic spaces by emphasizing coherent analytic sheaves, and is written by two major figures of the subject. It is a good place to learn about coherent sheaves and their applications to analytic spaces.

According to MathSciNet, "The book under review is one in which graduate students can learn all the basic material on coherent sheaves as well as a reference book in which a mathematician using methods of the theory of several complex variables can refresh his memory about those basic tools of coherent analytic sheaves."

According to Zentralblatt MATH, "Generally this book is written with great mastership. It comprises a very large scope of material, proofs are transparent, in the whole text the authors give numerous explanations and examples. In many places there are interesting historical notes. This book is indispensable for everyone interested in the theory of complex spaces."

M. Kashiwara, P. Schapira, *Sheaves on Manifolds*, Grundlehren der Mathematischen Wissenschaften, 292, Springer-Verlag, Berlin, 1990. x+512 pp.

Besides important applications of coherent sheaf theories in several complex variables and algebraic geometry, there are other aspects of sheaf theories and different applications. Since 1969, "the microlocal point of view" of sheaf theories was developed. One aim of this book is to study sheaf theory with this framework. They have played an important role in understanding analytic singularities of solutions of systems of linear differential equations, and D-modules and representation theories etc. This book gives a comprehensive and self-contained account of sheaf theory, with a tilt toward the microlocal point of view. It may not be an easy book, but it is a very reliable reference.

According to Bulletin of the L.M.S., "Clearly and precisely written, and contains many interesting ideas: it describes a whole, largely new branch of mathematics."

According to MathSciNet, "This book is devoted to the study of sheaves by microlocal methods which were first developed by the authors in a series of journal articles and then partly summarized elsewhere... The book may serve as a reference source as well as a textbook on this new subject. Houzel's historical overview of the development of sheaf theory will identify important landmarks for students and will be a pleasure to read for specialists."

R. Godement, *Topologie algébrique et théorie des faisceaux,* Actualit'es Sci. Ind. No. 1252, Publ. Math. Univ. Strasbourg, No. 13 Hermann, Paris, 1958. viii+283 pp.

This is a classic. It has served as a reference for sheaf theory for several decades after its publication. Unfortunately it has not been translated into English, though it was translated into Russian. The main purpose of this book is to study the cohomology of an arbitrary topological space with coefficients in a sheaf. The abundance of ideas, clarity and effectiveness of the presentation make it an excellent book. It is perhaps helpful to remember that the expository writing of Godement has had a lot of impact on mathematics, in particular, on the French school of automorphic forms. This book is probably his best known book.

3.2.8 H. Weyl, The Concept of a Riemann Surface, Translated from the third German edition by Gerald R. MacLane, ADIWES International Series in Mathematics, Addison-Wesley Publishing Co., Inc., Reading, Mass.-London, 1964. xi+191 pp.

This is a famous book where the first modern definition of a Riemann surface is given, i.e., as a 1-dimensional complex manifold. Its impact on modern mathematics is not restricted only to complex analysis. For example, the general definition of manifolds follows this book. As the book description says, "This classic on the general history of functions was written by one of the 20th century's best-known mathematicians. Weyl combined function theory and geometry in this high-level landmark work, forming a new branch of mathematics and the basis of the modern approach to analysis, geometry, and topology." This is a classic, and everyone can still read it with benefit.

According to the preface, "Klein had been the first to develop the freer conception of a Riemann surface, in which the surface is no longer a covering of the complex plane; thereby he endowed Riemann's basic ideas with their full power. It was my fortune to discuss this thoroughly with Klein in divers conversations. I shared his conviction that Riemann surfaces are not merely a device for visualizing the many-valuedness of analytic functions, but rather an indispensable essential component of the theory; not as a supplement, more or less artificially distilled from the functions, but their native land, the only soil in which the functions grow and thrive."

The review in MathSciNet is detailed and also explains well the history of abstract definition of Riemann surfaces: "It does not occur frequently that a book which has been a classic for four decades appears in a completely rewritten edition. The event is the more significant as it concerns a work which has undoubtedly had a greater influence on the development of geometric function theory than any other publication since Riemann's dissertation.

The book originated in Weyl's lectures at Göttingen in the winter semester 1911–1912...

3.2. COMPLEX ANALYSIS

It seems appropriate to summarize here the historical background of the new book... The reader's interest is, of course, focused on the question: how did the subject of the book, the idea of abstract Riemann surface, come into being? The author's reference to Klein as the originator suggests a closer scrutiny, which turns out to be rewarding and not devoid of surprises.

It is well known that Riemann introduced the surfaces now bearing his name as covering surfaces of the complex plane. Although various passages of his other writings can be interpreted as suggesting a more abstract approach, there is nowhere an explicit mention of it. Klein's first thought in this direction was inspired by Prym... In 1923, Klein supplements this acknowledgement with an interesting remark, which seems to have remained quite unnoticed. He says that he had sent his book to Prym and the latter had, in a letter of April 8, 1882, answered "ausdrücklich" [expressly] that, although he (Prym) did not exactly recall his conversation with Klein, he had certainly not referred to ordinary surfaces as carriers of complex functions. Thus we are faced with the unique situation that the less restrictive idea of Riemann surface is due to some unknown comment made by Prym and misunderstood by Klein.

The formal priority naturally remained with Klein... But there still was a long way to go from Klein's idea to that of Weyl. First of all, the former only considered closed surfaces... Furthermore, his reasoning was purely intuitive, based on physical illustrations, without any attempt at mathematical rigour. Weyl achieved perfection in both respects: he was the first to consider a Riemann surface in its full generality, and he gave the first axiomatic treatment of the concept.

The main advantage of the new notion was that it offered the natural carrier for analytic functions. Analyticity being locally defined, the globally imposed complex plane could only allow an artificially restricted theory of complex functions. Weyl's locally defined surface removed this obstacle and also permitted a full exploitation of Riemann's revolutionary idea of reversing the conventional process by giving the surface a priori and studying functions on it a posteriori.

Oddly enough, it took 35 years to find out whether the introduction of abstract surfaces had added any new specimens to conformal equivalence classes of covering surfaces of the plane...

Although Weyl restricted his attention to the theory of Riemann surfaces, the influence of his book has extended a great deal further. The work was written shortly after Brouwer's topological discoveries, and it was the author's ambition to give Riemann's theory the first presentation in which topological concepts and theorems received up-to-date rigorous treatment. The resulting axiomatic way of reasoning has inspired thinking in many doctrines quite remote from complex analysis..."

3.2.9 More books on Riemann surfaces

Riemann surfaces have becoming central objects in mathematics since it was introduced by Riemann. They correspond to algebraic curves over the complex number. They have been intensively studied in complex analysis, complex geometry and algebraic geometry. There are many books on Riemann surfaces from various points of view. In this subsection, we list some books on Riemann surfaces which emphasize the point of view of complex analysis (or geometric function theory). Within this framework, there are some more modern (sheaf theoretic) or classical (potential theory and analytic functions). We will discuss other books in algebraic geometry on algebraic curves over the complex numbers (or over any algebraically closed field) in §5.4.6.

G. Springer, *Introduction to Riemann Surfaces*, Addison-Wesley Publishing Company, Inc., Reading, Mass, 1957. viii+307 pp.

This is one of the first introductions to abstract Riemann surfaces after the foundational book of Weyl. When it was written, it was considered modern. In fact, according to the preface, "This book presents a self-contained, modern treatment of the fundamental concepts and basic theorems concerning Riemann surfaces." Certainly it is not modern anymore in comparison with the other books listed below it. It is indeed self-contained and makes it a very useful introduction to Riemann surfaces.

According to MathSciNet, "The book is written specifically with graduate (and advanced undergraduate) students in mind. There are no prerequisites beyond standard first courses in complex variables, real variables, and algebra. What is needed of topology and Hilbert space theory is derived from the beginning. Concepts and theorems are illuminated by examples and excellent figures, proofs are clarified by heuristic remarks, and the inventiveness of even the good student is challenged by a well chosen problem collection. The style, while very readable, never becomes 'insultingly simple' and even the specialist can derive pleasure from reviewing basic material in a well organized form."

R.C. Gunning, *Lectures on Riemann surfaces*, Princeton Mathematical Notes Princeton University Press, Princeton, N.J., 1966, iv+254 pp.

This is an introduction to Riemann surfaces from a modern point of view at the time when it was written. It is still a very good introduction to the subject by a leading figure on Riemann surfaces in the modern time.

According to MathSciNet, "This book consists of the lecture notes from a year course in the elementary theory of compact Riemann surfaces, from a modern point of view. There are two noteworthy aspects to the author's approach to the subject. First, the exposition is heavily sheaf-theoretic. The concepts of sheaves and Cech cohomology with values in a sheaf are used whenever possible. The sheaf theory used is developed from the beginning, and the prospective reader need not have any background in it. Second, the main analytic tool is the Serre duality theorem instead of the more usual theory of harmonic integrals... These notes are well written and quite readable. The heavily sheaf-theoretic approach may make this book seem too formal and algebraic for a classically oriented mathematician, but this may also make the book the ideal introduction to this beautiful subject (and its various generalizations) for the reader raised with the contemporary bias toward abstraction and formalization. This approach also makes the theories of one and of several complex variables look like part of the same subject."

O. Forster, *Lectures on Riemann Surfaces*, Translated from the German by Bruce Gilligan, Graduate Texts in Mathematics, 81, Springer-Verlag, New York-Berlin, 1981. viii+254 pp.

This is another good introduction to Riemann surfaces from a more modern point of view, and an introduction to the theory of complex manifolds using this special 1-dimensional case.

According to MathSciNet, "With the appearance of Serre's sheaf-theoretic treatment of the Riemann-Roch theorem [Comment. Math. Helv. 29 (1955), 9–26], it must have occurred to many that there was now an opportunity to present a course on Riemann surfaces that takes advantage of the many new tools developed in the theory of several complex variables, and at the same time can serve to introduce and motivate the use of those tools in a conceptually simple case.

A pioneering effort in this direction was R. C. Gunning's book *Lectures on Riemann surfaces*, 1966 ... The book under review is a very attractive addition to the list in the form of a well-conceived and handsomely produced textbook based on several years' lecturing experience. An initial chapter provides a good intuitive feeling for the subject by introducing Riemann surfaces as covering surfaces and algebraic curves... This book deserves very serious consideration as a text for anyone contemplating giving a course on Riemann surfaces."

H.M. Farkas, I. Kra, *Riemann Surfaces*, Second edition, Graduate Texts in Mathematics, 71, Springer-Verlag, New York, 1992. xvi+363 pp.

This is a comprehensive introduction to Riemann surfaces. Chapter 0 contains a good survey of three parts of the development of theory of Riemann surfaces: topological, algebraic and analytic. Its point of view and emphasis are different from the earlier books. For example, the uniformization theorem is an important part of this book. It is a good textbook and a useful reference. It can be read and enjoyed by readers from different backgrounds.

According to a review in Bulletin of AMS of the above two books, "Riemann surfaces, those old and venerated structures, show their smiling faces in many different connections, from the geometry of algebraic curves to the integration of nonlinear partial differential equations in mathematical physics. Even with all that is familiar, each generation finds frontiers beyond, exciting the explorers with a unique combination of explicitness and richness of technique: algebraic, analytic, and geometric. The great masters of the 19th century (Abel, Riemann,...) left a wealth of information and insight on algebraic and abelian functions. Around the turn of the century the uniformization theorem became firmly established by Poincaré and Koebe. (As Ahlfors says, 'This is perhaps the single most important theorem in the whole theory of analytic functions of one variable.') At about the same time, the theory of Fuchsian groups and automorphic functions, which through the uniformization theorem provides its own approach to the subject of Riemann surfaces, was formulated by Poincaré. The classic book of Hermann Weyl in 1913 put the abstract definition of Riemann surface on a solid foundation (still an excellent introduction in the English translation of its third (1955) edition)... The Farkas-Kra work adopts the classical approach and treats in detail compact surfaces while the Forster book uses the tools developed for complex manifold theory in higher dimensions to give a broad based introduction without putting special emphasis on any one area..."

3.2.10 Quasiconformal mappings

Quasi-conformal maps are natural extensions of conformal maps. They have been used to study deformation of conformal maps. After the initial work of Grötzsch in 1928, Teichmüller made essential use of them in his work on deformation spaces (Teichmüller spaces) of Riemann surfaces to solve the problem of moduli of Riemann surfaces and turned them into a major tool in geometric function theory.

L. Ahlfors, *Lectures on Quasiconformal mappings*, Second edition, With supplemental chapters by C.J. Earle, I. Kra, M. Shishikura and J. H. Hubbard. University Lecture Series, 38. American Mathematical Society, Providence, RI, 2006. viii+162 pp.

One can feel the freshness of lectures and informal style of lectures by reading this book. It starts out with seven reasons with justification or disapproval that state why quasi-conformal maps are important in the theory of analytic functions of a single complex variable. Then it proceeds directly to the problem and definition of Grötzsch. This is a classic and still remains a very suitable first introduction to quasiconformal mappings.

According to MathSciNet, "A reprinting of Lars Ahlfors' introductory lectures on quasiconformal mappings was long overdue. Most library copies of the first edition are now very well-thumbed from use by generations of mathematicians, so this second edition is very welcome. The original text from the 1964 lectures at Harvard does not seem dated, apart from one or two changes in terminology; this is testament both to the clarity of the exposition and to the major influence that Ahlfors had on the modern development of the subject. Much of the theory described in the lectures was created by Ahlfors himself, together with his collaborator Lipman Bers. The editors of the second edition provide their own review in their preface to the Ahlfors lectures, including some historical remarks and an appreciation of the influence of this book on generations of researchers. They clearly hold it in high esteem.

The second edition is made much more valuable by the addition of three new chapters by leading authorities in fields where quasiconformal mappings have proved especially fruitful. The supplement by the editors, Clifford Earle and Irwin Kra, is a survey of the developments of the last 40 years in the theory of quasiconformal mappings and Teichmüller spaces. Mitsuhiro Shishikura's chapter covers applications to complex dynamics, undreamt of when Ahlfors gave his lectures. The final chapter by John H. Hubbard describes and sketches a proof of one of the most spectacular applications of the foundations laid down by Ahlfors and Bers, the hyperbolization theorem for 3-manifolds that fiber over the circle. For the student using Ahlfors' lectures to learn about quasiconformal mappings these supplemental chapters will open up many topics for further study. Each includes an extensive bibliography and is sure to inspire further reading. But the new chapters will also be enjoyable reading for researchers already familiar with the topics covered."

L. Ahlfors, *Conformal Invariants: Topics in Geometric Function Theory*, McGraw-Hill Series in Higher Mathematics, McGraw-Hill Book Co., New York, 1973. ix+157 pp.

In his book, Ahlfors shared with the reader his excitement on geometric function theory. As he said, "Geometric function theory of one variable is already a highly developed branch of mathematics, and it is not one in which an easily formulated classical problem awaits its solution. On the contrary it is a field in which the formulation of essential problems is almost as important as their solution; it is a subject in which methods and principles are all-important, while an isolated result, however pretty and however difficult to prove, carries little weight." This book reflects his conviction well. It is a good introduction to geometric function theory written by a master of the subject.

According to MathSciNet, this book "encompasses a wealth of material in a mere one hundred and fifty-one pages. Its purpose is to present an exposition of selected topics in the geometric theory of functions of one complex variable, which in the author's opinion should be known by all prospective workers in complex analysis. From a methodological point of view the approach of the book is dominated by the notion of conformal invariant and concomitantly by extremal considerations..."

O. Lehto, K. Virtanen, *Quasiconformal Mappings in the Plane*, Second edition, Translated from the German by K. W. Lucas, Die Grundlehren der mathematischen Wissenschaften, Band 126, Springer-Verlag, New York-Heidelberg, 1973. viii+258 pp.

After the German version of this book was published, quasiconformal mappings were applied fruitfully to the theory of Kleinian groups and differential equations. Because of this, it was translated into English. This book has contributed to further developments of quasiconformal mappings. Recently, quasiconformal mappings and quasiconformal geometry have also been applied to more applied subjects such as applied mathematics, computer vision and medical imaging. This is a very good reference book for experts who are interested in quasiconformal mappings in the plane.

According to MathSciNet, "Quasiconformal mappings derive much of their fascination from the constant interplay between geometric and analytic methods. This is the leading point of view in this book, to the extent that the authors ignore all applications and even the higher-dimensional analogues. The book is thus intended for readers who are already convinced of the importance of quasiconformal mappings as an independent theory. It is written with a great deal of enthusiasm that will appeal to experts and non-initiates alike. The authors have a very fresh approach, and in almost all instances they improve on existing proofs. In this sense the book contains much original work.

The geometric definition requires merely that the ratio between the conformal moduli of a quadrilateral and its image be bounded. This means that the theory starts out without any differentiability conditions whatsoever. The authors make a point of showing, in the first two chapters, how far one can go in this purely geometric framework... After a review of the analytic tools that are needed (all very elementary), several variants of the analytic definition are introduced and compared with the geometric one."

K. Astala, T. Iwaniec, G. Martin, *Elliptic Partial Differential Equations and Quasiconformal Mappings in the Plane*, Princeton Mathematical Series, 48, Princeton University Press, Princeton, NJ, 2009. xviii+676 pp.

This is a natural sequel to the two books above on quasi-conformal mappings. It is both modern and comprehensive.

According to MathSciNet, "The aim is to present the latest achievements of the analytical theory and their consequences to plane PDE theory and nonlinear analysis in general. Here the authors are successful, although it requires 676 pages. Another aim is to present the optimal assumptions on the existence and regularity of quasiconformal mappings and related PDEs. These are not so apparent in the classical set up but the study of very weak solutions and quasiconformal mappings with finite dilatation opens new possibilities and applications. Much of the recent work has been done in this area where the authors have been the key persons. Many proofs in the book are new."

From the book description, "It gives a thorough and modern approach to the classical theory and presents important and compelling applications across a spectrum of mathematics: dynamical systems, singular integral operators, inverse problems, the geometry

3.2. COMPLEX ANALYSIS

of mappings, and the calculus of variations. It also gives an account of recent advances in harmonic analysis and their applications in the geometric theory of mappings..."

3.2.11 Teichmüller theory

The Teichmüller theory was developed by Teichmüller in 1930s in order to understand the moduli space of Riemann surfaces, which is an important space and has been pursued by many people after Riemann counted the number of parameters (called moduli) that determine Riemann surfaces. Besides applications to the moduli space of Riemann surfaces, the Teichmüller theory is now a vast subject with many unexpected applications from low dimensional topology to mathematical physics, complex dynamics and representation theories. The theory of quasi-conformal maps was developed by Teichmüller and used essentially in Teichmüller theory. Though there are several books on Teichmuller spaces, they have not exhausted many important topics.

O. Lehto, *Univalent Functions and Teichmüller Spaces*, Graduate Texts in Mathematics, 109, Springer-Verlag, New York, 1987. xii+257 pp.

This is not a book that gives comprehensive discussions of either univalent functions or Teichmüller spaces. Instead, it emphasizes the interplay between them. Teichmüller theory is a very well-developed subject now, and there are many different perspectives to study it. This well-written book is a very good introduction to Teichmüller space by emphasizing connection with univalent functions.

From the introduction of the book: "The development of the theory of Teichmüller spaces... gives rise to several interesting problems which belong to the classical theory of univalent analytic functions. Consequently, in the early seventies a special branch of the theory of univalent functions, often studied without any connection to Riemann surfaces, began to take shape... The interplay between the theory of univalent functions and the theory of Teichmüller spaces is the main theme of his monograph."

According to Zentralblatt MATH, "This book painstakingly develops the tools necessary to achieve its stated purpose. The first chapter deals with quasiconformal mappings, the second with univalent functions - specifically those aspects that developed in response to Teichmüller theory, as mentioned in the author's introduction... The book is the measured thoughtful product of an elder statesman in the field."

S. Nag, *The Complex Analytic Theory of Teichmüller Spaces*, Canadian Mathematical Society Series of Monographs and Advanced Texts, A Wiley-Interscience Publication, John Wiley & Sons, Inc., New York, 1988. xiv+427 pp.

There are many different aspects of Teichmüller theory. One important result is the existence of a complex structure which is invariant under the mapping class groups. Since the quotient of a Teichmüller space of a surface by the mapping class group of the surface is the moduli space of complex structures on the surface, this gives a complex structure on the moduli space. This is Teichmüller's approach to solve the moduli problem via Teichmüller spaces. There are several methods to put complex structures on Teichmüller spaces, and the method of Bers via Bers embedding seems to be the most efficient one and reminds one of realization of Hermitian symmetric spaces of noncompact type as bounded domains in $\mathbb{C}^n$. This book gives a comprehensive discussion of complex structure and the Bers embedding of Teichmüller spaces.

According to Zentralblatt MATH, "The book of S. Nag marks a new stage in the intensive development of the Teichmüller space theory. Indeed, it offers for the first time a systematic access way till the core of the most recent results in the complex analytic theory of these and other moduli spaces. It represents not only a very necessary but also very important achievement providing a self-contained, clear and attractive exposition in spite of the huge amount of concepts and facts which are presented of the complexity of the proofs and the variety of the utilized technics."

According to MathSciNet, "This book is a modern treatment of the complex analytic theory of Teichmüller spaces as developed by Ahlfors, Bers, Rauch, Royden and others. The emphasis is on the Bers embedding of Teichmüller space into the Banach space of bounded holomorphic quadratic differentials on a fixed base surface. Related questions on boundaries of moduli spaces and fiber spaces over these moduli spaces are studied in some detail."

W. Abikoff, *The Real Analytic Theory of Teichmüller Space*, Lecture Notes in Mathematics, 820, Springer, Berlin, 1980. vii+144 pp.

One of the major achievements of Teichmüller is that the Teichmüller space is homeomorphic to an Euclidean space by identifying extremal quasiconformal mappings through holomorphic quadratic forms. Another more intuitive way to see this is to use the Fenchel-Nielsen coordinates associated with pants decompositions. These coordinates are only real analytic with respect to the natural complex structure. In comparison with the previous book, this book emphasizes the real analytic aspects of Teichmüller spaces.

According to MathSciNet, "In this book the author gives a highly readable account of a large part of the theory of Teichmüller spaces, the real analytic theory. The aim of the book is to present the necessary background and a coherent part of the Bers theory of moduli to be able to state and prove the theorems of Bers and Thurston which classify the self-mappings of Teichmüller space. It complements the set of notes on Thurston's work edited by Fathi, Laudenbach and Poenaru who presented much of the same material from a topological viewpoint."

Y. Imayoshi, M. Taniguchi, *An Introduction to Teichmüller Spaces*, Translated and revised from the Japanese by the authors, Springer-Verlag, Tokyo, 1992. xiv+279 pp.

3.2. COMPLEX ANALYSIS

This is a comprehensive introduction to Teichmüller space and includes some topics not covered by other books, for example, the Weil-Petersson metric and a compactification of moduli space of Riemann surfaces. Two major figures in Teichmüller spaces are Ahlfors and Bers, and in some sense, this book follows the work of Ahlfors and Bers fairly closely.

According to MathSciNet, "In a clear and well written manner, the authors expose the reader to a wide portion of the theory of moduli of (compact) Riemann surfaces, covering many important topics for the first time in a textbook."

J. Hubbard, *Teichmüller Theory and Applications to Geometry, Topology, and Dynamics. Vol. 1. Teichmüller Theory*, With contributions by Adrien Douady, William Dunbar, Roland Roeder, Sylvain Bonnot, David Brown, Allen Hatcher, Chris Hruska and Sudeb Mitra, With forewords by William Thurston and Clifford Earle, Matrix Editions, Ithaca, NY, 2006. xx+459 pp.

This book starts out with basic materials and gives an accessible introduction to the foundation of Teichmüller theory, which includes both complex analysis and hyperbolic geometry aspects.

According to MathSciNet, "This volume is the first part of a book whose final purpose is to give complete and self-contained proofs of the following four theorems shown by W. Thurston between 1970–1980: (1) The classification of homeomorphisms of a surface. (2) The topological classification of rational maps. (3) A hyperbolization theorem for 3-manifolds that fiber over the circle. (4) A hyperbolization theorem for Haken 3-manifolds.

Although these theorems seem to be unrelated, the proofs of the theorems have a common stream, that is, Teichmüller theory. In the volume under review, the author presents Teichmüller theory in order to understand the proofs of Thurston's theorems, but as a result, the book becomes an excellent textbook on Teichmüller theory.

This volume begins with a foreword by Thurston in which the importance and the beauty of Teichmüller theory are discussed from the point of view of his mathematical experiences and his philosophy on mathematics. The foreword itself is worth reading... In conclusion, the reviewer is convinced that this well-written book will be very useful not only for people who are not familiar with Teichmüller theory, but also for the experts."

F. Gardiner, *Teichmüller Theory and Quadratic Differentials*, Pure and Applied Mathematics (New York), A Wiley-Interscience Publication, John Wiley & Sons, Inc., New York, 1987. xviii+236 pp.

Besides the more standard topics for introductory books on Teichmüller spaces, a unique feature of this book is that it also includes two chapters on quadratic differentials: the theory of Jenkins-Strebel differentials, and Thurston's theory of measured foliations.

According to MathSciNet, "This book is an important addition to the growing literature on the theory of moduli of Riemann surfaces. It is a scholarly work, well written and covering an important portion of Teichmüller theory. The level of exposition is quite good. The book is appropriate for an advanced graduate course."

F. Gardiner, N. Lakic, *Quasiconformal Teichmüller Theory*. Mathematical Surveys and Monographs, 76, American Mathematical Society, Providence, RI, 2000. xx+372 pp.

This book is different from a typical book on Teichmüller spaces as it is more oriented towards preparation for applications. From the preface, "The goal of this book is to provide background for applications of the Teichmüller theory to dynamical systems and in particular to iterations of rational maps and conformal dynamics, to Kleinian groups and three-dimensional manifolds, to Fuchsian groups and Riemann surfaces, and to one-dimensional dynamics. Although Teichmüller theory is a theory of two-dimensional objects, it naturally impinges on three-dimensional topology through its relationship to Kleinian groups and on one-dimensional dynamics through the quasisymmetric boundary action of a quasiconformal self-map of a disc... None of these topics is dealt with in this book. Rather we focus on new developments in the theory in both the finite and infinite dimensional cases. Our program is to give an exposition of the main theorems regardless of dimensionality, emphasizing techniques that apply generally, and hoping to provide background for more applications."

According to MathSciNet, "Gardiner and Lakic have produced a formidable treatise on the modern theories of quasiconformal mappings, Riemann surfaces and Teichmüller spaces. They have gathered, into a unified exposition, results which, for the most part, have not previously been found in book form or are otherwise scattered in the literature. Many of the approaches and results are new; others are more detailed than can be found elsewhere."

A. Fletcher, V. Markovic, *Quasiconformal maps and Teichmüller theory*, Oxford Graduate Texts in Mathematics, 11, Oxford University Press, Oxford, 2007. viii+189 pp.

Several books on Teichmüller spaces emphasize Teichmüller spaces of surfaces with finite topology, and hence the Teichmüller spaces have finite dimension. This book uses the framework of infinite dimensional Teichmüller spaces. It also includes some recent topics such as holomorphic motions, isomorphisms of Teichmüller spaces and local rigidity of Teichmüller spaces.

According to Zentralblatt MATH, "This is a very well written book on the complex analytic aspects of Teichmüller theory. It includes an exposition of important bases of the theory, in the tradition of Ahlfors and Bers, as well as some modern developments.

3.2. COMPLEX ANALYSIS

The book is written in the general framework of infinite-dimensional Teichmüller spaces and of the universal Teichmüller space."

K. Strebel, *Quadratic Differentials*, Ergebnisse der Mathematik und ihrer Grenzgebiete (3), 5, Springer-Verlag, Berlin, 1984. xii+184 pp.

Each quadratic differential on a Riemann surface gives a foliation of the surface. An important class of quadratic differentials consists of those whose associated horizontal trajectories are closed. These are called Jenkins-Strebel quadratic differentials. A major part of this book studies such quadratic differentials, and this book is a very good reference on this topic written by one of the main contributors.

According to Zentralblatt MATH, "This book deals with the theory of quadratic differentials initiated by Teichmüller and developed by the reviewer [Jenkins] with subsequent contributions being made by the author and others. It concentrates on providing a highly detailed presentation of the properties of quadratic differentials per se rather than treating the applications to geometric function theory and quasiconformal mappings or the more recently developed connections with foliations."

According to MathSciNet, "One of the most important notions in more recent geometric function theory is the notion of quadratic differential... But like some other important notions, quadratic differentials, with their connections to topology and differential geometry, soon came to have their own fascination. As a result many more or less scattered results on quadratic differentials exist in the literature, and it was desirable to collect and to set in order these results in a separate monograph. Exactly this has been done by the author in the book under review. The author is especially qualified for this work by his many important contributions to quadratic differentials. He provides in this monograph an excellent introduction to the topic, a book of reference, and a starting point for further research."

A closely related book on Riemann surfaces and Teichmüller spaces is the next one.

P. Buser, *Geometry and Spectra of Compact Riemann Surfaces*, Progress in Mathematics, 106, Birkhäuser Boston, Inc., Boston, MA, 1992. xiv+454 pp.

The geometry of Riemann surfaces (or rather hyperbolic surfaces) is explicit and also complicated. There are close relations between the geometry of hyperbolic surfaces and the spectra of the Laplace operator of the hyperbolic surfaces. One famous example is the Selberg trace formula. This book contains a comprehensive collection of results on the hyperbolic geometry of surfaces and their applications to the moduli spaces of hyperbolic surfaces and spectra of the Laplace operator. It is a unique reference on hyperbolic geometry of Riemann surfaces and its application to spectral theory.

According to MathSciNet, "This book consists of two parts. The first part is a very detailed presentation of the geometry of Riemann surfaces from a cutting-and-pasting approach. Starting with hyperbolic trigonometry, it develops Riemann surface theory through Teichmüller space... The second part is a self-contained exposition of the spectrum of the Laplacian of Riemann surfaces, with particular attention given to the problem of isospectral surfaces... Anyone familiar with the author's hands-on approach to Riemann surfaces will be gratified by both the breadth and the depth of the topics considered here. The exposition is also extremely clear and thorough... The book is pitched at the level of graduate students. The reviewer is unable to imagine anyone who can understand the title not benefiting greatly from the book."

According to a review in Bulletin of AMS, "This is a thick and leisurely book which will repay repeated study with many pleasant hours –both for the beginner and the expert. It is fortunately more or less self-contained, which makes it easy to read, and it leads one from essential mathematics to the 'state of the art' in the theory of the Laplace-Beltrami operator on compact Riemann surfaces. Although it is not encyclopedic, it is so rich in information and ideas..."

3.3 Harmonic analysis

Though Fourier made crucial contributions to Fourier analysis, there were many other contributors. For example, in 1754, Clairaut used series of cosine functions, and in 1759 Lagrange computed the coefficients of a trigonometric series for a vibrating string, and in 1805 Gauss used both sine and cosine functions for trigonometric interpolation of asteroid orbits. Several people including d'Alembert and Gauss used trigonometric series to study the heat equation, but the real breakthrough was the 1807 paper by Fourier titled "Mémoire sur la propagation de la chaleur dans les corps solides", where he made the crucial insight to model all functions by trigonometric series, introducing the Fourier series in the current sense.

Figure 3.3: Fourier's book.

Harmonic analysis studies the representation of functions as the superposition of basic functions. It corresponds to the representation of signals as superposition of basic waves. The basic waves are called "harmonics" in physics, hence the name "harmonic analysis" for the subject. Classically, it studies Fourier series and Fourier transforms. There has been many generalizations and it has become a vast subject with applications in areas as diverse as signal processing, quantum mechanics, and neuroscience. The classical Fourier analysis also has been generalized to analysis on locally compact topological groups, in particular, Lie groups. They are closely related to representation theories of Lie groups.

3.3.1 Y. Katznelson, An Introduction to Harmonic Analysis, 3rd Edition, Cambridge Mathematical Library, Cambridge University Press, 2004. xviii+314 pp.

This book studies harmonic analysis on commutative locally compact topological groups. The unit circle S^1 and the real line $\mathbb{R}$ are the two most basic and important commutative locally compact topological groups, and the book starts with detailed discussion of harmonic analysis on them, i.e., Fourier analysis and Fourier transformation. This combination of concrete examples and abstract theory makes this book accessible and attractive. This is a classic.

According to the citation of the Steele Prize for Mathematical Exposition, "Yitzhak Katznelson's book on harmonic analysis has withstood the test of time. Written in the sixties and revised later in the seventies, it is one of those classic Dover paperbacks that has made the subject of harmonic analysis accessible to generations of mathematicians at all levels. The book strikes the right balance between the concrete and the abstract, and the author has wisely chosen the most appropriate topics for inclusion. The clear and concise exposition and the presence of a large number of exercises make it an ideal source for anyone who wants to learn the basics of the subject."

According to MathSciNet, "Professor Katznelson starts the book with an exposition of classical Fourier series. The aim is to demonstrate the central ideas of harmonic analysis in a concrete setting, and to provide a stock of examples to foster a clear understanding of the theory. Once these ideas are established, the author goes on to show that the scope of harmonic analysis extends far beyond the setting of the circle group, and he opens the door to other contexts by considering Fourier transforms on the real line as well as a brief look at Fourier analysis on locally compact abelian groups."

The MathSciNet review of the first edition says, "In harmonic analysis, as in many other parts of mathematics, the beginning student, wishing to learn seriously about modern and classical aspects of his subject and to see how they fit together, has for a long time been handicapped by the lack of suitable books. The available books on harmonic analysis fall into two quite distinct camps, and within each camp there is further division... In respect of the classical theory of harmonic analysis on the line or on R^n, the situation is, if anything, worse. But perhaps most important is the serious general objection that the classical books on harmonic analysis do not direct the student's attention adequately towards the more "abstract" studies which have brought so much in the way of unity, perspective, new problems and new techniques to the subject. In the "modern" camp, the various presentations of the abstract theory ... presuppose, at least implicitly, a solid understanding of the central areas of the classical theory and are therefore to some extent inaccessible to the raw recruit. To be brief, there has been to date a serious gap in the textbook and treatise literature and a corresponding need for books which would 'span' the two camps. The recent appearance of two books, both by leading mathematicians, which help to fill this gap, therefore constitutes an event of major significance in the development of the subject... The present book, although only 264 pages long, contains a wealth of information, in a concise and highly polished form, about both classical and modern aspects of harmonic analysis... the author has written a book which is at once masterly, rich and stimulating. It will certainly inspire much further research in harmonic analysis, and will occupy an important place in the literature of the subject."

3.3.2 More books on harmonic analysis and Fourier series

Fourier series decomposes periodic functions or periodic signals into the sum of sine and cosine (or complex exponential) functions. They were introduced by Joseph Fourier in his famous book *Théorie analytique de la chaleur* in 1822. A closely related theory is the Fourier transformation, which can be thought of as the limit when the period goes to the infinity. Both Fourier series and Fourier integrals belong to Fourier analysis and are related to the more general harmonic analysis.

J. Duoandikoetxea, *Fourier Analysis*, Translated and revised from the 1995 Spanish original by David Cruz-Uribe. Graduate Studies in Mathematics, 29. American Mathematical Society, Providence, RI, 2001. xviii+222 pp.

The essential objects of classical Fourier analysis are Fourier series and Fourier integrals, but it is a huge subject with connections with many other subjects, and can be studied from different perspectives and by different methods. This book uses the real variable methods introduced by two major figures of the subject: Calderón and Zygmund. For example, it contains the following topics: the Hardy-Littlewood maximal function, the Hilbert transform, singular integrals, H^1, BMO, weighted inequalities, Littlewood-Paley theory and multipliers, etc. This book is a very good introduction to Fourier analysis on Euclidean spaces.

According to MathSciNet, "The book was originally written in Spanish by Javier Duoandikoetxea and it was a big success. This English version is based on the original Spanish version and was completed with the assistance of David Cruz-Uribe...

At the end of each chapter there are sections that discuss topics without proofs but which motivate further study and refer the reader to different sources in the literature. The references are very thorough and extensive."

3.3. HARMONIC ANALYSIS

H. Dym, H. McKean, *Fourier Series and Integrals*, Probability and Mathematical Statistics, No. 14, Academic Press, New York-London, 1972. x+295 pp.

This is an unusual book on Fourier series and integrals. Besides harmonic analysis on the unit circle and the real line, it also discusses harmonic analysis on finite commutative groups and a few important noncommutative groups. It avoids purely technical aspects of the subject such as the pointwise convergence of Fourier series of wild functions, but emphasizes many striking applications such as the isometric problem, Jacobi's identity for the theta function, wave motion and Huygens's principle, random walks, the central limit theorem, Gauss' law of quadratic reciprocity, etc. It is a good book where one can learn and appreciate the extraordinary power and flexibility of Fourier theory.

According to MathSciNet, "The emphasis is placed on the power and flexibility of Fourier methods in a variety of applications... the book should be suitable for advanced undergraduates and for graduates in mathematics and in particular for students in applied fields of mathematics (statistics, physics, engineering etc.). The treatment is in places formal... As a text on the senior/graduate level for students interested in applications of mathematics, this should be an inspiring and very interesting choice."

R. Paley, N. Wiener, *Fourier Transforms in the Complex Domain*, Reprint of the 1934 original, American Mathematical Society Colloquium Publications, 19, American Mathematical Society, Providence, RI, 1987. x+184 pp.

This book covers many topics, but a common thread is the Fourier transformation in the complex domain, in particular, the Paley-Wiener theorem. It "represents a definite statement of the results obtained by" the authors. This is a book written by the founders of a theory and is a classic. It is still a very good reference on the Paley-Wiener theory.

Paley is probably best known for the Paley-Wiener theorem, which relates decay properties of a function or distribution at infinity with analyticity of its Fourier transform. It is a fundamental tool in Fourier transformation and harmonic analysis. Wiener was the more senior of the two, but Paley died at a young age 26 in a skiing accident in the Canadian Rockies during a short vacation taken during his intense work with Wiener.

A.S. Besicovitch, *Almost Periodic Functions*, Dover Publications, Inc., New York, 1955. xiii+180 pp.

Periodicity of functions is needed in the theory Fourier series, and only periodic functions admit expanding into Fourier series. But there are many functions such as $\sin x + \sin \sqrt{2} x$, or $\sin x + \cos \pi x$, which are almost invariant under translation by a sequence of numbers which fill in the whole line $\mathbb{R}$ up to a fixed distance. These are called almost periodic functions. This book by Besicovitch is a **classic** on this topic.

Besicovitch made his first major contribution to mathematics in almost periodic functions, and a class of which is named after him. He also did important work on the Kakeya needle problem and the problem of computing the Hausdorff dimension, both of which are of current interests. Roughly speaking, a Kakeya set, also called a Besicovitch set, is any set of points in Euclidean space which contains a unit line segment in every direction. While some obvious objects satisfy this property, several interesting results and questions are motivated by the problem on how small such sets can be. Besicovitch showed that there are Besicovitch sets of measure zero. The notion of Hausdorff dimension is an important invariant of irregular subsets, and it is often not easy to compute. Many of the technical methods used to compute the Hausdorff dimension of highly irregular sets were developed by Besicovitch.

H. Bohr, *Almost Periodic Functions*, Chelsea Publishing Company, New York, N.Y., 1947. ii+114 pp.

The original book was written in German and was published in 1932. This book is a classic written by the founder of theory.

As the book description says, "Motivated by questions about which functions could be represented by Dirichlet series, Harald Bohr founded the theory of almost periodic functions in the 1920s. This beautiful exposition begins with a discussion of periodic functions before addressing the almost periodic case. An appendix discusses almost periodic functions of a complex variable." Besides the original applications to theories of functions, the notion of almost periodicity is also useful in dynamical systems and in communication and information theory.

K. Gröchenig, *Foundations of Time-frequency Analysis*, Applied and Numerical Harmonic Analysis, Birkhäuser Boston, Inc., Boston, MA, 2001. xvi+359 pp.

This book gives a unified approach to most of the modern theory for time-frequency analysis from a mathematician's point of view. It is written by one of the leading experts in this field.

According to MathSciNet, "Time-frequency analysis can be described as the study of the local frequency behavior of functions. As a tool for signal analysis, it has been used by engineers for decades. It has also played an important role in the early development of quantum mechanics. However, as a mathematical theory, specifically a branch of harmonic analysis, it has only recently come into its own. Foundations of time-frequency analysis provides a clear and thorough exposition of some of the fundamental results in the theory and gives some important perspectives on a rapidly growing field... The book is definitely suitable for a graduate level course in mathematical time-frequency analysis. It assumes a background in real analysis, Fourier analysis and Hilbert spaces. It is also suitable for self-study, as the exposition is superb."

A close relative of Fourier transform is the **Laplace transform**. The Fourier transform expresses a function as the superposition of modes of vibration (or exponential

3.3. HARMONIC ANALYSIS

functions with imaginary exponents), the Laplace transform resolves a function in terms of its moments. Both the Fourier transform and the Laplace transform can be used for solving differential and integral equations. They are also used in physics and engineering for systems such as electrical circuits, harmonic oscillators, optical devices, and mechanical systems.

D.V. Widder, *The Laplace Transform*, Princeton Mathematical Series, v. 6, Princeton University Press, Princeton, N.J., 1941. x+406 pp.

This is a classic and definite reference on the Laplace transform.

According to MathSciNet, "The present book is a significant contribution to a field of analysis whose importance becomes more and more obvious... The present book... will be particularly welcomed by anyone who needs a general introduction to this fascinating subject. It is very clearly written and can be easily understood by a student who has a knowledge of principles of functions of a single real or a single complex variable."

E. Hewitt, K. Ross, *Abstract Harmonic Analysis. Vol. I., Structure of Topological Groups, Integration Theory, Group Representations*, Second edition; *Vol. II: Structure and Analysis for Compact Groups. Analysis on Locally Compact Abelian Groups*, Grundlehren der Mathematischen Wissenschaften, 115, Springer-Verlag, Berlin-New York, 1979, ix+519 pp.; 1970, ix+771 pp.

These two volumes provide a comprehensive introduction to topological groups and harmonic analysis on them. They set out to write a single book which should be accessible to beginners, and also useful to experts. Since these two goals are not consistent, they divided the project into two volumes. They are useful references on abstract harmonic analysis.

According to Math Review of the first edition, "This volume [volume 1] is an exposition of the main body of knowledge concerning topological groups as it exists today. In outline the treatment follows the lines laid down by Weil and Pontryagin, with many additions and expansions. Especially valuable is the assemblage of facts concerning a large number of concrete examples. The book aims to be both useful to specialists and accessible to beginners. The beginner will most likely find the book accessible in detail but formidable in toto... In this second volume the authors have taken on a doubly difficult task. In the first place, they seek to cover an extensive range of material... In the second place, they have sought to present this material in such a way as to make the book useful to relative beginners and to relative experts alike."

L.S. Pontryagin, *Topological Groups*, Third edition. Translated from the Russian by Arlen Brown with additional material translated by P. S. V. Naidu. Classics of Soviet Mathematics Series, L. S. Pontryagin Selected Works, Vol. 2, Gordon and Breach Science Publishers, Switzerland, 1986, 544 pages.

Pontryagin made important contributions to harmonic analysis on abelian locally compact topological groups. For example, he introduced the dual group and proved the Pontryagin duality, which explains in a unified way some properties of the Fourier transform on locally compact groups, such as $\mathbb{R}$ and the unit circle. This book contains this and many other results. It is rather self-contained and starts with the definition of a group. This is a classic written by a giant of the twentieth century. This book is still a good introduction to and reference on topological groups, in particular, on the Pontryagin duality.

This book represents only one portion of the works of Pontryagin. In topology he posed the basic problem of cobordism theory. This led to the introduction around 1940 of a theory of certain characteristic classes, now called Pontryagin classes, which vanish on a manifold that is a boundary of another manifold.

According to MathSciNet review of the first edition, "This book is a lucid and comprehensive account of the present state of the subject. The notion of topological group arose out of that of Lie group by abstracting its algebraic, topological and continuity properties and ignoring the analytical features (that is, a topological group is a topological space whose points form the elements of a group in such a way that the group operations are continuous in the topology). The major development of the subject took place during the past ten years, chiefly through the efforts of van Dantzig, von Neumann and Pontryagin. The result has been the casting of new light on the older theory of Lie groups and the solution of some of its fundamental problems." The second edition is a vast revision of the first edition.

B.Sz.-Nagy, C. Foias, *Harmonic Analysis of Operators on Hilbert Space*, Translated from the French and revised North-Holland Publishing Co., Amsterdam-London; American Elsevier Publishing Co., Inc., New York, Akademiai Kiadó, Budapest 1970, xiii+389 pp.

B.Sz.-Nagy was a Hungarian mathematician. His father, G. Sz.-Nagy was also a famed mathematician, collaborated with A. Haar and F. Riesz, and together with Riesz wrote a famous book *Functional Analysis*. B.Sz.-Nagy contributed to the theory of Fourier series and approximation theory. This book is a very valuable reference for experts related to operator theory.

According to MathSciNet, "The theory of self-adjoint operators on Hilbert space reached a definitive form during the first third of the century through the work of Hilbert, von Neumann, Stone and others. A comparable completeness of the theory for unitary operators awaited the development of the seminal ideas of Beurling by Lax, Helson, Lowdenslager, and Halmos to mention a few. About twenty years ago M. S. Livic initiated a study of "almost unitary" operators and introduced the notion of characteristic operator function to study such operators. In the subsequent work of many mathematicians

3.3. HARMONIC ANALYSIS

including M. S. Brodsky and M. G. Krein, this theory has enjoyed rather striking accomplishments. About fifteen years ago the first author showed how to associate a unique unitary operator, the minimal unitary dilation, with each contraction operator and proceeded to develop a theory for contractions based on this correspondence. He was joined in this work by the second author... This book presents a detailed exposition of their theory for contractions along with its relation to the aforementioned theories stemming from Beurling and Livic..."

N.T. Varopoulos, L. Saloff-Coste, T. Coulhon, *Analysis and Geometry on Groups*, Cambridge Tracts in Mathematics, 100, Cambridge University Press, Cambridge, 1992, xii+156 pp.

This book started out with lecture notes from lectures by Varopoulos at Université Paris VI during the period 1982-87. It grew and became an advanced research monograph. It is not self-contained and can serve as a solid reference for experts.

According to MathSciNet, "The book is written in a very concise, clear and very elegant way. It presents an extremely interesting account of some of the most important developments of the last decade in analysis on groups. It will be used and cherished by anyone seriously interested in the subject."

According to Zentralblatt MATH, "This is a useful exposition of extensive work by the authors, scattered in many papers over the past decade, involving subjects of classical analysis such as Sobolev, Harnack and Dirichlet inequalities, heat diffusion semigroup, random walk, convolution operators, Dirichlet spaces in the sense of Beurling & Deny, ... in the context of (possibly discrete) groups... There are many open questions in the references and comments at the end of each chapter."

3.3.3 A. Zygmund, Trigonometric Series. 2nd ed. Vols. I, II. Cambridge University Press, New York, 1959, Vol. I., xii+383 pp.; Vol. II. vii+354 pp.

This book is the first classical book in mathematics I have heard of. Almost every professor in the mathematics department of my college talked about it and had a zerox copy of it. It is comprehensive and contains almost all results one wants to know about trigonometric series. This book has been very influential. It is still a very valuable, maybe the only authoritative, reference on trigonometric series. It is perhaps helpful to note that Zygmund has many, almost 1000, mathematical descendants, many of them are leading figures in the modern subject of Fourier theory and harmonic analysis!

From the foreword of the third edition by Fefferman: "Surely, Antoni Zygmund's Trigonometric series has been, and continues to be, one of the most influential books in the history of mathematical analysis... Its tremendous longevity is a testimony to its depth and clarity. Generations of mathematicians from Hardy and Littlewood to recent classes of graduate students specializing in analysis have viewed Trigonometric series with enormous admiration and have profited greatly from reading it..."

According to a review in Bulletin of AMS of the third edition, "J.E. Littlewood called it the Bible. After so many years, it is more the Bible than ever: it is a message of permanent value, an absolute chef-d'oeuvre, and a reference book for now and for years to come...

This book has to be considered as a piece of art as well as a source of information. It has a global structure, but its beauty has to be discovered in every chapter, page, or sentence. Everything is clear, condensed, and complete. Zygmund had very broad views, and they have a strong historical interest. Now, however, we may and should have other interests and perspectives, as shown in particular by all the books by Elias Stein and collaborators. Antoni Zygmund had also a personality as a human being: open to all problems of the world, warm with his friends, pleasant with his colleagues, and good to his students, and at the same time absolutely rigorous in all aspects of life. The style of his book, pure and simple, mirrors this personality. I am sure that it will remain as a model forever."

According to MathSciNet, "It is a book of almost overwhelming scholarship, and yet it is written in so lucid a style that reading it is a positive pleasure. The author indulges in occasional asides and footnotes which point out the ideas behind proofs and illuminate obscure technical points. He frequently gives two proofs of important theorems—a real variable and a complex variable proof, say, or a conformal mapping and a non-conformal mapping proof. All chapters but one end with a section called 'Miscellaneous theorems and examples', which range from simple exercises to quite difficult theorems (frequently hints are given). For every chapter, there are notes assembled at the end of the volume giving historical background, references for further reading, and often statements of results not included in the main text. There is a brief and probably inadequate index... Every analyst should be familiar with, and every harmonic analyst should carefully study, this book. In spite of the vast scope of the work, Professor Zygmund's complete mastery of his subject, and his charming style (reminiscent of Hardy's), have made it eminently readable. The beautiful presswork will add to the reader's enjoyment. It can be taken, if the extensive cross-references are used, in almost any order. The preface is an essay that will repay careful study.

It seems safe to predict, indeed, that as analysts for 25 years have been reared on the first edition of Zygmund's book, so analysts for another quarter century will be brought up on the second."

According to a review in Bulletin of AMS in 1936, "If one looks through the long list of books on Fourier series one can not help feeling that even the bulkiest of them are far from giving an adequate picture of the present status of the field. The non-existence of a monograph giving such a picture was very badly felt not only by the beginners but also by specialists, and the failure of so many attempts to write a real book on Fourier series created an impression that the task was almost hopeless. The author of the present monograph completely succeeded in dispelling this 'inferiority complex' and produced a book which not only introduces the reader into the immense field of the theory of Fourier series but at the same time almost imperceptibly brings him to the very latest achievements, many of them being due to the author himself. The style of the book is rigorous and vigorous and the exposition elegant and clear to the smallest details.

Without wasting his and the reader's time on unessential things the author endeavors to treat each special problem by methods throwing light on the problem from a general standpoint, and showing the place occupied by it in the whole structure. Such a method of exposition will prove to be extremely helpful to a neophyte and will delight a specialist."

3.3.4 Potential theory

Potential theory is an important subject of analysis. It studies solutions of the Laplace equation and generalizations and hence is closely related to theory of Laplace equation and differential equations. It is also closely related to the real and complex analysis and harmonic analysis.

M. Tsuji, *Potential Theory in Modern Function Theory*, Reprinting of the 1959 original. Chelsea Publishing Co., New York, 1975. x+590 pp.

This is a classic, and the modern function theory referred to in the title is not modern anymore. When it was written, it contained the cutting edge results. Due to its comprehensive nature, it is still a good introduction and reference.

According to MathSciNet, "Over a period of several decades the author has contributed to the theory of functions an impressive array of notes and papers, ranging from remarks on classical theorems to ingenious proofs of important new results. The present book serves first of all as a synthesis of the extensive literature of the author, some of which is virtually inaccessible, and second as a sketch of the background material needed to read the author's papers..."

N.S. Landkof, *Foundations of Modern Potential Theory*, Translated from the Russian by A. P. Doohovskoy, Die Grundlehren der mathematischen Wissenschaften, Band 180, Springer-Verlag, New York-Heidelberg, 1972. x+424 pp.

This book is also called modern potential theory, and is comprehensive in regards to potential theory related to concrete kernel functions. It was written as a textbook and can still serve as a good reference now.

According to MathSciNet, "The book deals with that part of potential theory which is connected with concrete kernels in Euclidean spaces... Although it is apparent that the book is primarily intended as a textbook, the author has succeeded in doing more; his book will also be of interest to specialists, and may serve as a useful reference on potential-theoretic results connected with Riesz and Green kernels in Euclidean spaces."

T. Ransford, *Potential Theory in the Complex Plane*, London Mathematical Society Student Texts, 28, Cambridge University Press, Cambridge, 1995. x+232 pp.

Dimension 2 is special in many ways. One is that in this case, there are close relations between potential theory and complex analysis. This book is a good introduction to interplay between these two topics. The author claims that this book is the one that he would have when he was learning the subject for the first time. It seems that he has achieved it.

According to MathSciNet, "The book is pitched at a level accessible to first-year graduate students. The main prerequisites are elementary complex analysis and basic integration theory (for Lebesgue measures), though the final chapter is more demanding... There are plenty of exercises, varying widely in difficulty. Notes at the end of each chapter cite original sources and indicate where developments beyond the scope of the book can be found.

Graduate students and researchers in complex analysis will find in this book most of the potential theory that they are likely to need. Others, primarily concerned to find out about potential theory in any of its many settings, would do well to see first how it works in the context of the plane; to them, too, this attractive book is recommended."

M. Klimek, *Pluripotential Theory*, London Mathematical Society Monographs, New Series, 6, Oxford Science Publications, The Clarendon Press, Oxford University Press, New York, 1991. xiv+266 pp.

Roughly speaking, potential theory studies subharmonic functions and in particular harmonic functions, and pluripotential theory studies plurisubharmonic functions. This book is the first comprehensive book on pluripotential theory. It is a valuable introduction and also a comprehensive reference.

According to MathSciNet, "This book is the first relatively comprehensive book ever written on the subject of the title. Pluripotential theory is, loosely speaking, the study of plurisubharmonic (psh) functions; in this text, all psh functions are defined on domains in $\mathbb{C}^n$... The beginner should find the book very readable, especially with the long introductory Chapter 2 on subharmonic and plurisubharmonic functions. The experts will be pleased to have the material in Section II of the book all in one place for the first time and will find this book to be a valuable reference tool, especially with a 12–13 page bibliography... All in all, this is a very welcome and much overdue book on an important topic in several complex variables."

3.3. HARMONIC ANALYSIS

E. Saff, W. Totik, *Logarithmic Potentials with External Fields. Appendix B by Thomas Bloom*, Grundlehren der Mathematischen Wissenschaften, 316. Springer-Verlag, Berlin, 1997. xvi+505 pp.

Given a compact subset E of $\mathbb{C}$, denote by $M(E)$ the set of all probability measures E. The equilibrium distribution μ_E of E is the unique element of $M(E)$ for which the energy $I(\mu) = \int \log(1/|z-t|) d\mu(z) d\mu(t)$ is minimal among all $\mu \in M(E)$. The measure μ_E arises naturally in several other areas of analysis, and the determination of μ_E is a classical problem in potential theory. The book studies a generalization of the extremal equilibrium distribution and discuss recent applications to various problems in analysis. This is a good reference.

According to MathSciNet, "In this outstanding monograph, E. B. Saff and V. Totik have done for 'weighted' potential theory what N. S. Landkof did for 'ordinary' potential theory in his monograph *Foundations of Modern Potential Theory*, 1972...

In the past two decades, potential theory has had a dramatic impact on a number of problems that involve some extremal property: these include the analysis of orthogonal polynomials for weights on the whole real line; rates of rational approximation of analytic functions; numerical conformal mapping; quadrature and interpolation; energy level of quantum systems; random matrices... These problems in turn led to the development of a new direction in potential theory, which some call weighted potential theory, and others call potential theory with external fields. This monograph contains a systematic and thorough development of that theory. While the foundations were laid by several researchers, including fundamental advances from the present authors, there are very many important new results that appear for the first time in this monograph. Moreover, many older results appear here with minimal assumptions for the first time.

The monograph is self-contained, well written, and should be accessible to graduate students with a reasonable background in analysis, measure theory and functional analysis. However, many researchers in a broad array of subjects will find this not only an essential reference, but will be able to learn a broad variety of techniques from this work..."

J. Heinonen, T. Kilpeläinen, O. Martio, *Nonlinear Potential Theory of Degenerate Elliptic Equations*, Unabridged republication of the 1993 original, Dover Publications, Inc., Mineola, NY, 2006. xii+404 pp.

Classical potential theory is associated with the Laplace equation and has had many applications in mathematics and physics. Recently, many nonlinear differential equations arise naturally from different contexts. Probably the best known is the p-Laplace operator, which is a quasilinear elliptic partial differential operator of 2nd order, and when $p = 2$, it is reduced to the usual Laplace operator. Therefore, there is a need to develop a corresponding potential theory for them. This book is one of the first systematic, self-contained introductions to nonlinear potential theory. It is suitable textbook for beginners and also a useful reference for others.

According to Zentralblatt MATH, "The book gives an introduction to nonlinear potential theory. Most of the considerations are done for the quasilinear possibly degenerate elliptic equation $-\text{div} A(x, \nabla u) = 0$, that should be regarded as a perturbation of the weighted p-Laplace equation $-\text{div}(w(x) |\nabla u|^{p-2} \nabla u) = 0$."

3.3.5 J. Garnett, Bounded Analytic Functions, Pure and Applied Mathematics, 96, Academic Press, Inc., New York-London, 1981. xvi+ 467 pp.

This is a modern classic and a popular book on bounded analytic functions on the unit disc and in particular on the Hardy spaces in one dimension. This book was awarded the 2003 Steele prize for mathematical exposition. The citation of the award says: "The book, which covers a wide range of beautiful topics in analysis, is extremely well organized and well written, with elegant, detailed proofs. The book has educated a whole generation of mathematicians with backgrounds in complex analysis and function algebras. It has had a great impact on the early careers of many leading analysts and has been widely adopted as a textbook for graduate courses and learning seminars in both the U.S. and abroad."

In complex analysis, the Hardy spaces are spaces of certain holomorphic functions on the unit disk (or equivalently upper half plane). They were introduced by Riesz and named them after Hardy because of a paper of Hardy in 1915. In real analysis, Hardy spaces are spaces of certain distributions on the real line, which are boundary values, in the sense of distributions, of the holomorphic functions of the complex Hardy spaces on the upper half plane, and they are related to the $L^p(\mathbb{R})$ spaces of the real line (or the unit circle if the unit disk is used). For $1 \leq p \leq \infty$, these real Hardy spaces are subspaces of $L^p(\mathbb{R})$, while for $p < 1$, the $L^p(\mathbb{R})$ spaces have some undesirable properties, and the Hardy spaces are much better behaved. There are also higher dimensional Hardy spaces consisting of certain holomorphic functions on tube domains in the complex case, or spaces of certain distributions on $\mathbb{R}^n$ in the real case. Hardy spaces have several applications in mathematical analysis and other applied subjects.

According to MathSciNet, "The motif of the theory of Hardy spaces is the interplay between real, complex, and abstract analysis. While paying proper attention to each of the three aspects, the author has underscored the effectiveness of the methods coming from real analysis, many of them developed as part of a program to extend the theory to Euclidean spaces, where the complex methods are not available... The author has not attempted to produce a compendium. Rather, he has selected a range of topics in a many-faceted theory and, within that range, penetrated to considerable depth. The book is written in the Satz-Beweis style, which is to say it does not make light reading. On the other hand, it should be extremely rewarding to the reader willing to put forth the requisite effort. The utmost care has gone into the presentation of each Satz and Beweis, the author having added his own touch in many places. Although, because of the heavy demands it puts on the reader, the book may not be the best starting place for the novice, it will be a gold mine for someone already acquainted with the subject who wants to delve deeper...

Finally, the reviewer wishes to say that the author has succeeded in bringing out the beauty of a theory which, despite its relatively advanced age—now approaching 80 years—continues to surprise and to delight its practitioners. The author has left his mark on the subject."

J. Garnett, D. Marshall, *Harmonic Measure,* New Mathematical Monographs, 2, Cambridge University Press, Cambridge, 2005. xvi+571 pp.

Harmonic measure is a measure on the boundary of a domain of $\mathbb{C}$ (or in $\mathbb{R}^n$, $n \geq 2$) which arises from the solution to the classical Dirichlet boundary value problem. It also describes the exit measure of the Brownian motion of the region. It is a relatively old notion, but a lot of progress has been made recently on harmonic measures for domains in $\mathbb{C}$, for example, results due to Bishop, Carleson, Jones, Makarov, Wolff et al. This book is a comprehensive introduction to harmonic measures for planar domains ranging from the classical to recent results.

According to a review in Bulletin of AMS, "Over the last 20 years I have often been asked to suggest a 'good place to learn about harmonic measure' and from now on the book of Garnett and Marshall will be my first suggestion. It is a great place for graduate students to learn an important area from the foundations up to the research frontier or for experts to locate a needed result or reference. Almost all the topics we have touched on, and many more, are discussed with greater depth and clarity than I have been able to achieve here, and since you are obviously quite interested in harmonic measure (after all, you have read this far), I recommend you get the book for the full story."

3.3.6 More books on Hardy spaces and function spaces

Hardy spaces are important for certain applications. For other applications, different function spaces such as Sobolev spaces and Besov spaces are needed.

P. Duren, *Theory of H^p Spaces,* Pure and Applied Mathematics, Vol. 38, Academic Press, New York-London, 1970, xii+258 pp.

This comprehensive introduction to Hardy spaces of planar domains cover both the classical theory such as the work of Hardy and Littlewood, and more modern results such as Beurling's theorem on invariant subspaces and Carleson's proof of the corona theorem.

It was written as a textbook for both graduate students and also as a source book for experts.

According to MathSciNet, "Till now there has been no single book in English offering a comprehensive introduction to the subject from the classical point of view... This situation has been remedied by the appearance of the book under review: it provides an efficient introduction to the basics of the subject, from the classical point of view, with a minimum of fuss. The book should thus be especially valuable to the student. It also goes into many of the deeper aspects of the subject, such as the work of Hardy and Littlewood."

K. Hoffman, *Banach Spaces of Analytic Functions*, Prentice-Hall Series in Modern Analysis Prentice-Hall, Inc., Englewood Cliffs, N.J., 1962, xiii+217 pp.

This is one of the first books that emphasize the interplay between functional analysis and the theory of analytic functions. According to the author, the reason for the lack of books is that "many of the techniques of functional analysis have a 'real variable' character and are not directly applicable to problems which belong intrinsically to analytic function theory, e.g., problems of conformal mapping and Riemann surfaces. But there are parts of this theory which blend beautifully with the concepts and methods of functional analysis. These are fascinating areas of study for the general analyst, for three principal reasons: (a) the point of view of the algebraic analyst leads to the formulation of many interesting problems concerned with analytic functions; (b) when such problems are solved by a combination of the tools from the two disciplines, the depth of each is increased; (c) the techniques of functional analysis often lend clarity and elegance to the proofs of classical theorems, and thereby make the results available in more general situations.

The main purpose of this monograph is to provide an introduction to the segment of mathematics in which functional analysis and analytic function theory merge successfully." This book has achieved its goal well.

According to MathSciNet, "This book deals with an area of analysis in which the modern methods of functional analysis and the classical methods of the theory of analytic functions are successfully employed side by side, to mutual advantage... The Banach spaces under consideration are Hardy's H^p-spaces."

H. Triebel, *Theory of Function Spaces*, Monographs in Mathematics, 78, Birkhäuser Verlag, Basel, 1983, 284 pp.

This book is useful as a reference for experts.

According to MathSciNet, "This is a third monograph by the author devoted to spaces of differentiable functions. Even though it takes up a considerable amount of material from its forerunners, *Fourier Analysis and Function Spaces*, 1977; *Interpolation Theory,*

Function Spaces, Differential Operators, 1978, differs from them by a considerable update of the content and of its presentation."

H. Triebel, *Interpolation Theory, Function Spaces, Differential Operators*, VEB Deutscher Verlag der Wissenschaften, Berlin, 1978. 528 pp.

Sobolev spaces are basic spaces in the study of differential equations. Sobolev spaces and Besov spaces are families of spaces depending on parameters. Different differential equations lead to weighted function spaces, for example, weighted Sobolev spaces. In order to study these function spaces systematically, the theory of interpolation is needed. This book gives a thorough discussion of the three topics in its title and is a useful reference.

According to MathSciNet, "This book examines the three topics of the title from the general viewpoint of interpolation theory; in particular, of interpolation theory in complex Banach spaces... This book contains a surprising amount of material, extensive references,... The style is clear but quite formal. Many proofs are long and technical; others are brief with many details omitted. No attempt is made to place definitions, theorems, or techniques in any historical or heuristic context. For these reasons (and even though the only prerequisite to formal understanding of the text is graduate level functional analysis) the book will be useful primarily to researchers already well familiar with at least one of its three main topics. Such experts, however, should find the book's comprehensiveness, precision, and unified point of view of great value."

K. Zhu, *Operator Theory in Function Spaces*, Second edition, Mathematical Surveys and Monographs, 138, American Mathematical Society, Providence, RI, 2007. xvi+348 pp.

This book studies three important classes of linear operators: Toeplitz operators, Hankel operators, and composition operators in various spaces of analytic functions on the unit disc. It is a useful reference for graduate students and specialists in complex analysis and operator theory.

According to MathSciNet, "This book deals primarily with three types of operators: Hankel operators, Toeplitz operators and composition operators, in the context of both the Hardy and the Bergman spaces on the unit circle **T** and the unit disk **D**, respectively. The common theme that links these operators is that they are parametrized by functions, called the symbols. The book emphasizes the development since the early 1980s, in particular the theory in the context of the Bergman space, an area in which the author has personally been involved."

C. Cowen, B. MacCluer, *Composition Operators on Spaces of Analytic Functions*, Studies in Advanced Mathematics, CRC Press, Boca Raton, FL, 1995. xii+388 pp.

The main topic of this book is the following: for an analytic function φ on the unit disk D (or a more general plane domain Ω), describe relations between the analytical and geometrical properties of the function φ and the analytical and topological properties of the composition operator $C_\varphi(f) = f \circ \varphi$ on the space of analytic functions on the disk D induced by φ. It lies on the borderline between complex function theory, functional analysis, and operator theory. It is a relatively new subject. This is a good textbook or introductory book.

From the book description, "It provides a comprehensive introduction to the linear operators of composition with a fixed function acting on a space of analytic functions. This new book both highlights the unifying ideas behind the major theorems and contrasts the differences between results for related spaces.

By including the theory for both one and several variables, historical notes, and a comprehensive bibliography, the book leaves the reader well grounded for future research on composition operators and related areas in operator or function theory."

According to MathSciNet, "Besides operator theory and complex analysis, composition operators play a role in subjects such as: functional equations, complex dynamical systems and even the theory of branching processes in probability theory. For example, the functional equation $f \circ \phi = \lambda f$, studied by Schroeder in an 1871 paper on iteration of analytic functions, is discussed here as an eigenvalue problem for the operator C_ϕ. This book, by two major contributors to the field, is about the interplay between the function-theoretic properties of ϕ and the operator theoretic-properties of C_ϕ. Clearly written, packed with examples and exercises and treating many topics that are of broad interest, it would be quite suitable for a second graduate course in complex analysis."

J. Bergh, J. Löfström, *Interpolation Spaces. An Introduction,* Grundlehren der Mathematischen Wissenschaften, No. 223, Springer-Verlag, Berlin-New York, 1976. x+207 pp.

Interpolation theory started in classical Fourier analysis. It is a basic theory in analysis and has been generalized to Banach spaces and more general spaces. This book is a classic on this general theory which has been applied to many different areas of mathematics. This is one of the highly cited books that has not been reprinted or revised.

According to the review in Bulletin AMS, "The subject has its origins in classical Fourier analysis, where it was conceived as an elementary means of finding L^p-estimates. The very nature of interpolation theory, however, is functional-analytic: typically, a linear operator T is bounded between spaces X_a and Y_a when $a = 0$ and $a = 1$, and one wants to conclude that T carries X_a to Y_a whenever $0 < a < 1$. Such problems arise in many areas of analysis, and the abstract theory has always been influenced, even guided, by the potential applications to such areas as harmonic analysis, approximation theory, and

3.3. HARMONIC ANALYSIS

the theory of partial differential equations. As a result, interpolation has no one place to call home; it is, quite simply, interesting mathematics. Its success, like that of a good executive, stems from its ability to handle specifics while operating on a generally higher plane...

The strength of the book lies in its elegant treatment of the theory of interpolation spaces and interpolation methods. Here, the authors have given us a beautiful account of some beautiful mathematics. It is for this that the book will be valued and, more importantly, used."

C. Bennett, R. Sharpley, *Interpolation of Operators,* Pure and Applied Mathematics, 129, Academic Press, Inc., Boston, MA, 1988. xiv+469 pp.

The book is an elementary, self-contained introduction to the real method of interpolation and some of its applications. It is a useful reference.

The purpose of this book was explained in the preface: "This is a book about the real method of interpolation. Our goal has been to motivate and develop the entire theory from its classical origin, that is, through the theory of spaces of measurable functions although the influence of Riesz, Thorin, and Marcinkiewicz is everywhere evident, the work of G. H. Hardy, J. E. Littlewood, and G. Pólya on rearrangements of functions also play a seminal role. It is through the Hardy-Littlewood-Pólya relation that spaces of measurable functions and interpolation of operators come together, in a simple blend which has the capacity for great generalization. Interpolation between L^1 and L^∞ is thus prototype for interpolation between more general pairs of Banach spaces. This theme airs constantly throughout the book. The theory and applications of interpolation are as diverse as language itself. Our goal is not a dictionary, or an encyclopedia, but instead a brief biography of interpolation, with a beginning and end, and (like interpolation itself) some substance in between."

3.3.7 E. Stein, Singular Integrals and Differentiability Properties of Functions, Princeton Mathematical Series, No. 30, Princeton University Press, 1970, xiv+290 pp.

Singular integrals are integral operators whose kernels are singular on the diagonal. They are central to harmonic analysis and are intimately connected with the study of partial differential equations. This book grew out of the author's course in Orsay in 1966–1967. His purpose was to present some of the required background and at the same time clarify the essential unity that exists between several related areas of analysis. The general Fourier theory that provides the background to most of the proofs in this book is contained in his book with G. Weiss mentioned below. These two books are companions in some sense, but can be read independently. Stein's influence on the subject of harmonic analysis and Fourier theory is immense, and his many well written books are part of the reason. This book is a classic.

From the book description, "One reason for its success as a text is its almost legendary presentation: Stein takes arcane material, previously understood only by specialists, and makes it accessible even to beginning graduate students. Readers have reflected that when you read this book, not only do you see that the great of the past have done exciting work, but you also feel inspired that you can master the subject and contribute to it yourself.

Singular integrals were known to only a few specialists when Stein's book was first published. Over time, however, the book has inspired a whole generation of researchers to apply its methods to a broad range of problems in many disciplines, including engineering, biology, and finance."

According to MathSciNet, "This book deals with several flourishing aspects of 'hard' analysis of the modern variety...

In respect of difficulty, the book falls roughly into two parts. The first consists of the first three chapters, the first two of which provide a very readable introduction to singular integral operators and related techniques, while the third offers ample motivation in the shape of important examples and applications. These first eighty pages give excellent insight into basic techniques and results, some study of which might well be expected of all graduate students in analysis...

The author has been successful in his attempt to make the book essentially self-contained. Only a very meagre acquaintance with harmonic analysis is needed; distributional derivatives are used, but only in the case of locally integrable functions, for which case the definitions are included and the basic properties sketched; and all necessary knowledge from functional analysis (mainly referring to separable Hilbert spaces) is adequately dealt with ad hoc...

To sum up, a welcome and expertly written account that should be a great help to those wishing to learn about several currently very active fields."

Stein has written and co-written several other important books which have been well-received. They include the following two.

E. Stein, G. Weiss, *Introduction to Fourier Analysis on Euclidean Spaces,* Princeton Mathematical Series, No. 32, Princeton University Press, Princeton, 1971. x+297 pp.

The idea of writing this book came to the authors in 1958–1959 when they taught a graduate course. It was written as an introduction to harmonic analysis on $\mathbb{R}^n$. But it is also a basic reference for experts. It shows rich structures of harmonic analysis on $\mathbb{R}^n$ in comparison with harmonic analysis on symmetric spaces, semisimple real Lie groups, and spaces and groups over local fields, or with abstract harmonic analysis. It is also a classic.

3.3. HARMONIC ANALYSIS

According to MathSciNet, "This book is a soundly conceived and brilliantly executed treatment of harmonic analysis on n-dimensional Euclidean spaces $\mathbb{R}^n$ and tori T^n. Although it is not encyclopedic, it is far more than the title indicates. The authors' intention was to select topics that show how both real and complex variable methods extend from the 1- to the n-dimensional case. They have also made judicious use of the special structure possessed by $\mathbb{R}^n$ and T^n. The groups $\mathbb{R}^n$ and T^n display features that intrinsically set them apart from general locally compact Abelian groups and compact groups: the existence of differential operators, of dilations, and of rotations, the truly special fact that the coordinate functions $(x_1, x_2, \cdots, x_n) \mapsto x_j$ can be used to generate spherical harmonics and hence to decompose $L^2(\mathbb{R}^n)$, and so on. The groups $\mathbb{R}^n$ and T^n are also the domains of classical special functions, whose lore has accumulated for centuries."

E. Stein, *Harmonic Analysis: Real-variable Methods, Orthogonality, and Oscillatory Integrals*, Princeton Mathematical Series, 43, Monographs in Harmonic Analysis, III, Princeton University Press, 1993. xiv+695 pp.

The book is a sequel to the previous two books by Stein and provides an updated description of the author's own contributions and a comprehensive, largely self-contained presentation of the work of a broad subject in the past 20 years. It is a very solid, comprehensive reference.

According to MathSciNet, "In the early 1970s, a number of seminal ideas in harmonic analysis on Euclidean spaces made their appearance. These included the use of real-variable methods to develop n-dimensional analogues of the Hardy spaces of one complex variable, the use of "almost orthogonality" (Cotlar's lemma) to prove L^2-boundedness of singular integral operators, and the use of the Kakeya construction to probe the geometry of the Euclidean plane. Interaction with partial differential equations led on the one hand to analysis on the Heisenberg group, used to tackle boundary value problems arising in several complex variables, and on the other to the appropriation and sharpening of ideas such as pseudodifferential and Fourier integral operators. The development of these ideas has occupied the attention of a substantial number of mathematicians over the last twenty years; amongst these, Stein and his collaborators and students have played leading roles. The book under review presents Stein's version of much of this activity. To complete a very broad description of its contents, it suffices to add that maximal functions and oscillatory integrals are emphasized, and that wavelets are barely mentioned..."

3.3.8 G. Folland, Harmonic Analysis in Phase Space, Annals of Mathematics Studies, 122, Princeton University Press, Princeton, N.J., 1989. x+277 pp.

This is a very good introduction and a valuable reference on harmonic analysis on phase space by an expert and a known expositor.

According to its preface, "The phrase 'harmonic analysis in phase space' is a concise if somewhat inadequate name for the area of analysis on $\mathbb{R}^n$ that involves the Heisenberg group, quantization, the Weyl operational calculus, the metaplectic representation, wave packets, and related concepts: it is meant to suggest analysis on the configuration space $\mathbb{R}^n$ done by working in the phase space $\mathbb{R}^n \times \mathbb{R}^n$. The ideas that fall under this rubric have originated in several different fields—Fourier analysis, partial differential equations, mathematical physics, representation theory, and number theory, among others. As a result, although these ideas are individually well known to workers in such fields, their close kinship and the cross-fertilization they can provide have often been insufficiently appreciated. One of the principal objectives of this monograph is to give a coherent account of this material, comprising not just an efficient tour of the major avenues but also an exploration of some picturesque byways."

According to Zentralblatt für Mathematik, "In the beginning was the Heisenberg group... And the Heisenberg group begat quantum mechanics, the orbit method of representation theory, the symplectic group, theta functions, the oscillator semigroup, the Hermite operator, the $\bar{\partial}_b$-problem, pseudoconvex domains, pseudodifferential operators, wavelets, and others. Actually, most of these descendants were alive and well before the Heisenberg group's role as a common ancestor was brought to light over the last twenty years. The book under review is a valiant attempt to present an account of (most of) these areas of mathematics, with an emphasis on the analysis - quantum mechanics and pseudodifferential operators... [This book] is a valiant attempt to present an account of [harmonic analysis in phase space], with an emphasis on the analysis-quantum mechanics and pseudodifferential operators... The author has taken great pains to express himself clearly and ... the notation is consistent throughout..."

3.3.9 I. Daubechies, Ten Lectures on Wavelets, CBMS-NSF Regional Conference Series in Applied Mathematics, 61, Society for Industrial and Applied Mathematics (SIAM), Philadelphia, PA, 1992. xx+357 pp.

This book is one of the early and successful books on wavelets, an important and relatively new subject. It is written by a major contributor and is a valuable introduction to the subject.

3.3. HARMONIC ANALYSIS

In plain English, wavelets means "small waves". Though that this name has been used in digital signal processing and exploration geophysics for decades, as a subject in mathematics, it is a relatively recent development in applied mathematics and has been growing rapidly. Wavelets are mathematically simple but can be effectively applied to many disciplines. The reason lies in the ability of wavelets to describe local phenomena more accurately than a traditional Fourier expansion. Therefore, wavelets are applied in many fields where an approach to transient behavior is needed.

From the book description, "Wavelets are a mathematical development that many experts think may revolutionize the world of information storage and retrieval. They are a fairly simple mathematical tool now being applied to the compression of data, such as fingerprints, weather satellite photographs, and medical x-rays — that were previously thought to be impossible to condense without losing crucial details."

According to MathSciNet, "This book is both a tutorial on wavelets and a review of the most advanced research in this domain... As mentioned in the introduction, this is a mathematics book that states and proves many theorems. In addition, it also gives many practical examples and describes several applications (in particular, in signal processing, image coding and numerical analysis)."

Other reviews include "a great part of the volume under review is dedicated to the engineering and physical origins of wavelets... The book style is alert and the interest of the potential reader is continuously kept alive. I think that this book is very useful to mathematicians as well as to people interested in the wavelets applications (engineers, physicists, etc.)."

Other important books on wavelets include the following:

Y. Meyer, *Wavelets and Operators. I.* Translated from the 1990 French original by D. H. Salinger, Cambridge Studies in Advanced Mathematics, 37, Cambridge University Press, Cambridge, 1992. xvi+224 pp. *Wavelets and Operators. II. Calderón-Zygmund operators.* Y. Meyer, R.R. Coifman, *Wavelets and Operators. III. Multilinear Operators*, Actualités Mathématiques. Hermann, Paris, 1990. pp. i–xii and 217–384.; 1991. pp. i–xii and 385–538.

This is the first comprehensive book on wavelets by a major contributor. It is a solid and systematic introduction to the subject.

According to MathSciNet, "Wavelets are not pursued as an end in themselves, but as a new and extremely useful tool for Fourier analysis. Thus the three volumes contain new and elegant approaches to topics in approximation theory, the theory of Hardy spaces or the theory of singular integral operators. The development of the tool takes up only half of Volume I, whereas the rest is concerned with the mathematical applications of wavelets. The first volume is devoted to multiresolution analysis, wavelet bases and the theory of classical function spaces, and addresses a wider audience of mathematicians, physicists, and engineers."

3.4 Functional analysis and operator theory

Banach was one of the founders of functional analysis and one of the original members, probably the most important one, of the Lwów School of Mathematics. The mathematicians in this group often met at the famous Scottish Cafe to discuss mathematical problems. This school was famous for its productivity and its extensive contributions to point-set topology, set theory and functional analysis. Banach's major work was the 1932 book, Théorie des opérations linéaires, *which was the first monograph on the general theory of functional analysis. It was translated into English in 1987 and reprinted by Dover in 2009.*

Figure 3.4: Banach's book.

Functional analysis mainly studies vector spaces endowed with suitable topologies such as those given by inner products and norms, and linear operators acting upon these spaces and respecting these structures in a suitable sense. Historically, functional analysis studied spaces of functions and formulated properties of transformations of functions such as the Fourier transform as transformations which define continuous, unitary etc. operators between function spaces. This point of view turned out to be particularly useful for the study of differential and integral equations. The word functional goes back to the calculus of variations and means a function whose argument is a function, and the name was first used in Hadamard's 1910 book on that subject. Hadamard also founded the modern school of linear functional analysis, which is further developed by Riesz and the group of Polish mathematicians around Stefan Banach.

Linear algebra deals mostly with finite dimensional vector spaces, and the topology is not explicitly used or emphasized. An important part of functional analysis is the extension of the theory of measure, integration, and probability to infinite dimensional spaces.

An important sub-area of functional analysis studies operator algebras. An operator algebra is an algebra of continuous linear operators on a topological vector space with the multiplication given by the composition of mappings.

3.4. FUNCTIONAL ANALYSIS AND OPERATOR THEORY

Since functional analysis is a basic subject, there are many books on its various aspects. They will be put into different groups. We start with a truly classic book by one of its founders.

3.4.1 S. Banach, Theory of Linear Operations, Translated from the French by F. Jellett, With comments by A. Peczyski and Cz. Bessaga, North-Holland Mathematical Library, 38, North-Holland Publishing Co., Amsterdam, 1987. x+237 pp.

This classic by Banach was originally published in 1938, and was reprinted by Dover Publications in 2009. It is the first book on functional analysis. Besides the great historical value, it is still a valuable book to read in order to feel the freshness of the subject when it was written by its founder.

Banach space theory was well-established in 1938 and it appeared as the title of a chapter. On the other hand, due to the clean and modern style, this book is still useful to the modern reader, and it has remained in print up to now. In the preface, Banach argued elegantly and briefly about the importance of the theory of operators, "This theory, therefore, well deserves, for its aesthetic value as much as for the scope of its arguments (even ignoring its numerous applications) the interest that it is attracting from more and more mathematicians. The opinion of J. Hadamard, who considers the theory of operators one of the most powerful methods of contemporary research in mathematics, should come as no surprise."

According to a review in Bulletin of AMS 1934, "The present book ... represents a noteworthy climax of long series of researches started by Volterra, Fredholm, Hubert, Hadamard, Frechet, F. Riesz, and successfully continued by Steinhaus, Banach, and their pupils... In conclusion the reviewer may express his conviction that Banach's monograph should occupy a permanent and honorable place on the desk of every one who is interested in the theory of linear operations, to be replaced only by subsequent, corrected and augmented editions, which undoubtedly will follow before long."

The chapter headings gives a good sense of the contents of this book. It also shows some unexpected topics. Introduction A. The Lebesgue-Stieltjes Integral B. (B)-Measurable sets and operators in metric spaces. Chapter I. Groups. Chapter II. General vector spaces. Chapter III. F-spaces. Chapter IV. Normed spaces. Chapter V. Banach spaces. Chapter VI. Compact operators Chapter VII. Biorthogonal sequences. Chapter VIII. Linear functionals. Chapter IX. Weakly convergent sequences. Chapter X. Linear functional equations. Chapter XI. Isometry, equivalence, isomorphism. Chapter XII. Linear dimension. Appendix. Weak convergence in Banach spaces.

3.4.2 More basic books on functional analysis

A. Kolmogorov, S.V. Fomin, *Elements of the Theory of Functions and Functional Analysis. Vol. 1. Metric and Normed Spaces,* Translated from the first Russian edition by Leo F. Boron, Graylock Press, Rochester, N. Y., 1957. ix+129 pp., *Vol. 2: Measure. The Lebesgue Integral. Hilbert Space,* Translated from the first (1960) Russian ed. by Hyman Kamel and Horace Komm Graylock Press, Albany, N.Y., 1961, ix+128 pp.

This book is a classical introduction to functional analysis and was based on lectures of two courses by the authors at Moscow State University. One course includes the theory of sets, measures, the Lebesgue integral, general theories such as functional analysis and applications to concrete problems in classical analysis, and many other topics. Another course is less comprehensive. It is still an excellent introduction to functional analysis.

According to MathSciNet, "This admirable book presents within its brief compass most of the elements of the theory of metric spaces and normed linear spaces. While rigor is maintained at a high level, the needs of mathematical physicists have evidently been kept in view. The present volume is elementary in the sense that Lebesgue measure is neither assumed nor discussed. Later fascicules are promised, in which Lebesgue integration, Hilbert space, integral equations with symmetric kernel, and applications to computational mathematics will be dealt with."

According to an Amazon review, "The authors introduce step by step all the key concepts needed to get a thorough understanding of the subject and proceed all the way long from set theory to Fredholm integral equations.

This book is appreciated not only because of the topics it includes but mostly because of the insight with which it was written. It is a pleasure to find through every page of the book the great genius of Kolmogorov who not only mastered most areas of mathematics but who also had an almost unparalleled understanding of what the trends of future mathematics would be."

F. Riesz, B. Sz.-Nagy, *Functional Analysis,* Translated from the second French edition by Leo F. Boron, Reprint of the 1955 original, Dover Books on Advanced Mathematics, Dover Publications, Inc., New York, 1990. xii+504 pp.

This is a very famous classical textbook on functional analysis. The original preface describes the content and the philosophy of this classic well, "The first part covers the modern theories of differentiation and integration and serves as an introduction to the second, which deals with integral equations and the theory of linear operators in Hilbert space. The two parts form an organic unit centered about the concept of the linear operator. This concept is reflected in the method by which the Lebesgue integral is constructed."

According to MathSciNet, "This classic textbook has gone through many printings and translations... This book shows the authors' skill in choosing for each result the proof best suited to give a student sufficient understanding to make further applications and extensions appear easy and natural. Careful references are made throughout, both for the material in the text and for related topics. This book covers so much material so well that it should be noted that it does not discuss rings of operators in Hilbert space nor spectral multiplicity. But a student interested in integration or spectral theory will find in this book much pertinent background material as well as a clear, concise discussion of these major topics."

According to a review in Bulletin of AMS, "This work is superb. For the field which it covers, it cannot be approached now nor will be soon by other books. It is not presented as a treatise for specialists, the essential purpose of which is to report advanced and complex results. Nor is it written as a textbook for the young student. Its aims are much higher and much more elegant. And in accomplishing these aims its authors have put together a magnificent advanced treatise and a most excellent though not elementary text. The purpose of the work is to set down, within the spirit and context of the undertaking, a certain coherent and central portion of mathematics in final and definite form. And within the spirit of the undertaking, this version is final and correct... The hallmark of the work is its balance and good taste: in the choice of subjects, in the extent and detail in which they are developed, in the methods used to present them, and in the critical question of style and exposition."

P. Lax, *Functional Analysis,* Pure and Applied Mathematics (New York), Wiley-Interscience, New York, 2002. xx+580 pp.

This book was written by a famous mathematician who is also an expert user of functional analysis. This book conveys the usefulness of functional analysis well.

As the author pointed out in the Foreword, there are now many good books on functional analysis, but this book is different, "the order in which the material is arranged, the interspersing of chapters on theory with chapters on applications, so that cold abstractions are made flesh and blood, and the inclusion of a very rich fare of mathematical problems that can be clarified and solved from the functional analytic point of view." Functional analysis does indeed give people the feeling of being too formal and abstract.

According to MathSciNet, "What is attractive about the book is the interspersing of chapters on theory with chapters on applications of the theory to a variety of problems arising in various branches of mathematics. The book is well suited for graduate courses in functional analysis and useful for research mathematicians."

According to a review in Bulletin of AMS, "The author also includes some first-hand historical notes: except for ones about Heisenberg (in chapters thirty-three, thirty-five and thirty-seven), most of them are stories about the tragic fate of many founding fathers of functional analysis in WW II, including Hausdorff, König, Banach, Tauber, Schauder and Hellinger."

R. Zimmer, *Essential Results of Functional Analysis*, Chicago Lectures in Mathematics, University of Chicago Press, Chicago, IL, 1990. x+157 pp.

The famous novelist Tolstoy wrote a famous short story "How much land does a man need". Functional analysis is such a basic subject in mathematics and probably every mathematician needs to know something about it. One can ask how much functional analysis does a working mathematician need? This slim book by Zimmer probably gives a good answer. This is also a book written by an expert user of functional analysis. It covers many major topics in functional analysis and illustrates them with real theorems. Everything is included in 157 pages!

According to MathSciNet, "This superb little book contains material for a one-semester course. It assumes familiarity with the rudiments of real-variable theory: integration and elementary facts about Banach and Hilbert spaces. The necessary background is summarized in 12 pages. The remaining 140 pages are packed with an impressive amount of high quality information... In spite of the brevity of the book the proofs are written out in detail and with utmost clarity. The author consistently finds the most elegant and streamlined treatment, thus, e.g., locally convex topologies are defined by families of seminorms, and convex balanced neighborhoods of the origin figure only in the exercises, of which there is an excellent collection at the end of each of the six chapters".

W. Rudin, *Functional Analysis*, Second edition, International Series in Pure and Applied Mathematics, McGraw-Hill, Inc., New York, 1991. xviii+424 pp.

In the preface, Rudin said that a good introductory book on functional analysis "should include a presentation of its axiomatics (i.e., of the general theory of topological vector spaces), it should treat at least a few topics in some depth, and it should contain some interesting applications to other branches of mathematics..." and set out to write one. The time has shown that he has achieved his goal. His criterion can probably be applied to any book on any subject, a text book or a research monograph.

According to Zentralblatt MATH, "By the richness of its contents, by the clearness and compactness of its exposition, by the wealth of applications indicated, the book will certainly have much success with readers."

According to MathSciNet, "This book is directed primarily towards the analytic rather than the algebraic or topological aspects of functional analysis. It is in some respects a sequel to the author's earlier book [Real and complex analysis, McGraw-Hill, New York, 1966; second edition, 1974], with whose more advanced parts there is some overlap. It is written in a very readable and not over-formal style; the author has avoided the temptation to organize his material so logically that it becomes thoroughly boring.

Its contents form an excellent basis for more advanced work in topics on the analytic side of functional analysis..."

J.B. Conway, *A Course in Functional Analysis,* Graduate Texts in Mathematics, 96, Springer-Verlag, New York, 1985. xiv+404 pp.

This is a good textbook written by an expert on functional analysis.

According to Zentralblatt MATH, "Functional Analysis means different things to different people and an author who decides to write a book of moderate size on this subject inevitably finds himself in a situation where he has to make a selection of topics which excludes many that are important both by virtue of their applicability and in their own intrinsic appeal. The book under review is no exception where Professor Conway has made his personal selection of topics which may not coincide with someone else's list. The author has chosen to write on linear functional analysis which may broadly and blandly be described as the study of linear spaces with topological structure and of linear mappings between such spaces with particular emphasis on normed spaces with or without an inner product. It goes to the credit of the author that in making his choice of topics from this rather wide area, he has strived to capture much of what is beautiful in linear functional analysis and to treat in a more or less balanced manner the different topics that are at the core of that subject."

According to MathSciNet, "This book is an excellent text for a first graduate course in functional analysis... This book is a fine piece of work. It includes an abundance of exercises, and is written in the engaging and lucid style which we have come to expect from the author."

3.4.3 More books on Banach spaces

J. Lindenstrauss, L. Tzafriri, *Classical Banach Spaces. I. Sequence spaces, II. Function spaces,* Ergebnisse der Mathematik und ihrer Grenzgebiete, vol. 92, 97. Springer-Verlag, Berlin-New York, 1977. xiii+188 pp., 1979. x+243 pp.

The point of this book is to study special and important classes of Banach spaces in some depth instead of the general Banach spaces. It is a very valuable reference.

From the preface, "The purpose of this book is to present the main results and current research directions in the geometry of Banach spaces with an emphasis on the study of the structure of the classical Banach spaces, that is $C(K)$ and $L_p(\mu)$ and related spaces. We do not attempt to write a comprehensive survey of Banach space theory, or even only of the theory of classical Banach spaces, since the amount of interesting results on the subject makes such a survey practically impossible... We therefore hope that it will be possible to use the present book as a textbook on Banach space theory and as a reference book for research workers in the area."

According to Zentralblatt für Mathematik, "The material it presents is hard to find in other books. For people working in the structure theory of Banach spaces it will be most valuable as a source of references and inspiration. For those who wish to learn the subject the book deserves a warm welcome too... The geometry of Banach lattices is a rich, beautiful, ... and rewarding subject. The proof is in the reading and perusing of the masterpiece."

J. Diestel, H. Jarchow, A. Tonge, *Absolutely Summing Operators,* Cambridge Studies in Advanced Mathematics, 43, Cambridge University Press, Cambridge, 1995. xvi+474 pp.

It is well-known that Grothendieck laid down the foundation and built up the modern algebraic geometry. But his work in 1950s planted the seeds of the theory of p-summing operators and absolutely summing operators. This book studies these operators and their generalizations, and is a valuable reference for experts.

From the book description, "Many fundamental processes in analysis are best understood by studying and comparing the summability of series in various modes of convergence. This text provides the beginning graduate student, one with basic knowledge real and functional analysis, with an account of p-summing and related operators. The account is panoramic, with detailed expositions of the core results and highly non-trivial applications to, for example, harmonic analysis, probability and measure theory and to operator theory..."

According to MathSciNet, "Today the theory of absolutely summing operators lies at the heart of modern Banach space theory–elegant in its own right, it has powerful applications in harmonic analysis, approximation theory, probability theory and operator theory, among others. In recent years several books have been published which–concentrating on special aspects–show that there is a deep and fruitful interaction between operator theory in Banach spaces and other well-established fields in analysis... The book under review is a very welcome addition to a rather long list of recently published monographs concerning deep mathematics in Banach spaces from the last three decades? and it is the first time that the important subject of 'summing operators' has been presented in almost complete detail in book form. Large parts of the book are within reach of graduate students, extensive 'notes and remarks' sections fill even sophisticated corners of the field with light..."

H. Schaefer, *Topological Vector Spaces,* Third printing corrected, Graduate Texts in Mathematics, Vol. 3, Springer-Verlag, New York-Berlin, 1971. xi+294 pp.

This is one of the first complete books in English treating topological spaces and grew out of courses which the author gave around 1960. It is both a good introduction and a valuable reference.

3.4. FUNCTIONAL ANALYSIS AND OPERATOR THEORY

According to Zentralblatt MATH, "The book has firmly established itself both as a superb introduction to the subject and as a very common source of reference. It is becoming evident that the book itself will only become irrelevant and pale into insignificance when (and if!) the entire subject of topological vector spaces does. An attractive feature of the book is that it is essentially self-contained, and thus perfectly suitable for senior students having a basic training in the area of elementary functional analysis and set-theoretic topology. My view—let even possibly biased for sentimental reasons—is that the book under review would make for a very practical and useful addition to every mathematician's personal office collection."

3.4.4 More books on Hilbert spaces

P. Halmos, *A Hilbert Space Problem Book,* Second edition, Graduate Texts in Mathematics, 19, Encyclopedia of Mathematics and its Applications, 17, Springer-Verlag, New York-Berlin, 1982. xvii+369 pp.

This is an unusual book. When it first came out, the approach was quite new as far as books were concerned. It corresponds to the Moore method (or Inquiry-based learning), which has becoming popular with more elementary level courses. It probably takes more efforts to read this book, but one also gains more.

From the Preface: "The only way to learn mathematics is to do mathematics. That tenet is the foundation of the do-it-yourself, Socratic, or Texas method, the method in which the teacher plays the role of an omniscient but largely uncommunicative referee between the learner and the facts. Although that method is usually and perhaps necessarily oral, this book tries to use the same method to give a written exposition of certain topics in Hilbert space theory... This book was written for the active reader. The first part consists of problems, frequently preceded by definitions and motivation, and sometimes followed by corollaries and historical remarks... The second part, a very short one, consists of hints... The third part, the longest, consists of solutions: proofs, answers, or constructions, depending on the nature of the problem... This is not an introduction to Hilbert space theory. Some knowledge of that subject is a prerequisite: at the very least, a study of the elements of Hilbert space theory should proceed concurrently with the reading of this book."

According to MathSciNet, "The book is divided into three parts: problems, hints and solutions. According to the instructions, if one is unable to solve a problem, even with the hint, he should, at least temporarily, grant it (if it is a statement) and proceed to another. He should be prepared, however, to use the as yet unproved statement in the hope of thereby getting some clues as to its solution. If he does solve a problem he should look at the hint and the solution anyway for perhaps some other variations on the theme. The rules are simple and the advantages of following them to the conscientious and diligent reader are surely obvious and incalculable. Nevertheless, there is the ever lurking temptation to combine each problem-hint-solution into a unified whole, to be read together and at one time. Although this procedure admittedly thwarts, at least partially, the aim of the book, it is certainly not without its own rewards. Thus, a researcher thereby may quickly find a compact presentation of a particular issue, often including some history, references, and enlightening side remarks...

The book will be of benefit not only to the serious student but also to the 'expert' who may very well find here the solution or counterexample to some nagging or persistently baffling question. The presentation is lively, often humorous and always interesting. The volume can be highly recommended to all Hilbert space fanciers."

N.I. Akhiezer, I.M. Glazman, *Theory of Linear Operators in Hilbert Space. Vol. I. Vol. II,* Translated from the third Russian edition by E. R. Dawson. Translation edited by W. N. Everitt, Monographs and Studies in Mathematics, 9. Pitman (Advanced Publishing Program), Boston, Mass.-London, 1981. xxxii+312 pp., pp. i–xii and 313–552 and xiii–xxxii.

For many people in geometric analysis and representation theories, one of the most important parts of functional analysis concerns Hilbert spaces and spectral theory of unitary and self-adjoint operators. This book is a classic textbook which discusses in detail the geometry of Hilbert space and the spectral theory of unitary and self-adjoint operators. It is also a valuable reference.

According to MathSciNet, "The first six chapters of this book present a clear and detailed exposition of the standard theory of linear operators in Hilbert space, culminating in the spectral theorems for self-adjoint and unitary operators. The seventh chapter and two appendices are concerned with the extension of symmetric operators and with the theory of differential operators. This material is less elementary and well known than that of the first six chapters and some of it represents quite recent work."

I.C. Gohberg, M.G. Krein, *Introduction to the Theory of Linear Nonselfadjoint Operators,* Translated from the Russian by A. Feinstein, Translations of Mathematical Monographs, Vol. 18, American Mathematical Society, Providence, R.I. 1969, xv+378 pp.

The theory of non-self-adjoint operators in Hilbert space is a relatively young branch of functional analysis. Completeness of eigenvectors were established around 1951. This book has an unusual history. It evolved from a rejected survey paper and has become a unique reference on non-self-adjoint operators.

According to MathSciNet, "This book represents a significant contribution to the literature concerning bounded linear operators on a separable Hilbert space. It collects, for the first time in book form, many results that have hitherto been relatively inacces-

3.4. FUNCTIONAL ANALYSIS AND OPERATOR THEORY

sible. Despite its sweeping title, the book is mostly confined to a discussion of certain classes of completely continuous (= compact) linear operators... However, it presents a large number of results which are not to be found in these two references, but which are only available in the recent research (and largely Soviet) literature. The authors have performed a great service in organizing and presenting this difficult material in such a splendid form."

3.4.5 L. Schwartz, Théorie des Distributions, Hermann, Paris, 1966, xiii+420 pp.

Schwartz is an unusual mathematician. He received a Fields Medal for his work on distributions, the subject of this classical book. He was also a well-known outspoken intellectual.

According to a description in Notices of AMS of his another book *A Mathematician Grappling with His Century*, "Laurent Schwartz is one of the most remarkable intellects of the 20th century. His discovery of distributions, one of the most beautiful theories in mathematics, earned him a 1950 Fields Medal. Beyond this formidable achievement, his love for science and for teaching led him to think deeply and lecture broadly to the general public on the significance of science and mathematics to the well-being of the world. At the same time, his commitment to the social good, even at the expense of his beloved research, proved a moral compass throughout his life. The fight for human rights and his major role in the battle against the wars in Algeria and Vietnam were typical of matters close to his heart. The story of his life in the context of his century provides for future generations an inspiring testimonial from an extraordinary mathematician and thinker. Laurent Schwartz is a strategist of ideas, within mathematics and without. He is a great communicator who has drawn huge audiences and conveyed to them the fragrance of research, or the joy of teaching, or the value of freedom. His is a mind whose company is never dull. He belongs to the great libertarian tradition of France."

According to MathSciNet, "This is a generally clear, carefully organized, and detailed account of the basic aspects of the theory of 'distributions' due to the author, and described by him in earlier publications... This theory provides a convenient formalism for many common situations in theoretical and applied analysis, but its greatest significance may be in connection with partial differential equations, particularly those of hyperbolic type, where its adaptability to local problems gives it an advantage over Hilbert space (and other primarily global) techniques. On the other hand, the latter, when applicable, are considerably more powerful than any techniques as yet provided by distributions."

The second volume "is a detailed treatment of the harmonic analysis of distributions on Euclidean spaces or toruses, and like volume I is unusually clearly written and well organized... The book has excellent summaries preceding each chapter, very good indexes of nomenclature and notations, and includes a bibliography pertinent to the first as well as to the present volume."

3.4.6 I.M. Gelfand, G.E. Shilov, Generalized Functions. Vol. 1. Properties and Operations, Vol. 2. Spaces of Fundamental and Generalized Functions, Vol. 3. Theory of Differential Equations, I.M. Gelfand, N.Ya. Vilenkin, Vol. 4. Applications of Harmonic Analysis, I.M. Gelfand, M.I. Graev, N.Ya. Vilenkin, Vol. 5. Integral Geometry and Representation Theory, Academic Press, New York-London, 1964, xviii+423 pp., x+261 pp., x+222 pp., xiv+384 pp., xvii+449 pp.

This is a very impressive series of books which cover different areas of modern mathematics. Gelfand is known for his broad view of mathematics, and this series of book reflects that. Usually it is assumed that this series consists of the above 5 volumes. But the authors also considered the following book as the sixth volume of the above series:

I.M. Gelfand, M. Graev, I.I. Pyatetskii-Shapiro, *Representation Theory and Automorphic Functions*, W. B. Saunders Co., Philadelphia, Pa.-London-Toronto, Ont., 1969, xvi+426 pp.

This sixth volume will be considered separately in the section on automorphic forms and automorphic representation later.

To understand the motivation and scope of this impressive series, and also the history of generalized functions, it is perhaps best to quote from the Foreword of the first Russian edition: "Generalized functions have of late been commanding constantly expanding interest in several different branches of mathematics. In somewhat nonrigorous form, they have already long been used in essence by physicists.

Important to the development of the theory have been the works of Hadamard dealing with divergent integrals occurring in elementary solutions of wave equations, as well as some work of M. Riesz. We shall not discuss here the even earlier mathematical work which could also be said to contain some groundwork for the future development of this theory.

The first to use generalized functions in the explicit and presently accepted form was L. Sobolev in 1936 in studying the uniqueness of solutions of the Cauchy problem for linear hyperbolic equations.

From another point of view Bochner's theory of the Fourier transforms of functions increasing as some power of their argument can also bring one to the theory of generalized functions. These Fourier transforms, in Bochner's work the formal derivatives of continuous functions, are in essence generalized functions.

3.4. FUNCTIONAL ANALYSIS AND OPERATOR THEORY

In 1950–1951 there appeared Laurent Schwartz's monograph *Théorie des Distributions*. In this book Schwartz systematizes the theory of generalized functions, basing it on the theory of linear topological spaces, relates all the earlier approaches, and obtains many important and far reaching results. Unusually soon after the appearance of *Théorie des Distributions*, in fact literally within two or three years, generalized functions attained an extremely wide popularity. It is sufficient just to point out the great increase in the number of mathematical works containing the delta function.

In the volumes of the present series we will give a systematic development of the theory of generalized functions and of problems in analysis connected with it. On the other hand our aims do not include the collation of all material related in some way to generalized functions, and on the other hand many of the problems we shall consider can be treated without invoking them. However, the concept is a convenient link connecting many aspects of analysis, functional analysis, the theory of differential equations, the representation theory of locally compact Lie groups, and the theory of probability and statistics. It is perhaps for this reason that the title generalized functions is the most appropriate for this series of volumes on functional analysis.

Let us briefly recount the contents of the first four volumes of the series.

The first volume is devoted essentially to algorithmic questions of the theory... The second volume develops the concepts introduced in the first, use topological considerations to prove theorems left unproved in the latter, and constructs and studies a large number of specific generalized function spaces... The third volume is devoted to some applications of generalized functions to the theory of differential equations, ... In the fourth and fifth volumes we consider problems in probability theory related to generalized functions (generalized random processes) and the theory of representation theory of Lie groups. The unifying concept here is that of harmonic analysis (the analog of Fourier integral theory) of generalized functions, in particular questions related to the representations of positive definite functions..."

The Foreword of the fourth volume says "This book is the fourth volume of a series of monographs on functional analysis appearing under the title 'Generalized Functions'. It should not, however, be considered a direct sequel to the proceeding volumes. In writing this volume the authors have striven for the maximum independence from the proceeding volumes..."

According to a review in Bulletin of AMS, "At the beginning of the 1950's the theory of generalized functions was in somewhat the same state that nonstandard analysis is in today. Mathematicians were by no means of one mind as regards the benefits of the theory. Critics felt that it was an overblown way of describing a modest but useful scheme for making computations in one area of harmonic analysis: the Heaviside calculus. Even those who were enthusiastic about the theory regarded distributions as shadowy entities like quarks or mirons. It was felt that to understand the theory one had first to become familiar with a formidable array of topics in abstract analysis: barreled topological vector spaces, Montel spaces and so on. Graduate students were discouraged from going into distribution theory and advised to do Schauder-Leray estimates instead.

By the end of the 1950s this situation had completely changed. Generalized functions had come to be viewed as an indispensable tool in almost every area of analysis. The reasons for this were not hard to account for. The novelty of the theory wore off and people gradually got used to thinking of distributions as house-and-garden variety objects. It turned out that the function-theoretic underpinnings of the theory could be reduced to standard facts about Sobolev spaces, so one did not need to know about espaces tonnelles. In fact to learn enough of the theory to be able to work with distributions, albeit nonrigorously, one could get by with a few elementary facts about the Fourier transform. This meant that distributions could be made a regular part of the graduate curriculum. Finally, two extremely important mathematical developments, both occurring in the middle 1950s, turned out to depend on the theory of distributions in an absolutely essential way. One of these was in the area of linear partial differential equations... The other development was in the area of group representations ...

The five volumes of Generalized functions were mostly written while these developments were taking place and, in fact, in no small way contributed to these developments."

3.4.7 K. Yosida, Functional analysis, reprint of the sixth (1980) edition. Classics in Mathematics, Springer-Verlag, Berlin, 1995. xii+501 pp.

This is a classic written by an expert on functional analysis and also an expert user of functional analysis. It has been both very successful as a textbook and as a reference on functional analysis.

As the author points out in the preface, "as a textbook to be studied by students on their own or to be used in a course on functional analysis... While the book is primarily addressed to graduate students, it is hoped it might prove useful to research mathematicians, both pure and applied."

According to MathSciNet of the first edition, "the book covers a great variety of topics in functional analysis. Besides, each chapter of the book contains a great deal of material, both classical and modern, as well as many useful examples in concrete spaces. Of course, the book also includes the results of the author, who has enriched functional analysis during the last 30 years with many valuable contributions. Actually, each chapter of the book deserves to be treated in a separate monograph and, in fact, as is well known, for most of them there do already exist such monographs... The book is a welcome contribution to the existing literature on functional analysis. The reviewer

3.4. FUNCTIONAL ANALYSIS AND OPERATOR THEORY

has no doubt that this book will be very useful for students and that it will also be read with interest by specialists, both in pure and in applied functional analysis."

3.4.8 **N. Dunford, J. Schwartz, Linear Operators, Part I, General Theory; Part II, Spectral theory. Selfadjoint operators in Hilbert space. Part III, Spectral operators, Wiley Classics Library, A Wiley-Interscience Publication, John Wiley & Sons, Inc., New York, 1988. xiv+858 pp., x+859–1923, xx+1925–2592.**

This is probably the longest and also the most famous book on functional analysis. It is a classic and has been very influential in many ways.

This book was written for "the student and the mature mathematician", and "the present treatise is relatively self-contained, and nearly everything in it can be read by one who has studied the elementary algebraic and topological properties of the real and complex number systems, and those basic results of the theory of functions of a complex variable which center around the Cauchy integral theorem", according to the preface of Part I. Many people have probably heard of these two huge volumes and some even attempted to read them in full. It is not clear how many people have actually finished reading them. They contain a huge amount of information.

According to a review in Bulletin of AMS, "With the long awaited appearance of the second volume, we now see before our eyes a panoramic view, rich in colorful detail, of the whole output of a school of mathematical analysis that started with the work of Volterra and Fréchet near the turn of the century, and reached its peak in Poland, Hungary and the Soviet Union as well as at Chicago and Yale in the thirties and forties... In these days of extreme fragmentation of fields, when some mathematicians take pride in not knowing the applications of their own work, the authors' constant concern with the 'Zusammenhang' between theory and application, and between distinct branches of mathematics, accounts for the unusual length of this treatise. ... To have a nearly complete view of an entire field, that of functional analysis, available in a coherent exposition (which, we hope, will eventually reach several thousand pages with the publication of the remaining 2 volumes), is a benefit that few mathematical disciplines have ever enjoyed, one thinks of Fricke-Klein or Russell-Whitehead as feeble precedents; inevitably it will have a decisive influence upon the future development of all analysis, as well as of much theoretical physics and other neighboring disciplines."

According to MathSciNet, "This is a comprehensive account of the modern theory of linear operators, mainly in Banach spaces, and of applications of the theory of such operators to other parts of mathematics..... The present book is an important contribution to the mathematical literature. The wealth of material and the extensive bibliography should make it a standard reference for the expert in the field. It should also prove valuable to the advanced graduate student seeking knowledge in one of the many subjects treated. The latter, however, should be warned that one important part of modern linear space theory is touched on only occasionally, namely, the theory of linear spaces which are not normed. If, e.g., he wants to be informed about the present state of the theory of linear convex topological spaces, he ought to turn to other sources."

3.4.9 Semigroups and functional analysis

A. Pazy, *Semigroups of Linear Operators and Applications to Partial Differential Equations*, Applied Mathematical Sciences, 44, Springer-Verlag, New York, 1983. viii+279 pp.

This is a valuable and popular textbook on semigroups of linear operators and their applications in differential equations.

From the book description, "Since the characterization of generators of C_0-semigroups was established in the 1940s, semigroups of linear operators and its neighboring areas have developed into an abstract theory that has become a necessary discipline in functional analysis and differential equations. This book presents that theory and its basic applications, and the last two chapters give a connected account of the applications to partial differential equations."

According to MathSciNet, "The few prerequisites needed and the clear exposition make this monograph an excellent textbook for a graduate course or for part of a functional analysis or a partial differential equations sequence (although the lack of exercises does not help here). It presents a reasonably complete and deep overview of an important theory and of some of its applications; thus it will be of interest to users of partial differential and other equations. Finally, it offers the tools necessary for access to areas of mathematics where important research is presently going on, chiefly the theory of nonlinear semigroups. On these and other counts, it is a welcome addition to the literature."

K. Engel, R. Nagel, *One-parameter Semigroups for Linear Evolution Equations*, With contributions by S. Brendle, M. Campiti, T. Hahn, G. Metafune, G. Nickel, D. Pallara, C. Perazzoli, A. Rhandi, S. Romanelli and R. Schnaubelt, Graduate Texts in Mathematics, 194, Springer-Verlag, New York, 2000. xxii+586 pp.

The theory of one-parameter semigroups of linear operators on Banach spaces started in the early 1900 and became mature in 1957. It was perfected in 1970s and 80s. Many applications to different fields were made after that, and this comprehensive book was written to reflect such an exciting state. One chapter title is "Semigroups Everywhere". It contains a lot of information. Later the authors wrote another shortened version of this book titled *A short course on operator semigroups*. This book is a bit unusual in the sense that it has ten additional contributors. It is a very good introduction to the theory of semigroups with many applications to various evolution problems, with some emphasis on motivation. It is also a valuable reference.

According to MathSciNet, "This book is a broad introduction to the theory of C_0-semigroups.... The steady pace of the text and the very high quality of the exposition make the material in this book easily accessible to anyone with a background knowledge of functional analysis. The length of the book may appear a little intimidating but there is clear guidance as to which topics can be omitted at first reading... the book's greatest impact is likely to be through tempting graduate students to learn about semigroups of operators and to appreciate a little of the applications. Nevertheless the book should also be of interest to experienced researchers. Most specialists in semigroup theory and applications will be glad to have it available for general reference, and it is also very suitable for non-specialists who wish to use just one book on basic semigroup theory for occasional consultation."

E. Hille, R. Phillips, *Functional Analysis and Semi-groups*, rev. ed., American Mathematical Society Colloquium Publications, vol. 31, American Mathematical Society, Providence, R.I., 1957. xii+808 pp.

This is a classic. The first edition of this book was published in 1948. Due to a lot of breakthroughs on the analytic theory and applications of semi-groups of linear operators, it needed to be revised substantially, and Phillips was asked to join in the revision process. It has been an important book since then.

According to MathSciNet, "This volume is a revised edition, with Phillips as co-author, of the well-known book of Hille under the same title in 1948. The revision is a thorough new one, involving rearrangement, the addition of new background material, and the incorporation of many new results." The Math Review of the first edition of this book says: "This book is devoted to the study of semi-groups (associative multiplicative systems for which no cancelation rules are postulated) and their linear representations in Banach spaces. Commutative semi-groups receive generous but not exclusive attention. Much important material contained in this volume has not been published elsewhere. Certain well-known results about groups and their representations in Banach and Hilbert spaces appear here as special cases of results about semi-groups... It is appropriate to note the systematic, incisive and polished character of the author's treatment. His scholarly handling of the subject will be appreciated by all users of the book, as will the clarity of his expository style. The general character of the work is analytical rather than algebraic, although algebraic concepts and methods are everywhere given due prominence (as they must be in any adequate discussion of semi-groups)."

A. Lunardi, *Analytic Semigroups and Optimal Regularity in Parabolic Problems*, Progress in Nonlinear Differential Equations and their Applications, 16, Birkhäuser Verlag, Basel, 1995. xviii+424 pp.

The purpose of this book is to give a systematic treatment of the basic theory of analytic semigroups and abstract parabolic equations in general Banach spaces, and to explain how such a theory can be used in studying parabolic partial differential equations. While focusing on the classical solutions, it also include some more recent developments. It gives a comprehensive overview of the status of the field, and also teaches the basic techniques of the subjects. This book is suitable for both students and mathematicians.

According to MathSciNet, "This very interesting book provides a systematic treatment of the basic theory of analytic semigroups and abstract parabolic equations in general Banach spaces, and how this theory may be used in the study of parabolic partial differential equations; it takes into account the developments of the theory during the last fifteen years. The book is focused on classical solutions, with continuous or Hölder continuous derivatives: this choice allows one to consider any kind of nonlinearities (possibly of nonlocal type), even involving the highest-order derivatives of the solution, in contrast with the growth limitations required by the L^p approach. Analytic semigroups are described independently of the general theory of semigroups, in order to emphasize their special properties among C^0-semigroups... The only necessary prerequisites are the fundamental notions of functional analysis and some familiarity with partial differential equations..."

3.4.10 A. Connes, Noncommutative Geometry, Academic Press, Inc., San Diego, CA, 1994. xiv+661 pp.

The preface of this book explains the meaning of noncommutative geometry well: "The correspondence between geometric spaces and commutative algebras is a familiar and basic idea of algebraic geometry. The purpose of this book is to extend this correspondence to the noncommutative case in the framework of real analysis. The theory, called noncommutative geometry, rests on two essential points: 1. The existence of many natural spaces for which the classical set-theoretic tools of analysis, such as measure theory, topology, calculus, and metric ideas lose their pertinence, but which correspond very naturally to a noncommutative algebra. Such spaces arise both in mathematics and in quantum physics and we shall discuss them in more detail below: examples include: a) The space of Penrose tilings, b) The space of leaves of a foliation, c) The space of irreducible unitary representations of a discrete group, d) The phase space in quantum physics, e) The Brillouin zone in quantum physics, f) Space-time. ... 2. The extension of the classical tools, such as measure theory, topology, differential calculus and Riemannian geometry, to the noncommutative situation..." This is a very important introduction to noncommutative geometry written by its founder.

According to Vaughan Jones, this book is "... A milestone for mathematics. Connes has created a theory that embraces most aspects of 'classical' mathematics and sets us out on a long and exciting voyage into the world of noncommutative mathematics. The book contains a colourful account of the meaning of the term 'non-commutative space' based on an extraordinary wealth of examples, including the set of all Penrose tilings, the space of leaves of a foliation, the quantum Hall effect and an intriguing non-commutative model of four-dimensional space-time that reproduces the standard model of elementary particles from quite general considerations... The reader of the book should not expect proofs of theorems. This is much more a tapestry of beautiful mathematics and physics which contains material to intrigue readers with any mathematical background. At the same time there is a comprehensive bibliography that will lead the reader straight to the sources and proofs of the results."

According to a review in Bulletin of AMS, "Abstract analysis is one of the youngest branches of mathematics, but is now quite pervasive. However, it was not so many years ago that it was considered rather strange. The generic attitude of mathematicians before World War II can be briefly evoked by the following story told by the late Norman Levinson. During Levinson's postdoctoral year at Cambridge studying with Hardy, von Neumann came through and gave a lecture. Hardy (a quintessential classical analyst) remarked afterwards, 'Obviously a very intelligent young man. But was that mathematics?'... The 'geometry' part of the title is a bit programmatic; a more descriptive title might be Topics in operator algebra and its applications. The approach is subjective and informal, although technically highly sophisticated. The author runs through the main features of his own and related works of the past three decades with minimal attention (or references) to earlier, including some ideational basic, works. The result appears as a brilliant and sustained tour de force that will fascinate sufficiently knowledgeable devotees of the author's main lines of investigation but may baffle non-specialists. This would be a pity, for the book goes into relations between diverse aspects of its subject that are otherwise not well represented in the literature, especially those in a topological or homological algebraic direction.

As an effervescent and yet sustained account of a wide range of advanced and refined aspects of abstract analysis, this book appears quite valuable and is certainly unique."

3.4.11 Operator algebras

Although the theory of operator algebras is usually classified as a branch of functional analysis, it has direct applications to representation theory, differential geometry, quantum statistical mechanics and quantum field theory. Such algebras can be used to study arbitrary sets of operators with little algebraic relation. Operator algebras can be regarded as a generalization of spectral theory of a single operator. In general operator algebras are non-commutative rings.

Though algebras of operators are studied in various contexts, they usually consist of bounded operators on a Banach space or on a separable Hilbert space, endowed with the operator norm topology. In the case of operators on a Hilbert space, the adjoint map on operators gives a natural involution which provides an additional algebraic structure which can be imposed on the algebra, and the best studied examples are self-adjoint operator algebras such as C^*-algebras and von Neumann algebras.

Commutative self-adjoint operator algebras can be regarded as the algebra of complex valued continuous functions on a locally compact space, or that of measurable functions on a standard measurable space. Thus, general operator algebras are often regarded as a noncommutative generalizations of these algebras, or the base space on which the functions are defined. This is the philosophy of noncommutative geometry, which tries to study various non-classical and pathological objects such as non-Hausdorff quotients by noncommutative operator algebras. The book above of Connes fits this framework. There are several other good books on operator algebras.

M. Takesaki, *Theory of Operator Algebras. I. II. III.*, Springer-Verlag, New York-Heidelberg, 1979. vii+415 pp., 2003. xxii+518 pp., xxii+548 pp.

Rings of operators, also called von Neumann algebras, were introduced by von Neumann in 1929 with "his grand aim of giving a sound foundation to mathematical sciences of infinite nature". Together with Murray, he laid down the foundation for this new subject in a series of papers near the end of 1930s. A lot of progress has been made by many people since then, and this book by one of the main contributors gives a comprehensive overview of the subject starting from the basics. It is valuable both as a textbook and as a reference.

According to MathSciNet, "The present volume is a detailed and comprehensive account of basic C^*-algebra and von Neumann algebra theory... Each chapter contains at least one section of historical notes, and there are many exercises giving useful complements to the text or applications of the theory..."

Volume II is mainly concerned with type III von Neumann algebras and their automorphism groups, and is a valuable reference. Volume III is a direct continuation of the second volume and is mainly concerned with the structure theory of approximately finite-dimensional factors. It is a reference for experts.

J. Dixmier, *von Neumann Algebras*, With a preface by E. C. Lance, Translated from the second French edition by F. Jellett, North-Holland Mathematical Library, 27, North-Holland Publishing Co., Amsterdam-New York, 1981. xxxviii+437 pp.

3.4. FUNCTIONAL ANALYSIS AND OPERATOR THEORY

This is a classic, and different editions of this book represent different historical periods of this subject.

According to MathSciNet of the original first French version, "This book presents a thorough reworking and systematic development of the theory of weakly-closed *-algebras of bounded operators on a Hilbert space, which the author renames von Neumann algebras. Although only twenty years old, this has been one of the most active and fruitful areas of modern functional analysis, as the author's bibliography of nearly two hundred items attests. The present exposition divides into three large chapters: the first on the global theory, the second on the direct integral reduction theory, and a final one presenting those standard but more technical topics which do not fit into the pattern of development chosen by the author. A valuable part of the book are the numerous exercises, which present illuminating examples and sketch additional bits of theory.

The first chapter is probably the most important, since the global theory is the principal aspect of the general theory to be developed since the original von Neumann and von Neumann-Murray papers. Its goal is the global decomposition of a von Neumann algebra according to type and finiteness, and its essential tool is the theory of traces, upon which the key definitions are based. This procedure avoids the more difficult global dimension theory of projections (which is here relegated to the third chapter), but it does not yield the complete classification analysis..."

J. Dixmier, C^*-*algebras*, Translated from the French by Francis Jellett, North-Holland Mathematical Library, Vol. 15, North-Holland Publishing Co., Amsterdam-New York-Oxford, 1977. xiii+492 pp.

This is a classical book on C^*-algebra. The theory of C^*-algebras was rooted in the work of Gelfand and Naimark which characterizes C^*-algebras among all involutive Banach algebra. It turned out that this class of Banach algebras plays a basic role in the study of representations of a very extensive class of involutive Banach algebras. A substantial part of this book studies C^*-algebras, including discussion of results due to Fell, Glimm, Kadison, Kaplansky, Mackey and Segal et al. The latter part of this book summarizes some aspects of unitary representations of groups, since the theory of groups provides some of the most interesting examples of C^*-algebras. The theory of C^*-algebras is still expanding and applied to new subjects, but this work remains a clear and accessible introduction to this subject.

Other related books on C^*-algebras include the following three.

S. Sakai, C^*-*algebras and* W^*-*algebras*, Ergebnisse der Mathematik und ihrer Grenzgebiete, Band 60, Springer-Verlag, New York-Heidelberg, 1971. xii+253 pp.

This book can be considered as an update of the classical book on C^*-algebra by Dixmier. It gives an excellent and comprehensive account of the theory of von Neumann algebras, and hence is a valuable reference for people at different levels.

According to MathSciNet, "This book is a treatise on the theory of von Neumann algebras. This book invites comparison with the older and better-known treatise of J. Dixmier... The two books cover very much the same material, ... Perhaps the best illustration of the philosophical differences between the two books is that Dixmier defines a von Neumann algebra as a *-subalgebra of $L(H)$ which equals its double commutant and then proceeds via the spectral theorem, whereas the author defines a W^*-algebra as a C^*-algebra which is a dual Banach space and then proceeds via the Gelfand theory of commutative C^*-algebras.

The present book does not include any exercises for the reader but does include many research questions, a goodly number of which have now been answered."

C. Pedersen, C^*-*algebras and Their Automorphism Groups*, London Mathematical Society Monographs, 14, Academic Press, Inc., London-New York, 1979. ix+416 pp.

As explained in the preface, the theory of C^*-algebra has grown to a size which cannot be covered comprehensively in a single book, and the author picked topics which might be limited by his "knowledge and prejudice to form a somewhat manageable version". He emphasized problems connected with groups of automorphisms of C^*-algebras. The good reception of this book proves his good choice. It is a valuable reference.

According to MathSciNet, "This book contains an up-to-date account of most of the basic material on C^*-algebras and von Neumann algebras, as well as a great deal of more advanced material. Many of the results in the book were unknown in 1965, and the proofs of many of the older results are different from the original proofs. In the reviewer's opinion, this book is destined to replace Dixmier's two books as the main reference book for researchers in operator algebras... The results are all presented in a very elegant and polished form. A one-semester course in general functional analysis is formally adequate training for reading this book. In fact, however, the proofs are so slick that the book may be more suitable for researchers than it is for students just learning the subject.

Each section of the book ends with a subsection of historical notes. These notes, as well as the many humorous remarks scattered throughout the book, are sufficient reason for anybody with even a passing interest in operator algebras to at least glance through the book."

R. Kadison, J. Ringrose, *Fundamentals of the Theory of Operator Algebras. Vol. I. Elementary Theory. Vol. II. Advanced Theory*, Pure and Applied Mathematics, 100, Academic Press, Inc., New York, 1983. xv+398 pp., 1986. pp. i–xiv and 399–1074.

Not all Banach algebras are the same, and two classes stand out. They are C^*-algebras and von Neumann algebras. These two substantial volumes give an introduction to the subjects up to the point where the reader can read research papers. It can also serve as a good reference.

According to MathSciNet, "The authors' purpose is clearly stated in the first paragraph of the preface: '[Our] primary goal is to teach the subject and lead the reader to the point where the vast recent literature, both in the subject proper and in its many applications, becomes accessible.' This point of view is to be contrasted with the many (excellent) existing books on the subject, which are primarily directed toward the research mathematician, or exist to provide background on operator algebras for applications in theoretical physics... Although the material in this first volume is standard and well known to experts, the authors have presented it in a fresh and attractive way which conveys the spirit and beauty of the subject. They are to be commended for writing a beautiful book which, in the reviewer's opinion, fulfills all of the promises made in the preface."

"As with the first volume, the primary goal [of the second volume] is to teach the subject rather than try to be encyclopedic. For this reason and also because of the enormous present-day size, scope, and complexities of the theory, the authors have made no attempt to include more recent developments involving the 'noncommutative aspects' of topology, differential geometry, shape theory, K-theory, etc. Lest one feel that this is cause for alarm let me hasten to say that what is presented here is an exciting and careful account of some very substantial mathematics!"

L. Gillman, M. Jerison, *Rings of Continuous Functions*, The University Series in Higher Mathematics, D. Van Nostrand Co., Inc., Princeton, N.J., 1960. ix+300 pp.

This book has an unusual preface: "This book is addressed to those who know the meaning of each word in the title: none is defined in the text." Its goal is to give a systematic study of the ring of real-valued continuous functions on a topological space and the subring of bounded functions, and the interplay between the algebraic properties of these rings and the topological properties of the space. The correspondence between algebraic properties of the ring of continuous functions and topological properties of the space is one of the motivations for the noncommutative geometry formulated by A. Connes. This book is rather complete and covers almost all known and important results on the topics under discussion and will be a valuable reference.

From the book description, "Great emphasis is placed on the study of ideals, and on the associated residue class rings. Questions of extending continuous functions from a subspace to the entire space arise as a necessary adjunct and are dealt with in considerable detail. Many problems provide additional information about the material covered in the text."

3.4.12 H. Brezis, Analyse Fonctionnelle, Théorie et Applications, Collection Mathématiques Appliquées pour la Matrise. Masson, Paris, 1983. xiv+234 pp.

Functional analysis and many books on them usually give people an impression that they are beautiful abstract theories. This book emphasizes applications. As the title indicates, half of the book is about concrete applications to differential equations, and the other half is about the more standard abstract theory of functional analysis. It is a very useful introduction to functional analysis and can complement other books on functional analysis well.

According to Zentralblatt MATH, "This is a very readable introduction to functional analysis, leading quickly to nontrivial applications, especially in ordinary and partial differential equations. More than one half of the book is devoted to Sobolev spaces and the solution of elliptic equations by means of the Lax-Milgram theorem, and to abstract evolution (the Hille-Yosida theorem) and their application to the heat and wave equations... The book is clearly well written and gives the reader a good impression of the applicability of functional analysis. This emphasis makes it an outstanding contribution to the vast literature on this field."

These two halves of the book are closely related. Indeed as the author explained in the preface, "Historically, 'abstract' functional analysis was developed to answer questions raised by the solution of partial differential equations. Conversely, progress in 'abstract' functional analysis has greatly stimulated the theory of partial differential equations."

According to an Amazon review, "Brezis has intelligently chosen several fundamental concepts of functional analysis, and has build the book around them and their applications. For this reason this book is not as comprehensive a source as, e.g., Dunford & Schwartz, Edwards or Yosida's classical texts, but for a newcomer who intends to become a user of functional analysis this book is an ideal place to start."

The book above has been revised and updated to

H. Brezis, *Functional Analysis, Sobolev Spaces and Partial Differential Equations*, Universitext, Springer, New York, 2011.

This book continues the feature of the previous French book by emphasizing fruitful interaction of abstract theories and concrete applications. It updates the French book and also includes many problems to make it more suitable as a textbook.

According to Zentralblatt MATH, "A previous version of this book, originally published in 1983 in French, became very popular worldwide, and was adopted as a textbook in many European universities. A deficiency of the French text was the lack of exercises. The present volume contains a wealth of problems. The author's aim is to give a systematic treatment of some of the fundamental abstract results in functional analysis and of their applications to certain concrete problems in linear differential and partial differential equations. Moreover, by interlacing extensive commentary and foreshadowing subsequent developments within the formal scheme of statements and proofs, by the inclusion of many apt examples and by appending interesting and challenging exercises, the author has written a book which is eminently suitable as a text for a graduate course. This volume is distinguished by the broad variety of problems which have been treated and by the abstract results which are developed."

H. Brezis, *Operateurs Maximaux Monotones et Semi-groupes de Contractions dans les espaces de Hilbert*, North-Holland Mathematics Studies, No. 5, Notas de Matematica (50), North-Holland Publishing Co., Amsterdam-London; American Elsevier Publishing Co., Inc., New York, 1973. vi+183 pp.

This book is one of the first systematic books on the theory of monotone operators on Hilbert spaces and the associated theory of nonlinear evolution equations and semigroups. Unlike the previous two books, it does not contain applications. Due to its comprehensive nature, it is suitable for both experts and beginners in this rapidly developing subject.

3.4.13 K. Deimling, Nonlinear Functional Analysis, Springer-Verlag, Berlin, 1985. xiv+450 pp.

It is well-known that the most basic object in linear algebra is a linear transformation. In functional analysis, the most basic objects in functional analysis are linear maps between Banach spaces (or more general topological linear spaces) satisfying certain boundedness conditions. Nonlinear functional analysis studies nonlinear mappings in the context of Banach spaces and includes topics such as generalizations of calculus to Banach spaces, implicit function theorems for infinite dimensional spaces, fixed-point theorems such as Brouwer fixed point theorem, fixed point theorems in infinite-dimensional spaces, topological degree theory, Morse theory etc. Since maps arising from applications are often nonlinear, and linear approximations have limited applicability, nonlinear functional analysis is becoming more important for applications.

This book is useful as an accessible introduction to beginners and also as a reference due to its comprehensive coverage.

According to MathSciNet, "Many problems in nonlinear functional analysis have their origin in the inspection of various types of equations or questions arising in nonlinear models for problems in, for instance, engineering and natural science. This leads to the study of certain classes of nonlinear mappings, solvability properties of equations involving nonlinear mappings, properties of their solutions, etc., in an abstract Banach space setting.

The aim of the book under review is to present a survey of the main ideas, concepts and methods constituting the field of nonlinear functional analysis... The presentation is very careful through and makes use only of elementary analysis and basic facts in functional analysis (e.g. normed linear spaces, bounded linear operators). Special emphasis is laid on examples illustrating abstract concepts and methods on the one hand, and showing how abstract results can be applied to special model problems or problems in other abstract contexts on the other. In addition, the presentation uses a uniform mathematical language and way of thinking which particularly facilitates reading of the material by beginners in this field. A special feature of the book consists in a large number of elaborated exercises accompanying each section... This is a fascinating book which presents a clear guide to the field of nonlinear functional analysis. Both printing and format are excellent, and the work can be thoroughly recommended."

E. Zeidler, *Nonlinear Functional Analysis and Its Applications. I. Fixed-point Theorems*, Translated from the German by Peter R. Wadsack, Springer-Verlag, New York, 1986. xxi+897 pp. *II/A. Linear Monotone Operators*, Translated from the German by the author and Leo F. Boron, 1990. xviii+467 pp. *II/B. Nonlinear Monotone Operators*, Translated from the German by the author and Leo F. Boron, 1990. pp. i–xvi and 469–1202. *III. Variational Methods and Optimization*, Translated from the German by Leo F. Boron, 1985. xxii+662 pp. *IV. Applications to Mathematical Physics*, Translated from the German and with a preface by Juergen Quandt, 1988. xxiv+975 pp.

This impressive and comprehensive series of books started from a not-so-modest goal. As the author explained in the preface of the first volume, there were good references to each important topic in nonlinear functional analysis, "What was missing was a comprehensive treatment of nonlinear functional analysis, accessible to a broader audience of mathematicians, natural scientist, and engineers, with a command of the basics of linear functional analysis only, which would provide a rapid survey of the subject. I attempted to close this gap with a five-part expansion of my lecture notes." The basic contents of these volumes can be seen from the subtitles: I) Fixed-point theorems, II) Linear and nonlinear monotone operators, III) Variational methods and optimization, IV-V) Applications in mathematical physics. It seems that the author has achieved his goal. This book is a very valuable guide and reference for everyone who is interested in nonlinear functional analysis.

According to Zentralblatt MATH, "All in all, this book is an excellent readable introduction in the prevalent classes of nonlinear operators. A large audience of mathematicians and natural scientists will find it a stimulating guide through the contemporary operator treatment of nonlinear partial differential and integral equations."

IV. The reading room of the Old Law Library of the University of Michigan

There is no question that the most important book on geometry is Euclid's *Elements*, though it is less emphasized by most people that Elements also includes important results in number theory. How about the most popular book or the longest lasting book on algebra? This honor might go to Euler's book *Elements of Algebra*. It is one of the first books that study algebra in the modern form. However, it is different from most modern approaches and is interesting to the contemporary reader, or even experts. Together with his other textbook *Introduction to Analysis of the Infinite*, they are two books written by Euler which are accessible and still used by many people now.

Tartaglia is best known for his solutions of cubic equations by radicals and his conflicts with Gerolamo Cardano. Though he published many books, his method for solving cubic equations was first published in a book by Cardano. Because of this, the formula to solve cubic equations is usually called the "Cardano-Tartaglia Formula". But del Ferro was the first person known to have solved cubic equations by radicals. Unfortunately, there are no surviving scripts from del Ferro, because of his resistance to communicating his works. He only showed them to a small, select group of friends and students. One reason might be a peculiar practice of mathematicians at that time: publicly challenging one another. When a challenge was accepted, each mathematician needed to solve the other's problems, and the loser was often disgraced and lost his university position. Del Ferro might be fearful of being challenged and likely kept his greatest work secret so that he could use it to defend himself in the event of a challenge. In spite of this, he kept a notebook which contained all his important results.

It is well-known that Abel proved that a general algebraic equation of degree 5 or above cannot be solved by radicals. But the credit should go to Ruffini. In 1799 he published a book containing this result, and its long title gives a good summary: *General theory of equations in which it is shown that the algebraic solution of the general equation of degree greater than four is impossible.* The introduction to the book is also informative: "The algebraic solution of general equations of degree greater than four is always impossible. Behold a very important theorem which I believe I am able to assert (if I do not err): to present the proof of it is the main reason for publishing this volume. The immortal Lagrange, with his sublime reflections, has provided the basis of my proof."

Emmy Noether is probably the best female mathematician in the history. She is well-known for her great contributions to abstract algebra. In some sense, she changed the whole landscape of modern algebra. For example, one important concept in algebra is the notion of Noetherian rings. The classical book *Modern Algebra* by van der Waerden was based on lectures of E. Artin and E. Noether. She also made fundamental contribution to the conservation laws in theoretical physics. It is probably less known that it was Noether who suggested that one should consider the homology groups of a space instead of only its Betti number. Of course, a group contains more information and this has far-reaching consequences.

Curriculum Vitae

I, Amalie Emmy Noether, Bavarian by nationality and Israelite by faith, was born in Erlangen on the 23rd of March, 1882, the daughter of Dr. Max Noether, Professor at the Royal University, and of his wife Ida, née Kaufmann. My schooling took place in Erlangen and Stuttgart, and in 1900 I passed both "examinations for teachers of French and English" in Ansbach, and in 1903, as a private student, I obtained the *absolutorium* of the Royal Realgymnasium in Nürnberg. From 1900–1903 I studied at the University of Erlangen, in the winter semester 1903/1904 in Göttingen, and since autumn of 1904 I have been enrolled at Erlangen for the study of mathematics. My teachers in Erlangen were Professors Gordan, Noether, Reiger, Wehnelt, Wiedemann, Pirson, Bulle, Fester, Fischer; in Göttingen Professors Blumenthal, Hilbert, Klein, Minkowski, Schwarzschild. To all of them I am indebted for promoting my scientific progress; I wish to express my particular gratitude to Herrn Geheimrat Gordan for encouraging me to write the present thesis, and for his constant interest during the time I worked on it.

Chapter 4

Algebra

Algebra studies rules of operations and relations between numbers and quantities, and the constructions and concepts arising from them, including polynomials, equations and algebraic structures. It is an essential part of mathematics.

The origin of algebra can be traced to the ancient Babylonians, who developed an advanced arithmetical system with which they were able to do calculations in an algorithmic fashion. The Babylonians developed formulas to calculate solutions for problems which are solved today by using linear equations, quadratic equations, and indeterminate linear equations. By contrast, most Egyptians of this era, as well as Greek and Chinese mathematicians in the first millennium BC, usually solved such equations by geometric methods. The Hellenistic mathematicians such as Diophantus as well as Indian mathematicians such as Brahmagupta continued the traditions of Egypt and Babylon. Diophantus' *Arithmetica* is on a higher level and the Greek mathematician Diophantus has traditionally been known as the "father of algebra". Viète's work at the end of the sixteenth century marks the beginning of the classical algebra. In 1637, Descartes invented analytic geometry and introduced modern algebraic notation. Abstract algebra was developed in the nineteenth century, initially motivated by the problem of solvability of algebraic equations, i.e., the Galois theory, and the problem of geometric constructibility. The modern algebra started with the work of Dedekind, Kronecker and others, and has deep connections with other branches of mathematics such as algebraic number theory, algebraic topology and algebraic geometry.

Algebra includes (1) abstract algebra, or modern algebra, in which algebraic structures such as groups, rings and fields are axiomatically defined and studied, (2) finite group theory, motivated by Galois groups of algebraic equations, (3) linear algebra, (4) commutative algebra, which is closely related to algebraic geometry, (5) homological algebra.

It is a trend of modern mathematics that many different subjects interact and intertwine, and new subjects often arise from such fusion. For example, algebraic geometry builds on both geometry and algebra. Banach algebra is an example of combining analysis and algebra. So is algebraic topology.

It is natural that infinite groups are also natural objects in algebra, and algebraic groups and Lie groups can be considered as notions in algebra.

In this chapter, we will deal with books in the 5 areas listed above.

4.1 Abstract algebras and finite groups

Weber worked in algebra, number theory, and analysis. He is best known for his three volume book Lehrbuch der Algebra *(Lectures on algebra), which was first published in 1895 and has been reprinted many times (for example, it was reprinted by AMS/Chelsea in 2000). Much of this book consists of his original research in algebra and number theory. His famous work with Dedekind* Theorie der algebraischen Functionen einer Veränderlichen *gave an algebraic foundation for Riemann surfaces, hence a purely algebraic formulation of the Riemann-Roch theorem. He loved books. His last position was at University of Strasbourg and he made the mathematics library there a great one.*

Figure 4.1: Weber's book.

4.1.1 Linear algebra

Linear algebra studies linear (or vector) spaces, and linear transformations between them. Since linear structures appear in many different contexts, for example, in approximating nonlinear structures, they are are easier to deal with. Linear algebra is central to modern mathematics and its applications.

There are also many books on linear algebra, and it is not easy to pick out the really good ones. We select several which have withstood the test of time or have been popular.

I.M. Gelfand, *Lectures on Linear Algebra*, with the Collaboration of Z. Ya. Shapiro, Translated from the second Russian edition by A. Shenitzer, Reprint of the 1961 translation, Dover Publications, Inc., New York, 1989. vi+185 pp.

This book was based on a course in linear algebra by Gelfand in Moscow State University around 1945 and S.V. Fomin contributed to the writing of this book. This book has a fast pace and gives people an impression of linear algebra in action. For example, the book starts out by motivating the definition of vector spaces, and it does so

4.1. ABSTRACT ALGEBRAS AND FINITE GROUPS

by discussing three important aspects from geometry, algebra and analysis in slightly less than one page and catching their common properties. The next page gives the definition of the dimension of a vector space, followed by four important examples. In the total of 185 pages, it covers the usual topics such as linear transformations and their canonical forms, self-adjoint operators and symmetric matrices etc. It also contains many other topics, for example, the last chapter has only twenty one pages but gives a self-contained introduction to tensors. This book shows the reader what is essential in elementary linear algebra and does so as quickly as possible. It is a compact and efficient book written by a master.

According to Amazon reviews, "I can only strongly recommend this somewhat forgotten textbook to anyone who needs a basic but otherwise thorough exposition to linear algebra. It conflates rigor, examples, and exercises very nicely, although in a very economic manner. Its coordinate-oriented presentation may not appeal to the mathematician in search of ultimate generality, but it is ideal for the physics and engineering-minded person."

"This is not a book to learn linear algebra from for the first time: this is an advanced book that is useful for graduate students who have already had a linear algebra course and who want to learn more topics, or understand topics on a deeper level. This is an excellent book."

P. Halmos, *Finite-dimensional Vector Spaces*, Reprinting of the 1958 second edition, Undergraduate Texts in Mathematics, Springer-Verlag, New York-Heidelberg, 1974. viii+200 pp.

Halmos is famous for his writing and his advice on writing, and this is the first book he wrote. It was first published in 1942 when he was only 26 years old. As he said in the preface, "My purpose in this book is to treat linear transformations on finite-dimensional vector spaces by the methods of more general theories. The idea is to emphasize the simple geometric notions common to many parts of mathematics and its applications, and to do so in a language that gives away the trade secrets and tells the student what is needed in the back of minds of people proving theorems about integral equations and Hilbert spaces." The Princeton University Press was hesitating to publish this elementary book. It turned out that it was a money maker for them. It is a good textbook.

According to MathSciNet, "The author exploits as completely as possible the methods and notions of the infinite in his presentation of the finite case; such a program has long been needed. Thus we find transformations studied with the help of adjoint spaces, reflexiveness, linear manifolds, disjointness, projections, etc. The book is written with great care; it should be much appreciated by the young graduate student who wishes to begin his studies of linear mathematics."

According to an Amazon review: "Besides the clarity which marks all of his books, this one has a pleasant characteristic: all concepts are patiently motivated (in words!) before becoming part of the formalism. It was written at the time when the author, a distinguished mathematician by himself, was under the spell of John von Neumann, at Princeton. Perhaps related to that is the fact that you find surprising, brilliant proofs of even very well established results (as, for instance, of the Schwartz inequality). It has a clear slant to Hilbert space, despite the title, and the treatment of orthonormal systems and the spectrum theorem is very good. On the other hand, there is little about linear mappings between vector spaces of different dimensions, which are crucial for differential geometry. But this can be found elsewhere. The problems are useful and, in general, not very difficult. All in all, an important tool for a mathematical education."

S. Axler, *Linear Algebra Done Right*, Undergraduate Texts in Mathematics, Springer-Verlag, New York, 1996. xviii+238 pp.

The unusual title of this book is explained in the preface to the instructor: "Almost all linear algebra books use determinants to prove that every linear operator on a finite-dimensional complex vector space has an eigenvalue. Determinants are difficult, nonintuitive, and often defined without motivation... In contrast, the simple determinant-free proofs presented here offer more insight. Once determinants have been banished to the end of the book, a new route opens to the main goal of linear algebra–understanding the structure of linear operators." Indeed, anyone who has tried to find formulas of (or evaluate) general determinants containing parameters can testify that that it is an art, much like finding antiderivatives (or integrals) is an art. On the other hand, determinant is an important notion, and for some applications, one still has to learn it. Learning it at a later stage as in this book might make it less painful. It is a good introduction to linear algebra from a slightly non-standard point of view.

According to MathSciNet, "This text for a second course in linear algebra is aimed at math majors and graduate students. The approach is novel, banishing determinants to the end of the book and focusing on the central goal of linear algebra: understanding the structure of linear operators on vector spaces. The author has taken unusual care to motivate concepts and to simplify proofs.

Though this text is intended for a second course in linear algebra—following one that focuses on matrices and computation—there are no prerequisites other than appropriate mathematical maturity. A variety of interesting exercises in each chapter helps students understand and manipulate the objects of linear algebra."

According to Zentralblatt MATH, "This second edition of an almost determinant-free, none the less remarkably far-reaching and didactically masterly undergraduate text on linear algebra has undergone some substantial improvements... However, no mitigation

4.1. ABSTRACT ALGEBRAS AND FINITE GROUPS

has been granted to determinants. Altogether, with the present second edition of his text, the author has succeeded to make this an even better book."

G. Shilov, *Linear Algebra*, Revised English edition, Translated from the Russian and edited by Richard A. Silverman, Dover Publications, Inc., New York, 1977. xi+387 pp.

As mentioned above, computing general determinants is difficult. But determinants are important. For example, in solving systems of linear equations whose number of variables is equal to the number of equations, whether the coefficient matrix is invertible or not is crucial. A natural question is whether there is a single invariant of the matrix which will tell us whether it is invertible. The determinant of the matrix is the invariant. In integration of functions of multi variables, change of variables leads to determinants. This book by Shilov starts with a careful discussion of determinants. It contains more materials than a standard textbook on linear algebra and is suitable for advanced undergraduate students and beginning graduate students. It is a solid book.

According to MathSciNet, "The present treatise gives a thorough and reliable exposition of material which is, in the main, classical... All in all, this is a useful and solid if not an enormously exciting book; but perhaps it would be unreasonable to look for excitement in a subject as mature as linear algebra."

According to an Amazon review, "I find it ironic that my two favourite Linear Algebra texts are this book and the Axler, for they are exact opposites: Axler shuns determinants, and Shilov starts with them and builds much of his theory off them. However, there is no book I have found that has such a deep and clear exposition of determinants. The first chapter alone makes this book worth buying.

However, there's an incredible amount of material in this book, and the later chapters are just as valuable. This is a dense book, but it is fairly easy to read once you get used to the style. I would recommend it to anyone learning linear algebra for the first time, as well as to people who want a deeper understanding or a different perspective.

Like I said before, this book is particularly useful when combined with a complementary text such as Axler, which provides a completely different approach to the subject. This book may come across as a bit old-fashioned, and some might say the material is obsolete, but I believe that everything contained in the book is useful, if only to give the reader a deeper understanding of the why's and how's of linear algebra."

4.1.2 Advanced linear algebra and matrix algebra

R. Horn, C. Johnson, *Matrix Analysis*, Cambridge University Press, Cambridge, 1985. xiii+561 pp.

This book can be considered a book on advanced linear algebra with special emphasis on matrices. It is a standard reference on matrix analysis.

From the book description, "Linear algebra and matrix theory have long been fundamental tools in mathematical disciplines as well as fertile fields for research. In this book the authors present classical and recent results of matrix analysis that have proved to be important to applied mathematics. Facts about matrices, beyond those found in an elementary linear algebra course, are needed to understand virtually any area of mathematical science, but the necessary material has appeared only sporadically in the literature and in university curricula. As interest in applied mathematics has grown, the need for a text and reference offering a broad selection of topics in matrix theory has become apparent, and this book meets that need. This volume reflects two concurrent views of matrix analysis. First, it encompasses topics in linear algebra that have arisen out of the needs of mathematical analysis. Second, it is an approach to real and complex linear algebraic problems that does not hesitate to use notions from analysis. Both views are reflected in its choice and treatment of topics."

According to MathSciNet, "Basic aspects of a topic appear in the initial sections of each chapter, while more elaborate discussions occur at the ends of sections or in later sections. This strategy has the advantage of presenting topics in a sequence that enhances the book's utility as a reference. It also provides a rich variety of options to the instructor. The book contains many exercises and problems which are essential to the development of an understanding of the subject and its applications."

R. Horn, C. Johnson, *Topics in Matrix Analysis*, Cambridge University Press, Cambridge, 1991. viii+607 pp.

This is a continuation of the book above and is a useful reference.

From the book description, "Building on the foundations of its predecessor volume, *Matrix Analysis*, this book treats in detail several topics with important applications and of special mathematical interest in matrix theory not included in the previous text. These topics include the field of values, stable matrices and inertia, singular values, matrix equations and Kronecker products, Hadamard products, and matrices and functions. The authors assume a background in elementary linear algebra and knowledge of rudimentary analytical concepts. The book should be welcomed by graduate students and researchers in a variety of mathematical fields both as an advanced text and as a modern reference work."

F.R. Gantmacher, *The Theory of Matrices. Vol. 1*, Translated from the Russian by K. A. Hirsch, Reprint of the 1959 translation, AMS Chelsea Publishing, Providence, RI, 1998. x+374 pp.

This is an excellent textbook on advanced topics of matrices and is a classic.

According to MathSciNet, "The work is an outstanding contribution to matrix theory and contains much material not to be found in any other text."

4.1. ABSTRACT ALGEBRAS AND FINITE GROUPS

P. Lancaster, M. Tismenetsky, *The Theory of Matrices*, Second edition, Computer Science and Applied Mathematics, Academic Press, Inc., Orlando, FL, 1985. xv+570 pp.

This is a classical book for students in applied mathematics and sciences. It also contains some important topics of advanced linear algebra not covered in standard books, and hence is also a valuable reference.

According to MathSciNet, "This is an introductory text book in matrix theory intended primarily for students of applied mathematics, engineering, or science. There is very little abstraction (linear transformations are apparently nowhere mentioned) but there are a number of topics not usually covered in elementary textbooks; these include the variational treatment of the eigenvalues of Hermitian matrices, functions of matrices, perturbation theory, bounds for eigenvalues (Gersgorin's theorem is included), stability problems (Lyapunov stability theory, Routh-Hurwitz criterion), and some of the main results concerning matrices with nonnegative elements (including the Perron-Frobenius theorem). More traditional topics contained in the book include the theory of linear equations, eigenvalue theory, and the elementary divisor theory. In the reviewer's opinion the book adequately meets the needs of its intended audience."

R. Bhatia, *Matrix Analysis*, Graduate Texts in Mathematics, 169, Springer-Verlag, New York, 1997. xii+347 pp.

This is a well-written, readable book on advanced matrix analysis with emphasis on matrix inequalities. It is both a valuable introduction and a reference.

According to MathSciNet, "According to the preface, many problems of operator theory present most of their complexities and subtleties in the finite-dimensional case; thus they are susceptible to the methods and results of this well-written advanced text.

The prerequisite assumption of a thorough familiarity with basic linear algebra is reinforced by the brisk pace of the first chapter, which in twenty pages touches upon Hadamard's inequality, the Cholesky decomposition, tensor product, and Schur's theorem. The remaining nine chapters cover enough material for several courses, especially perturbation theory of spectra and matrix inequalities... There is an ample selection of exercises carefully positioned throughout the text. In addition each chapter includes problems of varying difficulty in which themes from the main text are extended."

A. Berman, R. Plemmons, *Nonnegative Matrices in the Mathematical Sciences*, Computer Science and Applied Mathematics. Academic Press [Harcourt Brace Jovanovich, Publishers], New York-London, 1979. xviii+316 pp.

This is a classic book on nonnegative matrices.

According to MathSciNet, "As the title suggests, this book is devoted to nonnegative matrices and how they arise and are useful in the mathematical sciences. It is a particularly fine coverage of this whole area, complete with exercises, vast bibliographical references, and notes at the end of each chapter which supplement the text with historical remarks and connections. As such, this book can be highly recommended for a graduate course in numerical linear algebra. The book is truly outstanding in several regards. First, the treatment of the theory of M-matrices in Chapter 6 is given in great and careful detail and is one highlight of the book; there is simply no better book on the subject currently..."

P. Davis, *Circulant Matrices*, A Wiley-Interscience Publication, Pure and Applied Mathematics, John Wiley & Sons, New York-Chichester-Brisbane, 1979. xv+250 pp.

This is a valuable book on a special class of matrices, circulant matrices.

According to MathSciNet, "Circulant matrices are those in which a basic row of numbers is repeated again and again, but with a shift in position. They have many connections to problems in physics, image processing, probability and statistics, numerical analysis, number theory and geometry. The built-in periodicity means that circulants tie in with discrete Fourier analysis and group theory. As the author states in the preface of his book: 'Writers on matrix theory appear to have given circulants short shrift, so that the basic facts are rediscovered again and again. This book is intended to serve as a general reference on circulants as well as to provide alternate or supplemental material for intermediate courses in matrix theory. The reader will need to be familiar with the geometry of the complex plane and with the elementary portions of matrix theory up through unitary matrices and the diagonalization of Hermitian matrices. In a few places the Jordan form is used.' ...The book is well written; for people who are familiar with matrix theory, it can also be recreational reading."

4.1.3 van der Waerden, Algebra. Vol. I., Algebra. Vol. II., Based in part on lectures by E. Artin and E. Noether, Translated from the fifth German edition by John R. Schulenberge, Springer-Verlag, New York, 1991. xiv+265 pp., xii+284 pp.

The first volume of this classical book was first published in 1930, and the second volume in 1931. It has been revised and reprinted many times. Though this book is a bit out-dated, it is still beneficial to read this classic.

The introduction of this book explains the purpose of this book: "The most recent expansion of algebra far beyond its former bounds is mainly due to the 'abstract', 'formal', or 'axiomatic' school. This school has created a number of novel concepts, revealed hitherto unknown interrelations, and led to far-reaching results, especially in the theories of *fields* and *ideals*, of *groups*, and *hypercomplex numbers*. The chief purpose of this book is to introduce the reader into this whole world of concepts. Within the scope of these modern ideas classical results and methods will find their due place." It is the first book on abstract algebra and this goal has been well-achieved. In fact, it has been so successful that it has changed the world of abstract algebra so that it has now become obsolete. It is like that a grain that dies after new plants have risen from it. When it was first published, it was really modern. Now the new concepts discussed in the book are standard ones in mathematics. But it is still a very good book to read for both knowledge and pleasure.

van der Waerden is now mainly remembered for this book and his work on abstract algebra. Actually he was broad. He also worked on algebraic geometry, topology, number theory, geometry, combinatorics, analysis, probability and statistics, and quantum mechanics. In his later years, he studied the history of mathematics and science and wrote books and papers on them. It may not be well-known that he has a outstanding student Wei-Liang Chow, who is famous for Chow varieties and Chow's theorem on algebraicity of closed analytic subspace of the complex projective space.

From the book description,"This beautiful and eloquent text transformed the graduate teaching of algebra in Europe and the United States. It clearly and succinctly formulated the conceptual and structural insights which Noether had expressed so forcefully and combined it with the elegance and understanding with which Artin had lectured."

According to a review in Bulletin in 1932, "It will immediately occur to every reader of van der Waerden's new book that modern algebra is a subject quite different from the classical algebra built up in its last golden era by Dedekind, Weber, Frobenius, and Kronecker. It is true that a closer study often reveals that the main difference lies in the form of presentation, but it is equally true that in many instances the problems of modern algebra are broader and of a different character.

The new school of abstract algebra has developed into one of the strongest branches of present day mathematics in Germany. Its fundamental principles are closely related to Hilbert's ideas of a formal foundation of mathematics, reducing all theories to an axiomatic basis consisting of relational properties of undefined elements. It is of course nothing new to build up a mathematical theory from its axioms; the main problem of abstract algebra is however the determination of all systems with a given operational basis, i.e. to find the structural properties of all such systems. It is interesting to observe to what a remarkable extent this has been possible. An immediate consequence is an intimate knowledge of the fundamental assumptions of each theory and theorem; but still more important is the abstract identification of many mathematical investigations, also outside of algebra proper, which makes abstract algebra a unifying principle sorely needed in these times of specialization.

One of the central papers in abstract algebra is the well known analysis by Steinitz of the structure of fields. It has been followed by a vast number of investigations on the structure of groups, rings, ideals, hypercomplex systems, etc., works associated with the names of Artin, Krull, E. Noether, and others. For hypercomplex systems the investigations of Wedderburn and Dickson are outstanding.

This new Algebra proposes to be a guide to these investigations, and in many ways it is more than that; van der Waerden has coordinated the various investigations and he has tried as far as possible to consider them from the most general point of view. His book therefore gives considerably more than a summary of the previous theory, and it will certainly keep a prominent place among the books on algebra for many years to come. It is not a textbook in the ordinary sense; and it requires a wide preliminary knowledge on the part of the reader, even though every subject is worked up from its foundations. But I am certain that advanced students and mathematicians interested in algebra will study it with great pleasure."

4.1.4 More books on abstract algebra

After the appearance of the book on abstract algebra by van de Waerden, algebra has developed steadily, and new books are needed for each generation. There have been many books on algebra from different perspectives and at different levels. In this subsection, we list some of the most important and representative ones.

G. Birkhoff, S. Mac Lane, *A Survey of Modern Algebra,* Third edition, The Macmillan Co., New York; Collier-Macmillan Ltd., London, 1965. x+437 pp.

This book has had a huge impact on the education of algebra in USA. Indeed, as the book description say, "This classic, written by two young instructors who became giants in their field, has shaped the understanding of modern algebra for generations of mathematicians and remains a valuable reference and text for self study and college courses."

This was a standard text book for algebra. Its basic philosophy is well explained in its preface: "We have tried throughout to express the conceptional background of the various definitions used. We have done this by illustrating each new term by as many familiar examples as possible. This seems especially important in an elementary text because it serves to emphasize the fact that the abstract concepts arise from the analysis of concrete situations... Modern algebra also enables one to reinterpret the results of classical algebra, giving them far greater unity and generality."

When the first version appeared in 1941, Math Review says: "This is a text on modern algebra that is particularly suited for a first year graduate course or for an advanced undergraduate course. A very striking feature of the book is its broad point of view. There are contacts with many branches of mathematics and so it can serve as an introduction to nearly the whole of modern mathematics. Thus there is a careful development of real numbers, such as Dedekind cuts, and such set-theoretic concepts as

4.1. ABSTRACT ALGEBRAS AND FINITE GROUPS

order, countability and cardinal number are discussed. Throughout the study of matrices and quadratic forms the geometric point of view is emphasized. There is also contact with the field of mathematical logic in the chapter on the algebra of classes and with the ideas of topology in the proof of the fundamental theorem of algebra."

The book above by Birkhoff and Maclane was very successful and also became outdated after a few generations like van der Waerden's book. New concepts such as modules, categories and functors etc were developed. They wrote a more modern book on algebra for undergraduate and beginning graduate students in the framework of these new concepts. Note also that the order of the authors' names is swapped.

S. MacLane, G. Birkhoff, *Algebra*, The Macmillan Co., New York; Collier-Macmillan Ltd., London, 1967. xix+598 pp.

This is a classic and is still a very good introduction to algebra.

According to MathSciNet, "The concrete examples which underlie undergraduate algebra courses–integers, real and complex numbers, polynomials, vectors, matrices, determinants–are neither modern nor abstract. A textbook in which these objects are presented becomes a book on 'modern algebra' by the kind of scaffolding it erects to exhibit these objects. An earlier generation of mathematicians learned, for example in the authors' *A Survey of Modern Algebra*, that these objects are examples of groups, rings, fields, and vector spaces, and saw that the standard constructions in algebra often reduce to finding the right equivalence relation. It is refreshing to see in this book the standard material of undergraduate algebra presented systematically from a new point of view. This time the axiomatic systems and equivalence relations are there, but are governed in turn by the language of categories, functors, universal objects, and dualities... The book is clearly written, beautifully organized, and has an excellent and wide ranging supply of exercises, giving the student plenty of experience with the new ideas from category theory as well as traditional exercises on the concrete examples. The book contains ample material for a full year course on modern algebra at the undergraduate level. One subject sometimes included in undergraduate courses, and not in this book, is Galois theory."

S. Lang, *Algebra*, Addison-Wesley Publishing Co., Inc., Reading, Mass, 1965 xvii+508 pp.

Lang is famous for writing many books and other things. But this book is the longest of his books. The third edition in 2002 is at least twice long as most others of his books. If we want to describe this book in one sentence, this book can be called *Algebra for Working Mathematicians*. It is very broad and covers what one needs no matter if one works in almost every branch of mathematics. His perspective on algebra and the basic philosophy in writing this book is as follows: "As I see it, the graduate course in algebra must primarily prepare students to handle the algebra which they will meet in all of mathematics: topology, partial differential equations, differential geometry, algebraic geometry, analysis, and representation theory, not to speak of algebra itself and algebraic number theory with all its ramifications."

This book is well-known and has been successful and will probably continue to be so both as a textbook and a reference. One may not want to read it from cover to cover, but it is good to skip through to get an idea of some useful parts of the subject of algebra.

According to Zentralblatt MATH, "For the math graduate who wants to broaden his education this is an excellent account; apart from standard topics it picks out many items from other fields: Bernoulli numbers, Fermat's last theorem for polynomials, the Gelfond-Schneider theorem and (as an exercise, with a hint) the Iss'sa-Hironaka theorem. This makes it a fascinating book to read, but despite its length it leaves large parts of algebra untouched."

According to MathSciNet, "The author states that his aim in writing this book was to produce a modern basic text for a first year graduate course in algebra. He has succeeded very well indeed. In spirit and content, his textbook is similar to van der Waerden's classical *Moderne Algebra*... It is, however, much more up-to-date and complete..."

According to Notices of AMS, "Lang's *Algebra* changed the way graduate algebra is taught, retaining classical topics but introducing language and ways of thinking from category theory and homological algebra. It has affected all subsequent graduate-level algebra books."

According to The Mathematical Gazette, "Lang's *Algebra* has gained an iconic status, due both to the comprehensiveness of its coverage and its ability to be authoritative and lively at the same time... a revolutionary work, changing the way in which graduate algebra was taught... the author describes the book as "very stable", indicating that there is little that he has wished to change. This confidence is reflected in the wider mathematical community, and ... this new printing deserves a place in every university departmental library."

N. Jacobson, *Basic Algebra. I., II., Second Edition,* W.H. Freeman and Company, New York, 1985. xviii+499 pp,1989. xviii+686 pp.

This two volume set of books on algebra is a valuable introduction for both undergraduate and graduate students and a useful reference for others. Volume I covers all undergraduate topics in algebra, and volume II covers all of the subjects of first-year graduate algebra. They are written by one of the leading algebraists of the author's generation, who is also well-known for his writing skill and many other books.

According to Zentralblatt MATH, "This is an outstandingly well-written book, very clear and exceptionally well thought out from the point of view of the student... One very good feature is the introduction of each chapter, which gives an excellent background and overview of the material of the chapter. It is also good to give the student at least a small idea of the history of the subject as the author often does in both the text and in his many footnotes (also many references to the important sources)."

According to MathSciNet, "The guiding principles of the author in preparing this volume are forcefully stated as follows: 'Much of the material we present here has a classical flavor. It is hoped that this will foster an appreciation of the great contributions of the past and especially of the mathematics of the nineteenth century. In our treatment we have tried to make use of the most efficient modern tools. This has necessitated the development of a substantial body of foundational material of the sort that has become standard in textbooks on abstract algebra. However, we have tried throughout to bring to the fore well-defined objectives which we believe will prove appealing even to a student with little background in algebra.' ... All of this is given us with an intensification of the author's well-known expository power. This reader was struck especially by the vast number of historical notes that are cleverly inter-woven into the text and are of immense value in sustaining interest. The exercise sets are well constructed to provide some that are easy enough for all students, many which point out mild extensions of the basic theorems and quite a number that will challenge even the best students."

M. Artin, *Algebra*, Prentice Hall, Inc., Englewood Cliffs, NJ, 1991. xviii+618 pp.

This is an unusual book on algebra and covers many important topics which are not covered in a standard book on elementary algebra. For example, in the chapter on Galois theory, it has sections on Kummer extensions and cyclotomic extensions, and the chapter on bilinear forms contains a section on conics and quadric, and the chapter on symmetry contains a section on discrete groups of motions. This book shows many connections of algebras with other subjects, and geometry is one of them. This successful textbook will complement other books on algebra well. It might be helpful to keep in mind that the father of M. Artin wrote a famous nonstandard book on algebra, called *Geometric Algebra*, motivated by geometric applications.

According to MathSciNet, "This is a remarkable text designed for highly motivated undergraduates having some preparation in linear algebra and some other post-calculus mathematics. It is noteworthy for its contents and the style of presentation. In the preface, the author lists three principles that he followed (briefly: examples should motivate definitions, technical points are presented only if needed later in the book, topics should be important for the average mathematician) and takes pains to point out that 'Do it the way you were taught' is not one of them. The style throughout the text is to present basic concepts, give many nontrivial examples and present brief and understandable discussions of advanced material... The student who works through this book will gain knowledge of the topics in standard algebra and learn of the connections with much classical and advanced mathematics. The expert who reads this book will appreciate having in one source many connections between the various mathematical specialties."

I. Herstein, *Topics in Algebra*, Second edition, Xerox College Publishing, Lexington, Mass.-Toronto, Ont., 1975. xi+388 pp.

In this classical book, Herstein set out to write (or rather to experiment with) a book that is "a little beyond that which is usually taught at the junior-senior level" and the content and sophistication is halfway between "two great classics, *A Survey of Modern Algebra* by Birkhoff and MacLane and *Modern Algebra* by Van der Waerden." His experiment succeeded well by the test of time. Both the choices of topics and organization of materials are attractive and helpful to the reader.

According to MathSciNet, "This text covers most of the basic material on abstract algebra, with a more or less classical approach... The book starts with set theory, a smattering of number theory, and then plunges into a short course on finite groups, ... and the four squares theorem (beginning of Waring's problem). One gets the impression that in writing this text, the author was not interested in the content of the theorems alone, but also in the mathematical experience to which the student would be exposed, thus explaining several deliberate instances of inefficient proofs (the most significant was mentioned above: the independent reductions to Jordan and rational canonical form, without deriving either as a corollary of the other)."

J. Fraleigh, *A First Course in Abstract Algebra*, Historical notes by Victor Katz, 6th edition., Addison-Wesley, Reading, MA, 1999, xiii+536 pp.

Abstract algebra is a basic course for upper undergraduate students, and there are many introductory textbooks on it. Many of them, especially more modern ones, are probably simplified versions of the classics mentioned above by including more details but fewer topics. This book is one of the most popular recent textbooks on abstract algebra for undergraduate students. It contains a lot of details and applications to other subjects.

According to Zentralblatt MATH, "The textbook is intended for a first course in abstract algebra, and it covers topics such as Group theory (up to Sylow theorem), free groups and group presentation; Ring theory (incl. UFD and Gaussian integers), Fields and Galois theory. The book is traditionally considered as one of the most popular of the kind. The new edition retains all the features of the previous ones such as detailed proofs, informal style of explanations, carefully selected and organized exercise sets, answers to all odd-numbered exercises. The historical notes contributed by V. Katz add an excellent 'flavour' to the text."

One important branch of abstract algebra is lattice theory.

G. Birkhoff, *Lattice Theory*, Third edition, American Mathematical Society Colloquium Publications, Vol. XXV, American Mathematical Society, Providence, R.I., 1967, vi+418 pp.

Garrett Birkhoff is best known for his work on lattice theory, and this book was the culmination of his work, which had turned lattice theory into a major branch of abstract algebra.

This book is a classic. Since the original publication in 1940, it has been updated several times and is still one of the best books to learn lattice theory.

From the book description, "The purpose of the third edition is threefold: to make the deeper ideas of lattice theory accessible to mathematicians generally, to portray its structure, and to indicate some of its most interesting applications."

When people talk about lattices, they often mean lattices of the Euclidean spaces $\mathbb{R}^n$, and more generally locally compact topological groups, in particular lattices of Lie groups. The geometry of lattices of $\mathbb{R}^n$ is essentially the geometry of numbers. A lattice can also mean something completely different. In abstract algebra, a lattice is a partially ordered set in which any two elements have a unique least upper bound or join, and a unique greatest lower bound or meet. They can be characterized as algebraic structures satisfying certain axiomatic identities. Because of these two characterizations, lattice theory are connected to both order theory and universal algebra. They are closely related to Boolean algebras. Since partially ordered sets occur naturally from many different contexts, lattice theory can be broadly applied. For example, for any set A, the collection of all subsets of A forms a partially ordered set under the inclusion. By taking the union as join, and intersection as meet, we obtain a lattice. This is one of the simplest example of partially ordered set and lattices.

The first edition of the book by Birkhoff is the first comprehensive treatment of lattice theory and its applications. For the second edition, according to MathSciNet, "This expanded and thoroughly reorganized edition contains accounts of many important discoveries of the last twenty years, integrated with a streamlined and modernized treatment of the material of the previous editions. It amply attests to the growth and vitality of the subject during those years. With the size of the volume, the length of the list of unsolved problems continues to grow—166 in this edition. Both suggest that lattice theory will continue to be the subject of increasing research activity."

G. Grätzer, *General Lattice Theory*, Second edition, New appendices by the author with B. A. Davey, R. Freese, B. Ganter, M. Greferath, P. Jipsen, H. A. Priestley, H. Rose, E. T. Schmidt, S. E. Schmidt, F. Wehrung and R. Wille. Birkhauser Verlag, Basel, 1998. xx+663 pp.

This is a very important book in the development of lattice theory. The careful selection of topics and organization of this book has made it a valuable reference to many people. In the preface, Grätzer wrote: "I tried to include what I consider the most important results and research methods of all of lattice theory. To treat the rudimentary results in depth and still keep the size of the volume from getting out of hand, I had to omit a great deal. I excluded many important chapters of lattice theory that have grown into research fields on their own. Ordered algebraic systems and other applications have also been excluded. The reader will find appropriate references to these throughout this book. It is hoped that even those whose main interest lies in areas not treated here in detail will find this volume useful for obtaining the background in lattice theory so necessary in allied fields."

According to MathSciNet, "Since it was first published in 1978, George Grätzer's book General lattice theory has become the lattice theorist's Bible. Now, two decades on, we have the second edition, in which the old testament is augmented by a new testament that is epistolic. There is no need here to expound to the faithful on the old testament *General lattice theory*, 1978, save to say that there is an appendix by the author entitled "Retrospective" in which he discusses the major developments that have taken place since the first edition was published. In particular, for each of its chapters he reports on the remarkable progress achieved in relation to the 193 open problems posed therein. Here we find a myriad of results and references. The new testament gospel is provided by leading and acknowledged experts in their fields..."

H. Schaefer, *Banach Lattices and Positive Operators*, Die Grundlehren der mathematischen Wissenschaften, Band 215, Springer-Verlag, New York-Heidelberg, 1974. xi+376 pp.

An important source of lattices come from vector spaces and Banach spaces. A vector lattice is a real vector space with a partial order on its elements which defines the structure of a lattice and is compactible with the linear structure in a suitable way. If the vector lattice is a Banach space and the partial order is compatible with the Banach norm, then it is called a Banach lattice. The close connection of positive operators with Banach lattices makes both interesting and allows Banach lattices to arise from different contexts. The current book is an important reference on Banach lattices and positive operators.

According to MathSciNet, "The theory of vector lattices... was founded between 1935 and 1940 by F. Riesz, ..., by H. Freudentha... and by L. V. Kantorovic and his Leningrad school from 1935 on. A few years later, between 1940 and 1944, several Japanese mathematicians made important contributions to the subject (H. Nakano, T. Ogasawara, K. Yosida). At about the same time, in the U.S.A., S. Kakutani and H. F. Bohnenblust obtained fundamental results on abstract M-spaces and L_p-spaces and G. Birkhoff's book on general lattice theory (*Lattice Theory*... was published... The present book is the most recent addition to the list. In addition to an extensive account of the general theory, it contains, in particular, many results obtained by the active research group consisting of the author and H. Lotz, R. Nagel, U. Schlotterbeck and M. Wolff.... This is an important book, containing recent and very relevant information and showing, moreover, that vector lattice theory is closely tied up with analysis. The book is not easy to read; it will be difficult to use parts of it as a textbook."

S. Burris, H.P. Sankappanavar, *A Course in Universal Algebra*, Graduate Texts in Mathematics, 78, Springer-Verlag, New York-Berlin, 1981. xvi+276 pp.

Universal algebra studies algebraic structures themselves, not particular examples ("models") of algebraic structures. For example, instead of studying a particular group, universal algebra studies all groups at the same time, i.e., takes "the theory of groups" as one object of study. This explains the meaning of the word *universal* since it can be applied to all groups in this case. This book was written as an introduction after an extensive growth of the theory, and it has been an important textbook.

According to MathSciNet, "Algebra has had a long association with universality. Newton's lectures on algebra were published in 1707 as *Arithmetica Universalis*. However, the current meaning of the expression 'universal algebra' dates from the work of Birkhoff and Ore in the 1930s. The appeal of the subject in its early years was probably due to its universality, but the work of a few dozen people during the past two decades has added a dimension of depth to the breadth that was the original trademark of universal algebra. ... As a graduate textbook, the work is a sure winner. With its clear, leisurely exposition and generous selection of exercises, the book attains its pedagogical objectives stylishly. Moreover, the work will serve well as a research tool, even though it does not resemble an encyclopedia of universal algebra in any way. What this book offers is a rich assortment of significant new results that previously were scattered throughout technical literature...

In his recent review of the second edition of P. M. Cohn's *Universal Algebra*, 1981, W. Taylor pointed out that there are three viable views of universal algebra: it provides a common framework for the traditional topics of algebra; it is an important aspect of model theory; it is a legitimate domain of algebra, on a par with group theory, ring theory and other respected topics. Taylor observed that the existing books on universal algebra have slighted the last of these viewpoints. What about the present authors' work? The reviewer's opinion is that it does full justice to all three views of universal algebra."

A.H. Clifford, G.B. Preston, *The Algebraic Theory of Semigroups*, Vol. I. Mathematical Surveys, No. 7, American Mathematical Society, Providence, R.I., 1961, xv+224 pp.

The subject of semigroups is quite different from other areas of algebra, for example, group theory, and may not be so well-known to the general mathematics community, but is popular among some mathematicians. This is reflected by the high citation numbers of this book. It is also perhaps helpful to point out that sometimes people use semigroups without being too aware of them, for example, in the theory on the heat kernel of Riemannian manifolds.

This is the first English book on semigroups and has played a guiding role to many people who want to learn and work in semigroups.

One can get a glimpse of semigroup theory from the following chapter headings: Chapter 1, Elementary concepts; Chapter 2, Ideals and related concepts; Chapter 3, Representations by matrices over a group with zero; Chapter 4, Decompositions and extensions; Chapter 5, Representations by matrices over a field.

J. Howie, *Fundamentals of Semigroup Theory*, London Mathematical Society Monographs, New Series, 12, Oxford Science Publications, The Clarendon Press, Oxford University Press, New York, 1995. x+351 pp.

Motivated by the growth of the theories of semigroups and applications (in particular, to many parts of computer science), the author set out to write a book "offering both an overview of the subject for specialist and non-specialist alike, and an entrée for the graduate student." This book is a valuable updated introduction, and updates and complements the earlier book. This was intended for graduate students.

According to MathSciNet, review of an earlier edition (or rather book), "The book more than succeeds in this aim. First, it is a top-rate book for its main intended audience, students with some knowledge of algebra and mathematical sophistication, who are approaching semigroups for the first time. The material is made absolutely clear and the style is interesting and informative. Of particular interest and importance are the many discussions of how sections of the theory are related to one another. In places, an air of excitement is generated as the author takes the reader through the possible thought processes of the researcher trying to produce a result, a mere five years earlier. Second, a study of the entire book will provide for the potential researcher a strong background in material not previously available in book form..."

4.1.5 I.R. Shafarevich, Basic Notions of Algebra, Springer, Berlin, 1990; Springer-Verlag, Berlin, 1997. iv+258 pp.

Algebra studies the rules of operations and relations, and the constructions and concepts arising from them, including terms, polynomials, equations and algebraic structures such as groups, fields, rings and algebras. Together with geometry, analysis, topology, and number theory, algebra is one of the main branches of pure mathematics. This book by Shafarevich gives a comprehensive and accessible discussion of all major concepts and topics with brief deep insights.

This is a very enjoyable book. It is not often that one meets such a good book which one can open at any page to read and learn something. It is not difficult to read, and the fast pace motivates the reader to move on. In some sense, it is like a conversation with the author, a very knowledgable mathematician with a global view of mathematics.

From the book description, "Wholeheartedly recommended to every student and user of mathematics, this is an extremely original and highly informative essay on algebra and its place in modern mathematics and science. From the fields studied in every university maths course, through Lie groups to cohomology and category theory, the author shows how the origins of each concept can be related to attempts to model phenomena in physics or in other branches of mathematics. Required reading for mathematicians, from beginners to experts."

According to MathSciNet, "This book is a survey of algebra, emphasizing concepts, ideas, applications, and omitting all but the simplest proofs. The author devotes about 100 pages each to rings (including fields) and groups, and 50 to homological algebra; he has in mind a reader who is fairly ignorant in algebra but well versed in the rest of mathematics including theoretical physics... The author has accomplished the seemingly impossible task of giving a readable survey of a vast area, which should encourage algebra users to delve deeper into the subject."

4.1.6 R. Carter, Simple Groups of Lie Type, Pure and Applied Mathematics, Vol. 28, Reprint of the 1972 original, Wiley Classics Library, A Wiley-Interscience Publication, John Wiley & Sons, Inc., New York, 1989. x+335 pp

Classification of finite simple Lie groups was one of the central problems in mathematics, and its solution was a major achievement in the 20th century mathematics. The fact that many finite groups arise from classical Lie groups was observed as early as 1870 by Jordan and later by Dickson, Dieudonné and Artin, and was put into a uniform framework by Chevalley in 1955 by his fundamental work on Chevalley groups. This book by Cartier gives a systematic discussion of finite simple groups constructed from linear algebraic groups. It is made accessible by starting from basics of the Chevalley groups. Two thirds of this book are devoted to Chevalley groups. Even for people who have no interests in finite groups, this is a valuable book. It has become a standard reference on simple groups of Lie type.

According to MathSciNet, "É. Cartan, in a lecture on the topology of Lie groups, remarked that it is a kind of historical law that the general properties of the simple groups have been verified first in the various groups, and afterward one has sought and found general explanations that do not require the examination of special cases [Enseignement Math. 35 (1936), 177–200]. It would be difficult to trace the history of the classical groups back to the beginning since, for example, the orthogonal group was known, in a sense, to Euler. But one could make a good case for the assertion that Jordan, who defined the linear, orthogonal, and symplectic groups over finite fields, and proved their simplicity, was the first to study these groups from a standpoint akin to modern group theory. It is over a hundred years since Jordan began this work and now, after slow progress through a variety of special cases, we know how to construct and prove the basic properties of simple groups of Lie type over arbitrary fields. This is the subject of the book under review; it gives a good illustration of Cartan's law...

The book reviewed here gives a fine introduction to this central piece of mathematics. It is a clear and interesting account, accessible to a graduate student, involves a mild set of pre-requisites, and is complete enough to serve as a reference for the experts...

The Chevalley groups and their relatives will devour (are devouring) large parts of algebra, combinatorics, number theory, theory of special functions, and topology. This fine book tells us about the culmination of a hundred years of cooperative effort by many mathematicians. Buy it, read it, enjoy it, and work on some of the many problems accessible to its initiates."

A closely related book by Cartier is

R. Carter, *Finite Groups of Lie Type. Conjugacy Classes and Complex Characters*, Pure and Applied Mathematics (New York), A Wiley-Interscience Publication, John Wiley & Sons, Inc., New York, 1985. xii+544 pp.

It is well-known that there is a close relation between characters of representations of groups and conjugacy classes of the groups. For finite groups that arise from simple (or more general reductive algebraic groups), structures of the algebraic groups can be exploited fruitfully to understand representations of the finite groups in the fundamental works of Deligne and Lusztig, and Lusztig. This book is a very valuable reference on this important subject. It is also a good book showing how Tits buildings can be applied to understand representations of finite groups of Lie type.

According to MathSciNet, "In the book under review, the author has attempted two rather ambitious projects. The first, and in this he succeeds admirably, is to give a fairly comprehensive account, with relatively complete proofs, of the status of the complex character theory and conjugacy classes of finite groups of Lie type, as of around 1976. Thus, the present work contains a relatively detailed account of the Deligne-Lusztig theory, ... which has become the cornerstone in the understanding of the characteristic zero representation theory of the group under consideration.

The second project involves a sketchier account, largely without proofs, of some of the developments since 1976... Nevertheless the present book represents a marvelous piece of scholarship and service to the general mathematical community..."

4.1.7 Finite groups

Ever since groups were introduced by Galois for solutions of algebraic equations, they have been intensively studied for their own properties. There are many good books on finite groups.

4.1. ABSTRACT ALGEBRAS AND FINITE GROUPS

W. Burnside, *Theory of Groups of Finite Order*, 2d ed., Dover Publications, Inc., New York, 1955. xxiv+512 pp.

Originally published in 1897, this book was the first treatise on group theory in English. As he wrote in the introduction, "The present treatise is intended to introduce to the reader the main outlines of the theory of groups of finite order apart from any applications. The subject is one which has hitherto attracted but little attention in this country; it will afford me much satisfaction if, by means of this book, I shall succeed in arousing interest among English mathematicians in a branch of pure mathematics which becomes the more fascinating the more it is studied." This book has had a major influence in the development of group theory. It is a classic written by a major founder of theory of finite groups, and is still widely read today.

The second edition published in 1911 and contains a systematic development of the subject including Frobenius's character theory and Burnside's work using these methods. He wrote in the Preface to the second edition: "Very considerable advances in the theory of groups of finite order have been made since the appearance of the first edition of this book. In particular the theory of groups of linear substitutions has been the subject of numerous and important investigations by several writers; and the reason given in the original preface for omitting any account of it no longer holds good. In fact it is now true to say that for further advances in the abstract theory one must look largely to the representation of a group as a group of linear substitutions."

According to a review in Bulletin of AMS in 1900, "As may be inferred from the title, the author has practically confined his attention to the groups of a finite order. The known regions which are bounded by these restrictions are, at the present time, not too extensive to be described in one volume. Happily, our author is so familiar with these regions that he does not confine himself entirely to leading the reader by known roads to the many interesting objective points. He has pointed out many short routes as well as a number of new objects of interest.

The theory of groups of a finite order (under the form of substitution groups) was first developed as a branch of the theory of equations; and the works of Lagrange, Abel, and Galois on the solution of equations have contributed most powerfully towards its early development. In recent years it has been pointed out by Jordan, Klein, Cayley, and others that this subject has extensive geometrical applications. This has led to the study of groups in the abstract; i.e., independent of any particular mode of representation. The work before us aims to treat the subject in this modern spirit..."

H. Zassenhaus, *The Theory of Groups*, second edition, Chelsea Publishing Company, New York, 1958. x+265 pp.

The first edition was published in 1937 and has enjoyed and deserved its reputation as a classic.

According to editor's preface of the first English edition, "[it] has been the modern classic of discrete group theory ever since its first appearance twelve years ago. With this book now in English, it will be possible for the beginning student of groups to view them against the background of modern algebra, and to see how its spirit of generality applies, without stepping outside of his own language or rummaging through the journals to do so..."

The motivation for writing this book was clearly explained by the preface, "Investigations published within the last fifteen years have greatly deepened our knowledge of groups and have given wide scope to group-theoretic methods. As a result, what were isolated and separate insights before, now begin to fit into a unified, if not yet final, pattern. I have set myself the task of making this pattern apparent to the reader, a useful tool for the solution of mathematical and physical problems.

It was a course by E. Artin, given in Hamburg during the Winter Semester of 1933 and the Spring of 1934, which started me on an intense study of group theory. In this course, the problems of the theory of finite groups were transformed into problems of general mathematical interest..."

D. Gorenstein, *Finite Groups*, Second edition, Chelsea Publishing Co., New York, 1980. xvii+519 pp.

Though Gorenstein started out working on commutative algebra, for example, Gorenstein rings are named after him, he switched to work on finite simple groups and became a leading figure among the people who have finally classified finite simple groups. When he wrote this book in 1968, his motivation was clearly explained:"From the 1950's until 1968, the theory of finite groups underwent an intense period of growth, including the first major classification theorems concerning simple groups as well as the construction of the first new sporadic simple group in a hundred years. In writing this book, my aim was to describe that development in sufficient detail for the interested reader to reach the frontiers of the subject and thereby participate in the excitement that then surrounded the study of simple groups."

This book is a very valuable reference written by a major modern contributor. According to MathSciNet, "In recent years a vast amount of work has been done in the theory of finite groups. In particular, progress has been made in studying the structure of finite simple groups. This work has led to the intrinsic characterization of certain classes of finite groups as well as to the discovery of some new simple groups. This book constitutes the first serious attempt to present some of this material in a systematic fashion. As such it is required reading for anyone who wishes to study the subject. The aim of the book is not to present all the known results in the theory (which is virtually impossible because of the bulk of the material), but rather to bring the reader to a level from which he can begin to read the literature on finite simple groups."

4.1. ABSTRACT ALGEBRAS AND FINITE GROUPS

M. Hall, *The Theory of Groups*, Reprinting of the 1968 edition, Chelsea Publishing Co., New York, 1976. xiii+434 pp.

This is a classical textbook on groups. The initial goal of writing this book was explained in preface:"The present volume is intended to serve a dual purpose. The first ten chapters are meant to be the basis for a course in group theory, and exercises have been included at the end of each of these chapters. The last ten chapters are meant to be useful as optional material in a course or as reference material. When used as a text, the book is intended for students who have had an introductory course in modern algebra comparable to a course taught from Birkhoff and Mac Lane's A Survey of Modern Algebra. I have tried to make this book as self-contained as possible, but where background material is needed references have been given, chiefly to Birkhoff and Mac Lane." This goal has been well achieved.

Indeed, from the book description, "Perhaps the first truly famous book devoted primarily to finite groups was Burnside's book. From the time of its second edition in 1911 until the appearance of Hall's book, there were few books of similar stature. Hall's book is still considered to be a classic source for fundamental results on the representation theory for finite groups, the Burnside problem, extensions and cohomology of groups, p-groups and much more. For the student who has already had an introduction to group theory, there is much treasure to be found in Hall's Theory of Groups."

A.G. Kurosh, *The Theory of Groups*, Translated from the Russian and edited by K. A. Hirsch, second English edition, volumes Chelsea Publishing Co., New York, 1960. Vol. 1: 272 pp., Vol. 2: 308 pp.

Originally published in 1944, this is the first modern and high-level text on group theory. It is still a valuable reference. Unlike many other books, it contains substantial discussion of infinite groups, in particular nilpotent and solvable groups. The author explained the reason for not restricting to only finite groups: "The golden age of the theory of finite groups came at the end of the last century and the first decade of the present... But later it became clear that the finiteness of a group is a restriction that is too strong and not always natural. It was of particular importance that this restriction very soon led to conflicts with the need of neighboring branches of mathematics: in several parts of geometry, the theory of automorphic forms, topology, in all of these one again and again came across algebraic formulations similar to groups, but infinite, and so demands were made in the theory of groups that the theory of finite groups was not in a position to satisfy."

According to an Amazon review, "Kurosh was a remarkably clear thinker and writer, and K. A. Hirsch did an excellent job of translation, so this book is very readable for anyone who is already familiar with the rudiments of the subject. It's worth careful study; not many mathematics texts give so much insight into the thinking of a master of the topic. And, algebra in general as well as group theory in particular, is more about insight than about particular results, so reading this book can help a current student of mathematics to grow in professional skill."

More modern books on group theories include the next two.

D. Robinson, *A Course in the Theory of Groups*, Second edition, Graduate Texts in Mathematics, 80, Springer-Verlag, New York, 1996. xviii+499 pp.

This is comprehensive modern introduction to group theory including both finite and infinite groups. It is useful both as a textbook and reference.

From the preface: "This book is intended as an introduction to the general theory of groups. Its aim is to make the reader aware of some of the main accomplishments of group theory while at the same time providing a reasonable coverage of basic material. The book is addressed primarily to the student who wishes to learn the subject, but it is hoped that it will also prove useful to specialists in other areas as a work of reference. An attempt has been made to strike a balance between the different branches of group theory, abelian groups, finite groups, infinite groups, and to stress the unity of the subject. In choice of material I have been guided by its inherent interest, accessibility, and connections with other topics. No book of this type can be comprehensive, but I hope it will serve as an introduction to the several excellent research level texts now in print."

J. Dixon, B. Mortimer, *Permutation Groups*, Graduate Texts in Mathematics, 163, Springer-Verlag, New York, 1996. xii+346 pp.

Groups were first introduced as permutation groups. Until the early 1980, most study on permutation groups was restricted to the finite case. After the classification of finite simple groups, the subject was almost killed. By that time the study of infinite permutation groups picked up, and this book is a standard reference on the new results on finite permutation groups and infinite permutation groups. It is well-written and accessible and can serve as a valuable introduction.

B. Huppert, *Endliche Gruppen. I.*, Die Grundlehren der Mathematischen Wissenschaften, Band 134, Springer-Verlag, Berlin-New York, 1967, xii+793 pp.

B. Huppert, N. Blackburn, *Finite Groups. II. III*, Grundlehren der Mathematischen Wissenschaften, 242, 243. Springer-Verlag, Berlin-New York, 1982. xiii+531 pp., ix+454 pp.

This is a very valuable systematic and reliable reference on finite groups. Due to the rapid development and the amount of work on finite simple groups, this book can not cover all important topics.

From the preface to Volume II: "At the time of the appearance of the first volume, the tempestuous development of finite group theory had already made it virtually impossible to give a complete presentation of the subject in one treatise. The present volume and its successor have therefore the more modest aim of giving descriptions of the recent development of certain important parts of the subject, and even in these parts no attempt at completeness has been made."

According to MathSciNet, "The aim of this impressive book [Vol I] is to present in an organized manner a large part of the theory of finite groups. It should be regarded as a serious and valuable review of the subject as it is known today.

The author began his task in 1958, which is about the same time the present surge in activity in the field began. Thus his goal of a degree of comprehensiveness has become difficult to achieve; the author has met this problem by carefully selecting the material he presents, and the work is projected into two volumes, of which this is the first... The topics with which the author deals are ones with which he is thoroughly familiar, and every reader can gain from his insights. The book is developed with a strong emphasis on the means of working with particular examples.

Counter-examples of possible generalizations of theorems are also presented. A number of well chosen problems appear throughout the book. At the end of each chapter is a survey of the literature. These surveys are brief but useful."

According to a review in Bulletin of AMS, "The Huppert-Blackburn volumes consist of the continuation of the introductory first volume and are devoted to more advanced topics. The very high standards set in the first volume are maintained; these books are not only very careful and detailed, they are a pleasure to read. It is a real loss for the subject that there will be no further volumes in this series."

D. Gorenstein, R. Lyons, R. Solomon, *The Classification of the Finite Simple Groups*, Mathematical Surveys and Monographs, 40.1. American Mathematical Society, Providence, RI, 1994. xiv+165 pp.

A major achievement of mathematics in the 20th century is the completion of **classification of finite simple groups**. This is a huge project and involves many people and papers. A second generation proof is being prepared in a series of books. Many volumes are expected and the first volume is this book. As explained in the preface, "Though elated at the successful completion of the classification of finite simple groups in the early 1980s, Danny Gorestein nevertheless immediately appreciated the urgency of a 'revision' project, to which he turned without delay."

This is a valuable reference for many obvious reasons.

According to MathSciNet, "The Odd Order paper stunned mathematicians when it appeared over thirty years ago: it proved a single theorem and took over two hundred and fifty pages! Instead of being an anomaly it was a trendsetter as very long proofs became common in mathematics. A new kind of proof had been discovered. Moreover, this new phenomenon was to be followed by another, the compilation of dozens of such works in the twenty-year saga of the classification of finite simple groups, yet another new level of complexity of proofs. There were a number of good reasons to have confidence in this achievement, despite its enormity. Usually any simple group was classified a number of times in the overlapping theorems that made up the classification; the arguments were robust, in that errors were easily fixed and the failure of one approach simply gave more information that could be used in several other methods; the work on large simple groups became much more unified and clear-cut, much in the way higher-dimensional geometry can be compared to the low-dimensional cases; programs to find sporadic groups, which were independent of the classification and looked promising, led to only finitely many groups and no new programs. However, mathematics, for several millennia, has had a unique and demanding norm for lasting acceptance, a Euclidean Standard: to be of lasting value, mathematics must be readable, correct and alive. A body of discoveries must be accessible to readers, meticulous and of continuing interest. No amount of untransmitted insight, confidence or intuition can be a substitute; the passage of time enforces this over any force of personalities or general enthusiasms. Moreover, the burden is completely on the people who claim the results and not at all on the readers.

This volume is the first, of a number to follow, which mean to meet these standards for the classification, and, in so doing, set an example for all of mathematics. This new, second-generation proof is a great improvement while still being mostly based on the methods developed for the first proof. The exact background and assumed results (like the Odd Order result) are carefully spelled out and referenced. Each simple group is reached exactly once in the classification. Full use of the power of induction is made, giving more savings in length and simplified arguments. The entire organization of the proof is given in great detail. The wait for this project to reach this stage has been worth it.

Besides discussing these issues in detail, this volume contains a survey of the simple groups, a discussion of local and global methods and how they relate and a section introducing the basic ideas centering on the concepts of components. The bulk of this volume is devoted to a review of the whole series of books...."

After the **classification of finite simple groups** was finished, a very valuable book summarizing these groups, their structures and characters is the following unusual book, though this is not a book for people to read from cover to cover.

J.H. Conway, R.T. Curtis, S.P. Norton, R.A. Parker, R.A. Wilson, *Atlas of Finite Groups. Maximal Subgroups and Ordinary Characters for Simple Groups*, With computational assistance from J. G. Thackray, Oxford University Press, Eynsham, 1985. xxxiv+252 pp.

This book is an essential reference for experts and also a fun book for nonexperts to get some impression of finite simple groups. It has a very good introduction and contains many tables of characters of finite simple groups.

According to Zentralblatt MATH, "The ATLAS gives a kind of information, which in published papers on finite simple groups is hard to get, but which is so useful in a concrete situation. The ATLAS should provide for a large readership of mathematicians the quickest and most efficient way to get that kind of information. It also might be just fun (like for the reviewer) to run at random through the ATLAS and find new surprising details about groups which seemed familiar. In conclusion: This ATLAS will be of greatest importance for any mathematician who faces concrete problems in finite simple groups."

According to MathSciNet, "At last, an official collection of character tables and related information about many finite simple groups has appeared in book form. This information is important to specialists in finite group theory and the volume contains neatly presented instructional material which the nonspecialists can appreciate... The efforts of the last 25 years to classify finite simple groups created a greater need to have numerical and combinatorial information about the known groups... In sum, the five authors have collected some of this early and unpublished work, then greatly extended it and put it in a form suitable for easy modern applications..."

4.1.8 Finite groups and their representation theories

One important way to understand abstract groups is to study their representations, i.e., to understand them in terms of linear transformations or matrices. This is also a major reason why groups are useful. Though representation theories of groups enclose many different aspects, they started with representations of finite groups.

W. Feit, *The Representation Theory of Finite Groups*, North-Holland Mathematical Library, 25, North-Holland Publishing Co., Amsterdam-New York, 1982. xiv+502 pp.

Modular representation theory deals with linear representations of finite groups over a field of positive characteristic, in particular, when the characteristic divides the order of the group. It arises naturally from various contexts and has played an important role in the classification of finite simple groups. As explained in the preface, this book "is meant to give a picture of the general theory of modular representations as it exists at present." This has been an important reference since then. Feit is probably most famous for a long joint paper with G. Thompson which proved that all finite groups of odd order are solvable.

According to MathSciNet, "This book is the first comprehensive treatment of the general theory of modular representations of finite groups. Since the subject has been under steady development for nearly half a century the volume is very welcome, especially as the field is now extremely active... This book is a very important contribution."

According to a review in Bulletin of AMS, "The general modular theory of representations of finite groups ... began with the work of Richard Brauer in the 1940s. Its goals were expressed in Brauer's paper of 1944, On the arithmetic in a group ring: "We are far from knowing all important properties of group characters. In particular, we are interested in further results which connect the group characters directly with properties of the abstract group G. Any result of this kind means, in the last analysis, a result concerning the structure of the general group of finite order. One approach to our question is to study arithmetic properties ... " That motivation still holds today..."

I.M. Isaacs, *Character Theory of Finite Groups*, Corrected reprint of the 1976 original [Academic Press, New York], AMS Chelsea Publishing, Providence, RI, 2006. xii+310 pp.

Character theory is important in studying finite groups, for example, some important results such as Frobenius' theorem have no known proofs without using character theory. It is also an interesting branch of algebra in itself. This book provides a comprehensive description of major aspects of this theory.

According to MathSciNet, "This book introduces the reader to the theory of characters of finite groups, and develops a wide variety of applications (both classical and modern) to group structure theory. The book is designed to be suitable for use as a graduate-level text, and it is largely from this point of view that the reviewer will discuss it...."

4.1.9 Representation theories and associate algebras

An associative algebra is an associative ring that has a compatible structure of a vector space over a certain field or, more generally, of a module over a commutative ring. It arises naturally from representations of groups, in particular finite groups. In this case, the ring is the group ring.

C. Curtis, I. Reiner, *Representation Theory of Finite Groups and Associative Algebras*, Pure and Applied Mathematics, Vol. XI Interscience Publishers, a division of John Wiley & Sons, New York-London, 1962, xiv+685 pp.

Representations of finite groups are important for understanding and classifying finite groups. As explained in the preface, the scope of representation theories had been increasing, and this book was written to meet such a need: "During the past decade there has been increased emphasis on integral representations of groups and rings, motivated to some extent by questions arising from homological algebra. This theory of integral representations has been a fruitful source of problems and conjectures both in homological algebra and in the arithmetic of non-commutative rings.

4.1. ABSTRACT ALGEBRAS AND FINITE GROUPS

The purpose of this book is to give, in as self-contained a manner as possible, an up-to-date account of the representation theory of finite groups and associative rings and algebras." This is a classic and has been influential and introduced many researchers to representation theory. It is still a comprehensive introduction to representation theory of finite groups and associative algebras, and a very valuable reference.

According to MathSciNet, "The appearance of this masterful book is very timely, especially in view of the recent renewal of activity in the representation theory of finite groups. It provides a remarkably complete introduction to the subject and contains for the first time a large amount of material that has been available hitherto only in the original memoirs. It undoubtedly will be the canonical reference work on its subject for a long time, and should do much to revive interest in non-commutative algebras... In brief, the authors have written a book worthy of the subject and have rendered an important service to the mathematical community."

C. Curtis, I. Reiner, *Methods of Representation Theory. Vol. I., Vol. II., With Applications to Finite Groups and Orders*, Pure and Applied Mathematics. A Wiley-Interscience Publication, John Wiley & Sons, Inc., New York, 1981. xxi+819 pp., 1987. xviii+951 pp.

Though the previous book by the two authors had been a standard references for 20 years since its publication, it became outdated due to rapid development of the subject. These two books can be considered as a sequel to the previous book. They have also been influential in the further development of the subject.

The first volume gives a comprehensive survey of the established foundations of the representation theory of finite groups and of orders, and the second volume discusses a series of more recent topics. They are valuable references.

According to MathSciNet, "This is the first volume of a two-volume treatise on representation theory of groups and orders. After a long introduction with background material the book divides into two separate but related parts: two chapters on ordinary and modular representation theory of finite groups; two chapters on integral representations..." "This second volume, comprising Chapters 5 through 11, concludes a massive treatise on the representation theory of finite groups and related orders and algebras."

R. Pierce, *Associative Algebras*, Graduate Texts in Mathematics, 88. Studies in the History of Modern Science, 9, Springer-Verlag, New York-Berlin, 1982. xii+436 pp.

Probably the most basic and familiar concept of the abstract algebra is the notion of group. Since groups were introduced to study polynomial equations, it also connected well with the ancient algebra. But algebra is more than groups and the Galois theory. Indeed, as the author explained in the preface, "For many people there is life after 40;

for some mathematicians there is algebra after Galois theory. The objective of this book is to prove the latter thesis. It is written primarily for students who have assimilated substantial portions of a standard first year graduate algebra textbook, and who have enjoyed the experience." It is valuable both as a textbook and an accessible reference.

According to MathSciNet, "This book is a welcome addition to the literature. It brings together a wide spectrum of algebra. Any good fourth year honors course in algebra ought to form an adequate background.

The first half of the book covers the material standard to any beginning course in associative algebras, such as the Wedderburn structure theorem for semisimple algebras, discussions of the various chain conditions on modules, the Jacobson radical, Artinian algebras, the Krull-Shmidt theorem, the structure of projective modules over Artinian rings, and Wedderburn's principal theorem,... Chapter 12 begins the study of the Brauer group of a field... The book is very well written and has many exercises which enhance the text by either providing examples, giving a different proof of a theorem in the text, or generalizing a theorem in the text. The text also has many illustrative examples."

4.1.10 Rings and modules

Modules over rings are a generalization of the notion of vector spaces over fields. They also generalize the notion of abelian groups, which are modules over the ring of integers. Specifically, a module is an additive abelian group, and a product is defined between elements of the ring and elements of the module, and this multiplication is associative (when used with the multiplication in the ring) and distributive. Modules are very closely related to the representation theory of groups. They are also one of the central notions of commutative algebra and homological algebra, and are used extensively in representation theory, commutative algebra and algebraic topology.

F. Anderson, K. Fuller, *Rings and Categories of Modules*, Second edition, Graduate Texts in Mathematics, 13, Springer-Verlag, New York, 1992. x+376 pp.

This book was written as a self-contained textbook on a major portion of the theory of rings and modules for graduate students. It has achieved this purpose well and has also become one of the standard references on the approach to ring theory via the theory of modules and their categories.

The basic theme is to understand connection between one-sided ideals of rings and modules of the rings.

A Dedekind domain R is an integral domain in which every ideal is uniquely decomposed into a product of prime ideals. Given a Dedekind domain, let F be its quotient field. Let A be an algebra over F. A R-order of A is a R-submodule of A that is finitely generated, spans A over F, and contains R. A maximal order is an order that is not contained in any other order. An important example of order is as follows: F is a number field and R is its ring of integers.

Maximal orders are basic objects in the integral theory of semi-simple algebras and number theory.

I. Reiner, *Maximal Orders*, Corrected reprint of the 1975 original, With a foreword by M. J. Taylor, London Mathematical Society Monographs, New Series, 28, The Clarendon Press, Oxford University Press, Oxford, 2003. xiv+395 pp.

This has been a standard textbook and also a reference on maximal orders. As explained by the author, "The theory of maximal orders originated in the work of Dedekind, who studied the factorization properties of ideals of R, where R is a ring of algebraic integers in an algebraic number field K. As shown by Dedekind, the factorization is especially simple in the extreme case where R is the ring of all algebraic integers in K. This ring is in fact the unique maximal $\mathbb{Z}$-order in K. In this book we shall consider the generalization of Dedekind's ideal theory to the case of maximal R-modules in separable K-algebras."

According to MathSciNet, "This book is about the classical theory of maximal orders over a Dedekind ring in a separable algebra. The presentation and methods of proof are essentially the classical ones in a modernized version. The book has developed from a series of lectures for graduate students, and the author's intention has been to make the book—and the subject—easily accessible to a large variety of readers... The text is very well written and complete; all proofs are given in full detail. There are well-chosen exercises at the end of each chapter.

The book certainly fills a gap in the mathematical literature, since no modern textbook on maximal orders has been available."

A special class of rings consists of **group rings**. Given any ring R and any group Γ, there is a group ring $R\Gamma$. They appear naturally in group representations and topology. For example, representations of Γ correspond to modules of $R\Gamma$. In geometric topology, an important example of group ring is obtained by taking $R = \mathbb{Z}$ and Γ the fundamental group of a manifold, and $\mathbb{Z}\Gamma$ appears naturally in surgery theory of manifolds. Group rings are also interesting structures in their own right.

D. Passman, *The Algebraic Structure of Group Rings*, Pure and Applied Mathematics, Wiley-Interscience, New York-London-Sydney, 1977. xiv+720 pp.

Group rings of finite groups occur naturally in representations of finite groups, since representations of the groups can be identified with modules of the rings. Group rings of finite groups are markedly different from group rings of infinite groups, since the former case can be studied by trace functions and the strong structure theorems of finite dimensional algebras, and these methods are not available for the latter case. This book give the first coherent and reasonably complete account of infinite group rings which is also accessible to graduate students.

According to MathSciNet, "The group ring of a group G over a commutative ring K with identity is the ring $K[G]$ of all finitely-supported functions $G \to K$ with pointwise addition and convolution as multiplication of these functions. Convolution-type product operations are playing an increasing role in pure and applied mathematics: e.g., due to G.-C. Rota (1964) incidence algebras of posets and Möbius inversion are now main tools in combinatorial enumeration theory. So the theory of group rings can serve as an example of an advanced algebraic theory of 'linearization'. But first of all, group rings are important in their own right, as one of the few algebraic structures with a distinct analytic flavor. Also, they are a meeting place of ring theory questions with techniques from group theory, number theory and representation theory. The interplay between these disciplines for finite G is classic. But the last 30 years have demonstrated that it is also attractive and fruitful in the case of infinite groups G, where the old methods are no longer available.

The algebraists are fortunate to have now the first reasonably complete and up-to-date exposition of the theory of infinite group rings, written by an expert who has made valuable or decisive contributions to all aspects of the theory treated in this book..."

4.2 Commutative algebras

Figure 4.2: Abel's books.

Abel had a short life, but many things in mathematics are named after him. Probably the most basic definition related to Abel is that of abelian group, and the name came from his work on solutions of a polynomial equation by radicals. The adjective "abelian" in other objects such as abelian varieties, abelian categories also implies commutativity. Among mathematical adjectives derived from the name of a mathematician, the word "abelian" is rare in that it is often spelled with a lowercase a, rather than an uppercase A. This shows how ubiquitous the concept is in modern mathematics and should be considered as a great honor to Abel. (But this can be interpreted differently. One story says that when Weil made a comment on the fame in mathematics, he said that even the great Abel was often spelled with a lower case a).

The subject of commutative algebras studies commutative rings, their ideals, and modules over such rings. Important examples of commutative rings include polynomial rings, rings

4.2. COMMUTATIVE ALGEBRAS

of algebraic integers. It has played an essential role in algebraic geometry and algebraic number theory. In fact, it provides foundational tools in a similar way as differential analysis provides the basic tools for differential geometry and Riemannian geometry.

There are several good books on commutative algebras with different emphasis, which are suitable for people at different levels.

M. Atiyah, I. MacDonald, *Introduction to Commutative Algebra*, Addison-Wesley Publishing Co., Reading, Mass.-London-Don Mills, Ont. 1969. ix+128 pp.

This book was based on a course for undergraduate students at Oxford University with the goal of providing a rapid introduction to the theory of commutative algebra. This accessible book is also an excellent introduction for graduate students and others. Topics are well-chosen and arranged. Though it is a short book, it probably contains enough material for a general working mathematician who does not specialize in algebraic geometry or number theory.

From the introduction, "It is designed to be read by students who have had a first elementary course in general algebra. On the other hand, it is not intended as a substitute for the more voluminous tracts on commutative algebra such as those of Zariski-Samuel or Bourbaki. We have concentrated on certain central topics, and large areas, such as field theory, are not touched... we put more emphasis on modules and localization... The lecture-note origin of this book accounts for the rather terse style, with little general padding, and for the condensed account of many proofs. We have resisted the temptation to expand it in the hope that the brevity of our presentation will make clearer the mathematical structure of what is by now an elegant and attractive theory. Our philosophy has been to build up to the main theorems in a succession of simple steps and to omit routine verifications."

O. Zariski, P. Samuel, *Commutative Algebra.* Vol. I, Vol. II. Reprint of the 1960 edition, Graduate Texts in Mathematics, Graduate Texts in Mathematics, No. 28, Vol. 29, Springer-Verlag, New York-Heidelberg, 1975. xi+329 pp, 13-02 x+414 pp.

There are inseparable relations between commutative algebra and algebraic geometry. These two classical books give a systematic exposition on commutative algebra, and Zariski is one of the most influential algebraic geometers of the twentieth century. Though the original motivation was to provide the necessary background for Zariski's unwritten colloquium book on abstract algebraic geometry, they were written to include also results in commutative algebra beyond applications in algebraic geometry, i.e., the end products are independent books on commutative algebra. They have played an important role in the development of commutative algebra. They are still valuable references today.

According to MathSciNet, "Before the appearance of the present work, the only systematic account of commutative algebra was to be found in Krull's 'Idealtheorie'

[Springer, Berlin, 1935]... Motivation and examples are provided. When it seems appropriate, several proofs of a theorem are given. The authors have almost always resisted the temptation to quote 'well-known' facts without reference."

"It would be an under-statement to say that the second volume of this work lives up to the standards and expectations set by the first, because the scope and style of the final volume could not have been anticipated even though the first volume contained an outline of the authors' total program and ample evidence of their expository skill. Unlike the first, however, the second volume is concerned in large measure with those parts of commutative algebra that are the fruits of its union with algebraic geometry; and this fact has had a marked influence on the organization of the material and the character of the exposition. Throughout the work the algebro-geometric motivations and applications of the purely algebraic material are elaborated in such detail as to make the book (among other things) an excellent text for a course in the arithmetic foundations of algebraic geometry. This fact alone renders the book a welcome addition to the all too sparse list of expository treatises in this area."

N. Bourbaki, *Commutative Algebra. Chapters 1–7*, Translated from the French, Reprint of the 1972 edition, Elements of Mathematics (Berlin), Springer-Verlag, Berlin, 1989. xxiv+625 pp.

Bourbaki is famous for both the Bourbaki seminar and many books which they have produced. Not every Bourbaki book is good, but this is one of the good ones they produced. It is not an introduction to commutative algebra, but rather a comprehensive and valuable reference.

According to Zentralblatt MATH, "Ever since its first appearance, this volume has been considered one of the most complete, systematic and elegant reference books on modern commutative algebra. Without any doubt, this prominent role of Bourbaki's 'Commutative algebra' is wholly persisting, up to these days."

J.P. Serre, *Local Algebra*, Translated from the French by CheeWhye Chin and revised by the author, Springer Monographs in Mathematics, Springer-Verlag, Berlin, 2000. xiv+128 pp.

Like all good books by Serre, this one is precisely, clearly and elegantly written with sharp focus. Before being translated into English, it had already become a classic.

The original motivation of this relative short book on commutative algebra was explained in the preface of the English translation, "The original text was based on a set of lectures, given at the College de France in 1957–1958, and written up by Pierre Gabriel. Its aim was to give a short account of Commutative Algebra, with emphasis on the following topics:

a) *Modules* (as opposed to *Rings*, which were thought to be the only subject of Commutative Algebra, before the emergence of sheaf theory in the 1950s);

b) *Homological methods*, á la Cartan-Eilenberg;

c) Intersection multiplicities, viewed as Euler-Poincaré characteristics."

D. Eisenbud, *Commutative Algebra with a View Toward Algebraic Geometry*, Graduate Texts in Mathematics, 150. Springer-Verlag, New York, 1995. xvi+785 pp.

One important motivation of commutative algebra comes from algebraic geometry, and these two subjects have inseparable connections, but a gap has grown between them. This book tries to fill in this gap and is an excellent introduction and also a valuable comprehensive reference on commutative algebra with a view towards applications in algebraic geometry.

According to Zentralblatt MATH, "At present, commutative algebra is an independent, abstract, deep, and smoothly polished subject for its own sake, on the one hand, and an indispensable conceptual, methodical, and technical resource for modern algebraic and complex analytic geometry, on the other hand. Studying or actively pursuing research in geometry or number theory requires today a profound knowledge of commutative algebra; however, most textbooks on algebraic or complex analytic geometry usually assume such a knowledge from the beginning, often refer to the (undoubtedly excellent) great standard texts on abstract commutative algebra, or survey a minimal account of the basic results, mostly in a series of appendices... Conversely, most textbooks on commutative algebra present the material in its purely algebraic and perfectly polished abstract form, with at most a few elementary hints and applications to the related geometry. This situation really creates a dilemma for both students and teachers of algebraic geometry or commutative algebra."

Among many books on commutative algebra, this book is special in that it tries to a build a bridge over the gap. This is reflected in the subtitle of the book. As the author explained, "It has seemed to me for a long time that commutative algebra is best practiced with knowledge of the geometric ideas that played a great role in its formation: in short, with a view towards algebraic geometry."

According to MathSciNet, "this text has a distinctively different flavor than existing texts, both in coverage and style. Motivation and intuitive explanations appear throughout, there are many worked examples, and both text and problem sets lead up to contemporary research."

H. Matsumura, *Commutative Algebra*, Second edition, Mathematics Lecture Note Series, 56, Benjamin/Cummings Publishing Co., Inc., Reading, Mass., 1980. xv+313 pp.

This has been a standard textbook on commutative algebra and has been especially popular with students, and a second edition was prepared 9 years after the first edition.

According to MathSciNet, "This book goes rather far into the up-to-date theory of commutative rings. The author assumes that only the elements of algebra and homology are known, but I feel that a previous knowledge of elementary commutative algebra itself (e.g., part of the contents of M. F. Atiyah and I. G. Macdonald's book *Introduction to Commutative Algebra*, 1969, would be useful to many readers. For example, Chapter I deals in 16 pages with the Zariski topology, radicals, localization, globalization, Nakayama's lemma and the comparison between noetherian and Artinian rings. With this proviso, this book is very readable. "

"This revised edition of the author's successful monograph on commutative algebra is a welcome addition to the mathematical bookshelf. The first edition has served as a standard reference for several years, since it exposed commutative algebra as the subject is currently practiced. This revision serves the same function..."

H. Matsumura, *Commutative Ring Theory*, Translated from the Japanese by M. Reid, Cambridge Studies in Advanced Mathematics, 8, Cambridge University Press, Cambridge, 1986. xiv+320 pp.

This book of Matsumura is a more accessible and also updated version of his earlier book. It was written for graduate students and covers all the topics which are considered standard today. It is also a valuable reference.

According to MathSciNet, "This book is a welcome addition to a small but élite coterie of textbooks in commutative algebra: I. Kaplansky's *Commutative rings*, M. F. Atiyah and I. G. Macdonald's *Introduction to Commutative Algebra*, E. Kunz's *Introduction to Commutative Algebra and Algebraic Geometry*, and the author's own *Commutative Algebra*. For a field in a zestful state of development, these volumes, however distinctive, are just too few, particularly as they have different goals and with coverages that are not highly overlapping.

In at least two respects this effort by the author works better than his earlier book. Firstly, it gives a more systematic and smoother exposition of the basic notions and constructs of the field. It starts at a more elementary level and provides a fuller view of the subject; it also takes it further. This alone makes it an excellent basic introductory textbook. Secondly, its advanced topics focus on a large measure on Cohen-Macaulay phenomena. These are essential for understanding what has been the core of recent (and some not quite so recent) developments in commutative algebra. It has become too unwieldy for a student, or a worker in a related field, to have to learn the basic techniques of the area from its original sources.

Its other appealing features include: plenty of exercises, from the easy to the very hard; guides to the literature; a rich bibliographical listing that should be helpful to researchers; and an index that makes locating topics quite easy..."

M. Nagata, *Local Rings*, Interscience Tracts in Pure and Applied Mathematics, No. 13, Interscience Publishers, a division of John Wiley & Sons, New York-London, 1962, xiii+234 pp.

The theory of local ring is an important part of the theory of commutative algebra (or commutative ring) and has played an important role in both algebraic geometry and commutative algebra. For example, localization of a commutative ring at any prime ideal gives a local ring, and it allows one to study local properties of algebraic varieties. Nagata set out to write a book that "gives ... a systematic exposition of an up-to-date theory of local rings" and "to develop further the theory of local rings". This is a valuable reference.

According to MathSciNet, "The early part of the book contains basic results of commutative algebra, often with new proofs. The latter part of the book treats the advanced theory of local rings, reuniting most of the results of the author's many research papers, and containing some new results... On the whole, statements in the book are clear, and proofs easy to follow. However, there are exceptions..."

W. Bruns, J. Herzog, *Cohen-Macaulay Rings*, Cambridge Studies in Advanced Mathematics, 39, Cambridge University Press, Cambridge, 1993. xii+403 pp.

A local Cohen-Macaulay ring is a commutative noetherian local ring whose Krull dimension is equal to its depth. The depth is always bounded above by the Krull dimension, and the equality forces some regularity conditions on the ring, which allow some powerful theorems to be proved under this general assumption. A ring is a Cohen-Macaulay ring if its localization at every prime ideal is a local Cohen-Macaulay ring. The preface explains the purpose of this book: "The notion of a Cohen-Macaulay ring marks the cross-roads of two powerful lines of research in present-day commutative algebra. While its main development belongs to the homological theory of commutative rings, it finds surprising applications in the realm of algebraic combinatorics. Consequently this book is an introduction to the homological and combinatorial aspects of commutative algebra." Indeed, this is a very good introductory book and is also a valuable reference.

According to MathSciNet, "For years commutative algebra has been a field in which introductory texts are plentiful, and advanced texts rare, even though in many respects the research is mature and the trails to the frontier are well marked... The authors have chosen not to be encyclopedic, but rather to concentrate on an efficient presentation of fundamentals. There is a clear point of view: Cohen-Macaulay rings come up absolutely everywhere ... and the techniques for studying them are worthwhile since they do extend to more general settings...

This book is a genuine text, not a specialist's monograph... With its combination of various threads of the field, this book should be a valuable addition to the collection of anyone, beginner or expert, who is interested in commutative algebra."

A related, more computational book on commutative algebra is

D. Cox, J. Little, D. O'Shea, *Ideals, Varieties, and Algorithms. An Introduction to Computational Algebraic Geometry and Commutative Algebra*, Undergraduate Texts in Mathematics, Springer-Verlag, New York, 1992. xii+513 pp.

This elementary book on commutative algebra and algebraic geometry is quite different from the books above. It emphasizes algorithmic aspect of commutative algebra and algebraic geometry. Though this book was written as a textbook for undergraduate students, it is also a useful introduction for others.

According to MathSciNet, "Since the 1960s, a new field of mathematics has been steadily growing in significance. Broadly describable as computational commutative algebra, this field grew out of the Gröbner basis algorithm of Buchberger for computing in the ring of multivariate polynomials over a field modulo an ideal, coupled with the rapidly advancing level of computing power available for such computations. There have been some notable successes, such as the automatic proof using polynomial algorithms of a huge chunk of the Euclidean geometry theorems known to the world (as well as some not previously known), a feat only technically feasible in recent years. This dramatic success has stimulated a wide interest in the area, in disciplines such as engineering, where the theory of robotics has found a new tool. The ideas have been generalised in various ways, for example to noncommutative situations.

This book fills a niche by providing a comprehensive yet easily accessible introduction to the most important concepts of computational commutative algebra. Written as an undergraduate text, the book will probably become a standard reference work for students and, increasingly, workers whose main interest is not in mathematics proper."

All the books above deal with commutative rings and algebras, and the next one studies non-commutative Noetherian rings.

J. McConnell, J. Robson, *Noncommutative Noetherian Rings*, With the cooperation of L. W. Small, Pure and Applied Mathematics (New York), A Wiley-Interscience Publication, John Wiley & Sons, Ltd., Chichester, 1987. xvi+596 pp.

Though commutative rings are familiar to most people through the important examples of polynomial rings and rings of integers, noncommutative rings also occur abundantly, for example, from matrix rings, rings of differential operators, group rings, enveloping algebras of Lie algebras etc. General noncommutative rings are too difficult to study since there is no analogue of localization of commutative rings. For noncommutative Noetherian prime rings, localization at the zero ideal is possible. This book is the first book giving a systematic treatment of noncommutative Noetherian rings.

According to Zentralblatt MATH, "Noncommutative Noetherian Ring Theory emerged as a discipline in its own right with the publication of Goldie's Theorems for prime and semiprime rings in 1958-60. Goldie's Theorem occupies the same place in the noetherian theory as the Artin-Wedderburn Theorem in the Artinian theory. Indeed, Goldie's Theorem provides an important link between noetherian and Artinian methods. In the thirty years since 1958 the subject has developed both as an intrinsically interesting branch of algebra and as a tool in other branches of algebra. Until the publication of this book there has been no attempt to provide an overview of, and a general reference for, the most important developments in the theory. The present authors have set out to fill this gap and have succeeded admirably. The book is written in a style that makes it easy to read and use: the authors aim for clarity of expression rather than for the most general statement of results."

4.3 Homological algebra

The Homology group is an important invariant of a topological space and can be computed relatively easily. As pointed out before, it was Noether who emphasized the homology group of a space rather than its Betti numbers. The subject of homological algebra studies homology in a general algebraic setting. For example, the homological properties of groups have been extensively studied, and its development is intertwined with the emergence of category theory. It is a relatively young subject, motivated by results in combinatorial topology, algebraic topology, and abstract algebra such as theory of modules and syzygies at the end of the nineteenth century.

Figure 4.3: Noether's book.

Homological algebra studies homology and cohomology groups in a general algebraic setting and related algebraic structures. In fact, homological algebra provides general methods to extract homology and cohomology invariants of chain complexes which arise from algebric structures and topological spaces.

From the beginning, homological algebra was motivated by and has played an important role in algebraic topology. It has also been applied to many subjects including commutative algebra, algebraic geometry, algebraic number theory, representation theory, mathematical physics, operator algebras, complex analysis, the theory of partial differential equations.

4.3.1 H. Cartan, S. Eilenberg, Homological Algebra, With an appendix by David A. Buchsbaum, Reprint of the 1956 original, Princeton Landmarks in Mathematics, Princeton University Press, Princeton, NJ, 1999. xvi+390 pp.

This is the first book on homological algebra and is a classic. It was written when methods of algebraic topology had caused revolutions in the world of pure algebra, and it is the culmination of the first period in the development of homological algebra. It has greatly influenced further development of the subject itself and its far-reaching applications in other branches of mathematics.

The reason and purpose for writing this book was clearly explained in the preface: "During the last decade the methods of algebraic topology have invaded extensively the domain of pure algebra, and initiated a number of internal revolutions. The purpose of this book is to present a unified account of these developments and to lay the foundations of a full-fledged theory.

The invasion of algebra had occurred on three fronts through the construction of cohomology theories for groups, Lie algebras, and associative algebras. The three subjects have been given independent but parallel developments. We present herein a single cohomology (and also a homology) theory which embodies all three: each is obtained it by s suitable specialization. This unification possesses all the usual advantages. One proof replaces three. In addition an interplay takes place among the three specializations; each enriches the other two. The unified theory also enjoys a broader sweep. It applies to situations not covered by the specializations. An important example is Hilbert's theorem concerning chains of syzygies in a polynomial ring of n variables."

According to MathSciNet, "The title 'Homological Algebra' is intended to designate a part of pure algebra which is the result of making algebraic homology theory independent of its original habitat in topology and building it up to a general theory of modules over associative rings...

The appearance of this book must mean that the experimental phase of homological algebra is now surpassed. The diverse original homological constructions in various algebraic systems which were frequently of an ad hoc and artificial nature have been absorbed in a general theory whose significance goes far beyond its sources. The basic principles of homological algebra, and in particular the full functorial control over the manipulation of tensor products and modules of operator homomorphisms, will undoubtedly become standard algebraic technique already on the elementary level. It is probably with such expectations that the authors have put so much missionary zeal into the systematization of their approach and the cataloguing of the basic results."

According to a review in Bulletin of AMS, "Perhaps Mathematics now moves so fast—and in part because of vigorous unifying contributions such as that of this book—

4.3. HOMOLOGICAL ALGEBRA

that no unification of Mathematics can be up to date. The reviewer might also add his strictly personal opinion that the authors have not kept sufficiently in mind the distinction between a research paper and a book: a good research paper presents a promising new idea when it is hot— and when nobody knows for sure that it will turn out to be really useful; a good research book presents ideas (still warm) after their utility has been established in the hands of several workers. This book contains too large a proportion of shiny new ideas which have nothing to recommend them but their heat and promise..."

S. Mac Lane, *Homology*, Die Grundlehren der mathematischen Wissenschaften, Bd. 114, Academic Press, Inc., Publishers, New York; Springer-Verlag, Berlin 1963. x+422 pp.

This is another important book in the history of homological algebra and was written when the homological algebra had reached a more mature stage. Instead of being abstract, for every topic, this book emphasizes motivations, in particular motivations from topology, and usefulness of the results.

According to MathSciNet, "Since the appearance of the first definitive book on homological algebra by Cartan and Eilenberg eight years ago, the subject has had numerous applications and has undergone many changes. In this treatise on homology, the author not only incorporates his personal point of view in this field, but also includes many of these later developments both as additional material and as motivation for a good deal of the original subject matter of homological algebra.... It indicates, however, the attempt to develop things from the special to the general. There are helpful exercises at the end of each section, and historical notes are generally found at the end of each chapter."

According to a review in Bulletin of AMS, "Approximately seven years have elapsed between the appearance of the first book on homological algebra by Cartan and Eilenberg, and the publication of this book on homology by Mac Lane... Since that time, the methods of homological algebra have been applied extensively in many areas, and homological techniques have been refined, elaborated, and generalized. Moreover, the value of homological tools and of the functorial approach has been conceded by more than a dozen mathematicians. Thus Mac Lane approaches his subject not as a magician with a set of tricks, but as an expositor who intends to demonstrate how a viable coherent theory illuminates meaningful questions. His general pattern, then, is to go from the particular to the general, throwing as much light as possible along the way.

It must of course be borne in mind that the lapse of time and development of the subject are not solely responsible for the clarity of presentation of much of the material in this volume. The author evidently believes in concrete examples and concrete applications of general machinery and within the limited scope of the book, does a fine job in providing them."

J. Rotman, *An Introduction to Homological Algebra*, Pure and Applied Mathematics, 85, Academic Press, Inc., New York-London, 1979. xi+376 pp.

At the time when it was written, the first edition of this book was an updated textbook on homological algebra. The second edition was also an updated version of the book. Briefly, there have been three periods in the development of homological algebra. The first period started in 1940s and ended in the early 1960s, and the second stage continued through 1980s, which involved abelian categories and sheaf cohomology under the influence of the works of Grothendieck and Serre, and the stage involves derived categories and triangulated categories and is still ongoing. The first edition of this corresponds to the first stage, and the second edition introduces results of the second stage too. Two newer books by Gelfand & Manin, and Weibel discuss all three stages.

Rotman tried to make the homological algebra lovable by presenting the subject "in the context of other mathematics". It is a good introduction to homological algebra.

According to the MathSciNet review of the first edition, "This book gives a treatment of homological algebra which motivates the subject in terms of its origins in algebraic topology... the author has been very careful throughout to keep the subject in perspective in a number of ways which include, for example, tracing the development of terminology, supplying many examples where homological methods have been used to solve rather down-to-earth problems which are intractable from other points of view, and maintaining a level of generality which, while sufficient for the vast majority of applications, is not so great so as to overwhelm the novice. A great deal of solid, satisfying mathematics is presented. Applications to commutative algebra are particularly abundant..."

4.3.2 More books on homological algebra and related subjects

C. Weibel, *An Introduction to Homological Algebra*, Cambridge Studies in Advanced Mathematics, 38, Cambridge University Press, Cambridge, 1994. xiv+450 pp.

Homological algebra has progressed a lot since the books from 1950s to 1970s. According to Zentralblatt MATH, "Largely due to the influence of *A. Grothendieck, J.-P. Serre*, and their French school of algebraic geometry, homological algebra has been increasingly formulated in the general terms of Abelian categories and derived functors. This provided the necessary framework for the rigorous foundation of modern algebraic geometry by the theory of schemes and their morphisms, sheaves and their cohomology, local cohomology, and other central notions. Related areas, such as commutative algebra, algebraic topology, algebraic number theory, and algebraic analysis were just as influenced, for instance by homological concepts like spectral sequences, group cohomology, Galois cohomology, Hochschild homology, André-Quillen homology, cyclic homology, triangulated categories, derived categories, $\mathcal{D}$-modules in microanalysis, etc... In these

4.3. HOMOLOGICAL ALGEBRA

days, homological algebra is a fundamental tool, of obviously still increasing importance, for mathematicians and (nowadays also) theoretical physicists." On the other hand, according to the preface, "Unfortunately, many of these later developments are not easily found by students needing homological algebra as a tool. The effect is a technological barrier between casual users and experts at homological algebra."

The time is ready for new generations of textbooks on homological algebra. This book of Weibel is such a book "to break down that barrier by providing an introduction to homological algebra as it exists today." It is a very good introduction and can also serve as a valuable reference.

According to the preface "this book intends to paint a portrait of the landscape of homological algebra in broad brushstrokes". To achieve this aim, Weibel, on the one hand, provides a full and clear account of the fundamentals of the subject, and on the other hand, he also offers introductions to several other subjects which have made heavy use of, or have been strongly influenced by, homological algebra ... this book is aimed at a second- or third-year graduate student".

According to MathSciNet, "It is, ..., the ideal text for the working mathematician needing a detailed description of the fundamentals of the subject as it exists and is used today; the author has succeeded brilliantly in his avowed intention to break down the technological barrier between casual users and experts."

S. Gelfand, Y. Manin, *Methods of Homological Algebra*, Second edition, Springer Monographs in Mathematics, Springer-Verlag, Berlin, 2003. xx+372 pp.

Since the beginning in the 1940s, homological algebra has come a long way both internally and externally. This is the most modern book on homological algebra and is a valuable reference for students and experts.

Its Foreword explained the history of homological algebra well. According to Zentralblatt MATH, "In contrast to the existing textbooks on homological algebra, the authors provided an approach to the most recent viewpoint of the subject, from the beginning on, and that in a consequent, systematic way. At this time, and yet until now, their book represented the first attempt to develop Verdier's theory of derived categories, and some of its applications, in a comprehensive textbook-style, and to embed the basics of classical homological algebra into this approach. The outcome was a brilliant and unique treatise ..."

According to MathSciNet of the first edition, "This is long-awaited modern textbook on homological algebra. The exposition is centered around the notion of derived category, which is treated in full detail for the first time in the textbook literature... In fact, the book contains more than just an exposition of the subjects involved. It teaches the reader, via examples, how to 'think homologically'. The book has many 'levels of understanding' and can be read, with much benefit, on each of them. Thus, it can be used by students just beginning to study homological algebra, as well as by specialists who will find there some points which have never been clarified in the literature." For the revised second edition, "It is based on the systematic use of the language and technics of derived categories and derived functors. The reader has all the basic material and a lot of examples ... This book can be used by students just beginning to study homological algebra, as well as by specialists who will find there some points which have never been clarified in the literature."

S. Mac Lane, *Categories for the Working Mathematician*, Second edition, Graduate Texts in Mathematics, 5, Springer-Verlag, New York, 1998. xii+314.

A category is an algebraic structure that comprises "objects" that are linked by "arrows". A category has two basic properties: the ability to compose the arrows associatively and the existence of an identity arrow for each object. A simple example is the category of sets, whose objects are sets and whose arrows are functions. In general, the objects and arrows may be abstract entities of any kind, and the notion of category provides a fundamental and abstract way to describe mathematical entities and their relationships. This is the central idea of category theory, which seeks to generalize all of mathematics in terms of objects and arrows. Indeed, many branches of modern mathematics can be described in terms of categories, use of the category theory brings out deep insights and similarities between seemingly different areas of mathematics. The ideas and methods are most transparent in the homological algebra.

The aim of this book is "to present those ideas and methods which can now be effectively used by Mathematicians working in a variety of other fields of Mathematics research." This explains the attractive title of this book and probably convinces every working mathematician to take a look at it to see what is missing in his/hers education. It is an accessible introduction for beginners and also a very comprehensive reference for experts.

J. McCleary, *User's Guide to Spectral Sequences*, Mathematics Lecture Series, 12, Publish or Perish, Inc., Wilmington, DE, 1985. xiv+423 pp.

A spectral sequence is an effective method to compute homology and cohomology groups in homological algebra and algebraic topology by taking successive approximations. Spectral sequences are a generalization of exact sequences, and was introduced by Jean Leray in 1946. Its history is well explained in a review in Bulletin of AMS of this book, "Spectral sequences were developed and first used in France immediately after World War II. Apparently they were first introduced by J. Leray in some papers concerned with what would nowadays be called the sheaf-theoretic cohomology of locally compact spaces... Soon afterwards they were being used by H. Cartan and J.-L. Koszul. (The latter used them in his work on the cohomology of Lie algebras.) The reviewer recalls quite vividly the difficulties algebraic topologists on this side of the Atlantic had in trying to understand and digest these first papers about spectral sequences. Fortunately, the famous theses of J.-P. Serre and A. Borel appeared within a few years. These theses gave many interesting new results which were derived using spectral sequences, and the exposition was exceptionally clear. (Serre and Borel both had announced the principal results of their theses in brief Comptes Rendus notes which appeared a year or so before their respective theses.)

Probably the first application of spectral sequences to algebra was the above-mentioned work of Koszul on Lie algebras. The first explicit statement of the LHS spectral sequence was in a Comptes Rendus note by Serre in 1950, followed by a complete exposition by Serre and Hochschild in 1953. The earlier work of R. C. Lyndon, contained in his 1946 Harvard University Ph.D. thesis and published in 1948, was done before spectral sequences were known in the U.S.; hence he had to try to state his results without the use of spectral sequences.

After the publication of the theses of Serre and Borel, spectral sequences techniques gradually pervaded more and more of algebraic topology and homological algebra. They also came to be used in certain other parts of mathematics, such as algebraic geometry, category theory, and algebraic K-theory. To the best of the reviewer's knowledge, nobody has ever tried to catalog all the many different applications of spectral sequences in modern-day mathematics."

For non-specialists on spectral sequence, mentioning of its name brings only trepidation or confusion. This book provides a user-friendly manual to spectral sequence.

According to the author, "A 'good' user's manual for any apparatus should satisfy certain expectations. It should provide the beginner with sufficient details in exposition and examples to feel comfortable in starting to apply the new apparatus to his or her problems. The manual should also include enough details about the inner workings of the apparatus to allow a user to determine what is going on if it fails while in operation. Finally, a user's manual should also include plenty of information for the expert who is looking for new ways to use the device." So this is a book for everyone who is interested in spectral sequence.

According to MathSciNet of the first edition, "The book is written in a pleasantly discursive manner, mixing the memory of the pioneers of the subject with the desire to impart their hard-gained knowledge to us lesser mortals. The author is to be congratulated on having taken in a vast amount of research material and serving it up in a reasonably digested and digestible form."

J. Loday, *Cyclic Homology*, Appendix E by María O. Ronco, Second edition, Chapter 13 by the author in collaboration with Teimuraz Pirashvili, Grundlehren der Mathematischen Wissenschaften, 301, Springer-Verlag, Berlin, 1998. xx+513 pp.

Cyclic homology and **cyclic cohomology** are homology and cohomology theories for associative algebras. They were introduced by Alain Connes around 1980, and have played an important role in his noncommutative geometry as that played by homology and cohomology groups in the study of topological spaces. It arises from many sources such as index theorems for non-commutative Banach algebras, Lie algebra homology of matrices, algebraic K-theory, homology of S^1-spaces.

This book is a comprehensive introduction to cyclic homology and is also a valuable reference. According to European Mathematical Society Newsletter, "The book can be strongly recommended to anybody interested in noncommutative geometry, contemporary algebraic topology and related topics."

According to MathSciNet, "This book is written to introduce students and nonspecialists to the field of cyclic homology. As a prerequisite it requires only a rudimentary knowledge of algebraic topology and homological algebra. In the first half of the book the author meticulously outlines several approaches to cyclic homology, namely algebraic, geometric, and categorical. The second half contains more sophisticated results which use cyclic homology to express certain relative algebraic K-theory groups... A nice feature of the text is how the author interleaves introductory chapters with those containing recent results in cyclic homology... To conclude, Cyclic homology is written in a lucid expository style with careful, detailed proofs of most theorems. The text fills a gap in the mathematical literature."

V. The Science Library of the University of Michigan

Gauss initiated and made fundamental contributions to many branches of mathematics. His contribution to geometry includes the Gauss-Bonnet formula, Gauss curvature, and the Theorema Egregium (remarkable theorem), which says that the curvature of a surface can be determined intrinsically by measuring angles and distances on the surface. He is also one of the discovers of the non-Euclidean geometry. It should be mentioned also that he picked the last one from the three problems for Habilitation proposed by Riemann and hence forced Riemann to think about it and to introduce the notion of Riemannian manifolds, which has changed the mathematical world forever.

Riemann did not publish too many papers, but he changed every subject he touched, especially if the subject is in mathematics. Probably he is best known for Riemannian geometry, the Riemann zeta function, and Riemann surfaces. Things were not easy for Riemann. He had some difficulty in getting a permanent job and died seven years after he got the permanent chair in Gottingen. It is probably less known that after Gauss picked the Habilitation problem on geometry, Riemann sweated and struggled in order to prepare for it. His letters to his family during those months explained the situation well and are still preserved at the library of University of Gottingen.

Minkowski was a child prodigy. Before I learnt anything about mathematics of Minkowski, I learnt this from the biography *Hilbert* by C. Reid. His most original contribution is the creation of geometry of numbers, and his best known result is the Minkowski geometry, which put the relativity theory into the proper mathematical framework. He first became famous at the age of 18 when, in 1883, he was awarded the Mathematics Prize of the French Academy of Sciences for his work on the theory of quadratic forms. The co-winner of the prize was an old established mathematician Henry Smith, who passed away in 1883. In all works of Minkowski, the geometry is a underlying theme. In geometry of number, some very difficult problems in number theory can be reduced to elementary geometry facts. This is an excellent example of the power of combining two seemly different subjects.

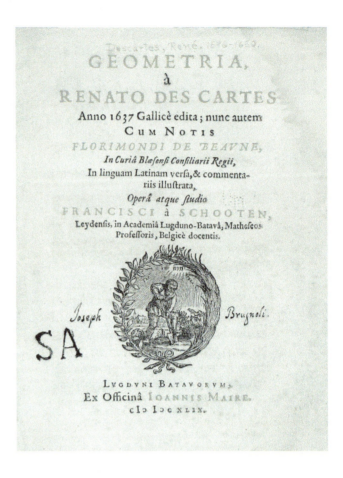

Descartes (or Cartesius) was a philosopher, mathematician, and writer. His most important contribution to mathematics was the Cartesian coordinate system, which established a bridge between algebra and geometry. Hence he is often called the father of analytical geometry. It is difficult to imagine mathematics, in particular calculus and geometric analysis, without the Cartesian coordinate system, but people often try to formulate results in a coordinate free way. His most famous philosophical statement is *I think, therefore I am*. Maybe mathematicians have something similar: they can define space and time into existence.

Chapter 5

Geometry

Geometry is one of the oldest branches of mathematics and is concerned with shape, size, relative position of geometric figures, and properties of space. Based on practical knowledge about lengths, areas, and volumes in space, geometry was put into an axiomatic form by Euclid in the third century BC. Euclidean geometry set a standard of geometry (and also of mathematics) for many centuries to follow. The introduction of coordinates by Descartes in the early seventeenth century allows one to represent geometric figures, such as plane curves, analytically, i.e., in terms of functions and equations, and this marked new stage for the development of geometry. The subject of geometry was further enriched by the study of intrinsic structure of geometric objects that originated with Gauss and led to the creation of differential geometry (or Riemannian geometry). Differential geometry is the foundation of modern mathematics, and every mathematician will need to use manifolds and analysis on them at one point or another. People in other fields, especially physics, also need differential geometry. Indeed, as is well-known, it is the work of Einstein on general relativity that brought Riemannian geometry to the central stage of mathematics.

Geometry includes (1) differential geometry and Riemannian, (2) complex geometry, (3) algebraic geometry, (4) convex geometry. It also includes projective geometry, analytic geometry, symplectic geometry etc. In this chapter, we will concentrate on books related to the 4 subjects listed above.

5.1 Differential geometry

Differential geometry uses the techniques of differential and integral calculus, as well as linear and multilinear algebra, to study problems in geometry of subspaces of the Euclidean spaces and differential manifolds. In some sense, it is the combination of differential calculus and differential manifolds. In the eighteenth and nineteenth centuries, the classical differential geometry studies plane and space curves and surfaces in the three-dimensional Euclidean space. Since the late nineteenth century, the focus has shifted to general geometric structures on differentiable manifolds which are not necessarily embedded into Euclidean spaces. It is closely related to differential topology, and to the geometric aspects of the theory of differential equations on manifolds.

The connection is one of the most basic notions in differential geometry, and the Christoffel symbols are crucial for explicit computation of connections in local coordinates. For example, the obvious notion of parallel transport does not make sense without the notion of connection. Connections are also crucial in the gauge theory in physics. It might be interesting to note that Christoffel was a student of the famous number theorist Kummer and wrote a thesis on the motion of electricity in homogeous bodies.

Figure 5.1: Christoffel.

5.1.1 H. Hopf, Differential Geometry in the Large, Notes taken by Peter Lax and John Gray, With a preface by S. S. Chern, Lecture Notes in Mathematics, 1000, Springer-Verlag, Berlin, 1983. vii+184 pp.

This is the first and an important book on the global aspects of differential geometry written by a master expositor. In spite of its old age, this book is still a very valuable introduction to the modern, global differential geometry for students.

This is a very special volume (1000th) of the prestigious book series from Springer, and they used this volume for the lecture notes by Hopf. According to the preface by Chern, "Heinz Hopf is a mathematician who recognized important mathematical ideas and new mathematical phenomena through special cases. In the simplest background the central idea or the difficulty of a problem usually becomes crystal clear. Doing geometry in this fashion is a joy. Hopf's great insight allows this approach to lead to serious mathematics, for most of the topics in these notes have become the staring-points of important developments... Hopf's mathematical exposition is a model of precision and clarity. His style is recognizable in these notes."

Though local geometry of submanifolds of the Euclidean space and more generally abstract Riemannian manifolds have been studied since Gauss, it is a relatively recent

5.1. DIFFERENTIAL GEOMETRY

trend to study the global aspects of geometry, i.e., relations between geometric properties and topological properties of Riemannian manifolds. The work of Hopf and his students such as Stiefel, Rinow, Hirzebruch etc. and this book have had a huge impact on the development of global differential geometry, or "differential geometry in the large".

According to MathSciNet, "To quote from the Introduction, 'The first chapter will be a review of classical differential geometry in the small, and the second will be devoted to some general remarks on closed, differentiable surfaces, not necessarily in E^3. These will be followed by a short chapter on the Riemannian geometry of closed surfaces in which will be considered the relation between the Gauss curvature of a surface, the singularities of fields of directions on the surface and the topological structure of the surface. The remainder of the lectures will deal with surfaces in E^3.' One might well enquire as to the relevance of material, which is over twenty-five years old, today. Has it not all been superseded by intervening work? In the reviewer's opinion, the answer is a resounding no! Some of the areas of differential geometry discussed in this book are still very much under investigation. Some of the questions raised in this book are still open. Some of the methods used have been elaborated and generalized. Some have not. It is pure pleasure to see a first class mathematical mind at work cobbling together a bit of calculus, a bit of combinatorics, a bit of algebra, a bit of topology, etc., into a beautiful and surprising result... The book should be a delight to anyone interested in differential geometry."

5.1.2 Introduction to differential geometry and Riemannian geometry

Differential geometry is a classical and large area. An important sub-field is Riemannian geometry. Riemannian geometry originated with the vision of Bernhard Riemann expressed in his inaugural lecture. It is a generalization of the differential geometry of surfaces in $\mathbb{R}^3$.

Riemannian geometry studies Riemannian manifolds, which are smooth manifolds with a Riemannian metric, i.e., with an inner product on the tangent space at each point which varies smoothly from point to point. This gives local geometric notions such as length and volume, and global quantities can be derived by integrating local contributions.

The book above by Hopf is one of the beginning ones on the global geometry of Riemannian manifolds. There are many books on differential geometry. We mention some of them and put them into three groups. The first group is more elementary, and the other groups are more advanced and specialized.

F. Warner, *Foundations of Differentiable Manifolds and Lie Groups*, Corrected reprint of the 1971 edition, Graduate Texts in Mathematics, 94, Springer-Verlag, New York, 1983. ix+272 pp.

This has been quite popular with graduate students. For many of them, this is probably the first book to learn basics in differential manifolds, Lie groups and also a bit about the important Hodge theorem and De Rham cohomology. The book is carefully written and enough details are provided. Unlike some other books on differential manifolds, Lie groups are emphasized here. Putting these important concepts and results together conveys a bigger perspective.

According to MathSciNet, "... Overall, the style is clear, clean, and formal, perhaps a bit thin on motivation, as is customary, except through examples and problems. The problems appearing after each chapter are elementary in the beginning, but after Chapter 2 the problems also contain significant major theorems. For example, the fundamental theorem on abelian Lie groups and the Peter-Weyl theorem are problems. Many problems are devoted to filling in necessary pieces of the text, e.g., a problem on the use of a partition of unity to prove the existence of a Riemannian metric is placed after Chapter 1, and problems on the star operator and the adjoint of left multiplication by a 1-form, both used in the Hodge theorem, are found after Chapter 2..."

J. Lee, *Riemannian Manifolds. An Introduction to Curvature*, Graduate Texts in Mathematics, 176, Springer-Verlag, New York, 1997. xvi+224 pp.

This is a new accessible introduction to Riemannian geometry. As the subtitle indicates, it focuses "on developing an intimate acquaintance with the geometric meaning of curvature. In so doing, it introduces and demonstrates the use of all the main technical tools needed for a careful study of Riemannian manifolds."

According to MathSciNet, "In Riemann's renowned inaugural lecture 'On the hypotheses which lie at the foundations of geometry', the concepts of n-manifolds and Riemannian metrics were introduced. In a later paper submitted to the Paris Academy, Riemann extended the study and introduced curvature into Riemannian geometry. Ever since, curvature has been at the center in the study of Riemannian geometry, although there are different viewpoints to the concept. In the book under review, the author focuses on developing the geometric meaning of curvature... The topics in the book are well-chosen and the content is well-organized. Arguments and proofs are written down precisely and clearly. The expertise of the author is reflected in many valuable comments and remarks on the recent developments of the subjects. Serious readers would have the challenges of solving the exercises and problems. The book is probably one of the most easily accessible introductions to Riemannian geometry."

R. Bishop, R. Crittenden, *Geometry of Manifolds*, Pure and Applied Mathematics, Vol. XV, Academic Press, New York-London, 1964, ix+273 pp.

When this book was published in 1964, the differential geometry was not so developed yet. There have been a lot of developments after that. But this well-written book is still a good reference that covers basic topics in differential geometry.

5.1. DIFFERENTIAL GEOMETRY

According to MathSciNet, "This book represents an excellent treatment of a wide section of modern differential geometry. As the authors acknowledge in the preface it owes much to the distinguished school of differential geometry at M.I.T. and, in particular, to W. Ambrose and I. Singer, who in turn draw upon the pioneering work of Cartan, Chern, and Ehresmann. The style is elegant and at the same time considerate for the needs of a beginner. Many comments are given to related questions as well as a great number of well chosen problems with pertinent references whenever desirable. This makes it possible to cover in considerable detail the foundations of the field. The authors seem to have been quite successful in avoiding serious errors without doing it the easy way by glossing over the details. In summa, the reviewer thinks that anybody who chooses to base his course on differential geometry at the graduate level on this book could not do better."

M. do Carmo, *Riemannian Geometry*, Translated from the second Portuguese edition by Francis Flaherty, Mathematics: Theory & Applications, Birkhäuser Boston, Inc., Boston, MA, 1992. xiv+300 pp.

This is a widely used textbook for graduate courses. The main texts are not long but cover many basic and important results. They are supplemented by many well-chosen exercises.

The book description describes its feature well: "The author's treatment goes very directly to the basic language of Riemannian geometry and immediately presents some of its most fundamental theorems. It is elementary, assuming only a modest background from readers, making it suitable for a wide variety of students and course structures. Its selection of topics has been deemed 'superb' by teachers who have used the text.

A significant feature of the book is its powerful and revealing structure, beginning simply with the definition of a differentiable manifold and ending with one of the most important results in Riemannian geometry, a proof of the Sphere Theorem. The text abounds with basic definitions and theorems, examples, applications, and numerous exercises to test the student's understanding and extend knowledge and insight into the subject."

M. do Carmo, *Differential Geometry of Curves and Surfaces*, Translated from the Portuguese, Prentice-Hall, Inc., Englewood Cliffs, N.J., 1976. viii+503 pp.

The geometry of curves and surfaces in $\mathbb{R}^3$ has been studied for a long time, especially the local properties. This book gives an accessible introduction to both the local and global geometry of curves and surfaces from the modern point of view. It is one of the most successful textbooks on this basic topic for undergraduate students.

I. Chavel, *Riemannian Geometry–A Modern Introduction*, Cambridge Tracts in Mathematics, 108, Cambridge University Press, Cambridge, 1993. xii+386 pp.

This is another new introduction to Riemannian geometry. Besides the usual topics, it includes a substantial discussion of isometric inequalities together with more advanced topics such as Cheeger's finiteness theorem.

According to MathSciNet, "This book provides a comprehensive introduction to Riemannian geometry, including a number of topics appearing in book form for the first time. The text assumes an introductory course on differentiable manifolds. The book is denser than, say, M. P. Do Carmo's *Riemannian geometry*, 1992, and contains an extensive bibliography as well as frequent notes to orient the reader to the literature. Thus the book can serve as a resource for experts in the field. An unusual feature of the book is the very extensive notes and exercises section at the end of each chapter. These are used to further develop the ideas in the main body of the text, to give references to the literature, and to introduce a smorgasbord of topics, including many recent results, not otherwise covered. Exercises vary from the routine to important results in the literature."

S. Gallot, D. Hulin, J. Lafontaine, *Riemannian Geometry*, Third edition, Universitext, Springer-Verlag, Berlin, 2004. xvi+322 pp.

This is an attractive introduction to global Riemannian geometry. It discusses many current and interesting topics with the help of many examples and exercises. For example, besides the standard topics of textbooks in Riemannian geometry, the first edition also contains results on interplay between topology and curvature, curvature and volume, curvature and growth of the fundamental group, curvature and topology, an entire chapter on analysis on manifolds, in particular spectral geometry, and also a chapter on Riemannian submanifolds. The latter editions include more current topics such as Levy-Gromov isoperimetric inequality, pseudo-Riemannian geometry etc. Being carefully written and well-arranged, this book conveys quite accurately a picture of the active field of Riemannian geometry.

B. Dubrovin, A. Fomenko, S. Novikov, *Modern Geometry—Methods and Applications. Part I. The Geometry of Surfaces, Transformation Groups, and Fields*, Second edition, *Part II. The Geometry and Topology of Manifolds*, Translated from the Russian by Robert G. Burns. Graduate Texts in Mathematics, 93, 104, Springer-Verlag, New York, 1992. xvi+468 pp., 1985. xv+430 pp.

This is a modern introduction to differential geometry and differential topology and a scope determined by the criterion "which parts of modern geometry should be regarded as absolutely essential to a modern mathematics education," or what a working mathematician outside geometry might need. It also contains substantial discussion of algebraic topology, for example, the Morse theory and smooth structures on the 7-spheres, and many applications not found in other books on differential manifolds. It is a comprehensive introduction and a valuable reference too.

5.1. DIFFERENTIAL GEOMETRY

According to MathSciNet, "This book is a result of several years of teaching experience, and of an ambitious project of collecting in a single text all of the geometry, the topology, the global analysis, and the physical applications which a mathematics student with inclinations to theoretical physics, and a student of theoretical physics should learn..."

According to a review in Bulletin of AMS, "The best way to describe this volume is to say that it is a contemporary treatise on modern methods in geometry with deep applications to the physical sciences."

In contrast to the above recent books on Riemannian geometry, the next is a classical one when the subject of Riemannian geometry was not so fashionable in mathematics or understood by the mathematics community.

L. Eisenhart, *Riemannian Geometry*, Eighth printing, Princeton Landmarks in Mathematics, Princeton Paperbacks, Princeton University Press, Princeton, NJ, 1997. x+306 pp.

This book was written when the usefulness of Riemannian geometry in the general relativity motivated people to study and develop the intrinsic properties of Riemannian geometry. As the author wrote, "Riemann proposed the generalization, to spaces of any order, of the theory of surfaces, as developed by Gauss, and introduced certain fundamental ideas in this general theory. From time to time important contributions to the theory were made by Bianchi, Beltrami, Christoffel, Schur, Voss and others, and Ricci coordinated and extended the theory with the use of tensor analysis and his Absolute Calculus. Recently there has been an extensive study and development of Riemannian Geometry, and this book aims to present the existing theory."

From the book description, "In this book, Eisenhart succinctly surveys the key concepts of Riemannian geometry, addressing mathematicians and theoretical physicists alike."

T. Frankel, *The Geometry of Physics. An Introduction*, Cambridge University Press, Cambridge, 1997. xxii+654 pp.

This book was written as a textbook on differential geometry for physics students. For mathematics students, it is not a standard text book, but it is very useful and enjoyable as a supplementary read (at a leisurely pace) because of its broad coverage of topics, discussion of physics, and interplay between mathematics and physics. Indeed, as the book description says, "This book is intended to provide a working knowledge of those parts of exterior differential forms, differential geometry, algebraic and differential topology, Lie groups, vector bundles and Chern forms that are essential for a deeper understanding of both classical and modern physics and engineering. Geometric intuition is developed through a rather extensive introduction to the study of surfaces in ordinary space; consequently, the book should be of interest also to mathematics students. This book will be useful to graduate and advanced undergraduate students of physics, engineering and mathematics. It can be used as a course text or for self study."

According to Zentralblatt MATH, "physicists and mathematicians are traditionally trained with different books, and this is certainly one of the main reasons why they often have problems to find a common language. The book under review may be viewed as a valuable contribution to bridge this gap. On the one hand, it meets the mathematical standards of a text-book on differential geometry; on the other hand, it strongly appeals to intuition and emphasizes applications to physics."

5.1.3 More advanced books on Riemannian geometry

H. Busemann, *The Geometry of Geodesics*, Academic Press Inc., New York, 1955. x+422 pp.

We all know from elementary school that the straight line is one of the most basic concepts of geometry. In Riemannian geometry, geodesics are equally basic. In comparison with the Euclidean space, their structures can be much more complicated, and they reflect the geometry of Riemannian manifolds.

This is a very original and visionary book. It propose a systematic method to study global geometry of Riemannian manifolds using behaviors of distance between geodesics and hence avoiding differentiation and curvature of the manifold. Such an approach can be applied to more general metric spaces and has been very influential in geometric group theory and related spaces, for example, the CAT(0)-spaces and the more general Busemann spaces. The Busemann function associated with a geodesic is also a fundamental object in differential geometry. Busemann received the Lobachevsky Medal he received in 1985 for this book.

According to a review in Bulletin of AMS, "This book represents a most interesting and successful effort to treat in a unified fashion the geometry of a class of metric spaces general enough to include Riemannian and Finsler spaces. Only the barest minimum is postulated; in particular, the spaces are metric; bounded infinite sets have limit points; any two points can be connected by a (geodesic) segment, i.e. a curve whose length equals the distance of the points; every point has a neighborhood in which a geodesic segment can be prolonged in a unique way. It is a consequence that a given segment can be prolonged indefinitely in both directions to yield a geodesic, i.e. a curve which is locally a segment. These are more or less paraphrases, the five axioms themselves being stated in terms of the metric, and in very simple form. The spaces satisfying them are called G-spaces and are the object of study throughout the book. They include the complete Riemannian and Finsler manifolds. It should be pointed out however that it is not assumed that the spaces are differentiable, or even topological manifolds. ... This is a

5.1. DIFFERENTIAL GEOMETRY

book with a very original method of approach to problems of what might be called 'general differential geometry' (with a slant toward the foundations of geometry). It assumes very little and proves a great many theorems, many of which were not known, certainly not in so general a form, before the work of the author. For this reason it is not a book that can be skimmed lightly, but rather it must be studied to be followed. The organization, however, is good and the proofs clearly written, their length in many instances being due to the extremely weak assumptions made. It must certainly be counted an important addition to the literature and will deserve the careful consideration of mathematicians interested in the geometry of Riemannian and, particularly, Finsler spaces."

P. Petersen, *Riemannian Geometry*, Second edition, Graduate Texts in Mathematics, 171, Springer, New York, 2006. xvi+401 pp.

This is a more advanced, comprehensive introduction to Riemannian geometry and also covers some topics that are close to the recent research topics. Some results on convergence of Riemannian manifolds and finiteness theorems appear for the first time in a book form.

According to Zentralblatt MATH, "it is extremely important to have a textbook like this available, giving introductions to the advanced achievements of modern Riemannian geometry in a comprehensible way, covering more or less all of the current mainstream of that area. In that sense, it will be a must for a student interested in the topic, because he will get an overview, will know precisely which preparatory texts should be consulted, and after having made his choice, he will get a detailed and motivating description of the topic, leading him to subjects of current research. Furthermore, the selection of results presented in the text is complemented by a rich problem section. There are added comments concerning the development of the subject and recommendations for further reading. These may not be shared by many geometers, because, according to the reviewer's opinion, they should have been more open to what has been achieved in the classical literature or in related areas of geometry. Moreover, the presentation of the material in this monograph is so convincing, that this kind of additional self-promotion of modern Riemannian geometry is not needed."

J. Cheeger, D. Ebin, *Comparison Theorems in Riemannian Geometry*, North-Holland Mathematical Library, Vol. 9., North-Holland Publishing Co., Amsterdam-Oxford; American Elsevier Publishing Co., Inc., New York, 1975. viii+174 pp.

Spaces of constant curvature are special and their geometry is easier to understand. One effective method in Riemannian geometry is to compare a Riemannian manifold with such a model manifold. Comparison theorems relate information on curvature to geometric and topological properties of Riemannian manifolds, and are fundamentals tools in global Riemannian geometry.

This is the first book in differential geometry that concentrates on comparison theorems and their applications. It is concisely and clearly written and has been influential in the development of Riemannian geometry. As the authors said in the preface to the second edition, "In the period since the first edition of this book appeared (1975), Riemannian geometry has experienced explosive growth." But the topics in this book are still essential for people who want to work in Riemannian geometry.

According to the review in Bulletin of AMS, "Differential geometry is an almost unique area within mathematics, since it involves both the old and the new in an essential way. Riemannian geometry itself has, of course, been around for over one hundred years: About twenty-five years ago geometers began to ask how the local curvature of Riemannian manifolds could influence their global properties. (There were clues that this was an interesting question, e.g., the theorem of Hadamard and Cartan that a complete simply connected Riemannian manifold of nonpositive sectional curvature was diffeomorphic to Euclidean space.) The major opening salvo in this campaign was Rauch's work, published in 1951, showing that a (positive definite) Riemannian manifold whose sectional curvature function is sufficiently close to the curvature of the usual metric on the sphere is, in fact, homeomorphic to the sphere. Rauch combined techniques whose roots lie in the classical work: Sturm-type theorems for the systems of linear ordinary differential equations which result from linearization of the geodesic equations, and the distance minimizing property of the geodesies. Berger, Klingenberg and Toponogov then developed the conditions on the curvature which assure that the manifold is homeomorphic to the sphere and analyzed what happens at the precise point that the conditions are violated. They also developed a refined and powerful methodology to deal with this type of problem. In the sixties the methods were successfully applied to two general problems: Find Rauch-type conditions on the curvature which would assure that the manifold is diffeomorphic to the sphere, and study general global properties of Riemannian manifolds whose sectional curvature has a fixed sign. This superb book gives us a masterful and definitive account of this work."

From the book description, "Much of the material, particularly the last four chapters, was essentially state-of-the-art when the book first appeared in 1975."

W. Klingenberg, *Riemannian Geometry* , de Gruyter Studies in Mathematics, 1. Walter de Gruyter & Co., Berlin-New York, 1982. x+396 pp.

Klingenberg has been very influential in the development of global differential geometry in Germany. He wrote a book with his students

D. Gromoll, W. Klingenberg, W. Meyer, *Riemannsche Geometrie im Grossen*, Lecture Notes in Mathematics, No. 55, Springer-Verlag, Berlin-New York, 1968. vi+287 pp.

5.1. DIFFERENTIAL GEOMETRY

It has also been influential in the development of global differential geometry.

Besides his work on the sphere theorem, Klingenberg is also well-known for his work on existence of closed geodesics (fortunately or unfortunately). In this book, he presented his approach to geodesics of Riemannian manifolds.

According to MathSciNet review for the first edition, "... This book is not really a textbook. Perhaps it could best be described as an excellent monograph on geodesics which has been expanded to include an introduction to the preliminary ideas which this requires. As each technique or idea is introduced it is preceded by a paragraph or two of history and motivation and followed by a discussion of particular examples. The main subject of this book is exciting and new results are constantly being obtained. Any differential geometer should know something about geodesics and this is a good place to read about it, as well as to find many useful ideas and examples for reference."

From the preface: "... there is one result of very recent origin, which I am quite happy to include in the second edition. I call it the main theorem for surfaces of genus 0: There always exist on such a surface infinitely many geometrically distinct closed geodesics.

It is a truly centennial result since, already in 1905, H. Poincaré showed that on a convex surface there exists a simple closed geodesic. His work was continued and extended by G. Birkhoff, L. Lyusternik, L. Shnirelman and many others, until finally, in 1993, the combined efforts of V. Bangert and J. Franks led to a proof of the main theorem in full generality. In the same year, N. Hingston published a paper which only uses methods which have been developed in this monograph and avoids the approach that Franks used. It therefore became quite natural for me to present in the final section of Chapter 3 a complete proof of the main theorem. I feel that this constitutes an important and beautiful finale to my work."

The problem whether a surface of genus 0 admits infinitely many geometrically distinct closed geodesics is still open. It seems that this problem consumed the rest of the academic life of Klingenberg.

According to a review in Bulletin of AMS, "In short, the present book presents a unique, and excellent, selection of topics, and it presents them well. In my view, it is the best available introduction to those topics in contemporary Riemannian geometry centering around the geometry of geodesics. No serious student or practitioner of geometry should be without it."

M. Spivak, *A Comprehensive Introduction to Differential Geometry. Vol. I.–Vol. V.*, Second edition. Publish or Perish, Inc., Wilmington, Del., 1979. xiv+668 pp., xv+423 pp., xii+466 pp., viii+561 pp., viii+661 pp.

This is indeed a very long and comprehensive introduction to differential geometry with an emphasis on historic roots. It is probably difficult to read it from cover to cover, but it is certainly valuable and educational to browse through it. The author received a Steele prize for mathematical exposition for this book.

According to MathSciNet, "This book is distinguished from other books of its kind by its historical approach to the subject. The connection is made between the classical works of Euler, Gauss, Riemann and other geometers and the newest investigations in differential geometry. In regards to methodology, a particularly useful comparison of classical and contemporary methods for the exposition of specific problems of differential geometry is frequently applied. The material is further elucidated by the large number of original drawings that are in the text of the book. All of this, together with a huge quantity of examples scattered throughout the text and problems that are suggested for the reader to solve, make this book extremely useful to everyone who studies or teaches differential geometry."

Indeed, according to a review in Bulletin of AMS, "The book takes as its theme the classical roots of contemporary differential geometry. Spivak explains his Main Premise (my term) as follows: 'in order for an introduction to differential geometry to expose the geometric aspect of the subject, an historical approach is necessary; there is no point in introducing the curvature tensor without explaining how it was invented and what it has to do with curvature'. His second premise concerns the manner in which the historical material should be presented: 'it is absurdly inefficient to eschew the modern language of manifolds, bundles, forms, etc., which was developed precisely in order to rigorize the concepts of classical differential geometry'. Here, Spivak is addressing to 'a dilemma which confronts anyone intent on penetrating the mysteries of differential geometry'. On the one hand, the subject is an old one, dating, as we know it, from the works of Gauss and Riemann, and possessing a rich classical literature. On the other hand, the rigorous and systematic formulations in current use were established relatively recently, after topological techniques had been sufficiently well developed to provide a base for an abstract global theory; the coordinate-free geometric methods of E. Cartan were also a major source. Furthermore, the viewpoint of global structure theory now dominates the subject, whereas differential geometers were traditionally more concerned with the local study of geometric objects.

Thus it is possible and not uncommon for a modern geometric education to leave the subject's classical origins obscure. Such an approach can offer the great advantages of elegance, efficiency, and direct access to the most active areas of modern research. At the same time, it may strike the student as being frustratingly incomplete. As Spivak remarks, 'ignorance of the roots of the subject has its price-no one denies that modern formulations are clear, elegant and precise; it's just that it's impossible to comprehend how any one ever thought of them.' ..."

5.1. DIFFERENTIAL GEOMETRY

S. Kobayashi, K. Nomizu, *Foundations of Differential Geometry. Vol I., Vol. II.*, Interscience Publishers, A division of John Wiley & Sons, New York-London, 1963. xi+329 pp., 1969, xv+470 pp.

This is an early comprehensive introduction to modern differential geometry. It was the standard reference before the next generation of textbooks on differential geometry appeared. It is not as easy or accessible as some of the more recent books, nor does it emphasize global differential geometry. Though there are many competing books now on this well-developed subject, this book still has a special place and is a very valuable reference. This is well-described by an Amazon review, "Volume 1 still remains unrivalled for its concise, mathematically rigorous presentation of the theory of connections on a principal fibre bundle—material that is absolutely essential to the reader who desires to understand gauge theories in modern physics. The essential core of Volume 1 is the development of connections on a principal fibre bundle, linear and affine connections, and the special case of Riemannian connections, where a connection must be 'fitted' to the geometry that results from a pre-existing metric tensor on the underlying manifold."

According to MathSciNet, "Conceptual material and technical detail alike are handled carefully and coherently..."

5.1.4 Special topics in differential geometry

J. Wolf, *Spaces of Constant Curvature*, Sixth edition, AMS Chelsea Publishing, Providence, RI, 2011. xviii+424 pp.

There are two types of questions and hence two approaches to differential geometry: either study general Riemannian manifolds, proving conclusions under various assumptions, or study some special classes of Riemannian manifolds. Riemannian manifolds of constant curvature are a distinguished class of Riemannian manifolds. They are even special among the important class of symmetric spaces. Though their universal covering spaces are simple, their geometry and topology are interesting and can be very complicated. For example, the seemingly simple question of classifying compact quotients of the Euclidean spaces (i.e., compact flat manifolds) is not simple at all. This comprehensive book on spaces of constant curvature is the book on a large class of special Riemannian manifolds of constant curvature. It also contains an effective and relatively short discussion of the theory of symmetric spaces. This is both a good introduction and a valuable reference.

According to MathSciNet, "This book is one of those studies that bridge two fields, in this case, group theory and differential geometry. The book serves as a serious application of the concepts in each field, and workers in either field should find it useful. It is not a book for beginners, but a serious graduate student with a little background in the fields mentioned, plus some topology, can find here ideas put to work... Strong points include the wealth of examples, the detailed exposition on classical groups in their various guises, crystallographic groups, Grassmann manifolds, and all the model spaces needed for classification theory. The book is an excellent view of the tactics of classification, and the author does not avoid the sticky details which such theory demands. He has collected and organized the work of many researchers and included much of his own work, including new results on symmetric spaces and indefinite metric manifolds of constant curvature. Needless to say, there is much work yet to be done, and the author helps point the way. A complete connected Riemannian manifold of constant curvature K is called a space form. The classification of space forms was formulated by W. Killing in 1891 and named the 'Clifford-Klein space form problem'. After the notions of completeness and covering manifold were well understood, H. Hopf formulated the problem in modern terminology in 1925. It became a question concerning fundamental groups since the simply connected covering must be either a sphere ($K > 0$), Euclidean space ($K = 0$), or hyperbolic space ($K < 0$)..."

B. O'Neill, *Semi-Riemannian Geometry*, With applications to relativity, Pure and Applied Mathematics, 103, Academic Press, Inc., New York, 1983. xiii+468 pp.

When differential geometry was used by Einstein to formulate the theory of general relativity, it was not Riemannian manifolds which were used to describe the space-time, but rather Lorentz manifolds. This book is probably the first successful book that combines Riemannian and Lorentz geometries under the general class of semi-Riemannian manifolds.

As the author explained in the preface, "For many years these two geometries have developed almost independently: Riemannian geometry reformulated in coordinate-free fashion and directed toward global problems, Lorentz geometry in classical tensor notation devoted to general relativity. More recently, this divergence has been reversed as physicists, turning increasingly toward invariant methods, have produced results of compelling mathematical interest."

According to a review in Bulletin of AMS, "An important feature of this book is the inclusion of two of the various singularity theorems of Hawking and Penrose. These theorems show that under certain conditions involving the Ricci curvature and the convergence of geodesies, the given space-time contains an incomplete causal geodesic... Another strong point of this book is the inclusion of a large number of important and useful problems. The book is written so that the readers with less time at their disposal may skip the exercise sections; however, most people will find these sections very helpful. This is an excellent book for anyone interested in semi-Riemannian geometry and/or General Relativity. It fills an important niche in the literature."

5.1. DIFFERENTIAL GEOMETRY

R. Montgomery, *A Tour of Subriemannian Geometries, Their Geodesics and Applications*, Mathematical Surveys and Monographs, 91, American Mathematical Society, Providence, RI, 2002. xx+259 pp.

Subriemannian geometries, or Carnot-Caratheodory geometries, are a certain type of generalization of Riemannian manifolds, and can be viewed as degenerated limits of Riemannian geometries. They occur naturally in constrained systems in classical mechanics, such as the motion of vehicles or robot arms, and the orbital dynamics of satellites. Roughly speaking, a subriemannian manifold is a manifold endowed with a distribution (i.e., a k-dimensional subbundle of the tangent bundle), called horizontal together with an inner product on that distribution. If k equals the dimension of the manifold, it is a usual Riemannian manifold. The Heisenberg group carries a natural sub-Riemannian structure determined by its center. For a subriemannian manifold, the distance between two points just as in the Riemannian case using lengths of paths between them, except the paths must be tangential to the distribution, i.e., it must travel along horizontal paths between the two points.

This book gives a panoramic guided tour of the subject and it is a valuable introduction.

H.B. Lawson, M. Michelsohn, *Spin Geometry*, Princeton Mathematical Series, 38, Princeton University Press, Princeton, NJ, 1989. xii+427 pp.

The spin group $Spin(n)$ is the double cover of the special orthogonal group $SO(n)$. When $n \geq 3$, $Spin(n)$ is simply connected and hence equal to the universal cover of $SO(n)$. It is a very important group in geometry and topology. One reason is that the possibility of reducing the structure group of the tangent bundle of a compact smooth manifold to spin allows one to define the Dirac operator, which is an elliptic differential operator on the manifold and can be used effectively to study the geometry and the topology of the manifold. For example, the Dirac operator was first used in the work of Atiyah and Singer on index theory in 1963 and has been used in many other contexts. This book discusses the spin geometry from that beginning up to recent results on the existence of Riemannian metrics with positive scalar curvature. It is both a good textbook and a valuable reference.

According to MathSciNet, "Spinors and the Dirac operator are concepts which are nowadays common currency in the global applications of differential geometry, index theory and gauge theory, yet they are nevertheless difficult to pinpoint in the basic literature. This is especially a problem for the graduate student who must either choose an exposition which is rooted in the geometry of four dimensions, or delve into the original papers of Atiyah, Bott and Shapiro or Cartan and Chevalley. The book by the authors is a godsend for this category of readers, as it deals with the algebra, geometry and analysis of spinors in a thorough yet readable manner... The sum total is a book of outstanding clarity. It is surely essential reading for any mathematician intent on learning how to use analytical methods in global Riemannian geometry."

A. Besse, *Einstein Manifolds*, Ergebnisse der Mathematik und ihrer Grenzgebiete (3) 10, Springer-Verlag, Berlin, 1987. xii+510 pp.

This is a very comprehensive reference on Einstein manifolds. Every smooth manifold admits many Riemannian metrics. A natural question is, "what are the best metrics for a given manifold?" In dimension 2, every smooth manifold admits a Riemannian metric of constant curvature, and they are the best in some sense and are relatively easy to be understood and to be used to understand the topology of the manifold. A direct generalization of this to higher dimension does not work for several reasons. First, there are at least three different notions of curvature: sectional curvature, Ricci curvature and scalar curvature. If we require constant sectional curvature, then here are many differential manifolds which do not admit Riemannian metrics of constant sectional curvature. If we require constant scalar curvature, the Riemannian metric does not reflect too much of the topology of the manifold, i.e., it is not a strong enough condition on the manifold. It turns out that the right generalization in the higher dimension is the constant Ricci curvature. This corresponds to the Einstein's vacuum field equations equation in the general relativity, and hence a metric with constant Ricci curvature is called an Einstein metric. It turns out that the structure of Einstein metrics are rich and Einstein manifolds have powerful applications in many other subjects in mathematics. This book tells a coherent story of such important manifolds, as the book description say, "It is an efficient reference for many fundamental techniques of Riemannian geometry as well as excellent examples of the interaction of geometry with partial differential equations, topology and Lie groups. Certainly the monograph provides a clear insight into the scope and diversity of problems posed by its title."

A. Besse, *Manifolds all of Whose Geodesics are Closed*, With appendices by D. B. A. Epstein, J.-P. Bourguignon, L. Barard-Bergery, M. Berger and J. L. Kazdan. Ergebnisse der Mathematik und ihrer Grenzgebiete, 93, Springer-Verlag, Berlin-New York, 1978. ix+262 pp.

If we ask the question, "what is so special about the unit sphere S^2 in $\mathbb{R}^3$ as a Riemannian manifold?", there are many different answers. One possible answer might be that all its geodesics are closed, and all simple geodesics have the same common length. A simple question whether there are any other surfaces having the same property turns out to be not so simple. Another question concerns the same property for higher dimensional Riemannian manifolds. The compact symmetric spaces of rank one enjoy this property. The property that all geodesics are closed and have the same length (or a slightly relaxed condition) has a natural interpretation in terms of the geodesic flow. These indicate that the subject is connected to many other topics of mathematics.

5.1. DIFFERENTIAL GEOMETRY

This book gives the first systematic discussion of known results and open problems regarding manifolds having only closed geodesics. This book is also a valuable reference for such special Riemannian manifolds. According to MathSciNet, "Although the title could indicate a rather narrowly specialized book, just the opposite is true. The author manages to touch on most of the major themes in modern differential geometry while presenting the current state of knowledge concerning Riemannian manifolds having certain restricted behaviors of their geodesics."

D. Joyce, *Compact Manifolds with Special Holonomy*, Oxford Mathematical Monographs, Oxford University Press, Oxford, 2000. xii+436 pp.

For any differential manifold with a connection, in particular a Riemannian manifold with its canonical Levi-Civita connection, there is a holonomy group, which measures the change of tangent vectors under parallel transport along loops based at a fixed point. For a generic Riemannian manifold of dimension n, its holonomy group is equal to $SO(n)$. If it is different from $SO(n)$, it is called a manifold with special hononomy. The holonomy group reflects the geometry of the manifold. This subject has attracted attention from both differential geometers and also string theorists. In the middle of 1990s, the author produced the first examples of compact manifolds with exceptional holonomy G_2 and $Spin(7)$, and is an expert in this subject.

This is the first systematic book that studies Riemannian manifolds with special holonomy. It contains many topics which are hard to be found in any other book such as a good introduction to Calabi-Yau manifolds and a proof of the Calabi conjecture.

From the book description, "The book starts with a thorough introduction to connections and holonomy groups, and to Riemannian, complex and Kähler geometry. Then the Calabi conjecture is proved and used to deduce the existence of compact manifolds with holonomy $SU(m)$ (Calabi-Yau manifolds) and $Sp(m)$ (hyperkähler manifolds). These are constructed and studied using complex algebraic geometry. The second half of the book is devoted to constructions of compact 7- and 8-manifolds with the exceptional holonomy groups G_2 and $Spin(7)$. Many new examples are given, and their Betti numbers calculated..."

According to MathSciNet, "The book is well written, with many references to the literature, helpful guidance to the reader, discussion of relationships to physics and pointers to directions for future research... This book is an important reference in the study of Ricci-flat geometry."

S. Kobayashi, *Differential Geometry of Complex Vector Bundles*, Publications of the Mathematical Society of Japan, 15, Kano Memorial Lectures, 5, Princeton University Press, Princeton, NJ; Iwanami Shoten, Tokyo, 1987. xii+305 pp.

Einstein metrics on Riemannian manifolds are a natural generalization of spaces of constant curvature and sought after for various reasons. For Hermitian complex vector bundles over complex manifolds, the existence of the Einstein-Hermitian metric on them is probably the right choice and corresponds to the stability of the vector bundles in algebraic geometry, which is needed to construct moduli spaces of vector bundles. The theory of complex vector bundles has attracted attention of algebraic and differential geometers, and also low-dimensional topologists and mathematical physicists working on gauge theory. The purpose of this book is to lay a foundation for the theory of Einstein-Hermitian vector bundles. This is a valuable introduction and a useful reference.

According to MathSciNet, "The central theme is the relationship between the fundamental notions of stability of a bundle on a compact Kähler manifold, on the one hand, and the Einstein-Hermitian condition on the curvature of the bundle in some Hermitian metric, on the other hand. Stability was introduced by Mumford for curves and later Takemoto in general, while the Einstein-Hermitian condition was introduced by the author as an attempt to interpret stability in terms of curvature conditions. This interface has had a remarkable flourishing over the last ten years, culminating in recent work of Donaldson, Uhlenbeck-Yau and others...... The author has provided a valuable introduction to a field that is evolving very rapidly..., this volume is highly recommended for making this area quickly accessible to students and interested researchers alike."

R. Sharpe, *Differential Geometry. Cartan's Generalization of Klein's Erlangen Program*, With a foreword by S. S. Chern, Graduate Texts in Mathematics, 166, Springer-Verlag, New York, 1997. xx+421 pp.

For school children, geometry means the Euclidean geometry. There have been several great generalizations of the Euclidean geometry, for example the hyperbolic geometry. These were unified by Felix Klein using homogeneous spaces of Lie groups through his famous Erlangen program. Another great generalization is Riemannian geometry. These two generalizations seemed to share little in common. In the early 1920s, Elie Cartan obtained a common generalization of these theories. These are called Cartan geometries. The purpose of this book is to explain why differential geometry is the study of a connection on a principal bundle and to introduce these Cartan geometries in a self-contained way to beginning graduate students.

According to MathSciNet, "This book discusses geometric structures infinitesimally and locally modeled on classical geometries. A 'classical geometry' is—according to Klein's Erlangen program—the geometry of a homogeneous space G/H of a Lie group G. For Klein, the fundamental entities of the geometry are just classes of objects on G/H invariant under G. The author calls such a geometry a Klein geometry. A natural question is how (and whether) a given smooth manifold M can be modeled (infinitesimally, locally, or globally) on a Klein geometry. For example, Euclidean geometry is the geometry of R^n invariant under the group G of rigid motions of $\mathbb{R}^n$, and a Euclidean structure is just a flat Riemannian metric. Of closed surfaces, only the torus and Klein bottle can be locally modeled (or uniformized) by $\mathbb{R}^2$ but every surface has an 'infinitesimal Euclidean'—that is, Riemannian—geometry. A Cartan geometry is a structure infinitesimally modeled on a Klein geometry. For example, Riemannian metrics are infinitesimally Euclidean geometries. A Riemannian structure is locally Euclidean if and only if the metric is flat—that is, if the Riemann curvature tensor vanishes. In a similar way, a general Cartan geometry possesses an infinitesimal invariant whose vanishing detects whether the infinitesimal model given by the Cartan geometry is an exact local model of the geometry... Valuable to a wide audience of mathematicians and physicists, this book should stimulate further research into hitherto unexplored applications of the ideas of Klein and Cartan."

L. Santaló, *Integral Geometry and Geometric Probability*, With a foreword by Mark Kac, Encyclopedia of Mathematics and its Applications, Vol. 1, Addison-Wesley Publishing Co., Reading, Mass.-London-Amsterdam, 1976. xvii+404 pp.

Integral geometry studies measures on sets of geometrical objects of a space such as points, lines, and motions, which is invariant under the symmetry group of that space. More recently, it also includes invariant (or equivariant) transformations from the space of functions on one geometrical space to the space of functions on another geometrical space, which often take the form of integral transforms such as the Radon transform and its generalizations.

Integral geometry first emerged as an attempt to refine certain statements of geometric probability theory, and the early contributors include Morgan Crofton, Luis Santaló and Wilhelm Blaschke. One example is the classic theorem of Crofton that expresses the length of a plane curve as an expectation of the number of intersections with a random line.

This book is a classic and gives a comprehensive and accessible introduction to both integral geometry and geometric topology.

According to MathSciNet, "In integral geometry, appropriate group invariance requirements determine the most interesting measures; these notions have been applied to geometric questions such as isoperimetric problems. Since the time of Blaschke's school, integral geometry has grown steadily. Separately, geometric probability and its relative, stochastic geometry, of the late R. Davidson (1968–1970), have resumed a flourishing development, often motivated by practical applications.

The author's monograph covers all these topics, in their current states, with encyclopedic fullness and pedagogic skill. By so doing, it promises to become both a standard textbook and the basic reference for years to come. Yet its greatest service may be the re-unification of integral geometry with modern geometric probability, to the benefit of both subjects."

According to the review in Bulletin of AMS by S.S. Chern, "Integral geometry was the name coined by Wilhelm Blaschke in 1934 for the classical subject of geometric probability. During that year the author came to Hamburg from Madrid and the reviewer from China, and we sat in Blaschke's course on geometric probability. The main reference was an 'Ausarbeitung' of a course by the same name given by G. Herglotz in Göttingen. At the end of 1934 the author found his now famous proofs of the isoperimetric inequality in the plane and Blaschke himself found the fundamental kinematic formula and started a series of papers under the general title of 'integral geometry'. It was a fruitful and enjoyable year for all concerned. Integral geometry is exactly 200 years old if we identify its birth with Buffon's solution in 1777 of the needle problem... As the first volume of an ambitious encyclopedia, the book sets a style. It is a combination of a lucid exposition of the introductory aspects and a complete survey of the area. No topic seems to have been left uncovered. There is also a very complete, but selective, bibliography. The author handled his material with great dexterity and ease. The book should serve as an excellent text for a graduate course on integral geometry. The encyclopedia and the author are to be congratulated for their success."

E. Giusti, *Minimal Surfaces and Functions of Bounded Variation*, Monographs in Mathematics, 80, Birkhäuser Verlag, Basel, 1984. xii+240 pp.

Minimal surfaces, or rather soap bubbles, are probably one of the earliest geometric shapes children play with. Mathematically, the problem of finding minimal surfaces bounded by a given curve in the space was one of the first problems considered after the foundation of the calculus of variations was established, but a satisfactory solution was obtained only relatively recently. After the initial works of Douglas and Rado, De Giorgi, Reifenberg, Federer and Fleming, Almgren et al made substantial contributions. This book is the first self-contained comprehensive introduction to this classical and rapidly developing subject.

According to MathSciNet, "This book deals with Plateau's problem, which is to find a hypersurface of least area spanning a given boundary. It was only in 1930/31 that a solution of this old problem was found by Douglas and Rado for surfaces in $\mathbb{R}^3$, and it took another 30 years until the higher-dimensional case was attacked by means of measure-theoretic methods... These remarkable results arise from a most fruitful period in the theory of higher-dimensional minimal surfaces. This book gives at first a comprehensive and self-contained representation of this important development. The reading requires only a "fairly good knowledge of general measure theory and some familiarity with the theory of elliptic partial differential equations'. It is facilitated by a well-balanced notation. In its entirety, the book is, in the opinion of the reviewer, a masterpiece of clear (and aesthetic) representation of highly nontrivial analysis."

5.1. DIFFERENTIAL GEOMETRY

5.1.5 M. Gromov, Metric Structures for Riemannian and Non-Riemannian Spaces, Based on the 1981 French original, With appendices by M. Katz, P. Pansu and S. Semmes, Translated from the French by Sean Michael Bates, Progress in Mathematics, 152, Birkhäuser Boston, Inc., Boston, MA, 1999. xx+585 pp.

Gromov has revolutionized the study of geometry and related topics by his "soft" or "coarse" viewpoint and approach. This book contains a lot of Gromov's original ideas to study geometry, topology of Riemannian manifolds and relations between them. The first French edition of this book was very influential, and the new and much longer edition contains a lot more material. One can read this book in order to become familiar with Gromov's approach to geometry, to get inspiration, gain new perspectives and learn open problems in geometry.

According to a review of the first French edition in Bulletin of AMS, "The influence of this 'little green book' can hardly be overestimated. Its 150 pages were packed with striking new concepts and ideas. In particular, it was this book that spread the idea of convergence of Riemannian manifolds to a larger audience..."

From the preface of the first edition: "Our purpose was to present some of the links that have been established between the curvature of a Riemannian manifold V and its global behavior. Here, the word 'global' applies not only to the topology of V but also to a family of metric invariants of Riemannian manifolds and mappings between these manifolds. The simplest metric invariants of V are, for example, its volume and its diameter; the dilatation is an important invariant for a mapping from V_1 into V_2. In fact, such metric invariants also appear in a purely topological context, and they provide an important link between infinitesimal data on V (generally expressed by a condition on the curvature) and the topology of V. For example, the now classical theorem of Bonnet gives an upper bound for the diameter of a manifold V with positive curvature, from which one can deduce the finiteness of the fundamental group of V. A deeper topological study of Riemannian manifolds requires finer metric invariants than diameter or volume; we have tried to present a systematic treatment of these invariants, but this study is far from being as comprehensive as we had hoped."

For the second edition, "Among the most substantial additions, each taking over a hundred pages, there is a chapter on convergence of metric spaces with measures, and an appendix on analysis on metric spaces written by Semmes."

W. Ballmann, M. Gromov, V. Schroeder, *Manifolds of Nonpositive Curvature*, Progress in Mathematics, 61, Birkhäuser Boston, Inc., Boston, MA, 1985. vi+263 pp.

This book gives a systematical study of manifolds of nonpositive curvature and then discusses a generalization of the Mostow strong rigidity and finiteness results for manifolds of finite volume. It puts the Mostow strong rigidity in the framework of manifolds of nonpositive curvature. Some related new and known results are also discussed such as the geodesic compactification and the compactification by horofunctions, structures of nonpositively curved manifolds of higher rank. It is based on lectures delivered by M. Gromov at the Collége de France in Paris.

This book has influenced the active study of Riemannian manifolds of nonpositive curvature and more generally geodesic spaces of nonpositive curvature, i.e., the so-called CAT(0)-spaces. It is still a very valuable introduction and reference on manifolds of nonpositive curvature.

According to a review in Bulletin of AMS, "Its purpose is twofold, namely to give an introduction to manifolds of nonpositive curvature and to give the proof of two outstanding results: the rigidity of locally symmetric spaces in the class of all manifolds of nonpositive curvature (in generalization of Mostow's rigidity theorem), as well as an estimate for the topology of nonpositively curved analytic manifolds of finite volume (for more precise statements see below)...

The book by M. Gromov, V. Schroeder, and W. Ballmann is an extraordinary one which contains a wealth of new ideas as well as plenty of inspiration for further work... It is of course not a textbook in the usual sense. Although it starts quite easy it ends more or less like a research paper. But due to the excellent work of V. Schroeder the presentation is always clear. I think the book will have a strong influence on the further development of the theory of manifolds of nonpositive curvature." This indeed true.

D. Burago, Y. Burago, S. Ivanov, *A Course in Metric Geometry*, Graduate Studies in Mathematics, 33, American Mathematical Society, 2001.

The theory of length spaces has grown rapidly in the past few decades and has been used effectively in differential geometry, geometric group theory, dynamical systems, and partial differential equations etc. The purpose of this book is to give an introduction to the theory of length spaces and many other topics in geometry that are related to the notion of distance, or the broad geometry in the style of Gromov. This is a textbook and the authors tried to "bridge the gap between students and researchers interested in metric geometry, and modern mathematical literature."

According to MathSciNet, "In recent years the area of synthetic methods in geometry has experienced an explosive growth. While the traditional approach to differential geometry is based on a build-up of heavy analytic machinery, synthetic methods have the advantage of easy accessibility based on simple geometric axioms that have an immediate appeal to geometric intuition. A textbook for students interested in entering this area was long overdue. The present volume fills this gap... It was one of the goals of the authors to make this book more accessible, and clearly they have accomplished this goal. For example, the reader will appreciate the detailed discussion of Gromov-Hausdorff con-

5.1. DIFFERENTIAL GEOMETRY

vergence in Chapter 7 which complements the rather brief treatment of this subject in Gromov's book."

5.1.6 M. Bridson, A. Haefliger, Metric Spaces of Non-positive Curvature, Grundlehren der Mathematischen Wissenschaften, 319, Springer-Verlag, Berlin, 1999. xxii+643 pp.

This is a carefully written comprehensive book on spaces of nonpositive curvature and groups that act properly on them. It is accessible and can be used as a book for beginners to learn the subject, and is also a valuable reference for experts.

According to MathSciNet, "This book constitutes an excellent introduction to the theory of metric spaces of non-positive curvature. The bases of the theory as well as the most important recent developments are treated. All the results are proved in full generality with complete details, and the most interesting examples are given.

The basic notion involved is that of a geodesic metric space with curvature bounded above. A. D. Aleksandrov, who laid the foundations of the theory, gave several equivalent definitions of this property, ... We note by the way that H. Busemann worked independently on a slightly different theory of curvature in metric spaces, based on the convexity of the distance function, and in fact his first paper on the subject is prior to the work of Aleksandrov... More recently, M. Gromov has revived the interest in Aleksandrov's theory... In conclusion, the book under review assembles several most interesting results in a very active subject which otherwise were spread over various articles and books."

The content of this book is well summarized by the book description, "A description of the global properties of simply-connected spaces that are non-positively curved in the sense of A. D. Alexandrov, and the structure of groups which act on such spaces by isometries. The theory of these objects is developed in a manner accessible to anyone familiar with the rudiments of topology and group theory: non-trivial theorems are proved by concatenating elementary geometric arguments, and many examples are given."

According to a review in Bulletin of AMS, "The study of nonpositively curved spaces goes back to the discovery of hyperbolic space, the work of Hadamard around 1900, and Cartan's work in the 20's. These spaces play a significant role in many areas: Lie group theory, combinatorial and geometric group theory, dynamical systems, harmonic maps and vanishing theorems, geometric topology, Kleinian group theory, and Teichmüller theory. In some of these contexts, for instance in dynamics and in harmonic map theory, nonpositive curvature turns out to be the right condition to make things work smoothly, while in others such as Lie theory, 3-manifold topology, and Teichmüller theory, the basic objects of study happen to be nonpositively curved spaces. With so many closely related interdependent fields, nonpositive curvature has been a very active topic in the last twenty years."

5.1.7 S. Helgason, Differential Geometry, Lie Groups, and Symmetric Spaces, Corrected reprint of the 1978 original, Graduate Studies in Mathematics, 34, American Mathematical Society, Providence, RI, 2001. xxvi+641 pp.

Though symmetric spaces and their quotients (locally symmetric spaces) have played a fundamental role in mathematics, there are not many books on them. This book was the first systematic introduction to the theory of symmetric spaces and has been the standard reference on symmetric spaces. It is an expanded version of an earlier book by the same author:

S. Helgason, *Differential Geometry and Symmetric Spaces*, Pure and Applied Mathematics, Vol. XII, Academic Press, New York-London, 1962, xiv+486 pp.

According to MathSciNet, "Symmetric spaces were discovered by Élie Cartan in 1926 and have been the object of extensive study first by him and since by many others. From a global viewpoint, a symmetric space is a Riemannian manifold which possesses a symmetry about each point, that is, an involutive isometry leaving the point fixed. This generalizes the notion of reflection in a point in ordinary Euclidean geometry. From this assumption it follows that these spaces are homogeneous in the sense that the group of isometries, which is a Lie group, is transitive on the manifold. This circumstance accounts for an intimate connection of the theory of these spaces with Lie groups. Symmetric spaces were actually classified by Cartan: they include the spaces of classical geometry (Euclidean, non-Euclidean, projective, etc.) and compact Lie groups, among others. They furnish a sufficiently varied class to be of great interest in differential geometry, algebraic topology, certain types of analytic or geometric number theory, and in certain areas of analysis. Moreover, the study of their geometry is useful in the study of semi-simple Lie groups. In spite of their usefulness and interest, there has been, to the reviewer's knowledge, no self-contained comprehensive book on symmetric spaces. For this reason alone, this book would be most welcome. But in fact, it has much more to commend it. It is a well-organized and very well-written text, which will certainly make this beautiful theory accessible to many mathematicians who would not otherwise study it and which, it may be predicted, will become a standard reference for those working in the subject.... this book is quite self-contained and thoroughly covers the basic results on symmetric spaces, without assuming a preliminary knowledge of either differential geometry or Lie group theory..."

According to some people, the earlier book published in 1962 might be better organized and more concise than the expanded version in 1978. Indeed, the citation of the Steele Prize to Helgason in 1988 says, "In 1962 Sigurdur Helgason published a book which has become a classic. The subject matter included central topics in geometry and

Lie group theory, with important ramifications for harmonic analysis. More recently this material has been revised and expanded into a two volume treatment... The exposition throughout is a model of clarity. Arguments in proofs are very clean, the organization is superb, and the material ranges over a wide vista of important topics of interest to a broad segment of the mathematical community."

Chapter X, "Functions on symmetric spaces", of the book in 1962 was expanded into another book.

S. Helgason, *Geometric Analysis on Symmetric Spaces*, Second edition, Mathematical Surveys and Monographs, 39, American Mathematical Society, Providence, RI, 2008. xviii+637 pp.

By geometric analysis in the title, the author meant integration and partial differential equations on symmetric spaces which enjoy some invariant properties with respect to the Lie groups that act on the symmetric spaces. In particular, it includes joint eigenfunctions and eigenspace representations of all invariant differential operators, generalization of Fourier decomposition and Fourier transformation, the Radon transformation etc. This is a book written by an expert on these topics. The writing is clear, and the style is a bit formal sometimes.

According to MathSciNet, "This volume is intended to serve both as a text and as a reference... This volume makes an excellent companion to the author's *Differential geometry, Lie groups, and symmetric spaces*, putting to work many of the abstract concepts developed in the earlier volume. The introductory material and large number of exercises (with answers!) will make the book quite appropriate for students. Researchers will find numerous useful references on geometric analysis, along with proofs, connections with other parts of mathematics, and valuable historical remarks."

J. Faraut, A. Korányi, *Analysis on Symmetric Cones*, Oxford Mathematical Monographs, Oxford Science Publications, The Clarendon Press, Oxford University Press, New York, 1994. xii+382 pp.

A particularly important example of symmetric spaces is the space of positive definite matrices $GL^+(n,\mathbb{R})/SO(n)$, which is a symmetric cone and has been extensively studied in number theory, in particular in connection with the geometry of numbers and the theory of quadratic forms. It has also been very useful in statistics. Another important symmetric cone is the Lorentz cone, which is played a crucial role in the wave equation. These examples are the most basic ones among the class of symmetric cones in Euclidean spaces, which give rises to *linear symmetric spaces*. This book is the first systematic exposition of the geometry and analysis of symmetric cones. It is an accessible introduction and a valuable reference.

According to Zentralblatt MATH, "The approach is based on the theory of Jordan algebras in the way worked out by M. Koecher and his school. This approach gives explicit formulas and also a classification of the symmetric cones relatively easily. Symmetric cones and tubes over them are examples of Riemannian symmetric spaces. The class of symmetric cones has special features which make it possible to go much further in explicit harmonic analysis than in the general case of arbitrary Riemannian symmetric spaces. This is due to the interplay of Euclidean harmonic analysis on the ambient vector space, and of the non-commutative harmonic analysis of the cone as a symmetric space."

5.2 Geometric analysis

Ricci's full name is Gregorio Ricci-Curbastro, and his last name should be Ricci-Curbastro. Since he wrote an important book on the calculus of tensors with his former student Levi-Civita and used his shortened name Ricci, he is usually called by Ricci now. Riemann did not have many formal students (only Roch was listed in the mathematics genealogy project), but he influenced and educated many Italian students where he was recovering in the warm Southern Italy, and Ricci was one such person. The Ricci flow is one of the most important topics in geometric analysis due to its success with a positive solution of the Poincaré conjecture.

Figure 5.2: Ricci's book.

Geometric analysis combines differential geometry and differential equations. It includes both the use of geometrical methods in the study of partial differential equations on $\mathbb{R}^n$ and general manifolds, and the application of the theory of partial differential equations to understand geometry and topology of manifolds. Besides the linear differential equations such as the Laplace equation, heat and wave equations, there are also important nonlinear differential equations such as the minimal surface equation and the Monge-Ampére equation. Geometric analysis has been applied to solve many outstanding problems in algebraic geometry via the Calabi conjecture and in topology such as the Poincaré conjecture.

5.2. GEOMETRIC ANALYSIS

5.2.1 R. Schoen, S.T. Yau, Lectures on Differential Geometry, International Press, 1994. v+235 pp.

Schoen and Yau have contributed significantly to geometric analysis on manifolds. They have introduced many fundamental methods and tools to solve important problems. This book was based on their lectures starting around 1984, and many results were published in book form for the first time. The excitement and freshness of the lectures are preserved in the book. This book gives an excellent account of differential geometry from an analytic point of view. Geometric analysis is a very active field of research nowadays, and the work of the authors had a great impact on the subject. Therefore, this is a book written by the main and active contributors. It contains a wealth of information and can serve both as a unique textbook for students and a very solid reference for all.

According to MathSciNet, "The book contains significant results in differential geometry and global analysis; many of them are the works of the authors. The main topics are differential equations on a manifold and the relation between curvature and topology of a Riemannian manifold. There are nine chapters in the book, with the last three chapters more like appendices, which focus on problems concerning different areas of differential geometry... The book under review is very well written. Readers will find comprehensive and detailed discussions of many significant results in geometric analysis. The book is both useful as a reference book for researchers and as a course book for graduate students. With details of proofs and background materials presented in a concise and delightful way, the book provides access to some of the most exciting areas in differential geometry."

Harmonic maps are generalizations of harmonic functions on Riemannian manifolds and geodesics in Riemannian manifolds (they are respectively harmonic maps to $\mathbb{R}$ and from $\mathbb{R}$). They are, in some sense, the best maps one can get between Riemannian manifolds. Harmonic maps has many applications in differential geometry, topology and mathematical physics. For example, several rigidity results of manifolds can be proved by harmonic maps.

R. Schoen, S.T. Yau, *Lectures on Harmonic Maps*, Conference Proceedings and Lecture Notes in Geometry and Topology, II. International Press, Cambridge, MA, 1997. vi+394 pp.

This book was also written by two experts with real insight and is a very valuable reference on harmonic maps. It is a sequel to the book above.

According to Zentralblatt MATH, "The monograph under review covers a series of recent publications on harmonic maps and their relationship with geometry and topology of Teichmüller spaces and Kähler manifolds. Consequently, it is a fundamental reference for those working on such subjects."

5.2.2 T. Aubin, Nonlinear Analysis on Manifolds, Monge Ampére equations, Grundlehren der Mathematischen Wissenschaften, 252, Springer-Verlag, New York, 1982. xii+204 pp.

This is probably the first systematic and advanced introduction to nonlinear differential equations on manifolds. It has been a valuable and influential reference for geometric analysts.

According to MathSciNet, "This book deals with nonlinear problems in geometry such as the problems of Yamabe, Calabi, Nirenberg and some other problems, Monge-Ampere equations on compact Kähler manifolds and on a bounded domain. The book presents a number of methods to deal with these problems, some basic results, and some open problems. The book is intended as an introduction to research... The book is a valuable contribution to the literature by one of the active researchers in the field."

T. Aubin, *Some Nonlinear Problems in Riemannian Geometry*, Springer Monographs in Mathematics, Springer-Verlag, Berlin, 1998. xviii+395 pp.

This is an expanded and updated version of the previous book.

According to MathSciNet, "This book is a significantly expanded revision of the author's classic 1982 book. It contains everything that was in the 1982 version, essentially unchanged, plus a nearly equal amount of new material.

The 1982 book began with a four-chapter overview of essential background material from Riemannian geometry, partial differential equations, and functional analysis. It then treated in depth three of the most significant geometric applications of nonlinear PDEs up to that time: the Yamabe problem, to find a metric of constant scalar curvature conformal to a given one; the Calabi conjecture, that every 2-form representing the first Chern class of a compact Kähler manifold is the Ricci form of some Kähler metric, along with the related problem of finding Kähler-Einstein metrics; and the Dirichlet problem for real or complex Monge-Ampère equations on bounded domains..."

5.2.3 More books on geometric analysis

J. Jost, *Riemannian Geometry and Geometric Analysis*, Fifth edition, Universitext, Springer-Verlag, Berlin, 2008. xiv+583 pp.

There have been many books on differential geometry and Riemannian geometry, and also some books on analysis and geometric analysis. This is probably the first book which gives a systematic introduction to both subjects in a single book. With the updates in these editions, this book is also becoming more comprehensive. As the author stated in the preface, "It attempts to a synthesis of geometric and analytic methods in the study of Riemannian manifolds."

5.2. GEOMETRIC ANALYSIS

According to MathSciNet, "This book provides a very readable introduction to Riemannian geometry and geometric analysis. The author focuses on using analytic methods in the study of some fundamental theorems in Riemannian geometry, e.g., the Hodge theorem, the Rauch comparison theorem, the Lyusternik and Fet theorem and the existence of harmonic mappings...

With the vast development of the mathematical subject of geometric analysis, the present textbook is most welcome. It is a good introduction to Riemannian geometry. The book is made more interesting by the perspectives in various sections, where the author mentions the history and development of the material and provides the reader with references."

I. Chavel, *Eigenvalues in Riemannian Geometry*, Including a chapter by Burton Randol, With an appendix by Jozef Dodziuk., Pure and Applied Mathematics, 115, Academic Press, Inc., Orlando, FL, 1984. xiv+362 pp.

Though the Laplace operator of a Riemannian manifold is the most basic one, and the spectral theory of the Laplace operator is of fundamental importance, there are not many books devoted to the spectral theory of the Laplace operator and its interplay with the geometry, i.e., the subject of spectral geometry. This book was written when the spectral geometry has reached one height, and it is a valuable reference.

According to a review in Bulletin of AMS. "The recent explosion of activity studying the relation between geometric and analytic properties of spaces has fused many areas of mathematics, such as the traditionally disparate fields differential geometry, partial differential equations, topology, mathematical physics, and number theory. One of the most popular topics in this study is the search for properties of the spectrum of the Laplace operator of a manifold in terms of its geometric invariants... Being invariantly defined, it is the simplest geometric elliptic operator which appears everywhere in geometry. It is the principal part of the expression for scalar curvature of a conformal factor in a metric as well as the mean curvature and stability form of a hypersurface. More importantly it is the linearization of the many nonlinear operators in geometry such as the Gauss curvature operator, the mean curvature operator and the Monge-Ampere operator. It is essential, therefore, to the understanding of nonlinear phenomena as well as the operator to study in the first model problems.

The early results were summarized fifteen years ago by Berger, Gauduchon, and Mazets in *Le spectre d'une variété riemannienne*, which at that time well represented the literature on the 'geometry of the Laplace operator.' Growing out of a set of lectures from the late sixties, the purpose of these notes was to acquaint graduate students or newcomers to the field of eigenvalues on manifolds with the basic background and results. One of the high points was the asymptotic expansion for the heat kernel which enabled one to read off geometric invariants of the manifold such as the volume and integral of scalar curvature from the ' high-end' behavior of the spectrum. There were also a few scattered results on the first nonzero eigenvalue estimates. However, with the currently vigorous activity, it is high time for another overview of newer results and, perhaps more importantly, of the techniques from the field. To this end, Isaac Chavel offers his *Eigenvalues in Riemannian Geometry*."

E.B. Davies, *Heat Kernels and Spectral Theory*, Cambridge Tracts in Mathematics, 92, Cambridge University Press, Cambridge, 1989. x+197 pp.

This is not an easy textbook, but rather an advanced monograph. Besides the Laplace operator of complete Riemannian manifolds, it also discusses differential operators with variable coefficients on regions in the Euclidean space. Due to this generality, new and general techniques are used, which sometime lead to better results on bounds of the heat kernel of the differential operator. The writing is sometimes dense or concise, but it is a valuable reference.

According to Zentralblatt MATH, "This advanced monograph investigates through a study of heat kernels the spectral properties of linear, selfadjoint elliptic differential operators and obtains pointwise bounds for associated eigenfunctions. The book is a first account of the new techniques which can be used as a result of the recent improvements in understanding of the heat kernels obtained by means of quadratic form technique and logarithmic Sobolev inequalities. In contrast to older theories the new method enables pointwise upper and lower bounds to be obtained in terms of constants which are computable and, in many cases, sharp. The adopted approach, which is entirely analytical, enables operators in divergence form with measurable second order coefficients to be dealt with in a straight forward way... The theory for operators with variable coefficients is developed in a region of Euclidean space whilst heat kernels of Laplace-Beltrami operators are studied on Riemannian manifolds."

N. Berline, E. Getzler, M. Vergne, *Heat Kernels and Dirac Operators*, Grundlehren der Mathematischen Wissenschaften, 298, Springer-Verlag, Berlin, 1992. viii+369 pp.

The Dirac operator has gained wide recognition in mathematics since the work of Atiyah and Singer on index theory, and the heat kernel has provided an important approach to the index theory. This book provides a systematic exposition of the two topics in the title by incorporating relatively recent results, some of which are due to the authors. It is a valuable reference.

According to a review in Bulletin of AMS, "Perhaps the most famous theorem of the 1960's, the Atiyah-Singer index theorem bears all the hallmarks of great mathematics: it draws on and relates several fields in mathematics, explicating and amplifying relationships between them, and in turn contributes to the internal structure of each... The book... is an outstanding monograph on this beautiful subject. As its title promises, the

5.2. GEOMETRIC ANALYSIS

viewpoint adopted here is the systematic use of heat-kernel techniques to study a variety of topics in index theory. The topics covered reflect the tastes of the authors; many results here are due to them, and many other important results are included, often with new proofs to fit the general flow of ideas... Although in principle self-contained, this book might be a bit heavy for the utter novice, but it treats all the background material thoroughly and well. In particular, a construction of the heat-kernel for second order geometric operators is included; although this appears in other places, this is a particularly readable and complete account. The section on geometry, including the discussion of characteristic classes, is probably too brief for someone who has never encountered them before, but the discussion of superconnections is good."

P. Gilkey, *Invariance Theory, the Heat Equation, and the Atiyah-Singer Index Theorem*, Second edition, Studies in Advanced Mathematics, CRC Press, Boca Raton, FL, 1995. x+516 pp.

This book discusses a heat kernel proof of the Atiyah-Singer index theorem and other related theorems. Though it is well-known that the heat kernel admits small time asymptotic expansions, explicitly identifying the coefficients is not easy. It turns out that for the integrand of the index theorem that comes from the heat kernel asymptotics, there are amazing cancelation, and these coefficients can be evaluated by using the invariance theory. This book presents a self-contained introduction to this approach to the index theory.

According to a review in Bulletin of AMS, "It is relatively self-contained and collects material previously scattered in research literature. It is very suitable to be a textbook for a graduate course, yet it takes the reader up to an area of active research. Gilkey's book will be very valuable to practitioners as well as students."

G. Pólya, G. Szegö, *Isoperimetric Inequalities in Mathematical Physics*, Annals of Mathematics Studies, no. 27, Princeton University Press, Princeton, N. J., 1951. xvi+279 pp.

The Laplace operator and its spectrum describe the vibration of membranes and have played an important role in mathematical physics. On the other hand, clamped plates are described by a higher order differential operator. Many inequalities concerning eigenvalues of the Laplace operator of domains in Euclidean space with either Dirichlet and Neumann boundary conditions, and other differential operators, are studied in this book, and many results and methods are due to the authors. Some problems on extremal eigenvalues raised in this book have been influential and have been intensely studied recently. This is a classic book on bounds of eigenvalues and other invariants of domains of Euclidean space.

According to MathSciNet, "The book brings together the known results, many of which are due to authors, by taking, perfecting, developing questions of geometric extrema, some old and more or less known or dealt with before: the old isoperimetric problem, maximum torsional rigidity of a cylinder section of area fixed (reached for the circular cross-section according to intuitive considerations of Saint-Venant), minimum electrostatic capacity of a body of given volume (performed according to the sphere of Poincaré incomplete information), fundamental frequency minimum of membranes or plates embedded in given area (Lord Rayleigh) and a more or less imposed, etc. These questions repeatedly in recent decades by various authors lead to many inequalities, called here by extension of the isoperimetric problem known and are, moreover, given the difficulty of exact calculations, useful in practice, compared with cases individuals for whom we can give explicit formulas used or numerical results and the authors detail in tables."

5.2.4 M. Gromov, Partial Differential Relations, Ergebnisse der Mathematik und ihrer Grenzgebiete (3), 9, Springer-Verlag, Berlin, 1986. x+363 pp.

This is an unusual book, full of new ideas and results. It is a far-reaching development of work in differential geometry and topology of Whitney, Nash, the author and many others. It studies many questions about immersions of Riemannian manifolds. Some of the results in this book may not be easy to follow, but this book has been influential.

The extensive review in MathSciNet gives a very good description of this book, "Around 1970, the world of differential geometry was astounded by the news that a young Russian by the name of Mikhael Gromov had proved that any noncompact differential manifold admits a Riemannian metric of positive sectional curvature, and also one of negative sectional curvature. We were also told that this was achieved by a "soft" method of topological sheaves. Moreover, in one and the same setting, Gromov also proved generalizations of both the Hirsch-Smale immersion theorem and the A. Phillips submersion theorems. Many more results were promised. Slowly, Gromov's papers (some in collaboration with Ya. M. Éliashberg and V. A. Rokhlin) filtered to the West in the early seventies... After a lapse of some fifteen years, the author has now presented what would appear to be his valedictory statement on the subject. Within the covers of the volume under review, he has deepened, generalized and synthesized the materials from the diverse earlier publications to arrive at a coherent account starting from first principles. The appearance of this book is a major event in geometry during the past decade.

The aim and scope of the book are succinctly set forth in the foreword: ' The classical theory of partial differential equations is rooted in physics, where equations (are assumed to) describe the laws of nature. Law-abiding functions, which satisfy such an equation, are very rare in the space of all admissible functions.... Moreover, some additional conditions often insure the uniqueness of solutions....We deal in this book with a completely different class of partial differential equations (and more general relations) which arise

5.2. GEOMETRIC ANALYSIS

in differential geometry rather than in physics. Our equations are, for the most part, under-determined (or, at least, behave like those) and their solutions are rather dense in spaces of functions. We solve and classify solutions of these equations by means of direct (and not so direct) geometric constructions. Our exposition is elementary and the proofs of the basic results are self-contained.' The partial differential relations alluded to above are usually either equations or inequalities. A typical example of the former is the system of partial differential equations arising from the isometric imbedding problem for Riemannian manifolds... This should not be taken literally. One should rather approach this as a research monograph where new ideas turn up almost in every page. Many of these ideas will undoubtedly inspire further developments."

5.2.5 H. Federer, Geometric Measure Theory, Die Grundlehren der mathematischen Wissenschaften, Band 153, Springer-Verlag New York Inc., New York, 1969, xiv+676 pp.

Geometric measure theory studies geometric properties of the measures of sets in Euclidean spaces and more generally manifolds, such as arc lengths and areas. It uses measure theory to generalize differential geometry of submanifolds to subspaces with mild singularities which are called rectifiable sets. These subspaces often arise from some extremal problems such as minimal surfaces and minimal submanifolds. It also studies less regular subspaces such as fractals which arise from dynamical systems.

This book is a classic on geometric measure theory written by one of the leading figures of the subject. The beginning chapters can be used as a textbook and the latter part can bring the reader to the research level. After all these years, its comprehensiveness still makes it a good reference.

According to the preface, "During the last three decades the subject of geometric measure theory has developed from a collection of isolated special results into a cohesive body of basic knowledge with an ample natural structure of its own, and with strong ties to many other parts of mathematics. These advances have given us deeper perception of the analytic and topological foundations of geometry, and have provided new direction to the calculus of variations. Recently the methods of geometric measure theory have led to very substantial progress in the study of quite general elliptic variational problems, including the multidimensional problem of least area.

This book aims to fill the need for a comprehensive treatise on geometric measure theory. It contains a detailed exposition leading from the foundations of the theory to the most recent discoveries, including many results not previously published. It is intended both as a reference book for mature mathematicians and as a text-book for able students... Study of the later chapters is suitable preparation for research."

According to MathSciNet, "The book has already become an invaluable resource for those doing research in geometric measure theory. Specialists in such fields as partial differential equations, the calculus of variations, differential geometry and differential topology will find it useful as an exposition of the powerful (and previously rather inaccessible) results of geometric measure theory. Discussions of open problems are scattered throughout the text... Also included are many historical comments and identification of the original sources of all relatively new and important material presented in the text."

According to an Amazon review, "This is absolutely essential material to those wishing to apply the techniques of metric measure theory but is not a book to sit down and read. Instead buy both this book and Frank Morgan's *Geometric Measure Theory* [see below]. Frank Morgan's book is an easy read for a graduate student with a semester of real analysis completed and is beautifully intuitive and has many illustrations. Even more convenient, it refers directly to the theorems in Federer's book by number so you can then go to Federer for the complete detailed proofs and the full general statements."

5.2.6 More books on geometric measure theory

P. Mattila, *Geometry of Sets and Measures in Euclidean Spaces, Fractals and Rectifiability*, Cambridge Studies in Advanced Mathematics, 44, Cambridge University Press, Cambridge, 1995. xii+343 pp.

This is a book on geometric measure theory but covers some topics different from the book by Federer. Geometric sets that arise from calculus of variation can be thought of as limits of submanifolds and often enjoy some properties of surfaces and submanifolds, for example, rectifiable properties. Besides these subsets with nice regularity properties, this book emphasizes less regular subspaces such as fractals, fractal-type sets which occur naturally in dynamical systems. It is a valuable reference.

According to MathSciNet, "It contains a wealth of material not in Federer, but with a style that places less emphasis on stating the most general result and more emphasis on giving instructive proofs. The table of contents is complete and helpful."

Comparison of this book with other important books on geometric measure theory and fractals is given in a review of this book in Bulletin of AMS, "**Federer's Book**. *Geometric Measure Theory*, the first monograph of the field, came out in 1969. It is a monumental treatise with a wealth of material on both parts of the field as well as on related questions. Geometric properties of measures are studied as a tool to define a homological integration theory, culminating finally in a very general measure-theoretic calculus of variations... Although still a valuable reference, [it] is not a textbook, however, due to its extremely condensed style. Various attempts were made to establish a more comprehensible presentation... **Falconer and Mattila.** The text under review [by

5.2. GEOMETRIC ANALYSIS 267

Mattila] concentrates on the geometric part of the field, indicated by the subtitle "Fractals and Rectifiability" – although "Rectifiability and Fractals" would be more appropriate, as we shall see below. Compared to [*Geometric measure theory*], this book reads like a novel, but it is definitely a graduate text. For an introduction, the successful books by Falconer are recommended. Falconer provides a clear understanding of the basic concepts; Mattila goes further and leads the reader straight to the edge of current research."

F. Morgan, *Geometric Measure Theory, A Beginner's Guide*, Fourth edition, Elsevier/Academic Press, Amsterdam, 2009. viii+249 pp.

As the subtitle indicates, this book is an accessible introduction to geometric measure theory. It can also serve as a preparation to read Federer's classic on geometric measure theory. It contains a large number of beautiful illustrations which help the reader to get the right intuition and understand better the complicated definitions.

5.2.7 Calculus of variation

Calculus of variations deals with extremal functionals, which are functions defined on a space of functions. A simple example of such a problem is to find a curve of shortest length, or a geodesic, connecting two points. If the space is the Euclidean space $\mathbb{R}^3$ and there are no constraints, then the solution is obviously a straight line between the points. However, if the curve is constrained to lie on a surface in space $\mathbb{R}^3$, then the solution is a geodesic, and many solutions may exist. A related problem is posed by Fermat's principle: light follows the path of shortest optical length connecting two points, where the optical length depends upon the material of the medium. One corresponding concept in mechanics is the principle of least action.

Many important problems in calculus of variation involve functions of several variables. For example, Plateau's problem requires finding a surface of minimal area that spans a given contour in the space $\mathbb{R}^3$. Though solutions can often be found by experiments with bubbles, rigorous mathematical treatment is not simple.

C. Morrey, *Multiple Integrals in the Calculus of Variations*, Die Grundlehren der mathematischen Wissenschaften, Band 130, Springer-Verlag, New York.

This is not an easy book, but it contains a lot of valuable information for researchers. Morrey was a real expert on the calculus of variations and the theory of partial differential equations and made fundamental contribution to them. It is a good reference for mature mathematicians.

According to Journal of Optimization Theory and Applications, "the book contains a wealth of material essential to the researcher concerned with multiple integral variational problems and with elliptic partial differential equations. The book not only reports the researches of the author but also the contributions of his contemporaries in the same and related fields. The book undoubtedly will become a standard reference for researchers in these areas. The book is addressed mainly to mature mathematical analysts. However, any student of analysis will be greatly rewarded by a careful study of this book."

I.M. Gelfand, S.V. Fomin, *Calculus of Variations*, Revised English edition translated and edited by Richard A. Silverman, Prentice-Hall, Inc., Englewood Cliffs, N.J. 1963. vii+232 pp.

This is a classic introduction to the calculus of variations. It was based on lectures of Gelfand and conveys the freshness and liveliness of the lectures. Its purpose is to introduce basics of the calculus of variations "in a form which is both easily understandable and sufficiently modern." Many physical applications of variational methods are also emphasized.

According to MathSciNet, "This book fills a long-standing need for a text on the calculus of variations which can be used at the advanced undergraduate level, which provides sufficient physical background, and which at the same time takes some account of 20th century mathematical developments in the subject. Through unerring good taste and elegance of style, a great amount of material has been put into a little over 200 pages..."

According to an Amazon review, "It contains, for instance, a wonderful treatment of Noether's theorem, hardly to be surpassed. The Hamilton-Jacobi equation is also treated with brilliance and clarity. Gelfand (and Fomin!) developed a style in which the precision of the mathematics does not interfere with the general panorama. The applications are very well selected and perfectly illustrate the theory."

B. Dacorogna, *Direct Methods in the Calculus of Variations*, Second edition, Applied Mathematical Sciences, 78, Springer, New York, 2008. xii+619 pp.

This is a good self-contained introduction to the calculus of variation. It can serve as a guide to nonspecialists to discover the most important problems, results and techniques of the subject.

According to MathSciNet, "The author deals with one of the central problems of the calculus of variations, which is to find among all functions with prescribed boundary conditions those which minimize a given functional. During the nineteenth century, new methods were derived by Lagrange, Riemann, Weierstrass, Jacobi and Hamilton, and these methods are called classical. With the beginning of the twentieth century, different techniques were used by Hilbert and Lebesgue in connection with the study of Dirichlet integrals. These methods were generalised by Tonelli and are now known as the direct methods. Since then, these methods have been extensively used and generalised. In this monograph, the author mentions some of the more recent developments."

5.2. GEOMETRIC ANALYSIS

M. Giaquinta, *Multiple Integrals in the Calculus of Variations and Nonlinear Elliptic Systems*, Annals of Mathematics Studies, 105, Princeton University Press, Princeton, N.J., 1983. vii+297 pp.

This is a very accessible introduction to an important aspect of the calculus of variation, the existence and differentiability of minimum points or more generally stationary points of regular functions in the calculus of variations.

According to MathSciNet, "We may say that the central problem in the calculus of variations is the existence and regularity of stationary points for the so-called regular functionals... Research in this area has been dominated by Problems 19 and 20 posed by Hilbert in the 1900 International Congress of Mathematicians, respectively, asking for analyticity and existence of solutions for regular problems."

H. Attouch, *Variational Convergence for Functions and Operators*, Applicable Mathematics Series, Pitman (Advanced Publishing Program), Boston, 1984. xiv+423 pp.

To approach the limit of sequences of variational problems, relatively new concepts of convergence of sequences of functions and operators appeared in the period 1964–1984. This is called the variational convergence. Each type of variational problem has a particular type of variational convergence. This book is the first systematic exposition of the variation convergence and applications, with emphasis on the convergence associated with minimization. It has been a useful reference.

G. Dal Maso, *An Introduction to Γ-convergence*, Progress in Nonlinear Differential Equations and their Applications, 8, Birkhäuser Boston, Inc., Boston, MA, 1993. xiv+340 pp.

This book is a very valuable introduction to a notion of "variational convergence", called Γ-convergence, which has been very useful in the study of variational problems involving severe perturbations of the data, such as phase transitions, singular perturbations, boundary value and control problems in widely perturbed domains and also various homogenization problems in mathematical theory of composite materials.

According to MathSciNet, "This book is the first one completely devoted to a detailed presentation of several basic aspects of Γ-convergence (often known as epi-convergence). The author has substantially contributed to many aspects of the subject.

The exposition is very detailed and full of notes, remarks, interesting examples (sometimes devoted to auxiliary topics as well). The level of the book is advanced and requires a considerable degree of mathematical maturity from the reader. However, several prerequisites playing a basic role in the topics treated are developed in considerable depth and detail, in particular, topological and functional-analytic tools. The book is very clearly written, and to some extent is self-contained. The mathematical theory is developed in a coherent way although nonmathematical motivations or applications are absent..."

O. Bolza, *Lectures on the Calculus of Variations*, Dover Publications, Inc., New York, 1961, xi+271 pp. second edition Chelsea Publishing Co., New York, 1961, ix+271 pp.

The purpose of this classic is to give a detailed account of the development of calculus of variation starting from Weierstrass, Hilbert et al. The author Bloza has also made fundamental contributions to the subject. This book has played an important role in the development of minimal surfaces, which are now actively pursued by many people. This book is of both historical and current interests.

5.2.8 Fractal geometry

Pictures of fractals are beautiful and familiar to many people, and their rise can partly explained by the easy access to computers or computing power. Actually, their roots were much earlier and went back to the seventeenth century. A mathematically rigorous treatment of fractals can be traced back to functions which were continuous but nowhere differentiable, studied by Karl Weierstrass, Georg Cantor and Felix Hausdorff in the nineteenth century. However, the term fractal was coined by Benoit Mandelbrot only in 1975. It was derived from the Latin *fractus* which means "broken" or "fractured." It is not easy to formally define fractals. There are several examples of fractals, which are defined as portraying exact self-similarity, quasi self-similarity, or statistical self-similarity. Mathematically, fractals arise as limits of certain points under iteration of maps. They also appear in nature and are useful in sciences and technology.

K. Falconer, *Fractal Geometry. Mathematical Foundations and Applications*, John Wiley & Sons, Ltd., Chichester, 1990. xxii+288 pp.

This has been an important and popular textbook on the mathematics of fractals since the first edition. Its aim is to provide an accessible treatment of the mathematics that characterize fractals and define and compute the fractal dimension (or Hausdorff dimension). This is a richly illustrated, accessible and fun to read book.

According to MathSciNet, "This marvelous textbook differs from its successful predecessor *The geometry of fractal sets*, 1986, both in content and in style. The first book was essentially an accessible introduction to geometric measure theory, with the last chapter devoted to fractal examples. The new text is definitely directed to a wide audience of mathematicians interested in fractals and their applications... The presentation is lucid... Avoiding any speculation, he concentrates on facts which can be stated and shown rigorously... Compared with other texts on fractals, the treatment is very careful...

It is impossible to deal with all aspects of 'fractals' in 300 pages. The author's emphasis is on Hausdorff measure, while numerical questions in calculating dimensions, computer algorithms, topological features and recursive structure are not treated systematically. On the whole, the book offers a lot of information and insight without requiring too much effort".

K. Falconer, *The Geometry of Fractal Sets*, Cambridge Tracts in Mathematics, 85, Cambridge University Press, Cambridge, 1986. xiv+162 pp.

This is an accessible, yet rigorous, self-contained introduction to geometric measure theory, with some emphasis on applications to fractals. It is also enriched by many examples of curves of fractional dimension, attractors, examples from number theory, convex analysis, Brownian motion etc.

According to MathSciNet, "The book is an excellent complement to C. A. Rogers' *Hausdorff Measures*, 1970. It restricts attention to subsets of Euclidean space, and gives a clear picture of the inherent difficulties and beautifully intricate arguments which are needed both to prove general results and to calculate the dimensions of specific sets... The minimum amount of measure theory is introduced at the beginning, and the first topic then discussed concerns density properties of s-sets (sets of finite positive Hausdorff s-measure). Naturally, this is an account based on the work of Besicovitch, clearly the inspiration for the whole book..."

B. Mandelbrot, *The Fractal Geometry of Nature*, Schriftenreihe für den Referenten, [Series for the Referee] W. H. Freeman and Co., San Francisco, Calif., 1982. v+460 pp.

Mandelbrot coined the term fractal and his book *The Fractal Geometry of Nature* brought fractals into the mainstream of professional and popular mathematics.

According to MathSciNet, "This book is a substantial revision of an earlier one *Fractals: form, chance and dimension, English translation*, 1977... Many people, scientists and others, were very much excited by these works. Many patterns of Nature are irregular and fragmented. There is a need for a mathematical theory of such forms. The author responds to this challenge by using fractal sets, sets whose Hausdorff-Besicovitch dimension is bigger than their topological dimension. Fractional dimension, self-similarity or statistical self-similarity and power laws are the main mathematical tools of this book. Using these devices, many physical or even biological systems are studied."

5.3 Complex geometry and complex analysis

Algebraic geometry over the field of complex numbers is closely related to complex geometry and complex analysis. One important example is the application of Hodge theory, in particular, the Hodge decomposition of compact Kähler manifolds, to complex projective varieties. This brings analysis to bear on the topology of algebraic varieties. The Hodge theory was initiated by W.V.D. Hodge in his book published in 1941 and completed by Kodaira and Weyl. Hodge also made the famous Hodge conjecture on the cohomology of algebraic varieties, which is the number one open problem in algebraic geometry.

Figure 5.3: Hodge's book.

Complex geometry and complex analysis are two closely related subjects, and domains in the complex space $\mathbb{C}^n$ and complex manifolds are the basic underlying spaces of these two subjects. By definition, complex geometry emphasizes the geometric aspect, and complex analysis emphasizes the analysis aspect. But it is often difficult to draw the line between them.

5.3.1 L. Hörmander, An Introduction to Complex Analysis in Several Variables, Third edition, North-Holland Publishing Co., Amsterdam, 1990. xii+254 pp.

This has been a very important textbook and a valuable reference on complex analysis of several variables and complex geometry since it was published. The analytic methods based on differential equations, in particular the L^2-method, used in this book have been very influential in the development and application of complex analysis of several variables. Though there are newer books now, probably they still can not replace this book.

According to MathSciNet, "This book develops the theory of analytic functions of several complex variables from the viewpoint of the theory of partial differential equations. In the preface, the author states: 'Two recent developments in the theory of partial differential equations have caused this book to be written. One is the theory of overdetermined systems of differential equations with constant coefficients, which depends very heavily on the theory of functions of several complex variables. The other is the solution of the so-called ∂ Neumann problem, which has made possible a new approach to complex analysis through methods from the theory of partial differential equations. Solving the

5.3. COMPLEX GEOMETRY AND COMPLEX ANALYSIS 273

Cousin problems with such methods gives automatically certain bounds for the solution, which are not easily obtained with the classical methods, and results of this type are important for the applications to overdetermined systems of differential equations.'

The treatment is not intended to achieve completeness in any direction, but rather to provide an introduction to the theory for those whose main interests are in analysis... The book is well written and very clear... For anyone wishing to obtain a grasp of the basic theory, this book is an excellent introduction."

5.3.2 More books on complex geometry

Complex geometry studies complex manifolds, Hermitian and Kähler metrics on them, and complex vector bundles and their holomorphic sections, which generalize functions of many complex variables. Complex manifolds are closely related to algebraic varieties over the field of complex numbers and complex geometry also provides important transcendental methods to study algebraic geometry. The Hodge decomposition of cohomology of Kähler manifolds is one such example. Though complex geometry is an important subject, there are not many books about it.

S.S. Chern, *Complex Manifolds Without Potential Theory*, With an appendix on the geometry of characteristic classes, Revised printing of the second edition, Universitext. Springer-Verlag, New York, 1995. vi+160 pp.

Chern is well-known for his work on Chern class of complex vector bundles. This book was written by a master of the subject. The appendix to the second edition gives a good summary of the Chern classes of complex vector bundles and is very useful. It might be the most interesting part of the whole book.

According to MathSciNet, "Despite its pocket size this booklet contains an impressive quantity of material. It starts assuming little more than the concept of holomorphic function and proceeds initially at a leisurely pace; further on it becomes rather fast-moving... This book is a welcome source of information for those mathematicians who have already some acquaintance with the field. Students may need additional sources on basic differential geometry, sheaves and possibly some other topics."

K. Kodaira, *Complex Manifolds and Deformation of Complex Structures*, Translated from the Japanese by Kazuo Akao, With an appendix by Daisuke Fujiwara, Grundlehren der Mathematischen Wissenschaften, 283, Springer-Verlag, New York, 1986. x+465 pp.

Though the deformation theory of Riemann surfaces, i.e., complex manifolds of 1 dimension, were developed and understood by Teichmüller, Ahlfors, Bers and others since 1930s, the higher dimensional cases are much more complicated. A coherent theory of deformations of general complex manifolds was developed by Kodaira and Spencer and has had many applications in various subjects. This book is the first, and the only, systematic introduction to complex manifolds and the Kodaira-Spencer deformation theory. As the author said in the preface, the development of the deformation theory was his most interesting mathematical experience and he tried to reproduce this experience in this book. So this is a book on an important subject written by its creator.

According to MathSciNet, it "will be of service to all who are interested in this by now vast subject. Although intended for a reader with a certain mathematical maturity, the author begins at the beginning, with a discussion of foundational material in several complex variables, complex manifolds, differential forms, bundles, sheaves, and cohomology.

The author honors us by recording some of his thoughts and feelings at the time that the theory was created,... This is a book of many virtues: mathematical, historical, and pedagogical. Parts of it could be used for a graduate complex manifolds course."

J. Morrow, K. Kodaira, *Complex Manifolds*, Holt, Rinehart and Winston, Inc., New York-Montreal, Que.-London, 1971. vii+192 pp.

This is a good introduction to complex manifolds, cohomology of sheaves, and their applications to geometry of complex manifolds, based on lectures of Kodaira, a master of the subject. For example, it gives the original proof of the Kodaira embedding theorem, which says that a Kahler manifold admitting a Hodge class can be embedded as an algebraic variety in the projective space. It can also serve as an introduction to the Kodaira-Spencer theory of deformations of complex structures.

D. Huybrechts, *Complex Geometry: An Introduction*, Universitext, Springer-Verlag, Berlin, 2005. xii+309 pp.

This is a modern book on complex geometry written by an algebraic geometer with some emphasis on Kaher manifolds and applications in algebraic geometry.

According to MathSciNet, "This is the book that a generation of complex geometers will wish had existed when they first learned the subject, and that the next generation of geometers will surely use. It is Griffiths-and-Harris-lite, a perfect modern update of the classic, *Principles of algebraic geometry*, 1978. It is easier to read, without the idiosyncracies, set at an easier level, and contains some beautiful recent developments in place of the more advanced topics in [op. cit.]. Importantly, it also blends in a little of [A. L. Besse, Einstein manifolds, Springer, Berlin, 1987]—to the analytic geometry are added the relevant parts of Riemannian geometry: holonomy, connections, gauge theory, Ricci curvature, etc. This differential geometry of course sheds a good deal of light on many of the results of complex geometry and makes the subject much easier to understand.

The author concentrates entirely on geometry, leaving the necessary analysis and topology to other texts; even sheaf cohomology is relegated to an appendix. This works very well.

5.3. COMPLEX GEOMETRY AND COMPLEX ANALYSIS

Inserted into the standard material are some excellent appendices to stimulate interest and further reading. In this way the reader learning the basic material is brought quickly and often to some fascinating areas of current research."

R. Wells, *Differential Analysis on Complex Manifolds*, Third edition, With a new appendix by Oscar Garcia-Prada, Graduate Texts in Mathematics, 65, Springer, New York, 2008. xiv+299 pp.

This is one of the first books that discusses basics of analysis and geometry of compact complex manifolds. It has been one of the standard sources for this material due to the several editions and updates.

According to Zentralblatt MATH, "The book has proven to be an excellent introduction to the theory of complex manifolds considered from both the points of view of complex analysis and differential geometry."

According to MathSciNet, "The book is rather concise in that it aims to prove the Hodge and Lefschetz decompositions for the cohomology of a compact Kahler manifold (Chapter V), and Kodaira's vanishing and projective embedding theorems for Hodge manifolds (Chapter VI). These topics are covered in full detail, and this dictates the extent of the foundational material in Chapters I-IV. In addition, there are brief allusions to other topics and problems which can be formulated in the language developed, for example, Hirzebruch's Riemann-Roch theorem, and deformation theory and the period mapping."

S. Kobayashi, *Hyperbolic Manifolds and Holomorphic Mappings*, Pure and Applied Mathematics, 2 Marcel Dekker, Inc., New York, 1970. ix+148 pp.

The following book is a revision and expansion of this book.

S. Kobayashi, *Hyperbolic Complex Spaces* , Grundlehren der Mathematischen Wissenschaften, 318, Springer-Verlag, Berlin, 1998. xiv+471 pp.

The notion of Kobayashi distance and Kobayashi hyperbolicity has been applied in different subjects ranging from number theory to dynamics. These concepts were first systematically discussed in the book in 1970, and the updated version also contains many new results obtained since then. It is an essential reference on the Kobayashi hyperbolicity.

According to MathSciNet, "the Poincaré metric has proven extremely useful in the study of hyperbolic Riemann surfaces and their associated function theory. In higher dimensions, there is no uniformization theorem, and so one would like an intrinsically defined metric or distance function on "most" complex spaces that has properties similar to those of the Poincaré metric... In 1967, the author of the book currently under review 'dualized' Carathéodory's construction and defined an intrinsic pseudodistance on complex spaces based on mappings from D. For hyperbolic Riemann surfaces, the

Kobayashi pseudodistance coincides with the metric of constant negative curvature inherited from the Poincaré metric. Thus, the author calls a complex space hyperbolic when the Kobayashi pseudodistance is an honest distance, i.e., the distance between distinct points is always positive.

The book under review can be regarded as an updated and much expanded version of the author's 1970 monograph on hyperbolic spaces, *Hyperbolic Manifolds and Holomorphic Mappings*, 1970... The author's book is exceptionally well organized, with an impressive collection of references to the literature... The author did an excellent job of selecting and treating examples that are essential for developing the reader's intuition about the subject and contented himself with citing the literature for technical examples that illustrate finer points..."

5.3.3 C.L. Siegel, Topics in Complex Function Theory. Vol. I. Elliptic Functions and Uniformization Theory. Vol. II. Automorphic Functions and Abelian Integrals. Vol. III. Abelian Functions and Modular Functions of Several Variables, Interscience Tracts in Pure and Applied Mathematics, No. 25, Wiley-Interscience, 1969. ix+186 pp., 1988. xii+193 pp., 1989. x+244 pp.

These three volumes correspond to three courses taught by Siegel in Göttingen. This "text written by a great mathematician of our time will present one of the most fascinating fields of mathematics", according to the preface by Magnus. Indeed, Siegel initiated several important methods in the field of modular forms in several variables. He was one of the most important mathematicians of the twentieth century.

According to a review in Bulletin of AMS, "It is that in these three books, the modern spirit of abstraction is given a definitely secondary role, consciously and deliberately pushed out of the way for the concrete development of the area which these books treat, that of algebraic functions over the complex numbers and of automorphic functions and forms. In the reviewer's opinion these books are an excellent treatment of that subject, well-seasoned with numerous examples and important special results. In spite of the fact that there may be slicker treatments of various parts of the subject on the market, the relatively uninitiated graduate student who wants an introduction, as well as the specialist who wants to look up complete proofs for 'well-known' facts whose proof is difficult to find elsewhere, would be well advised to turn to this set... the work as a whole stands out for its excellent treatment of a broad subject, and what, for obvious reasons of space, it lacks in completeness, it more than makes up for in inspirational quality."

According to MathSciNet, the first volume "Develops the higher parts of function theory in a unified presentation. Starts with elliptic integrals and functions and uniformization theory, continues with automorphic functions and the theory of abelian integrals and ends with the theory of abelian functions and modular functions in several variables. The last topic originates with the author and appears here for the first time in book form."

The second volume consists of two chapters: Automorphic functions, Abelian integrals. For the third volume, "it may be said that this text is still an optimal introduction to the topic of abelian and Siegel modular functions. Everyone interested in this still-growing topic should read this book before or alongside the now existing books of, which cover the same material with more detail and more of the modern developments. With this text one will learn how the fundamental notions in the theory came into being. And, by comparing, for instance, the meaning of a Picard variety or a Jacobi function in this book with the notions of Picard scheme or Jacobi form in more recent treatments, one will see how things are developing."

5.4 Algebraic geometry

Clebsch had a relatively short life and died at the age of 39. It is probably not so well-known that he co-founded the mathematical research journal Mathematische Annalen. He made important contributions to algebraic geometry and invariant theory, and his students include several important algebraic geometers such as Brill, Gordan, Max Noether etc. Together with Carl Neumann at Göttingen, he founded the journal Mathematische Annalen in 1868. Clebsch had a huge influence on Felix Klein as his postdoctoral mentor, and Klein inherited Mathematische Annalen and turned it into the leading journal in the world.

Figure 5.4: Clebech's book.

Algebraic geometry studies solutions to systems of polynomial equations, or more generally, algebraic varieties, by combining concepts, methods and techniques of abstract algebra, especially commutative algebra, with geometry. It occupies a central place in modern mathematics and has substantial overlapping and connections with such diverse fields as complex analysis, topology and number theory. Solving systems of polynomial equations in several variables is a classical problem. The subject of algebraic geometry starts when the emphasis is shifted towards understanding the intrinsic properties of the totality of solutions of system of equations. Plane algebraic curves include lines, circles and parabolas, and form one of the best studied classes of algebraic varieties. Indeed, they were studied intensively before and during the 19th century. One key distinction between classical projective geometry and modern algebraic geometry is that the former is concerned with the more geometric notions, while the latter emphasizes the more analytic concepts of regular functions, maps and sheaf theory. Another important difference lies in the broad scope of the modern algebraic geometry. Grothendieck's idea of scheme connects algebraic geometry with algebraic number theory with far reaching consequences for both subjects.

A brief history of algebraic geometry was given in the MathSciNet review of the book of Griffiths and Harris: "Algebraic geometry, as a mutually beneficial association between major branches of mathematics, was set up with the invention by Descartes and Fermat of Cartesian coordinates. Geometry was as old as mathematics; but it was not until the seventeenth century, more or less, that algebra had matured to the point where it could stand as an equal partner. Calculus too played a major role (tangents, curvature, etc.); in the early stages algebraic and differential geometry could be considered to be two aspects of 'analytic' (as opposed to 'synthetic') geometry.

During the nineteenth century the horizons of the subject were expanded (to ∞!) by the development of projective geometry and the use of complex numbers as coordinates. Gradually, out of an intensive study of special curves and surfaces, the idea emerged that algebraic geometry should deal with an arbitrary algebraic subset of n-dimensional projective space over the complex numbers (i.e. a set of points where finitely many homogeneous polynomials with complex coefficients vanish simultaneously). This was the proper context for the working out of concepts like transformation groups and their invariants, correspondences, and 'enumerative' geometry (how to count the number of solutions of a geometric problem).

In the middle of the nineteenth century, Riemann appeared on the scene like a supernova. His conceptions of intrinsic geometry on a manifold, topology, function theory on a Riemann surface, birational transformations, abelian integrals, and zeta functions, fueled almost all the subsequent developments. In the analytic vein, which is relevant to the book under review, some of the more prominent contributors have been Picard, Poincaré, Lefschetz, Hodge, Kodaira, and Hirzebruch. In particular Hodge and Kodaira used the theory of partial differential equations to establish basic results, some of which have not yet been proved otherwise.

It is not my purpose here to summarize the history of algebraic geometry, but rather to suggest that since it began algebraic geometry has been a prime exhibit of the unity of mathematics, an area where diverse methods from analysis, topology, geometry, algebra and even number theory have interacted in a marvellously fruitful way. Indeed, though

5.4. ALGEBRAIC GEOMETRY

the subject has sometimes grown in directions which seemed exclusively algebraic, geometric, or analytic, history teaches us that it will continue to flourish only if nourished by ideas from all the different fields.

It seems inevitable in mathematics that powerful methods will eventually be pushed to the point of excess, either of generality or complexity. In a period of less than fifteen years, Grothendieck, building on Serre's foundational work, introduced revolutionary concepts and techniques (enough, according to Dieudonné, to keep several generations of algebraic geometers occupied). In this comparatively brief time, the enormous energy of Grothendieck and his school resulted in a shelf-full of seminars and IHES publications whose sheer mass threatened to unbalance algebraic geometry. The inhuman task of working through thousands of pages made the essential contributions virtually inaccessible to anyone who was not in contact with Grothendieck, either first hand in Paris, or second hand in Cambridge, Princeton, or Moscow. This led to some resentment, grumblings about 'empty generality', 'fashions in mathematics', etc. The situation is now much improved. In the past five years a number of first-rate down to earth introductory texts have appeared, with the foreseeable consequence that algebraic geometry should once again occupy its proper place in the general mathematical consciousness..."

5.4.1 A. Grothendieck, The Éléments de géométrie algébrique, Inst. Hautes Études Sci. Publ. Math., 1960–1967.

Grothendieck changed the landscape of algebraic geometry. These works by him are probably still the best records of his revolutionary ideas and approaches to modern algebraic geometry. They are not easy to read, but of utmost importance.

One important feature of Grothendieck is that whenever possible, he will try to do the most general things and also isolate the most basic and essential parts of any theories.

According to a review in Bulletin of AMS, "The present work, of which Chapters 0 and 1 are now appearing together, is one of the major landmarks in the development of algebraic geometry. It plans to cover eventually everything that is known in algebraic geometry over arbitrary ground rings, and of course a lot more besides. In order to get a more specific idea of what is to come, one should consult first Grothendieck's address to the International Congress at Edinburgh, 1958, and also the whole series of talks at Bourbaki seminars given in the past two years...) in which he has given a sketch of the proofs of important results to appear in later chapters. These talks will provide the necessary motivation to the whole work. They are written concisely, directly, and excitingly. Such motivation could not be given in the actual text, which is written very lucidly, is perfectly organized, and very precise. Thanks are due here to Dieudonné, without whose collaboration the labor involved in writing and publishing the work would have been insurmountable...

To conclude this review, I must make a remark intended to emphasize a point which might otherwise lead to misunderstanding. Some may ask: If Algebraic Geometry really consists of (at least) 13 Chapters, 2,000 pages, all of commutative algebra, then why not just give up?

The answer is obvious. On the one hand, to deal with special topics which may be of particular interest only portions of the whole work are necessary, and shortcuts can be taken to arrive faster to specific goals. Thus one may expect a period of coexistence between Weil's Foundations and Elements. Only history will tell if one buries the other. Projective methods, which have for some geometers a particular attraction of their own, and which are of primary importance in some aspects of geometry, for instance the theory of heights, are of necessity relegated to the background in the local viewpoint of Elements, but again may be taken as starting point given a prejudicial approach to certain questions.

But even more important, theorems and conjectures still get discovered and tested on special examples, for instance elliptic curves or cubic forms over the rational numbers. And to handle these, the mathematician needs no great machinery, just elbow grease and imagination to uncover their secrets. Thus as in the past, there is enough stuff lying around to fit everyone's taste. Those whose taste allows them to swallow the Elements, however, will be richly rewarded."

According to Wikipedia, "EGA for short, is a rigorous treatise, in French, on algebraic geometry that was published (in eight parts or fascicles) from 1960 through 1967 by the Institut des Hautes Etudes Scientifiques. In it, Grothendieck established systematic foundations of algebraic geometry, building upon the concept of schemes, which he defined. The work is now considered the foundation stone and basic reference of modern algebraic geometry.

Initially thirteen chapters were planned, but only the first four (making a total of approximately 1500 pages) were published. Much of the material which would have been found in the following chapters can be found, in a less polished form, in the Seminaire de geometrie algebrique (known as SGA). Indeed, as explained by Grothendieck in the preface of the published version of SGA, by 1970 it had become clear that incorporation all of the planned material in EGA would require significant changes in the earlier chapters already published, and that therefore the prospects of completing EGA in the near term were limited. An obvious example is provided by derived categories, which became an indispensable tool in the later SGA volumes, was not yet used in EGA III as the theory was not yet developed at the time. Considerable effort was therefore spent to bring the published SGA volumes to a high degree of completeness and rigour.

Grothendieck nevertheless wrote a revised version of EGA I which was published by Springer-Verlag. It updates the terminology, replacing 'prescheme' by 'scheme' and

5.4. ALGEBRAIC GEOMETRY

'scheme' by 'separated scheme', and heavily emphasizes the use of representable functors. The new preface of the second edition also includes a slightly revised plan of the complete treatise, now divided into twelve chapters... The work on EGA was finally disrupted by Grothendieck's departure first from IHES in 1970 and soon afterwards from the mathematical establishment altogether...

In historical terms, the development of the EGA approach set the seal on the application of sheaf theory to algebraic geometry, set in motion by Serre's basic paper FAC. It also contained the first complete exposition of the algebraic approach to differential calculus, via principal parts. The foundational unification it proposed (see for example unifying theories in mathematics) has stood the test of time."

5.4.2 I. Shafarevich, Basic Algebraic Geometry. 1. Varieties in Projective Space. 2. Schemes and Complex Manifolds. Second edition, Springer-Verlag, Berlin, 1994. xx+303 pp., xiv+269 pp.

It is known that algebraic geometry, in particular, the modern algebraic geometry, is abstract, and it is not easy to understand the geometric intuition behind theories and concepts. This book emphasizes geometric intuition, motivations of problems and their solutions, and the complex algebraic varieties instead of the general abstract schemes. It is a book written by a master mathematician and master expositor. It discusses many advanced topics and constructions in an accessible way. This is a great book for people at all levels.

According to MathSciNet, "The first edition of this book appeared (in Russian) in 1972. At that time, this textbook was the first and the only one which built bridges between the geometric notions, the classical origins and achievements, the modern concepts and methods, and the complex-analytic aspects in algebraic geometry... In the meantime, it has become one of the most valuable, recommended and used textbooks on algebraic geometry worldwide, together with the other standard textbooks of R. Hartshorne, D. Mumford, and P. A. Griffiths and J. Harris. The special feature of the author's book, in comparison to the other textbooks, is provided by the fact that it really conveys the many different aspects of contemporary algebraic geometry, without focussing on any particular approach and without requiring any advanced prerequisites. In this sense, it has proved an extremely useful addition to the other (here and there) more thoroughgoing textbooks, a recommendable introduction to them and to current research, and in any case, an excellent invitation to algebraic geometry... The author manages, with his inimitable masterly skill, to insert more concrete, advanced and topical material, to present the interrelations, the complexity, and the vividness of algebraic geometry in a comprehensible way and to make his outstanding textbooks even more valuable, in particular for students and teachers."

According to an Amazon review, "In the 1950's algebraic geometry was tedious and hard to grasp because it was mostly commutative algebra, developed by Zariski and Weil and their schools to fill logical gaps in the Italian arguments of the previous half century. The rich geometric texture of the Italian school was lost. In the 1960's Serre and Grothendieck introduced homological algebra to the subject and greatly expanded and enhanced it to embrace also arithmetic, but the abstraction level went WAY up, so again it was hard to grasp and relate to geometry. Hartshorne is a member of both Zariski and Grothendieck's schools and appreciates down to earth objects like space curves, but his book has a long beginning section on schemes and cohomology that can definitely throw a beginner off the horse.

Pardon the delay in getting here, but the point is that Shafarevich's book has none of the tediousness of the previous generation, yet benefits from the rigorous foundations via commutative algebra of Zariski's works. I would say Shafarevich's book, is a geometrically oriented explanation of the material that can be explained using Zariski's methods. I.e., it has a rich geometric feel, is very well explained, includes many easy examples, and is rigorous in its use of commutative algebra. This book allowed many of us who were stymied by the huge amount of algebra needed for 1960's Grothendieck style AG, to finally gain admission to the subject."

Another classical introduction to algebraic geometry which emphasizes geometric aspect is the following.

W. Hodge, D. Pedoe, *Methods of Algebraic Geometry. Vol. I. Book I: Algebraic Preliminaries. Book II: Projective Space. Vol. II. Book III: General Theory of Algebraic Varieties in Projective Space. Book IV: Quadrics and Grassmann Varieties. Vol. III. Book V: Birational Geometry.* Reprint of the 1947 original, Cambridge Mathematical Library, Cambridge University Press, 1994. viii+440 pp., x+394 pp., x+336 pp.

When it was written in 1947, it aimed to "provide a convenient account of the foundations and methods of modern algebraic geometry." According to Atiyah, this was intended to update and replace H. F. Baker's *Principles of Geometry*. Of course, algebraic geometry has advanced since that time, but concrete discussion of the classical algebraic geometry makes this classic book still a relevant and useful reference.

From the book description, "This classic work (first published in 1947), in three volumes, provides a lucid and rigorous account of the foundations of modern algebraic geometry. The authors have confined themselves to fundamental concepts and geometrical methods, and do not give detailed developments of geometrical properties but geometrical meaning has been emphasized throughout. This first volume is divided into two parts. The first is devoted to pure algebra: the basic notions, the theory of matrices over a non-commutative ground field and a study of algebraic equations. The second

5.4. ALGEBRAIC GEOMETRY

part is in n dimensions. It concludes with a purely algebraic account of collineations and correlations."

H.F. Baker, *Principles of Geometry. Volume 1. Foundations.* Reprint of the 1922 original; *Volume 2. Plane Geometry; Volume 3. Solid Geometry; Volume 4. Higher Geometry; Volume 5. Analytical Principles of the Theory of Curves; Volume 6. Introduction to the Theory of Algebraic Surfaces and Higher Loci.* Cambridge Library Collection, Cambridge University Press, Cambridge, 2010. ii+xii+184 pp., ii+xvi+244 pp., ii+xx+228 pp., ii+xvi+250 pp., ii+x+247 pp., ii+x+308 pp.

This is another classic in algebraic geometry. This is a valuable reference on the algebraic geometry in the early twentieth century.

Baker was a British mathematician, working mainly in algebraic geometry, but was also remembered for contributions to partial differential equations (related to what would become known as solitons), and Lie groups. Perhaps his most important contribution is the set of six volume books. It was the first British work on geometry that uses axiomatic methods instead of coordinates. The first two volumes cover the foundations of Euclidean geometry and the introduction of a coordinate system, volume 3 studies solid geometry considering quadrics, cubic curves in space, and cubic surfaces. Volume 4 considers higher dimensions, mainly dimensions four and five, and the final two volumes study the analytical principles of the theory of curves and the theory of algebraic surfaces and higher loci. These books provide a detailed insight into the geometry which was developing at the time of publication.

5.4.3 D. Mumford, The Red book of Varieties and Schemes, Second, expanded edition. Includes the Michigan lectures (1974) on curves and their Jacobians, With contributions by Enrico Arbarello. Lecture Notes in Mathematics, 1358. Springer-Verlag, Berlin, 1999. x+306 pp.

The second edition of this book is the combination of two books. Both of them are classics and very valuable.

According to MathSciNet, "The book under review is a reprint of Mumford's famous Harvard lecture notes, widely used by the past few generations of algebraic geometers. Springer-Verlag has done the mathematical community a service by making these notes available once again. When the book first appeared, Grothendieck's ideas were not widely known or understood and the Red Book was the first approach to the precipitous multi-volume EGA. The book covers the basic theory of schemes and varieties. Considerable emphasis is placed on the study of morphisms using down-to-earth calculations in commutative algebra. A number of topics are covered which are omitted in the now standard text of R. Hartshorne [Algebraic geometry, Springer, New York, 1977]... The informal style and frequency of examples make the book an excellent text. The introduction includes the exciting announcement that a companion volume on cohomology is to appear shortly, written in collaboration with D. Eisenbud and J. Harris."

The second edition also contains a reprint of another classic by Mumford, *Curves and Their Jacobians*, 1975, which were also out of print.

According to MathSciNet review on *Curves and their Jacobians*, "This beautiful book is an exposition, starting from scratch, of the theory of algebraic curves (over C) and their Jacobians with emphasis on explicit realizations of curves and moduli questions. Most of the material is presented without proof; some special topics (either new or lesser known results) are sketched. The book is a slightly expanded version of four Ziwet lectures given by the author at the University of Michigan in the Fall of 1974; hence the (pleasantly) informal style. As the author explains in the introduction, the book is designed for any reader having the basic standard training in topology, complex analysis and algebra. No special knowledge of commutative algebra or the foundations of algebraic geometry is required."

D. Mumford, *Algebraic Geometry. I. Complex Projective Varieties*, Corrected reprint, Grundlehren der Mathematischen Wissenschaften, 221, Springer-Verlag, Berlin-New York, 1981. x+186 PP.

Together with his "red book of varieties and schemes", this classic is still one of the best introductions to algebraic geometry.

According to Zentralblatt MATH, "In the twentieth century, algebraic geometry has undergone several revolutionary changes with respect to its conceptual foundations, technical framework, and intertwining with other branches of mathematics. Accordingly the way it is taught has gone through distinct phases. The theory of algebraic schemes, together with its full-blown machinery of sheaves and their cohomology, being for now the ultimate stage of this evolution process in algebraic geometry, had created — around 1960 — the urgent demand for new textbooks reflecting these developments and (henceforth) various facets of algebraic geometry. The famous volumes *Éléments de géométrie algébrique* by **A. Grothendieck** and **J. Dieudonné** were entirely written in the new language of schemes, without being linked up with the classical roots, and the so far existing textbooks just dealed with classical methods. It was **David Mumford**, who at first started the project of writing a textbook on algebraic geometry in its new setting. His mimeographed Harvard notes *Introduction to algebraic geometry: Preliminary version of the first three chapters* (bound in red) were distributed in the mid 1960's, and they were intended as the first stage of a forthcoming, more inclusive textbook. For some years, these mimeographed notes represented the almost only, however utmost convenient and abundant source for non-experts to get acquainted with the basic new concepts and ideas of modern algebraic geometry. Their timeless utility, in this regard, becomes apparent from the fact that two reprints of them have appeared, since 1988, as a proper book under the title *The Red book of Varieties and Schemes* [cf. Lect. Notes Math. 1358 (1988)]. In the process of extending his Harvard notes to a comprehensive textbook, the author's teaching experiences led him to the didactic conclusion that it would be better to split the book into two volumes, thereby starting with complex projective varieties (in volume I), and proceeding with schemes and their cohomology (in volume II). – In 1976, the author published the first volume under the title *Algebraic geometry. I: Complex projective varieties* (1976; corrected second edition 1980), where the corrections concerned the wiping out of some misprints, inconsistent notations, and other slight inaccuracies.

The book under review is an unchanged reprint of this corrected second edition from 1980. Although several textbooks on modern algebraic geometry have been published in the meantime, Mumford's *Volume I* is, together with its predecessor *The Red Book of Varieties and Schemes*, now as before, one of the most excellent and profound primers of modern algebraic geometry. Both books are just true classics!"

5.4.4 R. Hartshorne, Algebraic Geometry, Graduate Texts in Mathematics, No. 52, Springer-Verlag, New York-Heidelberg, 1977. xvi+496 pp.

This is probably the most popular introduction to modern algebraic geometry. It has affected the education of generations of algebraic geometers and others who want to learn and use algebraic geometry. Due to the excellent index, it is also a very convenient reference.

Though the multivolume books by Grothendieck are systematic and profound, they are difficult to penetrate for most people. According to MathSciNet, "This text is intended to introduce graduate students to the methods and results of abstract algebraic geometry as practised today. No such exposition can succeed unless it enables the reader to make the drastic transition between the basic, intuitive questions about affine and projective varieties with which the subject begins, and the elaborate general methodology of schemes and cohomology employed currently to answer (or attempt to answer) these questions. The present text, notable for generality and depth, is also notable for its author's concern, throughout, to keep the important issues about varieties clearly in the foreground... Granting these necessary commitments, the present text succeeds admirably, in the reviewer's opinion, in introducing its difficult subject at a level appropriate for preparing future workers in the field."

As one Amazon review says, "This is THE book to use if you're interested in learning algebraic geometry via the language of schemes. Certainly, this is a difficult book; even more so because many important results are left as exercises. But reading through this book and completing all the exercises will give you most of the background you need to get into the cutting edge of AG. This is exactly how my advisor prepares his students, and how his advisor prepared him, and it seems to work."

5.4.5 P. Griffiths, J. Harris, Principles of Algebraic Geometry, Pure and Applied Mathematics, Wiley-Interscience, New York, 1978. xii+813 pp.

This classical book on algebraic geometry is different from other classical and popular books in the sense that it emphasizes the point of view of complex geometry and differential geometry. For example, it contains detailed discussion of such fundamental results as the Hodge decomposition, the Kodaira theorem, and a detailed discussion of the geometry of Riemann surfaces. It also covers classical topics in the projective and birational geometry of curves and surfaces. This is a very good book to learn algebraic geometry from a more differential geometric approach and is also a valuable reference on algebraic geometry and classical algebraic geometry.

According to MathSciNet, "This textbook on algebraic geometry may be regarded as a highly welcome complement to others which have appeared since A. Grothendieck's rigorous foundation of algebraic geometry from the 'scheme-theoretic' point of view. More than those others, this book reflects the intimate connection of algebraic geometry with complex analysis as one of the roots of algebraic geometry, in addition to emphasizing the detailed study of concrete classical and nonclassical geometric questions. Therefore, the ground field is always the field of complex numbers, and considerations are limited to projective varieties. As the authors point out in their preface, the methodological principles of this book are the following: (1) The general machinery and techniques should be developed only insofar as they are necessary to handle some concrete geometric questions and special classes of algebraic varieties. (2) There should be a balance between the general theory and the study of examples. (3) The book should be completely self-contained to facilitate the study and to avoid confusing cross-references, which are so typical of the literature in algebraic geometry... Altogether this voluminous book is written in a beautiful, clear, and intuitive style. It is a very nice geometric counterpart to most 'algebraic' books on algebraic geometry, and as well a good source as a reference book. The exposition is very inspiring for the reader and certainly should become a standard text."

This book "Establishes a geometric intuition and a working facility with specific geometric practices. Emphasizes applications through the study of interesting examples and the development of computational tools."

According to a review in Bulletin of AMS, "...But before Griffiths' and Harris' Principles of algebraic geometry, there was no systematic gathering together of the modern

5.4. ALGEBRAIC GEOMETRY

transcendental methods-basically those of differential geometry on complex manifolds, like Hodge theory on Kaähler manifolds, currents, Chern classes, residues etc.-together with extensive illustration of how they can be applied to particular situations. So this book could fill an essential gap in the presentation of algebraic geometry to the mathematical public. Overall, the book should be most successful. This is first of all because of the choice of topics-the authors have a firm grasp of what is fundamental. Secondly there is the underlying philosophy that, in the best tradition of the subject, emphasis must be placed on the interaction between general theory and particularly interesting examples. So it is that about half the book deals with the general theory of complex manifolds and algebraic varieties (interspersed with examples), while the other half has a wealth of applications to curves and their Jacobian varieties, surfaces, and finally (seventy pages!) to one example of a three-dimensional variety, the quadric line complex. There appears to be no mention of the general algebraic approaches of Weil, Zariski, or Grothendieck. The basic objects of study throughout are algebraic varieties of the most concrete kind-subvarieties of complex projective space... Globally, *Principles of algebraic geometry* is an impressive scholarly work, not only as a compendium of basic analytic methods, but also as a guidebook to the vital geometric core of the subject. My advice to a student of algebraic geometry would be: 'Start with one of the other introductory books, but have it in mind to get thoroughly familiar with this one. You will probably need a knowledgeable teacher to help you over the rough spots. If it makes you feel better, think of this book as a set of lecture notes, or even as a fantastic collection of exercises, with copious hints. This is very high quality mathematics; put forth the effort and learn as much of it as you can.' "

J. Harris, *Algebraic Geometry. A First Course*, Graduate Texts in Mathematics, 133, Springer-Verlag, New York, 1992. xx+328 pp.

This is a good introduction to algebraic geometry from the more geometric and elementary algebraic point of view.

According to MathSciNet, "It represents his solution to a basic dilemma: on the one hand, the modern approach to algebraic geometry using schemes needs to be learned by anyone wishing to work in the subject; on the other hand, the rather daunting technicalities of scheme theory may overshadow the more approachable classical theory and tend to put people off. As well, by now there are a number of excellent textbooks which mix the classical and modern in a variety of ways and from a variety of perspectives. Given this, the author is surely correct when he argues that the best way to approach the subject is to introduce suitable topics from 'elementary' classical algebraic geometry before going on to the modern theory. The classical theory is, in the author's words, a glorious subject; it is also needed if one is to understand the motivation for schemes and the insight the modern approach gives.

Of course everything depends on how this project is carried out. Here the book succeeds brilliantly; by concentrating on a number of core topics ... and by treating them in a hugely rich and varied way..., the author ensures that the reader will learn a large amount of classical material and, perhaps more importantly, will also learn that there is no one approach to the subject—the essence lies in the range and interplay of possible approaches. Only the minimum of prerequisites is required, and apart from relatively few unproved statements (and even then, these are usually put in context or supported by interesting examples) everything necessary is developed as the discussion proceeds.

The book demands active reading. The treatment is relatively informal; the main highlights are set out clearly in a coherent and persuasive manner, and the reader is asked (implicitly and very often explicitly) to fill in details or work out further examples or generalisations, hints for selected exercises being given at the end... the reader is encouraged to jump around in the text and to follow certain topics through, as they appear and reappear in the book."

5.4.6 Introduction to algebraic geometry and algebraic curves

W. Fulton, *Algebraic Curves. An Introduction to Algebraic Geometry*, Notes written with the collaboration of Richard Weiss, Reprint of 1969 original, Advanced Book Classics, Addison-Wesley Publishing Company, 1989. xxii+226 pp.

The purpose of this book is to introduce the reader to algebraic geometry by developing the theory of algebraic curves from the view point of modern algebraic geometry without too much prerequisites. It has succeeded well.

According to MathSciNet, "The author has the commendable purpose of introducing the reader to algebraic geometry —at the undergraduate or beginning graduate level— by treating the classical theory of plane curves by methods that are modern but do not demand a great deal of background (e.g., sheaves do not appear)..."

R. Walker, *Algebraic Curves*, Reprint of the 1950 edition, Springer-Verlag, New York-Heidelberg, 1978. x+201 pp.

This is another classical introduction to abstract algebraic geometry through algebraic curves. It has withstood the test of time. It shows how the more abstract algebraic concepts relate to traditional analytical and geometrical problems. The prerequisite is kept as elementary as possible.

K. Smith, L. Kahanpää, P. Kekäläinen, W. Traves, *An Invitation to Algebraic Geometry*, Universitext. Springer-Verlag, New York, 2000. xii+155 pp.

As the title says, this is an invitation to algebraic geometry and the reader can get a taste of some underlying principles, important developments, and some of the problems of algebraic geometry in the twentieth century.

According to MathSciNet, "This aptly-named volume could also be compared to a tourist guide: portable and invitingly illustrated, it points out the major landmarks, gives some sense of history, and suggests where one might go to learn more or even perhaps to stay and work in the area... By eschewing all but the simplest proofs, it manages to pack an impressive number of topics into fewer than 160 pages."

R. Bix, *Conics and Cubics. A Concrete Introduction to Algebraic Curves*, Second edition, Undergraduate Texts in Mathematics, Springer, New York, 2006. viii+346 pp.

By focusing on conics and cubics, this carefully written book is really a concrete and geometric introduction to abstract algebraic geometry. Though it was written as an undergraduate textbook, it is also useful for beginning graduate students.

According to MathSciNet, "The book therefore belongs in the admirable tradition of laying the foundations of a difficult and potentially abstract subject by means of concrete and accessible examples... Two major strengths of the book are its historical perspective, in the form of informative introductions to the chapters which give the main developments in non-technical language, and its exercises, which are numerous and interesting."

5.4.7 Topology of algebraic varieties

F. Hirzebruch, *Topological Methods in Algebraic Geometry*, Third enlarged edition, Die Grundlehren der Mathematischen Wissenschaften, Band 131, Springer-Verlag, New York, 1966, x+232 pp.

This classical book was not planned as a book. Instead, it was a detailed presentation of a major theorem by the author. So it is a mixture of a report, a textbook and a paper. It has been very influential in mathematics.

According to MathSciNet, "This monograph is an exposition of the main ideas and concepts centering around the fundamental Riemann-Roch theorem for algebraic manifolds of arbitrary dimension, a theorem whose proof has recently been found by the author and which is presented here in complete form for the first time..."

J. Milnor, *Singular Points of Complex Hypersurfaces*, Annals of Mathematics Studies, No. 61, Princeton University Press, 1968, iii+122 pp.

This well-written book has been influential in several ways. The fibration introduced near an isolated singular point in this book is now called the Milnor fibration and is an important invariant of the singular point.

The motivation of this book is well-explained in the preface, "The topology associated with a singular point of a complex curve has fascinated a number of geometers, ever since K. Brauner showed in 1928 that each such singular point can be described in terms of an associated knotted curve in 3-sphere. Recently E. Brieskorn has brought new interest to the subject by discovering similar examples in higher dimensions, thus relating algebraic geometry to higher dimensional knot theory and the study of exotic spheres.

This manuscript will study singular points of complex hypersurfaces by introducing a fibration which is associated with each singular point."

J. Bochnak, M. Coste, M. Roy, *Real Algebraic Geometry*, Translated from the 1987 French original, Revised by the authors, Ergebnisse der Mathematik und ihrer Grenzgebiete (3), 36, Springer-Verlag, Berlin, 1998. x+430 pp.

The basic geometric objects in real algebraic geometry are algebraic varieties, in particular polynomial equations, over real closed fields. It is a relatively new subject and did not arise as an independent field of research till the early seventies. This book was the first book that discusses basic problems, some of the most important concepts, and techniques in the field.

According to MathSciNet, "the book is wonderfully written, clear and precise. It is not often that these two qualities coincide. Here, even the few missing details no book can avoid are so accurately placed that they actually encourage the reader. On the other hand, the contents are a fantastic guide for several post-graduate courses, skipping some chapters and teaching word-by-word some others. In addition, each chapter finishes with a short historical report where every problem, solution, example and idea are properly fathered. This includes some surprises. The bibliography is excellent: 274 titles starting back in 1636(!)."

5.4.8 Symplectic geometry and symplectic topology

Symplectic manifolds are differentiable manifolds equipped with a closed, nondegenerate 2-form, in particular Kähler manifolds are symplectic manifolds. Symplectic geometry uses methods of differential geometry and differential topology to study symplectic manifolds, and has its origins in the Hamiltonian formulation of classical mechanics. Symplectic geometry has a number of similarities and differences with Riemannian geometry. A major difference is that symplectic manifolds have no local invariants such as the curvature of Riemannian manifolds. Symplectic topology is mainly concerned with global properties of symplectic manifolds.

V. Guillemin, S. Sternberg, *Symplectic Techniques in Physics*, Cambridge University Press, Cambridge, 1984. xi+468 pp.

5.4. ALGEBRAIC GEOMETRY

This book shows how symplectic geometry can be used to formulate physical laws and the solution of associated problems. It is another instance of fruitful interaction between mathematics and physics. This book is an important contribution of mathematicians to physics.

From the book description, "Symplectic geometry is very useful for formulating clearly and concisely problems in classical physics and also for understanding the link between classical problems and their quantum counterparts. It is thus a subject of interest to both mathematicians and physicists, though they have approached the subject from different viewpoints. This is the first book that attempts to reconcile these approaches..."

D. McDuff, D. Salamon, *Introduction to Symplectic Topology*, Oxford Mathematical Monographs, Oxford Science Publications, The Clarendon Press, Oxford University Press, New York, 1995. viii+425 pp.

Unlike Riemannian manifolds, symplectic manifolds are all locally equivalent. Symplectic topology studies the global properties of symplectic manifolds and corresponds to global differential geometry in some sense. It is a relatively new subject, and this is the first book written by two leading experts in the subject.

According to MathSciNet, "While the term itself is a relatively recent coinage, the roots of this field of research can be traced back to Poincaré's last geometric theorem, proved by Birkhoff in the 1920s, which asserts that an area-preserving twist map of the annulus has at least two distinct fixed points. In 1965 Arnold realized that the correct generalization of this theorem to higher dimensions had to be a statement about fixed points of symplectomorphisms rather than merely volume-preserving maps. This conjecture provided one of the most influential sources of inspiration in symplectic topology... the book succeeds at the difficult task of forming a genuine introduction on the one hand, while on the other hand being as up to date as possible and conveying the excitement of a very active area of research. This makes it not only compulsory but also compelling reading for anyone new to the field and more experienced readers alike and, apart from providing the ideal basis for a graduate course on symplectic topology, it also constitutes an important reference work."

According to a review in Bulletin of AMS, "This book has already earned its place as a basic reference for workers in the field; it will also make the task of beginning research substantially simpler for graduate students starting work in this area."

D. McDuff, D. Salamon, *J-holomorphic Curves and Symplectic Topology*, American Mathematical Society Colloquium Publications, 52, American Mathematical Society, Providence, RI, 2004. xii+669 pp.

This is a book about an important and active research topic as the title indicates. It is a timely contribution.

According to Zentralblatt MATH, "The theory of J-holomorphic curves was founded by M. Gromov around 1985. Its applications include many key results in symplectic topology and it was one of the main inspirations for the creation of Floer homology. It provides a natural context in which one can define Gromov-Witten invariants and quantum cohomology, which form the so-called A-side of the mirror symmetry conjectures. Insights from physics have themselves inspired many fascinating developments, for example the till now little understood connections between the theory of integrable systems and Gromov-Witten invariants. The book under review is a wonderful work about this topic."

H. Hofer, E. Zehnder, *Symplectic Invariants and Hamiltonian Dynamics*, Birkhäuser Advanced Texts: Basler Lehrbücher, [Birkhäuser Advanced Texts: Basel Textbooks] Birkhäuser Verlag, Basel, 1994. xiv+341 pp.

This is another recent book on symplectic topology written by two experts. This book shows how analysis, differential geometry, algebraic topology and physics can interplay in such a fruitful way.

According to MathSciNet, "This book is a beautiful introduction to one outlook on the exciting new developments of the last ten to fifteen years in symplectic geometry, or symplectic topology, as certain aspects of the subject are lately called. The types of results presented in this book fall naturally into three classes: the now 'classical' rigidity results of Gromov and Eliashberg, some results on the structure of the group of compactly supported diffeomorphisms of $\mathbb{R}^{2n}$, and results on the Arnold conjecture on the number of fixed points of symplectic diffeomorphisms..."

D. Blair, *Riemannian Geometry of Contact and Symplectic Manifolds*, Second edition, Progress in Mathematics, 203. Birkhauser Boston, Inc., Boston, MA, 2010. xvi+343 pp.

Contact geometry is in some sense an odd-dimensional counterpart of symplectic geometry, which is always even-dimensional. Both contact and symplectic geometry are motivated by the mathematical formalism of classical mechanics. When an earlier version of this book was written and published in Springer Lecture Notes in Mathematics, the subject of contact manifolds from the Riemannian point of view was not well-known. This book is an important contribution in this direction, and can serve either as a textbook and a reference.

According to MathSciNet, "In this book, contact and symplectic manifolds are studied from a Riemannian point of view; as such it unites two worlds which are too often separated: that of contact and symplectic geometry on one hand and that of Riemannian geometry on the other hand. As remarked by the author himself, the book contains more contact geometry than symplectic geometry."

5.4.9 D. Mumford, J. Fogarty, F. Kirwan, Geometric invariant theory, Third edition, Ergebnisse der Mathematik und ihrer Grenzgebiete (2), 34, Springer-Verlag, Berlin, 1994. xiv+292 pp.

This is a very influential book on the development and application of geometric invariant theory (GIT) to construct moduli spaces. It gives the first algebraic construction of the moduli space of algebraic curves (or Riemann surfaces) and abelian varieties. With the updates in the new edition it is still the best reference on this important topic.

GIT studies an algebraic action of an algebraic group G on an algebraic variety (or scheme) X and provides techniques for forming a quotient of X by G as a variety. One motivation was to construct moduli spaces of algebraic curves as quotients of moduli spaces of certain marked algebraic curves. It was developed by David Mumford in 1965 in this book, using ideas from a paper of Hilbert in classical invariant theory. In the 1970s and 1980s, GIT was related to symplectic geometry and equivariant topology, and was used to construct moduli spaces in differential geometry and mathematical physics, such as instantons and monopoles.

The MathSciNet review of the first edition says "It is a classical problem to construct moduli spaces for algebraic curves, abelian varieties, etc. This book expounds significant progress made in the field. Among the new concepts: 'coarse moduli scheme', and 'stable points'. It is not always possible to construct a universal object over a moduli scheme; hence one has to state, and to solve, a classification problem without the moduli functor being representable. A 'coarse moduli scheme' (cf. pages 99, 129) is the representable functor closest to the moduli functor such that its geometric points are in one-to-one correspondence with the objects to be classified. The famous obstacle to constructing orbit spaces is pinned down in the definition of 'stable points' (and in every problem one has to characterize these points). Thus the constructions reduce to existence theorems of certain orbit spaces under actions of reductive groups... The reader will find an ingenious mixture of quite general and deep methods on the one hand, and special cases on the other hand. Besides the quality of the mathematics to be found here, the easy style of the presentation is striking. The author makes clear his motives and intuitive arguments before giving precise definitions and proofs. Together with many instructive examples, it gives the reader the inside information that is sometimes lacking in treatises on modern mathematics."

Since the fundamental work of Mumford on moduli spaces, there have been several other books on moduli spaces of curves, other varieties and bundles etc.

P. Newstead, *Introduction to Moduli Problems and Orbit Spaces*, Tata Institute of Fundamental Research Lectures on Mathematics and Physics, 51, Tata Institute of Fundamental Research, Bombay; by the Narosa Publishing House, New Delhi, 1978. vi+183 pp.

This is a good accessible introduction to the geometric invariant theory of Mumford developed in his book. It does not use schemes, and uses varieties instead.

According to MathSciNet, "These lectures form an excellent introduction to quotient varieties and Mumford's theory of stability which avoids scheme-theoretic complexities without baulking at essential difficulties."

I. Dolgachev, *Lectures on Invariant Theory*, London Mathematical Society Lecture Note Series, 296, Cambridge University Press, Cambridge, 2003. xvi+220 pp.

This is a more recent introduction to geometric invariant theory with some emphasis on the classical explicit invariant theory.

According to MathSciNet, "The exposition is informal; the results and main steps of their proofs are clearly presented, but the reader may have to supply some assumptions and many details. The emphasis is on explicit methods, illustrated by numerous examples and exercises, and on connections with classical projective geometry. Recent developments concerning toric quotients and variations of geometric invariant theory quotients are also discussed. All of this makes the book a very worthy introduction to invariant theory."

5.4.10 Classification of varieties and moduli spaces

A moduli space is an algebraic variety, a scheme, or an algebraic variety whose points represent isomorphism classes of algebro-geometric objects of some fixed kind, for example, algebraic varieties, bundles over varieties, or sheaves over algebraic varieties. Moduli spaces often arise as solutions to classification problems. The notion of moduli was first introduced by Riemann and meant parameters that determine isomorphism classes of Riemann surfaces.

H. Nakajima, *Lectures on Hilbert Schemes of Points on Surfaces*, University Lecture Series, 18, American Mathematical Society, Providence, RI, 1999. xii+132 pp.

A Hilbert scheme is a scheme that is the parameter space for the closed subschemes of some projective space (or a more general variety). It refines the Chow variety. The basic theory of Hilbert schemes was developed by Alexander Grothendieck in 1961, motivated by the work of Teichmüller on the moduli space of marked Riemann surfaces, i.e., the Teichmüller space.

This book deals with the Hilbert scheme of points on a complex surface. This object is originally studied in algebraic geometry but is also related to many other branches of mathematics, such as singularities, symplectic geometry, representation theory, and even to theoretical physics, as this book shows. It is perhaps helpful to note that the author was a differential geometer by training and did early work in that subject.

5.4. ALGEBRAIC GEOMETRY

According to MathSciNet, "Moduli spaces of objects associated with a given space X carry interesting, at times surprising, structures. They reflect some of the properties of X. However, they carry more geometric structure and bring out seemingly hidden aspects of the geometry of X.

This beautifully written book deals with one shining example: the Hilbert schemes of points on algebraic surfaces."

D. Huybrechts, D. Lehn, *The Geometry of Moduli Spaces of Sheaves*, Aspects of Mathematics, E31, Friedr. Vieweg & Sohn, Braunschweig, 1997. xiv+269 pp.

This book can serve as a good introduction to the theory of semistable coherent sheaves over arbitrary algebraic varieties and to their moduli spaces. It also has the character of a research monograph and a comprehensive survey on some very recent results on those moduli spaces of stable sheaves over algebraic surfaces, and hence is a valuable reference for experts as well. It is a unique book of the above features which deals with an important active topic in algebraic geometry.

According to MathSciNet, "A more elementary introduction (limited to projective spaces) can be found in the book by C. Okonek, M. Schneider and H. Spindler [Vector bundles on complex projective spaces, 1980]... Some aspects of the theory of semistable sheaves are not treated in this book, in particular the case of projective spaces, where a lot of work has been done, Picard groups of moduli spaces, Donaldson polynomials and gauge theoretical aspects of moduli spaces (for this see the book by R. D. Friedman and J. W. Morgan [Smooth four-manifolds and complex surfaces, 1994)..."

S. Mukai, *An Introduction to Invariants and Moduli*, Translated from the 1998 and 2000 Japanese editions by W. M. Oxbury, Cambridge Studies in Advanced Mathematics, 81, Cambridge University Press, Cambridge, 2003. xx+503 pp.

This book provides an accessible introduction to both algebraic varieties and their moduli spaces. The emphasis is on algebraic moduli spaces, together with related classical and geometric invariant theory, and the necessary basic concepts of algebraic geometry are developed for this purpose.

According to MathSciNet, "It is a self-contained and very explicit book, which makes it excellent for its purpose as an introductory text on invariant theory and moduli. Since the classical theory of Mumford is treated in all generality, this book treats the theme in the most usual situations. It is fully possible to follow the constructions in the text without a deep knowledge of categories of schemes and sheaves of modules. The book starts off by defining invariants of groups and goes through the classical binary invariants. This is done for the group GL(n,k) and for the ordinary complex manifolds... The 'problem' of the invariants to separate orbits leads to the study of stability. One of the examples treated in full detail is the moduli space of hypersurfaces in P^n, treated as a geometric quotient. A really nice feature is the extension of the idea of a quotient, from values to ratios, making the construction of projective quotients more natural, and also introducing the theory of compactifications in a natural way... All in all, this book gives really good explanations to why invariant theory and moduli theory are as they are, and in an understandable manner."

J. Harris, I. Morrison, *Moduli of Curves*, Graduate Texts in Mathematics, 187, Springer-Verlag, New York, 1998. xiv+366 pp.

The moduli space of algebraic curves has been studied intensively since Riemann and there are several excellent books devoted to it. But none of them give an introductory, systematic and reasonably comprehensive account of this subject including a survey of the some current results in this active subject. This book is an attempt in this direction and the targeted reader includes non-experts and graduate students. It is a valuable reference.

According to MathSciNet, "It is safe to say that this book will be an indispensable resource for people working on algebraic curves. It is based on notes from a course taught at Harvard in 1990, and captures the enthusiastic lecture style of the first author. It is not simply a compilation of theorems, but rather stresses how to use modern tools and techniques to obtain results about curves and families of curves."

According to a review in Bulletin of AMS, "this is not an easy book; rather, it is one that, despite the friendly attitude of the authors, one has to fight with. However, if you accept the challenge, you can get a lot out of it; and when you, panting, reach the end, you become aware of the fact that you liked it so much that you're tempted to go back right away to page 1 and start again."

J. Kollár, *Rational Curves on Algebraic Varieties*, Ergebnisse der Mathematik und ihrer Grenzgebiete. 3. Folge. A Series of Modern Surveys in Mathematics [Results in Mathematics and Related Areas, 3rd Series, A Series of Modern Surveys in Mathematics], 32, Springer-Verlag, Berlin, 1996. viii+320 pp.

This book gives an excellent introduction to the geometry of the structure of algebraic varieties of dimensions 3 and up through the geometry of rational curves on them.

According to MathSciNet, "In the classification theory of algebraic varieties families of rational curves appear to be significant objects to study. Now, from the modern minimal model program point of view, it is clear that certain classes of algebraic varieties can be characterized by properties of these families. Many results in this direction were obtained in works of Mori, Miyaoka, the author and others. This book contains a good exposition of the theory of rational curves... The book is a very good introduction to the theory of rational curves. It will be very useful to a wide audience."

5.4. ALGEBRAIC GEOMETRY

J. Kollár, S. Mori, *Birational Geometry of Algebraic Varieties*, With the collaboration of C. H. Clemens and A. Corti, Translated from the 1998 Japanese original, Cambridge Tracts in Mathematics, 134, Cambridge University Press, Cambridge, 1998. viii+254 pp.

This book gives a reasonably self-contained introduction to the minimal model program of threefolds and is written by two of the most important figures of this subject.

According to MathSciNet, "The minimal model program, or Mori's program, was one of the great successes of algebraic geometry in the 1980s. The basic goal was to understand the birational geometry of threefolds in a way analogous to the birational theory of surfaces. This approach got its start with work of Mori, who began to study the role of curves on threefolds X which intersected the canonical class of KX negatively... The book under review, ..., is a comprehensive treatment of the minimal model program. The text strives to be self-contained and to give complete proofs; the level of knowledge needed is that of [R. Hartshorne, *Algebraic geometry*, 1977]... the text under review will prove invaluable for the more advanced student of the minimal model program, as well as researchers in the field."

5.4.11 Algebraic curves

Algebraic curves have been studied for a long time, and the theory of algebraic curve is well-understood, unlike varieties of higher dimension. Indeed, after many particular examples had been considered, starting with circles and other conic sections, the theory of general algebraic curves was fully developed in the nineteenth century already.

E. Arbarello, M. Cornalba, P. Griffiths, J. Harris, *Geometry of Algebraic Curves. Vol. I*, Grundlehren der Mathematischen Wissenschaften, 267, Springer-Verlag, New York, 1985. xvi+386 pp.

The main purpose of the book is to study special linear series on algebraic curves. It includes both classical results and new achievements in this field, in particular on the Brill-Noether theory. It is written by four experts on the theory of algebraic curves. This book is not an introduction to the geometry of algebraic curves, but rather a research monograph and can serve as a good reference.

According to MathSciNet, "The theory of algebraic curves and their Jacobians is the first—and in some respects the best understood and most beautiful—topic in algebraic geometry. The legacy left by the classical geometers of the 19th and early 20th century (such as Riemann, Clebsch, Noether, Halphen, Castelnuovo, Enriques, and Severi) is very rich in beautiful results, ingenious ideas, and challenging conjectures. However, many problems that arose in the classical theory of curves proved to be of extremely deep nature and resisted the efforts of several generations of geometers.

A decisive breakthrough in attacking, solving, or completing some of those long-standing classical problems (such as the Brill-Noether problem on special linear systems and many enumerative problems in algebraic curve theory) has been achieved only during the last two decades. A tremendous new activity in curve theory, based on the general and rigorous concepts and techniques of modern algebraic geometry, has led to a much deeper understanding of the geometry of algebraic curves...

The goal of the book under review... is to provide the first unified and systematic representation of the basic old and new ideas, techniques and results of this recent progress, with principal emphasis on the study of linear series on curves... It is addressed to those readers who have a working knowledge of basic algebraic geometry, in particular of the basics of curve theory itself, and wish to venture beyond them into more recently explored ground..."

E. Arbarello, M. Cornalba, P. Griffiths, *Geometry of Algebraic Curves, Vol. II*, With a contribution by Joseph Daniel Harris, Grundlehren der Mathematischen Wissenschaften, 268, Springer, Heidelberg, 2011. xxx+963 pp.

The first volume of *Geometry of Algebraic Curves* concentrates on the geometry (both external and intrinsic) of a fixed algebraic curve, this second volume gives a comprehensive treatment of the foundations of the theory of moduli of algebraic curves. Moduli spaces of curves have been intensively studied since Riemann and are expected to be pursued in many years to come. This book not only discusses algebraic approaches to moduli spaces, instead, it combines algebro-geometric, complex analytic and topological/combinatorial methods. It is a valuable introductory book for students to learn the subject and a comprehensive reference for experts.

E. Brieskorn, H. Knörrer, *Plane Algebraic Curves*, Translated from the German by John Stillwell, Birkhäuser Verlag, Basel, 1986. vi+721 pp.

This book is not a usual introductory book on algebraic curves. It was based on a course on plane algebraic curves for undergraduate students. The purpose was to show through beautiful, simple and concrete examples of curves in the complex projective plane the interplay between algebraic, analytic and topological methods in the study of algebraic varieties. Its emphasis on geometric intuitions and concrete examples differentiates from and complements other books on algebraic curves.

C.H. Clemens, *A Scrapbook of Complex Curve Theory*, Second edition, Graduate Studies in Mathematics, 55, American Mathematical Society, Providence, RI, 2003. xii+188 pp.

This book contains many elementary and classical facts and results on algebraic curves. As the author said, this is a book of "impressions" of a journey through the theory of complex algebraic curves. It is not self-contained or balanced, but one focal

5.4. ALGEBRAIC GEOMETRY

point is the theory of theta functions culminating with the Schottky problem. It is an enjoyable book to read and is also a useful reference.

From the book description, "The author's intent was to motivate further study and to stimulate mathematical activity. The attentive reader will learn much about complex algebraic curves and the tools used to study them. The book can be especially useful to anyone preparing a course on the topic of complex curves or anyone interested in supplementing his/her reading."

According to MathSciNet, "This book should be brought to the attention of anyone who wants an orientation in the modern theory of Jacobian varieties, who is preparing a course on the topic, or who is looking for collateral reading matter for such a course..."

5.4.12 Algebraic surfaces

The theory of algebraic surfaces is much more complicated than that of algebraic curves, i.e., compact Riemann surfaces of finite type. Many results on algebraic surfaces were obtained by the Italian school of algebraic geometry, though not rigorously. The birational geometry of algebraic surfaces is rich, because of blowing-up. Basic results on algebraic surfaces include the Hodge index theorem, and the classification of algebraic surfaces. Not much is known about moduli spaces of algebraic surfaces in general.

W. Barth, K. Hulek, C. Peters, A. Ven, *Compact Complex Surfaces*, Ergebnisse der Mathematik und ihrer Grenzgebiete. 3. Folge / A Series of Modern Surveys in Mathematics, Vol. 4, 2nd enlarged ed., 2004. XII+ 436 pp.

The classification of compact, complex surfaces is an important topic in algebraic geometry. A rough classification is given according to Kodaira dimension, and a refinement is the Enriques-Kodaira classification.

The first edition of this book was the first comprehensive, modern account of the Enriques-Kodaira classification, a deeper study of some of surfaces including K3-surfaces, Enriques surfaces and surfaces of general type. It has been a popular textbook and also a useful reference book.

According to Zentralblatt MATH, "the second, enlarged edition of this (meanwhile classic) textbook on complex surfaces has gained a good deal of topicality and disciplinary depth, while having maintained its high degree of systematic methodology, lucidity, rigor, and cultured style. No doubt, this book remains a must for everyone dealing with complex algebraic surfaces, be it a student, an active researcher in complex geometry, or a mathematically ambitious (quantum) physicist."

According to MathSciNet, "For the past 10 years there has been a bumper crop of books in algebraic geometry which deal with the classification of algebraic surfaces [A. Beauville, Complex algebraic surfaces, 1978: H. Kurke, Lectures on algebraic surfaces,

1982; P. A. Griffiths and J. Harris, Principles of algebraic geometry, 1978]. However, all of them served only as an introduction to the subject. The book under review goes far beyond this goal. For the first time, the Enriques classification of compact algebraic surfaces is complemented by the Kodaira classification of compact complex surfaces; also, many of the most recent results on surfaces of general type and surfaces of type $K3$ are fully represented. The title of the book emphasises the role of transcendental methods in the theory. This choice is certainly necessary for the treatment of nonalgebraic complex surfaces. It also makes this book more accessible to nonspecialists in algebraic geometry. The reader interested in learning the projective technique and the beauty of classic Italian style theory of algebraic surfaces should look for another book (those of Griffiths-Harris or Beauville give much better introductions into this subject). As an illustration of the spirit of this book, we should mention the absence in it of an example of a cubic surface with its 27 lines, the symbol of the theory of algebraic surfaces for more than a century... This book was the first one to put the classic theory of algebraic surfaces (the case of nonalgebraic surfaces was not considered) on a modern footing and appeared before or at the same time as most of Kodaira's papers on classification of complex surfaces were published..."

For the second edition, "The Enriques-Kodaira classification is carried out in the spirit of Mori theory and many new developments have been added, including new analytic tools as well as new algebraic methods... A new section is devoted to the stunning results achieved by the introduction of Donaldson and Seiberg-Witten invariants."

A. Beauville, *Complex Algebraic Surfaces*, Second edition, London Mathematical Society Student Texts, 34, Cambridge University Press, Cambridge, 1996. x+132 pp.

This is a relatively elementary and concise introduction to complex algebraic surfaces and their classification theory. It presents Enriques's birational classification of surfaces in a modern and rigorous way. This is a good textbook for students and non-experts interested in the subject, and has been a popular book on complex surfaces.

According to MathSciNet, "It should be noted that since the classic book of F. Enriques himself [*Le superficie algebriche*, Zanichelli, Bologna, 1949] there have been many attempts to rewrite this classification in a more modern and rigorous way. The first attempts are connected mainly with the names of J.-P. Serre, K. Kodaira, I. Shafarevich and O. Zariski. After a ten year break, new expositions appeared (E. Bombieri and D. Mumford, E. Bombieri and D. Husemoller), reflecting recent progress in the theory: the case of a ground field of positive characteristic (Mumford, Bombieri) and surfaces of general type (Bombieri, Bogomolov, Yau). The present book is rather close to a seminar of Shafarevich, which in turn follows Enriques' book closely and is written in a style that tempts any reader to rewrite the book..."

5.4. ALGEBRAIC GEOMETRY

O. Zariski, *Algebraic Surfaces*, Second supplemented edition, With appendices by S. S. Abhyankar, J. Lipman, and D. Mumford, Ergebnisse der Mathematik und ihrer Grenzgebiete, Band 61, Springer-Verlag, New York-Heidelberg, 1971. xi+270 pp.

This is a classical book on algebraic surfaces and their classification. When it was written, it served a very important purpose of putting the work of Italian school on algebraic surfaces on a more solid ground as explained by Lefschetz in a review. It is probably outdated now. But it is of historical interests and the appendices in the new edition are valuable.

According to a review in Bulletin of AMS by Lefschetz in 1936, "It is indeed the first time that a competent specialist, informed on all phases of the subject, has examined it carefully and critically. The result is a most interesting and valuable monograph for the general mathematician, which is, in addition, an indispensable and standard vade mecum for all students of these questions... Altogether we cannot recommend this monograph too strongly to all lovers of geometry in every form. They will leave it convinced of the beauty and vitality of this most attractive branch of modern mathematical research."

A closely related earlier book on algebraic surfaces written by a founder of the theory of algebraic surfaces is

F. Enriques, *Le Superficie Algebriche*, Nicola Zanichelli, Bologna, 1949. xvi+464 pp.

Enriques was the first person to give a classification of algebraic surfaces up to birational equivalence. He also made other fundamental contributions to algebraic geometry. For example, Enriques surfaces were named after him. His classification of surfaces was made rigorous and refined by later generations of algebraic geometers such as Oscar Zariski, Kunihiko Kodaira and Igor Shafarevich, and their schools. The work of Enriques has had a huge impact on the modern algebraic geometry.

According to Igor Dolgachev, "Enriques's book *Le Superficie Algebriche* was one of the most influential books in algebraic geometry". Indeed, all the books above on classification of algebraic surfaces are following the footsteps of this classic by a master.

A google translation of the Foreword of the book, "In this work we aim to expose the theory of algebraic surfaces, regarded as a particular aspect of algebraic-geometric, which has been developing, over a half century, especially in the Italian geometric school. For many years we have developed such exposure, reviewing and sometimes imitating, or at least rearranging the oldest demonstrations, so as to give the same theory of the structure more rigorous and easier.

This work of reviewing and sorting took place primarily through the courses of lectures held by us in the University of Rome... Today, emerging out of the stormy period of the war, in time to pass the manuscript to the printers, we can even now difficult to make exact notion of the scientific literature of the last three years, and in addition to new work commitments prevent us from returning work done to review it in light of some studies, more or less roughly, we came to know only later. So keep to the text of these lessons the form given, as mentioned, in 1942, except to finish a few bibliographic and to state, with some notes, the gaps that require further examination."

It contains the following: 1. Foreword by Guido Castelnuovo - 2. Author's Preface - 3. Introduction - Chapter I. Linear systems of curves - Fuel Systems Chapter Il. covariants and invariants - Chapter III. The area added - Chapter IV . The number and kind of Riemann-Roch theorem for surfaces - Chapter V. Numerical invariants and multiple planes - Chapter VI. regular size: Minimum of kinds and conditions of rationality - Chapter VII. Classification of surface usually linear (p. I) - Chapter VIII. regular size, and canonical pluricanoniche - Chapter IX. Surface irregular and continuous systems of curves disequivalenti - Chapter X. The surface of geometric genus zero - Chapter XI. General classification of the surface"

D. Mumford, *Lectures on Curves on an Algebraic Surface,* With a section by G. M. Bergman. Annals of Mathematics Studies, No. 59, Princeton University Press, Princeton, N.J., 1966, xi+200 pp.

This book consists of lecture notes from a course by the author in the exact form as they were written and distributed. The purpose is to give a complete clarification of one "theorem" on algebraic surfaces, "the so-called completeness of the characteristic linear system of a good complete algebraic systems of curves, on a surface". It has been very influential in the study of the Picard variety of a complete non-singular surface over an algebraically closed ground field of arbitrary characteristic.

5.4.13 W. Fulton, Intersection Theory, Second edition, Ergebnisse der Mathematik und ihrer Grenzgebiete. 3. Folge, Springer-Verlag, Berlin, 1998. xiv+470 pp.

Intersection theory of algebraic varieties has been studied for a long time and played a fundamental role in the development of mathematics. This awarding winning book gives a systematic, comprehensive and modern account of this important theory.

The citation of the Steele Prize for this book says "It introduced a new order into a field that had been in disarray, by introducing a new and simpler approach that gave all the old results and more. Moreover it gave clarifying expositions of many classical computations in intersection theory, often reducing lengthy old arguments to a few lucid paragraphs. By its very clear exposition and the high quality of its content, this book has had an enormous impact on the field."

According to MathSciNet, "In the development of algebraic geometry, enumerative problems have always played a decisive role. For more than 200 years, the elaboration,

5.4. ALGEBRAIC GEOMETRY

interpretation, and application of an appropriate framework for enumerative algebraic geometry has fascinated and occupied generations of mathematicians. Although, especially during the last century, many brilliant ideas and fruitful approaches for establishing a suitable notion of local intersection multiplicity as well as for constructing intersection products of algebraic cycle classes have been provided, the lack of generality, often of clarity and exactness, and the limited applicability to only special classes of algebraic varieties and subvarieties has been a serious embarrassment. In this connection, the intensified search for an appropriate, general, and rigorously founded intersection theory since the 1930s reflected the foundational crisis of algebraic geometry, prepared the way for the modern, well-established algebraic geometry of today, and served as a major testing ground for the new foundations. Inspired by, and simultaneously in altercation with the pioneering, but often incomplete and incorrect work of the old Italian school, the foundational work of van der Waerden, Weil, Chow, Chevalley, Samuel, Serre, Grothendieck, and many others was heavily based upon the attempt to make intersection theory rigorous and clear. Despite the progress in building rigorous concepts of intersection theory during the past decades, and despite the vast range of applications, the main problem consisted, now as before, in the lack of full generality, of comparability between different approaches, and in rather complicated prerequisites and set-ups. The present book is the first modern attempt at a general, complete, and self-contained presentation of intersection theory. The approach given here was developed by the author and R. MacPherson during the past 10 years, and has an extremely general, coherent, and definitive character. It provides, finally, a unified framework for enumerative algebraic geometry. The intersection theory developed here is not only more general, stronger, and of wider applicability than the previously available ones, but also is more foundational, in that it is developed from scratch with only a few prerequisites from commutative algebra and algebraic geometry...

Overall, the book is brilliant for many reasons. It introduces an epoch-making, revolutionary new approach to an old and important theory, with far-reaching prospects. It touches, reconsiders, clarifies, and strengthens many classical topics from a unified, general, and rigorous point of view. It introduces ideas and techniques that have a vast range of applicability in modern and classical geometry... This text, with its brilliant content and excellently arranged, is a prime mover in algebraic geometry, and a must for each algebraic geometer! It is an indispensable reference book, an outstanding textbook, and a great source for geometric research in the future."

A more expository book on intersection theory by Fulton is the following one. It can serve as an introduction to and a good summary of the previous monograph.

W. Fulton, *Introduction to Intersection Theory in Algebraic Geometry*, CBMS Regional Conference Series in Mathematics, 54, Published for the Conference Board of the

Mathematical Sciences, Washington, DC; The American Mathematical Society, Providence, RI, 1984. v+82 pp. Third printing with corrections. 1996.

According to Zentralblatt MATH, "In fact, these notes provide a broad survey on many topics in both classical and modern intersection theory, the details of which had been elaborated in that brilliant monograph of W. Fulton's. While the author's nearly encyclopaedic book 'Intersection theory' has quickly become the most important standard text on this subject, these notes have maintained their outstanding role as both a beautiful introduction and a masterly survey in this area of algebraic geometry... W. Fulton's introductory notes are an excellent invitation to this subject, and a valuable spring of information for any mathematician interested in the methods of algebraic geometry in general."

According to a review of the above two books in Bulletin of AMS, "The two books under review are complete up-to-date accounts of intersection theory. The Ergebnisse book offers a detailed treatment; the CBMS book, a general introduction. Both present a revolutionary new approach, developed by the author in collaboration with MacPherson, which is technically simpler and cleaner, yet much more refined and general. To better appreciate the subject of intersection theory and the contribution of these books, it is useful to know some history. Intersection theory was founded in 1720 by Maclaurin, 93 years after Descartes promoted the use of coordinates and equations. Maclaurin stated that two curves, defined by equations of degrees m and n, intersect in mn points. A proof was sought by Euler in 1748 and Cramer in 1750. Finally in 1764 Bezout introduced a more refined method of eliminating one of the two variables from the two equations, producing a polynomial in one variable of minimal degree, which he proved is equal to mn. (Euler did so independently the same year.) Bezout went on to treat the case of r equations in r unknowns, and so most any theorem of intersection theory about projective r-space is called Bezout's theorem.

Intersection theory remained centered around Bezout's theorem for a century and a half... A revolutionary change in intersection theory took place in 1879 with the appearance of Hermann Schubert's book, *Kalkül der abzählenden Geometrie*. Schubert introduced explicitly the first intersection rings and, implicitly the operations of pullback and pushout. He applied these tools to enumerative geometry with great success, systematized and simplified much earlier work, and solved problems that had previously defied attack. Schubert's work grew out of Chasles's. In 1864 Chasles gave the first theory of enumerative geometry. In particular, he found many valid formulas and correct numbers, including the correct number of conies tangent to five others, 3264... Schubert was not fully satisfied with the mathematical foundation available for his theory, but he forged ahead anyway... The revolutionary approach of the books under review has two major aspects: a suggestive point of view and a flexible technical device. The point of view is

5.4. ALGEBRAIC GEOMETRY

that intersection theory is primarily concerned with the construction and study of operators on cycle classes on singular schemes, ... This point of view of operators actually comes closer to Schubert's original one than does the traditional point of view..."

5.4.14 R. Lazarsfeld, Positivity in Algebraic Geometry. I. Classical Setting: Line Bundles and Linear Series; II. Positivity for Vector Bundles, and Multiplier Ideals, Ergebnisse der Mathematik und ihrer Grenzgebiete. 3. Folge, Springer-Verlag, Berlin, 2004. xviii+387 pp., xviii+385 pp.

The notion of positivity has played a fundamental role in algebraic geometry, for example, through various vanishing and embedding theorems. This book gives a comprehensive account of various results and applications of ampleness and positivity in complex algebraic geometry. This book motivates and explains important results through their applications, and explains historical developments and mathematical interplay between various results well. A lot of the material in the latter part appears in book form for the first time. Though this book contains a lot of results and many topics and is targeted at the reader in complex algebraic geometry, it is not too technical and others can enjoy reading it too. Volume I is more elementary and can be read independent of Volume II.

According to a review in Bulletin of AMS, "Parts one and two treat the traditional theory of positivity in algebraic geometry, most of whose basic definitions and results have been established for at least 20 years. Modern developments have led to some changes in proofs and emphasis, but the basic framework has been stable, and it is ready for a monograph that serves as the definitive summary of the field for decades to come. I believe that Lazarsfeld's volumes give this treatment. At the same time, the presentation is easy to follow, thus it can also be used in a second course of algebraic geometry. The theory is laid out clearly, emphasizing the main ideas and theorems, and a huge number of examples also show the many ramifications, applications and special cases that have been studied or are worth exploring further.

By contrast, part 3 is the first detailed treatment of a rapidly developing new area, thus it is likely to become dated sooner."

According to MathSciNet, "Positivity has long been a major theme in various branches of algebraic geometry, both as an object of study and as a technical tool. A first book attempting to treat positivity, under the heading of ampleness, appeared already three and a half decades ago. That is R. Hartshorne's book [Ample subvarieties of algebraic varieties, Springer, Berlin, 1970], where the main goal was to extend the notion of ample divisor to arbitrary subvarieties of a given variety. Since then, the subject has seen major developments in a large number of different directions: intersection theory, singularities, topology of algebraic varieties, vanishing theorems and their applications to linear series, and higher dimensional geometry, to give a certainly incomplete list. However, these developments have mostly remained scattered in the literature and some, maybe precisely for this reason, have not been worked out in a systematic fashion. In the book under review the author succeeds wonderfully in putting together under the same heading most of the areas of classical and modern complex algebraic geometry dedicated to, or influenced by, the study of positivity. The book is divided into three parts, each with a separate introduction. In addition, each chapter contains introductory remarks and concluding notes which emphasize the history of the topic, sources of inspiration, and further references... one interesting feature of the text is the fact that at times it provides new simplified proofs of, or new approaches to, well-established results... There are numerous examples scattered throughout the text (together with various applications, they form about a third of the book). Some are very explicit, making essentially every concept introduced in the book quite easy to grasp. Others serve as a guide to further literature and encourage independent study... The book under review is exceptionally well written. It treats a large number of concepts and topics, but always in a gentle and explicit manner. It can be used both as a textbook and as a source for current research problems. As such, it will be of great value to both students and experts in the field. It is also excellent as a guide to further literature."

5.4.15 Toric varieties

A toric variety or torus embedding is a normal variety containing an algebraic torus as a dense subset such that the action of the torus on itself extends to the whole variety. One of the origins of toric varieties is to construct toroidal embeddings of arithmetic Hermitian locally symmetric spaces. Since then, toric varieties have been developed independently and are becoming one of the most important classes of varieties in algebraic geometry. Besides their own beauty and importance, the rich supply of examples and possibility of explicit computation via connection with combinatorics make them the testing examples for various theories.

T. Oda, *Convex Bodies and Algebraic Geometry, An Introduction to the Theory of Toric Varieties*, Translated from the Japanese, Ergebnisse der Mathematik und ihrer Grenzgebiete (3), 15, Springer-Verlag, Berlin, 1988. viii+212 pp.

This book is one of the first and systematic introductions to this beautiful subject since its foundation was laid down at the early 1970's. It has been a very useful introduction and reference on toric varieties and their applications.

According to MathSciNet, "In spite of the fact that the theory of toric varieties has been around for about fifteen years, and many nice surveys of it have appeared [D.

Mumford et al., Toroidal embeddings, Vol. I, Lecture Notes in Math., 339, 1973; V. I. Danilov, Uspekhi Mat. Nauk 33 (1978), no. 2(200), 85–134;, T. Oda, Torus embeddings and applications, Tata Inst. Fund. Res., 1978], it is still practiced only by a small group of algebraic geometers and combinatorists. The present book should change this situation for the better. Together with fundamentals, it gives a very nice exposition of many applications of the theory to questions from algebraic geometry or combinatorial convex geometry. Omitted are most of the applications to singularities and to compactifications of quotient spaces of bounded homogeneous domains."

W. Fulton, *Introduction to Toric Varieties*, Annals of Mathematics Studies, 131, The William H. Roever Lectures in Geometry, Princeton University Press, Princeton, N.J., 1993. xii+157 pp.

This is a more recent introduction to toric varieties (the second one after Oda's book). It was based on the lecture notes of a minicourse by the author, but it can be used as a textbook for a course on toric varieties. Many well-chosen exercises make this a very good, but demanding textbook.

According to MathSciNet, "This book is a welcome addition to the basic comprehensive literature on the theory of toric varieties (which are also called torus embeddings).

The theory provides interesting interactions between two seemingly unrelated fields: the geometry of convex cones (together with lattice points) and algebraic geometry.

A basic dictionary relating these two fields was established already at the inception of the theory at the beginning of the 1970s. Nevertheless, not only have many new interesting applications been found, but even the basic theory itself has continued to grow...

The dictionary works both ways: in one direction, we can use convex geometry to construct many nice examples of varieties and phenomena in algebraic geometry which provide not only remarkably fertile testing grounds for general theories but serve as good material for an introduction to algebraic geometry as well. In the other direction, algebro-geometric theorems and theories, when applied to toric varieties, very often lead to new insight in convex geometry... Instead of trying to survey the subject, the author gives an efficient and insightful mini-course on toric varieties by first developing the foundational material with many examples and then concentrating on the topology, intersection theory and Riemann-Roch problems, on which the author is well known to be a leading expert..."

G. Kempf, F. Knudsen, D. Mumford, B. Saint-Donat, *Toroidal embeddings. I*, Lecture Notes in Mathematics, Vol. 339, Springer-Verlag, Berlin-New York, 1973. viii+209 pp.

This is the first book where toric varieties (or rather torus embeddings) were introduced in order to construct a class of compactifications with mild singularities of arithmetic locally symmetric varieties, which were given in a book by Mumford and his collaborators. In some sense, toroidal embedding originally served as a technical construction, but it has had unexpected influence on algebraic geometry. The authors would probably not expect that the theory of toric varieties developed independently and perpendicularly to this original motivation.

According to the introduction, "The goal of these notes is to formalize and illustrate the power of a technique which has cropped up independently in the work of at least a dozen people, but which does not seem to have been generally recognized as yet."

I.M. Gelfand, M.M. Kapranov, A.V. Zelevinsky , *Discriminants, Resultants, and Multidimensional Determinants*, Reprint of the 1994 edition, Modern Birkhäuser Classics, Birkhäuser Boston, Inc., Boston, MA, 2008. x+523 pp.

The elimination theory is a classical subject in commutative algebra and algebraic geometry that deals with algorithmic approaches to eliminating between polynomials of several variables, and resultants and discriminants appear naturally in the processes. However, elimination theory has been eliminated from algebra and geometry for several decades in favour of more abstract ideas. By presenting the recent important results of the authors in a uniform way together with some classical results, they started a renaissance of elimination theory and connected it with very recent developments in algebraic geometry, homological algebra, and combinatorial theory

According to MathSciNet, "This book revives and vastly expands the classical theory of resultants and discriminants. Most of the main new results of the book have been published earlier in more than a dozen joint papers of the authors. The book nicely complements these original papers with many examples illustrating both old and new results of the theory... This book can be considered as a first step towards exposition of the general theory of A-hyperelliptic functions developed by the first author and his numerous collaborators. Some of the remarks in the text point to this connection."

According to a review in Bulletin of AMS, "One basic question I would like to answer is whether this book is meant for (graduate) students. My feeling is that first of all it is very nice for students to see so many concrete examples and pictures. The reader here is well motivated by simple but illuminating examples... before being faced with general notions; this is done quite systematically in the book, where examples and pictures always precede the proofs and allow those to be easily stated and easily understood."

5.5 Convex geometry and discrete geometry

The major branches of pure mathematics consist of algebra, analysis, geometry and topology, combinatorics, logic and number theory. Geometry and number theory are probably the two oldest. Their combination, geometry of number, was a recent creation by a single person, Minkowski. Its basic result concerns the member of integral points inside a convex body in the Euclidean space. It turns out to have unexpected powerful applications in several other subjects.

Figure 5.5: Minkowski's book.

Convex geometry studies convex subsets of the Euclidean space, which arise naturally in many areas of mathematics. It is a relatively young mathematical discipline. Although the first known contribution to convex geometry dates back to antiquity and can be traced in the works of Euclid and Archimedes, it became an independent branch of mathematics at the turn of the nineteenth century, mainly due to the works of Hermann Brunn and Hermann Minkowski.

Convex analysis studies properties of convex functions and convex sets, and has important applications to convex minimization, which is a subdomain of optimization theory.

Discrete geometry is closely related to combinatorial geometry, and they study combinatorial properties and constructive methods of discrete geometric objects, such as points, lines, spheres, polygons. For example, it studies how lines and planes intersect, or how convex subsets and their translates may be arranged to cover a larger object. Discrete geometry has large overlap with convex geometry.

5.5.1 R.T. Rockafellar, Convex Analysis, Princeton Mathematical Series, No. 28, Princeton University Press, Princeton, N.J., 1970. xviii+451 pp.

This is a classic on convex analysis and it has applications to optimizations problems in computer sciences, economics and engineering. The exposition is self-contained and kept at an elementary level and hence makes it an accessible textbook. This book was based on the author's lectures at Princeton in 1966 and a course by Werner Fenchel at University of Coppenhagen 16 years before that. For people working in Teichmüller theory and

Fuchsian groups, Fenchel is famous for the Fenchel-Nielsen coordinates and also for his book on Fuchsian groups, and they would not expect his contribution to convex analysis. At the time when it was written, it also contained the most recent results, and it has also been a very useful reference.

According to MathSciNet, "In the study of extremum problems in many areas of applied mathematics, convexity plays an increasingly important role. This book is a self-contained account of the recent developments of the theory of convex sets and convex functions.

The book, divided into 8 parts, begins with the basic algebraic and topological properties of convex sets and convex functions. The emphasis is on the fact that the convex functions on $\mathbb{R}^n$ can be identified with certain convex sets in $\mathbb{R}^{n+1}$ (their epigraphs), while the convex sets in $\mathbb{R}^n$ can be identified with certain convex functions on $\mathbb{R}^n$ (their indicators). These identifications are frequently used in the book to switch back and forth between geometric and analytic approaches.... As a whole the book is well organized and unusually coherent. The author's approach, influenced by W. Fenchel [' Convex cones, sets and functions', mimeographed lecture notes, Princeton Univ., Princeton, N.J., 1953], is to a large extent, an outgrowth of his own research. Many results and proofs have their origin in the author's earlier papers. Much new material is also presented here for the first time."

I. Ekeland, R. Témam, *Convex Analysis and Variational Problems*, Translated from the French, Corrected reprint of the 1976 English edition, Classics in Applied Mathematics, 28, Society for Industrial and Applied Mathematics (SIAM), Philadelphia, PA, 1999. xiv+402 pp.

This is a classic on application of convex analysis to the calculus of variations and optimal control. This is the first general book that introduces convex analysis into the calculus of variations and that was written by the main contributors of the subject.

According to MathSciNet, "Not more than ten years have passed since the words 'convex analysis' appeared in the mathematical vocabulary but anyone working in optimization knows how deeply convex analysis has influenced all parts of the theory. There are two reasons why convex analysis has proved to be so effective in the calculus of variations and optimal control. The first is the abundance of meaningful convex problems, especially involving multiple integrals. The second, connected with the very nature of variational problems, is that a certain amount of convexity is inherent in any, even non-convex, variational problem."

C. Castaing, M. Valadier, *Convex Analysis and Measurable Multifunctions*, Lecture Notes in Mathematics, Vol. 580, Springer-Verlag, Berlin-New York, 1977. vii+278 pp.

5.5. CONVEX GEOMETRY AND DISCRETE GEOMETRY

This book studies convex analysis, measurable multi-functions and applications. It contains expositions, extensions and improvements of many important results in convex analysis. Though it is not comprehensive, it is a valuable reference.

J.P. Aubin, A. Cellina, *Differential Inclusions. Set-valued Maps and Viability Theory.* Grundlehren der Mathematischen Wissenschaften, 264, Springer-Verlag, Berlin, 1984. xiii+342 pp.

Differential inclusions are a generalization of the concept of ordinary differential equation. For a differential equation, the derivative of an unknown function is equal to something. For a differential inclusion, the derivative of an unknown function is required to belong to a given set. It arises in many situations such as differential variational inequalities, projected dynamical systems, dynamic Coulomb friction problems. When the book was first written, differential inclusions were newly developed, and this book was a comprehensive introduction to this subject. It has been a valuable reference.

According to MathSciNet, "This monograph presents a new, systematic and unified approach to the essential ideas of differential inclusions and related topics such as viability problems and optimal control. The text is intended for readers who are not necessarily mathematicians."

5.5.2 Convex functions and convex geometry

A.W. Roberts, D. Varberg, *Convex Functions*, Pure and Applied Mathematics, Vol. 57, Academic Press, New York-London, 1973. xx+300 pp.

Though the notion of convex functions is of fundamental importance, it was not recognized a long time ago, and there were not many books about it either. This book was written as an introduction for undergraduate students and graduate students. It is still a good introduction and a valuable reference.

According to MathSciNet, "It is quite surprising that the notion of convex functions is just 70 and not 170 or 270 years old and that its initiator, J. L. W. V. Jensen, could in 1906 state just as a hunch that: ' It seems to me that the notion of convex function is just as fundamental as positive function or increasing function. If I am not mistaken in this, the notion ought to find its place in elementary expositions of the theory of real functions.' While time has more than proved this conjecture, there still have been no books dealing exclusively with a comprehensive study of convex functions... The present book has been written for senior undergraduate students but certainly can be used also as a graduate textbook, and also the non-specialist mathematician or even the specialist may find interesting new information in it...."

P. Gruber, *Convex and Discrete Geometry*, Grundlehren der Mathematischen Wissenschaften, 336, Springer, Berlin, 2007. xiv+578 pp.

This is a comprehensive book on convex and discrete geometry, including main ideas, results and methods of the subjects. It is written by a leading expert, and is a good textbook for graduate courses and also a very valuable reference for experts.

According to MathSciNet, "Convex geometry is a classical area of mathematics that dates back to antiquity. There are many excellent books covering the area or its parts. What makes the book under review unique is that the author's main goal is to expose applications of convex geometry to other areas of mathematics and science, in particular number theory, mathematical physics, crystallography, tomography, optimization, and computational geometry. The coverage of convex geometry itself does not pretend to be complete, the author chooses different directions of the theory, presents the main results with proofs and gives references to further developments. Immediately after that or a little later one can find applications to other areas and open problems. The coverage of discrete geometry and the geometry of numbers also contains a wealth of applications..."

R. Gardner, *Geometric Tomography*, Second edition, Encyclopedia of Mathematics and its Applications, 58, Cambridge University Press, Cambridge, 2006. xxii+492 pp.

Tomography refers to a technology of imaging by sections or sectioning, through the use of any kind of penetrating waves such as X-ray. Mathematically, it means to reconstruct a set from measures of its sections by flats or its projections onto subspaces. This book is a first book devoted to geometric tomography. It is an accessible introduction enriched with beautiful figures and very informative notes, and is also a valuable reference.

E. Alfsen, *Compact Convex Sets and Boundary Integrals*, Ergebnisse der Mathematik und ihrer Grenzgebiete, Band 57, Springer-Verlag, New York-Heidelberg, 1971. x+210 pp.

It is intuitively clear that for any bounded closed convex subset of the plane, every point is a convex combination of the extremal points of the subset. Choquet theory is concerned with a similar representation by the extreme points of a convex set in an infinite-dimensional (locally convex Hausdorff) topological vector space, or equivalently about measures with support on such extreme points. When this book was written, it served as an up-to-date introduction to and also a complete reference on Choquet theory. This elegant short book is still a valuable reference for this basic topic in analysis.

R. Schneider, *Convex Bodies: the Brunn-Minkowski Theory*, Encyclopedia of Mathematics and its Applications, 44, Cambridge University Press, Cambridge, 1993. xiv+490 pp.

Brunn-Minkowski theory, or rather Brunn-Minkowski theorem, bounds the volume of the sum of two convex compact subsets of $\mathbb{R}^n$ from below by a suitable combination of the volumes of the two subsets. It is the classical core of the geometry of convex bodies. This book is a classical reference on the Brunn-Minkowski theory.

5.5. CONVEX GEOMETRY AND DISCRETE GEOMETRY

According to Zentralblatt MATH, "In recent years, convexity theory has been of increasing interest with connections to various other disciplines, and the Brunn-Minkowski theory has maintained its place at the centre of the field. The more recent developments include a solution of the Christoffel problem, local formulae of integral geometry, the theory of valuations, new inequalities for various functionals, zonoids and their applications in stochastic geometry, stability results improving classical inequalities... Over time, the necessity of a book, which includes the most important of these new developments, became clear. The monograph under review satisfies this need in an impressive manner. It is written by an author who has himself made important contributions to nearly all aspects of the theory and whose work has initiated many of the recent developments."

G. Pisier, *The Volume of Convex Bodies and Banach Space Geometry*, Cambridge Tracts in Mathematics, 94, Cambridge University Press, Cambridge, 1989. xvi+250 pp.

When it was written, there was a lot of progress on the local theory of Banach spaces, i.e., using finite dimensional tools to study infinite dimensional spaces. This book was written to give a self-contained presentation of a number of recent results, which relate the volume of convex bodies in n-dimensional Euclidean space and the geometry of the corresponding finite-dimensional normed spaces. It is still a very valuable reference on this growing subject.

V. Milman, G. Schechtman, *Asymptotic Theory of Finite-dimensional Normed Spaces*, With an appendix by M. Gromov, Lecture Notes in Mathematics, 1200, Springer-Verlag, Berlin, 1986. viii+156 pp.

This book deals with the local theory of Banach spaces, i.e., the geometrical structure of finite dimensional normed spaces as the dimension grows to infinity. Written by two experts on this subject, this book serves as a good introduction to some of the results, problems and major methods developed in the local theory.

According to MathSciNet, "this is an excellent book, which admirably achieves its goal of introducing the reader to the diverse methods of local theory. Although many questions remain open, this publication marks a first stage of maturity for this theory. It is clearly a 'must' for specialists, but it deserves a large diffusion among other mathematicians, and it can serve as the basis for an advanced graduate course."

VI. The Lurie Tower on the North Campus of the University of Michigan, which houses a 60-bell grand carillon

LEONARD EULER.

From a Medalion, as large as life, by Ruchotte, in the profession of John Wilmot Esq.

Topology has a relatively short history. Euler's paper in 1736 on the negative solution to the problem on the Seven Bridges of Konigsberg is often viewed as the beginning of topology. The title of his paper is "Solutio problematis ad geometriam situs pertinentis". It was a famous problem understood by everyone and solved elegantly by a new idea in mathematics, Euler's formula for polyhedra. This formula has been generalized to give the Euler characteristic (or Euler-Poincare characteristic) of a topological space and is a fundamental invariant in modern mathematics.

Though some important concepts in topology were already introduced by Enrico Betti and Bernhard Riemann, the general theory, or the subject of algebraic topology, was single-handedly created by Poincaré. His first article on topology appeared in 1894. He introduced the abstract topological definition of homotopy and homology and also the basic concepts and invariants of combinatorial topology, such as Betti numbers and the fundamental group, which led to the famous Poincare conjecture on characterization of the three dimensional sphere. He also proved a formula, called the Euler-Poincare theorem, relating the number of edges, vertices and faces of a polyhedron in any dimension and hence generalizing the Euler polyhedron formula.

L. E. J. BROUWER (1881–1966)

To most students in mathematics, Brouwer's best known result is the fixed point theorem on a continuous self-map of the unit disc. Probably his most important result is his proof of the topological invariance of dimension, which finally justified the continuity approach of Poincaré and Klein in their attempts to prove the uniformization theorem of Riemann surfaces. His simplicial approximation theorem is also a foundational result in algebraic topology. In spite of this, he might be better known for others for his role as the founder of the mathematical philosophy of intuitionism opposing the formalism of David Hilbert. He was kicked out of the editorial board of Mathematische Annalen because of this and established a new journal Compositio Mathematica, which was also taken away from him eventually.

Lobachevsky was mainly known for his work on hyperbolic geometry. His idea was so revolutionary and he was called "Copernicus of Geometry". It was also said that Gauss learned Russian in order to read the work of Lobachevsky. Due to the work of Poincaré, Klein, and later Thurston, hyperbolic geometry plays an essential role in the low dimensional topology, and almost all important problems in two and three dimensional topology are solved recently with the help of hyperbolic geometry. Though the three people, Gauss, Lobachevsky and Bolyai, are often credited with discovery of hyperbolic geometry, Lobachevsky deserved most credit both for his original work and multiple books. On the other hand, Gauss never published his ideas and Bolyai did not publish too much either or present his work on non-Euclidean geometry successfully to the world mathematical community. In fact, Bolyai published more than 24 pages as an Appendix to a mathematics textbook by his father, but he left more than 20,000 pages of mathematical manuscripts when he died.

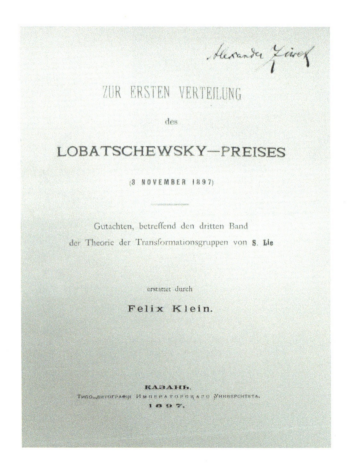

Chapter 6

Topology

Topology is a relative new branch of mathematics. It is now a major and very active branch of the modern mathematics. Topology studies properties of spaces and figures that are preserved under continuous deformations, such as deformations that involve stretching, but no tearing or gluing. One can also impose various conditions on spaces and maps between them. For example, smooth maps between smooth manifolds are the basic objects in differential topology. Another example is algebraic topology which studies topological spaces using algebraic methods.

The origin of topology can be traced to as early as 1736 when Euler wrote his paper on the seven bridges of Königsberg, but it was only near the end of the nineteenth century that a distinct discipline referred to in Latin as the geometria situs (geometry of place) or analysis situs (Greek-Latin for picking apart of place) was developed. One founding father is Poincaré. It was later given the modern name of topology. By the middle of the twentieth century, topology had become one of the most important subjects of mathematics. This can be confirmed by the number of Fields medals which have been given to people working in topology and related topics.

Topology includes (1) point-set topology, or general topology, (2) algebraic topology, (3) differential topology, and (4) geometric topology. In algebraic topology, homology, cohomology, and homotopy theories play an important role.

In this chapter, we discuss books related to the topics above.

6.1 More classical topology

In the definition of topology spaces, there are at least four conditions on separation of points. The most natural one is the Hausdorff property. In his classic book of 1914, Grundzüge der Mengenlehre, *Hausdorff defined and studied partially ordered sets abstractly, and proved that every partially ordered set has a maximal linearly ordered subset using the axiom of choice. He also axiomatized the topological concept of neighborhood of points and introduced the topological spaces that are now called Hausdorff spaces. Besides his mathematical work, he also wrote plays and philosophical works under a pseudonym.*

Figure 6.1: Hausdorff's book.

6.1.1 P. Alexandroff, H. Hopf, Topologie I. Berlin, Springer, 1935. xiii+ 636 pp.

A popular or elementary book on topology by Alexandroff was mentioned before. The current book is at a different level and has a different purpose. It is a classical book and played an important role in its time. In fact, it was considered as a Bible in topology for a period. It is outdated now.

According to MathSciNet, "There can be no better example than this book of the effect of the right people being together at the right time. The year 1935 was a watershed in the history of topology, not least for the appearance of this book, and, though the subject has changed enormously since then (so that, for example, our perspective is very different from that of Kuratowski's splendid review, cited above), 'Alexandroff-Hopf' is still useful and important today."

The history of this book is explained by MacTutor History of Mathematics archive as follows: "The atmosphere in Göttingen had proved very helpful to Alexandroff, particularly after the death of Urysohn, and he went there every summer from 1925 until 1932. He became close friends with Hopf and the two held a topological seminar in Göttingen...

From 1926 Alexandroff and Hopf were close friends working together. They spent some time in 1926 in the south of France with Neugebauer. Then Alexandroff and Hopf spent the academic year 1927–1928 at Princeton in the United States. This was an important year in the development of topology with Alexandroff and Hopf in Princeton

6.1. MORE CLASSICAL TOPOLOGY

and able to collaborate with Lefschetz, Veblen and Alexander. During their year in Princeton, Alexandroff and Hopf planned a joint multi-volume work on Topology the first volume of which did not appear until 1935. This was the only one of the three intended volumes to appear since World War II prevented further collaboration on the remaining two volumes."

According to an article titled "Moscow 1935: Topology moving toward America" by Whitney, "The international Conference in Topology in Moscow, September 4–10, 1935, was notable in several ways. To start, it was the first truly international conference in a specialized part of mathematics, on a broad scale. Next, there were three major breakthroughs toward future methods in topology of great import for the future of the subject. And, more striking yet, in each of these the first presenter turned out not to be alone: At least one other had been working up the same material.

At that time, volume 1 of P. Alexandroff / H. Hopf, *Topologie*, was about to appear. I refer to this volume as A-H. Its introduction gives a broad view of algebraic topology as then known; and the book itself, a careful treatment of its ramifications in its 636 pages. (It was my Bible for some time). Yet the conference was so explosive in character that the authors soon realized that their volume was already badly out of date; and with the impossibility of doing a very great revision, the last two volumes were abandoned..."

According to a review in Bulletin of AMS in 1936, "Books on topology are so few that the appearance of a new one is an important event. The volume now before us is especially impressive, not only for its own size and thoroughness, but because it is the first of three volumes which together are intended to give a detailed survey of topology as a whole. The authors are distinguished geometers who have been in close contact with the various topological centers in Europe and America." The time is different now.

The planned second part of this book never appeared. On the other hand, Alexandroff wrote a closely related book on homology. Some parts of the book with Hopf were revised and added into the new book.

P.S. Alexandrov (also Alexandroff), *Combinatorial Topology*, Vol. 1, 2 and 3, Translated from the Russian, Reprint of the 1956, 1957, and 1960 translations, Dover Publications, Inc., Mineola, NY, 1998. 650 pp.

This is more elementary than the book with Hopf, but is still very substantial. According to the preface, "This book is an introduction to modern homology theory. It can be understood by anyone familiar with general set-theoretic and algebraic concepts and has been designed for the reader striving to acquire a knowledge of topology through a systematic study of its essentials. Hence, in this book, a reader can become acquainted with the ideas of modern topology only by a detailed study of the fundamental topological facts...In writing this book I have made extensive use of *Topologie* I, a joint work of the well-known Swiss mathematician H. Hopf and myself ... "

According to MathSciNet, "... The author is at pains to disclaim completeness in the variety of topics covered, or the development of any topic, or in historical reference. The book is presented at every point as an introduction to its subject matter..."

6.1.2 K. Kuratowski, Topology. Vol. I., Vol. II., New edition, Revised and augmented, Translated from the French by J. Jaworowski Academic Press, New York-London; Państwowe Wydawnictwo Naukowe, Warsaw, 1966. xx+560 pp., 1968. xiv+608 pp.

This is a classical book on the general topology written by an authority. It is really a comprehensive reference. The two volumes of over 1100 pages cover all the important topics in the general topology, and it each topic is treated systematically. For example, the dimension theory is studied in both volume 1 and volume 2 in about 100 pages. Another 100 pages are devoted to connected spaces, which are usually treated in a few pages of a modern textbook on topology. This is followed by another 100 pages on locally connected spaces. In some sense, this book can be thought of as a collection of monographs on all important topics in the broad area of general topology. No wonder that it was called the "Bible" in the set-theoretic topology.

Besides the standard topics, it also covers other interesting topics, for example, applications of topology to mathematical logic. The prefaces of these two books also contain a list of major books on the general topology which had been written. They provide an interesting historical perspective.

The general topology is a foundational subject and is used by almost all mathematicians. It might not be a major research area as it was in the early part of the 20th century. This book provides a very good summary of important topics and results in the general topology.

This book is especially good for people who have had some introduction to topology and is a very valuable reference to look up some non-standard topics, for example, even a definition of a topological space in Chapter 1. The author has organized a huge amount of material in strict logical order, and the book is easy to use.

According to MathSciNet, "This translation of what is generally regarded as the set-theoretic topologist's 'Bible' is a most welcome addition to the mathematical literature in English."

6.1.3 More books on topology

H. Seifert, W. Threlfall, *Seifert and Threlfall: A Textbook of Topology*, Pure and Applied Mathematics, 89, Academic Press, Inc., New York-London, 1980. xvi+437 pp.

6.1. MORE CLASSICAL TOPOLOGY

This is a classic book on topology and emphasizes the geometric intuition. It has taught generations of students about the geometric intuition in topology.

According to a review in Bulletin of AMS in 1935, "The textbook of Seifert and Threlfall should do much to smooth the path of the student who wants to learn the fundamentals of (combinatorial) topology. The authors have concentrated on basic concepts and methods, avoiding generalizations; they have explained these concepts and methods in as simple and concrete a fashion as possible, yet one which is thorough and rigorous. The exposition proceeds by easy stages, with examples and illustrations at every turn. It presupposes nothing; even the necessary group theory is developed in a special supplementary chapter."

S. Lefschetz, *Algebraic Topology*, American Mathematical Society Colloquium Publications, v. 27, American Mathematical Society, New York, 1942. vi+389 pp.

This is a classic written by a master.

According to the MathSciNet review by Whitney, "The book will certainly be a fundamental source of knowledge for some time to come; it will be interesting to see how well topologists will be able to apply its methods to particular theories. The first half will be well repaying to any topologist; anyone studying the last half will find himself fully rewarded in the end."

J. Kelley, *General Topology*, Graduate Texts in Mathematics, Vol. 27, 1975. xiv+298 pp.

This classic book is a systematic exposition of general topology. It is especially intended as background for modern analysis. It covers some topics not covered in more modern books on general topology and hence is a good reference.

According to MathSciNet, " It is roughly similar in scope to the first volume of Bourbaki's 'Topologie générale', though with variations of emphasis... As with Bourbaki, a valuable feature of the book is the collection of exercises; besides routine exercises, these include further (often extensive) theoretical developments and applications (with suitable hints), and some useful counterexamples. There is very little on the geometrical aspects of the subject; local connectedness receives little more than mention, and arcs are not mentioned at all...."

R.L. Moore, *Foundations of Point Set Theory*, Revised edition, American Mathematical Society Colloquium Publications, Vol. XIII, American Mathematical Society, Providence, R.I. 1962. xi+419 pp.

This is a classic and has been influential. Moore was known for his work in general topology, but now probably much more for the Moore method of teaching university mathematics. His teaching method must work well since he has supervised 50 Ph. D students, almost all at Texas, including R.H. Bing, Mary Ellen Rudin, and Raymond Louis Wilder.

According to MathSciNet, "A whole generation of topologists has matured since this book first appeared (and mostly since it went out of print)...The book is concerned mainly with three large and closely related sectors of point-set topology: the theory of continuous curves (locally connected, connected spaces), the topology of the plane and 2-sphere, and upper semi-continuous collections and decompositions. While the methods are elementary (in the technical sense) and, strictly speaking, no knowledge of the literature is required or referred to, certainly some knowledge of general topology will be found helpful...As it stands, the work as a whole can only be described as monumental"

J. Dugundji, *Topology*, Allyn and Bacon, Inc., Boston, Mass., 1966. xvi+447 pp.

This is a comprehensive introduction to general topology. Its aim is to provide a foundation in general topology which will be sufficient for a broad variety of disciplines in mathematics. Its comprehensive coverage of topics makes this classic a very valuable introduction and a reference.

According to MathSciNet, "This book is designed to serve mainly as a text for a two-semester course in general topology. For this purpose it is considered by the reviewer to be one of the best among the numerous books on the subject now available.

The inclusion of material of a more specialized character... and the coverage in some depth of several topics ... render it also valuable as a reference... Several hundred problems are provided. Some of these are simple exercises; others are theorems complementing the text... The user of this book will find the exposition extremely lucid, concise and abundantly illustrated by concrete examples."

R. Engelking, *General Topology*, Translated from the Polish by the author, Second edition, Sigma Series in Pure Mathematics, 6, Heldermann Verlag, Berlin, 1989. viii+529 pp.

This book gives a reasonably complete and up-to-date account of general topology. It is a good textbook for people who wish to specialize in general topology, and is also a valuable reference for specialists in general topology.

According to MathSciNet, "this is a very good book. It is a comprehensive exposition of that part of topology that centers around compactness, metrizability, paracompactness and normality. It covers a vast amount of material in a coherent, well organized way. The choice of topics discussed is excellent (the reviewer feels that the two or three topics he misses only reflect personal taste) and guarantees that the book will retain its value for a long time to come."

S. Nadler, *Continuum Theory: An Introduction*, Monographs and Textbooks in Pure and Applied Mathematics, 158, Marcel Dekker, Inc., New York, 1992. xiv+328 pp.

A continuum is a compact connected Hausdorff or metric space. Besides the regular ones such as submanifolds, there are many irregular ones which appear naturally in

6.1. MORE CLASSICAL TOPOLOGY

dynamical systems and limiting sets. Though they are important for more advanced courses and research, results on continua are usually not found in standard textbooks on topology. This book serves an important purpose by filling in a gap. It is useful both as a textbook and as a reference book.

According to MathSciNet, "the book will be a tremendous help for anyone who wishes to learn continuum theory in an organized and pedagogical sequence and has had an upper-level undergraduate or beginning graduate course covering continuous functions, compactness, connectedness, the Tietze extension theorem and the Baire theorem."

6.1.4 Fixed point theory

There are several fixed point theorems, and they are used in many branches of mathematics, for example, the Lefschetz fixed-point theorem and the Brouwer fixed point theorem in topology, and the Schauder fixed point theorem and Schaefer's fixed point theorem in functional analysis, which are particularly useful for proving existence of solutions of nonlinear differential equations.

K. Goebel, W. Kirk, *Topics in Metric Fixed Point Theory*, Cambridge Studies in Advanced Mathematics, 28, Cambridge University Press, Cambridge, 1990. viii+244 pp.

Every mathematician knows some fixed point theorems and has used them in some ways. This book gives a systematic introduction to various fixed point theorems related to Banach spaces.

According to MathSciNet, "There are two classical theorems, one rigidly metric and one purely topological, enclosing this subject... The two proofs are totally unrelated, the first constructs a sequence converging to a fixed point while the second is a nonconstructive existence proof using the compactness. This book considers a number of situations where the hypotheses on the function and on the space are balanced somewhere between those of these classic cases. ... In short, everything anyone wants to know about metric fixed point theory is discussed somewhere, clearly and with recent proofs where there are any."

A. Granas, J. Dugundji, *Fixed Point Theory*, Springer Monographs in Mathematics, Springer-Verlag, New York, 2003. xvi+690 pp.

This is the most comprehensive book on the fixed point theorems that lie on the border-line of topology and nonlinear functional analysis. It is a useful reference for people who are interested in fixed point theorems.

According to MathSciNet, "This is the most comprehensive, well-written and complete book on fixed point theory to date. The book studies just about every aspect of fixed point theory in terms of classes of operators, spaces, and methodology... The book carries an extensive literature on the subject and many examples. Many of the interesting results, given as exercises, constitute an extension of the theory established in the main text. No results from the main text rely upon those of the exercises. I recommend this excellent volume on fixed point theory to anyone interested in this core subject of nonlinear analysis."

6.1.5 Dimension theory

Though the dimension of a vector space or a manifold is intuitively clear and easy to define, the dimension of a general subset of the Euclidean space is not so clear. The dimension theory is a branch of general topology which defines and studies dimensions and dimensional invariants of general topological spaces.

W. Hurewicz, H. Wallman, *Dimension Theory*, Princeton Mathematical Series, v. 4, Princeton University Press, Princeton, N.J., 1941. vii+165 pp.

This is a classic and has been a standard reference on dimension theory for many years. The authors aimed "to give a connected and simple account of the most essential parts of dimension theory" on topics "which are of interest to the general worker in mathematics as well as the specialist in topology." They have succeeded. It presents the theory of dimension for separable metric spaces with "depth, clarity, precision, succinctness, and comprehensiveness."

Hurewicz was a major topologist and is now remembered for his discovery of the higher homotopy groups in 1935–1936, and his discovery of exact sequences in 1941. His contributions in general topology are centered around dimension theory.

J. Nagata, *Modern Dimension Theory*, Revised edition. Sigma Series in Pure Mathematics, 2, Heldermann Verlag, Berlin, 1983. ix+284 pp.

The classical book *Dimension Theory* by W. Hurewicz and H. Wallman only studied separable metric spaces. The current book is a sequel in some sense. Besides including the development after that book, it also emphasized the dimension theory for general metric spaces.

According to MathSciNet, "The present monograph is a rich, reliable and well-written source of information on general dimension theory; in the opinion of the reviewer it will prove to be indispensable to all mathematicians working in dimension theory and will become the new standard reference work for this area."

According to a review of Bulletin of AMS, "Dimension theory is one of the triumphs of point-set topology. When Cantor showed that Euclidean spaces of different dimensions nevertheless admitted one-one correspondences, and Peano showed that this could even happen in a continuous way, the naive ideas about dimension were shattered. Was there even a topological invariant that could be called dimension? Brouwer showed that this was so, at least for Euclidean spaces; but his work did not lead to a satisfactory general theory. The key idea was contained in a remark of Poincaré: Euclidean space $\mathbb{R}^3$ is 3-dimensional because prison walls are 2-dimensional. This was developed, in the early 1920s, by Urysohn and independently by Menger, into a satisfactory theory of dimension...The first edition is much easier to read; but the present one is even more worth reading. It gives a very good account of its subject, and its title is well deserved."

Y. Pesin, *Dimension Theory in Dynamical Systems*, Contemporary views and applications, Chicago Lectures in Mathematics, University of Chicago Press, Chicago, IL, 1997. xii+304 pp.

Nonregular subsets such as Cantor set which do not share properties with submanifolds occur naturally in dynamic systems as invariant sets and they often have fractional dimensions. In fact, their dimensions (or rather Hausdorff dimensions) are the most important characteristics of the dynamical systems. The purpose of this book is to present a comprehensive and systematic account of dimension theory in dynamical systems. It has achieved its purpose.

According to MathSciNet, "This book is the first one to offer a unified approach to many different notions of dimension, as well as important methods and results previously scattered throughout the literature, at the interface between dimension theory and the theory of dynamical systems—an exciting area that has rapidly developed over the last decade. In fact, it is an excellent and highly enjoyable book, self-contained, and clearly written from the first to the last page. It puts many ideas and concepts in a higher and more general framework without introducing cumbersome technicalities, and still touches on the pathologies arising in connection with the dimensions and the tantalizing questions in the field. As a result, there is a smoother ride to the core of the subject than experienced by many of us when the area was in its early stages."

6.2 Algebraic topology

Algebraic topology uses tools from abstract algebra, for example, homological algebra, to study topological spaces. The purpose is to find algebraic invariants that distinguish and classify topological spaces up to homeomorphism, or up to homotopy equivalence. Although algebraic topology primarily uses algebra to study topological problems, topology can also be used to solve algebraic problems. For example, a famous result that any finite-dimensional real division algebra must be of dimension 1, 2, 4, or 8 was proved by using techniques of algebraic topology, in particular K-theory. (An earlier result that finite-dimensional real commutative division algebra is either 1 or 2 dimensional was also proved by methods of topology).

Cech was a Czech mathematician and worked in differential geometry and topology. He is best known for Cech cohomology. Though Cech was interested in geometry but he was appointed to the chair of analysis at Masaryk University, Bron in 1923, whose geometry chair was held by Seifert. So Cech had to teach courses in analysis and algebra, and he proceeded to master these two subjects. Cech also became interested in topology, and became one of the leading experts on combinatorial topology.

Figure 6.2: Cech.

6.2.1 J. Milnor, J. Stasheff, Characteristic Classes, Annals of Mathematics Studies, No. 76, Princeton University Press, 1974. vii+331 pp.

This is a classical book, one of the best books written by Milnor.

The book evolved from lectures by Milnor in 1957 at Princeton University. The lecture notes were written up by Stasheff and Milnor and were then circulated to many universities. They provided a foundation of their training in algebraic topology. The lecture notes were finally turned into a book and appeared in 1974.

This book discusses many characteristic classes including the work of the founders H. Whitney and E. Stiefel around 1935. After the initial work, in 1942 L. Pontryagin investigated the homology of the real Grassmann manifolds and introduced new characteristic classes for real vector bundles, and in 1846 S.S. Chern studied the cohomology of the complex Grassmann manifolds and showed that it is much easier to understand than the real case and accordingly defined the characteristic classes of complex vector bundles and showed that they enjoy particularly satisfying properties.

According to a Review in Bulletin of AMS, "In 1957 there appeared notes by Stasheff of lectures on characteristic classes by Milnor at Princeton University. These notes are a clear concise presentation of the basic properties of vector bundles and their associated characteristic classes. Since their appearance they have become a standard text regularly used by graduate students and others interested in learning the subject. The present, long-anticipated book is based on those notes. It follows the order of the notes but is considerably expanded with more detail and discussion. In addition, exercises have been added to almost each section, there are many useful references to the textbooks on algebraic topology that are available now, and there is an epilogue summarizing main developments in the subject since 1957. All of these strengthen the book and make it even more valuable as a text for a course as well as a book that can be read by students on their own. The material covered should be required for doctoral students in algebraic or differential topology and strongly recommended for those in differential geometry. ... The book under review is an elegant treatment of the subject. Prerequisite are the standard facts of algebra and point set and algebraic topology such as a typical student might have by the end of a year of graduate study. The main concepts and results are lucidly presented and easily accessible. It is an excellent source for anyone who wants to learn or review the essentials of the subject...there is a lot of mathematics included in the book. It is a valuable and welcome addition to the literature."

N. Steenrod, *The Topology of Fibre Bundles*,Reprint of the 1957 edition, Princeton Landmarks in Mathematics, Princeton University Press, Princeton, N.J., 1999. viii+229 pp.

This is a classic and the first book on fiber bundles. It was written when the theory of fiber bundles and applications had attracted the general attention but the literature was in a state of confusion. Its purpose was to organize systematically the known results. It is an important book written by an important topologist.

According to MathSciNet, "This is the first systematic account of fibre bundles, covering the development of the subject from its birth in 1935 until the most recent period. Most of the essential aspects are treated, the only major omission being the homology theory first developed by Gysin for sphere bundles and later generalized by Leray, Hirsch, and others to general fibre bundles."

6.2.2 S. Eilenberg, N. Steenrod, Foundations of Algebraic Topology, Princeton University Press, 1952. xv+328 pp.

This is an important classic written by two major figures. In this book, they gave the first axiomatic approach to homology and cohomology groups of topological spaces. Its impact on the development on algebraic topology is immense.

According to a review in Bulletin of AMS, "This book has had a profound influence on the development of topology both before and after its publication. In the five years since its first printing it has become a standard textbook and reference work for anyone interested in topology.

The first course in algebraic topology is usually a difficult one for the student. He faces a mass of unfamiliar algebraic machinery whose motivation is difficult to grasp and whose applicability is appreciated only much later. Realizing this, the authors have adopted an axiomatic approach to the subject of homology theory. Starting with seven easily stated axioms relating algebra and geometry (and assuming only the basic concepts of algebra and point set topology as prerequisites) they show how many important and interesting theorems can be proved directly from these axioms. The axioms themselves are presented without motivation, but their immediate application is intended to make it easier for the student to accept them. Only after the reader has seen the power of the theory is he led into the details of the existence and uniqueness of homology theories."

6.2.3 More books on algebraic topology

E. Spanier, *Algebraic Topology*, McGraw-Hill Book Co., New York-Toronto, Ont.-London, 1966. xiv+528 pp.

This is a very important textbook on algebraic topology. This systematic, comprehensive book has been the standard textbook for many years, and is still a very valuable reference. This is not an easy book, due to its emphasis on abstract and algebraic approach. Though there are many new books on algebraic books now, this is still the best source for some topics, especially for people who know some algebraic topology already.

According to MathSciNet, "This is a new attempt to squeeze a huge amount of algebraic topology into a single volume... The extensive coverage is achieved naturally by extreme condensation in both formulations and proofs. It is readable to those who are willing and able to verify all of the briefly sketched arguments and who can retain all the concepts and results that have been given previously. Other readers may easily get lost. Fortunately, no radical change of terminology is adopted in the present book... Because of this, it will serve as a very useful reference."

A. Hatcher, *Algebraic Topology*, Cambridge University Press, Cambridge, 2002. xii+544 pp.

This is a modern textbook on algebraic topology. It is very user-friendly and contains detailed proofs of theorems and discussions of results and extensions. One of the priorities of this book is readability. It is probably the most popular textbook on algebraic topology now.

According to MathSciNet, "The style is refreshingly informal, although precise definitions are always given. The author always prepares the reader for the next idea, and there are basic, simple black and white pictures which really help. He is not afraid to repeat ideas with increasing generality. For example, there are treatments of Postnikov towers which begin with a cellular version, go on to the usual tower of principal fibrations and end with Moore-Postnikov decompositions. The reviewer finds this a very 'student-friendly' approach (very different from the texts which always look for the greatest generality). There are many excellent examples."

6.2. ALGEBRAIC TOPOLOGY

G. Bredon, *Topology and Geometry*, Graduate Texts in Mathematics, 139, Springer-Verlag, New York, 1993. xiv+557 pp.

This book was written as a textbook for a graduate course in algebraic topology with some emphasis on smooth manifolds as the title suggests. It is comprehensive with good choices of topics and hence is also a useful book for many others as well.

According to MathSciNet, "It is a pleasure to review an interesting and original graduate text in topology and geometry. The topics covered include some general topology, smooth manifolds, homology and homotopy groups, duality, cohomology and products. The text is somewhat informal with an occasional offering of two proofs of the same result. Some optional topics are marked with a star, but a good lecturer can use this text to create a fine course at the appropriate level.

There are various innovative things, which are accessible but not traditionally covered. An example would be the Hopf maps, or the calculation of some cohomology of compact simple Lie groups."

J. Munkres, *Elements of Algebraic Topology*, Addison-Wesley Publishing Company, Menlo Park, CA, 1984. ix+454 pp.

This book was written as a textbook for a first year graduate course in algebraic topology. Because of great attention paid to details of proofs, clear expositions of the whole book and a good selection of exercises, this has been a popular textbook.

According to MathSciNet, "The author's general topology text [Topology: a first course, 1975] is so good that one brings high expectations to his latest book. This is a solid introduction to homology theory and, as in the first book, we find clear exposition supported by lots of helpful illustrations and excellent problems."

J. Munkres, *Topology: A First Course*, Prentice-Hall, Inc., Englewood Cliffs, N.J., 1975. xvi+413 pp.

This is another popular text on general topology with some discussion of the fundamental group and covering spaces. It was written as a first course in topology to serve as a foundation for future study in analysis, geometry and algebraic topology. As the author emphasized, there is no agreement on what topics can be considered appropriate for such a course. Its popularity might have proved the right choice in this book.

R. Switzer, *Algebraic Topology—Homotopy and Homology*, Die Grundlehren der mathematischen Wissenschaften, Band 212, Springer-Verlag, New York-Heidelberg, 1975. xii+526 pp.

This is a well-written advanced textbook on algebraic topology for a reader who has learned some basics of the subject. It is well organized, and for some topics, it leads the reader to the point where he could start to do research. Therefore, it is also a very valuable reference for experts.

According to MathSciNet, "The author has attempted an ambitious and most commendable project. He assumes only a modest knowledge of algebraic topology on the part of the reader to start with, and he leads the reader systematically to the point at which he can begin to tackle problems in the current areas of research centered around generalized homology theories and their applications... The author has sought to make his treatment complete and he has succeeded. The book contains much material that has not previously appeared in this format. The writing is clean and clear and the exposition is well motivated. There is, however, a regrettable lack of significant exercises and problems for the student to solve as an aid to and a test of his understanding...This book is, all in all, a very admirable work and a valuable addition to the literature and its value is not diminished by the somewhat minor flaws mentioned."

J.P. May, *Simplicial Objects in Algebraic Topology*, Van Nostrand Mathematical Studies, No. 11, D. Van Nostrand Co., Inc., Princeton, N.J.-Toronto, Ont.-London, 1967. vi+161 pp.

Simplicial sets first arose from earlier work in combinatorial topology and in particular from the notion of simplicial complexes and were introduced in the late 1940s. They are discrete analogs of topological spaces and have played an important role in algebraic topology, geometric topology and algebraic geometry. Formally, the homotopy theory of simplicial sets is equivalent to the homotopy theory of topological spaces, and one can apply discrete, algebraic techniques to perform and understand basic topological constructions. This book has been the standard reference for the theory of simplicial sets and their relationship to the homotopy theory of topological spaces.

According to MathSciNet, "This book, based on a second year graduate course given at Yale in 1965, presents much of the elementary material of algebraic topology from the semi-simplicial viewpoint. It should prove very valuable to anyone wishing to learn semi-simplicial topology. ... Although most of the material in this book was previously scattered through the literature, the author has provided new proofs for a number of the results. A good bibliography and various historical notes have been provided...."

6.2.4 D. Rolfsen, Knots and Links, Corrected reprint of the 1976 original, Mathematics Lecture Series, 7, Publish or Perish, Inc., Houston, TX, 1990. xiv+439 pp.

A knot is an embedding of a circle in the 3-dimensional Euclidean space $\mathbb{R}^3$. Two mathematical knots are equivalent if one can be transformed into the other via a deformation of $\mathbb{R}^3$ upon itself. A link is a collection of knots which do not intersect, but which may

6.2. ALGEBRAIC TOPOLOGY

be linked (or knotted) together. Clearly, knots and links are closely connected. They have played an important role in the study of 3-dimensional topology.

This is an unusual book. It does not follow the usual pattern of theorems and proofs. Instead it conveys spirit and the geometric flavor of topics around knots and links. Exercises are integral parts of the main text. This is a fun book to read and is also a very useful reference.

From the book description, "Besides providing a guide to understanding knot theory, the book offers 'practical' training. After reading it, you will be able to do many things: compute presentations of knot groups, Alexander polynomials, and other invariants; perform surgery on three-manifolds; and visualize knots and their complements. It is characterized by its hands-on approach and emphasis on a visual, geometric understanding.

Rolfsen offers invaluable insight and strikes a perfect balance between giving technical details and offering informal explanations. The illustrations are superb, and a wealth of examples are included."

According to a review in Bulletin of AMS, "It is charming and the most enjoyable mathematics book I have ever read. It is also a scholarly disaster. Ideas and theorems are usually unreferenced, leaving the unsophisticated reader to either assume the author as progenitor or categorize the result as not important enough for attribution (neither alternative will please the individual who first proved the theorem or introduced the idea). Occasionally, the author has selected the second person to prove a theorem as his reference; this is even more reprehensible, as we know how much easier it is to prove a known theorem than to do it first...The examples in the book are hard to find elsewhere, no one other book contains the quantity and timeliness of material it presents. I loved reading it, I learned a great deal from it and I recommend its purchase by anyone with even a passing interest in low dimensional geometric topology."

6.2.5 More knots and their invariants books

G. Burde, H. Zieschang, *Knots*, Second edition, de Gruyter Studies in Mathematics, 5, Walter de Gruyter & Co., Berlin, 2003. xii+559 pp.

This book is a good textbook and also a useful reference on knots. It concentrates on classical knots and emphasizes the combinatorial aspect. Results and their proofs are usually chosen for their simplicity rather than for their generality, but more general results are usually stated too.

According to MathSciNet,"When D. Rolfsen's book *Knots and links* appeared in 1976, it became the standard work on knot theory, and for many their introduction to geometric topology. Rolfsen viewed knots and links as key examples in 3-manifold topology, and emphasized the technique of surgery, which leads naturally to higher-dimensional considerations. The present authors have produced an excellent text which is quite different from that of Rolfsen, despite sharing much of its material. Their book concentrates on classical knots and is somewhat more combinatorial and less geometric in tone."

R. Crowell, R. Fox, *Introduction to Knot Theory*, Based upon lectures given at Haverford College under the Philips Lecture Program Ginn and Co., Boston, Mass., 1963. x+182 pp.

This is a classical introduction to knot theory. The senior author Fox is called "the father of modern knot theory". It was written at the time when the knot theory could be studied systematically from the algebraic point of view. Though it was based on a junior-senior level, it is also a good introduction for graduate students and everyone else.

According to MathSciNet, "This book is in many ways a very surprising work. In the first place it places topology, algebraic topology and knot theory in the position of solving a 'practical problem'... In the second place, for a very innocent task, namely distinguishing knots, a quite respectable amount of interesting and sophisticated mathematics is introduced... In the third place, while there is a great emphasis on practicality, and informal discussion prior to making a definition or proof is frequent, there is no sacrifice of rigor. This book is on the side of formality for necessity's sake, rather than formality for its own sake...The book closes with an interesting 'Guide to the Literature' intended to orient the student with respect to recent work in the field, and a fairly complete bibliography."

W. Lickorish, *An Introduction to Knot Theory*, Graduate Texts in Mathematics, 175, Springer-Verlag, New York, 1997. x+201 pp.

Knot theory has been developed and applied and connected to various subjects since the earlier books, for example, the quantum knot invariants are such examples. This book is a thorough, carefully written modern introduction to knot theory.

According to MathSciNet, "The venerable subject of knot theory was revitalized during the 1980's by the development of new polynomial invariants, evolving in the 1990's to new invariants for 3-manifolds. In this text, the author achieves a timely integration of the classical topics and the new developments... The author does a nimble job of developing the necessary background concepts from topology without becoming too enmeshed in details. Complete proofs are given for the major results on knots and invariants. The historical development of the topics is explained and credit is carefully assigned for the various developments of the theory."

6.2. ALGEBRAIC TOPOLOGY

V. Turaev, *Quantum Invariants of Knots and 3-manifolds*, Second revised edition, de Gruyter Studies in Mathematics, 18, Walter de Gruyter & Co., Berlin, 2010. xii+592 pp.

There have been a lot of results on invariants of knots and 3-manifolds defined through combinatorial-algebraic-topological methods, or rather topological quantum field theories (TQFTs). This book gives the first systematic account by one of the major players of the new development. It is a useful textbook for the beginners to learn the subject and also a useful reference for experts.

From the book description, "The monograph gives a systematic treatment of 3-dimensional topological quantum field theories (TQFTs) based on the work of the author with N. Reshetikhin and O. Viro. This subject was inspired by the discovery of the Jones polynomial of knots and the Witten-Chern-Simons field theory. On the algebraic side, the study of 3-dimensional TQFTs has been influenced by the theory of braided categories and the theory of quantum groups."

According to MathSciNet, "Since the early 1980s there has emerged an extraordinary new mathematical theory. It is called quantum groups by algebraists and quantum topology or topological quantum field theory by topologists. None of these terms quite does the subject justice. The present book provides one of the broadest introductions to the subject, emphasizing its roots in topology, Hopf algebras and fundamental categorical ideas.

The theory has been strongly influenced by ideas that arose in theoretical physics. In many cases, structures that had no original hint of topology turned out to be of key importance in the construction of new topological invariants of knots, links and 3-manifolds..."

6.3 Generalized cohomology theory and homotopy theories

When mentioning Cauchy, people think of analysis (real or complex). He was one of the early people who made use of the notion of homotopy as in the Cauchy residue formula and Cauchy integral formula. Though the notion of homotopy was old, it is not easy to compute homotopy groups. For example, computation of stable homotopy groups of spheres is still one of the most difficult problems in algebraic and differential topology.

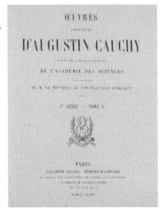

Figure 6.3: Cauchy's book.

6.3.1 J. Adams, Stable Homotopy and Generalised Homology, Chicago Lectures in Mathematics, University of Chicago Press, Chicago, Ill.-London, 1974. x+373 pp.

This is a classic by the founder and leading expert on stable homotopy theory. It is probably still the best introduction to the stable homotopy theory.

According to MathSciNet "This book brings together three sets of lecture notes by the author, all of which have been influential in guiding recent research in complex cobordism, Brown-Peterson cohomology, and their applications to stable homotopy..."

According to the cover description, "The three series focused on Novikov's work on operations in complex cobordism, Quillen's work on formal groups and complex cobordism, and stable homotopy and generalized homology. Adams's exposition of the first two topics played a vital role in setting the stage for modern work on periodicity phenomena in stable homotopy theory. His exposition on the third topic occupies the bulk of the book and gives his definitive treatment of the Adams spectral sequence along with many detailed examples and calculations in KU-theory that help give a feel for the subject."

D. Ravenel, *Complex Cobordism and Stable Homotopy Groups of Spheres*, Pure and Applied Mathematics, 121, Academic Press, Inc., Orlando, FL, 1986. xx+413 pp.

This book is the first readable and extensive book on computation of the stable homotopy groups of spheres by the Adams-Novikov spectral sequence. It also carefully treats the general definitions and methods.

According to the author, "The purpose of this book is threefold; (i) to make BP-theory and the Adams Novikov spectral sequence more accessible to nonexperts, (ii) to provide

6.3. COHOMOLOGY THEORY AND HOMOTOPY THEORIES

a convenient reference for workers in the field, and (iii) to demonstrate the computational potential of the indicated machinery for determining stable homotopy groups of spheres."

According to MathSciNet, "One of the nice features of this book is Chapter 1, ' An introduction to the homotopy groups of spheres' . It begins with a quick historical survey, starting with the Hurewicz and Freudenthal theorems and leading, via the Hopf map, to the Serre finiteness theorem, the Nishida nilpotence theorem, and the exponent theorem of Cohen, Moore, and the reviewer. Then results relating to the special orthogonal group are described, for example, Bott periodicity and the image of J. The history of computing homotopy groups is illustrated by a brief discussion of the Cartan Serre method of killing homotopy groups and of its descendent, the classical Adams spectral sequence. Some of the triumphs of this spectral sequence, or, more precisely, of the secondary cohomology operations related to it, are indicated; for example, the solutions to the classical and $\mod p$ Hopf invariant one problems. At this point, the author makes the transition to the main subject matter of this book by describing the complex cobordism ring, formal group laws, and the Adams Novikov spectral sequence."

6.3.2 K-theory

The K-theory in topology is a generalized (or extraordinary) cohomology theory, and there are different versions of K-theory for algebraic varieties, C^*-algebras, and rings.

The K-theory started with Alexander Grothendieck in 1957, who used it to formulate the Grothendieck-Riemann-Roch theorem. He defined a group using isomorphism classes of sheaves on an algebraic variety as generators modulo relations defined in terms of exact sequences. He called it a K-group of the algebraic variety because the German word for class is "Klasse".

By applying the same construction to vector bundles over topological spaces, Atiyah and Friedrich Hirzebruch defined topological K-groups in 1959.

M. Atiyah, *K-theory, Notes by D. W. Anderson*, Second edition, Advanced Book Classics, Addison-Wesley Publishing Company, 1989. xx+216 pp.

This is the first book that introduces topological K-theory. It was written by a co-invertor of the theory and master expositor.

Author's introduction: "These notes are based on the course of lectures I gave at Harvard in the fall of 1964. They constitute a self-contained account of vector bundles and K-theory assuming only the rudiments of point-set topology and linear algebra. One of the features of the treatment is that no use is made of ordinary homology or cohomology theory. In fact rational cohomology is defined in terms of K-theory.

The theory is taken as far as the solution of the Hopf invariant problem and a start is made on the J-homomorphism."

J. Rosenberg, *Algebraic K-theory and Its Applications*, Graduate Texts in Mathematics, 147, Springer-Verlag, New York, 1994. x+392 pp.

This is an accessible and comprehensive introduction to algebraic K-theory. It emphasizes motivations and applications to and connections with other subjects such as geometric topology, number theory, algebraic geometry, operator theory and functional analysis. It is both valuable as a textbook and a reference.

According to MathSciNet, "The development of algebraic K-theory has been spectacular during the last twenty-five years because of Quillen's work on the higher K-groups, and because of the important role of algebraic K-theory in many areas of modern mathematics, especially number theory, algebraic topology, algebraic geometry and functional analysis. However, there are not many books providing an introduction to algebraic K-theory. Furthermore, the subject has changed a bit since the publication of the excellent texts by H. Bass [Algebraic K-theory, 1968] and J. Milnor [Introduction to algebraic K-theory, 1971]. Therefore, it is sometimes hard for students or for anybody interested in this field to find references for the basic results and their proofs.

The book under review fulfills that need: it is a very nice textbook which presents all the classical topics, gives some information on the recent progress in higher K-theory and cyclic homology, and describes many applications. This book is self-contained in the following sense: all definitions are explained, the results concerning the subject of the book are proved (with some exceptions in the last two chapters), and the results from other areas of mathematics are mentioned without proof... Throughout the book, the author emphasizes the importance of the applications of algebraic K-theory and its interactions with other fields... This makes the reading of this text particularly attractive."

J. Milnor, *Introduction to Algebraic K-theory*, Annals of Mathematics Studies, No. 72, Princeton University Press, 1971. xiii+184 pp.

Algebraic K-groups are important invariants of rings, in particular, rings of integers of number fields. It is relatively easier to define the lower K-groups, and a difficult problem was to introduce higher algebraic K-groups. In this well-written book, Milnor introduced his Milnor K_2-group and studied its properties and gave explicit computations for some nice rings. It has a huge influence on the recent development of algebraic K-theory.

M. Karoubi, *K-theory. An Introduction*, Grundlehren der Mathematischen Wissenschaften, Band 226, Reprint of the 1978 edition, With a new preface by the author and a list of errata, Classics in Mathematics, Springer-Verlag, Berlin, 2008. xviii+308+e-7 pp.

This book can be considered as a sequel to the book by Atiyah on topological K-theory. It gives a comprehensive introduction to topological K-theory and some applications.

6.3. COHOMOLOGY THEORY AND HOMOTOPY THEORIES

According to MathSciNet, "This book represents only the second comprehensive text on topological K-theory. It extends the foundational text of M. F. Atiyah K-theory, 1967, while at the same time bringing to the material which is common to both the benefit of innovations which have come to light in the meantime... Incidentally, each chapter ends with a historical note, which should be very useful in enabling a nonexpert to pursue the literature..."

H. Bass, *Algebraic K-theory*, W. A. Benjamin, Inc., New York-Amsterdam, 1968. xx+762 pp.

This is the first book on algebraic K-theory written by one of the founders.

According to MathSciNet, "Algebraic K-theory flows from two sources. The (J.H.C.) Whitehead torsion, introduced in order to study the topological notion of simple homotopy type, leads to the groups K_1. The Grothendieck group of projective modules over a ring leads to K_0. The latter notion was applied by Atiyah and Hirzebruch in order to construct a new cohomology theory which has been enormously fruitful in topology.

The observation that these two ideas could be unified in a beautiful and powerful theory with widespread applications in algebra is due to the author, who is also responsible for a major portion of those applications. The author of this book is thus uniquely qualified to produce a fundamental document in this new and expanding field; he has also produced an admirable exposition of its methods and results."

6.3.3 K. Brown, Cohomology of Groups, Graduate Texts in Mathematics, 87, Corrected reprint of the 1982 original, Graduate Texts in Mathematics, 87, Springer-Verlag, New York, 1994. x+306 pp.

Cohomology and homology groups of a topological space are fundamental invariants of the space, and for any group, there are also cohomology and homology groups associated with it which can either be defined through the classifying space of the group or directly in terms of the group. Cohomology of groups has played an important role in homological algebra, algebraic topology and representation theories of groups, and they are also closely related to K-groups. For example, the algebraic K-groups of the ring $\mathbb{Z}$ of integers can be studied through the cohomology groups of the family of arithmetic groups $SL(n, \mathbb{Z})$.

This well written book by Brown is one of the most useful books on this important topic in homological algebra and algebraic topology. It is accessible and yet comprehensive. It is the standard textbook and reference for cohomology groups of infinite groups, especially those related to arithmetic subgroups of Lie groups.

According to a review in Bulletin of AMS, "The cohomology theory of abstract groups is a tool kit, in much the same way as is representation theory. One of its attractions is its breadth: the methods derive from algebra and topology and often connect with algebraic number theory. A good example is the theory of Euler characteristics, one of the highlights of the present book and one to which its author has made brilliant contributions...The book fills an important gap in the literature. I hope it will be studied by all who wish to understand group cohomology; they will be richly rewarded."

D. Benson, *Representations and Cohomology. I. Basic Representation Theory of Finite Groups and Associative Algebras, II. Cohomology of Groups and Modules* , Cambridge Studies in Advanced Mathematics, 30, 31, Cambridge University Press, Cambridge, 1991. xii+224 pp., x+278 pp.

Though cohomology appears in the title, the emphasis of these two books by Benson is on representation theory of finite groups, and cohomology groups of finite groups provide an effective approach. Therefore, they are very different from the book by K. Brown, which emphasizes cohomology of infinite groups, in particular, important groups in geometric group theory. These two volumes by Benson contain a wealth of information and are valuable introductions to representation theories of finite groups, from the perspective of cohomology groups.

According to MathSciNet, "This first volume is a very welcome addition to the existing classical literature on representation theory and cohomology of finite groups... This book cannot be compared to any of them—not only because of the author's individual choice of topics, but also because he sets up his own standards... This book is fascinating reading for someone who already has a basic knowledge of representation theory and cohomology, since only he can appreciate the proofs, which are very elegant and often new. These proofs are stripped of anything superfluous, stress the essential ideas and demonstrate the author's deep insight and masterly knowledge of the subject. This also makes it sometimes hard to follow the arguments. To quote the author: 'The pace is brisk.' I can recommend this book highly to anyone who works with representation theory and cohomology of finite groups."

Another book on cohomology of finite groups is

A. Adem, J. Milgram, *Cohomology of Finite Groups*, Second edition, Grundlehren der Mathematischen Wissenschaften, 309. Springer-Verlag, Berlin, 2004. viii+324 pp.

Cohomology groups of groups are important invariants of the groups and have been applied to many different subjects. But they are usually very difficult to compute, and this books give a comprehensive summary of methods and results on computation of cohomology of finite groups. This book provides an important and convenient source for this difficult problem.

According to MathSciNet, "The emphasis is on computational aspects of the cohomology of finite groups with coefficients in a field, and here is where it differs from

6.3. COHOMOLOGY THEORY AND HOMOTOPY THEORIES

previous treatments. Nowhere else will you find separate chapters on the cohomology of the symmetric groups, the groups of Lie type, and the sporadic simple groups."

According to Zentralblatt MATH, "In large parts the present book is devoted to a presentation of methods and results that bear on the calculation of cohomology groups. Numerous nontrivial examples of explicit computations are described in full detail."

6.3.4 G.W. Whitehead, Elements of Homotopy Theory, Graduate Texts in Mathematics, 61, Springer-Verlag, New York-Berlin, 1978. xxi+744 pp.

The notion of homotopy is really basic in topology and other areas of mathematics. For example, homotopy between two loops is also used in calculus. Homotopy equivalence between functions is also well-known and ubiquitous. An important family of homotopy invariants of a topological space consists of the homotopy groups, and the first of which is the fundamental group.

The current book is an authoritative overview of the elementary part of homotopy theory. This comprehensive book was written by a leading expert on homotopy theory in a leisurely way. Applications are also emphasized. Its elementary nature makes a good textbook, and its comprehensive coverage makes it a good reference.

From the book description, "The writing bears the marks of authority of a mathematician who was actively involved in setting up the subject. Most of the papers referred to are at least twenty years old but this reflects the time when the ideas were established and one imagines that the situation will be different in the second volume. Because of the length, it is unlikely that many people will read this book from cover to cover, but it will be used for reading up on a particular topic or dipping into for sheer pleasure - and the fact that the details may not be quite as expected should add to the enjoyment."

According to the review in Bulletin of AMS by Adams, "To conclude, I think George Whitehead proves at least one thing we already knew: in calibre as a mathematician, and for qualities of taste, style and judgement, he rates a lot higher than many who have written books on algebraic topology. I welcome this book most warmly, and I look forward to the second volume."

M. Hovey, *Model Categories*, Mathematical Surveys and Monographs, 63, American Mathematical Society, Providence, RI, 1999. xii+209 pp.

This book is a good introduction to and reference on Quillen's model categories, which has been becoming better known and more popular. The exposition is modern and accessible.

According to MathSciNet, "This book provides a thorough and well-written guide to Quillen's model categories. Model categories form the foundation of homotopy theory and play an important role not only in algebraic topology, but also in homological algebra and algebraic K-theory. To read this book one requires only a basic knowledge of category theory and some familiarity with chain complexes and topological spaces. This makes the text not only a volume for experts, but also usable in a classroom setting....To conclude, Model categories is written in a very clear style. The text fills a gap in the mathematical literature."

6.4 Differential topology

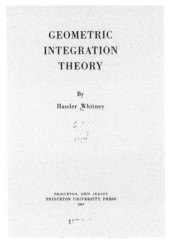

Figure 6.4: Whitney's book.

Whitney was one of the founders of differential topology. He obtained basic results on manifolds, embeddings, immersions, and characteristic classes. He was also one the founders of singularity theory. These two subjects are closely related, for example, submanifolds are crucial in the Whitney stratification of singular varieties. His book Geometric Integration Theory *connects algebraic topology with analysis and differential geometry. But it is probably less known that he worked on graph theory at the beginning of career, and his results contributed to the ultimate computer-assisted solution to the four-color problem. He laid the foundations for matroids, which is a fundamental notion in modern combinatorics and representation theory.*

Differential topology studies differentiable functions on differentiable manifolds, and is closely related to differential geometry. Differential topology studies the properties and structures that depend only on a smooth structure on a manifold, i.e., it is concerned with properties and invariants of smooth manifolds which are invariant under diffeomorphisms of manifolds, from the point of view of the Erlangen program. Topics in differential topology include the study of special kinds of smooth mappings between manifolds, namely immersions and submersions, and the intersections of submanifolds via transversality. Morse theory is a major branch of differential topology, in which topological information about a manifold is deduced from structures of singularities of a Morse function on the manifold, where a function is called Morse if at every critical point, its Hessian is non-degenerate. A generic function is Morse, but useful Morse functions come from explicit constructions.

6.4. DIFFERENTIAL TOPOLOGY

6.4.1 J. Milnor, Morse Theory, Based on lecture notes by M. Spivak and R. Wells, Annals of Mathematics Studies, No. 51, Princeton University Press, Princeton, N.J., 1963. vi+153 pp.

Morse theory gives a very direct way of analyzing the topology of a manifold by studying singularities of differentiable functions on that manifold. According to the basic insights of the founder Marston Morse, a differentiable function on a manifold will, in a typical case (i.e., when all singular points are non-degenerate), give a CW decomposition of the manifold according to the signature of the singular points.

The foundational book on Morse theory by Morse is the following

M. Morse, *The Calculus of Variations in the Large*, Reprint of the 1932 original, American Mathematical Society Colloquium Publications, 18, American Mathematical Society, Providence, RI, 1996. xii+368 pp.

It seems to be difficult to many people. The book of Milnor plays an important role in popularizing the Morse theory. Most people probably learned Morse theory from this book.

According to a review in Bulletin of AMS, "The theory of Marston Morse deals with the topological analysis of a manifold or a function space together with a real function on this space. Calculus of variations in the large was originally the main purpose of the theory. A typical subject was the study of the geodesics connecting two given points in a complete Riemannian n-manifold. In recent years a number of important applications have considerably increased the interest in Morse theory...Milnor's book is a lucid rapid introduction to the subject, with a highly geometrical flavour. It is well written, and points to many subjects of current research."

According to MathSciNet, "This book is devoted to an exposition of Morse theory. Starting from scratch, it goes through the proofs of the periodicity theorems of Bott for the unitary and orthogonal groups. The path taken to these theorems is by no means a minimal geodesic but the detours along the way serve to display the power of the theory as it is developed ..."

In some sense, the book above on Morse theory provides the foundation for the next book of Milnor.

J. Milnor, *Lectures on the h-cobordism Theorem*, Notes by L. Siebenmann and J. Sondow, Princeton University Press, Princeton, N.J., 1965. v+116 pp.

This is the authoritative detailed account of the h-cobordism theorem of Smale, which is of central importance in differential topology and was the culmination of Smale's work on the generalized Poincaré conjecture in high dimensions. This book is also an excellent introduction to differential topology by a master expositor who first became famous for his work in differential topology, i.e., the existence of exotic spheres in dimension 7.

According to MathSciNet, "These notes are devoted to an exposition of the differentiable h-cobordism theory entirely from the viewpoint of Morse theory. Although the basic facts of the study of critical points of functions as developed by Morse are clearly stated, it would be in order to suggest that the reader have a somewhat more expanded background in Morse theory.

The object is the following theorem, due originally to Smale. Let W be a compact, smooth manifold with two boundary components V and V'. If V is simply connected, has dimension greater than 4 and if V and V' are both deformation retracts of W, then W is diffeomorphic to $V \times [0, 1]$ ..."

6.4.2 Differential topology

J. Milnor, *Topology from the Differentiable Viewpoint*, Based on notes by David W. Weaver, Revised reprint of the 1965 original, Princeton Landmarks in Mathematics, Princeton University Press, Princeton, N.J., 1997. xii+64 pp.

This is a very beautiful, quick introduction to differential topology by a master expositor and one of the best differential topologists. It is very clearly written and is also very brief. The reader can see both the beauty and power of this important subject from reading this slim book.

From the author's preface: "These lectures were delivered at the University of Virginia in December 1963...They present some topics from the beginnings of topology, centering about L. E. J. Brouwer's definition, in 1912, of the degree of a mapping. The methods used, however, are those of differential topology, rather than the combinatorial methods of Brouwer. The concept of regular value and the theorem of Sard and Brown, which asserts that every smooth mapping has regular values, play a central role."

M. Hirsch, *Differential Topology*, Corrected reprint of the 1976 original, Graduate Texts in Mathematics, 33, Springer-Verlag, New York, 1994. x+222 pp.

After the rapid development of differential topology in the later 50s and 60s, there has been a need for introductory books on differential topology. This is one of the first books which filled this need. Its informal style emphasizing motivations and applications is both a disadvantage and advantage. It has been a valuable reference.

According to MathSciNet, "Viewed as a whole, this is a very valuable book. In little over 200 pages, it presents a well organized and surprisingly comprehensive treatment of most of the basic material in differential topology, as far as is accessible without the methods of algebraic topology. Newly introduced concepts are usually well motivated,

6.4. DIFFERENTIAL TOPOLOGY

and often the historical development of an idea is described. There is an abundance of exercises which supply many beautiful examples and much interesting additional information, and help the reader to become thoroughly familiar with the material of the main text.

V. Guillemin, A. Pollack, *Differential Topology*, Prentice-Hall, Inc., Englewood Cliffs, N.J., 1974. xvi+222 pp.

This is an effective, concise introduction to differential topology, and is a popular text book.

From the book description, it "provides an elementary and intuitive introduction to the study of smooth manifolds. In the years since its first publication, Guillemin and Pollack's book has become a standard text on the subject. It is a jewel of mathematical exposition, judiciously picking exactly the right mixture of detail and generality to display the richness within. The text is mostly self-contained, requiring only undergraduate analysis and linear algebra. By relying on a unifying idea–transversality–the authors are able to avoid the use of big machinery or ad hoc techniques to establish the main results. In this way, they present intelligent treatments of important theorems, such as the Lefschetz fixed-point theorem, the Poincaré-Hopf index theorem, and Stokes theorem. The book has a wealth of exercises of various types... The book is suitable for either an introductory graduate course or an advanced undergraduate course."

M. Golubitsky, V. Guillemin, *Stable Mappings and Their Singularities*, Graduate Texts in Mathematics, Vol. 14, Springer-Verlag, New York-Heidelberg, 1973. x+209 pp.

Smooth maps between smooth manifolds are fundamental in differential topology. Stable maps between them have been studied by Whitney, Morse and others from different perspectives. The Whitney embedding theorem asserts that certain stable maps are smooth, while the Morse theory uses singularities of stable functions (or Morse functions) to understand the topology of manifolds. This book provides an accessible introduction to this important topic including both points of view. This carefully written book will be a useful textbook for students and a reference for nonexperts.

6.4.3 R. Bott, L. Tu, Differential Forms in Algebraic Topology, Graduate Texts in Mathematics, 82, Springer-Verlag, New York-Berlin, 1982. xiv+331 pp.

This is an excellent introduction to algebraic topology using differential forms by a master lecturer and expositor Bott and his former student Tu. "The guiding principle in this book is to use differential forms as an aid in exploring some of the less digestible aspects of algebraic topology." de Rham cohomology is not the only cohomology considered, and the book covers a broad range of topics.

According to the book cover, "Developed from a first-year graduate course in algebraic topology, this text is an informal introduction to some of the main ideas of contemporary homotopy and cohomology theory. The materials are structured around four core areas: de Rham theory, the Cech-de Rham complex, spectral sequences, and characteristic classes. By using the de Rham theory of differential forms as a prototype of cohomology, the machineries of algebraic topology are made easier to assimilate. With its stress on concreteness, motivation, and readability, this book is equally suitable for self-study and as a one-semester course in topology."

6.4.4 W.V.D. Hodge, The Theory and Applications of Harmonic Integrals, Reprint of the 1941 original, With a foreword by Michael Atiyah, Cambridge Mathematical Library, Cambridge University Press, Cambridge, 1989. xiv+284 pp.

Hodge is a very well-known name in mathematics due to the Hodge decomposition, the Hodge structure, the Hodge numbers and the Hodge conjecture. It is a book written by the creator of the theory and contains his fundamental contribution to the Hodge theory.

From the book description, "First published in 1941, this book, by one of the foremost geometers of his day, rapidly became a classic. In its original form the book constituted a section of Hodge's essay for which the Adam's prize of 1936 was awarded, but the author substantially revised and rewrote it."

According to MathSciNet, "The theory of harmonic integrals on a Riemannian manifold was developed by Hodge in the course of his investigations on the integrals attached to an algebraic variety. It is possible indeed to put a finger on the exact spot of the transcendental theory of algebraic varieties which, in last analysis, is responsible for the appearance of this stimulating book: it is the question of whether or not there exist double integrals on an algebraic surface (or, more generally, m-fold integrals on an m-dimensional algebraic variety) which are everywhere finite and which have all their periods equal to zero. This question remained unsolved for a long time until Hodge proved that such integrals do not exist (subsequently it has been recognized that a similar theorem holds, more generally, for integrals of closed forms on a topological manifold, as a corollary of results obtained previously by de Rham). However, the new theory, although an outgrowth of specific problems in algebraic geometry, goes far beyond algebraic geometry proper. Indeed a glance at the table of contents shows that of the five chapters which make up the book, only one deals with algebraic varieties. The rest is an ingenious blend of analysis pure and simple, topology, tensor calculus, differential geometry and continuous groups. The book therefore should also prove of considerable interest to specialists in the fields just mentioned. Since the original investigations of Hodge are scattered in many notes in which the topic is at times presented in a somewhat condensed fashion, the present book will be welcome by those who wish to gain a better understanding of the central points of the theory of harmonic integrals. The book is exceedingly well written, there is great emphasis on rigor, and the original material is presented anew in a systematic and unified form."

According to a review in Bulletin of AMS, "This is one of those books which everyone who specializes in a particular branch of group theory, of the theory of algebraic surfaces, of the theory of Riemann surfaces, of topology or of the tensor analysis should consult. It shows how all these different fields are connected, and not connected in some superficial way or in the form of an analogy, but in an essential manner, so that interesting and profound theorems in one field cannot be understood without a thorough knowledge of other fields. In reading this book one is reminded of books like Klein's 'Ikosaeder', which is also a blend of several important fields."

6.4.5 M. Goresky, R. MacPherson, Stratified Morse theory, Ergebnisse der Mathematik und ihrer Grenzgebiete (3), 14, Springer-Verlag, Berlin, 1988. xiv+272 pp.

A natural generalization of the Euclidean space $\mathbb{R}^n$ and the sphere consists of smooth manifolds, and many results on topology and analysis on $\mathbb{R}^n$ can be generalized. On the other hand, singular spaces do occur and are often more interesting. (Recall that in the Morse theory, only the singular points count!) Singular spaces can often be decomposed into unions of smooth manifolds and give rise to stratified spaces. Since the Morse theory is a fundamental tool to understand topology of manifolds, a corresponding theory for stratified spaces will also be important for singular spaces, in particular, singular varieties.

This is an important book on understanding the topology of singular varieties, and more generally stratified spaces, by a generalization of the Morse theory for smooth manifolds. It is both an introduction to and a research monograph on this important subject.

According to MathSciNet, "The first and main part of the book 'contains a systematic exploration of the natural extension of Morse theory to include singular spaces' which are Whitney stratified. Many definitions and simpler results have obvious generalizations from classical to stratified Morse theory. But it is not immediately clear how to generalize the main result [of the Morse theory]... the book is a most important contribution in understanding the topology, especially the homology and homotopy of singular spaces and their smooth but noncompact complements. It not only presents the long-awaited complete proofs of results that the authors have been announcing since 1980 but also contains many historical and informal comments about contributions of other mathematicians."

According to a review in Bulletin of AMS, "Summing up, the book will certainly very soon prove an indispensable reference for everybody working in this field, and it seems not unjustified to expect that the book might quickly become a classic of modern Morse theory."

S. Weinberger, *The Topological Classification of Stratified Spaces*, Chicago Lectures in Mathematics, University of Chicago Press, Chicago, IL, 1994. xiv+283 pp.

Stratified spaces occur naturally from many different contexts such as geometric topology (surgery theory, controlled topology, rigidity etc.), quotients of group actions, limits of manifolds and varieties. They are important but complicated objects. This book is a very advanced introduction to the topology of stratified spaces, together with wide ranging applications in manifold theory, algebraic varieties, orbit spaces of group actions. The writing is clear and also informal at places.

P. Orlik, H. Terao, *Arrangements of Hyperplanes*, Grundlehren der Mathematischen Wissenschaften, 300, Springer-Verlag, Berlin, 1992. xviii+325 pp.

An arrangement of hyperplanes is a finite collection of codimension-one affine subspaces in a finite-dimensional vector space. It has a humble beginning in recreational mathematics on how to cut cheese, but it has been developed from different perspectives and is connected to deep and beautiful results in many subjects such as algebraic topology, combinatorics, algebraic geometry, algebra, group actions. This book is the first comprehensive account of hyperplane arrangements using all these subjects. It is self-contained and can serve as an introduction to this beautiful subject and also a useful reference.

According to MathSciNet, "In this book the authors have collected and presented, in a particularly clear and fun to read exposition, the many results of the last 20 years (and remaining open problems) in the study of hyperplane arrangements. This book will certainly become an indispensable reference for researchers in the field and for graduate students wishing to learn the subject as well as for mathematicians working in many related fields."

6.5 Geometric topology

There are several mathematicians named Jordan, and a famous one is Camille Jordan. Besides his extensive work on groups theory and linear algebra and his famous book on Galois groups of algebraic equations, he is also known for the basic Jordan curve theorem. A higher dimensional generalization is called Schoenflies theorem. It states that if an $(n-1)$-dimensional sphere S is embedded into the n-dimensional sphere in a locally flat way, then the embedding is homeomorphic to the standard embedding as an equator. It was a major achievement in geometric topology in the 1960s.

Figure 6.5: Jordan's book.

6.5.1 W. Thurston, The Geometry and Topology of Three-Manifolds, Lecture notes at Princeton University, 1978–1980.

These are notes for a graduate course which Thurston taught at Princeton University between 1978 and 1980. Thurston shared his notes, duplicating and sending them to people who requested them. It has a wide circulation and has become the most important and influential text in the low dimensional topology.

According to the MathSciNet review of a related book by Thurston, "It is natural to start this review by recalling the history of the book. In 1978, W. Thurston gave a course at Princeton University, whose subject was the geometry and topology of three-dimensional manifolds. He wrote notes for that course, and the notes immediately circulated all over the world. It is probably the opinion of all the people working in low-dimensional topology that the ideas contained in these notes have been the most important and influential ideas ever written on the subject. These notes created a new circle of ideas, and the expression 'Thurston type geometry' has become very common. The 1978 Princeton lecture notes, although written in an informal style, are self-contained and accessible to graduate students in topology or geometry. At some places the proofs are only sketched, but for most of the important new results, the arguments in the proofs are given completely. Besides the fact that many of these ideas are completely new, the notes present the subject matter in a great coherent expository (although special) style. Thurston's style of exposition is special in that it asks the reader to participate actively in what's going on by providing room for mental images, and this is one of the reasons why it is easy to get stuck if one tries to read these notes in a linear manner."

Though there have been a lot of progress on problems and questions raised in the original lecture notes of Thurston, reading them is still very valuable and gives people the excitement and unique insights.

Detailed comments on later developments are given in the following paper of Canary in 2006, *A new foreword for Notes on Notes of Thurston* of the next paper:

R. Canary, D. Epstein, P. Green, *Notes on notes of Thurston, With a new foreword by Canary*. London Math. Soc. Lecture Note Ser., 328, *Fundamentals of hyperbolic geometry: selected expositions*, pp. 1–115, Cambridge Univ. Press, Cambridge.

"The paper 'Notes on notes of Thurston' was intended as an exposition of some portions of Thurston's lecture notes *The Geometry and Topology of Three-Manifolds*. The work described in Thurston's lecture notes revolutionized the study of Kleinian groups and hyperbolic manifolds, and formed the foundation for parts of Thurston's proof of his Geometrization theorem. At the time, much of the material in these notes was unavailable in a published form."

There was a plan to expand and polish the lecture notes above of Thurston for publication. Only the first volume has appeared. It contains much of the material in Chapter 1, 2, and 3 of the original lecture notes, and some of sections 5.3 and 5.10. But the most exciting and creative portions of his original notes have not been published yet, and probably never will.

W. Thurston, *Three-dimensional Geometry and Topology. Vol. 1*, Edited by Silvio Levy, Princeton Mathematical Series, 35, Princeton University Press, Princeton, N.J., 1997. x+311 pp.

This is not a standard book on low dimensional topology. It is full of new ideas, nonstandard examples, pictures, exercises and problems. There are also many interesting remarks and comments on further developments and connections with other fields. The writing is dense and visual in the Thurston style and depends often on intuition. This is the only book written by Thurston which is related to (or rather serve as the baby foundation of) the famous geometrization program of Thurston for three manifolds, and one can read and enjoy it, or be amazed by mathematics of his style.

According to MathSciNet, "... here is a book which is, fortunately, still written in Thurston's style, demanding the participation of the reader's imagination, but with many more details than the chapters of the Princeton 1978 notes from which it grew. In particular, the book contains hundreds of exercises, problems and challenging questions which constitute an integral part of the logical progression.

The subject matter is therefore that of geometric structures on low-dimensional manifolds. It is important to point out that although the title of the book mentions only

6.5. GEOMETRIC TOPOLOGY

dimension three, the exposition also covers the theory of geometric structures on surfaces, and there are many ideas and results which apply to all dimensions.

The word 'geometry' that is used here can be given several meanings (according to Thurston himself) and, indeed, is used with at least three different meanings. Some sections of the book treat geometry in the very classical sense, that is, from the point of view of the axioms on points, lines, planes, angles and so on in a given space. In this sense of 'basic geometry' the book contains a complete exposition of hyperbolic geometry, but there are also several sections which concern Euclidean geometry, spherical geometry and projective geometry.

Another sense to the word 'geometry' that also applies here is that of the study of G-manifolds... Finally, the word 'geometry' is also used in a sense proper to Thurston. This involves the definition of a model geometry...

It is in the sense of the last definition above that there are precisely three two-dimensional model geometries (spherical, Euclidean and hyperbolic), and eight three-dimensional model geometries, and we recall that the main conjecture (Thurston's geometrization conjecture) states that the interior of every compact three-manifold has a canonical decomposition into pieces each of which is modelled on one of these eight geometries....There is no doubt that every geometer will look to this book for inspiration, for results and for problems. The book can also be used as a textbook for an elementary or an advanced course (depending on the audience). There is a lot of beautiful mathematics here."

Given a manifold, the **homotopy classes of homeomorphisms (or diffeomorphisms)** of the manifold define the mapping class group of the manifold. Besides their importance for understanding the topology and geometry of manifolds and related spaces, the mapping class groups also give rise to interesting, special groups in geometric group theory. Thurston's work has established close relations between automorphisms (or mapping class groups) of surfaces and geometric topology of three dimensional manifolds. One component of his fundamental work on Teichmüller spaces and mapping class groups of surfaces is a classification of automorphisms of surfaces. The next book deals with this work of Thurston.

A. Casson, S. Bleiler, *Automorphisms of Surfaces After Nielsen and Thurston*, London Mathematical Society Student Texts, 9, Cambridge University Press, Cambridge, 1988. iv+105 pp.

This is a very concise relatively accessible introduction to part of Thurston's reworking and extension of Nielsen's classic work on the structure and dynamics of diffeomorphisms of 2-manifolds, which is an important ingredient of Thurston's geometrization program.

According to MathSciNet, "The style of the book is close to that of the original lecture notes. There are many assertions which the careful reader will want to verify, but they are usually 'routine', and the verification is usually illuminating. Though there are few formal prerequisites, a high level of mathematical maturity on the part of the reader is advisable. It would be an excellent choice for a graduate seminar."

There are several books on or related to the Thurston program on geometrization of 3-manifolds.

M. Kapovich, *Hyperbolic Manifolds and Discrete Groups*, Progress in Mathematics, 183, Reprint of the 2001 edition, Modern Birkhäuser Classics, Birkhäuser Boston, Inc., Boston, MA, 2009. xxviii+467 pp.

Thurston did not write down any detailed proof of his hyperbolization theorem: Suppose that M is a compact atoroidal Haken 3-manifold that has zero Euler characteristic. Then the interior of M admits a complete hyperbolic structure of finite volume.

According to MathSciNet, "In the book under review the author presents a complete proof of the hyperbolization theorem in the generic case and an outline of Otal's approach [see J.-P. Otal, Astérisque No. 235 (1996), x+159 pp.] for the proof in the remaining case of manifolds which fiber over the circle."

A. Marden, *Outer Circles. An Introduction to Hyperbolic 3-manifolds*, Cambridge University Press, Cambridge, 2007. xviii+427 pp.

There has been a lot of progress on the theory of Kleinian groups and associated hyperbolic 3-manifolds, and several major conjectures were proved recently. This book presents a comprehensive introduction to hyperbolic 3-manifolds and an overview of the most recent developments in the subject. The purpose is to convey a big picture of hyperbolic 3-manifolds.

According to MathSciNet, "The present book is a far-reaching introduction and report on the development and the recent progress in the field of Kleinian groups and their associated hyperbolic 3-manifolds: it is a mixture of a textbook, a survey and a handbook/guide to the literature."

R. Benedetti, C. Petronio, *Lectures on Hyperbolic Geometry*, Universitext, Springer-Verlag, Berlin, 1992. xiv+330 pp.

This book is an advanced introduction to hyperbolic 3-manifolds and the work of Thurston, and can serve as a preparation for Thurston's famous lecture notes. It is a valuable introduction.

J. Ratcliffe, *Foundations of Hyperbolic Manifolds*, Second edition, Graduate Texts in Mathematics, 149, Springer, New York, 2006. xii+779 pp.

This book is a comprehensive introduction to hyperbolic geometry and leads to some exposition of topics in Thurston's notes.

6.5. GEOMETRIC TOPOLOGY

According to MathSciNet, "This book has a tremendous amount of depth. In addition to the careful and complete exposition, each chapter ends with a fascinating section containing historical notes, putting many of the ideas into context. This volume will play an important role in the continuing development of this fascinating field."

J. Hempel, *3-manifolds*, Reprint of the 1976 original, AMS Chelsea Publishing, Providence, RI, 2004. xii+195 pp.

This is an important book that gives a self-contained account of the status of 3-dimensional topology before the era of Thurston. In order to state the Thurston geometrization program, results of this book are needed. It is still a very valuable reference on 3-dimensional topology.

According to MathSciNet, "This book is a careful and systematic development of the theory of the topology of 3-manifolds, focusing on the critical role of the fundamental group in determining the topological structure of a 3-manifold. It is directed at an audience with a knowledge of basic algebraic topology and infinite group theory (i.e., those aspects of the subject that deal with free groups, free products, free products with amalgamation, etc.) and also some familiarity with the techniques of pl topology (which could profitably be studied simultaneously with the monograph under review). For the fairly sophisticated reader with such a background the book is for the most part self-contained, so that one can learn the subject from it...In this reviewer's opinion the book would be very appropriate as a text for an advanced graduate course, or as a basis for a working seminar. The reader will find it helpful to have at his fingertips a good supply of concrete examples of all types since the author is very sparing with examples such as these. A complementary text, which deals with another aspect of 3-manifold topology, is D. Rolfsen's book [Knots and links, 1976]. The latter is filled with beautiful examples and there is almost no overlap between the two books, which supplement one another perfectly. Thus, it is now possible for the nonspecialist to get into the heart of current research in 3-manifolds with a minimum of effort."

W. Jaco, *Lectures on Three-manifold Topology*, CBMS Regional Conference Series in Mathematics, 43, American Mathematical Society, Providence, R.I., 1980. xii+251 pp.

This book gives an overview of the status of the 3-dimensional topology before the Thurston era and integrates the classical results with the then new results and directions. Results discussed in this book are foundation for the Thurston geometrization program.

According to MathSciNet, "The author's theme is the key role played by incompressible surfaces and hierarchies in the study of 3-manifolds."

6.5.2 Four dimensional manifolds

Though all smooth manifolds are locally of similar nature in all dimensions, they have different global properties depending crucially on the dimension. The topology of manifolds of dimension 2, i.e., surfaces, is relatively simple and well-understood, and the topology of manifolds of dimensions 5 and up share many common properties (i.e., independent of dimension) and can be studied by some common methods, for example, surgery theory, and is in some sense easier: for example, the Poincaré conjecture in higher dimension was solved a long time ago (in 1960s). But the topology of manifolds of dimension 3 and 4 are very different from 5 or higher, and there are also marked differences between dimension 3 and dimension 4. The difference between topological and differential structures is probably most dramatic in dimension 4.

S. Donaldson, P. Kronheimer, *The Geometry of Four-manifolds*, Oxford Mathematical Monographs, Oxford Science Publications, The Clarendon Press, Oxford University Press, New York, 1990. x+440 pp.

This is a systematic account of the Donaldson theory written by its creator and a top expert. It was written at the height of the Donaldson theory, and was a fundamental reference on 4-manifolds at that time. But the emergence of Seiberg-Witten theory put the Donaldson theory into shadow. This carefully written book is still a very valuable reference on 4-manifolds.

According to MathSciNet, "This book successfully gives a self-contained and comprehensive treatment of the applications of Yang-Mills theory to 4-manifold topology that were pioneered and developed by Donaldson and brings together the internal developments of Yang-Mills theory within the framework of contemporary differential and algebraic geometry."

R. Gompf, A. Stipsicz, *4-manifolds and Kirby Calculus*, Graduate Studies in Mathematics, 20, American Mathematical Society, Providence, RI, 1999. xvi+558 pp.

The Kirby calculus is a method to modify 3-manifolds in geometric topology. This book describes in detail topological techniques and ideas related to the Kirby calculus over the last decades in the theory of compact 4-manifolds. These topological methods complement other approaches such as the gauge theory to construct new manifolds with certain properties, to show that some 4-manifolds constructed differently are homeomorphic etc. Besides being a very valuable reference on the Kirby calculus and its many applications to 4-manifolds, this book is also a good reference on many important topics on 4-manifolds related to algebraic geometry and symplectic geometry.

From the book description, it "offers an exposition from a topological point of view. It bridges the gap to other disciplines and presents classical but important topological techniques that have not previously appeared in the literature." According to MathSciNet, "it is almost unique in that it does so from the point of view of differential topology."

6.5.3 Geometric topology and surgery theory

J. Birman, *Braids, Links, and Mapping Class Groups*, Annals of Mathematics Studies, No. 82, Princeton University Press, Princeton, N.J.; University of Tokyo Press, Tokyo, 1974. ix+228 pp.

This book is the first one that systematically introduces Artin's braid group and its many applications in lower dimensional topology such as links and knots and automorphism groups of surfaces. This book is carefully written and readable and hence is a valuable textbook. Its comprehensive coverage of topics related to Artin's braid group also makes it an important reference.

According to MathSciNet, "this thorough, skillfully written monograph, the first devoted entirely to the theory of braids, covers each of its topics—roughly, one for each of five chapters—from its historic beginnings... The book is a pleasure to read."

C.P. Rourke, B.J. Sanderson, *Introduction to Piecewise-linear Topology*, Reprint, Springer Study Edition, Springer-Verlag, Berlin-New York, 1982. viii+123 pp.

This is a basic textbook on piecewise-linear topology. According to MathSciNet, "The book covers a lot of material in few pages (123). A mature mathematician will appreciate the shorter proofs, unencumbered with too many details, but a student may have to consult papers or one of the other PL books."

An important theory in understanding structures and classification of manifolds of higher dimensions is the surgery theory. It consists of a collection of techniques which allow one to produce one manifold from another in a controlled way, such as cutting out parts of the manifold and replacing it with a part of another manifold, matching up along the cut or boundary. It was developed in the early 1960s. The following two books have played an important role in the development of this powerful subject.

C.T.C. Wall, *Surgery on Compact Manifolds*, Second edition, Edited and with a foreword by A. A. Ranicki, Mathematical Surveys and Monographs, 69, American Mathematical Society, Providence, RI, 1999. xvi+302 pp.

This is a difficult book, since it was written in the complete generality. But this is an important book in the surgery theory of manifolds of higher dimensions. It gives a complete treatment of the author's famous obstruction to surgery on non-simply connected manifolds and discusses many applications, computations of the surgery groups and examples. The surgery theory is known for its difficulty, and this book is still a very important and valuable reference.

Indeed, according to the book description, "The publication of this book in 1970 marked the culmination of a particularly exciting period in the history of the topology of manifolds. The world of high-dimensional manifolds had been opened up to the classification methods of algebraic topology by Thom's work in 1952 on transversality and cobordism, the signature theorem of Hirzebruch in 1954, and by the discovery of exotic spheres by Milnor in 1956. In the 1960s, there had been an explosive growth of interest in the surgery method of understanding the homotopy types of manifolds... The original edition of the book fulfilled five purposes by providing: · a coherent framework for relating the homotopy theory of manifolds to the algebraic theory of quadratic forms; · a surgery obstruction theory for manifolds with arbitrary fundamental group; · the extension of surgery theory from the differentiable and piecewise linear categories to the topological category; · a survey of most of the activity in surgery up to 1970; · a setting for the subsequent development and applications of the surgery classification of manifolds."

W. Browder, *Surgery on Simply-connected Manifolds*, Ergebnisse der Mathematik und ihrer Grenzgebiete, Band 65, Springer-Verlag, New York-Heidelberg, 1972. ix+132 pp.

This book gives a comprehensive introduction to the surgery theory on simply connected manifolds. Together with the book on surgery on non-simply connected manifolds by Wall, it gives a complete account of surgery of manifolds. Due to the simply connected assumption, this book is easier than Wall's book, but it is not a easy book either.

According to MathSciNet, "It has a number of features that make it especially useful for anyone wishing to learn surgery theory. First, the author takes special pains with the basic algebraic topology and algebra involved in 1-connected surgery... Secondly, the book collects a great deal of diverse background information from the literature, giving relatively complete expositions... This makes the book relatively self-contained. Of course, the abundance of the material needed forces the style to be somewhat compact; the book requires and merits careful reading. Finally, and perhaps most important is the way in which the author has organized the theory. After a chapter on preliminaries, he lists seven basic results of surgery theory and then proceeds to derive many important applications of the theory from these. The remainder of the book is devoted to a proof of the seven basic results. This procedure allows the student to arrive at interesting applications before becoming engulfed in the many technical details of the main proofs. It also focuses attention properly on the most significant tools provided by the theory for the study of 1-connected manifolds."

R. Kirby, L. Siebenmann, *Foundational Essays on Topological Manifolds, Smoothings, and Triangulations*, With notes by John Milnor and Michael Atiyah, Annals of Mathematics Studies, No. 88, Princeton University Press, Princeton, N.J.; University of Tokyo Press, Tokyo, 1977. vii+355 pp.

6.5. GEOMETRIC TOPOLOGY

This is not a regular book, but is rather a collection of five essays which are of fundamental importance in topological manifolds. They represented major achievements near the end of 1960s and early 1970s. This book is the only source for these important results.

According to MathSciNet, "From the days of Poincaré a major goal of topologists has been to understand the 'naked' homeomorphism and its natural home—topological (TOP) manifolds. The lack of techniques to deal with these homeomorphisms forced Poincaré and early topologists to consider manifolds with a richer structure, namely infinitely differentiable (DIFF) or piecewise-linear (PL) manifolds, so that arguments of a combinatorial nature could be employed...

The five essays under review, which have been in existence in various forms since 1970, represent the first full publication of these new results...An all-out effort was made to make this book self-contained. The authors present the material both expertly and well....This book is indispensable for anyone interested in topological manifolds. The neophyte should keep in mind what is set out to be proved; he should follow a line of least resistance, keeping alternative approaches and enlightening generalizations for later readings. Before a framework can be developed, a foundation must be laid. The book is demanding, but self-contained, well written, and its readers will be rewarded with a full and open understanding."

VII. The University of Michigan Museum of Art

It is probably true that most mathematicians do not like to write books, especially when they are young. This is also the advice given to people who have not gotten tenure yet. But Hilbert faced the challenge to write a report on the up-to-dated status of algebraic number theory when he was relatively young at the age of 31 and succeeded beyond expectation. His book Zahlbericht (report on numbers) in 1897 unified the field of algebraic number theory and provided a guide to research in algebraic number theory for the next half century. There are also a lot of writings about the impact of this book.

It might be better to treat Hardy and Littlewood as a single mathematician and say a few words about them. It is often quoted that during their life time, there were three great English mathematicians: Hardy, Littlewood, and Hardy-Littlewood. Their long time cooperation and their huge amount of joint work have had a long lasting impact on analyltic number theory and analysis. For example, the Hardy-Littlewood circle method and the Hardy-Littlewood maximal function are some names which might sound familiar to many mathematicians.

"*This is Very Different from Twickenham.*"
In the Members' Stand everybody is an Old Blue, an Old International or an Expert Critic.
(See overleaf)

Hua Loo-Keng was probably the most famous mathematician in China in the twentieth century and had the greatest impact on the development of mathematics in China. He was a well-rounded mathematician. Besides his important work in number theory, he also contributed significantly to complex geometry, harmonic analysis, algebra, optimization operations research and many other subjects. He had written multiple books at all levels, and his books have been popular and influential. It is even more amazing that he achieved all these without getting a proper education in his youth.

Kronecker is probably most famous for the Kronecker delta and for the quote "God made natural numbers; all else is the work of man". He made many important contributions to number theory. For example, he formulated the Kronecker-Weber theorem in Galois theory, and his conjecture "liebster Jugendtraum" (dearest dream of youth) was put by Hilbert in a modified form as the twelfth problem in his famous list of 23 problems. On the negative side, Kronecker's harsh public opposition and personal attacks on Cantor and his work made Cantor deeply depressed.

Chapter 7

Number theory

Number theory is one of the oldest branches of mathematics and studies properties of numbers, and in particular, integers, and problems that arise from their study. Though counting started from the beginning of civilization, the foundation of number theory was treated by Euclid in his famous book *Elements*.

Number theory was a favorite subject for the Greek mathematicians of the late Hellenistic period. They were aware of the Diophantine equation concept in numerous special cases. Diophantus was probably the first Greek mathematician to study these equations systematically.

Diophantine equations were extensively studied by mathematicians in several different cultures. In the thirteenth century, Fibonacci wrote one of his greatest works, *the Liber Quadratorum*. In this work he dealt with Pythagorean triples.

Further advances were made in the sixteenth and seventeenth centuries. In particular, infinite descent approach of Fermat was an important general method of solving diophantine questions.

In the eighteenth century, Euler and Lagrange made important contributions to number theory. For example, Euler created analytic number theory, and Euler and Lagrange studied Pell equations.

The modern number theory started with the work of Legendre and Gauss, for example, on the law of quadratic reciprocity, and the famous book of Gauss *Disquisitiones Arithmeticae* in 1801 marked the beginning of the modern theory of numbers.

Number theory includes (1) elementary number theory, (2) algebraic number theory, (3) geometry of numbers, (4) analytic number theory, (5) transcendental number theory, (6) arithmetic algebraic geometry, (7) modular forms and automorphic forms, etc. Though number theory is one of the purest branches of mathematics, it has been applied to vast areas of sciences and practical situations. In this chapter, we will discuss books related to the above 7 subjects.

7.1 Number theory

Legendre was a great mathematician who did not get proper credit for some of his work. For example, least squares method was first published in 1806 as an appendix to his book on the paths of comets, but most people attributed it to Gauss. Abel's work on elliptic functions and Galois' work on polynomial equations also built on Legendre's work. He conjectured the quadratic reciprocity law, which was proved by Gauss. But he did get credit for this since the Legendre symbol is named after him. To many people, he is probably best known for the Legendre polynomials and Legendre transformation. He is also well known for his book Éléments de géométrie *in 1794, which greatly rearranged and simplified many of the propositions from Euclid's Elements to create a more effective textbook, and hence was the leading textbook on geometry for 100 years.*

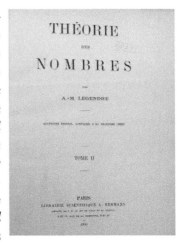

Figure 7.1: Legendre's book.

7.1.1 G.H. Hardy, E. Wright, An Introduction to the Theory of Numbers, Sixth edition, Revised by D. R. Heath-Brown and J. H. Silverman, With a foreword by Andrew Wiles, Oxford University Press, Oxford, 2008. xxii+621 pp.

Hardy is best known to the public for his book *A Mathematician's Apology*. To number theorists, he is probably better known for the Hardy-Littlewood circle method. But for many students, he is probably best known for this book on number theory.

This important and successful book, which first appeared more than 40 years ago and is now a well-known classic. It is still a very readable and useful book.

According to Nature, "Mathematicians of all kinds will find the book pleasant and stimulating reading, and even experts on the theory of numbers will find that the authors have something new to say on many of the topics they have selected... Each chapter is a model of clear exposition, and the notes at the ends of the chapters, with the references and suggestions for further reading, are invaluable."

According to a review by Mordell of the first edition (1938) in Mathematical Gazette, "It contains a wealth of interesting and often unexpected material which has not yet found its way into other so easily accessible books. The authors have spread their nets far and wide in gathering interesting material and their haul is a very nice one indeed. It is really surprising what an immense storehouse they have filled. There is sufficient variety in it to satisfy the most catholic taste and to cater for the reader in all his moods.

He may go through the book from cover to cover, or study a chapter here and there, or dip in now and then for a pleasant morsel. In lighter moments he may turn to the theory of the game of Nim, while on more austere occasions he may study the question of Euclidean algorithms in algebraic fields, or the Rogers-Ramanujan identities in the theory of partitions."

According to a review in Bulletin of AMS in 1939, "They have presented these introductions in a manner that should stimulate a reader to continue beyond some of them; and it seems safe to say a great deal more than what they themselves say, 'we can hardly have failed completely, the subject-matter being so attractive that only extravagant incompetence could make it dull.' The book is anything but dull; in fact it is as lively as the proverbial (not the English) cricket."

According to an Amazon review, "I feel the book is strongest in the area of elementary –not necessarily easy though – analytic number theory (Hardy was a world class expert in analytic number theory). An elementary, but difficult proof of the Prime number Theorem using Selberg's Theorem is thoroughly covered in chapter 22.

While modern results in the area of algorithmic number theory are not presented nor is a systematic presentation of number theory given (it is not a textbook), it contains a flavor, inspiration and feel that is completely unique. It covers more disparate topics in number theory than any other n.t. book I know of. The fundamental results in classical, algebraic, additive, geometric, and analytic number theory are all covered. A beautifully written book."

7.1.2 Books on basic number theory

Number theory is both an ancient and modern subject. Since Riemann introduced the method of complex analysis via Riemann zeta function into number theory, the theory of modular forms has become an integral part of number theory.

The ring of integers, the field of rational numbers and their finite extensions (number fields) are certainly basic objects in number theory. Other fields such as finite fields and functions fields over finite fields share many common properties. The analogy between number fields and functions fields is still an unfolding story in mathematics.

There are many books dealing with different aspects of number theory. In this subsection, we list some introductory books on basic number theory, modular forms, finite fields and function fields. We also include a book on applications of number theory in sciences and engineering, and a book on open problems in number theory.

These books are followed by some classics in number theory.

A popular book by the famous Chinese mathematician Hua is the following

L.K. Hua, *Introduction to Number Theory*, Translated from the Chinese by Peter Shiu, Springer-Verlag, Berlin-New York, 1982. xviii+572 pp.

This is a high quality, comprehensive introduction to number theory. The original Chinese version has affected so many students, given Hua's reputation and influence in China.

His philosophy and advice on learning in the preface is sound and helpful: "Apart from giving a broad introduction to number theory and some of its fundamental principles the author has also tried to emphasize several points to its readers.

First there is a close relationship between number theory and mathematics as a whole. In the history of mathematics we often see the various problems, methods and concepts in number theory having a significant influence on the progress of mathematics. On the other hand there are also frequent instances of applying the methods and results of the other branches of mathematics to solve concrete problems in number theory. However it is often not easy to see this relationship in many existing introductory books. Indeed many "self-contained" books for beginners in number theory may give an erroneous impression to their readers that number theory is an isolated and independent branch of mathematics. In this book the author tries to highlight this relationship within the scope of elementary number theory... Secondly an important progression in mathematics is the development of abstract concepts from concrete examples.... Thirdly, for beginners engaging in research, a most difficult feature to grasp is that of quality — that is the depth of a problem... We have to play often with the masters (that is, try to improve on the results of famous mathematicians); we must learn the standard works of the game (that is, the 'well-known' results). If we continue like this our progress becomes inevitable. This book attempts to direct the reader to work in this way..."

According to MathSciNet, "This is a valuable and important textbook on number theory, somewhat on the lines of Hardy and Wright [*An Introduction to the Theory of Numbers*, Oxford, 1954], but going far beyond it in scope."

According to a review in Bulletin of AMS, "One of the most striking features of the book is its economy of style, incorporating an astonishing number of results and topics. It achieves this efficiency, in part, by avoiding the temptation of stating and proving results in their greatest generality. To be sure, proofs of some theorems are left as exercises which supplement the, unfortunately, relatively few exercises to be found in the book.

As an example of this characteristic we single out especially the chapter on algebraic number theory. In the compass of 50 pages, the author takes us from the very beginnings of the subject through ideals, class numbers etc., and ends with applications to Mersenne primes and diophantine equations. This is achieved with no sacrifice of lucidity.

What is true of this chapter holds to a greater or lesser extent for most chapters. The interested mathematician may approach the material with minimal prior knowledge. The

7.1. NUMBER THEORY

language is classical and the reader will not be impeded by the necessity of having a large mathematical vocabulary. On the other hand, the reader will be amply rewarded with beautiful results of considerable depth and can come away with a sense of satisfaction.

In one of his letters to Sophie Germain, Gauss, referring to number theory, wrote that 'the enchanting charms of this sublime science are not revealed except to those who have the courage to delve deeply into them.' This book provides an admirable vehicle for so delving."

M. Rosen, *Number Theory in Function Fields*, Graduate Texts in Mathematics, 210, Springer-Verlag, New York, 2002. xii+358 pp.

The analogy between number fields and function fields has been fruitfully explored since Hasse, Weil, Deuring and Chevalley. Important recent results have also been obtained for function fields, for example, the Langlands conjectures over function fields by L. Lafforgue. Basically all important conjectures in number theory are now solved for algebraic function fields. This book gives a comprehensive account of the arithmetic of function fields. It is a valuable text book and also a convenient reference.

According to a review in Bulletin of AMS, "Rosen's book is perfect for graduate students, as well as other mathematicians fascinated by the amazing similarities between number fields and function fields."

R. Lidl, H. Niederreiter, *Finite Fields*, With a foreword by P. M. Cohn, Second edition, Encyclopedia of Mathematics and its Applications, 20, Cambridge University Press, Cambridge, 1997. xiv+755 pp.

The theory of finite fields is rooted in the work of Gauss and Galois. It is becoming more important with the emergence of discrete mathematics both in theory and practice. This is the first book entirely devoted entirely to finite fields. It is a reference for all users of finite fields.

Finite fields are usually treated as a minor topic in most books on abstract algebra and number theory. According to MathSciNet, "Yet there actually exists a large body of theory at various levels, from several viewpoints and with all kinds of contemporary applications. Hence this first-ever encyclopedic treatment of finite fields is a timely enterprise... Undoubtedly, this volume (though expensive) must become the handbook for finite fields. It should prove to be time- and energy-saving for the nonspecialist to consult as the need arises and could also be the starting point for the theorist to unify and develop areas of which he was perhaps previously unaware."

K. Ireland, M. Rosen, *A Classical Introduction to Modern Number Theory*, Second edition, Graduate Texts in Mathematics, 84, Springer-Verlag, New York, 1990. xiv+389 pp.

This book gives an introduction to modern developments in algebraic number theory and arithmetic algebraic geometry by showing their close relationship with classical, i.e. 19th century, number theory. The topics are well-chosen and reflect current interests of these subjects. Results and proofs are well illustrated by concrete examples. It is a very readable and interesting book for students and others as well.

According to MathSciNet, "The overall point of view of the book is very much that of the first half of the 19th century—almost a homage to Gauss and Eisenstein... But such a classical point of view gives the book a distinctive niche in the expository literature. Many mathematicians of this generation have reached the frontier of research without having a good sense of the history of their subject. In number theory this historical ignorance is being alleviated by a number of fine recent books. This work stands among them as a unique and valuable contribution."

"The charm of number theory derives from many sources: its long history, the simple formulations of some of its most difficult problems, the variety and intricacy of its techniques, and the unity which it lends to mathematics. This book offers a charming entryway to number theory and arithmetic geometry which is accessible to beginning graduate students. The book will be useful to mathematicians who seek an introduction to the subject which leads to some of its most important recent advances."

N. Koblitz, *p-adic Numbers, p-adic Analysis, and Zeta-functions*, Second edition, Graduate Texts in Mathematics, 58, Springer-Verlag, New York, 1984.

This book gives a self-contained elementary introduction to p-adic analysis and then discusses two important applications of p-adic analysis: construction of p-adic zeta function by p-adic integration and the rationality of the zeta function of algebraic varieties over finite fields. It has become a standard introductory book on p-adic numbers and their applications.

Though mathematics was motivated by applications, mathematics has developed on its own for its beauty and internal structures, and number theory is one such subject of the pure mathematics. Hardy believed that pure mathematics, in particular number theory, has no application: "Nothing I have ever done is of the slightest practical use." But the opposite is true. The next book is probably one of the best disproofs of Hardy's assertion.

M. Schroeder, *Number Theory in Science and Communication*, With applications in cryptography, physics, digital information, computing, and self-similarity, Fourth edition, Springer Series in Information Sciences, 7, Springer-Verlag, Berlin, 2006. xxvi+367 pp.

We all have heard that number theory is useful, especially in the current information age. This book will substantiate such a claim, and the reader can see many concrete well-explained applications.

7.1. NUMBER THEORY

According to MathSciNet, "It is hardly necessary to introduce this popular book to the mathematical public. Since its first edition in 1984 it has gained a large audience and also a wide range of critics. The contents of the book may be described as showing many applications of number theory to the 'real world', some well known (like the RSA cryptosystem), some less well known (freight classification yards, p. 80) and some quite unexpected (concert hall acoustics, p. 168). In this respect the book is a veritable source of information and an interesting collection of examples for the layman... Clearly, the present book is intended for the non-expert reader who is simply interested in number theory and some of its astounding applications. In this respect it provides a wealth of material and it is without doubt an interesting work. But even if intended for the layman a certain level of accuracy must be maintained. According to the preface 'number theory has not rested on its laurels since the appearance of the second edition in 1985.' It would be appropriate if the same could be said about the contents of the book. Then it would truly be an enrichment to the scientific literature."

7.1.3 H. Davenport, The Higher Arithmetic. An Introduction to the Theory of Numbers, Eighth edition, With editing and additional material by James H. Davenport, Cambridge University Press, Cambridge, 2008. x+239 pp.

This is a classic. It is a very accessible introduction to number theory. Now into its Eighth edition, "The Higher Arithmetic" introduces the classic concepts and theorems of number theory in a way that does not require the reader to have an in-depth knowledge of the theory of numbers.

According to MathSciNet, " This book is an extraordinarily lucid account of elementary number theory. Although a few of the common topics such as the Möbius inversion formula are omitted, there is more than ample compensation in the fresh style which enables the author to shed much light on topics of considerable depth that are here only touched on. Throughout the book one encounters the results and spirit of modern research as well as references to the literature. As a result, this work can serve not only as an account of elementary results but also as a guide to further reading. Unfortunately, the suitability of this book as a text is diminished by the fact that there are no problems... It is interesting to note that a factor contributing to the fine quality of the exposition is the author's rejection of the not uncommon practice of attempting to catch the reader's eye by frivolous anecdotes and number tricks. The result is a serious work of great charm."

7.1.4 H. Hasse, Number Theory, Translated from the third German edition and with a preface by Horst Günter Zimmer. Grundlehren der Mathematischen Wissenschaften, 229. Springer-Verlag, Berlin-New York, 1980. xvii+638 pp.

This is a classic written by a master. It is a definite account of number theory from the "divisor-theoretic" point of view.

According to a review in Bulletin of AMS MathSciNet, it is "a fine book ... [that] treats algebraic number theory from the valuation-theoretic viewpoint. When it appeared in 1949 it was a pioneer. Now there are plenty of competing accounts. But Hasse has something extra to offer. This is not surprising, for it was he who inaugurated the local-global principle (universally called the Hasse principle). This doctrine asserts that one should first study a problem in algebraic number theory locally, that is, at the completion of a valuation. Then ask for a miracle: that global validity is equivalent to local validity. Hasse proved that miracles do happen in his five beautiful papers on quadratic forms of 1923-1924... The exposition is discursive... It is trite but true: Every number-theorist should have this book on his or her shelf."

7.1.5 A. Borevich, I. Shafarevich, Number Theory, Translated from the Russian by Newcomb Greenleaf, Pure and Applied Mathematics, Vol. 20, Academic Press, New York-London, 1966, x+435 pp.

This is a beautifully written introduction to basic topics in number theory. It was co-written by a real expert and expositor.

According to the translator's note, "This book was written as a text for the learning of number theory, not as a reference work, and we have attempted to preserve the informal, slow-placed style of the original. The emphasis of the book is on number theory as a living branch of modern mathematics, rather than as a collection of miscellaneous results.

This book should prove accessible to any advanced undergraduate in mathematics, or to any graduate student..."

According to the Foreword, "This book is written for the student in mathematics. Its goal is to give a view of the theory of numbers, of the problems with which this theory deals, and of the methods that are used.

We have avoided that style which gives a systematic development of the apparatus and have used instead a freer style, in which the problems and the methods of solutions are closely interwoven. We start from concrete problems in number theory. General theories arise as tools for solving these problems. As a rule, these theories are developed sufficiently far so that the reader can see for himself their strength and beauty, and so that he learns to apply them.

7.1. NUMBER THEORY

Most of the problems that are examined in this book are connected with the theory of diophantine equations — that is, with the theory of the solutions in integers of equations in several variables..."

7.1.6 A.Y. Khinchin, Three Pearls of Number Theory, Translated from the Russian by F. Bagemihl, H. Komm, and W. Seidel, Reprint of the 1952 translation, Dover Publications, Inc., Mineola, NY, 1998. 64 pp.

This is an unusual classic book. This short book concentrates on three beautiful theorems and presents them as "some little mathematical pearls" to occupy the mind of an injured solider recovering in a hospital. The common things about these theorems are that they were proved by elementary arithmetic methods and by very young mathematicians at the age of the solider. It is a joy to read such a beautiful short book written with such an unusual motivation.

According to an Amazon review, "It is truly a pearl, and pearls have permanence; – they retain their beauty, from one generation to the next.

So too is the case for this little book. Measured in mathematical generations, you must count back a few;– back to the last year of the Second World War, and in what was then The Soviet Union; now Russia. The author, A. Y. Khinchin was (and is) a mathematical physicist of World Renown. He has seminal contributions to number theory, to statistics, to information theory, and to statistical physics.

The book is unique in many ways; for one, I believe it is for everyone, – even if you do not know math. But readers with math background will know that it is possible for writing in math to be both moving and beautiful. This is the case for this little classic. Both the historical background and the subject are unique.

The nature of the book (64 pages in all!) is almost like a personal letter written by a loving teacher to one of his students, but it is much more than that.

At the time, the War had devastated Russia, and almost everyone from the young generation, including students of the sciences was at the front. The casualties everywhere in The Soviet Union were staggering; many had lost parents and relatives during 4 long years of destruction.

Khinchin's student Seryozha was recovering (at the time of the letter) in an army hospital, and he had written his former teacher, asking for math problems to work on. We can't begin to imagine the terrible conditions of army hospitals on the front at this time. The care Khinchin took in responding is moving. In fact Seryozha had only taken one or two beginning classes at university, before being sent to war. And even though Khinchin had only a vague recollection of Seryozha from a class, he truly wanted to send him something he could use, – something that would make him happy. Students at the front were giving their lives for the rest of the country, and we must remember that this was a war where the difference between good and evil was crystal clear. Khinchin's students were heroes. The book opens with a moving and personal letter, full of empathy, gratitude and love.

As for the mathematics, Khinchin had carefully selected problems of great beauty, problems that can be stated and appreciated with little specialized knowledge; – in modern lingo, with very few prerequisites. And at the same time, they are problems Seryozha can work on in his hospital bed. They are profound, and they can be attacked with elementary means. Naturally, since 1945, there have been a lot of advances on all three. The problems are from arithmetic (or number theory), and they go under the names: (a) van der Waerden's theorem on arithmetic progressions, (b) Landau's hypothesis and Mann's theorem, and (c) an elementary solution of Waring's problem.

By now these three problems take a different form in modern math books, but none as beautiful, in my opinion as Khinchin's in his loving letter to his student written toward the end of the war."

A. Khinchin, *Continued Fractions*, With a preface by B. V. Gnedenko, Translated from the third (1961) Russian edition, Reprint of the 1964 translation, Dover Publications, Inc., Mineola, NY, 1997. xii+95 pp.

This is also a classic by a master. It was written in 1935 as a textbook to fill in a gap between standard textbooks on algebra and number theory, since neither discuss continued fractions much. The author tried to make this book as elementary and accessible as possible. It has been in print and read by many people since then.

According to Amazon reviews, "readable (no more mathematic needed than basics of analysis), complete (all fundamental conceptual aspects dealt with, included measure theory and implications on irrational numbers), brief (less than a hundred pages with virtually no applications - not even to Pell's equation!) and LIVELY in style. All in all a very good start for understanding this profound mathematical tool."

7.1.7 E. Artin, Galois Theory, Notre Dame Mathematical Lectures, No. 2, Edited and with a supplemental chapter by Arthur N. Milgram, Reprint of the 1944 second edition, Dover Publications, Inc., Mineola, N.Y., 1998. iv+82 pp.

This is a classic. It gives a very efficient introduction to Galois theory.

Galois used permutation groups to describe how the various roots of a given polynomial equation are related to each other, and these results were in turn used to address the basic question on how to solve polynomial equation in terms of the coefficients of the polynomial, using only the usual algebraic operations (addition, subtraction, multiplication, division) and application of radicals (square roots, cube roots, etc)?

Briefly, Galois theory provides a connection between field theory and group theory, and one modern aspect involves studying automorphisms of field extensions. For example, a major problem is to determine and understand all groups which occur as automorphism groups of extensions of the field of rational numbers.

Besides its importance to algebraic equations and number theory, the concept of group introduced by Galois has everlasting impact on mathematics. Group is the language to describe symmetry and is used almost everywhere in mathematics and sciences.

According to the MathSciNet review of the 2004 edition, "Artin's book is a gem and it is a pleasure to see it republished by the AMS. It remains one of the most elegantly presented texts on Galois theory and the exposition is as clear as if it were written yesterday... In summary, this book is a valuable supplement for a course on Galois theory. It may require a higher level of mathematical expertise than most undergraduate texts but is rewarding for its focused exposition and interesting examples..."

7.2 Algebraic number theory

To many calculus students, Dedekind is best known for the Dedekind cut, one approach to define real numbers from the rational numbers. But it is probably less known that he introduced this notion when he first taught calculus at ETH Zurich. Dedekind has the special honor of editing the collected works of Dirichlet, Gauss, and Riemann, the three famous people at Göttingen. Dedekind's study of Dirichlet's work motivated him to study algebraic number fields and ideals. In 1863, he published Dirichlet's lectures on number theory as Vorlesungen über Zahlentheorie. Actually, most part of the book itself was entirely written by Dedekind after Dirichlet's death.

Figure 7.2: Dedekind's book.

7.2.1 D. Hilbert, The Theory of Algebraic Number Fields, Translated from the German and with a preface by Iain T. Adamson, With an introduction by Franz Lemmermeyer and Norbert Schappacher, Springer-Verlag, Berlin, 1998. xxxvi+350 pp.

It is a classic written by a master at a young age. It is not easy to name other books which have such huge impact on the subjects they describe.

This book is a report on the status of algebraic number theory near the end of the 19th century written by Hilbert upon a request of the German mathematical society. Hilbert willingly accepted this in his early career. This book has influenced the direction of research in algebraic number theory in years after its publication. After all these years, it is still a very valuable reference. Writing this book had positive impact on his mathematics and this might be a counter example to the usual suggestion that young people should not spend time on writing books.

According to MathSciNet, "This is an English translation of Hilbert's famous *Zahlbericht* ... What makes this book so important is of course its weighty impact on the number theory of our century. Yet the book also contains some material of immediate interest to present-day students and researchers."

According to the introduction of the English translation, "Hilbert's so-called *Zahlbericht*, which appears for the first time in English, was the principal textbook on algebraic number theory for a period of at least thirty years after its appearance. Emil Artin, Helmut Hasse, Erich Hecke, Hermann Weyl and many others learned their number theory from this book. Even beyond its impact Hilbert's *Zahlbericht* has served as a model for many standard textbooks on algebraic number theory through the present day... As a matter of fact, except for minor details ... at least the first two parts of Hilbert's text can still today pass for an excellent introduction to classical algebraic number theory."

7.2.2 E. Hecke, Lectures on the Theory of Algebraic Numbers, Translated from the German by George U. Brauer, Jay R. Goldman and R. Kotzen, Graduate Texts in Mathematics, 77, Springer-Verlag, New York-Berlin, 1981. xii+239 pp.

The original German edition appeared in 1923, Akademische Verlag, Leipzig, 1923; Jbuch 49, 106.

Algebraic number theory is a major branch of number theory which studies algebraic structures related to algebraic integers, such as factorization of ideals. One of the classical results in algebraic number theory is that the ideal class group of an algebraic number field is finite, whose order is called the class number of the field and is computed by the class number formula via a special value of its Dedekind zeta function. Another is Dirichlet's unit theorem which describes the multiplicative group of units of the ring of integers of a number field. There are also various versions of reciprocity law.

There are many excellent books on algebraic number theory. This book is a classic by a master. The original German edition appeared in 1923.

According to the translator's preface, "Hecke is certainly one of the masters, and in fact, the study of Hecke L-series and Hecke operators has permanently embedded his name in the fabric of number theory. It is a rare occurrence when a master writes a basic

7.2. ALGEBRAIC NUMBER THEORY

book, and Hecke's *Lectures on the Theory of Algebraic Numbers* has become a Classic. To quote another master, Andre Weil: 'To improve upon Hecke, in a treatment along classical lines of the theory of algebraic numbers, would be a futile and impossible task.'
...

We have tried to remain as close possible to the original text in preserving Hecke's rich, informal style of exposition... One problem for a student is the lack of exercises in the book. However, given the large number of texts available in algebraic number theory, this is not a serious drawback. In particular we recommend *Number Fields* by D. A. Marcus (Springer-Verlag) as a particularly rich source..."

D. Marcus, *Number Fields*, Universitext, Springer-Verlag, New York-Heidelberg, 1977. viii+279 pp.

This can serve as a good supplement to the book above of Hecke by providing exercises.

According to MathSciNet, "This book gives a thoroughly delightful introduction to algebraic number theory. The author, who obviously enjoys the subject matter, assumes that the reader knows the basics of undergraduate linear and abstract algebra. The exposition is leisurely and 'down-to-earth', as the author puts it. He gives complete proofs of most of the results, although many of the exercises are of the 'fill in the details of Theorem X' or 'Prove Corollary Y' variety... The chief virtue of this book is the staggering collection of exercises."

7.2.3 J.W.S. Cassels, A. Fröhlich, Algebraic Number Theory, Proceedings of the instructional conference held at the University of Sussex, Brighton, September 1–17, 1965, Academic Press, London; Thompson Book Co., Inc., Washington, D.C., 1967, xviii+366 pp.

There are many conferences now and many conference proceedings. Most volumes are collecting dust on shelves in libraries or being disposed due to the advance of digital technology and shortage of library space. This book is an exception and it is a dream conference proceedings volume every conference organizer would like to edit and produce. It contains both well written expository papers and original papers. It is a very valuable introduction to number fields since its publication. It is probably still the best place to learn the difficult class field theory.

According to the book description. "First printed in 1967, this book has been essential reading for aspiring algebraic number theorists for more than forty years. It contains the lecture notes from an instructional conference held in Brighton in 1965, which was a milestone event that introduced class field theory as a standard tool of mathematics.

There are landmark contributions from Serre and Tate. The book is a standard text for taught courses in algebraic number theory."

From the editors' preface: "this volume also contains exercises compiled by J. T. Tate with J.-P. Serre's help, and above all Tate's doctoral thesis, which is for the first time published here after it had over many years had a deep influence on the subject as a piece of clandestine literature."

7.2.4 S. Lang, Algebraic Number Theory, Second edition, Graduate Texts in Mathematics, 110, Springer-Verlag, New York, 1994. xiv+357 pp.

This is a classical book written by Lang, among many books written by him. According to the preface, it "gives an exposition of the classical basic algebraic and analytic number theory." It is a very valuable introduction to algebraic number theory.

According to MathSciNet, "This book is the second edition of Lang's famous and indispensable book on algebraic number theory...

This book should be of great value to modern students of the subject because it contains clear, complete, and readable expositions of important earlier results that are in some danger of being forgotten. The author includes an excellent brief history of class field theory and explains the various approaches to it."

The MathSciNet review of the first edition says, "This book should be of great value to modern students of the subject because it contains clear, complete, and readable expositions of important earlier results that are in some danger of being forgotten. The author includes an excellent brief history of class field theory and explains the various approaches to it."

7.2.5 More books on algebraic number theory

A. Weil, *Basic Number Theory*, Reprint of the second (1973) edition, Classics in Mathematics, Springer-Verlag, Berlin, 1995. xviii+315 pp.

This is an important book written by a master. Though it treats "basic number theory", it is not an easy book.

According to MathSciNet, "This is an exposition of algebraic number theory and class field theory, handling in a unified manner all fields, called by the author **A**-fields, which are finite algebraic over the rational field or over a field of rational functions of one variable with a strictly finite field of constants. The author states in the foreword that he has tried 'to draw the conclusions from the developments of the last thirty years, whereby locally compact groups, measure and integration have been seen to play an increasingly

7.2. ALGEBRAIC NUMBER THEORY

important role in classical number theory' and to show that from the point of view which he has adopted one could give a coherent account, logically and aesthetically satisfying, of the topics he was dealing with.

The main features of the exposition of class field theory are the full use of analytic methods applied not only to commutative fields but also to simple algebras."

J. Neukirch, *Algebraic Number Theory*, Translated from the 1992 German original and with a note by Norbert Schappacher, With a foreword by G. Harder, Grundlehren der Mathematischen Wissenschaften, 322, Springer-Verlag, Berlin, 1999. xviii+571 pp.

This is a very good introduction to modern number theory. It can also serve as a modern reference on number theory. This book is a modern classic.

According to MathSciNet, "The book covers all standard topics in local and global algebraic number theory, and much more. It confesses to the policy of building the theory for its own sake, not avoiding abstraction and generality. This it does in a very efficient way.... Even though Neukirch's renowned book probably needs no further recommendation, the reviewer would like to recommend it anyway, hoping that curiosity, the reputation of the book, or even perhaps this review will induce many people to look at it; they will be richly rewarded."

L. Washington, *Introduction to Cyclotomic Fields*, Second edition, Graduate Texts in Mathematics, 83, Springer-Verlag, New York, 1997.

It is a standard introduction and reference on cyclotomic fields.

According to MathSciNet, "This book is intended as an introduction to the cyclotomic theory underlying the modern research in this area. Compared with S. Lang's recent book *Cyclotomic Fields*, having similar aims, the present book is written on a more elementary level. In fact, it should be accessible to any mathematician having had a one-semester course in algebraic number theory (including the basic facts about local fields). In the reviewer's opinion the author has succeeded well in choosing and organizing the material and in keeping the exposition on the intended level. The argumentations are simple and elegant, often showing great originality. The text is clearly and carefully written and the proofs are suitably detailed. In particular, the reader is pleased with the many passages in which the results and proofs are motivated or explained, frequently with small numerical examples. At the end of each chapter there is also a special section devoted to historical comments and references to the literature; these are highly welcome as well as the exercises (usually far from routine) appended to each chapter."

S. Lang, *Cyclotomic Fields I and II*, Combined second edition, With an appendix by Karl Rubin, Graduate Texts in Mathematics, 121, Springer-Verlag, New York, 1990. xviii+433 pp.

This book may not be easy as an introduction but is a useful reference.

According to MathSciNet, "The theory of cyclotomic fields has received a new life from recent work of Iwasawa, Leopoldt, Coates and Wiles, and Kubert and the author. The present book is intended as an introduction to the cyclotomic theory underlying their work. As such, it is weighted heavily toward modern developments while omitting several classical topics which one might expect in a book of its title, for example, the Kronecker-Weber theorem and Fermat's last theorem."

7.2.6 Computational algebraic number theory

H. Cohen, *A Course in Computational Algebraic Number Theory*, Graduate Texts in Mathematics, 138, Springer-Verlag, Berlin, 1993. xii+534 pp.

This book was written with two purposes in mind: a reasonably comprehensive introduction to computational number theory, and practicality of the implementation of the algorithms. It is a valuable introduction and reference.

According to MathSciNet, "Throughout history number theorists have been prone to computation, and the advent of computers has extended the reach of such calculations enormously. In recent years this natural proclivity has been encouraged by the appearance of number-theoretic ideas in computer science, cryptology, and other areas of mathematics. Computational algebraic number theory has had an especially vigorous growth in the last decade, and the book under review aims at providing a thorough introduction to this field that has a strong focus on 'practicality' ...

The book under review covers a vast amount of ground. In addition to giving numerous explicit algorithms, it presents the underlying mathematics in considerable detail... The combined attempts to present serious mathematics in tandem with detailed descriptions of algorithms that have been implemented by the author give this volume a special flavor; to the reviewer, it is a marvelous book that should prove genuinely useful to a broad audience."

H. Cohen, *Advanced Topics in Computational Number Theory*, Graduate Texts in Mathematics, 193, Springer-Verlag, New York, 2000. xvi+578 pp.

This book is a continuation of the earlier book by Cohen by generalizing algorithms to global function fields and relative extensions of number fields. It is a very valuable reference.

According to MathSciNet, "This book together with [the earlier one] will become the standard and indispensable reference on computational algebraic number theory and both books are strongly recommended"

7.2. ALGEBRAIC NUMBER THEORY

7.2.7 J.P. Serre, Local Fields, Translated from the French by Marvin Jay Greenberg, Graduate Texts in Mathematics, 67, Springer-Verlag, New York-Berlin, 1979. viii+241 pp.

This is a classic written by a master. It is an advanced introduction and a valuable reference.

From the preface, "The goal of this book is to present local class field theory from the cohomological point of view, following the method inaugurated by Hochschild and developed by Artin-Tate. The theory is about extensions-primarily abelian-of 'local' (i.e., complete for a discrete valuation) fields with finite residue. For example, such fields are obtained by completing an algebraic number field; this is one of the aspects of 'localisation'. "

A local field is a locally compact topological field with respect to a non-discrete topology. Given such a field, an absolute value can be defined on it. There are two basic types of local field: those in which the absolute value is archimedean and those in which it is not. In the first case, one calls the local field an archimedean local field, in the second case, one calls it a non-archimedean local field. Local fields arise naturally in number theory as completions of global fields.

Every local field is isomorphic (as a topological field) to one of the following: Archimedean local fields (characteristic zero): the real numbers $\mathbb{R}$, and the complex numbers $\mathbb{C}$, or non-archimedean local fields of characteristic zero: finite extensions of the p-adic numbers $\mathbb{Q}_p$ (where p is any prime number).

Non-archimedean local fields of characteristic p (for p any given prime number): the field of formal Laurent series $F_q((T))$ over a finite field F_q (where q is a power of p). There is an equivalent definition of non-archimedean local field: it is a field that is complete with respect to a discrete valuation and whose residue field is finite.

The local class field theory, introduced by Hasse in 1930, is the study of abelian extensions of local fields. It is the analogue for local fields of global class field theory. Local class field theory gives a description of the Galois group G of the maximal abelian extension of a local field K via the reciprocity map which acts from the multiplicative group $K^\times = K \setminus \{0\}$.

There is a wikipedia article on this book (note that very few books have wikipedia articles devoted to them).

J. Cassels, *Local Fields*, London Mathematical Society Student Texts, 3, Cambridge University Press, Cambridge, 1986. xiv+360 pp.

This is a self-contained introduction to local fields for beginners. It contains a lot of applications to motivate the results.

According to MathSciNet, "Upon an initial scanning of this book, one is impressed with the fact that is it so well-organized and very easy to read. However, the author indicates in the preface that the goal is 'not to bring the reader to the frontiers of knowledge but, rather, to illustrate the versatility, power and naturalness of the approach'. Moreover, there are sections marked with an asterisk which may be omitted at a first reading. But if these sections are skipped then the author cautions that 'you miss some of the lollipops' . Indeed, this is the case... There are many such palatable asterisked sections which the reader will enjoy. ... In the preface the author makes the following comment: 'The author will have failed if he does not persuade the reader that the p-adic numbers are every bit as natural and worthy of study as the reals and complexes.' In the opinion of this reviewer, Professor Cassels has succeeded magnificently."

7.2.8 Galois cohomology

Galois cohomology applies methods of homological algebra to study the group cohomology of Galois modules, i.e., modules for Galois groups. It started around 1950, when it was realized that the Galois cohomology of idele class groups in algebraic number theory was one way to formulate the class field theory.

J. Neukirch, A. Schmidt, K. Wingberg, *Cohomology of Number Fields*, Second edition, Grundlehren der Mathematischen Wissenschaften [Fundamental Principles of Mathematical Sciences], 323, Springer-Verlag, Berlin, 2008. xvi+825 pp.

The book is a comprehensive account of cohomology of number fields, or rather, advanced algebraic number theory. It is a very valuable reference.

According to MathSciNet, "The purpose of this volume is to provide a textbook for students, as well as a reference book for mathematicians interested in cohomology topics in number theory. Galois modules over local and global fields are the main subjects of this work, Galois cohomology being the basic technique used, which is essential for class field theory. The book is not a basic textbook in the sense of being completely self-contained... One of the most important achievements in this monograph is the providing of complete proofs of central and frequently used theorems such as the global duality theorem of Poitou and Tate and the celebrated theorem of Shafarevich on the realization of solvable groups as Galois groups over global fields. These results have been part of algebraic number theory for a long time, but their proofs are spread over many original articles, some of which contain mistakes and some of which even remain unpublished. The initial motivation of the authors was to fill these gaps and they definitely succeed in their goal..."

J.P. Serre, *Galois Cohomology*, Translated from the French by Patrick Ion and revised by the author, Springer-Verlag, Berlin, 1997. x+210 pp.

7.2. ALGEBRAIC NUMBER THEORY

This is a standard reference on Galois cohomology written by a master. It was one of the first books on this subject and has been updated in more than five editions.

According to MathSciNet, "This book surveys an elegant new subject which has developed out of the cohomological treatment of class field theory by E. Artin and J. Tate. The bulk of the early contributions were by Tate, and we are greatly indebted to the author for publishing them in his very lucid style. Many others have made impressive discoveries in the field since, and the copious results will now be summarized."

7.2.9 Geometry of numbers

The geometry of numbers is a relatively new subject in number theory, and studies convex bodies and integer vectors in n-dimensional space. It was initiated by Hermann Minkowski in 1910. Geometric methods have often led to effective and efficient proofs of results in algebraic numbers and other subjects.

P. Gruber, C. Lekkerkerker, *Geometry of Numbers*, Second edition. North-Holland Mathematical Library, 37, North-Holland Publishing Co., Amsterdam, 1987. xvi+732 pp.

This is a classic and has been the standard introduction to geometry of numbers.

According to the review of MathSciNet of the first edition, "This book is the product of a great deal of labour. The author describes almost all work available up to 1964 in the classical geometry of numbers over the n-dimensional Euclidean space. All basic results are proved. Others are described lucidly and references are given to the sources where complete proofs are available. The book deals with the homogeneous problem, in its various formulations, arithmetical and geometrical in terms of critical determinants as well as packings, and with both types of non-homogeneous problems, those corresponding to non-homogeneous determinants and those corresponding to coverings."

J.W.S. Cassels, *An Introduction to the Geometry of Numbers*, Corrected reprint of the 1971 edition, Classics in Mathematics, Springer-Verlag, Berlin, 1997. viii+344 pp.

This is a useful reference, but not an easy introduction.

According to Mathematical Gazette, "The work is carefully written. It is well motivated, and interesting to read, even if it is not always easy... historical material is included... the author has written an excellent account of an interesting subject."

C.A. Rogers, *Packing and Covering*, Cambridge Tracts in Mathematics and Mathematical Physics, No. 54, Cambridge University Press, New York, 1964, viii+111 pp.

One important aspect of geometry of numbers studies sphere packings and coverings on lattices. This is an accessible introduction to general sphere packings and coverings.

According to MathSciNet, "This monograph, based on the work of the author and his collaborators, gives a very elegant and readable account of many important results in the theory of packings and coverings, lattice as well as non-lattice. The author has managed in most cases to include proofs of best known results without affecting the readability of this book, which is entirely self-contained. Although the main text deals with estimates from above and below for the densities of best lattice and general packings and coverings for a set, ... the introduction contains a lucid and complete survey of the present knowledge of many other problems also. Because of the many unsolved problems in the field, this introduction is very valuable."

7.3 Analytic number theory

Littlewood was a great mathematician with a good sense of humor. His collaboration with Hardy was so successful that he was put into shadow sometimes. Several stories tell people's doubts on the existence of Littlewood due to the influence and fame of Hardy, and Littlewood responded with laughter. According to Hardy, in their collaboration, Littlewood was the stronger of the two. Besides his joint work with Hardy, he also contributed substantially to other subjects, for example, the "Littlewood-Paley theory in Fourier theory. He is also remembered for a unique book "A Mathematician's Miscellany".

Figure 7.3: Littlewood's book

Analytic number theory uses analytic methods to solve problems about the integers (or algebraic integers). Understanding distribution of primes is a major problem in analytic number theory, and the Riemann zeta function plays a pivotal role in attacking this problem and other problems in analytic number theory. For example, the Riemann hypothesis on the nontrivial zeros of the Riemann zeta function has motivated a lot of important work in number theory. It also has applications in physics, probability theory, and applied statistics.

First results about this zeta function were obtained by Leonhard Euler in the eighteenth century. It is named after Riemann, who established a relation between its zeros and the distribution of prime numbers in the paper "On the Number of Primes Less Than a Given Magnitude" in 1859.

7.3. ANALYTIC NUMBER THEORY

7.3.1 E.C. Titchmarsh, The Theory of the Riemann Zeta-Function, Second edition. Edited and with a preface by D. R. Heath-Brown, The Clarendon Press, Oxford University Press, New York, 1986. x+412 pp.

The Riemann zeta function plays a fundamental role in analytic number theory and has applications in many subjects in mathematics and sciences. It has guided the research direction of analytic number theory via the problem on zeros of the Riemann zeta function and related problems.

This is the classic on Riemann zeta-function!

According to MathSciNet, "the first edition of this classical treatise appeared in 1951, thus quite long ago. The editor of the present second edition, Heath-Brown, writes in the preface: 'Since the first edition was written, a vast amount of further work has been done. This has been covered by the end-of-chapter notes. In most cases, restrictions on space have prohibited the inclusion of full proofs, but I have tried to give an indication of the methods used whenever possible ...' this new edition makes very interesting reading even for those who are familiar with the original."

7.3.2 More books on the Riemann zeta function

A. Ivic, *The Riemann Zeta-function. The Theory of the Riemann Zeta-function with Applications*, A Wiley-Interscience Publication, Reprint of the 1985 original, Dover Publications, Inc., Mineola, NY, 2003. xxii+517 pp.

This is an updated version of E. Titchmarsh's classic *Theory of the Riemann Zeta-function*, 1951. It gives a comprehensive and accessible account of most of important results on the Riemann zeta-function and is a valuable reference.

H. Edwards, *Riemann's Zeta Function*, Pure and Applied Mathematics, Vol. 58, Academic Press, New York-London, 1974. xiii+315 pp.

This is an introduction to Riemann's zeta function with a historical perspective (or emphasis on historic roots).

According to MathSciNet, "it is not the author's purpose to give us an up-to-date account of all that is known about Riemann's zeta function. Instead, the author wishes to describe the principal ideas in the theory of the zeta function beginning with Riemann's epoch-making eight page memoir in which the Riemann hypothesis was first stated. In fact, it can strongly be argued that everything in the present book germinates from Riemann's classic paper and an unpublished manuscript of Riemann which contains the Riemann-Siegel formula. These two works are thoroughly discussed, ... For most readers, the most interesting feature of the book is the attention given to the calculation of the zeros of the Riemann zeta function $\zeta(s)$ on the critical line... A prevailing sense of history and excitement of mathematical discovery permeates the book. One gets the feeling of being with Riemann, Hadamard, Gram, and others as they struggle toward and achieve their discoveries... Historical conjecture, historical sidelights, speculation, interpretation, and motivation are happily a part of the book's style. Mistakes of the masters are pointed out. We are strongly reminded that doing mathematics is a truly human endeavor with pitfalls and faulty reasoning as well as with profound insights and inspiration..."

A. Karatsuba, S.M. Voronin, *The Riemann Zeta-function*, Translated from the Russian by Neal Koblitz, de Gruyter Expositions in Mathematics, 5, Walter de Gruyter & Co., Berlin, 1992. xii+396 pp.

This is also a more modern introduction to the theory of Riemann zeta function.

According to MathSciNet, "The aims of this book are twofold: first, to serve as an introduction to the theory of the Riemann zeta-function with its number-theoretic applications, and second, to acquaint readers with certain advances of the theory not covered by previous comprehensive treatises of the zeta-function such as the classic of E. C. Titchmarsh, *The theory of the Riemann zeta-function*, or a more recent monograph of A. Ivić *The Riemann zeta-function*, 1985."

According to a review in Bulletin by Sarnak, "This work is a valuable addition to the monographs on the zeta function. Other monographs on $\zeta(s)$ include the book by Edwards, which gives a nice historical account, and the book by Ivic, which covers in depth the Density Hypothesis and Density Theorems as well as Mean Value Theorems. For accounts of the important developments in the theory of Dirichlet L-functions, see the monographs by Bombieri, Davenport, Huxley, and Montgomery. If I were limited to having just one book on my shelves on the Riemann zeta function, I would opt for the 1986 edition of Titchmarsh's monograph."

Y. Motohashi, *Spectral Theory of the Riemann Zeta-function*, Cambridge Tracts in Mathematics, 127, Cambridge University Press, Cambridge, 1997. x+228 pp.

The spectral theory of the Riemann zeta-function is a relatively new field, and this book gives a comprehensive account of results of the field. It is a valuable reference.

According to MathSciNet, "This book is designed to show that the Riemann zeta function has deep connections with the theory of automorphic forms and then to exploit these connections to delve deeper into the theory of the zeta function itself... it might be noted that while viewing the Riemann zeta function through the theory of automorphic functions may bring better results, it has not made things easier. As Motohashi's book clearly demonstrates, analytic number theory, no matter how you slice it, is a very computational subject."

7.3.3 Analytic number theory

Analytic number theory uses methods from real analysis and complex analysis to solve problems about the integers (or algebraic integers). One early major result is Dirichlet's proof of Dirichlet's theorem on arithmetic progressions, i.e., the existence of infinitely many primes in every arithmetic progression, by L-functions, which is a generalization of the Riemann zeta function. Another major result is the prime number theorem.

Analytic number theory can be split up into two major parts: multiplicative and additive number theory, depending the type of problems they study rather than differences in technique. Multiplicative number theory deals with the distribution of the prime numbers, such as estimating the number of primes in an interval, and includes the prime number theorem and Dirichlet's theorem on primes in arithmetic progressions. On the other hand, additive number theory is concerned with the additive structure of the integers, such as Goldbach's conjecture that every even number greater than 2 is the sum of two primes. One of the main results in additive number theory is the solution to Waring's problem on representation of every natural number as the sum of at most s k^{th}-powers of natural numbers for suitable s and k.

G. Tenenbaum, *Introduction to Analytic and Probabilistic Number Theory*, Translated from the second French edition (1995) by C.B. Thomas, Cambridge Studies in Advanced Mathematics, 46, Cambridge University Press, Cambridge, 1995. xvi+448 pp.

This has proved to be a good introduction to analytic number theory and a valuable reference for experts as the author planned: "it was written with the double purpose of providing younger researchers with a self-contained account of analytic methods in number theory, and their elders with a source of references for certain basic questions".

According to MathSciNet, "The author has made important contributions to number theory and his mastery of the material is reflected in the exposition, which is lucid, elegant, and accurate. In its earlier version, the book had established itself as a basic work in number theory. Now in English and with wider distribution, this book should attract much attention as both a text and a reference."

T.M. Apostol, *Introduction to Analytic Number Theory*, (Undergraduate Texts in Mathematics), Springer, 1998.

This is an accessible, self-contained introduction to analytic number theory. Originally written for undergraduate students, it has been useful to others as well. It has been a successful book.

According to MathSciNet, "This book is the first volume of a two-volume textbook for undergraduates and is indeed the crystallization of a course offered by the author at the California Institute of Technology to undergraduates without any previous knowledge of number theory. For this reason, the book starts with the most elementary properties of the natural integers. Nevertheless, the text succeeds in presenting an enormous amount of material in little more than 300 pages... Each chapter is followed by a large and varied set of exercises; a student capable of handling the more ambitious ones is well on his way to independent research accomplishments. The presentation is invariably lucid and the book is a real pleasure to read."

H. Iwaniec, E. Kowalski, *Analytic Number Theory*, American Mathematical Society Colloquium Publications, 53, American Mathematical Society, Providence, RI, 2004. xii+615 pp.

This is a very comprehensive account of analytic number theory written by experts. It gives a global picture of the modern analytic number theory. It is both a substantial introduction for graduate students and a valuable reference for others.

According to MathSciNet, "This outstanding book makes accessible a substantial portion of analytic number theory. The book leads the reader from classical techniques and results in analytic number theory to the frontiers of research. At least as far as multiplicative number theory is concerned, the scope of this book is almost encyclopedic and the book will be both useful for students, and a valuable reference for researchers... this book covers a vast amount of analytic number theory. It will be an invaluable resource for researchers in the area. It is perhaps too advanced to serve as a first introduction to the subject, but students with some familiarity of analytic number theory will find much here that is deep and beautiful".

According to a review in Bulletin of AMS, "It is written in a relaxed, personal style with graduate students in mind, but there is plenty in it for the expert. The book is often witty, and in a long and unusually useful introduction there is a panoramic survey of the subject in which the authors set out their reasons for the choices of topics they have made... They demonstrate how the huge diversity of analytic methods have made an impact on the field. They write with authority and show their mastery of all the material on every page. One might say that they do for modern theory what Landau did for it in the first half of the last century, no mean achievement given all that has been accomplished since Landau's time."

H. Montgomery, *Ten Lectures on the Interface Between Analytic Number Theory and Harmonic Analysis*, CBMS Regional Conference Series in Mathematics, 84. the American Mathematical Society, Providence, RI, 1994. xiv+220 pp.

This is not a textbook or a comprehensive survey. But it is a very valuable reference.

According to MathSciNet, "These lectures give a lively, often original, look at a selection of topics from the area described by the title... The reviewer thinks this book deserves the attention of every analytic number theorist. Although it probably shouldn't

7.3. ANALYTIC NUMBER THEORY

be one's first book in the subject, it adds considerably more to the literature than a number of books which could serve that function."

According to Zentralblatt MATH, "The book is a masterpiece of exposition and can be highly recommended to anybody interested in the connections of analysis and number theory."

H. Halberstam, H. Richert, *Sieve Methods*, London Mathematical Society Monographs, No. 4, Academic Press, London-New York, 1974. xiv+364 pp.

Sieve methods are really basic in number theory and have a long history. This is the first comprehensive account of the modern sieve methods and has been a very valuable reference.

According to MathSciNet, "A notable gap in the literature of number theory has been the lack of any complete and connected account of the small sieve method and its applications. It is the aim of this book to fill this gap. The volume has been long awaited by sieve theorists and has attained the unusual distinction of having been referred to in several research papers that appeared some years before the book itself was actually published."

According to the review in Bulletin of AMS, "The modern theory of sieve method has developed gradually, with fits and starts, over the past sixty years. From the outset the literature was hard to read because of the complicated nature of the arguments, while in recent times many of the most important results have remained unpublished, making it almost impossible to be well informed. Moreover, the literature has become tangled, and fragmented by a lack of unifying perspective. Expository accounts of the subject have usually been restricted to specific aspects, and in many cases even these have made difficult reading... *Sieve methods* is the first exhaustive account of this important topic... For years to come, *Sieve methods* will be vital to those seeking to work in the subject, and also to those seeking to make applications. The heavy notation in the book seems to be essential in formulating such general methods. Some parts of the book are much more difficult to read than others, but generally the text is lively and conversational. In concept and execution this is an excellent, long-needed work."

R.C. Vaughan, *The Hardy-Littlewood Method*, Cambridge Tracts in Mathematics, 80, Cambridge University Press, Cambridge-New York, 1981. xi+172 pp.

This book describes the classical forms of the Hardy-Littlewood method and recent developments. This is the first modern general introduction and is valuable for both graduate students and experts.

According to MathSciNet, "The admirable book under review makes a notable introduction to the Hardy-Littlewood method, which sometimes is called the 'circle method', and its more recent developments in analytic number theory.

The genesis of the method is to be found in a paper of Hardy and Ramanujan in 1918 and the early essential work on the method appeared in a series of papers entitled "Some problems of Partitio Numerorum" by Hardy and Littlewood published during the period 1920–1928. Historical progress shows that during the past almost sixty years the method has had very striking and successful applications to Waring's problem, Goldbach's problem and many other additive problems in number theory. One of the remarkable examples is that Vinogradov refined the method to show that every large odd number is the sum of three odd primes...

The book will certainly interest both specialists and postgraduate students and become an excellent standard reference on the Hardy-Littlewood method."

7.3.4 Additive number theory

Additive number theory studies subsets of integers and their behavior under addition. Two classical problems are the Goldbach conjecture and Waring's problem.

L.K. Hua, *Additive Theory of Prime Numbers*, Translations of Mathematical Monographs, Vol. 13, American Mathematical Society, Providence, R.I., 1965, xiii+190 pp.

This is a classic. It gives a self-contained account of the additive number theory, in particular of Waring-Goldbach problem concerning representation of positive integers as sums of a given number of kth-powers of primes and includes results up to the point of publication. It was written as an introduction to modern additive number theory and served as a bridge to research literature. As Hua said, "After reading this book and a few other contemporary publications, ..., the reader will understand the basic principles of the modern additive theory of numbers and will be able to embark upon research."

M. Nathanson, *Additive Number Theory. Inverse Problems and the Geometry of Sumsets*, Graduate Texts in Mathematics, 165. Springer-Verlag, New York, 1996. xiv+293 pp.

Many classical problems in additive number theory are direct problems. Inverse problems in additive number theory are more recent and there have been a lot of progress. This book gives a self-account of such results on the inverse problems. It is a good introduction and a valuable reference too.

From the book description: "Inverse problems are a central topic in additive number theory. This graduate text gives a comprehensive and self-contained account of this subject. In particular, it contains complete proofs of results from exterior algebra, combinatorics, graph theory, and the geometry of numbers that are used in the proofs of the principal inverse theorems. The only prerequisites for the book are undergraduate courses in algebra, number theory, and analysis."

7.3.5 Multiplicative number theory

Multiplicative number theory is a subfield of analytic number theory that deals with prime numbers and with factorization and divisors of integers. The focus is usually on developing approximate formulas for counting these objects in various contexts. The prime number theorem is a key result in this subject.

H. Davenport, *Multiplicative Number Theory*, Third edition. Revised and with a preface by Hugh L. Montgomery, Graduate Texts in Mathematics, 74, Springer-Verlag, New York, 2000. xiv+177 pp.

This is a classic written by a real expert. The revision was also done by an expert.

According to MathSciNet, "This book originated from a lecture series [Multiplicative number theory, 1967] of the equally great mathematician and teacher given at the University of Michigan in 1966. Then its main merits were, aiming for the results on the (global) distribution of the primes and the primes in an arithmetic progression (with uniformity with respect to the modulus), to give a straightforward and very clear presentation of the necessary tools and techniques, and making the then most recent form of the large sieve and Bombieri's theorem readily accessible... This second edition has been revised with both great expertise and care by Montgomery... Altogether this is an important book which should be a must for anyone who is interested in analytic number theory."

H. Montgomery, R. Vaughan, *Multiplicative Number Theory. I. Classical Theory*, Cambridge Studies in Advanced Mathematics, 97, Cambridge University Press, Cambridge, 2007. xviii+552 pp.

This is a very valuable introduction for students and also a comprehensive reference for others. Its purpose "is to introduce the interested student to the techniques, results, and terminology of multiplicative number theory. It is not intended that our discussion will always reach the research frontier. Rather, it is hoped that the material here will prepare the student for intelligent reading of the more advanced research literature."

According to MathSciNet, "The monograph is a very readable, concise presentation of classical prime number theory, giving techniques as well as the underlying ideas, and describing an incredibly large range of topics. A study of this monograph seems to be a must for every number theorist. Many exercises, more or less difficult, challenge the reader. ('Some exercises are designed to illustrate the theory directly whilst others are intended to give some idea of the ways in which the theory can be extended, or developed, or paralleled in other areas.')"

7.4 Transcendental number theory

Though most (real or complex) numbers are transcendental, it is not easy to write down explicit ones. We all know two important transcendental numbers π and e. The name "transcendental" comes from a paper of Leibniz in 1682, where he proved $\sin x$ is not an algebraic function, and Euler was probably the first one who defined transcendental numbers. Joseph Liouville first proved the existence of transcendental numbers in 1844, and Hermite proved in 1873 that e is transcendental. Hermite's method was generalized next year by Lindemann to show that π is transcendental in his thesis (unfortunately this was basically the only work of Lindemann). Work on transcendental numbers continued partially due to its listing as Hilbert's seventh problem.

Figure 7.4: Leibniz's book.

A transcendental number is a number that is not algebraic, that is, it is not a root of a polynomial equation with rational coefficients. The most prominent examples of transcendental numbers are π and e. Transcendental number theory studies transcendental numbers. One question is how well can they be approximated by rational numbers. In 1850s, Joseph Liouville gave a necessary condition for a number to be algebraic, and thus a sufficient condition for a number to be transcendental. It says roughly that algebraic nonrational numbers cannot be very well approximated by rational numbers. In the twentieth century work by Axel Thue, Carl Siegel, and Klaus Roth, called Thue-Siegel-Roth theorem, improved the criterion of Liouville. Other major results due to Alexander Gelfond, Theodor Schneider and Alan Baker are related to a solution to the seventh of the Hilbert problems.

A. Baker, *Transcendental Number Theory*, Cambridge University Press, Cambridge, 1990. x+165 pp.

This is a classic by a real expert.

Its purpose was explained in the preface: "The study of transcendental numbers... has now developed into a fertile and extensive theory, enriching widespread branches of mathematics. My aim has been to provide a comprehensive account of the recent major discoveries in the field. Classical aspects of the subject are discussed in the course of the narrative."

According to MathSciNet, "This book gives a survey of the highlights of transcendental number theory, in particular of the author's own important contributions for which he was awarded a Fields medal in 1970. It is a very useful publication for mathematicians

7.4. TRANSCENDENTAL NUMBER THEORY

who want to obtain a general insight into transcendence theory, its techniques and its applicability. The style is extremely condensed, but there are many references for more detailed study. The presentation is very well done."

A. Gelfond, *Transcendental and Algebraic Numbers*, Translated from the first Russian edition by Leo F. Boron Dover Publications, Inc., New York, 1960, vii+190 pp.

It is a valuable reference by an expert. According to the author, the aims of this book are to show the contemporary state of the theory of transcendental numbers, to exhibit the fundamental methods of this theory, to present the historical course of development of these methods, and to show the connections which exist between this theory and other problems in the theory of numbers.

J.W.S. Cassels, *An Introduction to Diophantine Approximation*, Cambridge Tracts in Mathematics and Mathematical Physics, No. 45, Cambridge University Press, New York, 1957. x+166 pp.

This was written as an introduction for undergraduate students about basic techniques and some of the most striking applications of Diophantine approximation. It is also a useful reference.

According to MathSciNet, "It contains a remarkable wealth of material, and most of the major theories of the subject are treated fairly fully. The author has taken much trouble to present the proofs in small compass, though in consequence it is not always possible to follow the general principle of a proof without mastering the details... Everyone interested in the subject has reason to be grateful to the author for this valuable work."

M. Waldschmidt, *Diophantine Approximation on Linear Algebraic Groups. Transcendence Properties of the Exponential Function in Several Variables*, Grundlehren der Mathematischen Wissenschaften, 326, Springer-Verlag, Berlin, 2000. xxiv+633 pp.

This book gives a comprehensive account on the present state of the art of some of the most important developments in modern transcendence theory. It is a very valuable reference.

According to MathSciNet, "The present book is very nice to read, and gives a comprehensive overview of one wide aspect of Diophantine approximation. It includes the main achievements of the last several years, and points out the most interesting open questions... This book is of great interest not only for experts in the field; it should also be recommended to anyone willing to have a taste of transcendental number theory. Undoubtedly, it will be very useful for anyone preparing a post-graduate course on Diophantine approximation."

7.5 Arithmetic algebraic geometry

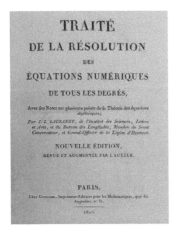

Lagrange was a great French mathematician born in Italy. He contributed enormously to many subjects in mathematics. For example, Lagrange's treatise Mécanique Analytique *in 1788 gave the most comprehensive treatment of classical mechanics since Newton had formed a basis for future development of mathematical physics. In number theory, he was the first to prove existence of a nontrivial solution to Pell's equation and that every positive integer is the sum of four squares, and he developed in 1775 a general theory of binary quadratic forms (including the reduction theory or binary quadratic forms) regarding the problem of representing integers by them.*

Figure 7.5: Lagrange's book.

Arithmetic algebraic geometry is one branch of number theory which solves number theoretic questions using powerful methods and techniques from algebraic geometry. Fermat's last theorem is one such example. Other examples include the Mordell theorem stating that for every elliptic curve defined over the field $\mathbb{Q}$ of rational numbers, its group of rational points is finitely generated, and the Mordell conjecture stating that an algebraic curve of genus greater than 1 defined over the field $\mathbb{Q}$ of rational numbers has only finitely many rational points.

7.5.1 G. Shimura, Introduction to the Arithmetic Theory of Automorphic Functions, Iwanami Shoten, Publishers, Tokyo; Princeton University Press, Princeton, N.J., 1971. xiv+267 pp.

This is a classic. It is both a valuable introduction for students and a useful reference for experts.

According to MathSciNet, "As the author notes in the Preface, there are two major topics treated in this volume: complex multiplication of elliptic or elliptic modular functions and applications of the theory of Hecke operators to the zeta-functions of algebraic curves and abelian varieties... The book contains many new points of view and deep results. It is written in a clear style and may also be used as a course textbook."

M. Vignéras, *Arithmétique des algèbres de quaternions*, Lecture Notes in Mathematics, 800, Springer, Berlin, 1980. vii+169 pp.

This book is the first modern introduction to quaternion algebras and their arithmetic applications. It is a useful reference.

According to MathSciNet, "The author's introduction informs us that this book has its origins in a course of lectures given in 1976 at the University of Paris (XI, Orsay) on the arithmetic of quaternion algebras. As these algebras are embraced as special cases within the corresponding theory of central simple algebras, the aim has been to present in detail those aspects which are typically quaternion."

7.5.2 J.P. Serre, A Course in Arithmetic, Translated from the French, Graduate Texts in Mathematics, No. 7, Springer-Verlag, New York-Heidelberg, 1973. viii+115 pp.

This is a classic by a master. Though it is short, it contains a wealth of information on modern number theory organized in a unique way.

According to the book description, Serre's "A Course in Arithmetic" is a concentrated, modern introduction to basically three areas of number theory, quadratic forms, Dirichlet's density theorem, and modular forms. The first edition was very well accepted and is now one of the leading introductory texts on the advanced undergraduate or beginning graduate level. "... The book is carefully written - in particular very much self-contained. As was the intention of the author, it is easily accessible to graduate or even undergraduate students, yet even the advanced mathematician will enjoy reading it. The last chapter, more difficult for the beginner, is an introduction to contemporary problems."

According to an Amazon review, "Serre's work could best be summarized in one word—Elegance... Due to the extreme elegance, the book is sometimes hard to read. This might sound like a paradox, but it's not and I'll explain why. The book takes some effort to read because it's terse and it often takes a while to figure out why something is 'obvious'. However, once you see it all, you'll realize that a great mind was guiding you through the pursuit. The choice of topics is just right to achieve the goals that the author sets out for himself. Also, I'd rather think for myself and read a smaller book than be given a huge fat tome where the author details his own thought process..."

7.5.3 J. Silverman, The Arithmetic of Elliptic Curves, Graduate Texts in Mathematics, 106, Second edition, Graduate Texts in Mathematics, 106, Springer, Dordrecht, 2009. xx+513 pp.

This is a popular and valuable introduction to and reference on arithmetic of elliptic curves.

According to the MathSciNet review of the first edition, "It meets the needs of at least three groups of people: students interested in doing research in Diophantine geometry, mathematicians needing a reference for standard facts about elliptic curves, and computer scientists interested in algorithms and needing an introduction to elliptic curves. For a long time one of the standard references for elliptic curves has been the survey article of J. W. S. Cassels, *Diophantine equations with special reference to elliptic curves*, J. London Math. Soc. 41 (1966), 193–291. In its choice of topics this book may be viewed as an amplification of Cassels' article, with technical details filled in, much more motivation, and an excellent set of exercises."

According to the MathSciNet review of the second edition, "It was one of the first modern textbooks on the arithmetic of elliptic curves written with emphasis on the relationships with algebraic geometry and algebraic number theory. Since then many nice textbooks on elliptic curves have appeared, in particular, those written by J. W. S. Cassels, D. Husemoller, J. Milne, and other authors. Nevertheless, ...[it] remains one of the better textbooks on elliptic curves".

J. Silverman, *Advanced Topics in the Arithmetic of Elliptic Curves*, Graduate Texts in Mathematics, 151, Springer-Verlag, New York, 1994. xiv+525 pp.

This is more advanced than the previous book and is a valuable reference.

According to MathSciNet, "Since its publication almost 10 years ago, Silverman's book *The arithmetic of elliptic curves*, 1986, has become a standard reference, initiating thousands of graduate students (the reviewer among them) to this exciting branch of arithmetic geometry. The eagerly awaited sequel,..., lives up to the high expectations generated by the first volume..."

S. Lang, *Elliptic Functions,* With an appendix by J. Tate, Second edition, Graduate Texts in Mathematics, 112, Springer-Verlag, New York, 1987. xii+326 pp.

This is a concrete, modern introduction to elliptic curves, which are the most basic objects in number theory.

According to MathSciNet, "The theory of elliptic functions is quite an old branch of mathematics, with origins in the work of Jacobi, Gauss and Abel, to mention only a few of the classical contributors to the subject. In recent years, the subject has again come to the forefront, especially in connection with algebraic geometry and arithmetic. The present volume provides an excellent introduction to the subject and carries the reader from the classical origins of elliptic function theory to some very modern topics, developed only in the last decade or so."

7.5.4 D. Mumford, Abelian Varieties. With appendices by C. P. Ramanujam and Yuri Manin, Corrected reprint of the second (1974) edition, Tata Institute of Fundamental Research Studies in Mathematics, 5, Hindustan Book Agency, New Delhi, 2008. xii+263 pp.

This is the first book that studies both the analytic and the algebraic theories of abelian varieties. This classical book is an excellent modern introduction to abelian varieties.

According to MathSciNet, "This book provides a modern, i.e., scheme-theoretic, treatment of most of the basic theory of abelian varieties... The exposition is clear, and often elegant and original. The book gives the impression that the author began with an idea of the most important results and set about proving them in the cleanest and quickest way (a method of exposition which could be recommended to other authors). There are aspects of the book one could niggle at (the lack of an index, the paucity of references) but one's final impression is of a beautiful book that will at last allow everyone to learn about abelian varieties without first having to grapple with Weil's language of algebraic geometry."

7.5.5 Abelian varieties and theta functions

D. Mumford, *Tata Lectures on Theta. I.*, With the assistance of C. Musili, M. Nori, E. Previato and M. Stillman, *II. Jacobian theta functions and differential equations*, With the collaboration of C. Musili, M. Nori, E. Previato, M. Stillman and H. Umemura, *III. With collaboration of Madhav Nori and Peter Norman*, Progress in Mathematics, Modern Birkhäuser Classics. Birkhäuser Boston, Inc., Boston, MA, 2007. xiv+235 pp., xiv+272 pp., viii+202 pp.

This three-volume book describes large parts of the classical theory of theta functions from a modern point of view. It starts from basics and combines the classic analytic and the modern geometric aspects. The writing is friendly and not formal. It is a valuable introduction.

According to MathSciNet, "The book will be profitable to anybody motivated to understand the deeper nature of thetas as functions (as it were) on the various moduli spaces obtained by acting on their arguments. Precisely these interpretations, which go a long way toward explaining the mystery of the theta identities, give the Tata books their unequalled role."

G. Faltings, C.L. Chai, *Degeneration of Abelian Varieties*, With an appendix by David Mumford, Ergebnisse der Mathematik und ihrer Grenzgebiete (3), 22, Springer-Verlag, Berlin, 1990. xii+316 pp.

There has been a lot of work on compactifications of arithmetic Hermitian locally symmetric spaces, in particular the Baily-Borel compactification, which is a normal projective variety defined over a number field. On the other hand, there are not many cases for which the Baily-Borel compactification is defined over the ring of integers. This book is an important book on explicit compactifications of the Siegel modular varieties which are defined over integers.

From the book description, "A new and complete treatment of semi-abelian degenerations of abelian varieties, and their application to the construction of arithmetic compactifications of Siegel moduli space, with most of the results being published for the first time. Highlights of the book include a classification of semi-abelian schemes, construction of the toroidal and the minimal compactification over the integers, heights for abelian varieties over number fields, and Eichler integrals in several variables, together with a new approach to Siegel modular forms. A valuable source of reference for researchers and graduate students interested in algebraic geometry, Shimura varieties or diophantine geometry."

H. Lange, C. Birkenhake, *Complex Abelian Varieties*, Second edition. Grundlehren der Mathematischen Wissenschaften, 302, Springer-Verlag, Berlin, 2004. xii+635 pp.

This book is a comprehensive account of complex abelian varieties and their applications. It is a valuable reference and introduction for many people, especially those who are less algebraically oriented. It fills in a gap in the literature on abelian varieties.

According to MathSciNet, "As its title indicates, this book studies complex abelian varieties and uses analytic rather than algebraic methods whenever possible.

Inevitably, a book on such a venerable subject is bound to overlap substantially with the many other existing texts, but the authors have somehow managed to make it unique in several respects: not only is it far more readable than most other books on the subject, but it is also much more complete. It is, in my opinion, a very valuable reference book more than a textbook.

The vast amount of results covered is probably the strongest feature of this book... Another advantage of having all this material put together in one book is that one can hope that the numerous normalizations and conventions involved will be consistent (which is definitely not the case in the existing literature). Another nice feature of this book is that prerequisites are kept to a minimum. Very little background in algebraic geometry is necessary, almost all proofs are complete, and accessible references are provided whenever they are not.

One could question the point of view taken here, and prefer the more algebraic methods of Mumford's book, for example, which often give more general results. But this would have meant writing another book, intended for another public, and also trying to do better than Mumford, which would have been difficult. On the contrary, because of

7.5. ARITHMETIC ALGEBRAIC GEOMETRY

the choices made by the authors, this book will be accessible to a larger audience, and really fills a gap in the existing literature. "

7.5.6 Diophantine geometry

A Diophantine equation is a polynomial equation where the variables can only take integers. More generally, the variables can take values in the field of rational numbers and other non-algebrically closed fields, and rings. Then solutions of Diophantine equations correspond to integral (or rational) points of algebraic varieties defined by the equations. Diophantine geometry studies Diophantine equations by formulating questions about such equations in terms of problems of algebraic varieties over a ground field that is not algebraically closed, such as the field of rational numbers or a finite field, or more general commutative ring such as the integers. Diophantine geometry has been studied for thousands of years, since the time of Pythagoras, and has continued to be an active subject pursued by number theorists such as the relatively recent proof of Fermat's Last Theorem and recent work on the ABC conjecture. We list three relatively recent books on this important topic.

S. Lang, *Fundamentals of Diophantine Geometry*, Springer-Verlag, New York, 1983. xviii+370 pp.

This is a good introduction to the classical theory of Diophantine geometry. It is one of the excellent books written by Lang.

According to MathSciNet, "The book is well written, and so far almost the only available introduction to the topic. Unfortunately, its limited size has not allowed the inclusion of very many newer results."

For the first edition, "This book is concerned with the interactions of diophantine analysis and algebraic geometry. It is up to the author's usual high standard, and should do much to renew profitable interest in the subject it treats... The historical notes throughout the book are of very high order. These include conjectures as to the future history as well, which possibly constitute one of the more valuable features of the book."

M. Hindry, J. Silverman, *Diophantine Geometry. An Introduction*, Graduate Texts in Mathematics, 201, Springer-Verlag, New York, 2000. xiv+558 pp.

This is a valuable advanced introduction to and a reference on modern Diophantine geometry.

According to MathSciNet, "In this excellent 500-page volume, the authors introduce the reader to four fundamental finiteness theorems in Diophantine geometry. After reviewing algebraic geometry and the theory of heights in Parts A and B, the Mordell-Weil theorem (the group of rational points on an abelian variety is finitely generated) is presented in Part C, Roth's theorem (an algebraic number has finitely many approximations of order 2+ ε) and Siegel's theorem (an affine curve of genus $g \geq 1$ has finitely many integral points) are proved in Part D, and Faltings' theorem (a curve of genus $g \geq 2$ has finitely many rational points) is discussed in Part E...

This volume will not only serve as a very useful reference for the advanced reader, but it will also be an invaluable tool for students attempting to study Diophantine geometry."

E. Bombieri, W. Gubler, *Heights in Diophantine Geometry*, New Mathematical Monographs, 4, Cambridge University Press, Cambridge, 2006. xvi+652 pp.

This book gives a comprehensive and self-contained account of basic and advanced topics in Diophantine geometry. It can serve as a systematic introduction to modern Diophantine geometry.

From the book description, "This monograph is a bridge between the classical theory and modern approach via arithmetic geometry. The authors provide a clear path through the subject for graduate students and researchers. They have re-examined many results and much of the literature, and provide a thorough account of several topics at a level not seen before in book form."

According to MathSciNet, "The present book also contains proofs of the theorems of Mordell-Weil, Roth and Faltings, but the emphasis is different. Now the heights are not a tool but, rather, the main object of interest, and the finiteness theorems occur as powerful demonstrations of the universality of the notion of height in modern Diophantine geometry."

7.5.7 Étale cohomology

The étale cohomology groups of an algebraic variety or scheme are algebraic analogues of the usual cohomology groups with finite coefficients of a topological space.

J. Milne, *Étale Cohomology*, Princeton Mathematical Series, 33, Princeton University Press, Princeton, N.J., 1980. xiii+323 pp.

This is an accessible excellent introduction to Étale cohomology and related topics.

According to MathSciNet, "Étale cohomology theory is a major contribution of Grothendieck and his co-workers to algebraic geometry. Its principal aim was achieved with Deligne's completion of the proof of the Weil conjectures in 1973. Though the main underlying idea was simple, its development involved considerable technical complications which frequently led in new directions. Because of this the chapter on Weil cohomology was planned by Grothendieck to be the conclusion of his Éléments and the only accounts of étale cohomology have been reports of the Bois-Marie seminars (SGA4) and Harvard lecture notes of M. Artin ['Grothendieck topologies', Lecture Notes, Harvard Univ., Cambridge, Mass., 1962]. The former are now widely circulated as Springer Lecture Notes. They demand a lot of the reader and are particularly unsatisfactory with regard to the

7.5. ARITHMETIC ALGEBRAIC GEOMETRY

category-theoretic foundations, especially to those impatient for geometry. All is put to rights by this book. It presents the main general features of étale cohomology with minimal demands as to category theory... The development of the theory is interwoven with important applications... The account does not aim to be self-contained, but rather to enable the reader to learn the subject in detail as well as in its overall perspective without the need to reconstruct proofs or consult papers of limited circulation. The choice of what to prove and what to refer to other sources is well done and the exposition flows smoothly."

According to a review in Bulletin of AMS by Bloch, "the author has done a tremendous service by organizing the material in a careful and united way which makes it possible for serious students to learn. In its way, the terse and brilliant account of the theory by Deligne (who disposes of the whole business in 65 pages) in SGA $4\frac{2}{2}$ is unexcelled."

P. Deligne, *Cohomologie étale*, Séminaire de Géométrie Algébrique du Bois-Marie SGA $4\frac{1}{2}$, Avec la collaboration de J. F. Boutot, A. Grothendieck, L. Illusie et J. L. Verdier, Lecture Notes in Mathematics, Vol. 569, Springer-Verlag, Berlin-New York, 1977. iv+312 pp.

This is a valuable reference written by a top expert.

According to MathSciNet, "This volume contains eight articles, mainly written by Deligne, on a variety of topics in étale cohomology. Its publication completes that of all the material required for the Grothendieck-Deligne proof of the Weil conjectures. It is written in a refreshingly direct and concise style."

7.5.8 Quadratic forms

A quadratic form is a homogeneous polynomial of degree two in a number of variables. Binary quadratic forms, i.e., quadratic forms in two variables, were systematically studied in the famous book *Disquisitiones Arithmeticae* by Gauss in 1801. Quadratic forms occupy a central place in various branches of mathematics, including number theory, linear algebra, linear algebraic groups, Lie group theory (by defining orthogonal groups), differential topology (as intersection forms of four-manifolds), etc.

T. Lam, *Introduction to Quadratic Forms Over Fields*, Graduate Studies in Mathematics, 67, American Mathematical Society, Providence, RI, 2005. xxii+550 pp.

This book is a very valuable reference for experts and also a good advanced introduction.

According to MathSciNet, "The author's book *The algebraic theory of quadratic forms* (ATQF, for short) appeared in 1973, and then a second printing with revisions was issued in 1980. The book was a great success and turned out to serve as the Bible for all quadratic form practitioners. The new book under a new title expands and updates ATQF.

The author confesses that writing ATQF took the young Ph.D. only one year, and rewriting the book after thirty years took six years! We should be grateful to the author for his effort. The result is a truly wonderful book. The story is told with amazing charm, clarity and precision. If ATQF won the Steele Prize in Mathematical Exposition, what kind of honors should be awarded to this perfect book?"

The Math Review of the first edition says "In 1937 E. Witt published a fundamental paper [J. Reine Angew. Math. 176 (1936), 31–44] on the general theory of quadratic forms over an arbitrary field... With all this activity in the last decade and every indication of even greater intensity of interest in the present decade at least, the time is very much ripe for a coherent compilation of results known to date. The mathematical community is very fortunate to have this need almost fully satisfied by the appearance of this remarkably well-written book."

W. Scharlau, *Quadratic and Hermitian Forms*, Grundlehren der Mathematischen Wissenschaften, 270, Springer-Verlag, Berlin, 1985. x+421 pp.

This book gives a comprehensive account of the algebraic theory of quadratic and Hermitian forms. It is a valuable reference and a good introduction.

According to MathSciNet, "The theory of quadratic forms, together with its companion theory of Hermitian forms, is a large subject, encompassing parts of number theory, algebra, and analysis, as well as interacting with a variety of other disciplines, including topology, real algebraic geometry, and model theory. The present text is especially rich in material and could be used to advantage in a number of essentially different graduate level courses in algebra. The first six chapters cover topics connected with the algebraic and arithmetic theories of quadratic forms. In the final four chapters the author turns his attention to Hermitian forms and simple algebras. There, the notion of semisimple algebra with involution can be viewed as the unifying concept. For a good overview of the book the reader is urged to read the author's foreword..."

O. O'Meara, *Introduction to Quadratic Forms*, Reprint of the 1973 edition, Classics in Mathematics, Springer-Verlag, Berlin, 2000. xiv+342 pp.

This is a classic and is still a very readable introduction to quadratic forms.

According to a review in Bulletin of AMS, "Anyone who has heard O'Meara lecture will recognize in every page of this book the crispness and lucidity of the author's style... The organization and selection of material is superb... deserves high praise as an excellent example of that too-rare type of mathematical exposition combining conciseness with clarity."

7.5. ARITHMETIC ALGEBRAIC GEOMETRY

J. Cassels, *Rational Quadratic Forms*, London Mathematical Society Monographs, 13, Academic Press, Inc., London-New York, 1978. xvi+413 pp.

This is a good elementary introduction to quadratic forms over rational numbers and integers.

According to MathSciNet, " According to the author, ' This (book) is not a treatise of the high tradition of Bachmann, Eichler, Watson and O'Meara. Its aims are much more modest: to indicate some of the major themes of classical arithmetical theory of quadratic forms in light of our present knowledge but from a totally elementary point of view.' There is no doubt that these goals are achieved and in a very elegant manner."

7.5.9 More books on number theory

S. Bosch, U. Güntzer, R. Remmert, *Non-Archimedean Analysis. A systematic Approach to Rigid Analytic Geometry*, Grundlehren der Mathematischen Wissenschaften, 261, Springer-Verlag, Berlin, 1984. xii+436 pp.

A rigid analytic space is an analogue of a complex analytic space over a non-archimedean field. This book is the first systematic introduction to rigid analytic spaces, or rigid analytic geometry. It is a valuable as an introduction and also as a reference.

According to a review in Bulletin of AMS, "Tate's rigid analytic geometry ... was created in 1961 by John Tate, who at that time had given a seminar on it at Harvard and written a manuscript entitled *Rigid analytic spaces*. These notes by Tate were distributed in Paris by the IHES in the spring of 1962 with(out) his permission and published as late as 1971 in Inventiones Mathematicae, whose editors thought it necessary to make these available to everyone. It is strange that the man who created this beautiful theory did nearly nothing to make it known. Further, to my knowledge he has never taken up research on the foundations of rigid analytic geometry which he has laid. I cannot guess for what reason he did not like his child later on."

According to MathSciNet, "The book is a broadly designed systematic presentation primarily of the 'local' theory of analytic spaces over a non-Archimedean valued field... The book will be welcomed by the expert who until now often had to conduct a painstaking search of the original literature, as well as by those who wish to familiarize themselves with the subject; no comparable presentation has existed until now."

S. Bosch, W. Lütkebohmert, M. Raynaud, *Néron Models*, Ergebnisse der Mathematik und ihrer Grenzgebiete (3), 21, Springer-Verlag, Berlin, 1990. x+325 pp.

This is a book on a specialized topic and is a useful reference for experts: the construction of the Néron models and to the study of their properties.

According to MathSciNet, "Although this [Néron model] is a very basic and widely used result, very few mathematicians have ever read Neron's proof. The authors' main purpose in writing their book has been to give a modern scheme-theoretic treatment of Neron's theorem. At the same time they have simplified his proof, and included some more recent work on Neron models."

M. Fried, M. Jarden, *Field Arithmetic*, Third edition, Revised by Jarden. Ergebnisse der Mathematik und ihrer Grenzgebiete, 3, Folge, A Series of Modern Surveys in Mathematics, 11, Springer-Verlag, Berlin, 2008. xxiv+792 pp.

This is a valuable and standard reference on a new combination of logic with other more mathematical areas such as field theory, number theory and arithmetic.

According to MathSciNet, "In recent years, starting from work of James Ax and Abraham Robinson, methods from 'logic' emerged into other areas such as field theory, number theory and arithmetic. Some of these developments gave valuable new insights into old questions of number theory and arithmetic; some of them led to new concepts. Most of the results so far have been scattered through the literature; no coherent treatment of the subject was available. Obviously the authors' intention is a detailed and self-contained presentation of part of these developments. The main focus and the central topic of the book is the theory of elementary statements in the sense of first-order logic of certain classes of fields, a program which, partly developed by the authors themselves, provides interesting links between the two mathematical disciplines of logic and arithmetic. As the authors themselves claim, the book should serve as a bridge between the two fields. Hence much stress is laid on thorough foundations. It is not an easy task to give comprehensive introductions to both subjects in one book. The result, however, should not only attract specialists from one of these fields. It should serve as well as a textbook for graduate courses or as an excellent survey suitable for personal studies."

Y. Manin, *Cubic Forms. Algebra, Geometry, Arithmetic*, Translated from the Russian by M. Hazewinkel, Second edition, North-Holland Publishing Co., Amsterdam, 1986. x+326 pp.

This is a very valuable book on many rich structures associated with cubic forms. Fermat's theorem on the sum of two squares of natural numbers is a classical result in number theory. The purpose of this book is "to find out what happens in the case of sums of three rational cubes... The author has generalized the problem along all the lines which occurred to him, and has used all technical resources known to him." It is a very original and interesting book.

According to MathSciNet, "This book is mainly concerned with the extent to which various results for elliptic curves have analogues for cubic surfaces (algebraic, geometrical and arithmetical). First and foremost, is it possible to define something like a group

7.6 Modular forms and automorphic representations

structure on the group of rational points $V(k)$ of a cubic surface V that is similar to the one for elliptic curves? It is clear that this cannot be done on $V(k)$, but it is possible on suitable quotients of $V(k)$ and there is a largest quotient admitting a composition law based on lines. The resulting structure is not a commutative group but a commutative Moufang loop..."

7.6 Modular forms and automorphic representations

A modular form is a complex analytic function on the upper half-plane satisfying some functional equations with respect to the action of the modular group, and also satisfying a growth condition. The name modular comes from the fact that the quotient of the upper half-plane by the modular group is the moduli space of elliptic curves. Hecke made deep contributions to the theory of modular forms and is honored with Hecke correspondences, Hecke L-functions etc. Special values of L-functions, in particular, Riemann zeta functions play an important role in the theory.

Figure 7.6: Hecke.

A modular form is a holomorphic function on the upper half-plane satisfying a certain kind of functional equation with respect to a finite index subgroup Γ of the modular group $SL(2,\mathbb{Z})$ (for example a congruence subgroup) and some growth condition at infinity, i.e., near the cusps of the fundamental domain of the group Γ. The theory of modular forms therefore belongs to complex analysis but the main application and importance of the theory has traditionally come from its connections with number theory. Modular forms appear in other areas, such as algebraic topology and string theory.

Automorphic forms are extension of modular forms. They are real analytic functions of several variables defined on a semisimple Lie group G and satisfying some invariance conditions with respect to an arithmetic subgroup Γ and growth conditions at infinity. The Lie group G generalizes the groups $SL(2,\mathbb{R})$ or $PSL(2,\mathbb{R})$ which act on the upper half-plane, where modular forms are defined, and the arithmetic subgroup Γ generalizes the modular group $SL(2,\mathbb{Z})$, or its finite index subgroups. Reformulated in terms of representations of G, automorphic forms become automorphic representations. They the starting point of the Langlands program.

7.6.1 R. Fricke, F. Klein, Vorlesungen uber die Theorie der automorphen Funktionen. Band 1: Die gruppentheoretischen Grundlagen. Band II: Die funktionentheoretischen Ausfhrungen und die Andwendungen. Bibliotheca Mathematica Teubneriana, Bande 3, 4, Johnson Reprint Corp., New York; B. G. Teubner Verlagsgesellschaft, Stuttg art 1965. Band I: xiv+634 pp.; Band II: xiv+668 pp.

This is a classical book on discrete subgroups and arithmetic subgroups of $SL(2,\mathbb{R})$. It is also a classic book in the sarcastic sense: everyone knows about it but very few people have read it. According to some German speaking mathematicians, it is not the issue with the language, but rather with the non-standard, less modern style of writing. However, it contains a wealth of information and a lot of buried treasure.

It is the first systematic exposition of discrete subgroups of $SL(2,\mathbb{R})$ and automorphic functions with respect to them. It contains detailed study of structures of arithmetic subgroups and lattices of $SL(2,\mathbb{R})$, and the geometry of their actions on the upper half plane, in particular geometry of Riemann surfaces and hyperbolic surfaces. It seems to be very comprehensive. For example, reduction theory of arithmetic groups and compactifications of noncompact quotients are addressed. Noncongruence arithmetic subgroups of $SL(2,\mathbb{Z})$ are also constructed. (One motivation for this problem is that structure of congruence subgroups is simpler and more methods can be applied to study them. For example, the adelic method developed after Fricke-Klein can only be applied to congruence subgroups, but not to general arithmetic subgroups.) The idea of Fenchel-Nielsen coordinates of Teichmüller spaces (or rather Fricke space, i.e., representation space of Fuchsian groups in $SL(2,\mathbb{R})$) is also described. Basically almost everything one wants to know about arithmetic subgroups of $SL(2,\mathbb{Z})$ and properties of the associated hyperbolic spaces is discussed or hinted at. Its influence on discrete subgroups of Lie groups is huge. Many results on discrete subgroups of general semisimple Lie groups and their associated locally symmetric spaces were motivated by results in these books.

But it is not read by people now. For example, according to a paper by Abikoff titled "The uniformization theorem", "To those of us trained in the Satz-Beweis [Theorem-Proof] school of mathematical exposition and discourse, reading Klein is often a mystical experience. In fact, many a mathematician has proved and published a deep and elegant result, later to discover with chagrin that there is a casual and vague reference to the result in Klein. Most of Klein's writing on areas related to uniformization are collected in two books written jointly with Fricke: those comprise over 2,000 pages without an index, and often without definitions or theorem statements. The mathematical insight contained therein is astounding, but it often seems that one can only appreciate a part of it after having independently rediscovered the results."

7.6. MODULAR FORMS AND AUTOMORPHIC REPRESENTATIONS

Since Klein is one of the founders of the discrete subgroups of Lie groups (another is Poincaré), it is not surprising that his book contains highly original materials. It is a pity that there is no English translation or modern update of this classic.

A brief outline of the contents of the above two books is as follows. The first volume is the natural continuation of their books on elliptic modular functions. Its preface gives a good summary of the historical development of the subject. The book is divided into three parts. The first part deals with basics of discontinuous groups of partial linear fraction transformations; the second part studies geometric aspects, in particular, polygon groups, and the last part studies arithmetic discontinuous groups. The second volume studies automorphic functions and their applications.

W. Magnus, *Noneuclidean Tesselations and Their Groups*, Pure and Applied Mathematics, Vol. 61. Academic Press [A subsidiary of Harcourt Brace Jovanovich, Publishers], New York-London, 1974. xiv+207 pp.

This is a useful reference that is also fun to read. According to the author, "The pictures that motivated the writing of the book are mainly those appearing in the mathematical works of Felix Klein and Robert Fricke. Most of them show tesselations of the non-Euclidean plane. Comments on their group theoretical, geometric, and function-theoretical meaning are, of course, available in the more than two thousand pages published by Klein and Fricke on this subject. But these comments are not easily accessible."

Klein had a broad vision of mathematics and had worked on several subjects, in particular function theory. More specifically, Klein showed that the modular group moves the fundamental region of the complex plane so as to tessellate that plane; he considered at the action of $PSL(2,7)$, thought of as an image of the modular group, and obtained an explicit representation of a Riemann surface today called the Klein quartic; Klein also considered equations of degree > 4, and solved them using the icosahedral group. This work led him to write a series of papers on elliptic modular functions.

Indeed, in his 1884 book on the icosahedron, Klein set out a theory of automorphic functions, connecting algebra and geometry. However Poincaré published an outline of his theory of automorphic functions in 1881, which led to a friendly rivalry between the two men. Both sought to state and prove a uniformization theorem for Riemann surfaces that would serve as a capstone to the emerging theory. Klein succeeded in formulating such a theorem and in sketching a strategy for proving it. But his health collapsed while doing this work and he never recovered from this in terms of original research.

Klein summarized his work on automorphic and elliptic modular functions in the following four volume book, written with Robert Fricke over a period of about 20 years."

R. Fricke, F. Klein (1890), *Vorlesungen über die Theorie der elliptischen Modulfunctionen* (Volume 1), B. G. Teubner, Leipzig.

R. Fricke, F. Klein (1892), *Vorlesungen über die Theorie der elliptischen Modulfunctionen* (Volume 2), B. G. Teubner, Leipzig.

R. Fricke, F. Klein (1897) (in German), *Vorlesungen über die Theorie der automorphen Functionen, Erster Band; Die gruppentheoretischen Grundlagen.*, Leipzig: B. G. Teubner.

R. Fricke, F. Klein (1912) (in German), *Vorlesungen über die Theorie der automorphen Functionen. Zweiter Band: Die funktionentheoretischen Ausführungen und die Anwendungen. 1. Lieferung: Engere Theorie der automorphen Funktionen*, Leipzig: B. G. Teubner.

The book on the icosahedron is the following.

F. Klein, *Lectures on the Icosahedron and the Solution of Equations of the Fifth Degree*, Translated into English by George Gavin Morrice, Second and revised edition, Dover Publications, Inc., New York, N.Y., 1956. xvi+289 pp.

This is a classic and is still widely read by people, unlike Klein's other more advanced books.

From the book description, "In this classic of mathematical literature, first published in 1884, Felix Klein elegantly demonstrates how the rotation of icosahedron can be used to solve complex quintic equations... The Icosahedron will be valued by experts in higher mathematics and students of algebra alike..."

J. Lehner, *Discontinuous Groups and Automorphic Functions*, Mathematical Surveys, No. VIII American Mathematical Society, Providence, R.I., 1964. xi+425 pp.

This is a classical introduction to automorphic forms in the classical style.

According to MathSciNet, "In view of the importance of the subject and the amount of work published in the last thirty years on automorphic functions and their groups, it is surprising that no comprehensive treatise in any language has appeared since the books by L. R. Ford [Automorphic functions, 1929] and P. Fatou [Fonctions automorphes (in Théorie des fonctions algébriques d'une variable, Tome II, 1930. Despite the excellence of these works and of their eminent precursor, namely, the treatise of R. Fricke and F. Klein [Vorlesungen ber die Theorie der automorphen Funktionen, Band 1, 1897; 1912], there has been for many years a great need for a book dealing with the subject in a modern way, using modern algebraic and topological concepts, particularly where Riemann surfaces and vector spaces are involved. In addition, the inclusion in book form of a concise and fairly comprehensive account of the important new ideas and techniques introduced in the last three decades by H. Petersson and others has for long been overdue. This book fills these gaps in an excellent manner and will be indispensable to the student of the subject and provide the basis from which he can begin his study of the extensive literature. It is also valuable since it provides proofs of many 'known' facts which cannot, or only with

7.6. MODULAR FORMS AND AUTOMORPHIC REPRESENTATIONS

difficulty, be found in the published literature. Although the scope of the book is wide, it naturally reflects the author's own special interests and there are several areas that are scarcely touched; this is particularly the case with regard to some parts of elliptic modular function theory.

In his first chapter the author gives a general historical account of the whole subject from its origin as an offshoot of elliptic function theory to the present day. The motivational advantage of such an introduction is considerable, and it is a feature that might well be copied more generally..."

7.6.2 I.M. Gelfand, M. Graev, I.I. Pyatetskii-Shapiro, Representation Theory and Automorphic Functions, W. B. Saunders Co., Philadelphia, Pa.-London-Toronto, Ont. 1969. xvi+426 pp.

This is a classical book and has been influential. For many people, this is the first book to see adeles in action. It contains a wealth of information about discrete subgroups and automorphic representations. It is still a very good introduction and a valuable reference.

From the Note at the beginning of the book (book description), "The theory of group representations has given us a new understanding of classical results in the theory of automorphic functions and has made it possible to attack the problems of this theory on a wider scale and obtain a number of new and profound results. The language of the theory of adeles—a recently developed branch of mathematics—plays an important role. The book contains many new ideas and results that have so far been accessible only in various circles of readers interested in contemporary mathematics. It may be recommended to students in advanced courses, to Ph.D. candidates and to research workers in pure mathematics."

The preface explains the history and relation of automorphic forms and representation theory well: "The classical theory of automorphic functions, created by Klein and Poincaré, was concerned with the study of analytic functions in the unit circle that are invariant under a discrete group of transformations. Since the unit circle can be regarded as a Lobachevskii plane in the Poincaré model, we may say that the classical theory of automorphic functions dealt with the study of functions analytic on the Lobachevskii plane and invariant under a discrete group of motions of the plane.

In the subsequent development of the theory of automorphic functions the papers of Hecke, Siegel, Selberg, and a number of other investigators played an essential part. In particular, papers by Godement, Maass, Roelcke, Peterson, and Langlands cover one or another aspect of the connection between automorphic functions and the theory of groups. Another very interesting direction in the theory of automorphic functions can be found in the works of Ahlfors and Bers.

The whole development of the theory of automorphic functions pointed forcely to the necessity of group-theoretical approach. Recently many of the ideas of the theory have been linked with arbitrary Lie groups and their discrete subgroups.

The connection between the theory of group representations and the theory of automorphic functions was made particularly precise in the last ten or twenty years, in the context of development of the theory of infinite-dimensional representations of groups. Although this connection was perceived much earlier (for example, in papers of Klein and Hecke), a true understanding was achieved only after the construction of the theory of infinite-dimensional representations of Lie groups.

One of the first papers to establish this relationship was by Gelfand and Fomin, in which the concepts of representation theory were linked with the theory of dynamical systems and the theory of automorphic finctions. The connection of automorphic functions with dynamical systems already occurs, in essence, in earlier papers of Hopf on dynamical systems.

Apart from the theory of infinite-dimensional representations of Lie groups, which had received a strong impetus in the last twenty years (in papers of Gelfand and Naimark, Harish-Chandra, Gelfand and Graev, and others), an important part in the construction of the modern theory of automorphic functions was the creation of the theory of algebraic groups by Chevalley, Borel, Harish-Chandra, Tits, and others.

Perhaps one of the most remarkable ideas that have arisen in the recent years is that of the group of adeles. In the process of writing this book the authors have convinced themselves how natural many concepts become when they are applied to the group of adeles and its discrete subgroup of principal adeles..."

7.6.3 H. Jacquet, R. Langlands, Automorphic Forms on $GL_(2)$, Lecture Notes in Mathematics, Vol. 114, Springer-Verlag, Berlin-New York, 1970. vii+548 pp.

This is an important research monograph and has been an essential reference. The famous Jacquet–Langlands correspondence in automorphic forms was introduced in Chapter 16 of this book and is a correspondence between automorphic forms on GL_2 and its twisted forms, proved using the Selberg trace formula. It was one of the first examples of the Langlands philosophy which states that maps between L-groups should induce maps between automorphic representations.

According to MathSciNet, "This important work is most easily understood in terms of the classical theory of automorphic forms which it so dramatically generalizes and revitalizes. ... The main contribution of the present work is that it not only generalizes these classical results but also reinterprets them from the point of view of group representation theory. In their work L-functions are defined over arbitrary global fields (not just

$\mathbb{Q}$) and the relevant group is the general linear group ($GL(2)$—not $SL(2)$)... As already implied, numerous works have appeared since its publication concerning applications, extensions, and refinements of its basic results. Because much still remains to be done in more general settings, and because this book's novel point of view is already amply justified, the authors' pioneering work is must reading for anyone interested in the theory of automorphic forms."

S. Gelbart, *Automorphic Forms on Adéle Groups*, Annals of Mathematics Studies, No. 83, Princeton University Press, 1975. x+267 pp.

This book can serve as an introduction to the book of Jacquet and Langlands. To many people, this is also a good first introduction to automorphic representations discussed in adelic languages.

According to MathSciNet, "The theory of automorphic forms on the upper half plane is classical. With the contributions of E. Hecke, the theory became one of the richest branches of mathematics. A basic theme in the work of Hecke is that deep arithmetic information is somehow wrapped up in natural analytic objects. It was later realized that these analytic objects could be profitably studied through another very rich area, the theory of group representations, developed by Harish-Chandra, Gelfand, Langlands and others. However, the connections between these seemingly unrelated subjects have not always been made explicit in the literature. The present book is an excellent introduction to the interplay between the two theories. Particular emphasis is given to the representation theory of GL_2."

7.6.4 More books on modular forms and automorphic forms

T. Miyake, *Modular Forms*, Translated from the Japanese by Yoshitaka Maeda, Springer-Verlag, Berlin, 1989. x+335 pp.

The book gives an efficient introduction to this vast subject of modular forms, leading the reader quickly to independent research. It is also a valuable reference.

According to MathSciNet, "The author offers a complete collection of definitions, formulas, and proofs as required for modular forms in one variable, particularly for Hecke operators (where the trace formulas are painstakingly derived). This book serves as a valuable source and handy secondary reference for results. Indeed, the methods and terminology are of a (minimal) handbook style, without excessive abstraction."

S. Lang, *Introduction to Modular Forms*, With appendixes by D. Zagier and Walter Feit, Corrected reprint of the 1976 original, Grundlehren der Mathematischen Wissenschaften, 222, Springer-Verlag, Berlin, 1995. x+261 pp.

This is a good introduction to modular forms with different emphasis from other books.

According to MathSciNet, "This book gives a thorough introduction to several theories that are fundamental to recent research on modular forms. Most of the material, despite its importance, had previously been unavailable in textbook form. Complete and readable proofs are given. Especially valuable and clearly written are the treatments of the Eichler-Shimura isomorphism on $SL_2(\mathbb{Z})$, the Atkin-Lehner theory of new forms, and the relation between congruences and Galois representations."

H. Iwaniec, *Topics in Classical Automorphic Forms*, Graduate Studies in Mathematics, 17, American Mathematical Society, Providence, RI, 1997. xii+259 pp.

This is a very valuable reference for experts and can also serve as an advanced introduction to some important topics of the vast classical automorphic forms.

According to MathSciNet, "The present book by the author is meant for advanced researchers in this field... This book differs from other texts in many aspects. For example, one can see from the contents that the author discusses many important topics in the theory of automorphic forms which are rarely seen in the textbooks available on the subject. Another aspect is the presentation of the proofs, which is also unusual, and this may give the reader a different flavor of the subject... Graduate students will certainly benefit from this book."

P. Lax, R. Phillips, *Scattering Theory for Automorphic Functions*, Annals of Mathematics Studies, No. 87, Princeton Univ. Press, Princeton, N.J., 1976. x+300 pp.

This is a self-contained introduction to automorphic forms from the point of view of scattering theory, on which both authors are experts. This will supplement other more standard books on automorphic forms.

According to MathSciNet, "The monograph under review is dedicated to a detailed and consistent account of the application of the Lax-Phillips scattering theory to the problem of the spectral decomposition of the operator L in $L_2(\Gamma\backslash H)$. The book is written by the two prominent specialists in functional analysis and presents a new application of the Lax-Phillips theory..."

7.6.5 R. Langlands, On the Functional Equations Satisfied by Eisenstein Series, Lecture Notes in Mathematics, Vol. 544, Springer-Verlag, Berlin-New York, 1976. v+337 pp.

This is truly a fundamental book and has had a huge influence on mathematics. For example, the celebrated Langlands program comes from the work in this book. It gives a fairly complete description of the spectral decomposition of locally symmetric space

7.6. MODULAR FORMS AND AUTOMORPHIC REPRESENTATIONS

of finite volume. (Note that there is almost no result on the spectral theory of general noncompact Riemannian manifolds of finite volume.) The discussion on structures of lattices of semisimple Lie groups also allows one to appreciate the famous work of Margulis on arithmeticity of lattices of semisimple Lie groups. But this is not an easy book at all, especially Chapter 7, even for experts.

This is reflected by the subtitle of the following book.

C. Moeglin, J.L. Waldspurger, *Spectral Decomposition and Eisenstein Series. Une paraphrase de l'Écriture [A paraphrase of Scripture]*, Cambridge Tracts in Mathematics, 113, Cambridge University Press, Cambridge, 1995. xxviii+338 pp.

This is a useful comprehensive reference on a central subject in modern mathematics, but it is not an easy book. The intertwined treatment of both number fields and function fields makes the reading even more difficult.

According to MathSciNet, "This is essentially a reworking of R. P. Langlands' classic book, with some interesting digressions and additions. It is motivated by a desire to understand more fully the construction of square integrable automorphic forms as residues of Eisenstein series. In places the techniques have been modified to reflect more recent work, but the overall outline and flavour of the original are clearly recognizable."

Another fundamental book written by Langlands is the next.

R. Langlands, *Base Change for* $GL(2)$, Annals of Mathematics Studies, 96. Princeton University Press, Princeton, N.J.; University of Tokyo Press, Tokyo, 1980. vii+237 pp.

This is also a very important book by Langlands. It is a valuable reference on a more specialized but essential topic.

According to MathSciNet, "This book is one of the most important works to appear in the area of automorphic forms and representation theory since the publication ten years ago of H. Jacquet and the author's *Automorphic forms on* $GL(2)$. Its results are of great interest, the trace formula methods used to prove them are exemplary, and the applications to Artin's L-functions are nothing short of spectacular..."

7.6.6 A. Borel, W. Casselman, Automorphic Forms, Representations and L-functions. Part 1, Part 2, Proceedings of Symposia in Pure Mathematics, XXXIII. American Mathematical Society, Providence, R.I., 1979. x+322 pp., vii+382 pp.

This is one of the best conference proceedings that has been published and has been a standard reference on the Langlands program and automorphic representations. It has been very influential.

From the foreword of the books: "A main goal of the Institute was the discussion of the L-functions attached to automorphic forms on, or automorphic representations of, reductive groups, the local and global problems pertaining to them, and of their relations with the L-functions of algebraic number theory and algebraic geometry, such as Artin L-functions and Hasse-Weil zeta functions. Besides seminars and lectures on recent and current work and open problems, the Institute also featured lectures (and even series of lectures) of a more introductory character, including background material on reductive groups, their representations, number theory, as well as an extensive treatment of some relatively simple cases."

7.6.7 More books on modular forms, automorphic representations and cohomology of arithmetic groups

A. Borel, N. Wallach, *Continuous Cohomology, Discrete Subgroups, and Representations of Reductive Groups*, Second edition, Mathematical Surveys and Monographs, 67, American Mathematical Society, Providence, RI, 2000. xviii+260 pp.

This is the first comprehensive book on cohomology of arithmetic subgroups of reductive Lie groups and its connection with automorphic representations of reductive Lie groups. It is a very valuable reference.

According to MathSciNet, "Let G be a semisimple Lie group, $K \subset G$ the maximal compact subgroup, and $\Gamma \subset G$ a discrete subgroup. The double coset space $K\backslash G/\Gamma$ is known as a locally symmetric space. Among such spaces are the Riemann surfaces of genus greater than one, and all complete Riemannian manifolds of constant negative curvature. This book, the product of a seminar given by the authors at the Institute for Advanced Study in 1976–77, is an account of practically all that was known at that time concerning the problem of computing the cohomology of $K\backslash G/\Gamma$, including some important results proved in the course of the seminar... As a thorough and careful presentation of basic machinery and important results in this interesting area of research, this book will be a valuable reference."

D. Bump, *Automorphic Forms and Representations*, Cambridge Studies in Advanced Mathematics, 55, Cambridge University Press, Cambridge, 1997. xiv+574 pp.

Though the subject of automorphic representations is a crucial one in modern mathematics, there are not many books about them. This is one of the successful ones.

From the book description, "This book covers both the classical and representation theoretic views of automorphic forms in a style that is accessible to graduate students entering the field. The treatment is based on complete proofs, which reveal the uniqueness principles underlying the basic constructions. The book features extensive foundational

7.6. MODULAR FORMS AND AUTOMORPHIC REPRESENTATIONS

material... Researchers as well as students in algebra and number theory will find this a valuable guide to a notoriously difficult subject."

According to MathSciNet, "This outstanding book covers, from a variety of points of view, the theory of automorphic forms on GL(2). Along the way the reader will encounter a host of information and techniques which go beyond GL(2). It comprises a solid foundation for the serious student or the researcher interested in automorphic forms, and contains a wealth of information which is not readily accessible in textbook form elsewhere. The book also features numerous and well-thought-out exercises, at a variety of levels...."

M. Eichler, D. Zagier, *The Theory of Jacobi Forms*, Progress in Mathematics, 55, Birkhäuser Boston, Inc., Boston, MA, 1985. v+148 pp.

This is the only book devoted to a systematic account of Jacobi forms. It is a valuable reference on and an introduction to this more specialized topic.

According to MathSciNet, "Jacobi's use of theta series as a foundation for the theory of elliptic functions is among the most fruitful ideas in the history of arithmetic. The authors have therefore chosen the term 'Jacobi form' to denote holomorphic (or just meromorphic) functions $f(\tau, z)$ of two variables, τ in the upper half-plane H and z in the complex plane $\mathbb{C}$, that enjoy the transformation properties of theta series and have Fourier expansions involving only the characters $(\tau, z) \to \exp(2\pi i(n\tau + rz))$ with rational numbers r, n satisfying $r^2 \leq 4nm$ for an integer m called the index (and intervening in the transformation laws just mentioned).

Reviewer's remarks: the material is of such intrinsic beauty and importance that no serious student of arithmetic should miss this book."

7.6.8 Hypergeometric series and theory of partitions

It is well known that a geometric series is a series for which the ratio of two successive terms is equal to a constant or a variable. A hypergeometric series, first used by John Wallis (Arithmetica Infinitorum, 1655), refers to a series for which the ratio of two successive terms is a simple function of the index.

G. Gasper, M. Rahman, *Basic Hypergeometric Series*, With a foreword by Richard Askey, Encyclopedia of Mathematics and its Applications, 35, Cambridge University Press, Cambridge, 1990. xx+287 pp.

As the author explains in the preface, hypergeometric functions are not discussed in standard courses in college or graduate schools. But they turn out to be very useful and important, arising from different subjects such as combinatorics, orthogonal polynomials, number theory and modular forms, Lie algebras etc. This comprehensive book on basic hypergeometric functions or series gives a systematic account of the theory and applications of the basic hypergeometric series and is a valuable reference.

N. Fine, *Basic Hypergeometric Series and Applications*, With a foreword by George E. Andrews, Mathematical Surveys and Monographs, 27, American Mathematical Society, Providence, RI, 1988. xvi+124 pp.

This is another good introduction to basic hypergeometric series. It complements the previous book.

According to MathSciNet, "For far too long, there has been a dearth of good references on basic hypergeometric series. The present book and *Basic hypergeometric series* by G. Gasper and M. Rahman have appeared in the past two years to greatly rectify this situation. It is a measure of the breadth of this field that after the respective first chapters there is virtually no overlap between these books.

Fine writes from the viewpoint of a number theorist, and his slim volume is rich with examples and results from the theory of partitions, the study of Ramanujan's mock theta functions, and modular equations. This is a very personal book, a distillation of those results in basic hypergeometric series which hold the most appeal to its author. Extensive notes and references by George Andrews connect Fine's text to the related research literature."

G. Andrews, *The theory of Partitions*, Encyclopedia of Mathematics and its Applications, Vol. 2, Addison-Wesley Publishing Co., Reading, Mass.-London-Amsterdam, 1976. xiv+255 pp.

A partition of a positive integer n is a way of writing n as a sum of positive integers. Many mathematicians starting with Euler have contributed to and used the theory of partitions, and the generating function of partition functions share some properties with modular forms, for example, they have similar product formulas, but this is the first book entirely devoted to the theory of partitions. It is well-written and is a good introduction to this subject. It is also a valuable reference due to its wealth of information.

From the book description, "This book considers the many theoretical aspects of this subject, which have in turn recently found applications to statistical mechanics, computer science and other branches of mathematics. With minimal prerequisites, this book is suitable for students as well as researchers in combinatorics, analysis, and number theory."

VIII. The Michigan Union of the University of Michigan

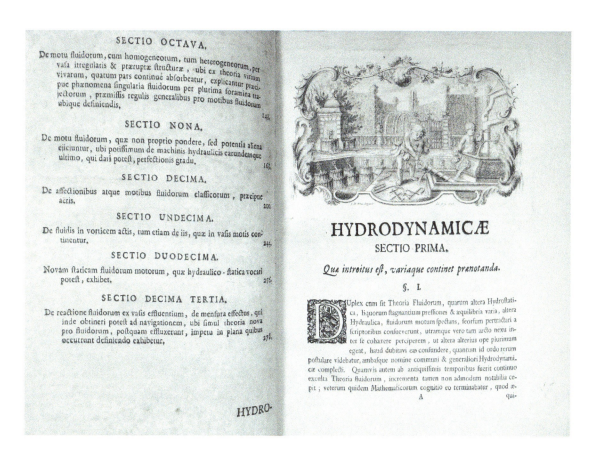

Daniel Bernoulli is one prominent member of the famous Bernoulli family, and was a contemporary and close friend of Leonhard Euler. One of his main contributions is his application of mathematics to mechanics, especially to fluid mechanics. His major work is summarized in his book *Hydrodynamique* (Hydrodynamica), published in 1738. This book imitated Lagrange's *Mecanique Analytique* by making all the results as consequences of a single principle: conservation of energy. He also wrote a memoir on the theory of the tides, which together with memoirs of Euler and Colin Maclaurin, contain all that was done on this subject between the publication of Newton's classic and the work of Pierre-Simon Laplace. He also wrote a large number of papers on various mechanical questions, especially on problems connected with vibrating strings.

Liouville is special among mathematicians in the sense that besides his contributions to mathematics, he also founded a high quality journal *Journal de Mathematiques Pures et Appliquees*. To most students, his best known theorem is called Liouville's theorem which states that bounded holomorphic function on the complex plane is constant. He was also the first to prove the existence of transcendental numbers by constructing Liouville numbers. His work on differential equations includes the Sturm-Liouville theory.

ŒUVRES DE JACQUES HADAMARD

Tome I

Editions du Centre National de la Recherche Scientifique
15, quai Anatole-France — Paris-VII
1968

Hadamard was a universal mathematician and made major contributions in various subjects such as number theory, complex function theory, differential geometry and partial differential equations. For example, he and Poussin independently proved the prime number theorem. His many important results in partial differential equations include the concept of a well-posed initial value and boundary value problem. When he visited Tsinghua University, China in the 1930s, he lectured on partial differential equations. He also wrote a popular book *Psychology of Invention in the Mathematical Field*, which is still popular now.

Most people have heard of Navier in the context of Navier-Stokes equations, which describe the motion of fluid substances. These equations arise from applying Newton's second law to fluid motion. Besides their interests as nonlinear differential equations, they are also useful because they can be used to model the weather, ocean currents, water flow in a pipe and air flow around a wing. On the other hand, existence of solutions are very difficult. Though Navier was usually considered as a engineer and physicist, his main contribution lies in Navier-Stokes equations.

Chapter 8

Differential equations

A differential equation is an equation for an unknown function of one or several variables that relates the values of the function itself and its derivatives of various orders. If there is only one variable, it is called an ordinary differential equation. Otherwise, it is called a partial differential equation. There could be more than one unknown functions, and also more than one equations, which give rise to systems of differential equations.

Differential equations do not have a long history and can be considered as a subject in analysis. Besides their own interests, differential equations have played a fundamental role in sciences such as physics and mechanics, engineering, economics, biology and many other disciplines. The basic reason is that things are changing, and laws governing them are often given by differential equations.

There are many different types of differential equations due to extensive application of differential equations. An important characteristic of a differential equation concerns whether it is linear or not. In this chapter, we classify books on differential equations broadly into three types: (1) ordinary differential equations and related dynamic systems, (2) linear partial differential equations, (3) nonlinear differential equations.

8.1 Ordinary differential equations

The theory of ordinary differential equations is a basic tool in mathematics, and there are many books on ordinary differential equations, especially at a more elementary levels. Much work has been done to find explicit solutions to ordinary differential equations. When the differential equation is linear, it can be solved by analytical methods in terms of elementary functions and integration. Unfortunately, most of the interesting ordinary differential equations are nonlinear and, with a few exceptions, cannot be solved exactly. Many important special functions arise from solutions of particular differential equations.

Instead of finding explicit solutions, another approach is to study qualitative behaviors of solutions of ordinary differential equations, which are complicated and interesting and related to dynamical systems.

The theory of ordinary differential equations is closely related to motions of bodies, especially with celestial bodies. Two prominent people with serious interests in such problems were Kelper and Galileo. Their work motivated Newton to develop calculus. Galileo's experiments on motion had also been very influential. Because of these works, Galileo is called the "father of modern observational astronomy", the "father of modern physics", and the "the father of modern science".

Figure 8.1: Galieo's book.

8.1.1 E. Coddington, N. Levinson, Theory of Ordinary Differential Equations, McGraw-Hill Book Company, Inc., New York-Toronto-London, 1955. xii+429 pp.

This is a comprehensive book on ordinary differential equations. Most textbooks on ordinary differential equations contain much less material than this one does. This is a very valuable reference.

According to MathSciNet, "This excellent book contains many topics which at present are of interest, such as asymptotic behavior of linear and nonlinear systems, boundary-value problems, both on finite and infinite intervals, stability, questions of perturbations, and the Poincaré-Bendixson theory of two-dimensional autonomous systems. Besides classical theorems there are to be found many results published during the last ten years, particularly results of the authors. In addition, new results of the authors are presented. The presentation, characterized by the exclusive use of vector and matrix notations, is elegant and lucid, even if brief...

The elegant manner of presentation and the choice of material, most of which is not to be found in standard text-books on differential equations, make the reading of the book a pleasure."

8.1.2 More books on ordinary differential equations

L. Cesari, *Asymptotic Behavior and Stability Problems in Ordinary Differential Equations*, Third edition, Ergebnisse der Mathematik und ihrer Grenzgebiete, Band 16, Springer-Verlag, New York-Heidelberg, 1971. ix+271 pp.

This book is a classic. It is a valuable reference on one of the central problems of ordinary differential equations: asymptotic behavior and stability problems.

According to MathSciNet, "According to the author 'the purpose of the present volume is to present many viewpoints and questions in a readable short report...' ; this objective has been accomplished admirably. In fact, this comprehensive work together with the recent books of R. Bellman, *Stability theory of differential equations*, 1953, E. A. Coddington and N. Levinson, *Theory of ordinary differential equations*, 1955, and S. Lefschetz *Differential equations: Geometric theory*, 1957, provides a rather complete picture of the state of the art up to the present time.... he feels that this excellent book should be in the hands of everyone doing research in this and related fields."

P. Hartman, *Ordinary Differential Equations*, Corrected reprint of the second (1982) edition, With a foreword by Peter Bates, Classics in Applied Mathematics, 38, Society for Industrial and Applied Mathematics (SIAM), Philadelphia, PA, 2002. xx+612 pp.

This book was based on lecture notes of courses, but it contains much more material than an one year course. Given its broad coverage, it is a valuable reference on the qualitative theory of ordinary differential equations.

According to MathSciNet, "This is an excellent book on the qualitative theory of ordinary differential equations. It presents a modern treatment of all the standard topics and also covers in depth much recent research that has not been available in book form until now. The author not only proves theorems but emphasizes techniques as well...."

I.G. Petrovski, *Ordinary Differential Equations*, Revised English edition, Translated from the Russian and edited by Richard A. Silverman, Prentice-Hall, Inc., Englewood Cliffs, N.J., 1966. x+232 pp.

This is a classic book by an expert on ordinary differential equations. There had been five Russian editions before an English translation in 1966. This slim book is concise yet comprehensive. It covers many major topics in ordinary differential equations. Besides the standard topics, Chapter 7 studies autonomous systems of differential equations, in particular asymptotics and the existence of limit cycles. This is not covered in usual textbooks. We note that the second part of Hilbert's XVI problem is concerned with limit cycles of autonomous systems with polynomial vector fields. This is a book for serious study of ordinary differential equations.

E.L. Ince, *Ordinary Differential Equations*, Dover Publications, New York, 1944. viii+558 pp.

This is a very valuable reference on ordinary differential equations. When it was written, it covered both classical results and the most recent developments. The first part studies differential equations in the real domain, and the second part studies differential equations in the complex domain. Topics include transformation group (Lie group) methods of solution, boundary eigenvalue problems, nature and methods of solution of regular, singular and nonlinear equations in the complex plane.

According to an Amazon review, "This classic combines readability with a vast wealth accurately presented material (much of which can still only be found in research papers and certainly can nowhere else be found in a single reference). Most aspects of theory are illustrated by examples... This is an essential reference for anyone working with ordinary differential equations."

8.1.3 V.I. Arnold, Ordinary Differential Equations, Translated from the third Russian edition by Roger Cooke, Springer Textbook, Springer-Verlag, Berlin, 1992. 334 pp.

This is an unusual book on ordinary differential equations. Though the topics it contains are elementary, it emphasizes geometric and qualitative aspects, and applications to mechanics and relations to dynamics. It is an elementary book written by a master and will supplement well other books on ordinary differential equations.

According to the MathSciNet review of the first Russian edition, "it is a most unusual book and, though most of the standard topics are covered, a great deal of routine and computational material is omitted and there is heavy emphasis on geometric qualitative ideas, on the one hand, and on applications to mechanics (both for their own sake and for a motivating source of ideas), on the other. Although the formal prerequisites are slight, some parts of the book (such as the last chapter) do require considerable mathematical maturity."

The book description says "Few books on Ordinary Differential Equations (ODEs) have the elegant geometric insight of this one, which puts emphasis on the qualitative and geometric properties of ODEs and their solutions, rather than on routine presentation of algorithms."

V.I. Arnold, *Geometrical Methods in the Theory of Ordinary Differential Equations*, Translated from the Russian by Joseph Szücs, Translation edited by Mark Levi, Second edition, Grundlehren der Mathematischen Wissenschaften, 250, Springer-Verlag, New York, 1988. xiv+351 pp.

This is really an advanced introduction to dynamical systems. It maybe not be easy as a textbook, but is a valuable reference.

According to MathSciNet, "This is a very attractive book on differentiable dynamical systems on compact manifolds as well as vector fields around their critical points. However, contrary to the author's previous book *Ordinary Differential Equations*, this book contains rather advanced material and consequently seems to be less easy to read..."

8.1.4 More books on ordinary differential equations and dynamical systems

M. Hirsch, S. Smale, *Differential Equations, Dynamical Systems, and Linear Algebra*, Pure and Applied Mathematics, Vol. 60, Academic Press, New York-London, 1974. xi+358 pp.

This is an attractive introduction to qualitative behaviors of ordinary differential equations. It is elementary, and many applications make this a very informative and interesting book. It is an excellent introduction for students and non-experts.

From the preface, "This book is about dynamical aspects of ordinary differential equations and the relations between dynamical systems and certain fields outside pure mathematics. A prominent role is played by the structure of linear operators on finite-dimensional spaces."

According to an Amazon review, "This book (the original version) has all the basics to introduce the future differential equations/dynamical systems researchers into the field. Written by authorities in the field (Hirsch and Smale,) this text offers a wide variety of topics, including linear systems, local and global stability theory for non-linear systems, and applications to physics and biology. As an added treat, the inclusion of basic linear algebra and operator theory makes this a rather self-contained work. The dedicated reader will not be disappointed - the material is well organized with sufficient level of detail, illustration, and exercises."

S.N. Chow, J. Hale, *Methods of Bifurcation Theory*, Grundlehren der Mathematischen Wissenschaften, 251, Springer-Verlag, New York-Berlin, 1982. xv+515 pp.

This book describes the bifurcation phenomena in ordinary and partial differential equations, together with related topics in nonlinear functional analysis and qualitative theory of differential equations. It is a valuable reference and can also serve as an advanced introduction.

According to a review in Bulletin of AMS, "When a system of equations—algebraic, differential, functional,... —depends on a parameter, it frequently happens that there are certain values of the parameter with the property that small variations in the parameter lead to significant changes in the qualitative behavior of the solutions of the corresponding system of equations. Loosely speaking, such values of the parameter are called bifurcation values and the general goal of bifurcation theory is to identify such points, to be as informative as possible about the nature of the solutions of the system of equations near such points and also to study the relations of such bifurcations with the local and global structure of the set of solutions. The importance of such a program stems from the ubiquity of examples of bifurcation occurring in systems of equations that arise naturally. Indeed a case can be made that a student's first systematic contact with nonlinear analysis should be bifurcation theory; that bifurcation is the central phenomenon of nonlinear behavior. It is safe to say that the number of interesting nonlinear equations for which one can find explicit solutions is exceedingly small. On the other hand advances in computational capability have significantly broadened our understanding of nonlinear systems of equations, revealing the astounding complexity of even the most innocent-looking systems. The goal is to obtain a coherent body of results which will lead to an understanding of the onset and progress of this complex behavior... The book, which was published in 1982, has become one of the standard references in the research literature on the subject. Since bifurcation theory is a field which is rapidly developing and also one which has many contact points with diverse areas of mathematics and applied science, it is no small task to present a treatment which is at once broad and coherent."

C. Conley, *Isolated Invariant Sets and the Morse Index*. CBMS Regional Conference Series in Mathematics, 38, American Mathematical Society, Providence, R.I., 1978. iii+89 pp.

This book is a valuable reference on the qualitative theory of ordinary differential equations from the topological viewpoint.

According to MathSciNet, "Every once in a while in mathematics, a strikingly original work appears which radically changes a particular field. The reviewer believes that the book under review fits this description, the area being the qualitative theory of nonlinear differential equations.

This book is basically a compilation and extension, with completely 'streamlined' proofs, of the researches of the author and his school over the past few years. By reading the book, one is shown how ideas of algebraic topology are introduced into the field of differential equations in a new and exciting way."

8.2 Linear differential operators

When students learn linear differential equations, they are usually taught three basic examples: the Laplace equation, the heat equation, and the wave equation. In some sense, the wave equation is the most difficult and powerful. One basic principle in the theory of waves is the Huygens principle on wave propagation, which was proposed by Huygens in 1678 and used by him to show that light consists of waves. He also contributed substantially to probability, astronomy and other subjects in physics.

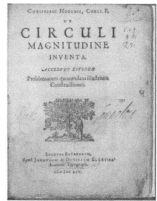

Figure 8.2: Huygens' book.

Linear differential operators have been studied extensively for a long time. Probably the three best known linear differential operators are the Laplace operator, the heat operator, and the wave operator on $\mathbb{R}^n$. These are special cases of linear differential operators with constant coefficients. The above three special differential operators can be generalized to smooth manifolds.

There are many good books on linear differential operators and linear differential equations. It is not easy to classify them and pick out the best ones.

8.2.1 L. Evans, Partial Differential Equations, Graduate Studies in Mathematics, 19, Second edition, Graduate Studies in Mathematics, 19, American Mathematical Society, Providence, RI, 2010. xxii+749 pp.

This is a modern classic and has become a standard textbook for graduate courses on partial differential equations. It is also a valuable reference.

According to MathSciNet, "After its appearance in 1998, the first edition of the book became an instant classic and the new standard for textbooks on PDEs. The innovative choice of contents and the mastery in their exposition by one of the outstanding experts in the area made it so. The first edition of the book has been broadly adopted by instructors in countless Math departments across the U.S. and also overseas."

According to a review in Bulletin of AMS, "Evans' tightly written book has excellent balance between linear and nonlinear equations. It has careful proofs which are sometimes real improvements on those available elsewhere... It shares with Courant's classic the properties that one can dive in at almost any topic, and that topics are often treated once and then treated later in more depth. The fact that the book can be read locally permits its use as a text even though there is way more than one could imagine covering in any one-year course. The same quality makes the book a valuable reference. It is one of the most consulted volumes on my shelves. It introduces you to topics and refers you to specialized works for the last word."

8.2.2 L. Hörmander, Linear Partial Differential Operators, Springer Verlag, Berlin-New York, 1976. vii+285 pp.

Hörmander has written many influential books on differential equations and related topics. This book *Linear Partial Differential Operators* is short and is more accessible than the more recent four volume set. It is a classic. It can serve as an advanced textbook and also a useful reference.

According to MathSciNet, "This book contains some of the recent developments in the theory of linear partial differential equations, in particular, those topics close to the interests of the author... The book will be invaluable to many. The researcher in the field will find it a ready reference. For others it will serve as a readable account of current research. It certainly does not cover all major trends in partial differential equations, but comes as close as any book can. On the other hand, it can hardly be recommended for the novice. The introduction to each topic is good, but little is done in the way of motivation. The exposition is essentially of the form: definition, lemma, theorem. References for the methods employed in the book are very good. However, it is difficult to trace the origin of a theorem. In content and exposition the author has done an excellent job. The book cannot be praised too highly."

L. Hörmander, *The Analysis of Linear Partial Differential Operators. I. Distribution Theory and Fourier Analysis. II. Differential Operators with Constant Coefficients, III. Pseudodifferential Operators. IV. Fourier Integral Operators,* Springer-Verlag, Berlin, 1983. ix+391 pp., viii+391 pp., 1985. viii+525 pp., vii+352 pp.

This comprehensive four volume set can be considered an expanded and updated version of the previous book. It contains an unusually large amount of information and is a reliable and valuable reference.

The MathSciNet review of Vol I and II of this series says, "Needless to say, these volumes are excellently written and make for greatly profitable reading. For years to come they will surely be a main reference for anyone wishing to study partial differential operators."

The review of Vol III and IV says, "No doubt, in the years to come, this monograph will be a classic in this very important area of mathematics."

8.2. LINEAR DIFFERENTIAL OPERATORS

L. Hörmander, *Lectures on Nonlinear Hyperbolic Differential Equations,* Mathématiques & Applications, 26, Springer-Verlag, Berlin, 1997.

This book is a valuable reference, and the first part can also serve as a good introduction to nonlinear hyperbolic problems.

It describes many recent aspects in the study of nonlinear hyperbolic differential equations by concentrating on three themes: (1) theorems of existence and uniqueness of the Cauchy problem for nonlinear first-order hyperbolic systems, (2) the global existence of the solutions for nonlinear perturbations of the wave equation and Klein-Gordon equation, (3) the paradifferential calculus, with application to propagation of singularities for fully nonlinear equations. The first part of the volume will be very useful for a beginner on nonlinear hyperbolic problems, whereas the subsequent two parts are mainly addressed to experts in the field.

L. Hörmander, *Notions of Convexity,* Progress in Mathematics, 127, Birkhäuser Boston, Inc., Boston, MA, 1994. viii+414 pp.

This book shows how various notions and results from complex analysis (one and several variables) and partial differential equations follow from the general notion of convexity. It also discusses some examples from geometry, game theory, and special functions. This comprehensive book is very rich in content and is a very valuable reference.

According to MathSciNet, "The notion of convexity is an old one. It was used by Archimedes in his axiomatic treatment of arc length. Later, it was used by Fermat, Cauchy, Minkowski, and others as a tool ancillary to other studies. Not until the 1930s did Bonneson and Fenchel formalize the notion of convexity and do a systematic study in a monograph.

Nowadays, convexity is an essential tool in analysis and geometry. Nonetheless, most treatments of the subject are specialized. They are usually directed to functional analysis considerations, minimax problems, or to classical geometry. There are few, if any, systematic treatments of the hard analysis aspects of convexity."

According to a review in Bulletin of AMS, "I hope students in complex analysis read this book to find that the theory of partial differential equations is closely related to the theory of analytic functions of several complex variables, and still more, I wish they would try to attack this adjacent field. *An introduction to complex analysis in several variables* [CASV] is an excellent book also for this purpose, but this new book is probably better suited for this purpose, as it contains much more material which is peculiar to the theory of partial differential equations. As one of the old students of [CASV], I believe the new generation will benefit much from this new book by Hörmander."

8.2.3 More books on linear differential equations

V. Guillemin, S. Sternberg, *Geometric Asymptotics*, Mathematical Surveys, No. 14, American Mathematical Society, Providence, R.I., 1977. xviii+474 pp.

This book gives a comprehensive account of symplectic geometry, Fourier integral operators, and applications to asymptotics of solutions of partial differential equations. It can serve as an excellent reference and an advanced introduction.

According to MathSciNet, "The topic of this nice book can be defined as a geometric approach to the investigation of some analytic problems, especially to the study of Fourier integral operators. These operators are now widely used for the analysis of singularities of solutions of linear partial differential equations and for the study of the spectra of the corresponding operators. The natural language for the theory of Fourier integral operators is that of differential geometry. Application of the method of stationary phase allows one to obtain asymptotic expansions for these operators and these are very important for mathematical physics and in particular for the wave theory of light... In general the book is very interesting and useful for specialists both in analysis and in differential geometry. The theory developed here seems to be very perspicuous, and the possibilities for its application are far from being exhausted."

V.P. Maslov, M.V. Fedoriuk, *Semiclassical Approximation in Quantum Mechanics*, Translated from the Russian by J. Niederle and J. Tolar, Mathematical Physics and Applied Mathematics, 7, Contemporary Mathematics, 5, D. Reidel Publishing Co., Dordrecht-Boston, Mass., 1981. ix+301 pp.

This is a valuable reference and important book on asymptotic solutions of linear partial differential equations.

According to MathSciNet, "The first author's 1965 book *Perturbation Theory and Asymptotic Methods*, 1965, has been a major source of ideas for recent research on asymptotic solutions of linear partial differential equations. Unfortunately, that book is notoriously difficult to read, and so Maslov's own presentation has been inaccessible to all but the most determined student. The book under review should remedy this situation. It is a 'textbook' presentation of the basic ideas contained in the 1965 book...

The presence of many worked out examples, as well as the very detailed exposition, make this book an excellent introduction (or companion) to V. Guillemin and S. Sternberg's book *Geometric Asymptotics*, 1977, in which global geometric aspects of the subject merely hinted at by Maslov and Fedoriuk are developed and applied to a wide range of areas in contemporary mathematics."

F. John, *Partial Differential Equations*, Fourth edition, Applied Mathematical Sciences, 1, Springer-Verlag, New York, 1982. x+249 pp.

This is a classic and is still an excellent introduction to partial differential equations.

8.2. LINEAR DIFFERENTIAL OPERATORS

From the book description, "the author identifies the significant aspects of the theory and explores them with a limited amount of machinery from mathematical analysis."

According to MathSciNet, "This introductory treatment of partial differential equations will be well suited to students having had a course in ordinary differential equations. Particular stress is laid on the geometrical interpretation, which is important for a later understanding of a more abstract formulation of these topics. Examples from natural phenomena make the book valuable to physicists and other users of mathematics."

D. Kinderlehrer, G. Stampacchia, *An Introduction to Variational Inequalities and Their Applications*, Reprint of the 1980 original. Classics in Applied Mathematics, 31. Society for Industrial and Applied Mathematics (SIAM), Philadelphia, PA, 2000. xx+313 pp.

As the title says, this book is a good introduction to variational inequalities and their applications, and it has played an important role in the further development of the subject. Variational inequalities provide an effective and powerful tool to study very complicated and complex problems in pure and applied sciences, and they also provide us with a natural and unified technique for developing efficient numerical techniques for solving equilibrium problems.

From the book description, it "is a resource for many important topics in elliptic equations and systems and is the first modern treatment of free boundary problems. Variational inequalities (equilibrium or evolution problems typically with convex constraints) are carefully explained. They are shown to be extremely useful across a wide variety of subjects, ranging from linear programming to free boundary problems in partial differential equations. Exciting new areas like finance and phase transformations along with more historical ones like contact problems have begun to rely on variational inequalities, making this book a necessity once again."

A. Bensoussan, J. Lions, G. Papanicolaou, *Asymptotic Analysis for Periodic Structures*, Corrected reprint of the 1978 original, AMS Chelsea Publishing, Providence, RI, 2011. xii+398 pp.

This is a valuable reference on boundary value problems for partial differential equations with periodic coefficients which depend on parameters. These equations appear naturally in material sciences, physics, and chemistry.

According to MathSciNet, "In the study of composite materials with a periodic structure (the period being small with respect to the size of the region in which the system is to be studied) an important problem consists in going back from the microscopic quantities to the macroscopic ones which are measurable."

V. Lakshmikantham, D.D. Bainov, P.S. Simeonov, *Theory of Impulsive Differential Equations*, Series in Modern Applied Mathematics, 6, World Scientific Publishing Co., Inc., Teaneck, NJ, 1989. xii+273 pp.

This is a useful reference on the theory of impulsive differential equations, i.e., differential equations involving impulse effects which give accurate description of observed evolution phenomena.

According to MathSciNet, "the book is divided into four chapters. As described in the preface, Chapter 1 introduces the impulsive evolution processes and offers examples for motivation. In Chapter 2 fundamental properties of solutions, variation of parameters formulae, upper and lower solutions, simple stability criteria, are discussed. Chapter 3 is devoted to the investigation of stability by means of discontinuous Lyapunov functions, impulsive differential inequalities and vector Lyapunov functions."

8.2.4 Inverse problems

An inverse problem is concerned with how to convert observed measurements or data into information about a physical object or system that we are interested in. In many cases, the systems are controlled by differential equations.

V. Isakov, *Inverse Problems for Partial Differential Equations*, Applied Mathematical Sciences, 127, Springer-Verlag, New York, 1998. xii+284 pp.

This book is a valuable reference on the current state of the theory and some numerical aspects of inverse problems in particular differential equations.

From the book description, "Applications include recovery of inclusions from anomalies of their gravity fields, reconstruction of the interior of the human body from exterior electrical, ultrasonic, and magnetic measurement. By presenting the data in a readable and informative manner, the book introduces both scientific and engineering researchers as well as graduate students to the significant work done in this area in recent years, relating it to broader themes in mathematical analysis."

According to MathSciNet, "Most mathematical problems in science, technology and medicine are inverse problems. Studying such problems is the only complete way of analyzing experimental results. In order to solve an inverse problem the following points have to be studied: (1) mastery of the special process both experimentally and theoretically; (2) possibility of mathematical modeling of the process; (3) mastery of the direct problem both theoretically and numerically; (4) studying the information content of the inverse problem, i.e., to find out which internal parameters of a system inaccessible to measurement can be determined in a stable and unique manner; (5) development of algorithms for the numerical solution of the inverse problem...

Mathematically, these problems are relatively new and quite challenging due to the lack of conventional stability and to nonlinearity and non-convexity. Currently, there are hundreds of publications containing new and interesting results. A purpose of the book is to collect and present many of these in a readable form..."

8.2. LINEAR DIFFERENTIAL OPERATORS

H. Engl, M. Hanke, A. Neubauer, *Regularization of Inverse Problems*, Mathematics and its Applications, 375, Kluwer Academic Publishers Group, Dordrecht, 1996. viii+321 pp.

From the book description, "This book is devoted to the mathematical theory of regularization methods and gives an account of the currently available results about regularization methods for linear and nonlinear ill-posed problems."

Inverse problems typically lead to ill-posed mathematical models. In particular, their solutions are unstable under data perturbations, and special numerical methods that can cope with these instabilities have to be developed. These are regularization methods.

According to MathSciNet, "The book not only contains recent results for ill-posed and inverse problems, but also surveys basic trends of recent decades... It can be recommended not only to mathematicians interested in this field, but also to students with a basic knowledge of functional analysis, and to scientists and engineers working in this field."

D. Colton, R. Kress, *Inverse Acoustic and Electromagnetic Scattering Theory*, Second edition, Applied Mathematical Sciences, 93, Springer-Verlag, Berlin, 1998. xii+334 pp.

This is a valuable reference on the mathematical basis of inverse scattering theory.

According to MathSciNet, "This monograph presents the state of the art on three-dimensional inverse scattering for acoustic and electromagnetic waves for obstacles and media. The inverse scattering problem is central in the areas such as radar, sonar, geophysical exploration and medical applications of computerized tomography... This volume is recommended as good introduction to non-quantum-mechanical scattering theory."

8.2.5 Critical point theory and minimax methods

Critical points of functions and critical functions of functionals often carry useful information about problems under study. In theories of variation of calculus and differential equations, one effective method to find critical functions is the minimax method.

P. Rabinowitz, *Minimax Methods in Critical Point Theory with Applications to Differential Equations*, CBMS Regional Conference Series in Mathematics, 65, The American Mathematical Society, Providence, RI, 1986. viii+100 pp.

This is a very valuable introduction to and reference on minimax methods in critical point theory with applications to differential equations.

From the book description, "The book provides an introduction to minimax methods in critical point theory and shows their use in existence questions for nonlinear differential equations... this volume is the first monograph devoted solely to these topics... It is addressed to mathematicians interested in differential equations and/or nonlinear functional analysis, particularly critical point theory."

M. Willem, *Minimax Theorems*, Progress in Nonlinear Differential Equations and their Applications, 24, Birkhäuser Boston, Inc., Boston, MA, 1996. x+162 pp.

This is an interesting introduction to minimax type theorems and theorems on minimizing and Palais-Smale sequences.

From the book description, "this text presents these theorems in a simple and unified way, starting from a quantitative deformation lemma. Many applications are given to problems dealing with lack of compactness, especially problems with critical exponents and existence of solitary waves."

J. Mawhin, M. Willem, *Critical Point Theory and Hamiltonian Systems*, Applied Mathematical Sciences, 74, Springer-Verlag, New York, 1989. xiv+277 pp.

This is a valuable reference on recent developments in minimax methods for semilinear differential equations, motivated by the development of a general theory of periodic solutions of Hamiltonian systems.

From the book description, "The last decade has seen a tremendous development in critical point theory in infinite dimensional spaces and its application to nonlinear boundary value problems. In particular, striking results were obtained in the classical problem of periodic solutions of Hamiltonian systems. This book provides a systematic presentation of the most basic tools of critical point theory... Application of those results to the equations of mechanical pendulum, to Josephson systems of solid state physics and to questions from celestial mechanics are given. The aim of the book is to introduce a reader familiar to more classical techniques of ordinary differential equations to the powerful approach of modern critical point theory."

K.C. Chang, *Infinite-dimensional Morse Theory and Multiple Solution Problems*, Progress in Nonlinear Differential Equations and their Applications, 6, Birkhäuser Boston, Inc., Boston, MA, 1993. x+312 pp.

This is a valuable reference on infinite-dimensional Morse theory and the critical point theory as a way of studying multiple solutions of differential equations in the calculus of variations.

According to MathSciNet, "This volume forms a treatise on modern Morse theory understood as a method of studying multiple solutions of ordinary and partial differential equations arising in the calculus of variations... In conclusion, the reviewer can recommend this book highly to experts as well as to beginners."

8.2. LINEAR DIFFERENTIAL OPERATORS

8.2.6 D. Gilbarg, N. Trudinger, Elliptic Partial Differential Equations of Second Order, Reprint of the 1998 edition, Classics in Mathematics, Springer-Verlag, Berlin, 2001. xiv+517 pp.

This is a classic and the standard book on elliptic partial differential equations of the second order. Starting from the classical Laplace equation and concluding with fully nonlinear equations, this book covers many major results of both linear and nonlinear elliptic equations. It is a very good introduction for beginners. Indeed, according to the preface, "This volume is intended as an essentially self-contained exposition of portions of the theory of second order quasilinear elliptic partial differential equations, with emphasis on the Dirichlet problem in bounded domains... By including preparatory chapters on topics such as potential theory and functional analysis, we have attempted to make the work accessible to a broad spectrum of readers. Above all, we hope the readers of this book will gain an appreciation of the multitude of ingenious barehanded techniques that have been developed in the study of elliptic equations and have become part of the repertoire of analysis." The goals set out by the authors are well-achieved.

This book can also serve as a very valuable reference to all users of elliptic partial differential equations. Everyone who is interested in or working on differential equations should have this book. Its popularity is reflected by its citation numbers. Indeed, it has one of the highest citation numbers of all math books.

8.2.7 More books on elliptic differential equations

J. Necas, *Les méthodes directes en théorie des équations elliptiques*, Masson et Cie, Éditeurs, Paris; Academia, Éditeurs, Prague 1967, 351 pp.

This is a classic and has been a standard reference for linear elliptic equations and systems. It gives a self-contained account of the elliptic theory based on the "direct method", i.e., the variational method. Because to its universality and close connections to numerical approximations, the variational method has become one of the most important approaches to the elliptic equations. One important point is that the method does not rely on the maximum principle or other special properties of the scalar second order elliptic equations, and it is better for handling systems of differential equations.

According to MathSciNet, "This is an up-to-date comprehensive treatment of the most important modern methods in the theory of linear elliptic partial differential problems... As a whole, the book is a good and sound introduction into the contemporary methods in the subject at an intermediate level between an advanced textbook and an encyclopedic monograph."

M. Protter, H. Weinberger, *Maximum Principles in Differential Equations*, Prentice-Hall, Inc., Englewood Cliffs, N.J., 1967. x+261 pp.

The **maximum principle** is one of the most basic techniques in differential equations. This book is a classical introduction to this important topic. It was written for undergraduate and graduate students in 1967, but it is still a valuable book for everyone today.

According to MathSciNet, "This book is at once an introduction to some aspects of partial differential equations suitable for an advanced undergraduate, and a book to be studied by a graduate student in search of a topic for his dissertation. At both levels, it is an unqualified success. The book begins with the maximum principle in one dimension. This allows the authors to present simple, clear arguments and to give a number of illustrations of the utility of maximum principles. After this introductory chapter, there are three chapters on maximum principles for elliptic, for parabolic, and for hyperbolic equations... The book is clearly written and pedagogically sound. The exercises are adequate."

S. Agmon, *Lectures on Elliptic Boundary Value Problems*, Prepared for publication by B. Frank Jones, Jr. with the assistance of George W. Batten, Jr. Van Nostrand Mathematical Studies, No. 2, D. Van Nostrand Co., Inc., Princeton, N.J.-Toronto-London, 1965. v+291 pp.

This is a classical introduction to the problem of the existence and regularity of solutions of higher-order elliptic boundary value problems, of spectral properties of operators associated with these problems, and of the asymptotic distribution of their eigenvalues.

According to MathSciNet, "this book gives a very clear and thorough introduction to the theory. Many results of fundamental importance for the subject are here presented in a readily accessible form, often for the first time with detailed proofs and under minimal assumptions..."

P. Grisvard, *Elliptic Problems in Nonsmooth Domains*, Monographs and Studies in Mathematics, 24, Pitman (Advanced Publishing Program), Boston, MA, 1985. xiv+410 pp.

The book gives the first comprehensive account of elliptic boundary value problems in domains with nonsmooth boundaries and problems with mixed boundary conditions. Such domains occur naturally, for example, domains with corners in the plane. It is a classic and is still a very valuable reference.

According to MathSciNet, "The monograph is the first comprehensive treatment of the subject, parts of which were until now available in scattered form. Most of the results are due to the author, but credit is given to other contributors too numerous to list in a review. Many results are new, and there are also new proofs of results published previously. It seems certain that this ground-breaking work will become an indispensable reference and an inspiration for everybody interested in the subject."

8.2. LINEAR DIFFERENTIAL OPERATORS

O. Ladyzhenskaya, N. Uraltseva, *Linear and Quasilinear Elliptic Equations*, Translated from the Russian by Scripta Technica, Inc, Translation editor: Leon Ehrenpreis, Academic Press, New York-London, 1968. xviii+495 pp.

This is a valuable reference for experts on the solvability of boundary value problems of linear and quasilinear equations of second order of elliptic type, and differential dependence of solutions on the boundary value data.

According to MathSciNet, "The book represents an exposition of investigations of the authors on (a) solvability of boundary value problems (primarily the Dirichlet problem in a finite region Ω of an n-dimensional space) for linear and quasilinear equations of second order of elliptic type, and (b) clarification of the dependence of the differential properties of solutions on analogous properties of the given data... The book contains solutions of a series of basic problems the study of which began at the beginning of the century, starting with the works of S. N. Bernstein. These results were obtained during the last few years and are presented in a monograph for the first time."

8.2.8 Pseudodifferential operators

A pseudo-differential operator is an extension of the concept of a differential operator. It is known that the Fourier transformation changes a differential operator with constant coefficients into a multiplication operator by a polynomial. If we localize this and replace a polynomial by certain infinite power series, we roughly obtain pseudo-differential operators. Pseudo-differential operators are used extensively in the theory of partial differential equations, index theorems, and harmonic analysis, etc.

M. Taylor, *Pseudodifferential Operators and Nonlinear PDE*, Progress in Mathematics, 100. Birkhäuser Boston, Inc., Boston, MA, 1991. 213 pp.

The close connection between the theory of pseudodifferential operators and linear differential equations is well known and has been well established. This book is an accessible account of developments in the theory of pseudodifferential operators and recent applications to regularity of nonlinear PDE.

According to MathSciNet, "In this book the author presents an account of the interplay between the theory of pseudodifferential operators and the theory of nonlinear partial differential equations. He uses the theory of paraproducts initiated by J.-M. Bony [Ann. Sci. École Norm. Sup. (4) 14 (1981), no. 2, 209–246] and the smoothing of symbols to give existence, uniqueness and regularity results in a variety of situations.

Taylor also wrote several other books. The next one is the most systematical.

M. Taylor, *Partial Differential Equations. I. Basic Theory. II. Qualitative Studies of Linear Equations. III. Nonlinear Equations*, Texts in Applied Mathematics, 23, Springer-Verlag, New York, 1996. xvi+563 pp., xxii+528 pp., 1997. xxii+608 pp.

These three books give an overview of the vast subject of partial differential equations. They are valuable references for many readers of different backgrounds.

According to MathSciNet, the review of the first volume says, "On the whole, the project of the author is ambitious, much more than how it may appear from the introduction, consisting of a short list of contents of chapters. Let us begin by observing that a replacement for the classical book of Courant and Hilbert seems missing in the contemporary literature, whenever intended as a work covering the general theory of PDE, starting from the basic results and leading up to current relevant contributions. There are of course many books, in fact hundreds of them, addressed to elementary approaches or to special topics in PDE, but a comprehensive treatment gets into evident difficulties, because of the vastness of the field. A possible answer to such difficulties is found in the so-called Encyclopedias of Mathematics, due to the enterprise of some publishers, which cover the PDE area by gluing together the contributions of several authors, with an inevitable mixture of different styles. No doubt, Taylor intends to give his personal courageous answer, by collecting in three volumes several course notes, the fruit of 20 years of teaching in the field."

M. Taylor, *Pseudodifferential Operators*, Princeton Mathematical Series, 34, Princeton University Press, Princeton, N.J., 1981. xi+452 pp.

Though there are several books on pseudodifferential operators, this book is a valuable reference on the theory and many applications.

According to MathSciNet, "The book under review proposes an introduction to the theory of ps.d.o., combined with many interesting applications concerning different fields of current research."

M. Shubin, *Pseudodifferential Operators and Spectral Theory*, Translated from the Russian by Stig I. Andersson, Springer Series in Soviet Mathematics, Second edition, Springer-Verlag, Berlin, 2001. xii+288 pp.

This is an accessible introduction to pseudodifferential operators and spectral theory of elliptic and hypoelliptic operators.

According to MathSciNet, "The aim of the book is to give a systematic treatment of the theory of pseudodifferential operators and the theory of Fourier integral operators with respect to the spectral theory of elliptic and hypoelliptic operators. The book is primarily addressed to graduate students in various fields of research who are interested in the subject... The book contains many illustrative examples throughout the text; these help the reader to gain a proper understanding of the ideas presented."

R. Melrose, *The Atiyah-Patodi-Singer Index Theorem*, Research Notes in Mathematics, 4, A K Peters, Ltd., Wellesley, MA, 1993. xiv+377 pp.

8.2. LINEAR DIFFERENTIAL OPERATORS

This is a very valuable introduction to and a reference on b-calculus of pseudodifferential operators on manifolds with boundary developed by Melrose. He illustrates the power of this calculus by treating the Atiyah–Patodi-Singer index theorem in his framework.

According to MathSciNet, "The beautiful index theorem due to Atiyah, Patodi, and Singer has held a central position of interest since its discovery in the early 1970s. It equates the analytic index of a Dirac operator on a manifold with boundary, where the metric and other data are assumed to be of product-type near the boundary, with the sum of an integral of a local expression, identical to the one in the boundaryless case, and the eta invariant, which is a global spectral invariant of the boundary Dirac operator. In this book, Melrose uses this index formula as a motivating context for a thorough discussion of his b-calculus of pseudodifferential operators on manifolds with boundary. Indeed, he gives a very direct proof, one particularly appealing for the naturality with which the eta invariant appears, right at the outset, and then regards the rest of the book as a 'filling in of details'. This proof is significant in that its formal structure generalizes to some much less trivial cases, such as the families index theorem and beyond. Of course, the book is much more than this... This book contains a complete and quite readable introduction to Melrose's pseudodifferential theory..."

8.2.9 Parabolic equations

A parabolic partial differential equation is a type of second-order partial differential equation with a time variable. The best example is the heat equation. Many problems in science such as heat diffusion and ocean acoustic propagation are described by parabolic equations.

D. Henry, *Geometric Theory of Semilinear Parabolic Equations*, Lecture Notes in Mathematics, 840, Springer-Verlag, Berlin-New York, 1981. iv+348 pp.

To most people, ordinary differential equations are much easier than partial differential equations. The basic point of this book is to study a qualitative theory of parabolic partial differential equations as ordinary differential equations in Banach spaces. It is a very valuable reference on this geometric theory of partial differential equations.

According to MathSciNet, "The author's point of view is to regard parabolic partial differential equations as ordinary differential equations in Banach spaces and to adapt standard ODE techniques to the PDE situation. Thus major concerns include the geometry of the flow, questions of stability and instability, invariant manifolds, the invariance principle, periodic solutions, bifurcation, and so on. The book contains a large number of interesting examples and applications. It is written in a lively style. It should be useful both for beginners and for experts."

A. Friedman, *Partial Differential Equations of Parabolic Type*, Prentice-Hall, Inc., Englewood Cliffs, N.J., 1964. xiv+347 pp.

This is a standard book on parabolic differential equations. It is suitable both as an introduction for beginners and as a reference for experts.

According to MathSciNet, "The author has succeeded in this book in providing for the reader whose background includes standard material on the theory of real variables and elementary knowledge of Hilbert spaces, an introduction to the subject of parabolic differential equations which brings him to the position of being able to read most papers in the field and to begin to consider research problems of his own.

The book largely consists of a harmonious collection of previously published results of the author and others. Expectedly, and deservedly, the author's own results take up a substantial portion of the book..."

O. Ladyenskaja, V. Solonnikov, N. Uralceva, *Linear and Quasilinear Equations of Parabolic Type*, Translated from the Russian by S. Smith, Translations of Mathematical Monographs, Vol. 23, American Mathematical Society, Providence, R.I., 1967. xi+648 pp.

This is a very valuable reference on linear and quasilinear equations of parabolic type. It is well-written with good choices of problems and results.

According to MathSciNet, "The book deals almost entirely with second order parabolic operators of one of the forms $u_t - \sum_{i=1}^n \partial a_i(x,t,u,u_x)/\partial x_i + a(x,t,u,u_x)$ (divergence form) or $u_t - \sum_{i,j=1}^n a_{ij}(x,t,u,u_x)u_{x_i x_j} + a(x,t,u,u_x)$ ('general' form), where $u_x = (u_{x_1}, \cdots, u_{x_n})$. It is a comprehensive survey of the results which can be achieved from the basic L^2 estimates and the refinements which follow from various definitions of generalized solutions of various problems associated with these operators. More than that, it is a manual of the techniques involved in these problems..."

A particularly important parabolic equation is the heat equation, which is useful both for theoretical considerations and practical applications. The following is a classic on heat conduction.

H.S. Carslaw, J.C. Jaeger, *Conduction of Heat in Solids*, Reprint of the second edition, Oxford Science Publications, The Clarendon Press, Oxford University Press, New York, 1988. x+510 pp.

This classic book describes many known exact solutions of problems of heat conduction or heat flow under all the most important boundary value problems. The wealth of beautiful solutions and methods makes it a very important, almost essential, reference for all people who are interested in heat conduction or the heat equation.

8.2.10 R. Adams, John J.H. Fournier, Sobolev Spaces, Second edition, Pure and Applied Mathematics (Amsterdam), 140, Elsevier/Academic Press, Amsterdam, 2003. xiv+305 pp.

This is a classic and a standard textbook and reference on the basic Sobolev spaces. It gives a systematic discussion of important results about Sobolev spaces motivated by applications.

A Sobolev space is a vector space of functions equipped with a norm that is a combination of L^p-norms of the function itself as well as its derivatives up to a given order. They are named after the Russian mathematician Sergei Sobolev. Their importance comes from the fact that solutions of partial differential equations are naturally found in Sobolev spaces, rather than in spaces of continuous functions and with the derivatives understood in the classical sense.

According to MathSciNet, "The monograph under review is probably the first book which is devoted to both extensive and intensive study of properties of Sobolev spaces. Although the book is a ' theoretical' one and no specific applications of the spaces mentioned are discussed, it intends to serve as a textbook for graduate students and research workers in differential equations... The book offers a considerable amount of material, preserving a strict and concise form of exposition.

The book can be highly recommended to every reader interested in functional analysis and its applications."

8.2.11 More books on Sobolev spaces

V. Maz'ja, *Sobolev Spaces*, Translated from the Russian by T.O. Shaposhnikova, Springer Series in Soviet Mathematics, Springer-Verlag, Berlin, 1985. xix+486 pp.

This book is a valuable reference. A more recent expanded version of this book is the following book by the author.

Sobolev Spaces with Applications to Elliptic Partial Differential Equations, Second, revised and augmented edition, Grundlehren der Mathematischen Wissenschaften, 342, Springer, Heidelberg, 2011. xxviii+866 pp.

According to MathSciNet, "This is a slightly reorganized and somewhat expanded English translation of three books of the author published previously in German [Imbedding theorems for Sobolev spaces, Part I, Teubner, Leipzig, 1979; Part II, 1980; On the theory of Sobolev spaces, Teubner, Leipzig, 1981]. As the author himself observes in the preface, the book is not a monograph on Sobolev spaces but mostly a report on his own work in the field."

W. Ziemer, *Weakly Differentiable Functions. Sobolev Spaces and Functions of Bounded Variation*, Graduate Texts in Mathematics, 120, Springer-Verlag, New York, 1989. xvi+308 pp.

Weakly differentiable functions are defined to integrable functions defined on an open subset of $\mathbb{R}^n$ whose partial derivatives in the sense of distributions are either L^p-functions or signed measures with finite total variation. This book studies pointwise behavior of Sobolev and bounded variation functions. This is a valuable reference and will supplement other books on Sobolev spaces.

According to MathSciNet, "The main purpose of this book is to study local behavior of these classes of functions. The central part of the book consists of Chapters 3 (for Sobolev spaces) and 5 (for the space BV). They concern the following problems: continuity properties of functions in terms of Lebesgue points, behavior of integral averages, density of smooth functions in corresponding Banach spaces, approximate and fine continuity, description of dual spaces in terms of measures and others."

8.2.12 T. Kato, Perturbation Theory for Linear Operators, Reprint of the 1980 edition, Classics in Mathematics, Springer-Verlag, Berlin, 1995.

This book is a classic. It is an authoritative and comprehensive book on perturbation theory for linear operators.

According to the review in Zentralblatt MATH, this is "An excellent textbook in the theory of linear operators in Banach and Hilbert spaces. It is a thoroughly worthwhile reference work both for graduate students in functional analysis as well as for researchers in perturbation, spectral, and scattering theory... I can recommend it for any mathematician or physicist interested in this field."

According to MathSciNet, "This is a treatise on linear transformations in Hilbert space as seen from the point of view of perturbation theory, as opposed to commutator theory or invariant subspaces. Transformations in Banach spaces are also considered when they can be treated as generalizations of self-adjoint transformations. The tone of the book is set in the first two chapters, which are concerned with transformations in finite-dimensional spaces and can be read with no prior knowledge of operator theory..."

8.3 Nonlinear differential equations

There are big differences between linear and nonlinear differential equations. Besides being more difficult, each nonlinear differential equation is different and requires special care. Probably the most famous, difficult and useful nonlinear differential equation is the Navier-Stokes equation. Both Navier and Stokes were outstanding mathematicians and did many other things. For example, Stokes was also a physicist, politician and theologist. He made seminal contributions to fluid dynamics, optics, and mathematical physics. He was secretary, then president, of the Royal Society.

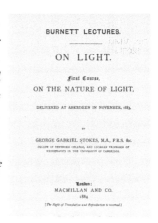

Figure 8.3: Stokes' book.

There are significant differences between linear and nonlinear differential equations. There are general theories for large classes of linear differential equations, but each nonlinear differential equation is different in a special way. A lot of work has been done in some important nonlinear differential equations that arise from mathematical physics and geometry such as the Navier-Stokes equations in fluid dynamics and the Monge-Ampére equations in differential geometry.

Given the vast numbers of interesting and important nonlinear differential equations, it is almost impossible to be comprehensive in discussing books on them. Only several books are listed here.

8.3.1 More geometric nonlinear differential equations

L. Caffarelli, X. Cabré, *Fully Nonlinear Elliptic Equations*, American Mathematical Society Colloquium Publications, 43, American Mathematical Society, Providence, RI, 1995. vi+104 pp.

This book gives a self-contained account of recent results on the regularity theory for viscosity solutions of fully nonlinear elliptic equations. It is a valuable reference for experts and an introduction to graduate students in nonlinear partial differential equations.

According to MathSciNet, "The theory of viscosity solutions for nonlinear partial differential equations has been developed by P.-L. Lions, M. G. Crandall, H. Ishii, L. C. Evans and many others. By now, it has reached a fairly complete stage of development and is mature enough to be treated in a textbook... This interesting and well-written book contains material selected with good taste. However, it is not exactly an introduction to the whole field of viscosity methods... The book is likely to be highly appreciated both by researchers and advanced students."

C. Villani, *Topics in Optimal Transportation*, Graduate Studies in Mathematics, 58, American Mathematical Society, Providence, RI, 2003. xvi+370 pp.

This book is the first comprehensive introduction to the theory of optimal transportation with its many striking applications. A special feature of this book is an entire chapter of open problems. It is a very valuable introduction for graduate students and beginners.

From the preface, "These notes are definitely not intended to be exhaustive, and should rather be seen as an introduction to the subject. Their reading can be complemented by some of the reference texts which have appeared recently... I have tried to keep proofs as simple as possible throughout the book, keeping in mind that they should be understandable by non-expert students. I have also stated many results without proofs, either to convey a better intuition, or to give an account of recent research in the field. In the end, these notes are intended to serve both as a course and as a survey."

C. Villani, *Optimal Transport. Old and New*, Grundlehren der mathematischen Wissenschaften, Vol. 338, 2009. XXII+978 pp.

This book is more comprehensive and emphasizes probability, geometry, and dynamical systems, and less on analysis and physics. According to the author's recommendation, both books can be read independently, and they enhance each other. It is a valuable introduction and reference.

According to MathSciNet, "At the close of the 1980s, the independent contributions of Yann Brenier, Mike Cullen and John Mather launched a revolution in the venerable field of optimal transport founded by G. Monge in the 18th century, which has made breathtaking forays into various other domains of mathematics ever since. The author presents a broad overview of this area, supplying complete and self-contained proofs of all the fundamental results of the theory of optimal transport at the appropriate level of generality. Thus, the book encompasses the broad spectrum ranging from basic theory to the most recent research results.

Ph.D. students or researchers can read the entire book without any prior knowledge of the field. A comprehensive bibliography with notes that extensively discuss the existing literature underlines the book's value as a most welcome reference text on this subject."

J.L. Lions, *Quelques méthodes de résolution des problèmes aux limites non linéaires*, Dunod; Gauthier-Villars, Paris, 1969. xx+554 pp.

This comprehensive book on methods to solve nonlinear boundary value problems for partial differential equations is a valuable reference written by an expert.

8.3. NONLINEAR DIFFERENTIAL EQUATIONS

According to MathSciNet, "In this book the author reports on methods of solving nonlinear boundary value problems for partial differential equations, on a theoretical and functional analysis basis. The vast material is organized into four chapters, according to methods. In all cases the methods are shown to yield existence theorems.

(1) Methods based on compactness... (2) Methods based on monotonicity. (3) Methods of regularization, viscosity methods and penalty methods. (4) Methods of successive approximation, Newton's method, discretization, decomposition methods... The presentation, although certainly adequate for the connoisseur, may not be so for students. The book should be thought of as the last step in a series of presentations, probably lectures, which may not be available to the reader. The author's classification of the vast material is a well conceived device for running the maze of technical and conceptual interrelations. However, the details and inherent difficulties may be found to strain the line of presentation."

M. Struwe, *Variational Methods. Applications to Nonlinear Partial Differential Equations and Hamiltonian Systems*, Fourth edition. Springer-Verlag, Berlin, 2008. xx+302 pp.

This book gives an accessible account of variational methods and their applications to global analysis, in particular nonlinear partial differential equations and Hamiltonian systems. A good selection of topics on both abstract results and applications and clear presentation make the book a good introduction to variational problems. It is also a very valuable reference.

According to MathSciNet, "This book presents an overview of some areas in the calculus of variations, with special emphasis on critical point theory, direct and minimax methods and limit cases... The material is presented at an advanced graduate level. The book includes several applications, with frequent comments and references to the recent literature"

V. Jurdjevic, *Geometric Control Theory*, Cambridge Studies in Advanced Mathematics, 52, Cambridge University Press, Cambridge, 1997. xviii+492 pp.

This book gives a good introduction to geometric control theory, which combines differential geometry, mathematical physics, and optimal control. It covers a wide range of mathematical topics and provides good geometric insights. This book can also serve as an important reference.

According to MathSciNet, "This textbook is mainly intended for use by graduate students and researchers in related fields who want to become familiar with fundamental results and problems in control theory. At the same time the book also provides many interesting and stimulating discussions for the expert. Much of the theoretical material treated in the first part of the book has been published in the scientific literature only and an introduction to these important concepts on a textbook level has long been overdue and is very welcome... The multitude of examples, many of them from classical problems, which are included and fully discussed in the text make this book a worthwhile introduction to the topic."

L. Ambrosio, N. Fusco, D. Pallara, *Functions of Bounded Variation and Free Discontinuity Problems*, Oxford Mathematical Monographs, The Clarendon Press, Oxford University Press, New York, 2000. xviii+434 pp.

This well-written book is a good introduction to student and also a valuable reference for experts. According to Bulletin of the London Mathematical Society, "This book provides an excellent account of the theory of BV functions (apparently for the first time in book form), and a nice introduction to geometric measure theory, as well as a rigorous survey of results for 'free discontinuity' problems modeled to the Mumford-Shah problem."

According to the Society for Industrial and Applied Mathematics, "it offers a comprehensive, unified, and self-contained treatment of contemporary BV theory that nicely complements the existing literature... This is a book with staying power, an invaluable source for experts in the calculus of variations, geometric measure theory, and partial differential equations, and a graduate textbook for advanced Ph.D. students ..."

8.3.2 Nonlinear differential equations and fluid mechanics

C. Canuto, M.Y. Hussaini, A. Quarteroni, T. Zang, *Spectral Methods in Fluid Dynamics*, Springer Series in Computational Physics, Springer-Verlag, New York, 1988. xiv+557 pp.

This book is a very valuable up-to-date reference on spectral methods in computational fluid dynamics and rigorous analysis in numerical analysis.

From the preface: "This is a book about spectral methods for partial differential equations: when to use them, how to implement them, and what can be learned from their rigorous theory... This book pays special attention to those algorithmic details which are essential to successful implementation of spectral methods. The focus is on algorithms for fluid dynamical problems in transition, turbulence, and aerodynamics. Specific applications in meteorology are not addressed.

We present a unified theory of the mathematical analysis of spectral methods and apply it to many of the algorithms in current use. We focus mainly on those aspects of the techniques which are typical of the analysis of spectral methods but which do not reduce to the straightforward application of well-known results from the analysis of finite difference methods...

8.3. NONLINEAR DIFFERENTIAL EQUATIONS

No numerical method is suitable for all problems. This book aims to guide the reader towards those applications for which spectral methods are preferable and not just workable. It is addressed both to computational fluid dynamicists who wish to use spectral methods and to numerical analysts who are interested in their rigorous analysis. It is directed towards research workers in these fields and presumes an elementary knowledge of linear algebra, differential equations, numerical analysis, and fluid dynamics."

P. Lax, *Hyperbolic Systems of Conservation Laws and the Mathematical Theory of Shock Waves*, Conference Board of the Mathematical Sciences Regional Conference Series in Applied Mathematics, No. 11, Society for Industrial and Applied Mathematics, Philadelphia, Pa., 1973. v+48 pp.

This is a good short introduction to hyperbolic systems of conservation laws and the mathematical theory of shock waves.

According to MathSciNet, "The author deals with all the phenomena that occur for systems of conservation laws, including various aspects of qualitative behavior such as formation of shocks, the entropy condition, etc. He also considers existence and uniqueness theorems, and the decay and asymptotic behavior of solutions, as well as difference scheme approximations. This monograph should prove useful to anyone who would like to work in this interesting field."

J. Smoller, *Shock Waves and Reaction-diffusion Equations*, Second edition, Grundlehren der Mathematischen Wissenschaften, 258, Springer-Verlag, New York, 1994. xxiv+632 pp.

This is a valuable and modern introduction to hyperbolic conservation laws and reaction-diffusion equations. It is also a reliable reference.

From the book description, "The purpose of this book is to make easily available the basics of the theory of hyperbolic conservation laws and the theory of systems of reaction-diffusion equations, including the generalized Morse theory as developed by Charles Conley. It presents the modern ideas in these fields in a way that is accessible to a wider audience than just mathematicians."

According to MathSciNet, "The book provides a valuable contribution to the expository literature on the subject."

R. Courant, K. Friedrichs, *Supersonic Flow and Shock Waves*, Interscience Publishers, Inc., New York, N.Y., 1948. xvi+464 pp.

This is a classic written by major contributors to its subjects.

From the book description, "Courant and Friedrichs' classical treatise was first published in 1948 and the basic research for it took place during World War II. However, many aspects make the book just as interesting as a text and a reference today. It treats the dynamics of compressible fluids in mathematical form, and attempts to present a systematic theory of nonlinear wave propagation, particularly in relation to gas dynamics. Written in the form of an advanced textbook, it should appeal to engineers, physicists and mathematicians alike."

According to MathSciNet, "The book is written in the clear and vivid style for which the authors are known. Open problems and unresolved difficulties are carefully noted, and the reader is never left in doubt as to whether he is presented with a mathematical theorem or with a conjecture based on physical experience."

P. Constantin, C. Foias, *Navier-Stokes Equations*, Chicago Lectures in Mathematics, University of Chicago Press, Chicago, IL, 1988. x+190 pp.

This book is a good introduction to relations between the solutions of the Navier-Stokes equations and finite-dimensional phenomena. It is not intended as a comprehensive book course on the Navier-Stokes equations, but it contains a lot of general material and is a valuable reference.

According to MathSciNet, "The authors provide a compact and self-contained course on Navier-Stokes equations... Throughout the book, the authors present simple, elementary proofs wherever possible, and they emphasize nondimensionality..."

M. Bardi, I. Capuzzo-Dolcetta, *Optimal Control and Viscosity Solutions of Hamilton-Jacobi-Bellman Equations*, With appendices by Maurizio Falcone and Pierpaolo Soravia, Systems & Control: Foundations & Applications, Birkhauser Boston, Inc., Boston, MA, 1997. xviii+570 pp.

This book gives a self-contained, up-to-date, and comprehensive account of the theory of viscosity solutions and applications to optimal control and differential games. It is a very valuable as a solid introduction and as a reference.

According to MathSciNet, "The book is devoted to the theory of viscosity solutions to Hamilton-Jacobi (H-J) type partial differential equations and its relation with Bellman's dynamic programming approach to optimal control and differential games... The book may be used by graduate students and researchers in control theory both as an introductory textbook, and as an up-to-date reference book, outlining the main approaches, results and problems of the H-J equations in the context of optimal control"

G. Galdi, *An Introduction to the Mathematical Theory of the Navier-Stokes Equations. Vol. I. Linearized Steady Problems*, Springer Tracts in Natural Philosophy, 38, Springer-Verlag, New York, 1994. xii+450 pp.

Besides being a good introduction, these two books can also be used as a valuable reference on an important and complicated subject.

According to MathSciNet, "The emphasis of this book is on an introduction to the mathematical theory of the stationary Navier-Stokes equations. It is written in the style of a textbook and is essentially self-contained. The problems are presented clearly and in an accessible manner. Every chapter begins with a good introductory discussion of the problems considered, and ends with interesting notes on different approaches developed in the literature. Further, stimulating exercises are proposed... It is an important book for it is very comprehensive on the topics considered..."

P.L. Lions, *Mathematical Topics in Fluid Mechanics. Vol. 1. Incompressible Models. Vol. 2. Compressible Models*, Oxford Lecture Series in Mathematics and its Applications, 3, 10, Oxford Science Publications, The Clarendon Press, Oxford University Press, New York, 1996. xiv+237 pp., 1998. xiv+348 pp.

These two books give an up-to-date account of various mathematical results on incompressible fluid models including density-dependent Navier-Stokes and Euler equations. They are valuable references.

According to MathSciNet, "This book is the first volume of a series of two, devoted to the mathematical issues of classical fluid mechanics... The two main equations for incompressible fluids are the Euler equation, for an inviscid fluid, and the Navier-Stokes equation, for a viscous one... In conclusion, this monograph does not pretend to be exhaustive, since each chapter would then be developed as a separate volume. It merely describes in brief the state of the art of the main problems, and gives a lot of new theorems. These are, at the date of publishing, the best results in the field. Its audience will thus be mainly the research community of the field, for which this book is certainly a milestone."

"This book is the second of two volumes ... [and] emphasizes the compressible side, with various contexts: viscous isentropic flows, viscous non-isentropic flows and inviscid flows... In conclusion, one can say that this book is a milestone in the mathematics of fluid dynamics. Perhaps no other before it offered so many novelties in one handful. The advances it makes in the field are such that specialists will need a decade to exploit them in their full range. It opens so many questions that the field will be active for a while. Certainly, every researcher on compressible fluids will wish to add this text to his private bookshelf."

O.A. Ladyzhenskaya, *The Mathematical Theory of Viscous Incompressible Flow*, Second English edition, revised and enlarged., Translated from the Russian by Richard A. Silverman and John Chu, Mathematics and its Applications, Vol. 2, Gordon and Breach, Science Publishers, New York-London-Paris, 1969. xviii+224 pp.

This is a valuable reference on the mathematical theory of viscous incompressible flow.

According to MathSciNet, "The philosophical and conceptual underpinnings of this work were initiated by Leray in a series of remarkable papers on solvability of initial-

and boundary-value problems for the Navier-Stokes equations, which appeared just 30 years ago; they were later developed and elaborated from a different point of view by E. Hopf in his paper of 1950 on the initial-value problem [Math. Nachr. 4 (1951), 213–231]. The present volume reflects the continuation of these developments which has been undertaken in recent years by various authors; as is appropriate, the emphasis is on the work of the writer and of her colleagues in the Institute Division with which she is associated. ..."

C. Dafermos, *Hyperbolic Conservation Laws in Continuum Physics*, Third edition, Grundlehren der Mathematischen Wissenschaften, 325, Springer-Verlag, Berlin, 2010. xxxvi+708 pp.

This book gives a modern introduction to hyperbolic conservation laws. It was written by a major contributor and is also a valuable reference on this active topic.

According to MathSciNet, "This book has several qualities which make it the best book ever written on the subject, in the reviewer's opinion. First, the style is so clear that some parts, especially the first few, can be read like a novel. Next, the choice of topics is so clever that the book will survive at least several decades, or even forever. At a finer level, Dafermos has chosen as often as possible to present theorems with physically motivated assumptions. This will be strongly appreciated by people looking for references. Last, the historical notes which follow each chapter, and the exhaustive bibliography, form a priceless database which does not exist elsewhere.

In conclusion, I strongly recommend this book to every researcher or graduate student in the field of conservation laws. Naturally, it will also be of interest to many scientists in related areas."

G.B. Whitham, *Linear and Nonlinear Waves* , Reprint of the 1974 original, Pure and Applied Mathematics (New York), A Wiley-Interscience Publication, John Wiley & Sons, Inc., New York, 1999. xviii+636 pp.

This is a classic written by an expert. It was the first book to give a comprehensive account of the many types of waves that occur both with and without dispersion and has become a standard book in the subject.

This book motivates problems and explains solutions to them well. One not only learns methods from this book, one also understands why one studies these problems. As the author explains in the preface, "The aim is to cover all the major well-established ideas but, at the same time, to emphasize nonlinear theory from the onset and to introduce the very active research areas in this field." "It was designed for applied mathematics students in the first and second years of graduate study; it appears to have been equally useful for students in engineering and physics."

8.3. NONLINEAR DIFFERENTIAL EQUATIONS

From the book description, "It covers the behavior of waves in two parts, with the first part addressing hyperbolic waves and the second addressing dispersive waves. The mathematical principles are presented along with examples of specific cases in communications and specific physical fields, including flood waves in rivers, waves in glaciers, traffic flow, sonic booms, blast waves, and ocean waves from storms."

According to an Amazon review, "Linear and Nonlinear Waves provides an excellent treatment of the wave phenomena. This is a very complex and mathematically intensive subject, which Whitham conveys in an understandable and followable manner. He takes the approach that the reader has some knowledge of the subject, but writes at an introductory graduate level. This book will give any mathematician or engineer interested in waves an excellent introduction and bring them to a superior understanding."

H. Lamb, *Hydrodynamics*, Reprint of the 1932 sixth edition, With a foreword by R. A. Caflisch, Cambridge Mathematical Library, Cambridge University Press, Cambridge, 1993. xxvi+738 pp.

This is a classic book on fluid dynamics. Since its first publication in 1879, it has remained as a standard classic text in hydrodynamics.

From the foreword, "The value of Lamb's Hydrodynamics today is first as a storehouse of exact solutions for fluid dynamic problems... In this aspect it is unequaled by modern texts. There are also certain fluid dynamic topics that are still best expressed in Lamb's book... Less tangible but equally important is the contact that Hydrodynamics provides with an earlier era of fluid dynamics. Lamb gave careful attribution to original sources, which is of great help to anyone interested in the history of fluid mechanics. More important to most readers is the perspective conveyed from a crucial period in the development of this subject. Since the final revision of Hydrodynamics in the 1930s, great progress has been made in fluid dynamics. Lamb's treatment of nonlinear water waves, shock waves, fluid dynamic stability, boundary layers and turbulence, for example, suggests many problems, of which a large number have since been solved but many others remain open. Thus Hydrodynamics provides us with a valuable measure of the past progress of fluid dynamics and with a compelling challenge for its future."

IX. The door of the Old Law Library of the University of Michigan

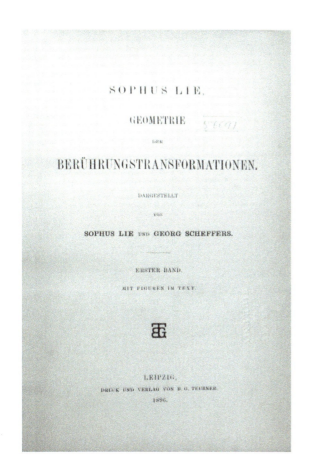
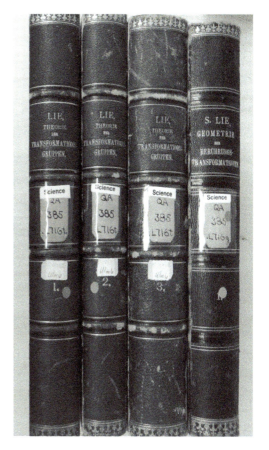

Lie theory was initiated by Lie in order to understand solutions of differential equations. Quickly, Lie theory followed its own course and found other striking applications in mathematics and sciences, especially in physics. One of his deep insights was to study Lie groups by their associated Lie algebras. Lie was a close friend of Klein and contributed enormously to the success of the Erlangen program proposed by Klein. Their friendship was beneficial to both of them, but they fell out due to the success (or the fame) of the Erlangen program and the poor health of Lie. They became friends again but Lie died shortly after that. It should be mentioned that E. Cartan and H. Weyl developed the global theory of Lie groups and made Lie groups central objects in mathematics.

Plucker started out as an experimental physicist and later turned to geometry. Though there is close connection between theoretical physics and mathematics, such a transition from experimental physics to pure mathematics seems to be nonexistent or rare now. He invented what was known as line geometry in the nineteenth century. He was the teacher of Klein in college and passed away before Klein finished college. But his impact on Klein was huge and the line geometry also has a definite influence on the joint work of Lie and Klein. Plucker was best known for Plucker coordinates in projective geometry.

Among all Lie groups, there are two most important classes: (1) semisimple Lie groups, in particular SL(2), and (2) nilpotent Lie groups, in particular the Heisenberg group. Heisenberg received a Nobel prize for the creation of quantum mechanics. Together with Max Born and Pascual Jordan, he gave the matrix formulation of quantum mechanics. Matrix algebras form a basic class of Lie algebras, and matrix groups (or linear algebraic groups) are also the most important Lie groups.

Chapter 9

Lie theories

In mathematics, there are not many subjects which are initiated and built up by a single man and their importance was recognized right away. Lie theory is one of them. With the passing of time, it is becoming more and more important. The basic reason is that the mathematical language of symmetry is the group theory, and continuous and multi-dimensional symmetry is crucial in solving many problems in sciences. In mathematics, Lie groups and spaces associated with them are some of the most important objects. Some major areas of application have been found, for example in automorphic representations and in mathematical physics, and the subject has become a busy crossroads.

Lie was motivated by the Galois theory of algebraic equations and introduced Lie groups in order to build a corresponding theory for differential equations. It did not turn out as successfully as Lie hoped, and the connection of Lie groups with differential equations does not dominate the subject of Lie theory. Instead, it is the direct application of Lie groups to the underlying geometric symmetry of many problems in mathematics and sciences that has made Lie theory a central subject of modern mathematics. The fact that there is a good structure theory for Lie groups and their representations has made them integral to large parts of the broad subject of algebra.

The Lie theory includes (1) Lie groups, Lie algebras and their interconnections, (2) algebraic groups, (3) transformation groups, (4) discrete subgroups of Lie groups, and automorphic forms, (5) representation theory of Lie groups and automorphic representation theories, (6) applications of Lie groups to differential equations, (7) generalizations of Lie groups and Lie algebras, (8) applications of Lie groups in sciences, especially in physics.

In this chapter, we discuss books related to these subjects.

9.1 Lie groups and Lie algebras

A Lie group is a group which has also a compatible differentiable manifold. Lie groups are the mathematical tool and language to describe continuous symmetry of mathematical objects and structures, which makes them indispensable tools for many parts of contemporary mathematics, as well as for modern theoretical physics.

One reason for the power of Lie groups is that they can be approximated well and effectively by their Lie algebras, which are the tangent spaces at the identity and hence are linear objects and can be handled relatively easily. In fact, many structures and properties of Lie groups are determined by and described explicitly in terms of Lie algebras. This is a deep insight of Sophus Lie. (Such effective linear approximation is not true with most smooth manifolds).

Lie groups are important and there are many books on them, but it is not easy to make a list of them. Another interesting, or rather unfortunate, thing is that, in spite of the importance and usefulness of Lie groups, most mathematical books on Lie groups do not discuss applications of Lie groups. Maybe the structure of Lie groups is rich enough and it also takes space and time to learn and clarify Lie groups.

After the concept of Lie groups and Lie algebras were introduced by Lie, two people, Weyl and E. Cartan, made crucial contributions to make this into the central tool in both mathematics and physics. In some sense, Weyl's impact was greater than E. Cartan. He developed the theory of compact groups, and proved a fundamental character formula. These results are foundational in understanding the symmetry structure of quantum mechanics. Non-compact groups and their representations, particularly the Heisenberg group, were also streamlined in his 1927 Weyl quantization, which still is the best bridge between classical and quantum physics to date.

Figure 9.1: Weyl's book.

9.1.1 C. Chevalley, Theory of Lie Groups. I, Fifteenth printing, Princeton Mathematical Series, 8, Princeton Landmarks in Mathematics, Princeton University Press, Princeton, NJ, 1999. xii+217 pp.

This book is a classic written by a major figure of the subject. It gives an accessible introduction to basics of Lie groups. It is a popular classic and has been reprinted over 15 times.

9.1. LIE GROUPS AND LIE ALGEBRAS

According to MathSciNet, "This book forms a companion to the well-known volumes by Weyl [The Classical Groups. Their Invariants and Representations, 1939] and Pontryagin [Topological Groups, 1939] in the same series."

According to a review in Bulletin of AMS, "In this masterpiece of concise exposition, the concept of Lie group is put together with all the craftsmanship of an expert. The finished product is a fascinating thing to contemplate, equipped as it is with three interrelated structures: algebraic, topological and analytic. It has no loose ends, no doubtful regions: it can be explored freely without the usual necessity of having to stay within a safe distance of the identity.

The reviewer was particularly struck that so much has been accomplished in a volume of such modest size. One of the things which makes this possible is the effectiveness of the definitions. Invariably they emphasize the property which can be most quickly put to work and which is most suitable to the logic of the situation. This insistence on calling things by their right names can be disconcerting at times since it tends to ignore intuitive meanings... But in the end, this book should be easier for most readers, and far more satisfying, than any exposition of the same material proceeding on a less exact level."

Claude Chevalley is a major mathematician of the 20th century. Probably his best known work concerns Chevalley groups, which are roughly linear algebraic groups defined over $\mathbb{Z}$ and give rise to all finite simple groups of Lie type. He also made major contributions to number theory, algebraic geometry. For example, in his thesis, he developed a new, more algebraic approach to class field theory.

9.1.2 J.P. Serre, Complex Semisimple Lie Algebras, Translated from the French by G.A. Jones, Reprint of the 1987 edition, Springer Monographs in Mathematics, Springer-Verlag, Berlin, 2001. x+74 pp.

This is a very valuable and influential book on complex semisimple Lie algebras written by a master expositor. In some sense, whenever Serre touches a subject, he makes it better. The famous Serre relations (or Chevalley-Serre relations) which give a presentation of complex semisimple Lie algebras are given in this book. They are of foundational importance, for example, in the construction of Kac-Moody algebras.

Complex semisimple Lie algebras were first classified by Wilhelm Killing in 1888-1890. His proof was made rigorous by Élie Cartan in 1894 in his Ph.D. thesis, who also classified real semisimple l Lie algebras. This was subsequently refined, and the present classification by Dynkin diagrams was given by then 22-year old student Eugene Dynkin in 1947. J.P. Serre also made some minor but important modifications, which are contained in this book of Serre.

According to MathSciNet, "This book consists of notes of the author's lectures at Algiers in 1965 and provides a concise presentation of the theory of complex semisimple Lie algebras. Actually, the book is intended for those who have an acquaintance with the basic parts of the theory, namely, with those general theorems on Lie algebras which do not depend on the notion of Cartan subalgebra. The author begins with a summary of these general theorems and then discusses in detail the structure and representation theory of complex semisimple Lie algebras. One recognizes here a skillful ordering of the material, many simplifications of classical arguments and a new theorem describing fundamental relations between canonical generators of semisimple Lie algebras. The classical theory being thus introduced in such modern form, the reader can quickly reach the essence of the theory through the present book..."

According to an Amazon review, "The core of ... is the existence and uniqueness proofs of semisimple Lie algebras corresponding to a root system. As an appendix, a theorem showing how to construct semisimple Lie algebras from root systems by means of generators and relations [that is, using presentations]. This result is of extreme importance, and constitutes one of the germs that lead to the notion of Kac-Moody algebras in 1968."

9.1.3 N. Bourbaki, Lie Groups and Lie Algebras. Chapters 4–6, Translated from the 1968 French original by Andrew Pressley. Elements of Mathematics (Berlin), Springer-Verlag, Berlin, 2002. xii+300 pp.

This is a very useful and important reference produced by Bourbaki.

According to MathSciNet, "This is a rich and useful volume. The material it treats has relevance well beyond the theory of Lie groups and algebras, ranging from the geometry of regular polytopes and paving problems to current work on finite simple groups having a (B, N)-pair structure, or 'Tits systems'. A historical note provides a survey of the contexts in which groups generated by reflections have arisen. A brief introduction includes almost the only other mention of Lie groups and algebras to be found in the volume. Thus the presentation here is really quite independent of Lie theory. The choice of such an approach makes for an elegant, self-contained treatment of some highly interesting mathematics, which can be read with profit and with relative ease by a very wide circle of readers (and with delight by many, if the reviewer is at all representative). Such drawbacks as it may have as part of a treatise on Lie theory, starting with the previous volume (Chapter I), arise out of the fact that the present considerations are inadequately motivated from the point of view of Lie groups or algebras. By reading Chapters IV, VII and VIII of N. Jacobson's Lie algebras, the student can easily and

9.1. LIE GROUPS AND LIE ALGEBRAS

smoothly close this motivational gap, and he should find his appreciation of this volume considerably intensified thereby... Previous Bourbaki volumes have led one to expect exercises in considerable quantity and depth, challenging even the reader who finds the text quite familiar. The volume in hand does full honor to this excellent tradition."

9.1.4 More books on Lie algebras and Lie groups

A.L. Onishchik, E.B. Vinberg, *Lie Groups and Algebraic Groups*, Translated from the Russian and with a preface by D.A. Leites, Springer Series in Soviet Mathematics, Springer-Verlag, Berlin, 1990. xx+328 pp.

This is a very valuable book on Lie groups and algebraic groups. It can be used as a very efficiently organized introductory book on Lie groups with an emphasis on algebraic groups, and as a convenient reference. It contains a wealth of information, which is achieved by formulating almost every statement as a problem, or as an exercise for the reader.

From the book description, "This is a quite extraordinary book on Lie groups and algebraic groups. Created from hectographed notes in Russian from Moscow University, which for many Soviet mathematicians have been something akin to a 'Bible', the book has been substantially extended and organized to develop the material through the posing of problems and to illustrate it through a wealth of examples. Several tables have never before been published, such as decomposition of representations into irreducible components. This will be especially helpful for physicists... The literature on Lie group theory has no competitors to this book in broadness of scope. The book is self-contained indeed. This distinguishes it favorably from other books in the area. It is thus not only an indispensable reference work for researchers but also a good introduction for students."

According to MathSciNet, "This book originated in the authors' 1967–1968 Moscow seminar, and was subsequently updated and expanded. As expressed in the preface, the "guiding idea was to present in the most economical way the theory of semisimple Lie groups on the basis of the theory of algebraic groups". This is in the spirit of what Chevalley set out to do in his never completed series of books on Lie groups.

Apart from its emphasis on algebraic groups, the book differs most radically from other accounts of the structure of Lie groups by presenting the results in Texas style as a sequence of problems (though proofs of some of the more difficult theorems are given directly in the text)... this book will be (for many readers of an algebraic disposition) an attractive alternative to the other large introductory treatises on Lie groups."

J. Humphreys, *Introduction to Lie Algebras and Representation Theory*, Graduate Texts in Mathematics, Vol. 9, Second printing, revised, Graduate Texts in Mathematics, 9, Springer-Verlag, New York-Berlin, 1978. xii+171 pp.

This is a standard textbook on Lie algebra. It is well-organized and clearly written.

From the author's preface: "This book is designed to introduce the reader to the theory of semisimple Lie algebras over an algebraically closed field of characteristic 0, with emphasis on representations... The first four chapters might well be read by a bright undergraduate; however, the remaining three chapters are admittedly a little more demanding..."

According to Math Review, "The reviewer would like to add that he has found this book to be exceptionally well written and ideally suited either for independent reading or as a text for an introduction to Lie algebras and their representations."

A. Knapp, *Lie Groups Beyond an Introduction*, Second edition, Progress in Mathematics, 140, Birkhäuser Boston, Inc., Boston, MA, 2002. xviii+812 pp.

This is a comprehensive book written by an expert on Lie groups. It is a big book and can serve as a valuable reference and an advanced introduction.

From the book description, "This book takes the reader from the end of introductory Lie group theory to the threshold of infinite-dimensional group representations. Merging algebra and analysis throughout, the author uses Lie-theoretic methods to develop a beautiful theory having wide applications in mathematics and physics. The book initially shares insights that make use of actual matrices; it later relies on such structural features as properties of root systems."

According to MathSciNet, "The book is eminently suitable as a text from which to learn Lie theory."

V.S. Varadarajan, *Lie Groups, Lie Algebras, and Their Representations*, Prentice-Hall Series in Modern Analysis, Prentice-Hall, Inc., Englewood Cliffs, N.J., 1974. xiii+430 pp.

This book was written as a systematic introduction to Lie groups, Lie algebras and their representations for graduate students and non-experts. Its combination of both the algebraic and analytic aspects of the Lie theory and discussion of finite dimensional representations also makes it a valuable reference for many people.

From the preface, "There are a number of books on the subject currently available—most notably those of Chevalley, Jacobson and Bourbaki—which present various aspects of the theory in great depth. However, I feel there is a need for a single book in English which develops both the algebraic and analytic aspects of the theory and which goes into the representation theory of semisimple Lie groups and Lie algebras in detail. This book is an attempt to fill this need."

According to MathSciNet, "The book is especially suitable for the graduate student as well as the nonspecialist mathematician having some acquaintance with topological

9.1. LIE GROUPS AND LIE ALGEBRAS

groups and differentiable manifolds who wishes to study Lie theory from the beginning in considerable depth. "

N. Jacobson, *Lie Algebras*, Republication of the 1962 original, Dover Publications, Inc., New York, 1979. ix+331 pp.

This is a classical book on Lie algebras written by an expert algebraist. It was written as a textbook and is still a very good introduction.

According to MathSciNet, "This book contains a very clear presentation of the theory of Lie algebras, from basic concepts to the deepest part of the theory. It is gratifying to see that the theory on the classification of irreducible modules is now available in such a simplified and elementary form. The author's work on the classification of simple Lie algebras over an arbitrary field is also presented in a unified manner."

9.1.5 A. Borel, Linear Algebraic Groups, Second edition, Graduate Texts in Mathematics, 126, Springer-Verlag, New York, 1991. xii+288 pp.

There are close relations between Lie groups and linear algebraic groups. In some sense, the most direct and accessible way to construct Lie groups is to use linear algebraic groups (or matrix groups). Of course, other groups such as finite groups of Lie type and p-adic groups can also be constructed from linear algebraic groups. One of the major contributions of Borel is to linear algebraic groups. For example, the notion of Borel subgroups has played a fundamental role.

This book is both a good introduction and a useful reference on linear algebraic groups written by one of the founders. It is well-organized and precise.

According to MathSciNet, "For over ten years the role of standard text on algebraic groups over a field has been played by the mimeographed notes of C. Chevalley [Séminaire C. Chevalley, 1956–1958. Classification des groupes de Lie algébriques, Secrétariat mathématique, Paris, 1958]. The present book, grown out of notes taken by H. Bass of a graduate course by the author at Columbia University in the spring of 1968, more or less covers the material treated in the first thirteen 'exposés' of Chevalley's seminar. Of course, the book shows the traces of developments in the field since the appearance of the seminar notes... What one should not expect in this book is an account of the results on algebraic group schemes by Grothendieck and his school. Borel and Bass have produced a very beautiful, outstanding book, clear and well organized. No doubt it will excellently serve two purposes: to give an introduction to the field, and to be a standard text which may be used for references..."

"Since the first edition of this book appeared in 1969, two more books with the same title (sic!) have appeared, viz., those by J. E. Humphreys, 1975, and T. A. Springer,

1981. Here is your consumer consultant's advice. Springer offers a fairly elementary introduction, not supposing preliminary knowledge of algebraic geometry and avoiding commutative algebra as much as possible. Except for a brief discussion of groups over finite fields, fields are always assumed algebraically closed, and representations are not dealt with. Humphreys needs more algebraic geometry, which is treated almost in full in a first chapter. He gives a quite complete account of the algebraically closed field case, also covering linear representations. Rationality questions are only surveyed. The author of the book under review is very complete, too, and puts heavy emphasis on rationality questions, but only lightly touches on linear representations. The reader interested in a first introduction may well use this book and skip the rationality questions in a first reading."

9.1.6 More books on algebraic groups, algebraic geometry and number theory

For many users of Lie groups in geometry and topology, when they consider linear algebraic groups **G**, it is the real and rational loci of algebraic groups and subgroups such as arithmetic groups which are important and useful (or relevant). But for more delicate questions in representation theories of algebraic groups, algebraic groups should be treated as group schemes and more refined structures are needed.

The MathSciNet review below of the book *Linear Algebraic Groups* by Humphreys explains well relations between several books on linear algebraic groups.

J. Humphreys, *Linear Algebraic Groups*, Graduate Texts in Mathematics, No. 21, Springer-Verlag, New York-Heidelberg, 1975. xiv+247 pp.

This is a well-organized and well-written introduction to modern algebraic groups.

According to MathSciNet, "The theory of linear groups in algebraic geometry over fields of arbitrary characteristic was originated by C. Chevalley, 1951, and strengthened by A. Borel, 1956. The famous Séminaire Chevalley, which Godement called the 'Bible' *Classification des groupes de Lie algébriques*, Vols. 1–2, 1958, was the completion of this theory over algebraically closed fields... Later, a decade ago, the 'Bible' of 288 pages was swallowed up in the lecture notes of 1742 pages by M. Demazure and A. Grothendieck (SGAD)... On the other hand, Borel produced a very well organized book of 398 pages *Linear Algebraic Groups*, 1969, which covered and improved Vol. 1 of the 'Bible'.

Now, in the present book of 247 pages, the author succeeds in reproducing the whole 'Bible' by following mostly Borel's book in the basic part and by adopting the method of SGAD whenever convenient. The reviewer has found this book to be exceptionally well-written and ideally suited either for independent reading or as a graduate level text for an introduction to everything about linear algebraic groups. It also contains a rich bibliography which is missing in the 'Bible'."

9.1. LIE GROUPS AND LIE ALGEBRAS

T. Springer, *Linear Algebraic Groups*, Reprint of the 1998 second edition, Modern Birkhäuser Classics, Birkhäuser Boston, Inc., Boston, MA, 2009. xvi+334 pp.

This second edition of this book is a completely new book. The old book presented the theory of linear algebraic groups over an algebraically closed field, but the new book studies the theory of linear algebraic groups over arbitrary fields. This book is a good introduction to linear algebraic groups written by an expert. The old book is elementary, and the new book is much more advanced.

According to MathSciNet, "it introduces the reader to the theory of affine algebraic groups over an algebraically closed field; this is done in nine chapters with an introductory chapter on algebraic geometry..., it covers the theory of algebraic groups over an arbitrary field in six chapters, again with an introductory chapter on algebraic geometry. Algebraic groups in this volume are viewed as groups in the category of (affine) algebraic varieties..."

M. Demazure, P. Gabriel, *Groupes algébriques. Tome I: Géométrie algébrique, généralités, groupes commutatifs. Avec un appendice Corps de classes local par Michiel Hazewinkel*, Masson & Cie, Éditeur, Paris; North-Holland Publishing Co., Amsterdam, 1970. xxvi+700 pp.

The first two chapters have been translated into English and appeared in the next book.

M. Demazure, P. Gabriel, *Introduction to Algebraic Geometry and Algebraic Groups*, Translated from the French by J. Bell, North-Holland Mathematics Studies, 39, North-Holland Publishing Co., Amsterdam-New York, 1980. xiv+357 pp.

This is an advanced modern introduction to linear algebraic groups, especially from the perspective of Grothendieck.

This book is suitable for readers who are interested in modern aspects of the theory of algebraic groups, but without enough knowledge of modern algebraic geometry to read the book SGA 3, which consists of three volumes: *Group Scheme, I: General Properties of Diagrams Groups*, Springer, Berlin, 1970, *Group Schemes, II: Groups of Multiplicative Type, and Structure Patterns in General Groups*, 1970, *Group Schemes, III: Structure of Reductive Group Schemes*, 1970.

Demazure was a student of Alexandre Grothendieck, and obtained his Ph.D. in 1965. Together with Grothendieck, he ran and edited the Séminaire de Géométrie Algébrique du Bois Marie on group schemes at the Institut des Hautes Études Scientifiques near Paris from 1962 to 1964. He was also one of the core members of the Bourbaki group from 1965 to 1985. In SGA3, Demazure introduced the definition of a root datum, a generalization of root systems for reductive groups that is central to the notion of Langlands duality.

V. Platonov, A. Rapinchuk, *Algebraic Groups and Number Theory*, Translated from the 1991 Russian original by Rachel Rowen. Pure and Applied Mathematics, 139, Academic Press, Inc., Boston, MA, 1994. xii+614 pp.

This book gives a comprehensive account of algebraic groups, their connections with and applications to number theory. Many results discussed here appear in book form for the first time. Written by two experts, the book contains a wealth of information and covers a broad range of topics, and is a very valuable reference for both students and experts.

According to MathSciNet, "The book is the first systematic exposition of the arithmetic theory of algebraic groups. In some sense it sums up a certain period of intensive treatment of algebraic groups over number fields that was started in the early 1960s and is still continuing. Before this book the only sources of information on the subject were numerous journal publications, different in their approaches, style and even terminology. It seems that the authors have succeeded in correcting this situation, giving a uniform and relatively self-contained exposition of such important topics as arithmetic groups and reduction theory, adéle groups, Galois cohomologies and Hasse principle, weak and strong approximation in algebraic groups, class numbers, and normal structure of the groups of rational points. It should be noted that a number of the results included in the book are original and many of them are due to the authors. In addition, the book contains a number of open problems which can be stimulating for new research in this direction. In general, it embraces all main results in the arithmetic theory of algebraic groups and so will be useful for mathematicians working in such areas as algebraic number theory, theory of algebraic groups, algebraic geometry, theory of discrete groups and automorphic functions."

R. Steinberg, *Lectures on Chevalley Groups*, Notes prepared by John Faulkner and Robert Wilson, Yale University, New Haven, Conn., 1968. iii+277 pp.

This is an important and original book. It has some impact on the theory of Chevalley groups.

According to MathSciNet, "These notes, based on the author's 1967–68 lectures at Yale, have been for a decade the standard reference for Chevalley groups over arbitrary (but especially finite) fields. In addition to the simple (adjoint type) groups treated by C. Chevalley [Tohoku Math. J. (2) 7 (1955), 14–66], the various central extensions and twisted analogues of these groups are described here.... In his concise style, the author has compressed an enormous amount of useful material into relatively few pages; this of course makes some demands on the reader..."

Though algebraic groups are important and closely related to algebraic geometry and Lie theories and have played an increasing important role in mathematics, there are not

9.1. LIE GROUPS AND LIE ALGEBRAS

many books on them. Besides the ones listed above, we mention two more which might be more accessible and recent.

P. Tauvel, R. Yu, *Lie Algebras and Algebraic Groups*, Springer Monographs in Mathematics, Springer-Verlag, Berlin, 2005. xvi+653 pp.

This book is a largely self-contained introduction to Lie algebra and algebraic groups. It contains a wealth of information on many related topics and is also a useful reference.

According to MathSciNet, "The theory of algebraic groups and Lie algebras is a deeply advanced and developed area of modern mathematics. It employs at the same time algebra, analysis and geometry, and contains numerous fundamental results which have found applications in many fields of mathematics and natural science. On the other hand, a researcher who wants to study this theory in detail or even only to enter into the subject, will collide with a huge amount of notation, notions and preliminary facts that are required in order to formulate and to prove central results. So it is very important to have a book which provides a self-contained introduction to the theory. Such an introduction is given in the present book. Moreover, in addition to basic facts on Lie algebras and algebraic groups, it covers a wide variety of deep results and even presents very recent developments on symmetric Lie algebras... The text is clearly written and the material is well organized and considered, so the present book may be strongly recommended both to a beginner looking for a self-contained introduction to the theory of algebraic groups and Lie algebras, and to a specialist who wants to have a systematic presentation of the theory."

M. Geck, *An Introduction to Algebraic Geometry and Algebraic Groups*, Oxford Graduate Texts in Mathematics, 10, Oxford University Press, Oxford, 2003. xii+307 pp.

Though algebraic geometry appears in the title, this is really an introduction to algebraic groups. It is in fact a very accessible introduction. It is easy to follow and uses classical examples to illustrate the general theories. In comparison with the other more systematic books on algebraic groups, this book can serve as a quick introduction to students and nonexperts who only want a taste or a quick view of algebraic groups.

According to MathSciNet, "The book under review can be viewed as an attempt to give a quick introduction to the theory of finite groups of Lie type... Taking into account his goals, the author restricts himself to considering affine algebraic varieties...

The style of exposition in the book is very reader-friendly. Each chapter contains bibliographic remarks and exercises (usually accompanied by hints). The proofs are clear and complete. The author luckily avoids overloading the text with technical details which might be frightening to a newcomer. All in all, I can agree with the author, who says in the introduction that among the readers one can imagine a student wishing to get a starting point before going over to more advanced material, as well as a lecturer or researcher with a specific interest in finite simple groups."

9.1.7 Algebraic invariant theories and representations of algebraic groups

Invariant theory studies actions of finite groups and algebraic groups on algebraic varieties by understanding their effect on functions. Classically, the theory was concerned with explicit description of polynomial functions that are invariant under the transformations from a given linear group. Invariant theory of finite groups is closely related to the Galois theory. One of the first major results was the theorem on the symmetric functions which describes the ring of polynomials in n variables which are invariant under the symmetric group S_n.

David Hilbert's positive answer to the question on finite generation of the algebra of invariants in 1890 opened a new direction in algebraic invariant theory. A later paper of Hilbert in 1893 dealt with the same questions in more constructive and geometric ways, but it remained virtually unknown until David Mumford brought these ideas of Hilbert back to life in the 1960s in his geometric invariant theory.

D. Hilbert, *Theory of Algebraic Invariants*, Translated from the German and with a preface by Reinhard C. Laubenbacher. Edited and with an introduction by Bernd Sturmfels, Cambridge University Press, Cambridge, 1993. xiv+191 pp.

The book is based on the handwritten course notes taken by Hilbert's student Sophus Marxen in 1897. Given Hilbert's revolutionary contribution to the invariant theory, it is a very valuable reference not only for historians of mathematics, but also for mathematicians.

According to MathSciNet, "The necessary explanation of the appearance in the 1990s of a review of an original text by D. Hilbert is given in the preface of the translator, R. C. Laubenbacher: 'Around the turn of the century, the University of Göttingen was a Mecca for mathematicians and students from around the world, including the United States. The visitors took back with them a large number of handwritten lecture notes. The present notes of Hilbert's 1897 course on invariant theory comprise 527 handwritten pages, taken by Hilbert's student Sophus Marxsen'... The significance of the time frame is emphasized in the introduction by the editor, B. Sturmfels: 'When Hilbert gave his course in 1897, his research in invariant theory had been completed. In particular, Hilbert's famous finiteness theorem had been proved and published in two striking papers... Thus 1897 was a perfect time for Hilbert to give an introduction to invariant theory, taking into account both the old approach of his predecessors and his new ideas.'... The short introduction highlights a few of the important developments in invariant theory during the century that has elapsed since Hilbert's original publications. Occasional editorial footnotes place some results in contemporary context. Two things make reading this book a particularly enjoyable experience. This is probably the closest one could get today to actually attending a course by Hilbert. And, as Sturmfels notes, 'It is this

9.1. LIE GROUPS AND LIE ALGEBRAS

bridge from nineteenth-century mathematics into twentieth-century mathematics which makes these course notes so special and distinguishes them from other treatments of invariant theory.'"

P. Olver, *Classical Invariant Theory*, London Mathematical Society Student Texts, 44, Cambridge University Press, Cambridge, 1999. xxii+280 pp.

The book gives an accessible yet in-depth introduction to the classical invariant theory, which attained its peak during the nineteenth century.

According to a review in Bulletin of AMS, "As a modern in-depth study of binary forms, this book is one of a kind. The book uses a rather ad-hoc approach to binary forms. This keeps the abstraction as low as possible. If one is more interested in invariant theory and representation theory for arbitrary classical groups, centered around Weyl's *The Classical Groups*, then *Representations and Invariants of the Classical Groups* by Goodman and Wallach is probably a better reference. The book is written in a pleasant style. The author cares a great deal about the subject. The many historical comments make it a quite enjoyable read."

According to MathSciNet, "The book is mainly oriented towards the applications of classical invariant theory, and certainly does not pretend to give the latest word on the present mathematical status of this beautiful subject... To summarize, this is a book that you can start reading knowing nothing whatsoever about the subject and that will get you to the point where you can start your own research. In fact, when you read it, you may want to start working immediately."

H. Derksen, G. Kemper, *Computational Invariant Theory*, Invariant Theory and Algebraic Transformation Groups, I, Encyclopaedia of Mathematical Sciences, 130, Springer-Verlag, Berlin, 2002. x+268 pp.

The book gives an accessible, comprehensive and up-to-date account of the algorithmic methods of invariant theory with respect to finite and linearly reductive groups. One basic question concerns how the invariant ring of a given group action can be computed. It is a valuable introduction to computational invariant theory for students and others.

According to Zentralblatt MATH, "The main aims of the book are centered around algorithmic methods of invariant theory, mainly with invariants of finite and linearly reductive groups. In case of finite groups the primary interest is devoted to the study of the modular case, where the characteristic of the ground field divides the order of the group. Of central interest is the question how the invariant ring of a given group action can be computed. Besides of the authors' interest in algorithmic approaches there are investigations about structural properties of invariant rings, such as the homological dimensions or the Noether bound for the degree of generators of the invariant rings... The text covers a lot of illustrating and instructing examples. It is intended as a companion for the study of rings of invariants with a view towards an algorithmic, computational approach."

J. Jantzen, *Representations of Algebraic Groups*, Second edition, Mathematical Surveys and Monographs, 107, American Mathematical Society, Providence, RI, 2003. xiv+576 pp.

This is a classical book on representations of algebraic groups by an expert.

According to MathSciNet, "This book is very well written and the author has taken great care over accuracy, both of mathematical details and in references to the work of others. The discussion is well motivated throughout. A fair amount of knowledge is assumed on the part of the reader in such subjects as algebraic geometry, homological algebra and the structure theory of algebraic groups. Some readers will no doubt find it convenient to have reference books on such subjects at hand when reading this volume...

This impressive and wide ranging volume will be extremely useful to workers in the theory of algebraic groups."

9.1.8 E. Artin, Geometric Algebra, Reprint of the 1957 original, Wiley Classics Library, A Wiley-Interscience Publication, John Wiley & Sons, Inc., New York, 1988. x+214 pp.

By the title "Geometric Algebra", it means the algebraic foundation or treatment of classical geometry, or rather classical groups. It explains how to characterize and understand difficult problems of these geometries or classical groups by elementary algebraic objects and properties. It is a classic on an important topic written by a master.

From the book description, "This classic text was written by one of the foremost mathematicians of the 20th century. Exposition is centered on the foundations of affine geometry, the geometry of quadratic forms, and the structure of the general linear group. Context is broadened by the inclusion of projective and symplectic geometry and the structure of symplectic and orthogonal groups."

According to a review in Bulletin of AMS, "When Hilbert's *Grundlagen der Geometrie* and other texts on the foundations of geometry appeared around the turn of the century, the approach was almost purely geometric. It is typical of the development of mathematics in the intervening years that, in this latest book on geometry, the approach is almost entirely algebraic. In the preface the author states that his aim is to offer a text (based on lecture notes of a course he has given at New York University) which would be of a geometric nature yet distinct from a course in linear algebra, topology, differential geometry, or algebraic geometry. This aim then accounts for the several different topics discussed in the book. For the student whose knowledge of modern algebra is meagre,

9.1. LIE GROUPS AND LIE ALGEBRAS

the first chapter offers a compilation of algebraic theorems (and their proofs) which he is most likely to need... The text is very well printed and the exposition is clear. The author makes every effort to encourage the reader by pointing out to him the easier parts of the book and by suggesting spots he may skip on a first reading. The text contains a number of exercises. The beginning graduate student, or very advanced undergraduate, will find this book an admirable introduction to material which is treated from a more advanced point of view and more extensively in Baer's Linear Algebra and Projective Geometry and Dieudonné's *La géométrie des groupes classiques*. Mathematicians will find on many pages ample evidence of the author's ability to penetrate a subject and to present material in a particularly elegant manner."

9.1.9 J. Tits, Buildings of Spherical Type and Finite BN-pairs, Lecture Notes in Mathematics, Vol. 386, Springer-Verlag, Berlin-New York, 1974. x+299 pp.

This is an important book on an important theory, Tits buildings, written by its inventor.

A building (also Tits building, Bruhat–Tits building) is a combinatorial and geometric structure which simultaneously generalizes certain aspects of flag manifolds, finite projective planes, and Riemannian symmetric spaces. Initially introduced by Jacques Tits as a means to understand geometrically the structure of exceptional groups of Lie type and hence to construct them over any field, the theory has also been used to study the geometry and topology of homogeneous spaces of both real and p-adic Lie groups and their discrete subgroups of symmetries. When a p-adic reductive Lie group has rank 1, then its associated building is a tree. As it is known, trees have been used to study free groups acting on them, and hence buildings are also important in understanding structures of the Lie groups and their subgroups. This book by its originator has played an important role in establishing Tits building as an important branch of modern mathematics.

Besides spherical Tits buildings discussed in this book, there are also Euclidean buildings, in particular, Bruhat-Tits buildings associated with reductive algebraic groups over local fields. Both types of buildings have been applied to various subjects of mathematics including differential geometry, topology, number theory, representation theory, harmonic analysis, algebraic geometry etc. For example, one first striking application of classifications of spherical Tits buildings in this book is the celebrated Mostow strong rigidity of compact locally symmetric spaces of higher rank. One reason is that geometry at infinity of symmetric spaces of noncompact type and noncompact semisimple Lie groups are often described by Tits buildings.

According to the review in Bulletin of AMS, "The relationships between certain algebraic, analytic and geometric structures and root systems in Euclidean spaces have been a source of methods and ideas that have had a profound impact on various parts of mathematics. Some particularly fruitful instances of this interaction are E. Cartan's classification of semisimple Lie algebras over the complex field, H. Weyl's papers on the representations of semisimple Lie algebras and Lie groups, E. Cartan's work on symmetric spaces, H. S. M. Coxeter's enumeration of finite groups generated by reflections, C. Chevalley's Tohoku paper and his classification of semi-simple algebraic groups, and Harish-Chandra's contributions to Fourier analysis on semisimple Lie groups. There are many others. It is not my purpose to give here a historical survey of this subject, but simply to remind the reader of some high points in the development of the theme, and to state that the volume which is the subject of this review belongs in this distinguished company.

The present volume is primarily a research article, devoted to the proofs of two main results. The mathematical community has been aware of the existence of these theorems for some time, and the author has announced parts of the work on several previous occasions. These notes contain the first detailed proofs of the main results. Their importance is difficult to overemphasize. They provide a common combinatorial foundation for several topics mentioned in the first paragraph, and are remarkable both for the originality and beauty of their conceptual framework, and for the depth and power required for the complete analysis of the possibilities.

There is no point for me to hide my enthusiasm for this work. It remains to express the hope that these notes will become familiar to a wide circle of readers, who will profit as much from the author's imagination, insight, and clear style, as I have."

9.1.10 More books on buildings and finite geometries

Due to wide and many unexpected applications of buildings in geometry, topology and algebra, there have been several good expositions of more topological aspects of them. We list several more recent books.

K. Brown, *Buildings*, Springer-Verlag, New York, 1989. viii+215 pp.

This is probably the most accessible introduction to buildings.

According to MathSciNet, "It is a pleasure to have, at last, books that can be used as introductions to this area. The emphasis on examples in the present well-written book, as well as its leisurely pace, make it very suitable for an introductory course. Moreover, the last two chapters are a valuable introductory source concerning affine buildings and their applications to arithmetic groups."

M. Ronan, *Lectures on Buildings,* Updated and revised, University of Chicago Press, Chicago, IL, 2009. xiv+228 pp.

9.1. LIE GROUPS AND LIE ALGEBRAS 455

This well-written book is a valuable introduction to buildings. Features and differences of the book by Brown and Ronan are well described in the following review in Bulletin of AMS: "The books of Brown and Ronan are the first (apart from the research-level monograph of Tits) to attempt a systematic development of the subject. They have some superficial resemblance, having similar titles and being of similar length and provenance (graduate courses given around 1987). Both are carefully written and liberally supplied with exercises. But in fact their aims and scope differ markedly. One symptom of this (mentioned above) is Brown's more leisurely build-up to the definition of buildings. Another symptom: Each book includes roughly 60 references, but only a quarter of these occur in both books. In spite of some inevitable overlap in coverage, the books are in fact largely complementary. Roughly speaking, Brown is more interested in describing special cases of buildings (spherical and Euclidean), in order to show how the applications to group cohomology come about, whereas Ronan is more concerned with classification problems, emphasizing group actions on buildings... Both of these fine books belong in every research library. Brown's approach will appeal to those looking for a gentle introduction to the 'classical' theory of buildings (and to those who share his interest in cohomology of discrete groups), but Ronan's book will do more to initiate the already motivated reader into current research on buildings and groups."

P. Abramenko, K. Brown, *Buildings. Theory and Applications*, Graduate Texts in Mathematics, 248. Springer, New York, 2008. xxii+747 pp.

There are several different ways to define and study buildings. This comprehensive book, a much expanded update of the earlier book by Brown, is the first one which presents all different approaches to buildings. It also contains a wealth of other information, and it a very valuable reference on buildings and applications of buildings.

According to MathSciNet, "The book under review, however, is the first encyclopedic treatment of buildings made available in the literature, so that with writing this book the authors manage to close a serious gap in the mathematical literature: the non-existence of an easily accessible reference for buildings beyond an introduction."

P. Garrett, *Buildings and Classical Groups*, Chapman & Hall, London, 1997. xii+373 pp.

This book is another textbook on buildings. In order to describe the structure of the classical groups such as general linear groups, symplectic groups and orthogonal groups over general fields and over p-adic fields, it gives a systematic study of buildings, both spherical and affine. This is a valuable introduction and complements the other books on buildings.

According to MathSciNet, "this is another introductory textbook to buildings after those of K. Brown and M. Ronan... It is intended to be and in fact is intelligible after completion of a basic course in algebra and contains accounts of the necessary facts about geometric algebra, reflection groups, p-adic numbers (and other valuation rings) and simplicial complexes and their geometric realizations. In particular, the reader is not expected to have any knowledge about algebraic groups."

Tits buildings is one kind of incidence geometry. A subject related to incidence geometry is finite geometry.

J.W.P. Hirschfeld, *Projective Geometries Over Finite Fields*, Second edition, Oxford Mathematical Monographs, The Clarendon Press, Oxford University Press, New York, 1998. xiv+555 pp.

This is the first systematic book devoted to finite projective spaces and their geometry and written by an expert. It is a very valuable reference.

According to MathSciNet, "The study of finite projective planes has made substantial progress in recent years. There are various applications in many fields of mathematics as well as in other sciences. The aim of this book is to give a comprehensive collection of our knowledge about projective planes over finite fields. Special attention is given to the properties of algebraic curves and of the combinatorial properties of subsets of points...This book is a very useful reference for all of those who are interested in finite Desarguesian planes and their applications."

J.W.P. Hirschfeld, J. Thas, *General Galois Geometries*, Oxford Mathematical Monographs, Oxford Science Publications, The Clarendon Press, Oxford University Press, New York, 1991. xiv+407 pp.

This book is the third and last volume of a treatise on projective spaces over a finite field. Together with *Projective Geometries Over Finite Fields*, and *Finite Projective Spaces of Three Dimensions*, they serve as a comprehensive and essential reference on projective geometry over finite fields.

According to MathSciNet, it "considers the general Galois geometry $PG(n,q)$, and thus encompasses the development in the two earlier volumes. However, the authors of the present volume have taken great care to ensure that it is largely independent of the earlier volumes, for example by including comprehensive and well-organized lists of notation in one of the appendices..."

S. Payne, J. Thas, *Finite Generalized Quadrangles*, Second edition, EMS Series of Lectures in Mathematics, European Mathematical Society (EMS), Zürich, 2009. xii+287 pp.

This is a valuable reference on generalized quadrangles, a special topic in finite geometry.

According to MathSciNet, "The book under review is a compendium of geometric and combinatorial results concerning generalized quadrangles... The authors were determined

9.1. LIE GROUPS AND LIE ALGEBRAS

to avoid all but the most elementary group theory in their presentation. This was in accordance both with the geometric and combinatorial nature of their own research, and with an attempted proof of the most important characterization theorem concerning generalized quadrangles... This book is a significant addition to the literature on finite geometries. It should stimulate further additions to the rapidly increasing literature on generalized polygons."

P. Dembowski, *Finite Geometries*, Ergebnisse der Mathematik und ihrer Grenzgebiete, Band 44, Reprint of the 1968 original, Classics in Mathematics, Springer-Verlag, Berlin, 1997. xii+375 pp

This is a comprehensive and systematic exposition of finite incidence structures. It is a very valuable reference.

According to MathSciNet, "Of the 1181 references in the bibliography of this book, 603 have been published since 1959. This fact alone suggests the value of the work to researchers in the field of finite geometry. The additional fact that this is the only attempt to survey such a vast amount of literature makes it almost indispensable... Such a vast amount of information as this book contains can only be accomplished in 375 pages by a very economical style of writing. Thus (in the usual spirit of the 'Ergebnisse' series), proofs which are readily available in the literature are omitted, while others are sketched in. Far from detracting from the value of the work, however, this style enhances it. For it enables one to have a good look at the forest without being too detracted by the individual trees. Having then obtained an understanding of the forest, the reader can determine the constituent nature of the trees to his heart's content through the use of any reasonably adequate library."

9.1.11 Applications of Lie theories to differential equations

As it is well known, finite groups were introduced in order to solve polynomial equations. Lie introduced Lie groups to find solutions of ordinary differential equations. This origin is gradually forgotten by some people who use Lie groups in spite of the fact that Lie groups are becoming more important. Algebraic groups are also closely connected with algebraic differential equations with coefficients that are meromorphic functions in domains of $\mathbb{C}^n$.

Applications of Lie groups to differential equations are well-explained by a review of Bulletin of AMS of a book of Peter Olver below: "A standard introductory textbook on ordinary or partial differential equations presents the student with a maze of seemingly unrelated techniques to construct solutions. Usually these unmotivated and boring techniques constitute the total experience with differential equations for an undergraduate. Faced with a given differential equation which is not a textbook model, one is hopelessly lost without 'hints!

In the latter part of the 19th century Sophus Lie introduced the notion of continuous groups, now known as Lie groups, in order to unify and extend these bewildering special methods, especially for ordinary differential equations. Lie was inspired by lectures of Sylow given at Christiania, present-day Oslo, on Galois theory and Abel's related works. [In 1881 Sylow and Lie collaborated in editing the complete works of Abel.] He aimed to use symmetry to connect the various solution methods for ordinary differential equations in the spirit of the classification theory of Galois and Abel for polynomial equations. Lie showed that the order of an ordinary differential equation can be reduced by one if it is invariant under a one-parameter Lie group of point transformations. His procedures were both constructive and aesthetic.

For ordinary differential equations Lie's work systematically and comprehensibly related a miscellany of topics including: integrating factors, separable equations, homogeneous equations, reduction of order, the method of undetermined coefficients, the method of variation of parameters, Euler equations, and homogeneous equations with constant coefficients. Lie also indicated that for linear partial differential equations, invariance under Lie groups leads directly to superpositions of solutions in terms of transforms. Why has Lie's approach not been adopted in standard textbooks?

A symmetry group of a system of differential equations is a group which maps solutions to other solutions of the system. In Lie's framework these groups consist of point transformations depending on continuous parameters, acting on the space of independent and dependent variables. Elementary examples include translations, rotations, and scalings. More generally, a solvable autonomous system of first-order ordinary differential equations essentially defines a one-parameter Lie group of transformations. Lie showed that, unlike discrete groups, for example reflections, the continuous group of point transformations admitted by a differential equation can be found by an explicit computational algorithm.

The applications of continuous groups to differential equations make no use of the global aspects of Lie groups. The applications use connected local Lie groups of transformations. Lie's three fundamental theorems showed that such groups are completely characterized in terms of their infinitesimal generators, which form a Lie algebra determined by its structure constants. Lie groups, and hence their infinitesimal generators, are naturally extended or ' prolonged' to act on the space of independent variables, dependent variables and derivatives of the dependent variables. Consequently, the nonlinear conditions of invariance of a given system of differential equations under Lie groups of transformations reduce to linear homogeneous conditions in terms of the infinitesimal generators of the group. These conditions are called the determining (or defining) equations of the group. Since the determining equations form an over-determined system of linear homogeneous partial differential equations, one can usually determine the infinites-

9.1. LIE GROUPS AND LIE ALGEBRAS

imals in closed form. For a given differential equation the entire procedure to determine the infinitesimal generators of its invariance group is routine; symbolic manipulation programs [2] have been developed to implement the scheme.

Accomplishments in applying Lie groups to differential equations also include: (1) The use of differential invariants, resulting from the infinitesimal (Lie group) approach, to find the most general differential equation invariant under a given symmetry group. (2) If an ordinary differential equation is derivable from a variational principle through a Lagrangian or Hamiltonian formulation, then invariance under a one-parameter 'variational' symmetry leads to a reduction of order by two. (3) If a system of partial differential equations is invariant under a Lie group of transformations, one can find constructively special classes of solutions, called similarity or invariant solutions, which are invariant under some subgroup of the full group admitted by the system. These solutions correspond to a reduction in the number of independent variables... (4) Without explicitly solving the determining equations for its infinitesimal generators, one can determine constructively whether or not a given nonlinear system of differential equations can be mapped into a system of linear differential equations by some one-to-one transformation. Moreover, one can find explicitly the transformation if such a mapping is possible. In general, Lie groups can be used to find mappings between equations provided the target equation can be characterized completely in terms of invariance under a Lie group of transformations. Most of the work in this direction has been developed in the past decade. (5) In 1918 Emmy Noether showed how the symmetry group of a variational integral (variational symmetry) leads constructively to a conservation law for the corresponding Euler-Lagrange equations. For example, conservation of energy follows from invariance under translation in time; conservation of linear and angular momenta respectively from translation in space and rotational invariances. Such variational symmetries leave invariant the Euler-Lagrange equations but the converse is false. (6) Recently the use of generalized symmetries defined by infinitesimal generators, including derivatives of the relevant dependent variables, has extended further the applicability of Lie groups to differential equations. The possibility of the existence of such transformations (which, according to Olver, are mistakenly called Lie-Backlund transformations in the modern literature) was recognized by Noether in her celebrated paper and came to fruition in the works of Kumei and other authors. These generalized symmetries cannot be represented in closed form from the integration of a finite system of ordinary differential equations as is the case for Lie groups. Generalized symmetries can be computed for a given differential equation by a simple extension of Lie's algorithm. They can be shown to account for the conserved Runge-Lenz vector for the Kepler problem and the infinity of conservation laws for the Korteweg-de Vries equation and other nonlinear partial differential equations exhibiting soliton behavior. Furthermore, multi-soliton solutions are similarity solutions for corresponding multiparameter generalized symmetries. The invariance of a partial differential equation under a generalized symmetry usually leads to invariance under an infinite number of generalized symmetries. The means of constructing an infinite number of such symmetries through the use of recursion operators was ingeniously shown by Olver in his famous paper."

E.R. Kolchin, *Differential Algebra and Algebraic Groups*, Pure and Applied Mathematics, Vol. 54, Academic Press, New York-London, 1973. xviii+446 pp.

This is a systematic self-contained book on differential algebras and algebraic groups written by a leading expert. It is a valuable introduction and reference.

Differential algebras are algebras equipped with a derivation. According to the preface of this book, "It is common knowledge that algebra, including algebraic geometry, historically grew out of the study of algebraic equations with numerical coefficients. In much the same way, differential algebra sprang from the classical study of algebraic differential equations with coefficients that are meromorphic functions in a region of some complex space $\mathbb{C}^m$. As a consequence, differential algebra bears a considerable resemblance to the elementary parts of algebraic geometry. Indeed, since an algebraic equation can be considered as a differential equation in which derivatives do not occur, it is possible to consider algebraic geometry as a special case of differential algebra."

According to MathSciNet, "This book, published after years of careful preparation, is a tour de force of the highest proportions. The author, as is well known, is the leading authority in the field of differential algebra. There are few people working in this area who have not benefitted enormously through personal contact with him and none who have not been influenced by his publications. His goal here is to present a unified exposition of the subject, in an algebraic setting, presuming no more than a standard first year graduate course in algebra."

An earlier, closely related book is the next one.

J. Ritt, *Differential Algebra*, American Mathematical Society Colloquium Publications, Vol. XXXIII, American Mathematical Society, New York, 1950. viii+184 pp.

This is a classic on differential algebra by one of its founders. It is still a valuable reference.

From the book description, "A gigantic task undertaken by J. F. Ritt and his collaborators in the 1930's was to give the classical theory of nonlinear differential equations, similar to the theory created by Emmy Noether and her school for algebraic equations and algebraic varieties. The current book presents the results of 20 years of work on this problem."

According to an Amazon review, "The book is still widely cited and can still be read profitably, despite its date of publication. The author's work was inspired by the work of

9.1. LIE GROUPS AND LIE ALGEBRAS

the mathematicians Emily Noether and van der Waerden, and is now referred to as the Ritt theory of characteristic sets of prime differential ideals. The author's work has been developed considerably further than what he envisaged, to subfields of mathematics now known as differential algebraic geometry and differential algebraic groups."

M. van der Put, M. Singer, *Galois Theory of Linear Differential Equations*, Grundlehren der Mathematischen Wissenschaften, 328, Springer-Verlag, Berlin, 2003. xviii+438 pp.

This book was written as an accessible textbook for first-year graduate students, but is a definitive, modern and comprehensive account of the Galois theory of linear differential equations. It is a very valuable reference.

According to MathSciNet, "This is a great book, which will hopefully become a classic in the subject of differential Galois theory. Although there is a large bibliography on the subject (which is very much en vogue today), it is almost exclusively composed of papers, and the specialist, as well as the novice, have long been missing an introductory book covering also specific and advanced research topics. This gap is filled by the volume under review, and more than satisfactorily. We delay the detailed comments on each chapter to state some general impressions. The book is carefully written: the authors have made a great effort to state the results in a language as common as possible, making use of specialized terminology only when strictly necessary. The material is introduced step by step and with a clear distinction of what is 'common knowledge' (like the Jordan Classification Theorem) and what is 'specifically required' (a bit of sheaf theory, some results on linear algebraic groups, etc.). Actually, the latter topics are briefly treated in the Appendices. The content is didactically presented. Almost each theorem is followed or preceded by one or several particular cases and examples. The reader finds numerous remarks on why a hypothesis is necessary, what kind of examples are not covered by a result, possible extensions of the theorems, and the like. Rather than presenting the more general and categorial proofs, the authors have taken care to make them understandable by people not familiar with the subject. In this aspect, one cannot help comparing this book (with advantage for the present volume) to Kolchin's and Ritt's treatises on differential algebra. The above qualities show not only the mathematical ability of the authors but also their competence as book writers. "

P. Olver, *Applications of Lie Groups to Differential Equations*, Second edition, Graduate Texts in Mathematics, 107, Springer-Verlag, New York, 1993. xxviii+513 pp.

This is an excellent introduction to applications of Lie groups to differential equations, which is a classical and yet new subject to many people. It is also a very valuable reference.

According to MathSciNet, "Since the discovery of the beautiful structure of the Korteweg-de Vries equation (KdV for short) in the 1960s, new frontiers in the treatment of nonlinear partial differential equations have been reached and numerous important techniques and methods have been developed. Apart from this – at least in this field – a new consciousness about historical perspectives has arisen. Many monographs treating recent developments in this field can be found but yet still missing was a book taking account of the historical context while reflecting the general spirit of research done in this area in a comprehensive way and at the same time presenting a broad view of some major parts of the recent developments. Ideally, the book should even do this in such a way that it may serve as a textbook for beginners.

Here is such a book, or at least a book which comes as close to these aims as possible at the present time... the present book is the second attempt to give an introduction to the classical treatment of applications of Lie groups to differential equations on one side and to serve on the other side as a textbook giving the beginner access to some beautiful mathematics developed over the last two decades."

According to a review in Bulletin of AMS, "Olver discusses effectively most of the accomplishments in applying Lie groups to differential equations. He has unearthed important results unknown to many specialists. Olver does not discuss connections with separation of variables nor applications to boundary value problems and relegates his limited presentation on mappings mostly to the exercises. These omissions in no way reflect negatively on this exciting text."

P. Olver, *Equivalence, Invariants, and Symmetry*, Cambridge University Press, Cambridge, 1995. xvi+525 pp.

The book studies the equivalence problem, i.e., to decide when two mathematical objects are the same under a change of variables. The symmetries of a given object are defined to be the group of self-equivalences, and conditions guaranteeing equivalence are expressed in terms of invariants, which are unaffected by the changes of variables. These problems arise from many different situations such as differential equations, variational problems, manifolds, Riemannian metrics, polynomials, and differential operators. This is a valuable reference on these important concepts, problems and solutions.

According to MathSciNet, "The book may profitably be used as the basis of a graduate course in which the central point is the study of differential invariants, with a detailed account of the equivalence problem for G-structures as formulated by E. Cartan. At any rate, it represents an important effort to present the theory of the equivalence problem in a modern language with many new applications to concrete situations; in the author's own words, 'it is a provocative blend of mathematical flavors'."

9.1. LIE GROUPS AND LIE ALGEBRAS

9.1.12 Discrete subgroups of Lie groups and algebraic groups

Discrete groups are groups with the discrete topology and carry the same information as the correspond abstract groups. A subgroup Γ of a Lie group G is called a discrete subgroup if the induced topology is the discrete topology. An important class of discrete subgroups of Lie groups consists of arithmetic subgroups of Lie groups. Though groups isomorphic to arithmetic subgroups of Lie groups are also called arithmetic groups, it is the embedding into Lie groups and actions on homogeneous spaces of the Lie groups that make arithmetic subgroups really interesting and important. For example, embedding of the infinite cyclic group $\mathbb{Z}$ into $\mathbb{R}$ produces a lot of deep mathematics such as the Poisson summation formula and the meromorphic continuation of the Riemann zeta function.

Discrete subgroups of semisimple Lie groups produce locally symmetric spaces, and the study of discrete subgroups of Lie groups are intertwined with the study of locally symmetric spaces.

A particularly important class of discrete subgroups are lattices of semisimple Lie groups. They correspond to locally symmetric spaces of finite volume. Locally symmetric spaces associated with the Lie group $SL(2,\mathbb{R})$ are hyperbolic surfaces, and their hyperbolic metrics can be continuously deformed and hence are not rigid. The beautiful Teichmüller theory has been developed to understand the collection of all hyperbolic metrics on surfaces. On the other hand, other locally symmetric spaces of noncompact type of finite volume enjoy many rigidity properties, for example, the Mostow strong rigidity.

Though discrete subgroups of Lie groups and locally symmetric spaces are central objects in mathematics, there are not many easily accessible books about them.

A. Borel, *Introduction aux groupes arithmétiques*, Publications de l'Institut de Mathématique de l'Université de Strasbourg, XV, Actualités Scientifiques et Industrielles, No. 1341, Hermann, Paris, 1969, 125 pp.

A crucial part of the theory of arithmetic groups is the reduction theory, i.e., finding good fundamental domains of arithmetic subgroups of algebraic groups. Borel made fundamental contribution towards this, and this is a classical introduction to arithmetic subgroups and the reduction theory and applications. It is still probably the best consist introduction to this important subject.

According to MathSciNet, "This book, based on lectures given by the author in 1964 at the Institut Henri-Poincaré, is devoted mainly to the study of fundamental domains in a reductive algebraic group G defined over $\mathbb{Q}$, relative to an arithmetic subgroup $\Gamma \subset G_{\mathbb{Q}}$... This book is a valuable addition to the literature of arithmetic groups; although the style is concise and the proofs (in later sections) are often demanding of the reader, the author's introductory remarks and bibliographical notes serve well to place the main theorems in historical perspective and to indicate their logical interdependence."

M.S. Raghunathan, *Discrete Subgroups of Lie Groups*, Ergebnisse der Mathematik und ihrer Grenzgebiete, Band 68, Springer-Verlag, New York-Heidelberg, 1972. ix+227 pp.

This is a standard reference on discrete subgroups of Lie groups written by an expert. Arithmetic subgroups of semisimple Lie groups are lattices, but the class of lattices is strictly bigger. This book gives a systematic discussion of lattices of all types of Lie groups. It may not be the easiest introduction, but it is certainly a very valuable and unique reference on discrete subgroups of general Lie groups.

According to MathSciNet, "The content of the book pretty much reflects the 'state-of-the art' at the time it was written (based on lectures given by the author at Yale in 1968/69 and at the Tata Institute in 1969/70)... Throughout the book many useful remarks and examples are given. The proofs of many of the known results are new. Unfortunately the case of SL(2,R), a great exception in the whole theory, is not discussed at all, nor is a reference given... Also not treated are questions of nonarithmeticity... Finally, the book was written before the important results of Margulis were available in print."

G. Mostow, *Strong Rigidity of Locally Symmetric Spaces*, Annals of Mathematics Studies, No. 78, Princeton University Press, Princeton, N.J.; University of Tokyo Press, Tokyo, 1973. v+195 pp.

This is a classic and a fundamental book. It is also quite readable. Mostow's strong rigidity proved in this book has established new subjects in mathematics and has a huge influence on both geometric group theory, geometry of manifolds of non-positive curvature and other areas. For example, it gave the first major application of the rigidity of Tits buildings and helped bring broader recognition to the theory of buildings. This is indeed a great book written by a master with great insight and vision.

According to MathSciNet, "It has been known since Riemann that the conformal equivalence classes of compact Riemann surfaces of genus g make up a manifold of dimension $6g - 6$. In the language of Lie groups, this means that the fundamental group Γ of one such surface considered as a subgroup of $PSL(2,\mathbb{R})$ (via the uniformization theorem) admits a continuous family of deformations not conjugate to Γ. A. Selberg, in 1960, made the remarkable discovery that certain higher-dimensional symmetric compact manifolds of negative curvature behave very differently in that the corresponding fundamental groups of such manifolds treated as discrete subgroups of the Lie groups G of isometries of the universal covering could not be deformed nontrivially...

These developments led the author to raise the following question (ten years ago): Let G and G' be connected semisimple Lie groups with no compact factors and with trivial centers; let Γ [Γ'] be a discrete 'irreducible uniform' subgroup of $G[G']$; assume that G is not locally isomorphic to $SL(2,\mathbb{R})$; then any isomorphism f of Γ on Γ' extends

9.1. LIE GROUPS AND LIE ALGEBRAS

to a (unique) continuous isomorphism of G on G'. (Roughly speaking, the question can be reformulated as follows: does the fundamental group of a compact locally symmetric space of higher dimension determine that space uniquely as to isometry? The work under review is the culmination of the author's profound investigations into this question; it answers the questions raised there in the affirmative..."

R. Zimmer, *Ergodic Theory and Semisimple Groups*, Monographs in Mathematics, 81, Birkhäuser Verlag, Basel, 1984. x+209 pp.

This is an influential book. Among many other things, it contains the first lucid exposition of the super-rigidity theorem and arithmeticity theorem of Margulis of lattices of semisimple Lie groups. It is well-written and has been a very good introduction to many important topics related to these major results of Margulis. It is also a very valuable reference.

According to MathSciNet, "A class of discrete subgroups of Lie groups, called lattices (namely, those for which the corresponding homogeneous space admits a finite invariant measure), play an important role in various branches of mathematics such as number theory, geometry, dynamics, ergodic theory, etc. In recent years there has been remarkable progress in the understanding of these subgroups, thanks to the profound work of G. A. Margulis on 'irreducible' lattices in semisimple Lie groups with trivial center and $\mathbb{R}$-rank at least 2. The theorems on superrigidity (about extending a finite-dimensional representation of the lattice to the whole group), arithmeticity (showing that the lattice can be viewed as arising from a well-known arithmetical construction) and on normal subgroups of such lattices are some of the high points.

To put this in a proper perspective for a reader not familiar with the area, it may be noted that various aspects of lattices in solvable Lie groups have been satisfactorily understood for quite some time, through the work of L. Auslander, G. D. Mostow and various other authors...

Ergodic theory plays a role in some of Margulis' proofs and especially in a simplification made by H. Furstenberg. On the other hand, the present author discovered that the rigidity theory could be recast in a wider framework to study ergodic actions with a finite invariant measure, of the semisimple Lie groups as above, proving in particular, an interesting and rather unexpected rigidity property of such actions (showing that under certain natural conditions 'orbit equivalence' of such actions, of possibly different groups, already implies conjugacy of the actions).

The book under review gives a lucid account of the work of Margulis and the subsequent developments. There is a major hurdle involved in such a task. The work involves ideas both from the theory of algebraic groups and ergodic theory, areas which are traditionally pursued by almost disjoint sets of mathematicians. The author has made an attempt to be accessible to both groups..."

G. Margulis, *Discrete Subgroups of Semisimple Lie Groups*, Ergebnisse der Mathematik und ihrer Grenzgebiete (3), 17, Springer-Verlag, Berlin, 1991. x+388 pp.

This is a very reliable and authoritative book written by the absolute leading expert on rigidity properties of lattices of Lie groups. It gives a coherent account of lattices of semisimple Lie groups, their structure and their classifications. It might not be easy as an introduction for students or nonexperts, but it is a very valuable and reliable reference for everyone.

According to MathSciNet, "Since the appearance of Raghunathan's book, there have been several very profound developments in the theory of lattices in semisimple groups and most of these developments are due to the author of the book under review. Among these are the proof of super-rigidity and S-arithmeticity of irreducible lattice in any group of the form $G = \coprod_{i=1}^{n} G_i$, where G_i is the group of rational points of a semisimple algebraic group G_i defined over a nondiscrete locally compact field k_i and $\sum k_i$-rank $G_i \geq 2$, and also the proof of finiteness of the index of any noncentral normal subgroup of a lattice in such a G. The author set himself the task of presenting a complete and reasonably self-contained account of these results in the book under review. He has admirably succeeded in his task. The proofs of results on super-rigidity, arithmeticity and the noncentral normal subgroups of lattices in real semisimple Lie groups are also given in a monograph by R. J. Zimmer [Ergodic theory and semisimple groups, 1984]. However, in the present book these results are proved in the most natural general setting..."

9.1.13 **J.P. Serre, Trees, Translated from the French by John Stillwell, Springer-Verlag, Corrected 2nd printing of the 1980 English translation, Springer Monographs in Mathematics, Springer-Verlag, Berlin, 2003. x+142 pp.**

This is a very well-written and influential book written by a master. It created the *Bass–Serre theory*, which is concerned with the algebraic structure of groups acting by automorphisms on simplicial trees. The theory relates group actions on trees with decomposing groups as iterated applications of the operations of free product with amalgamation and HNN extension, via the notion of the fundamental group of a graph of groups. This book is a very good introduction to people in geometric group theory, topology, algebraic group theory etc.

From the book description, "The seminal ideas of this book played a key role in the development of group theory since the 70s. Several generations of mathematicians learned geometric ideas in group theory from this book. In it, the author proves the fundamental theorem for the special cases of free groups and tree products before dealing with the proof of the general case."

9.1.14 Discrete subgroups of low rank Lie groups and algebraic groups

A. Lubotzky, *Discrete Groups, Expanding Graphs and Invariant Measures*, With an appendix by Jonathan D. Rogawski, Progress in Mathematics, 125, Reprint of the 1994 edition, Modern Birkhäuser Classics, Birkhäuser Verlag, Basel, 2010. iii+192 pp.

This is the first systematic exposition of expander graphs, their connections with number theory and several striking applications. There has been a lot of progress on expanders since the publication of this book, but this book is still probably the best introduction to the subject.

According to the book description, "The book presents the solutions to two problems: the first is the construction of expanding graphs – graphs which are of fundamental importance for communication networks and computer science; the second is the Ruziewicz problem concerning the finitely additive invariant measures on spheres. Both problems were partially solved using the Kazhdan property (T) from representation theory of semi-simple Lie groups. Later, complete solutions were obtained for both problems using the Ramanujan conjecture from analytic number theory. The author, who played an important role in these developments, explains the two problems and their solutions from a perspective which reveals why all these seemingly unrelated topics are so interconnected. The unified approach shows interrelations between different branches of mathematics such as graph theory, measure theory, Riemannian geometry, discrete subgroups of Lie groups, representation theory and analytic number theory.

Special efforts were made to make the book accessible to graduate students in mathematics and computer science. A number of problems and suggestions for further research are presented."

P. Sarnak, *Some Applications of Modular Forms*, Cambridge Tracts in Mathematics, 99, Cambridge University Press, Cambridge, 1990. x+111 pp.

This book gives a brilliant exposition of striking applications of modular forms to solve three famous open problems: Ruziewicz's problem on invariant measures. Ramanujan graphs in expander graph theory, Linnik's Problem on equidistribution of certain points on the sphere S^2. It is a very valuable reference and will complement well the above book of Lubotzky.

According to MathSciNet, "This book is concerned with the application of the theory of modular forms to the solution of three problems, each of which seems at first unrelated to the theory of modular forms. The three problems can each be reduced to the problem of estimating the size of the Fourier coefficients of modular forms, of either integral or half-integral weight... This book treats in detail a remarkable range of ideas and beautiful mathematics. It is highly recommended to everyone interested in modular forms."

A. Beardon, *The Geometry of Discrete Groups*, Graduate Texts in Mathematics, 91, Corrected reprint of the 1983 original, Graduate Texts in Mathematics, 91, Springer-Verlag, New York, 1995. xii+337 pp.

This is a popular book on Fuchsian groups and their higher dimensional generalizations acting on real hyperbolic spaces. This is comprehensive and is a very valuable introduction to the subject. It is also a convenient reference.

From the book description, "This text is about the geometric theory of discrete groups and the associated tessellations of the underlying space... A detailed account of analytic hyperbolic trigonometry is given, and this forms the basis of the subsequent analysis of tessellations of the hyperbolic plane. Emphasis is placed on the geometrical aspects of the subject and on the universal constraints which must be satisfied by all tessellations."

According to MathSciNet, "This book is suitable as a textbook for a course and is a valuable resource for graduate students or advanced undergraduates... It leads towards but does not enter the realm of hyperbolic three-manifolds, which has expanded rapidly since Thurston's work began to appear in 1977..."

W. Fenchel, J. Nielsen, *Discontinuous Groups of Isometries in the Hyperbolic Plane*, Edited and with a preface by Asmus L. Schmidt, Biography of the authors by Bent Fuglede, de Gruyter Studies in Mathematics, 29, Walter de Gruyter & Co., Berlin, 2003. xxii+364 pp.

This was a famous manuscript before it was finally published in 2003. It is a classic and has had a huge influence.

According to MathSciNet, "This monograph is based on what became known as the famous Fenchel-Nielsen manuscript. Nielsen (1890–1959) started this project well before World War II, and his interest arose through his deep investigations on the topology of Riemann surfaces and from the fact that the fundamental group of a surface of genus greater than one is represented by such a discontinuous group. Fenchel (1905–1988) joined the project later and took over much of the preparation of the manuscript. The present book is special because of its very complete treatment of groups containing reversions and because it avoids the use of matrices to represent Möbius maps... The monograph is intended for students and researchers in the many areas of mathematics that involve the use of discontinuous groups."

W. Goldman, *Complex Hyperbolic Geometry*, Oxford Mathematical Monographs, Oxford Science Publications, The Clarendon Press, Oxford University Press, New York, 1999. xx+316 pp.

Real hyperbolic spaces are well-known and there are many books about them. Complex hyperbolic spaces are a close cousin, but there are subtle differences them. This book is the first one which gives a comprehensive account of the geometry of the complex

9.1. LIE GROUPS AND LIE ALGEBRAS

hyperbolic space H_c^n, its boundary ∂H_c^n and isometric group actions. It is a valuable introduction and reference.

According to MathSciNet, "An interesting feature of complex hyperbolic space is that, unlike real hyperbolic space, it contains no totally geodesic real hypersurfaces (when the complex dimension is at least 2). This increases the difficulty of constructing polyhedra (for example fundamental polyhedra for discrete groups of complex hyperbolic isometries). One of the major themes of this book is the study of a particular class of real hypersurfaces which are a good substitute for totally geodesic ones. These hypersurfaces are bisectors: the locus of points equidistant from a particular pair of points... In addition to being well written, the book is profusely illustrated."

B. Maskit, *Kleinian Groups*, Grundlehren der Mathematischen Wissenschaften, 287, Springer-Verlag, Berlin, 1988. xiv+326 pp.

This is a relatively modern, systematic account of Kleinian groups, but does not include the work of Thurston on hyperbolic 3-manifolds and related topics. This is a valuable reference on Kleinian groups.

9.1.15 Combinatorial groups and geometric group theory

Geometric group theory studies finitely generated groups by exploring connections between algebraic properties of such groups and topological and geometric properties of metric (or more generally topological) spaces on which these groups act. When a group is not a discrete subgroup of a topological group or defined via a transformation group, then a basic idea is to endow a finitely generated group with word metrics and the group multiplication becomes an isometric action of the group, and the group can be viewed as a metric space (or as embedded into its Cayley graph). Geometric group theory is relatively new and has been actively pursued since late 1980s. Geometric group theory closely interacts with low-dimensional topology, hyperbolic geometry, algebraic topology, discrete subgroups of Lie groups, computational group theory, dynamical systems, probability theory, K-theory, and other areas of mathematics.

Combinatorial group theory is closely related to geometric group theory, and is often considered to be a part of geometric group theory. It is concerned the theory of free groups, and the concept of a presentation of a group by generators and relations. It is much used in geometric topology, since the fundamental group of a simplicial complex has in a natural and geometric way such a presentation. Consequently it is also concerned with group actions on geometric-topological objects. Combinatorial group theory also studies a number of algorithmically insoluble problems, most notably the word problem for groups, and the classical Burnside problem. Since combinatorial methods are used frequently, it is called "Combinatorial group theory".

H.S.M. Coxeter, W.O.J. Moser, *Generators and Relations for Discrete Groups*, Fourth edition, Ergebnisse der Mathematik und ihrer Grenzgebiete, 14, Springer-Verlag, Berlin-New York, 1980. ix+169 pp.

This is a classical book on combinatorial group theory written by experts. It is not a systematic book on combinatorial group theory, but contains a large collection of very valuable examples. It has been useful and will continue to be so in the future,

According to MathSciNet, "The scope and the flavour of this work are perhaps alike best illustrated by quoting the first two sentences of the preface. 'When we began to consider the scope of this book, we envisaged a catalogue supplying at least one abstract definition for any finitely-generated group that the reader might propose. But we soon realized that more or less arbitrary restrictions are necessary, because interesting groups are so numerous.' In fact, the book brings together a great deal of information on generators and relations for a wide variety of groups. For instance, it deals with the crystallographic groups, with the fundamental groups of surfaces, with the symmetric and related groups, with the projective linear groups over finite fields, and with groups on real Euclidean space generated by reflections. There is also a systematic method for checking the sufficiency of a set of relations, and considerable attention to graphical representations. The book will be invaluable to anyone who believes, as I do, that progress in Group Theory depends primarily on an intimate knowledge of a large number of special groups."

R. Lyndon, P. Schupp, *Combinatorial Group Theory*, Ergebnisse der Mathematik und ihrer Grenzgebiete, Band 89, Reprint of the 1977 edition, Classics in Mathematics, Springer-Verlag, Berlin, 2001. xiv+339 pp.

This is a standard reference and also an introduction to combinatorial group theory.

According to MathSciNet, "This book has the same title as the one by W. Magnus, A. Karrass and D. Solitar [Combinatorial group theory: Presentations of groups in terms of generators and relations, 1966], and like that book, defines the boundaries of the subject now called combinatorial group theory... the book is a valuable and welcome addition to the literature, containing many results not previously available in a book. It will undoubtedly become a standard reference."

P. de la Harpe, *Topics in Geometric Group Theory*, Chicago Lectures in Mathematics, University of Chicago Press, Chicago, IL, 2000. vi+310 pp.

Though geometric group theory is very actively pursued by many people, there are not too many books on the theory. This book is the first systematic account of this important new subject.

According to MathSciNet, "The main theme of geometric group theory is therefore that a finitely generated group Γ has a well-defined notion of geometry up to quasi-isometry, and to study this geometry one is free to pick as a model any proper, geodesic metric space on which Γ acts properly discontinuously and cocompactly. Intelligent choice of a model geometry, influenced by the problem at hand, is very important.

The author's introduction states quite nicely: 'The purpose of this book is to provide an introduction to this point of view, with emphasis on finitely-generated versus finitely-presented groups, on growth of groups, and on examples.'

The book functions on two levels. On one level the book gives an introduction to a topic, sometimes quite brief and sometimes very thorough, usually including enough definitions and proofs of elementary results so that a dedicated graduate student can follow what is about to come, with plenty of revealing examples and exercises. Ramping up to the next level, the book gives a rapid and broad overview of recent research, with sketches of ideas, and sprinkled with open problems. This combination makes for fun reading, stoking a reader's interest in the topic."

D. Epstein, J. Cannon, D. Holt, S. Levy, M. Paterson, W. Thurston, *Word Processing in Groups*, Jones and Bartlett Publishers, Boston, MA, 1992. xii+330 pp.

This is a very useful reference (and probably the only book) entirely devoted to theory of automatic groups, one important class of groups in geometric (or combinatorial) group theory.

According to Zentralblatt MATH, "The authors of this book have created the new theory of automatic structures on groups (automatic groups), a branch of combinatorial group theory. This book is the first to develop this new theory, and in fact it has been cited prior to publication as a reference in many papers... A major class of automatic groups (to geometers, this class is probably the most important general class) is the class of word hyperbolic or negatively curved groups of M. Gromov."

M. Gromov, *Asymptotic Invariants of Infinite Groups*, In *Geometric Group Theory*, Vol. 2 (Sussex, 1991), 1–295, London Math. Soc. Lecture Note Ser., 182, Cambridge Univ. Press, Cambridge, 1993.

Strictly speaking, this is not a monograph, but rather a long paper. This paper has had a huge impact on the geometric (or metric) aspect of infinite discrete groups. It has attracted and introduced many people into the subject of geometric group theory and large scale geometry.

According to a review in Bulletin of AMS, "It is most important, in reading this book, to take the author literally in his whole warning paragraph (section 0.4, page 10), including its concluding sentence: 'The readers of this paper should not expect new theorems (not even half proved ones), but they may come across some amusing problems.' The book is not meant to contain any theorems. It is not a monograph developing a concrete and well-defined new theory, as was the author's famous monograph on hyperbolic groups. It is clear to the reviewer that this book is written solely for the purpose of stimulating thought. The author is trying to convey, in very general terms, his overall vision of the subject. The best way to explain this is that the book is itself written in the asymptotic spirit! The author is trying to convey a look of the subject that fits together very well when looked at from afar or in the large scale, but where many details have not yet been worked out (and where in many cases the author has no intention of working them out) and may not yet fit together at all. Indeed, many of the 'amusing problems' in the book consist of an invitation (or perhaps a challenge) by the author to the reader to formulate and prove precise statements 'in the small scale' that the author can only glimpse 'asymptotically'... In summary, this book is really a collection of problems, presented in an unconventional way. The problems are built from a large collection of instructive and stimulating examples and are held together by a remarkable asymptotic view of the subject. Some problems may have a better-defined direction than others... The number of problems in the book is immense and impossible to review in a meaningful way... The reviewer can recommend to any mathematician picking up this book, glancing through it until he or she finds an area of mathematics dear to his or her heart, then reading that part to see if it makes sense. Once it makes some sense, read it again to see if it generates any new ideas."

B.A.F. Wehrfritz, *Infinite Linear Groups. An Account of the Group-Theoretic Properties of Infinite Groups of Matrices,* Ergebnisse der Matematik und ihrer Grenzgebiete, Band 76, Springer-Verlag, New York-Heidelberg, 1973. xiv+229 pp.

Linear groups (or matrix groups) provide some of the most important infinite discrete groups, or infinite abstract groups. This book is the first systematic book on different aspects of linear groups as discrete groups. It is a valuable reference.

According to MathSciNet, "A linear group is a group isomorphic to a subgroup of $GL_n(K)$ for a (commutative) field K. The study of such groups, especially the finite ones, is an established part of group theory... The book is '... an attempt to give an account of those properties of linear groups that seem relevant to infinite group theory'. Its author has played a substantial part in the development of these properties, and it is not surprising that the attempt is successful... the linear groups themselves will prove more useful than the more general infinite groups to which they may be applied."

According to a review in Bulletin of AMS, "Linear groups arise in many situations. The classical linear groups over finite fields are vital in the classification of finite simple groups. Linear groups occur as groups of rotations of lattices in three dimensions in a study of crystallographic groups. In the study of groups in general, linear groups occur very naturally as groups of automorphisms of certain classes of abelian groups or of groups with abelian factors. Free groups of countable rank and polycyclic groups have faithful linear representations over the integers, and simple proofs of the residual properties of these groups can be based on this fact. Often linear groups serve as readily constructed examples and counterexamples in group theory. The subject matter and techniques of the theory of group representations and the study of linear groups have much in common, but the emphases are different. In the former case one is trying to say something about the class of all representations of a given group, whilst in the second case one is usually studying a single representation or a small subset of representations of a group. Since very little seems to be known about the class of all representations of an infinite (discrete) group, the second point of view is of particular interest in the theory of infinite groups, and in this connection the linearity of a group is used as a kind of finiteness condition... In the book under review, Wehrfritz makes a systematic survey of what is now known about the answer to the question: If an infinite group has a faithful linear representation, what does this tell us about the structure of the group? It is a worthwhile piece of work because in the last fifteen to twenty five years a large number of results have been discovered, and much of this material was only available in scattered papers; often isolated results proved by people who needed them for some particular application..."

L. Fuchs, *Infinite Abelian Groups. Vol. I., Vol. II.* Pure and Applied Mathematics, Vol. 36, 36-II. Academic Press, New York-London 1970 xi+290 pp., 1973. ix+363 pp.

Abelian groups form a very special class of groups. They are relatively simple compared with noncommutative groups, and they provide a good motivating source for module theory. These two volumes provide a valuable and comprehensive introduction to infinite abelian groups. They have been a standard reference.

According to MathSciNet, "There are only three other books that deal seriously with abelian groups: I. Kaplansky's *Infinite Abelian Groups*, 1954, A. G. Kuros' *The Theory of Groups*, 1953, and the author's *Abelian groups*, 1958. In the dozen years since the last of these appeared, many important theorems have been discovered, usually by homological techniques. Indeed, the subject might be called "modules over a principal ideal domain" to emphasize that it is a chapter of that part of homological algebra dealing with modules over arbitrary rings. Many group-theoretic results are best understood in this broader context. Of course, the deepest and most interesting theorems about groups are those that do not hold for all R-modules.

This book, the first of two volumes, gives the basic definitions, constructions, and first theorems about abelian groups; the deeper results will appear in the second volume... The author's stated aims are 'to introduce graduate students to the theory of abelian groups and to provide a young algebraist with a reasonably comprehensive summary of the material on which research in abelian groups can be based'."

9.1.16 Coxeter groups

Coxeter groups are certain groups generated by reflections. A discrete subgroup of the isometry group of the Euclidean space generated by reflections are examples of Coxeter groups, called Euclidean reflection groups, which include the symmetry groups of regular polyhedra, and the Weyl groups of simple Lie algebras are such groups. The finite Coxeter groups are exactly the finite Euclidean reflection groups. Examples of infinite Coxeter groups include the triangle groups corresponding to regular tessellations of the Euclidean plane and the hyperbolic plane, and the Weyl groups of infinite-dimensional Kac–Moody algebras.

J. Humphreys, *Reflection Groups and Coxeter Groups*, Cambridge Studies in Advanced Mathematics, 29, Cambridge University Press, Cambridge, 1990. xii+204 pp.

This is an advanced introduction to the theory of Coxeter groups. It is both an introduction to the Bourbaki book on Lie groups and algebras (chapters 4-6) and an updating of it. It is also a valuable reference.

According to MathSciNet, "The symmetry groups of the regular polyhedra in $\mathbb{R}^n$ may be generated by reflections $s_1, \ldots, s_n$ and have a presentation with defining relations of the form $s_i^2 = (s_i s_j)^{k_{ij}} = 1$. In 1935, H. S. M. Coxeter enumerated all finite groups W generated by reflections and found that they are precisely the groups with presentations of this type, now called (finite) Coxeter groups. These groups and their infinite generalizations, some of which arise in elementary contexts as the symmetry groups of regular tessellations of Euclidean space, are the subject of this book. They have had extraordinary influence on geometry, Lie theory, finite groups, combinatorics and more distant parts of mathematics, for example singularity theory.

This is a useful book. The style is informal and the arguments are clear... It is the unique graduate level text on this subject and most of what it does is important for various aspects of Lie theory... The book is more ambitious than the undergraduate text *Finite Reflection Groups* by L. C. Grove and C. T. Benson and is, of course, not as formidable as Bourbaki. It may attract browsers, but it does not convey the excitement of the geometry in Coxeter's *Regular polytopes*, 1948."

H. Hiller, *Geometry of Coxeter Groups*, Research Notes in Mathematics, 54, Pitman (Advanced Publishing Program), Boston, Mass.-London, 1982. iv+213 pp.

This is an accessible introduction to an important subject: geometry and topology of flag varieties, and their connection with algebraic and invariant-theoretic properties of Coxeter groups.

According to MathSciNet, "The goal of this very readable book is to give an introduction, with minimal prerequisites, to the modern theories which reveal the beautiful

9.1. LIE GROUPS AND LIE ALGEBRAS

interactions between geometry and topology of flag varieties, algebraic and invariant-theoretic properties of Coxeter groups, representation theory and combinatorics."

According to a review in Bulletin of AMS, "Quotients of Lie groups appear throughout mathematics and physics. To an algebraist, the nicest Lie groups are the simple Lie groups over the complex numbers. The most interesting quotients formed from these groups are the flag manifolds G/P obtained by dividing out by a 'parabolic' subgroup. The algebraic geometric study of flag manifolds reveals several interesting interactions between algebraic geometry, representation theory, and combinatorics. The main topic of the book under review is the description of the cohomology rings of complex flag manifolds with an accent on the combinatorial aspects thereof. By far the most famous of the flag manifolds are the Grassmannians."

M. Davis, *The Geometry and Topology of Coxeter Groups*, London Mathematical Society Monographs Series, 32, Princeton University Press, Princeton, NJ, 2008. xvi+584 pp.

This book systematically studies Coxeter groups from the point of view of geometric group theory. It is a valuable introduction and a reference and will complement other more algebraic books on Coxeter groups well.

According to MathSciNet, "Books on Coxeter groups come in many flavours. When trying to get a feeling for the direction this book is heading, one should consider the tessellation of the hyperbolic plane by right-angled pentagons depicted on the front page and look for its re-occurrence in the book. It is impossible to describe the contents of this impressive book in one sentence; maybe it is fair to say that this book presents Coxeter groups in their natural environment of metric geometry.

This book is one of those that grows with the reader: A graduate student can learn many properties, details and examples of Coxeter groups, while an expert can read about aspects of recent results in the theory of Coxeter groups and use the book as a guide to the literature. I strongly recommend this book to anybody who has any interest in geometric group theory."

9.1.17 Transformation groups

Groups were first defined as transformation groups. Understanding of properties of groups also depends on actions on suitable spaces, and the importance of groups also comes from group actions. In some sense, a group is more interesting and useful if it admits more actions on natural spaces. A group together with an action on a space is called a transformation group.

G. Bredon, *Introduction to Compact Transformation Groups*, Pure and Applied Mathematics, Vol. 46, Academic Press, New York-London, 1972. xiii+459 pp.

This was written as the first comprehensive introduction to compact transformation groups. It has also been a useful reference since then.

According to MathSciNet, "This is a successful introduction to compact Lie groups of transformations on topological spaces and manifolds. A first course in algebraic topology is assumed, but the background demands on the reader are gradually increased throughout the book. The book contains a large portion of the subject, and touches on many interesting nontrivial aspects of mathematics of current interest. The reviewer has found this book to be exceptionally well written and ideally suited for research purposes or as a text for an introduction to the theory of compact transformation groups."

T. tom Dieck, *Transformation Groups*, de Gruyter Studies in Mathematics, 8, Walter de Gruyter & Co., Berlin, 1987. x+312 pp.

This book is an accessible modern introduction to the theory of transformation groups. It gives a good account of equivariant topology. It can supplement the book above by Bredon well.

According to MathSciNet, "The subject of transformation groups has undergone major developments over the past several years and is one of the most active areas of current research in topology...

The book under review is a well-thought-through, elegant introduction to equivariant homotopy theory and especially the role of localization of homology theories, induction theory and the Burnside ring. The material on localization and fixed point sets was already considered in great detail in the book of

W.Y. Hsiang, *Cohomology Theory of Topological Transformation Groups*, 1975. x+164 pp.

The remaining material, as well as this part, is very clearly presented and will be of great value to students as well as researchers who have not yet availed themselves of these tools."

According to MathSciNet, "The basic theme of this book [Hsiang's book] is that the understanding of topological transformation groups can proceed along lines analogous to the theory of linear representations of compact connected Lie groups."

Hsiang's book is a good introduction to equivariant cohomology theory.

S. Kobayashi, *Transformation Groups in Differential Geometry*, Ergebnisse der Mathematik und ihrer Grenzgebiete, Band 70, Springer-Verlag, New York-Heidelberg, 1972. viii+182 pp.

This is a valuable reference on automorphism groups of Riemannian and complex manifolds. As the author said, "The object of this book is to give a biased account of automorphism groups of differential geometric structures. All geometric structures are not created equal; some are creations of gods while others are products of lesser human minds. Amongst the former, Riemannian and complex structures stand out for their beauty and wealth."

According to MathSciNet, "The style is ... clear and concise, the notation is carefully chosen, and there are few, if any, errors. The author states that he has confined himself to basic results, referring to the bibliography for more difficult theorems. However, in many places the theory is carried far enough to serve as a starting point for new research and he frequently indicates open questions. This is an excellent selection of material by a fine expositor."

D. Montgomery, L. Zippin, *Topological Transformation Groups*, Interscience Publishers, New York-London, 1955. xi+282 pp.

This is a classical book on locally compact topological groups and their actions written by leading experts. It is a very valuable reference.

According to MathSciNet, "Almost two decades have passed since publication of the first edition of Pontryagin's *Topological Groups*, 1938, which has been since considered as one of the standard reference books in the field. In the meantime, the theory of topological groups has made outstanding progress, culminating in the solution of Hilbert's fifth problem by Gleason and by the authors of the present book. The authors give here a detailed account of those important results on locally compact topological groups obtained in this period, suggesting at the same time further future developments in the theory."

According to a review in Bulletin of AMS, "In writing this book the authors have had two purposes: (1) to present, in connected form, the recent solution by the authors and A. Gleason, based on the work of many mathematicians, of the famous 5th problem of Hubert, stating that a topological group which is locally homeomorphic to Euclidean space E^n is a Lie group; and (2) to report on the work done during the past twenty years on transformation groups, the emphasis being on the way a group acts on a space as a group of transformations. The first topic represents the final step in a long development to which many outstanding mathematicians have contributed: Introduction of the idea of Lie groups by Lie, via sets of transformations which depend (differentiably) on a number of parameters; intensive study of Lie groups, in particular the classification of all semisimple Lie groups (Killing, Cartan); representation theory of Lie groups (Cartan, Weyl); introduction of the concept of topological group (O. Schreier, 1925); analysis of compact groups (von Neumann, Pontryagin, 1933–1934); analysis of locally compact abelian groups (Pontryagin, 1934); invariant measure in locally compact groups (Haar, 1933); and then about twenty years of study of general locally compact groups (the compactness restriction seemed necessary to get significant results) by many mathematicians; finally, the solution of Hilbert's 5th problem, which actually involves a much more general statement —it might be formulated, somewhat loosely, as saying that in a sense (namely up to certain limit processes) there are no other locally compact groups but our old friends, the Lie groups. —We have sailed across a wide ocean; many times the course was not clear at all; we finally arrived at our own shores; from a different direction to be sure, but it is the old country. One might almost be a bit disappointed at the outcome; but one should remember that in the years of work on these questions many and very fruitful ideas have been discovered which have enriched all of mathematics tremendously, in addition to the many direct applications of the results."

W. Lück, L^2-*invariants: Theory and Applications to Geometry and K-theory*, Ergebnisse der Mathematik und ihrer Grenzgebiete, 3, Folge, A Series of Modern Surveys in Mathematics, 44, Springer-Verlag, Berlin, 2002. xvi+595 pp.

This is a user friendly introduction to L^2-invariants and their applications in geometry and topology. Its comprehensive coverage makes it also a very valuable reference.

According to MathSciNet, "L^2-invariants were introduced into topology by Atiyah in 1976, along with the L^2-index theorem and the definition of analytic L^2-Betti numbers... This book is an excellent survey of many up-to-date results about L^2-Betti numbers, Novikov-Shubin invariants, L^2-torsion, and the K- and L-theory of operator algebras needed in the definition of these L^2-invariants... The book combines features of a textbook and a reference work..."

From the book description, "In algebraic topology some classical invariants - such as Betti numbers and Reidemeister torsion - are defined for compact spaces and finite group actions. They can be generalized using von Neumann algebras and their traces, and applied also to non-compact spaces and infinite groups. These new L^2-invariants contain very interesting and novel information and can be applied to problems arising in topology, K-Theory, differential geometry, non-commutative geometry and spectral theory. The book, written in an accessible manner, presents a comprehensive introduction to this area of research, as well as its most recent results and developments."

G. Whyburn, *Analytic Topology*, American Mathematical Society Colloquium Publications, v. 28, American Mathematical Society, New York, 1942. x+278 pp.

By *Analytic Topology*, the book means transformation groups. In contrast with the earlier books, this book contains a lot of discussion of topological spaces which are not manifolds. These spaces often appear in geometric group theory and other contexts. This book is a very valuable reference on topics, some of which are not found in many other more standard topology books, in spite of that fact that it was published a long time ago.

9.1.18 V. Kac, Infinite-dimensional Lie Algebras, Third edition, Cambridge University Press, Cambridge, 1990. xxii+400 pp.

This is a classical book on Kac-Moody algebra written by one of its founders. It is a valuable introduction and reference.

A Kac–Moody algebra is a Lie algebra, usually infinite-dimensional, that can be defined by generators and relations through a generalized Cartan matrix, by generalizing the Serre relations for complex semisimple Lie algebras. Hence, they are a generalization of finite-dimensional complex semisimple Lie algebras, and many properties related to the structure of a Lie algebra such as its root system, irreducible representations can be generalized to Kac–Moody algebra. A class of Kac–Moody algebras called affine Lie algebras is of particular importance, especially in theoretical physics, for example in conformal field theory and the theory of exactly solvable models. An important application in math is an elegant proof by Kac of Macdonald identities in combinatorics by using the representation theory of affine Kac–Moody algebras.

According to MathSciNet, "Given the current interest in the subject of Kac–Moody algebras, a book on the topic is overdue. The author, drawing largely on his own contributions to the field, has filled the void with this concisely written volume which will provide a solid foundation for anyone wishing to begin working the area."

9.1.19 Loop groups, quantum groups, Hopf algebras and vertex operator algebras

Lie groups have played a fundamental role in mathematics and sciences. There are many generalizations of Lie groups. Given a Lie group or more generally a topological group, there is an associated loop group. A loop group is the group of loops in a group with multiplication defined pointwise.

As mentioned before, there are inseparable relations between Lie groups and Lie algebras. Generalization of Lie algebras or a closely related algebras include quantum groups and Hopf algebras. Another related algebra is a vertex operator algebra. In this subsection, we list some books on these topics.

A. Pressley, G. Segal, *Loop Groups*, Oxford Mathematical Monographs, Oxford Science Publications, The Clarendon Press, Oxford University Press, New York, 1986. viii+318 pp.

This is a classical and standard book on loop groups.

According to MathSciNet, "The loop group LG is the group of smooth maps from the circle S^1 to a group G, the multiplication being defined pointwise. The book unfolds the representation theory of LG when G is a compact finite-dimensional Lie group. The group LG is a special case in the class of infinite-dimensional groups $Map(X; G)$ consisting of smooth maps from a manifold X to the group G. However, in spite of their interesting geometrical structure and important applications in quantum field theory, relatively little is known about the representations of $Map(X;G)$ except in the case when X is the circle. ... The book contains a detailed treatment of the geometry of the Grassmannian $Gr(H)$ (stratification, Plücker coordinates) and the structure of LG in terms of various factorizations...

The authors have done an important service in giving the first coherent account of the geometric representation theory and structure of loop groups and their central extensions. The book is clearly written... Besides mathematicians working on infinite-dimensional groups and manifolds, the book is to be recommended for theoretical physicists working in quantum field theory, completely integrable systems and string theory."

G. Lusztig, *Introduction to Quantum Groups*, Progress in Mathematics, 110, Birkhäuser Boston, Inc., Boston, MA, 1993. xii+341 pp.

This is a very important reference on quantized enveloping algebras (or quantum groups) and their representations by a major contributor.

According to MathSciNet, "the book presents many topics, together with detailed arguments for most of them, in quantum group theory. It is written to a very high standard and perhaps is an introductory book for experienced researchers and experts".

C. Kassel, *Quantum Groups*, Graduate Texts in Mathematics, 155, Springer-Verlag, New York, 1995. xii+531 pp.

This is a comprehensive introduction to quantum groups and emphasizes both the purely algebraic and physical aspects.

According to MathSciNet, "The book provides an introduction to the algebra behind the words 'quantum groups' with emphasis on the fascinating and spectacular connections with low-dimensional topology and on Drinfeld's fundamental contribution. Despite the complexity of the subject the author makes the exposition accessible to a large audience. The book does not seek to address the problem of mentioning all the various aspects of quantum group theory. But, starting with elementary considerations, the author leads the reader to selected advanced results, providing detailed proofs of all necessary facts. A great many exercises are proposed to the diligent reader. Bibliographical and historical notes are not extensive but are very convenient for the beginner."

S. Majid, *Foundations of Quantum Group Theory*, Cambridge University Press, Cambridge, 1995. x+607 pp.

This is a self-contained introduction to the quantum group theory.

According to MathSciNet, "The book under review was written by one of the principal developers of the subject. It is mainly based on the author's own research papers and contains many of his results, presented now in a self-contained form... In summary, this is a remarkable book, written in a rigorous style (all necessary statements are proved explicitly) by an acknowledged leader in the field. This book should be strongly recommended both to students and to professionals."

S. Montgomery, *Hopf Algebras and Their Actions on Rings*, CBMS Regional Conference Series in Mathematics, 82, The American Mathematical Society, Providence, RI, 1993. xiv+238 pp.

This book is a good introduction to the algebraic structure of Hopf algebras. It is also a valuable introduction to quantum groups as special examples.

According to MathSciNet, "The point of view is the algebraic structure of Hopf algebras and their actions and co-actions. There has been an increased interest in Hopf algebras in recent years because of their appearance in statistical mechanics as quantum groups. In these notes, quantum groups are treated as important examples, but this still serves as a useful introduction to quantum groups."

M. Sweedler, *Hopf Algebras*, Mathematics Lecture Note Series, W. A. Benjamin, Inc., New York, 1969. vii+336 pp.

This is a valuable introduction to Hopf algebras and some applications.

According to MathSciNet, "The book is mainly concerned with Hopf algebras which are not graded... It begins with basic notions, develops at a leisurely pace and gives a clear and up-to-date presentation of the theory of ungraded Hopf algebras. A substantial part of the material consists of results obtained by the author and others during recent years. There are a good number of examples."

I. Frenkel, J. Lepowsky, A. Meurman, *Vertex Operator Algebras and the Monster*, Pure and Applied Mathematics, 134, Academic Press, Inc., Boston, MA, 1988. liv+508 pp.

This is an important book on vertex operator algebras and applications to the Monster, the largest sporadic finite simple group. It was written by major contributors and is a valuable reference.

From the book description, "The first part of the book presents a new mathematical theory of vertex operator algebras, the algebraic counterpart of two-dimensional holomorphic conformal quantum field theory. The remaining part constructs the Monster finite simple group as the automorphism group of a very special vertex operator algebra, called the 'moonshine module' because of its relevance to 'monstrous moonshine.'"

According to MathSciNet, "This book is a detailed research monograph which contains the complete proofs of the authors' previously announced results. It is mostly self-contained and the exposition, given the technicalities involved, could not have been any better."

According to American Scientist, "The present book shows how this group arises as the symmetry group of a certain vertex-operator algebra...'One fact, however, is undeniable. As the automorphism group of a distinguished conformal field theory, the Monster is fundamentally related to one of the most spectacular chapters of modern theoretical physics–string theory.' "

9.1.20 Applications of Lie groups in sciences

There are many applications of Lie groups in sciences. But some of the most striking applications are in physics, in particular, in elementary particles. One application concerns the Eightfold Way through representations of unitary group and the subsequent development of the quark model.

S. Sternberg, *Group Theory and Physics* ,(English summary), Cambridge University Press, Cambridge, 1994. xiv+429 pp.

Most mathematicians who are interested in Lie groups have heard of beautiful and powerful applications of Lie groups in physics, especially in elementary particles. This is a very good introduction to Lie groups and applications to physics.

According to MathSciNet, "The book contains a fresh approach to many topics and shows the highest degree of mathematical competency. The text seems to be very friendly to physicists though written in terms of modern mathematics (morphisms, orbits, vector bundles, etc.). In addition there are interesting and valuable excursions into the history of groups and spectroscopy and citations of classical works which make the reading of the book a real pleasure."

According to a review in Bulletin of AMS, "The urge to justify mathematics to students by pointing to 'real world' applications has led to more frequent inclusion of such topics as coding theory in recent textbooks on linear algebra and abstract algebra. But older books on algebra, or those written for graduate students, generally suppress any hint of applicability outside mathematics itself. In particular, most accounts of abstract group theory or group representations written by mathematicians ignore the rich applications of these subjects to twentieth century physics and chemistry... Symmetry is an ever-present consideration in the study of natural phenomena. Group theory provides a precise language with which to describe the possible symmetries of a physical system. In the case of crystallographic groups, the successful classification of relevant symmetry groups in the nineteenth century made possible many predictions of what may actually be observed in nature. But modern physics sometimes requires more subtle data coming not just from the obvious geometric actions of groups but also from the possible representations of the groups by linear operators (typically acting on spaces of functions). This is especially apparent in the study of elementary particles since the 1960s, leading to what is now known as the Standard Model.... Sternberg's book is neither a technical treatise nor a textbook (there are no exercises). It is written in an informal conversational style, with only a few results stated formally... the book is the next best thing to having a long chat with the author about subjects which are obviously near to his heart."

J. de Azcárraga, J. Izquierdo, *Lie Groups, Lie Algebras, Cohomology and Some Applications in Physics*, Cambridge Monographs on Mathematical Physics, Cambridge University Press, Cambridge, 1995. xviii+455 pp.

This book gives a self-contained account of modern cohomology theory of Lie groups and Lie algebras as well as some applications in physics. It contains many interesting examples.

According to MathSciNet, "This book is a useful contribution to the constantly growing literature that is devoted to the mathematics-physics interface. It presents a cohomological approach to the subject of anomalies in quantum gauge theories, one of the most important topics of modern quantum theory. The book is a welcome relief to the physicist searching for a physically motivated yet mathematically rigorous and systematic exposition of gauge anomalies. The emphasis is on the Lie algebraic method and stresses the importance of the cohomological point of view. Many excellent books and monographs have been written on the subject of differential geometry intended for a physics audience. The book under review reserves for itself a special place among other presentations by successfully trying to be self-contained. Its strong point is a systematic exposition of the mathematical background in an accessible form. In the reviewer's opinion, the book is most useful for physicists who want to go beyond the standard physics review type introduction to differential geometry. The book offers them a quite rigorous but at the same time lucid and self-contained presentation."

9.2 Representation theory

Usually mathematicians are famous for either theorems they proved or conjectures they made. But Schur is probably most famous for the Schur Lemma in representation theory. It is an elementary but extremely useful result in representation theory of groups and algebras. He belonged to a good academic family of representation theorists: his advisor was Frobenius, and his students included Richard Brauer. The subject of representation is beautiful and useful. According to Gelfand, "All of mathematics is some kind of representation theory". Maybe the difficulty is to how to recognize it and use it.

Figure 9.2: Schur's book.

Linear representations provide important means to understand structures of groups. They can be used to represent group elements as matrices so that the group operation can be represented by matrix multiplication. Representations of groups are important because they allow many group-theoretic problems to be reduced to problems in linear algebra, which is well-understood (or easier to be understood). Representation theories of groups were first developed for finite groups.

9.2.1 J.P. Serre, Linear Representations of Finite Groups, Translated from the second French edition by Leonard L. Scott, Graduate Texts in Mathematics, Vol. 42, Springer-Verlag, New York-Heidelberg, 1977. x+170 pp.

This is a classical textbook on representations of finite groups. It is very well organized and clearly written. It is suitable for both beginners and experts.

As one Amazon review says, "This is an excellent introduction to the subject. The book really breaks into 3 distinct parts. The first 5 chapters are a rapid introduction to the basics, similar to what one would get from any introductory text. They are most notable for actually going through the details on D_n, S_n cyclic groups... The second section (chapters 6-13) gives a more graduate level presentation of the material. Starting with a discussion of group algebras, moving onto inducted representations Artin's theorem (the existence of virtual characters) The third section is Brauer Theory. The book is by Serre so it goes without saying it one of the best if not the best book on the market... The index of notation is a fantastic asset for a subject where notation plays such a large role."

9.2.2 I.G. Macdonald, Symmetric Functions and Hall Polynomials, Oxford Mathematical Monographs, The Clarendon Press, Oxford University Press, New York, 1979. viii+180 pp.

This is a classic and has been the standard source and reference book for many aspects about symmetric functions.

According to MathSciNet, "Symmetric functions arise in several areas of mathematics, such as combinatorics, algebraic geometry, and the representation theory of symmetric and general linear groups. In particular, they are closely connected with Hall polynomials, which arise in the study of abelian p-groups... The purpose of the book is to give a self-contained account of Hall polynomials and related topics, especially symmetric functions..."

According to a review of this book in Bulletin, "The primary reason for the ubiquity of Hall polynomials lies in their close connection with symmetric functions. For this

9.2. REPRESENTATION THEORY

reason the author devotes about half of his book (Chapter I) to the theory of symmetric functions, without reference to Hall polynomials. Just this one chapter is a valuable source of information for anyone working in such fields as combinatorics, algebraic geometry, and representation theory, which frequently impinge on the theory of symmetric functions..."

Despite the amount of material of such great potential interest to mathematicians in so many diverse areas (to say nothing of physicists, chemists, et al., who deal with group representations), the theory of symmetric functions remains all but unknown to the persons it is most likely to benefit... Hopefully this beautifully written book will put an end to this state of affairs."

According to MathSciNet of the second edition, "Almost immediately after appearing in 1979, the first edition of this monograph became the standard reference for Schur functions (the irreducible characters of $GL_n(\mathbb{C})$), their relatives, and the rings of symmetric polynomials they generate. In 1984, A. V. Zelevinsky translated the first edition into Russian], and in the process, added several remarks, examples, and alternative proofs. This new edition incorporates the additions to the translated version, as well as a tremendous amount of new material. To call it a 'second edition' is an understatement—it has nearly tripled in length!

Readers familiar with the first edition will find that the author's uniquely brief but lucid style has not changed. Following the narrative portion of nearly every section is an extensive list of 'Examples' ('Excursions' would be more accurate), which in fact form a catch-all category of extended remarks, exercises (sometimes together with a solution), and digressions."

9.2.3 Representation theory of the symmetric group

The representation theory of the symmetric group is a particular case of the representation theory of finite groups. Due to special nature of the symmetry group and explicit information about its conjugacy classes of elements, a concrete and detailed theory of representation of the symmetric group can be obtained. For example, each such irreducible representation over the complex numbers can in fact be realized over the integers, i.e., every permutation acting by a matrix with integer coefficients, and it can be explicitly constructed by computing the Young symmetrizers acting on a space generated by the Young tableaux of shape given by the Young diagram.

The representation theory of the symmetric group has many applications, ranging from symmetric function theory to problems of quantum mechanics for a number of identical particles.

We list three books on symmetric groups and their representations.

W. Fulton, *Young Tableaux. With Applications to Representation Theory and Geometry*, London Mathematical Society Student Texts, 35, Cambridge University Press, Cambridge, 1997. x+260 pp.

This is a modern introduction to the Young tableaux and their applications to representation theory and algebraic geometry. It contains a wealth of information and is a valuable reference as well.

According to a review in Bulletin of AMS, "In conclusion, this book is a book that every student and researcher in algebraic combinatorics, algebraic geometry, and/or representation theory should have on their shelves. The author has organized a wealth of information that was previously available only spread out in many parts of the literature. The book does not really make an effort to indicate the directions in which this field is growing, but it will be of great use to those who want a brief and well organized treatment of any of the topics included. It will be particularly useful for graduate students: the author has that magical ability for getting to the good stuff without getting the reader mired in preliminary 'basics'."

According to MathSciNet, "The book develops the basic combinatorics of Young tableaux and their applications to representation theory and the geometry of flag varieties. For the first time readers can enjoy all three topics in one book, presented in an informal and lively way. The combinatorial first part is essentially self-contained, whereas the representation-theoretic second part requires at some points knowledge of certain general facts concerning algebraic groups; finally, the third, geometric, part requires selected topics from algebraic geometry and topology.... This masterfully written book presents clearly and attractively the interplay among selected topics in algebraic combinatorics, representation theory and geometry; I highly recommend it to everybody who wants to savour the beauty of mathematics."

G. James, A. Kerber, *The Representation Theory of the Symmetric Group*, With a foreword by P. M. Cohn. With an introduction by Gilbert de B. Robinson, Encyclopedia of Mathematics and its Applications, 16, Addison-Wesley Publishing Co., Reading, Mass., 1981. xxviii+510 pp.

This book gives a comprehensive account of the representation theory of the symmetric group. It is a very valuable introduction and a reference.

According to a review in Bulletin of AMS, "There are many excellent texts on the representation-theory of finite groups, ... The representation-theory of the symmetric groups S_n cannot, at present, be considered merely a specialization of the more general theory; there is a rich accumulation of concepts and theorems in the theory of S_n, whose analogs for arbitrary finite groups do not exist (or have not yet been found). The results of this special theory, not only have an intrinsic interest and beauty, but have applications in chemistry, physics and areas of mathematics as diverse as algebraic geometry

9.2. REPRESENTATION THEORY

(via flag manifolds, determinantal varieties and the Schubert calculus), classical invariant theory, rings with polynomial identity, multivariate statistics ... and, of course, combinatorics (in particular, via the Robinson-Schensted correspondence and the Redfield-Pólya enumeration-theory)... James and Kerber have, in the text under review, made accessible to the mathematical community a vast body of information and techniques concerning the representation theory of S_n and $GL(n)$. Great labor and thought have gone into the systematization of this material. The reviewer believes the subject-matter of their text is at present undergoing an intensive period of development; the James-Kerber text should contribute substantially to the acceleration of this development."

B. Sagan, *The Symmetric Group. Representations, Combinatorial Algorithms, and Symmetric Functions*, The Wadsworth & Brooks/Cole Mathematics Series, Wadsworth & Brooks/Cole Advanced Books & Software, Pacific Grove, CA, 1991. xviii+197 pp.

This book is a very accessible introduction to algebraic combinatorics, in particular representation theory, combinatorial algorithms, and symmetric functions through symmetry groups. It is well-written and clearly presented.

According to MathSciNet, "The symmetric group has motivated much of the increased interest in algebraic combinatorics in the last decade. The interplay among representation theory, symmetric function theory and combinatorics within the symmetric group has driven the development of this interplay within other Coxeter groups. The symmetric group also forms the background for the exciting research on Hall, Jack and Schubert polynomials. Indeed, it would be difficult to do research in algebraic combinatorics today without a good working knowledge of the symmetric group.

Previous to the publication of this book, an interested researcher had to learn about the symmetric group from a variety of sources, and from a variety of points of view. This book brings together, for the first time in one place, these three approaches to the symmetric group."

9.2.4 H. Weyl, The Classical Groups. Their Invariants and Representations, Fifteenth printing, Princeton Landmarks in Mathematics, Princeton University Press, Princeton, NJ, 1997. xiv+320 pp.

This is a classic, a popular classic, written by a master.

From the book description, "In this renowned volume, Hermann Weyl discusses the symmetric, full linear, orthogonal, and symplectic groups and determines their different invariants and representations. Using basic concepts from algebra, he examines the various properties of the groups. Analysis and topology are used wherever appropriate. The book also covers topics such as matrix algebras, semigroups, commutators, and spinors, which are of great importance in understanding the group-theoretic structure of quantum mechanics... Hermann Weyl was among the greatest mathematicians of the twentieth century. He made fundamental contributions to most branches of mathematics, but he is best remembered as one of the major developers of group theory... In *The Classical Groups*, his most important book, Weyl provided a detailed introduction to the development of group theory, and he did it in a way that motivated and entertained his readers. Departing from most theoretical mathematics books of the time, he introduced historical events and people as well as theorems and proofs. One learned not only about the theory of invariants but also when and where they were originated, and by whom... Weyl believed in the overall unity of mathematics and that it should be integrated into other fields. He had serious interest in modern physics, especially quantum mechanics, a field to which *The Classical Groups* has proved important, as it has to quantum chemistry and other fields."

According to a review in Bulletin of AMS in 1940, "It is a curious fact that while almost all the textbooks on higher algebra written prior to 1930 devote considerable space to the subject of invariants, the recent ones written from the axiomatic point of view disregard it completely. Because of this neglect the phrase 'invariant theory' is apt to suggest a subject that was once of great interest but one that has little bearing on modern algebraic developments. For this reason it is an important and original accomplishment that Professor Weyl has made here in connecting the theory of invariants with the main stream of algebra and in indicating that the subject has a future as well as a distinguished past.

In his treatment the theory of invariants becomes a part of the theory of representations. The most natural way to begin the study of invariants of a particular group is therefore to determine its representations. A large part of the book is concerned with this problem as it applies to the 'classical' groups $GL(n)$, the full linear group; $O(n)$, the orthogonal, and $S(n)$, the symplectic (complex or abelian) group.

In discussing the representations of an abstract group $\mathfrak{g}$ one finds it convenient to adjoin to the set of representing matrices their linear combinations. The resulting set is an algebra, the enveloping algebra of the original set, and defines a representation of a certain abstract algebra, the group algebra, that is completely determined by $\mathfrak{g}$ and the field of the coefficients. In this way the theory of algebra is applicable. The author has gone somewhat beyond his immediate needs in discussing this domain. Consequently his book may serve also as an excellent introduction to this theory."

R. Goodman, N. Wallach, *Representations and Invariants of the Classical Groups*, Encyclopedia of Mathematics and its Applications, 68, Cambridge University Press, Cambridge, 1998. xvi+685 pp.

This is a modern introduction to representations and invariants of the classical groups based on the principle: presenting the principal theorems of representation theory for the

9.2. REPRESENTATION THEORY

classical matrix groups as motivation for the general theory of reductive groups. It is a very valuable introduction and a reference.

According to MathSciNet, "As the title suggests, this book is inspired by H. Weyl's classic *The Classical Groups. Their Invariants and Representations...* here the authors focus primarily on the classical groups (over $\mathbb{R}$ and $\mathbb{C}$): general and special linear, unitary, orthogonal, symplectic. This allows them to keep prerequisites to a minimum, while covering all of the standard representation theory together with a modern treatment of classical invariant theory (in the spirit of Howe). At the same time, their proofs usually extend to the general setting of semisimple groups. Extensive appendices fill in much of the broader framework involving algebraic geometry and Lie groups. One novelty of the exposition, suggested by D.-N. Verma, is a reversal of the usual approach to relating tensor representations of symmetry and general linear groups: here the character results for symmetric groups are derived from those for the general linear group, based on Weyl's character formula..."

R. Goodman, N. Wallach, *Symmetry, Representations, and Invariants*, Graduate Texts in Mathematics, 255, Springer, Dordrecht, 2009. xx+716 pp.

This book is a revised and expanded version of the above book.

9.2.5 Representation theories of Lie groups

One important approach to describe *continuous* symmetry in mathematics and theoretical physics uses representations of Lie groups. Representation theories of Lie groups started with applications in physics and have been developed to a very high degree. Since they require advanced and complicated structures of Lie groups and Lie algebras, they present themselves as some of the most complicated topics in modern mathematics.

A lot of work has been done and a lot is known about representations of Lie groups, especially representations of compact Lie groups. One major open problem is the classification of unitary representations of noncompact semisimple Lie groups. A basic tool in representations of Lie groups is to use the corresponding 'infinitesimal' representations of Lie algebras.

There are many books on representations, and we have selected few books studying representations from different perspectives.

W. Fulton, J. Harris, *Representation Theory. A first Course*, Graduate Texts in Mathematics, 129, Readings in Mathematics, Springer-Verlag, New York, 1991. xvi+551 pp.

This is a very enjoyable and accessible book on finite dimensional representations of Lie groups by two major mathematicians who are not representation theorists.

From the book description, "The primary goal of these lectures is to introduce a beginner to the finite-dimensional representations of Lie groups and Lie algebras. Intended to serve non-specialists, the concentration of the text is on examples. The general theory is developed sparingly, and then mainly as useful and unifying language to describe phenomena already encountered in concrete cases. The book begins with a brief tour through representation theory of finite groups, with emphasis determined by what is useful for Lie groups. The focus then turns to Lie groups and Lie algebras and finally to the heart of the course: working out the finite dimensional representations of the classical groups. The goal of the last portion of the book is to make a bridge between the example-oriented approach of the earlier parts and the general theory."

According to MathSciNet, "The authors take great care to motivate everything by working through numerous examples and by providing informal proof sketches as well as geometric interpretations and applications wherever possible. General abstract theorems are postponed, sometimes to the appendices, in order to keep the treatment concrete and informal. There is an attractive blend of the Schur-Weyl methods for classical groups and the general Cartan-Weyl theory of highest weight representations. This should make the book especially appealing to mathematical physicists, combinatorists, and others who work concretely with tensor products or symmetric and exterior powers of various natural representations... Prospective readers should not be put off by the length of the book. Though the trees occasionally obscure the view of the forest, there is something for everyone here—even the expert in Lie theory..."

A. Knapp, *Representation Theory of Semisimple Groups. An Overview Based on Examples*, Princeton Mathematical Series, 36, Princeton University Press, Princeton, NJ, 1986. xviii+774 pp.

This is a comprehensive book on representations of semisimple Lie groups. It tries to alleviate the difficulty of this complicated but important subject by working on many examples instead of the general setup. It is a very valuable introduction to and an overview of the huge subject of representation theory of Lie groups.

From the book description, "In this classic work, Anthony W. Knapp offers a survey of representation theory of semisimple Lie groups in a way that reflects the spirit of the subject and corresponds to the natural learning process. This book is a model of exposition and an invaluable resource for both graduate students and researchers. Although theorems are always stated precisely, many illustrative examples or classes of examples are given. To support this unique approach, the author includes for the reader a useful 300-item bibliography and an extensive section of notes."

According to MathSciNet, "This book is the latest addition to the distinguished series which began with Hermann Weyl's *Classical Groups*. As the author puts it, 'the intention with this book is to give a survey of the representation theory of semisimple

9.2. REPRESENTATION THEORY

Lie groups, including results and techniques, in a way that reflects the spirit of the subject, corresponds more to a person's natural learning process, and stops at the end of a single volume'. The sheer size of the subject would seem overwhelming: from the Cartan-Weyl theory of finite-dimensional representations ... to the 'unitarity problem', still unsolved, perhaps the main problem of Harish-Chandra's legacy. Included are further the Langlands classification of irreducible admissible representations...

To be able to present all this material within one volume, the author chose an unusual device. As he says, 'Our approach is based on examples and has unusual ground rules. Although we insist (at least ultimately) on precisely stated theorems, we allow proofs that handle only an example. When the style of the proof is atypical of the subject matter of the book, we omit the proof altogether.'...

The book under review fills a gap in the literature which had seemed unfillable. It will immensely facilitate access to the representation theory of semisimple groups, as anyone will appreciate who had to learn the subject from the sources which had been available up to now."

N. Wallach, *Real Reductive Groups. I,* Pure and Applied Mathematics, 132, Academic Press, Inc., Boston, MA, 1988. xx+412 pp.

These two volumes study the representation theory of real reductive Lie groups with some emphasis on analytic aspects, and Harish-Chandra's Plancherel theorem. Volume 1 is the foundation for volume 2. Together they give a self-contained account of important parts of representation theory and harmonic analysis on a real reductive Lie group, an edifice erected by Harish-Chandra. They are important and reliable references on this important but difficult subject.

N. Chriss, V. Ginzburg, *Representation Theory and Complex Geometry*, Birkhäuser Boston, Inc., Boston, MA, 1997. x+495 pp.

This book studies representation theories through the geometry of varieties associated with a complex semisimple Lie group, such as flag varieties, nilpotent conjugacy classes, Springer fibers. It is a valuable introduction to and a reference on this geometric approach to representations.

According to a review in Bulletin of AMS, "An attractive feature of the book is the attempt to convey some informal 'wisdom', rather than only the precise definitions. As a number of results are due to the authors, one finds some of the original excitement. This is the only available introduction to geometric representation theory. The best recommendation is the fact that it has already proved successful in introducing a new generation to the subject."

According to MathSciNet, "Its main task is to produce a uniform geometric classification of the irreducible finite-dimensional representations of three different objects: Weyl groups, the Lie algebra $sl_n(\mathbb{C})$ (by an approach which may generalize to any complex semisimple Lie algebra), and affine Hecke algebras."

L. Corwin, F. Greenleaf, *Representations of Nilpotent Lie Groups and Their Applications. Part I. Basic Theory and Examples*, Cambridge Studies in Advanced Mathematics, 18, Cambridge University Press, Cambridge, 1990. viii+269 pp.

This book gives a self-contained modern account of basic representation theory and harmonic analysis for nilpotent Lie groups. It is a valuable introduction. For example, it also contains a nice chapter on cocompact discrete subgroups of nilpotent Lie groups.

According to MathSciNet, "One of the great successes of modern harmonic analysis is the representation theory of nilpotent Lie groups. Building on the pioneering work of Dixmier and using the Mackey machine, Kirillov gave a beautiful description of the unitary dual of a connected simply connected nilpotent Lie group. This description has proved vital for studying a number of problems in harmonic analysis on nilpotent Lie groups and has had a profound effect on the development of harmonic analysis for Lie groups in general... The field has grown tremendously since the appearance of these works, and the subject needed a comprehensive introduction to harmonic analysis on nilpotent Lie groups.

The authors have written a beautiful introduction to Kirillov theory for nilpotent Lie groups. A graduate student with basic training in analysis will find this a very accessible introduction to many aspects of harmonic analysis in a very concrete setting..."

T. Bröcker, T. tom Dieck, *Representations of Compact Lie Groups*, Graduate Texts in Mathematics, 98, Springer-Verlag, New York, 1985. x+313 pp.

This is an accessible introduction to the beautiful representation theory of Cartan-Weil of compact Lie groups written by two topologists. The approach is more geometric and analytic than algebraic. It is especially suitable to students and nonexperts. It is also a good reference because of many explicit computations.

According to MathSciNet, "this book is a text for a second-year graduate course on compact Lie groups and their representations. The treatment of the latter is quite thorough, and the most important examples arising from the classical groups are presented in considerable detail. This material and other classical topics are developed in modern terminology and in a clear, methodical, and accessible manner... The exercises in this book are excellent, and in particular they contain a wealth of very helpful information about specific examples. Aside from its usefulness as a text, the book is also an extremely good reference for a large body of important classical material translated into current terminology."

M. Auslander, I. Reiten, S. Smalo, *Representation Theory of Artin Algebras*, Corrected reprint of the 1995 original. Cambridge Studies in Advanced Mathematics, 36, Cambridge University Press, Cambridge, 1997. xiv+425 pp.

9.2. REPRESENTATION THEORY

This book is one of the first textbooks on the representation theory of Artin algebras, which is an important part of the module theory. It has been developed rapidly into an independent subject having many important applications in module theory, ring theory, representation theories, abelian group theory, the theory of quantum groups etc. It is a very valuable introduction to this developing subject.

From the book description, "The authors develop several foundational aspects of the subject. For example, the representations of quivers with relations and their interpretation as modules over the factors of path algebras is discussed in detail. Thorough discussions yield concrete illustrations of some of the more abstract concepts and theorems."

According to MathSciNet, "the book has an elementary and general character, and touches on only the simplest topics of the modern representation theory of Artin algebras... the book is addressed only to novices. It does not lead the reader to the most recent investigations in the representation theory of Artin algebras."

X. The Duderstadt Center on the North Campus of the University of Michigan

This original draft of a letter by Galileo on January, 1610 is the most treasured item in the large special collection at University of Michigan. After obtaining a description of a recently developed telescope in the Dutch town of Middelburg, Galileo built a telescope for himself, and in this letter offered it to the addressee, pointing out its potential use in warfare. The lower part of this sheet shows how Galileo used this telescope a few months later. As he viewed the skies on successive evenings, he had noticed several bright objects around Jupiter that changed position with time and he plotted their positions during one week. When he imagined how these movements would look if they were viewed from above Jupiter and drew the diagram in the lower left, he realized that the objects were moons of that planet. This was the first time that objects orbiting a body other than the earth were observed.

Fourier is best known for initiating the theory of Fourier series and their applications to problems of heat transfer and vibrations. Now the Fourier transform has been applied to a wide range of subjects in mathematics and sciences. Besides his highly original contribution to mathematics, Fourier was also a very successful politician. He went with Napoleon on his Egyptian expedition in 1798, and was made the Governor of Lower Egypt and the Secretary of the Institut d'Egypte. In this sense, he is a rare combination of a real mathematician and a successful politician.

Hamilton was a child prodigy and was appointed a professor of astronomy at the age of 22 before he graduated from college. He made important contributions to classical mechanics, optics, and algebra. He is probably best known for the discovery of quaternions. Quaternions and octonions often explain exceptional structures in Lie theory and topology, and several results related to them have the quality of exceptional beauty. His work on mechanical and optical systems led him to discover new mathematical concepts and techniques, and his greatest contribution is his reformulation of Newtonian mechanics, now called Hamiltonian mechanics.

> Clarendon Press Series
>
> AN
>
> ELEMENTARY TREATISE
>
> ON
>
> ELECTRICITY
>
> BY
>
> JAMES CLERK MAXWELL, M.A.
> LL.D. EDIN., D.C.L., F.R.SS. LONDON AND EDINBURGH
> HONORARY FELLOW OF TRINITY COLLEGE,
> AND PROFESSOR OF EXPERIMENTAL PHYSICS IN THE UNIVERSITY OF CAMBRIDGE
>
> EDITED BY
>
> WILLIAM GARNETT, M.A.
> FORMERLY FELLOW OF ST. JOHN'S COLLEGE, CAMBRIDGE
>
> Oxford
> AT THE CLARENDON PRESS
> 1881
> [All rights reserved]

Maxwell contributed significantly to several branches of physics, and his most important achievement was a new formulation of classical electromagnetic theory. This unites all previously seemingly unrelated observations, experiments, and equations of electricity, magnetism, and optics, and shows that they all satisfy Maxwell's equations. Hence electricity, magnetism and light are all manifestations of the electromagnetic field. In other words, all other classical laws or equations of these disciplines are special cases of Maxwell's equations. In spite of such important contributions to science, public recognition of Maxwell in Edinburgh was only a recent phenomenon in comparison with his fellow countryman Walter Scott, a novelist.

Chapter 10

Mathematical physics, dynamical systems and ergodic theory

Mathematical physics is both an ancient and young subject of the mathematics. For example, Newton developed calculus in order to solve problems in physics and astronomy. Recently, work on string theory has generated many unexpected theories and methods which can be used to solve longstanding problems in mathematics.

Roughly speaking, mathematical physics deals with "the application of mathematics to problems in physics and the development of mathematical methods suitable for such applications and for the formulation of physical theories."

In this chapter, we broadly classify books on mathematical physics into two types: classical and more modern. It is a vast subject and we have only selected a few books. We have also included dynamical systems and ergodic theory in this chapter.

10.1 Classical mathematical physics

The study of the motion of bodies is an ancient subject, and classical mechanics tries to understand the motion of various objects including projectiles and astronomical objects such as planets, stars, and galaxies. It is concerned with set of physical laws which describes the motion of bodies under the action of a system of forces. Hamiltonian mechanics is a reformulation of classical mechanics and uses a different more abstract mathematical formulation of the theory, which contributed to the formulation of the quantum mechanics.

Figure 10.1: Hamilton's book.

10.1.1 R. Courant, D. Hilbert, Methods of Mathematical Physics. Vol. I., Vol. II., Interscience Publishers, Inc., New York, 1953. xv+561 pp., 1962. xxii+830 pp.

This is a classic. As a standard book on mathematical physics, it has educated many generations of mathematicians. It is still a very valuable reference and introduction to classical analysis and applications to classical physics.

From the book description, "Since the first volume of this work came out in Germany in 1924, this book, together with its second volume, has remained standard in the field. Courant and Hilbert's treatment restores the historically deep connections between physical intuition and mathematical development, providing the reader with a unified approach to mathematical physics. The present volume represents Richard Courant's second and final revision of 1953."

According to a review in Bulletin of AMS by Weyl in 1938, "Thirteen years have elapsed between the publication of the first volume of Courant-Hilbert, Methoden der mathematischen Physik, and this concluding second volume. The two volumes are a beautiful, lasting, and impressive monument of what Courant, inspired by the example of his great teacher Hilbert and supported by numerous talented pupils, accomplished in Göttingen, both in research and advanced instruction. Courant came to Göttingen at a time of enormous political and economic difficulties for Germany, on a difficult inheritance, with the day of the heroes, Klein, Hilbert, and Minkowski drawing to a close. But by research and teaching, by personal contacts, and by creating and administering in an exemplary manner the new Mathematical Institute, he did all that was humanly possible to propagate and develop Göttingen's old mathematical tradition. How his fatherland rewarded him is a known story. The publication of the present volume seems to the reviewer a fitting occasion for expressing the recognition his work has earned him in the rest of the mathematical world.

The first volume treated a closed, relatively un-ramified field: the doctrine of eigenvalues and eigen functions. It met with enormous success, in particular among the physicists, because shortly after its publication these matters, due to Schrödinger's wave equation, gained an unexpected importance for quantum physics. This second volume covers the theory of partial differential equations in all of its aspects which are of importance for the problems of physics."

10.1.2 H. Weyl, The Theory of Groups and Quantum Mechanics, from the 2d rev., German ed., by H. P. Robertson, Dover Publications, 1949. 448 pp.

This is a classic. As Weyl wrote, "The importance of the standpoint afforded by the theory of groups for the discovery of the general laws of quantum theory has of late become more and more apparent. Since I have for some years been deeply concerned with the theory of the representation of continuous groups, it has seemed to me appropriate and important to given an account of the knowledge won by mathematicians working in this field in a form suitable to the requirements of quantum physics." This book has played a fundamental role in making group theory a basic, essential part of quantum physics.

According to an Amazon review, it is "One of the two great classics on group theory in physics". "The other one is Wigner's 'Group Theory and Quantum Mechanics'. As it is true of the other great books by Weyl, this is not an easy book, but it is, by all means, accessible... Written in the early years of the quantum theory, the author of this book foresaw the importance of considering symmetry in physics, the use of which now pervades most of theoretical high energy physics. Indeed, with the advent of gauge theories, and their experimental validation, it is readily apparent that symmetry principles are here to stay, and are just not accidental curiosities. A reader of the book can still gain a lot from the perusal of this book, in spite of its date of publication and its somewhat antiquated notation. Older books also have the advantage of discussing the material more in-depth, and do not hesitate to use hand-waving geometrical pictures when appropriate. This approach results in greater insight into the subject, and when coupled with eventual mathematical rigor gives it a solid foundation. One example where the discussion is superior to modern texts is in the author's discussion of group characters and their application to irreducible representations and spectra in atomic systems.

The reader will no doubt probably want to couple the reading of this book with a more modern text so as to alleviate the notational oddities in this book. The author's presentation is clear enough though to make an appropriate translation to modern notation. The reader will then be well prepared to tackle more advanced material in mathematical and theoretical physics that make use of the group-theoretic constructions that take place in this book."

G. Mackey, *Unitary Group Representations in Physics, Probability, and Number Theory*, Second edition, Advanced Book Classics, Addison-Wesley Publishing Company, Advanced Book Program, Redwood City, CA, 1989. xxviii+402 pp.

This is a useful reference on applications of unitary group representations in physics and other subjects of mathematics written by a major expert on representation theories.

According to MathSciNet, "This book is neither an ordinary text nor a mathematical monograph; rather, it is an expository treatment (at a rather advanced level) of group representations, physics, probability, ergodic theory and number theory, together with an account of how the first of these topics illuminates the other four. While a few theorems are proved in the text, most of the results are stated without proofs. In addition, many theorems are given in a technically incomplete state, in that the hypotheses necessary to make the theorem valid are not stated in detail. (In such cases, the author does state that there are further technical hypotheses, and he gives references for the statement and proof.) As compensation, the book is very clearly written, and the author does an excellent job of organizing the various topics so that each becomes coherent."

According to a review in Bulletin of AMS, "Even to one who does not wish to 'buy' group representations as the end all or be all that it sometimes pretends to be, this is a very nice book. In particular, with Jauch's *Foundations of quantum mechanics* on the one side and the present *Unitary Group Representations* by Mackey on the other, one has some forceful and interesting arm-chair reading in store. One should also keep Dirac's *Principles of quantum mechanics* close at hand... this is an extremely good book, written by a mathematician who is also a scientist and who is willing to make subjective statements to keep the theory alive and growing. It fills the bill in our current battles to revive the philosophy of mathematics as a part of a general scientific consciousness. It even passes the additional test of stating clearly certain open questions which remain in the theory and in the larger scientific investigations on which the theory may bear."

J. von Neumann, *Mathematical Foundations of Quantum Mechanics*, Translated from the German and with a preface by Robert T. Beyer, Twelfth printing, Princeton Landmarks in Mathematics, Princeton University Press, Princeton, NJ, 1996. xii+445 pp.

This is a classic written by a great mathematician. It is a good reference for understanding better the conceptual and mathematical foundation of quantum mechanics.

According to MathSciNet, "How does this classic text now stand in the light of modern developments? One would expect that by now it is hopelessly out-of-date. Of course, more modern treatments have appeared... These books are more concise, more direct, and contain simplifications in notation, methods and terminology. They each emphasize different aspects of the subject and they each carry these aspects to more distant frontiers. However, their core material and the germs of most of their ideas are contained in von Neumann's text. Moreover, they do not contain the personality, the motivating thoughts and the seminal calculations of the master himself."

F.W. Byron, R.W. Fuller, *Mathematics of Classical and Quantum Physics*, Dover Publications, 1992. x+661 pp.

10.1. CLASSICAL MATHEMATICAL PHYSICS

This is textbook for physics students and can complement physics textbooks in classical mechanics, electricity, magnetism, and quantum mechanics. It contains a wealth of information and is well-organized.

According to MathSciNet, "Students of theoretical physics have long been looking for a suitable mathematical textbook. So far they have had to be contented with one of two alternatives: (1) Use one or several good mathematical books but having little or no contact with physical problems. (2) Use a book on mathematical physics which is abundant in physical motivations, intuitions, and problems but lacks the elegant presentation and systematic organization usually associated with mathematical books. The authors of this book admirably combine the advantages of the above alternatives into one volume... Most of the motivations are taken from physical grounds. The examples and problems lean heavily on the applications to classical and quantum mechanics. Above all, the readers are not just presented with the mathematical techniques themselves. Instead, they are made familiar with the language, the style and the way of thinking characteristic of modern mathematics and physics, so that they may be able to keep up with the rapid growth of these two fields in the future."

10.1.3 L.D. Landau, E.M. Lifshitz, Course of Theoretical Physics. Vol. 1. Mechanics, Third edition, Pergamon Press, Oxford-New York-Toronto, Ont., 1976. xxvii+169 pp.

This is a all time classic and has had a huge influence on generations of physicists and mathematicians. All the information contained in these 10 volumes was considered by Landau to be the "minimum" that a physicist should know. It is a pity that there is no such an equivalent book (or a set of books) in mathematics, i.e., a book that contains the minimum materials (or a list of topics) a working mathematician should know. It is probably impossible due to the width of mathematics.

The Course of Theoretical Physics is a ten-volume series of books covering the basic topics of theoretical physics, in particular all major aspects of mechanics. The idea for this comprehensive book series started in 1935 when Landau became Professor of General Physics at Khar'kov University. He was thinking and carrying out programs for the "the theoretical minimum", which is the basic knowledge in theoretical physics needed by experimental physicists and by those who want to become theoretical physicists. In connection with this, he conceived the idea of this book series and began to write it. It was written in collaboration with his student Lifshitz. Almost all of the actual writing of the early volumes was done by Lifshitz, giving rise to the often repeated witticism, "not a word of Landau and not a thought of Lifshitz". The first eight volumes were finished in the 1950s, written in the Russian language, and translated into English by the late

1950s. The last two volumes were written in the early 1980s. Berestetskii and Pitaevskii also contributed to the series. The series is often referred to as "Landau and Lifshitz" in informal settings. The books are well known for their concise and accurate formulation of the laws of physics. Generations of physicists and applied mathematicians around the world have learnt physics from this series. The presentation of material is advanced and is suitable for graduate-level study.

Landau was not only a great physicist but also a great educator, and had dreamed of writing books on physics at every level. By the time of his almost fatal car accident in 1962, he had drafted plans for textbooks on mathematics for physicists, which would deal with practical applications of mathematics to physics and be free of rigors and complexities which are not necessary for applications.

From the preface by Lifshitz, "Landau's work in science was always such as to display his striving for clarity, his effort to make simple what was complex and so to reveal the laws of nature in their true simplicity and beauty. It was this aim which he sought to instil into his pupils, and which has determined the character of the *Course*."

According to an article on Landau by Lifshitz, "The striving for simplicity and order was an inherent part of the structure of Landau's mind. It manifested itself not only in serious matters but also in semi-serious things as well as in his characteristic personal sense of humour. Thus, he liked to classify everyone, from women according to the degree of their beauty, to theoretical physicists according to the significance of their contribution to science. This last classification was based on a logarithmic scale of five... On this scale, Einstein occupied the position $\frac{1}{2}$, while Bohr, Heisenberg, Schrödinger, Dirac and certain others were ranked in the first class. Landau modestly ranked himself for a long time in class $2\frac{1}{2}$ and it was only comparatively late in his life that he promoted himself to the second class."

The titles of all 10 volumes in this series are as follows:

1. Mechanics, 2. The Classical Theory of Fields, 3. Quantum Mechanics: Non-Relativistic Theory, 4. Relativistic Quantum Theory, 5. Statistical Physics, 6. Fluid Mechanics, 7. Theory of Elasticity, 8. Electrodynamics of Continuous Media, 9. Statistical Physics, Part 2., 10. Physical Kinetics.

According to an Amazon review, "If physicists could weep, they would weep over this book. The book is brief whilst deriving, in its few pages, all the great results of classical mechanics. Results that in other books take up many more pages...

The reason for the brevity is that, as pointed out by previous reviewers, Landau derives mechanics from symmetry. Historically, it was long after the main bulk of mechanics was developed that Emmy Noether proved that symmetries underly every important quantity in physics. So instead of starting from concrete mechanical case-studies and generalizing to the formal machinery of the Hamilton equations, Landau starts out from the most generic symmetry and derives the mechanics. The 2nd laws of mechanics, for example, is derived as a consequence of the uniqueness of trajectories in the Lagrangian. For some, this may seem too 'mathematical' but in reality, it is a sign of sophistication in physics if one can identify the underlying symmetries in a mechanical system. Thus this book represents the height of theoretical sophistication in that symmetries are used to derive so many physical results.

The difficulty with this approach, and the reason why this book is not a beginner's book, is that to the follow symmetric arguments, one really has to have already mastered vector calculus. Ideally, you should be able to transform coordinate in your sleep, perform integrals without missing a beat, whether they be line, area, or path, and differentiate functions in many dimensions. The arguments are not sloppy, as some have claimed - it only seems so if you have not mastered vector calculus.

Tradition says that in Plato's academy was engraved the phrase, ' 'Let no one ignorant of geometry enter here', so should the modern theoretical physicist, with Landau's Bible in hand, march under the arches engraved with the words 'Let no one ignorant of symmetry enter here'."

E. Lifshitz, L. Pitaevskii, *Course of Theoretical Physics*, Physical Kinetics. Vol. 10, Translated from the Russian by J. B. Sykes and R. N. Franklin, Pergamon International Library of Science, Technology, Engineering and Social Studies, Pergamon Press, Oxford-Elmsford, N.Y., 1981. xi+452 pp.

According to MathSciNet, "This is the closing volume of a series of ten volumes, which actually represent a general course of theoretical physics as was planned by L. D. Landau and the first author more than thirty years ago. However, the original scheme of this course was intended to have the ... nine volumes only."

Symmetry plays in a fundamental role in the above series of books by Landau and Lifshitz. A more mathematical book on mechanics and symmetry is the following one.

J. Marsden, T. Ratiu, *Introduction to Mechanics and Symmetry. A Basic Exposition of Classical Mechanical Systems* , Second edition, Texts in Applied Mathematics, 17, Springer-Verlag, New York, 1999. xviii+582 pp.

This book is a valuable and attractive introduction to some important topics in mechanics and symmetry such as reduction, stability and bifurcation. It is also a useful reference.

From the preface, "Symmetry and mechanics have been close partners since the time of the founding masters: Newton, Euler, Raglange, Laplace, Poisson, Jacobi, Hamilton, Kelvin, Routh, Riemann, Noether, Poincaré, Einstein, Schrodinger, Cartan, Dirac, and to his day, symmetry has continued to play a strong role, especially with the modern work of Kolmogorov, Arnold, Moser, Kirillov, Kostant, Smale, Souriau, Guillemin, Sternberg, and many others. This book is about these developments, with an emphasis on concrete applications..."

According to MathSciNet, "This book is another exposition of the basic ideas in Lagrangian and Hamiltonian mechanics. It does, however, manage to produce a distinctive and fresh outlook different from the well-known texts of Abraham and Marsden and of Arnold...

As the name of the book implies, a consistent theme running through the book is that of symmetry. Indeed the latter half of the book focuses on Poisson manifolds, momentum maps, Lie-Poisson reduction, co-adjoint orbits and the integrability of the rigid body... A pleasant feature of the book is that most of the theory that relates to finite-dimensional mechanical systems is illustrated concretely in terms of local coordinates, thereby making the book accessible even to beginners in the field."

A popular book on physics at the more elementary level is the following book.

R. Feynman, R. Leighton, M. Sands, *The Feynman Lectures on Physics. Vol. 1: Mainly Mechanics, Radiation, and Heat. Vol. 2: Mainly Electromagnetism and Matter. Vol. 3: Quantum Mechanics*, Addison-Wesley Publishing Co., Inc., Reading, Mass.-London, 1963. xii+513 pp., 1964. xii+569 pp., 1965. x+365 pp.

This is a classic on physics written by one of the most famous physicists and best lecturers. It is not only valuable to physics students, but also to mathematics students and mathematicians who are interested in physics.

From the book description, " 'The whole thing was basically an experiment,' Richard Feynman said late in his career, looking back on the origins of his lectures. The experiment turned out to be hugely successful, spawning a book that has remained a definitive introduction to physics for decades. Ranging from the most basic principles of Newtonian physics through such formidable theories as general relativity and quantum mechanics, Feynman's lectures stand as a monument of clear exposition and deep insight."

According to MathSciNet, "The text is based on lectures delivered by the first author to first year university students in an attempt to break out of the straitjacket of conventional introductory physics courses. On the evidence of this first volume of lectures, the attempt has been well worthwhile. The book is full of illustrations of basic principles drawn from modern physics; to give but one example, resonant oscillations are discussed not simply from the usual elementary mechanical standpoint but also in terms of electric circuits, oscillations in a crystal and the Mössbauer effect. It may be argued that some bits of this course are far beyond the capabilities of most first year students, but a touch of intellectual indigestion now and again is surely no great price to pay to escape the tedium of the conventional introductory course. In cutting away the deadwood that has been part of such courses for too long, the lectures have succeeded admirably."

10.2 More modern mathematical physics

Boltzmann was an Austrian physicist, best known for his deep work in the development of statistical mechanics, which explains and predicts how the properties of large numbers of atoms rather than individual ones determine the visible properties of matter such as viscosity, thermal conductivity, and diffusion. One street in Vienna is named after him, and Erwin Schrödinger International Institute for Mathematical Physics is located on this street.

Figure 10.2: Boltzmann's book.

10.2.1 General relativity and gravitation

A. Einstein, *The Meaning of Relativity*, 5th ed., Princeton University Press, Princeton, N. J., 1955. vi+169 pp.

This is a classic on the relativity theory written by its founder.

From Scientific American, "The present book is intended", Einstein wrote in 1916, "as far as possible, to give an exact insight into the theory of Relativity to those readers who, from a general scientific and philosophical point of view, are interested in the theory, but who are not conversant with the mathematical apparatus of theoretical physics... In the interest of clearness, it appeared to me inevitable that I should repeat myself frequently, without paying the slightest attention to the elegance of the presentation. I adhered scrupulously to the precept of that brilliant theoretical physicist L. Boltzmann, according to whom matters of elegance ought to be left to the tailor and to the cobbler." But it is elegant, in part because of the 1920 translation, by Robert W. Lawson, a British physicist who had polished his German while a prisoner of war in Austria. The introduction, by science writer Nigel Calder, guides the reader through the work section by section, even giving advice on which sections to skip, or at least not to worry about, if you can't "accompany Einstein through the forest of tricky ideas contained in this slim volume." Okay, this book isn't easy–again, in the master's elegant words, it "lays no small claims on the patience and on the power of abstraction of the reader"–but it is well worth the try.

From an Amazon review, "Many people feel frustrated because when they try to understand relativity, they find some authors that expound in their books a complex arrangement of equations referring to the mathematical part of the theory, namely, the books are accessible for people with certain levels of knowledge (that is the case of engineers, physicists, mathematicians, among others). Nevertheless, perceiving and anticipating this situation, Albert Einstein wrote this book (more than fifty years ago) whit the purpose of exposing the special and the general theory of relativity in such a way that anyone can understand it. In this sense, I think, Einstein succeeded because despite the shortness of the book, the same covers the most important aspects of relativity in a clear and concise form. Moreover, the book has appendixes where the author makes reference to some interesting subjects like the problem of space and relativity, the experimental confirmation of the theory, to name a few. If you have decided to learn something about relativity, and you do not have vast knowledge in physics and mathematics, I sincerely recommend you this book. On the other hand, if you were a reader looking for more technical information (mathematical foundation of general relativity), I would choose the book 'Gravitation' written by Misner, Thorne, Wheeler. This text represents an encyclopedia about general relativity."

P. Dirac, *General Theory of Relativity*, Reprint of the 1975 original, Princeton Landmarks in Physics, Princeton University Press, Princeton, NJ, 1996. viii+71 pp.

This is a classic on the relativity theory and the mathematics behind it written by a master physicist.

From the book description, "Einstein's general theory of relativity requires a curved space for the description of the physical world. If one wishes to go beyond superficial discussions of the physical relations involved, one needs to set up precise equations for handling curved space. The well-established mathematical technique that accomplishes this is clearly described in this classic book by Nobel Laureate P.A.M. Dirac. Based on a series of lectures given by Dirac at Florida State University, and intended for the advanced undergraduate, General Theory of Relativity comprises thirty-five compact chapters that take the reader point-by-point through the necessary steps for understanding general relativity."

According to MathSciNet, "This small (71 page) book has the aim of presenting the indispensible material that a student who wishes to understand Einstein's theory of relativity must master. The material presented does not require previous knowledge beyond the basic ideas of special relativity and the handling of differentiations of field functions. The material in the book is presented in the clear, direct and precise form for which Professor Dirac is noted. The contents of the book are divided into thirty-five short sections. Section 18 entitled the 'Schwarzschild solution' and Section 19 entitled 'Black holes' are of particular interest."

According to an Amazon review, "In his inimitable concise style, with not a word out of place, Dirac offers a 60 page sketch of the classical theory of general relativity.

10.2. MORE MODERN MATHEMATICAL PHYSICS

Doesn't cover any of the modern theoretical developments and offers not a single figure, but if you have a strong math background and very little time to spare, this is probably the book for you. No problems included, so serious autodidacts should supplement this with another text such as d'Inverno which does have problems."

H. Weyl, *Space, Time, Matter*, Dover, 1952. 330 pp.

This is a systematic exposition of the theory of relativity by a master mathematician. It contains many original ideas both in physics and mathematics.

From the book description: "A classic of physics ... the first systematic presentation of Einstein's theory of relativity." "Long one of the standard texts in the field, this excellent introduction probes deeply into Euclidean space, Riemann's space, Einstein's general relativity, gravitational waves and energy, and laws of conservation."

According to an Amazon Review, "This book bewitched several generations of physicists and students. Hermann Weyl was one of the very great mathematicians of this century. He was also a great physicist and an artist with ideas and words. In this book you will find, at a deep level, the philosophy, mathematics and physics of space-time. It appeared soon after Einstein's famous paper on General Relativity, and is, in fact, a magnificent exposition of it, or, rather, of a tentative generalization of it. The mathematical part is of the highest class, striving to put geometry to the forefront. Actually, the book introduced a far-reaching generalization of the theory of connections, with respect to the Levi-Civita theory. It was not a generalization for itself, but motivated by the dream (Einstein's) of including gravitation and electromagnetism in the same (geometrical) theory. The result was gauge theory, which, slightly modified and applied to quantum mechanics resulted in the theory which dominates present particle physics... Weyl's book is most famous for introducing gauge theory, which was later reborn in the form of phase transformations in quantum theory. Weyl did not live quite long enough to hear of the latter being applied by Yang and Mills, though he socially interacted with Yang in his last year at Princeton... Weyl also introduces what he calls 'tensor densities' which Shouten called 'Weyl tensors' and Synge and Schild call oriented tensors, often called twisted tensors. These are analogous to and include 'axial vectors'.... Weyl's introduction of the 'affine connection' after criticism of Levi-Civita's notion of parallelism led the way to further notions of connections and generalization of the notion of connection as such by Elie Cartan and others. These are but a few of the intellectual gems in this work."

R. Sachs, H.S. Wu, *General Relativity for Mathematicians*, Graduate Texts in Mathematics, Vol. 48, Springer-Verlag, New York-Heidelberg, 1977. xii+291 pp.

This is a very well-written book on general relativity for mathematicians as the title says. All concepts are precisely defined within the framework of modern differential geometry, and all assertions are proved rigorously. This is a valuable book for mathematicians who want to learn or to work on general relativity.

According to MathSciNet, "With the increased interest of late in general relativity, the time could not have been more opportune for a volume on this subject for mathematicians. The first author (a physicist), and the second author (a mathematician) acknowledge that presenting honest mathematics and honest physics together, while simultaneously preserving the distinction between the two, is a formidable task. To achieve such a balance requires a delicate touch, because it is difficult for mathematicians to steel themselves against their traditional methods. This quotation found in the book and attributed to Einstein contains the essential point: "As far as the laws of mathematics refer to reality, they are not certain; and as far as they are certain, they do not refer to reality."

Although the authors had moments of doubt during the writing of this book, and appear pessimistic about future efforts in explaining physics to mathematicians, they have produced a work of singularly stellar quality.

In summary, a mathematician with little or no physics background can obtain a leisurely introduction to present-day cosmology, both physical and theoretical, by reading the book...

The book should be on the desk of every mathematician who has wondered about the red shift or the microwave radiation background."

10.2.2 S. Hawking, G. Ellis, The Large Scale Structure of Space-time, Cambridge Monographs on Mathematical Physics, No. 1, Cambridge University Press, London-New York, 1973. xi+391 pp.

This is a classic written by a master. It gives a comprehensive account of the structure of space-time from the radius of an elementary particle to the radius of the universe, in particular singularities of the universe.

From the book description, "Einstein's General Theory of Relativity leads to two remarkable predictions: first, that the ultimate destiny of many massive stars is to undergo gravitational collapse and to disappear from view, leaving behind a 'black hole' in space; and secondly, that there will exist singularities in space-time itself. These singularities are places where space-time begins or ends, and the presently known laws of physics break down. They will occur inside black holes, and in the past are what might be construed as the beginning of the universe. To show how these predictions arise, the authors discuss the General Theory of Relativity in the large."

According to the review in Bulletin of AMS, "This is an exciting and important volume since it is the first comprehensive presentation of a theory of cosmology taking into

10.2. MORE MODERN MATHEMATICAL PHYSICS

account the discoveries of the past quarter century in particle physics, radio astronomy, and differential topology. The astronomical universe or cosmos is examined within the framework of general relativity and global differential geometry. The exposition is authoritative and painstaking, although in the search for logical completeness sometimes a bewildering tangle of alternatives and complexities is introduced (see, for instance, Chapter 6 on causal structure). The authors assume a basic knowledge of the physical aspects of general relativity theory, and write for the reader who is skilled in tensor calculus but who wishes to see the appropriate concepts defined in an intrinsic coordinate-free manner suitable for a global geometry."

According to an Amazon review, "The early seventies saw a revolution in cosmology; for the first time, modern mathematical methods were applied to the discipline, with intriguing results. This book was (along with Penrose's articles) the seminal work in global general relativity... The meat-and-potatoes of the book is the discussion of gravitational collapse, and the singularity theorems. They provide us with intuitively good reasons for believing in some very strange phenomenon. If you're interested in the frontiers of modern science, and have the appropriate mathematical background, this book cannot be recommended too highly. The little yellow book stands supreme in the hierarchy of works of modern physics."

According to MathSciNet, "Despite its imposing title, this book is a text on general relativity with a very mathematical orientation. It is an excellent introduction to the subject for a mathematician interested in relativity, because it is much more rigorous and uses a language much more familiar to the mathematician than that found in the usual texts. The thrust of the book is toward proving the 'singularity theorems', stating that large classes of solutions of Einstein's equations with reasonable equations of state reach a singularity (of some sort) in a finite time."

W. Misner, K. Thorne, J. Wheeler, *Gravitation*, W. H. Freeman and Co., San Francisco, Calif., 1973. ii+xxvi+1279 pp.

This book is a classic and is a manual on the modern general relativity theory. It can be used as an accessible introduction and an essential reference for people who are interested in mathematical astronomy.

From the book description, "This landmark text offers a rigorous full-year graduate level course on gravitation physics, teaching students to: • Grasp the laws of physics in flat spacetime, • Predict orders of magnitude, • Calculate using the principal tools of modern geometry, • Predict all levels of precision, • Understand Einstein's geometric framework for physics, • Explore applications, including pulsars and neutron stars, cosmology, the Schwarzschild geometry and gravitational collapse, and gravitational waves, • Probe experimental tests of Einstein's theory, • Tackle advanced topics such as superspace and quantum geometrodynamics."

According to Amazon reviews, "This book can be divided into three logical parts... This first part is the best introduction to the theory of relativity I have ever read. The mathematics is introduced in a very comprehensive manner, there are lots of exercises where the reader can get used to the tensor calculus. The physical explanations are just brilliant and what is more important general relativity is introduced in the manner Einstein itself viewed it: as a geometric representation of gravity!... The second part starts with the application of general relativity to stars (stars and relativity), goes on to the universe and to black holes (gravitational collapse and black holes, and describes finally gravitational waves and experimental methods.... The third part finally describes the frontiers of general relativity... Like part two it gives a good overview not showing many computational details."

R. Penrose, *The Road to Reality. A complete Guide to the Laws of the Universe,* Alfred A. Knopf, Inc., New York, 2005. xxviii+1099 pp.

This is an ambitious book by a famous person as the title suggest. It gives an overview of the contemporary mathematical physics. It is a serious book for the educated public, or at least the mathematically oriented.

From the weekly publisher, "Unlike a textbook, the purpose of which is purely to impart information, this volume is written to explore the beautiful and elegant connection between mathematics and the physical world. Penrose spends the first third of his book walking us through a seminar in high-level mathematics, but only so he can present modern physics on its own terms, without resorting to analogies or simplifications (as he explains in his preface, 'in modern physics, one cannot avoid facing up to the subtleties of much sophisticated mathematics')... Penrose transcends the constraints of the popular science genre with a unique combination of respect for the complexity of the material and respect for the abilities of his readers."

According to MathSciNet, "Do not be misled by the subtitle of Penrose's book to expect a detailed exposition of the current theories of modern physics. This book will not acquaint the reader intimately with the details of string theory, for example, nor even of general relativity!... In short, this book provides the opportunity to glimpse Penrose's fascinating vision of the physical world and how to describe it. And whether one agrees or not with his particular views, they are intriguing and provocative...

Another remarkable feature of this book is that Penrose has made an ambitious attempt to engage as wide an audience as possible in this very serious discussion of the critical issues of contemporary physics... What of the book's enduring qualities? Is it a book to read through (or dip into) and then convert into a structural component of a shaky bookcase? Penrose asserts in the Preface that he intends the book to serve pedagogical purposes; the book is peppered with exercises. While some are immediately

10.2. MORE MODERN MATHEMATICAL PHYSICS

intriguing, many are of the 'fill in the details' variety, and sometimes those details go well beyond the text. A few are even acausal, depending explicitly on material treated at a later stage of the book. To my mind, this book, whether one agrees with Penrose's assessments of the likes of string theory or not, can serve brilliantly as a companion to study for many years to come for those who are not already experts in most of the topics covered... This book conveys the diversity and power of mathematical thought and its impact on physical theory. Penrose's enthusiasm helps carry the reader through this difficult material. But, in essence, this book is an extraordinary invitation to share a little of Penrose's intellectual adventure along the road to reality, or at least in search of that road..."

10.2.3 Statistical mechanics

Statistical thermodynamics applies probability theory to study the thermodynamic behavior of systems composed of a large number of particles. It provides a connection between the microscopic properties of individual atoms and molecules and the macroscopic properties of materials that can be easily observed, hence explaining thermodynamics in terms of classical and quantum-mechanical description of statistics and mechanics at the microscopic level. It also provides a microscopic interpretation of macroscopic thermodynamic quantities such as work, heat, free energy, and entropy.

Statistical mechanics was initiated by Ludwig Boltzmann in 1870. His original papers on various aspects of statistical mechanics total about 2,000 pages. The term "statistical thermodynamics" was proposed by J. Willard Gibbs in 1902 (according to him, the term "statistical", in the context of statistical mechanics, was first used by James Clerk Maxwell in 1871.)

D. Ruelle, *Statistical Mechanics: Rigorous Results*, Reprint of the 1989 edition, World Scientific Publishing Co., Inc., River Edge, NJ; Imperial College Press, London, 1999. xvi+219 pp.

This is a classic on rigorous results about equilibrium statistical mechanics written by an authority. The emphasis is on general methods, and special models are not discussed.

According to MathSciNet, "it was a seminal opus which crucially contributed to the birth and growth of the field (and of related fields). Furthermore, for over three decades it has been an indispensable tool for the formation of successive generations of mathematical physicists. And, moreover, despite the enormous development of the area, it continues to be a classical reference where researchers and students find information, enlightenment and inspiration.

The main reason for the lasting value of this book is, of course, its focus on the building blocks of the theory..."

D. Ruelle, *Thermodynamic Formalism. The Mathematical Structures of Classical Equilibrium Statistical Mechanics*, Second edition, Cambridge Mathematical Library, Cambridge University Press, Cambridge, 2004. xx+174 pp.

This book studies certain mathematical aspects of equilibrium statistical mechanics, i.e., the theory of thermodynamic formalism which was developed to describe the properties of certain physical systems consisting of a large number of subunits. It is more mathematical and advanced than the previous book by the same author. It is a classical and valuable reference for both mathematicians and physicists working on related areas.

H. Georgii, *Gibbs Measures and Phase Transitions*, Second edition, de Gruyter Studies in Mathematics, 9, Walter de Gruyter & Co., Berlin, 2011. xiv+545 pp.

This book is an in-depth introduction to Gibbs measures and phase transitions in statistical mechanics. It is also a valuable reference due to its broad coverage.

According to MathSciNet, "This book is concerned with the description of the equilibrium states of infinite classical lattice models in statistical mechanics. The book is an excellent basis for a serious study. It is carefully written, the proofs are detailed and clear, the examples are illuminating. Over 650 annotated references provide a useful orientation in the huge existing literature."

C. Cercignani, *The Boltzmann Equation and Its Applications*, Applied Mathematical Sciences, 67, Springer-Verlag, New York, 1988. xii+455 pp.

This is an advanced, modern introduction to the Boltzmann equation in kinetic theory. It is an essential reference.

According to MathSciNet, "The author has written what will surely become one of the major references on the Boltzmann equation; for many areas this book will be 'the' reference. Successfully writing on the subject of the Boltzmann equation is not an easy task. The heterogeneous nature of the audience (and its expectations) requires the author to steer a careful path lest he exclude or, worse yet, bore many of his readers. These pitfalls have been adroitly avoided here without any loss in content. Rigorous mathematical techniques are used where required, but these are introduced in such a way that the non-mathematician should be able to follow closely and appreciate what is happening, while the mathematician will not find his attention wandering. The physical side of the theory is developed with similar care and is on an equally high level. This blending of mathematics and physics produces its own synergism and is responsible for the unique overall tone of the book. The author's subject is the modern theory of Boltzmann's equation, primarily as it relates to dilute monatomic gases..."

C. Cercignani, R. Illner, M. Pulvirenti, *The Mathematical Theory of Dilute Gases*, Applied Mathematical Sciences, 106, Springer-Verlag, New York, 1994. viii+347 pp.

10.2. MORE MODERN MATHEMATICAL PHYSICS

This book is a good introduction to and an overview of the mathematical results on the Boltzmann equation. It provides a unified presentation of the known mathematical results and can serve as a very useful reference on the literature about the Boltzmann equation.

According to MathSciNet, "This excellent book covers the modern mathematical theory of the Boltzmann equation in the kinetic theory of rarefied gases, and contains most of the significant results in the field of the last twenty years... The book, which is addressed to postgraduate students familiar with advanced calculus and functional analysis, gives a systematic treatment of the main mathematical results on the Boltzmann equation, a discussion of open problems, and a guide to the existing literature."

R. Baxter, *Exactly Solved Models in Statistical Mechanics*, Reprint of the 1982 original, Academic Press, Inc, [Harcourt Brace Jovanovich, Publishers], London, 1989. xii+486 pp.

This is a classic written by a master. It describes beautifully two-dimensional lattice models in statistical mechanics.

According to MathSciNet, "This is an excellent book which stands out by the brilliance of the work described in it, the clarity of its presentation and the fame of its author. Strictly speaking the book's short title is slightly too general. It may be useful to supplement it with the following more restrictive outline or abstract: 'Two-dimensional lattice models in equilibrium statistical mechanics, solved by means of the Bethe ansatz method and related methods—with an introduction into the basic concepts of equilibrium statistical mechanics and four chapters on the most important exactly solved models of a more elementary nature' ... The style of writing of this book is, in a sense, rather down-to-earth, with an emphasis on calculations and derivations of specific results, rather than on discussing general questions."

10.2.4 Quantum field theory

J. Glimm, A. Jaffe, *Quantum Physics. A Functional Integral Point of View*, Second edition, Springer-Verlag, New York, 1987. xxii+535 pp.

This is a classic written by two major contributors. It studies a broad range of problems in quantum field theory and statistical physics. It is a valuable references to both mathematicians and physicists working on related areas.

According to MathSciNet, "Quantum field theory is the most basic and most true branch of theoretical physics... Constructive Quantum Field Theory is the mathematical (rigorous) study of quantum field theories... Quantum physics provides a view of Constructive Quantum Field Theory, and related fields in Statistical Mechanics, by the two researchers with the greatest insights into the theory... This book may be read through as a thorough introduction to a large segment of modern mathematical physics. Mathematicians with a background in analysis will find the book self-contained. There is a grand scope of covered material, from non-relativistic quantum mechanics and scattering theory, through statistical mechanics, and most importantly to quantum field theory."

10.2.5 V.I. Arnold, Mathematical Methods of Classical Mechanics, Second edition, Graduate Texts in Mathematics, 60, Springer-Verlag, New York, 1989. xvi+508 pp.

This is a classic written by a leading expert. It is an excellent textbook for students to learn classical mechanics in the framework of modern mathematics. It can also serve as a valuable reference.

According to MathSciNet, "This beautifully written book has much to offer to students (in the extended sense) of both mathematics and physics. The author frequently drops terms and makes approximations like a physicist, but as a mathematician he never neglects to investigate the consequences of approximating a complicated system by a simpler one. In fact, the question of the validity of approximation methods forms one of the recurring themes of the book; another theme is the role of symplectic geometry in the formal structure of mechanics.

An attractive feature of the book is the frequent use of concrete examples to illustrate physical principles. These examples, as part of an overall attempt to present mechanics as it is understood by physicists, distinguish the book from more abstract recent works on mechanics [such as the books of R. Abraham, *Foundations of mechanics*, 1967. On the other hand, the mathematical content of the book makes it quite unlike the standard texts written by physicists such as the books by L. D. Landau and E. M. Lifshitz, *Course of Theoretical Physics, Vol. 1, Mechanics*; and H. Goldstein, *Classical Mechanics*, 1951."

Reviews from Amazon include the following: "Written by a great mathematician of our time, Vladimir Arnol'd, this truly outstanding book represents classical mechanics from a unifying geometrical point of view and is a 'must-to-read' book for any graduate student working in the field.!"

"Some readers might see this as a book of math rather than physics, but that would not be fair: Arnold always stresses the geometrical meaning and the physical intuition of what he states or demonstrates."

R. Abraham, J. Marsden, *Foundations of Mechanics*, Second edition, revised and enlarged, With the assistance of Tudor Ratiu and Richard Cushman, Benjamin/Cummings Publishing Co., Inc., Advanced Book Program, Reading, Mass., 1978. xxii+m-xvi+806 pp.

The first edition of this book provided the first rigorous foundation of mechanics on manifolds. The substantially expanded second edition marked the beginning of a rapid

10.2. MORE MODERN MATHEMATICAL PHYSICS

development of geometric methods in mathematical physics, in particular in mechanics and dynamics. This book has been a standard reference in the subject.

According to MathSciNet, "the book can be recommended as a basic reference work for the foundations of differentiable and Hamiltonian dynamics. For proofs of the deeper results in the Smale school of dynamics or, for example, the results of Kolmogorov, Arnold, and Moser, one must still refer to other texts and the original papers. A great effort has been made to make the book as up-to-date as possible prior to its actual publication in terms of results and modern proofs. Though occasionally it is profitable to first consult another book, such as V. I. Arnold's *Mathematical methods of classical mechanics*, 1978, for a simple geometric discussion of symmetries in Hamiltonian systems, there is no other published account that can match the present text in detail and wealth of material. For a classroom text the book by Arnold would probably be more appreciated by the average graduate student, especially for the elementary classical mechanics. And for the physics, treated from a rigorous mathematical point of view, the series by W. Thirring, *A Course in Mathematical Physics*, could be used. But this new edition of the text under review will be a leader in the field for quite a few years to come."

W. Thirring, *A Course in Mathematical Physics. Vol. I. Classical Dynamical Systems*, Translated from the German by Evans M. Harrell, Springer-Verlag, New York-Vienna, 1978. xii+258 pp.

This is a classical book. It is the first volume of a set of four volumes originally published in German, which has also been translated into Russian.

According to Zentralblatt MATH, "The author presents Hamiltonian mechanics as a geometry of phase flows on a manifold. This approach is modern and attractive from the mathematical point of view. It was used by some other authors (Abraham and Marsden, Arnold)... The main difference between this book and above mentioned books by Abraham and Marsden and Arnold is that this book is aid more at theoretical physics and is a base for other three volumes of the course."

H. Goldstein, C. Poole, J. Safko, *Classical Mechanics*, 3rd Edition, Addison Wesley, 2001. 680 pages.

This is a classic and a standard textbook on classical mechanics of particles and systems.

From the book description of the new edition in 2001, "For thirty years this has been the acknowledged standard in advanced classical mechanics courses. This classic book enables readers to make connections between classical and modern physics - an indispensable part of a physicist's education. this new edition, ... introduces readers to the increasingly important role that nonlinearities play in contemporary applications of classical mechanics."

According to an Amazon review, "Goldstein's *Classical Mechanics* appeared at the right time. The development of quantum mechanics demanded familiarity with methods of advanced mechanics that no student of physics had been introduced to. Dirac told in a seminar that he didn't know what a Poisson bracket was, when he was constructing his version of quantum mechanics (where Poisson brackets play a fundamental role). Heisenberg didn't know matrices, in similar circumstances. Max Born did know these things, and actually wrote a superb book on mechanics using them, but it was in German, at an advanced level and called Mechanics of the Atom. The book then available in English was the formidable Whittaker 'Analytical Dynamics', whose exercises took sometimes a whole page just to be stated! In this panorama, in the fifties, Addison-Wesley published the beautifully produced Goldstein. It was an instant sensation. In the introduction the author candidly confessed that, in his opinion, a course in mechanics justified itself only as a preparation for quantum mechanics, and that was clearly the slant of the book. It was extremely well written, except for a disastrous chapter on the Hamilton-Jacobi equation. The exercises were not at the level of the text: you found much better ones in Slater, Frank's 'Mechanics', for instance. The references were excellent, commented, and gave the reader a sense of perspective (and of awe, in the company of men like Riemann, Born, Weber...). Later on the slim book by Landau, Lifshitz, 'Mechanics', entered the scene and showed that Goldstein's program could be made better, briefer, and that the Hamilton-Jacobi equation, clearly and sensibly derived, was the jewel of the crown. Not only, in the subsequent volumes of their Theoretical Physics course, they showed how invaluable this Hamilton-Jacobi was, by applying it with great skill in all kinds of problems! Then, finally, it became clear that mechanics was not dead: the whole affair of stability, chaos, etc, exploded, and it became impossible to consider mechanics just as a ladder to quantum mechanics. So, even the philosophy of the venerable Goldstein had to be forgotten. Still, Goldstein's Classical Mechanics is alive, possibly now more Classical than Mechanics."

10.2.6 M. Reed, B. Simon, Methods of Modern Mathematical Physics. I. Functional Analysis. Second edition; II. Fourier Analysis, Self-adjointness; III. Methods of Modern Mathematical Physics; IV. Analysis of Operators, Academic Press, Inc., New York, 1980. xv+400 pp., 1975. xv+361 pp., 1979. xv+463 pp., 1978. xv+396 pp.

This four volume set contains a wealth of information on methods of modern mathematical physics. They are well-written and well-organized. Each volume can serve a good introduction to the subject under discussion and the whole set is a very valuable and comprehensive reference on many aspects of modern mathematical physics.

10.2. MORE MODERN MATHEMATICAL PHYSICS

According to MathSciNet, "It is hard for the reviewer to judge how suited the authors' mathematical banquet is to the appetite of the average theoretical physicist. For the mathematician who wants a guide to some of the most interesting recent developments in mathematical physics it is invaluable—the most important series of texts in functional analysis since Dunford-Schwartz."

According to an Amazon review, "Books on mathematical methods 'for physicists' are often criticized by their superficiality, a sacrifice deemed necessary for achieving completeness. This one is a glaring exception: the first of a set of 4 (!) volumes dealing with the finest tools for dealing with the delicate mathematical questions in quantum theory - namely, functional analysis..."

10.2.7 Scattering theory

Scattering theory studies the scattering of waves and particles in mathematics and physics. For example, wave scattering corresponds to the collision and scattering of a wave with some material object, for instance sunlight scattered by rain drops to form a rainbow. Scattering also includes the interaction of billiard balls on a table and the Rutherford scattering of alpha particles by gold nuclei etc. Mathematically, scattering theory studies how solutions of partial differential equations, propagating freely "in the distant past", come together and interact with one another or with a boundary condition, and then propagate away "to the distant future". There are two types of problem: the direct scattering problem and the inverse scattering problem.

S. Novikov, V. Manakov, L. Pitaevskii, V. Zakharov, *Theory of Solitons. The Inverse Scattering Method*, Translated from the Russian, Contemporary Soviet Mathematics, Consultants Bureau [Plenum], New York, 1984. xi+276 pp.

This is a classic written by leading experts.

According to MathSciNet, "This is the first book on the famous inverse scattering method. The authors are well-known physicists and mathematicians who have made a number of crucial contributions to the subject. .... The book requires very little preliminary knowledge from the reader. It includes pertinent material from such far-apart subjects as scattering theory and Riemann surfaces. The exposition is very informal, which is a big advantage for the reader learning such an immense subject as the inverse scattering method."

J. Dereziński, C. Gérard, *Scattering Theory of Classical and Quantum N-particle Systems*, Texts and Monographs in Physics, Springer-Verlag, Berlin, 1997. xii+444 pp.

The book gives a comprehensive and largely self-contained introduction to scattering theory of classical and quantum N-particle systems. This well-written book is suitable as an introduction for students and a valuable reference for researchers.

From the book description, "It is a modern presentation of time-dependent methods for studying problems of scattering theory in the classical and quantum mechanics of N-particle systems... The book is self-contained and explains in detail concepts that deepen the understanding. As a special feature of the book, the beautiful analogy between classical and quantum scattering theory (e.g., for N-body Hamiltonians) is presented with deep insight into the physical and mathematical problems."

According to MathSciNet, "... this book, written by two active authors, intends to summarize complete achievements, novel results and basic techniques related to the multi-particle systems from the viewpoint of the time-dependent method, and its significance would be properly understood in terms of recognition of the history of scattering theory..."

D.R. Yafaev, *Mathematical Scattering Theory. General Theory*, Translated from the Russian by J. R. Schulenberger, Translations of Mathematical Monographs, 105, American Mathematical Society, Providence, RI, 1992. x+341 pp.

This book gives a systematic account of the stationary, i.e. time-independent, mathematical scattering theory and fills a gap in the literature. It is both a good introduction for students and a valuable reference for researchers.

From the book description, "Scattering theory presents an excellent example of interaction between different mathematical subjects: operator theory, measure theory, the theory of differential operators and equations, mathematical analysis, and applications of these areas to quantum mechanics. Because of the interplay of these fields, a deep understanding of scattering theory can lead to deep insights into the developing world of modern mathematics. Yafaev's book provides such an understanding of scattering theory, starting with basic principles and extending to current research. He presents a comprehensive and systematic exposition of the theory, covering different methods (of trace class and smooth perturbations) and approaches (time dependent and stationary) and discussing the relationships among them."

V. Marchenko, *Sturm Liouville Operators and Applications*, Translated from the Russian by A. Iacob, Operator Theory: Advances and Applications, 22, Birkhäuser Verlag, Basel, 1986. xii+367 pp.

This is a classical book on inverse spectral problems for Sturm-Liouville operators using the transformation operator method. It is a useful reference on this basic subject.

According to MathSciNet, "As pointed out in the author's preface, the main goal of this monograph is to show the importance of transformation operators in spectral theory (these are operators that can be used to provide a complete solution to the problem of recovering a Sturm-Liouville equation from its spectral function). Both traditional and new applications of such operators are discussed, the latter in connection with the use of spectral theory in the study of nonlinear equations... The monograph is not intended to be an exhaustive account of spectral-theoretic methods."

L.D. Faddeev, L.A. Takhtajan, *Hamiltonian Methods in the Theory of Solitons*, Translated from the Russian by A. G. Reyman, Springer Series in Soviet Mathematics. Springer-Verlag, Berlin, 1987. x+592 pp.

The book gives a systematic account of the inverse scattering method and its application to soliton theory. Though it deals with the classical part of the subject only, it is a milestone in dynamical systems literature. It is a classic written by real experts.

From the book description, "The main characteristic of this now classic exposition of the inverse scattering method and its applications to soliton theory is its consistent Hamiltonian approach to the theory."

From the preface, "This book presents the foundations of the inverse scattering method and its applications to the theory of solitons in such a form as we understand it in Leningrad.

The concept of soliton was introduced by Kruskal and Zabusky in 1965. A soliton (a solitary wave) is localized particle-like solution of a nonlinear equation which describes excitations of finite energy and exhibits several characteristic features: propagation does not destroy the profile of a solitary wave; the interaction of several solitary waves amounts to their elastic scattering, so that their total number and shape are preserved. Occasionally, the concept of the soliton is treated in a more general sense as a localized solution of finite energy. At present this concept is widely spread due to its universality and the abundance of applications in the analysis of various processes in nonlinear media. The inverse scattering method which is the mathematical basis of soliton theory has developed into a powerful tool of mathematical physics for studying nonlinear differential equations, almost as vigorous as the Fourier transform.

The book is based on the Hamiltonian interpretation of the method, hence the title..."

D. Colton, R. Kress, *Integral Equation Methods in Scattering Theory*, Pure and Applied Mathematics (New York), A Wiley-Interscience Publication, John Wiley & Sons, Inc., New York, 1983. xii+271 pp.

This is a valuable reference on integral equation methods in scattering theory, especially for people interested in wave scattering by obstacles.

According to MathSciNet, "According to the authors, the following topics are emphasized in their presentation: (1) the regularity properties of acoustic and electromagnetic potentials, (2) the close relationship between Maxwell's equations, the vector Helmholtz equation, and the scalar Helmholtz equation, (3) the reformulation of the boundary value problems of scattering theory as integral equations that are uniquely solvable for all values of the wave number, (4) the low frequency behavior of solutions to the boundary value problems of scattering theory, (5) the use of function-theoretic methods to study the inverse scattering problem, (6) the role of compactness in stabilizing the inverse scattering problem, and (7) the use of integral equations to reformulate the inverse scattering problem as a problem in constrained optimization."

10.3 Dynamical systems

The theory of dynamical system has its origins in Newtonian mechanics. A dynamical system on a space, which can be a topological space, a smooth manifold or a measure space, is a rule which describes the time dependence of a point in the space. For example, one important example of a dynamical system is a smooth manifold together with a family of diffeomorphisms of the manifold, and one basic problem is to understand structures of orbits of this family of maps and their large time asymptotic behaviors. Dynamical systems occur naturally in many different practical situations such as the swinging of a clock pendulum and the flow of water in a pipe.

Lagrangian mechanics formulates the classical mechanics using Hamilton's principle of stationary action. It can be applied to systems whether or not they conserve energy or momentum and was introduced by Lagrange in 1788. The core element of Lagrangian mechanics is the Lagrangian function, which summarizes the dynamics of the entire system in a very simple expression. In Lagrangian mechanics, the trajectory of a system of particles is derived by solving the Lagrange equations of either the first or second kinds (the second kinds are also called the Euler-Lagrange equations).

Figure 10.3: Lagrange's book.

10.3.1 Dynamics and celestial mechanics

Celestial mechanics studies motions of celestial objects. It applies principles of physics, in particular classical mechanics, to astronomical objects such as stars and planets. It is an important branch of dynamical systems.

    C. Siegel, J. Moser, *Lectures on Celestial Mechanics*, Translation by Charles I. Kalme, Die Grundlehren der mathematischen Wissenschaften, Band 187, Springer-Verlag, New York-Heidelberg, 1971. xii+290 pp.

10.3. DYNAMICAL SYSTEMS

This is a classic written in a clear and straightforward way by two great mathematicians. It is a valuable reference.

According to MathSciNet, "This book is essentially a translation of a book by the first author *Vorlesungen über Himmelsmechanik*, 1956, with the addition of several sections... The flavor of the original edition is unchanged by the translation into a clearly written text... The importance of the sections added to the original edition should be emphasized. Those added to Chapter III represent the introduction into celestial mechanics of powerful techniques that allow the problem of stability of equilibria in Hamiltonian systems of two degrees of freedom to be resolved. Applications of these methods to prove the stability of Lagrange's equilibria in the restricted problem of three bodies are explored thoroughly. The explosion of interest in celestial mechanics during the late 1950's that has continued to the present has sustained an audience that welcomes this translation of a fine book on the foundations of celestial mechanics".

J. Moser, *Stable and Random Motions in Dynamical Systems: With Special Emphasis on Celestial Mechanics*, Reprint of the 1973 original, With a foreword by Philip J. Holmes, Princeton Landmarks in Mathematics, Princeton University Press, Princeton, NJ, 2001. xii+198 pp.

This book grew out of the Herman Weyl lectures by the author at IAS, Princeton. It gives a selected survey of recent progress in celestial mechanics, in particular with the stability problem, and the opposite case of random orbits. This book is accessible and is a valuable reference.

According to MathSciNet, "This monograph describes how stable behavior (given by quasi-periodic motion on invariant tori) and statistical behavior (given by shifts on symbol spaces) take place side by side in analytic conservative systems of differential equations. Historical background on the significance of these questions, many examples where these phenomena occur, as well as an introduction to the results are given. Because the harder proofs are placed in separate chapters, this presentation is accessible to the reader without specialized background in dynamical systems. The book by V. I. Arnold and A. Avez *Ergodic Problems of Classical Mechanics*, 1968, is similar in that it treats both stable behavior and statistical or ergodic behavior. The book by C. L. Siegel and the author *Lectures on Celestial Mechanics*, 1971, is a more complete introduction to celestial mechanics. For an alternate approach see the book by R. Abraham and J. Marsden *Foundations of Mechanics*, 1967."

10.3.2 Dynamical systems

G. Birkhoff, *Dynamical Systems*, With an addendum by Jurgen Moser, American Mathematical Society Colloquium Publications, Vol. IX, American Mathematical Society, Providence, R.I. 1927. xii+305 pp. Reprinted in 1966.

This book is a classic written by a leading expert. Since it was first published in 1927, it has always been in print. It has been a very influential book on the development of dynamics and many other branches of mathematics.

According to Zentralblatt Math, "The author's great book ... is well known to all, and the diverse active modern developments in mathematics which have been inspired by this volume bear the most eloquent testimony to its quality and influence."

From the book description, "In 1927, G. D. Birkhoff wrote a remarkable treatise on the theory of dynamical systems that would inspire many later mathematicians to do great work. To a large extent, Birkhoff was writing about his own work on the subject, which was itself strongly influenced by Poincaré's approach to dynamical systems. With this book, Birkhoff also demonstrated that the subject was a beautiful theory, much more than a compendium of individual results. The influence of this work can be found in many fields, including differential equations, mathematical physics, and even what is now known as Morse theory. The present volume is the revised 1966 reprinting of the book, including a new addendum, some footnotes, references added by Jurgen Moser, and a special preface by Marston Morse. Although dynamical systems has thrived in the decades since Birkhoff's book was published, this treatise continues to offer insight and inspiration for still more generations of mathematicians."

A. Katok, B. Hasselblatt, *Introduction to the Modern Theory of Dynamical Systems*, With a supplementary chapter by Katok and Leonardo Mendoza, Encyclopedia of Mathematics and its Applications, 54, Cambridge University Press, Cambridge, 1995. xviii+802 pp.

This is a very comprehensive introduction to the modern theory of dynamical systems and can serve as a valuable reference for people at all levels.

According to MathSciNet, "Most introductory texts in dynamical systems concern somewhat limited systems, such as homeomorphisms of the interval, or only particular techniques, such as symbolic dynamics or simulation of bifurcation. The book under review is an introduction to differentiable dynamical systems and all that is connected to their analysis. Thus it must include thorough treatments of topological dynamics, symbolic dynamics, and ergodic theory.

In order to begin a comprehensive exposition without sacrificing motivation, the authors use examples interlaced with definitions and propositions in the first chapter. Later chapters are organized by topic, providing easier reference and some independence among chapters.

This book emphasizes topological, measure-theoretic, and number-theoretic invariants associated to dynamical systems, and methods for deciding a system's asymptotic behavior, including many local-to-global results.

10.3. DYNAMICAL SYSTEMS

The motivation is evident throughout the book... The audience for this book consists of graduate students or researchers in mathematics or related fields who have the 'first year' background in measure theory, functional analysis, topology (with some differential geometry), and algebra, or are willing to learn it quickly from the appendix. The book is a pleasure to read."

K. Alligood, T. Sauer, J. Yorke, *Chaos. An Introduction to Dynamical Systems*, Textbooks in Mathematical Sciences, Springer-Verlag, New York, 1997. xviii+603 pp.

This is a good introduction, written by experts, to dynamical systems with particular emphasis on chaos for undergraduate students and other beginners.

According to MathSciNet, "With regard to both style and content, the authors succeed in introducing junior/senior undergraduate students to the dynamics and analytical techniques associated with nonlinear systems, especially those related to chaos.

There are several aspects of the book that distinguish it from some other recent contributions in this area, such as one by R. Devaney, *A First Course in Chaotic Dynamical Systems*. First, the treatment of discrete systems here maintains a balanced emphasis between one- and two- (or higher-) dimensional problems. This is an important feature since the dynamics for the two cases and methods employed for their analyses may differ significantly. Also, while most other introductory texts concentrate almost exclusively upon discrete mappings, here at least three of the thirteen chapters are devoted to differential equations... In short, the book is a significant contribution to the increasing collection of texts on this topic. The authors have succeeded in taking the most important ideas from dynamical systems and chaos, and presenting them to undergraduates in a serious but accessible manner. The work should be a definite consideration for anyone contemplating an introductory course in this area."

R. Devaney, *An Introduction to Chaotic Dynamical Systems*, Reprint of the second (1989) edition, Studies in Nonlinearity, Westview Press, Boulder, CO, 2003. xvi+335 pp.

The book gives an excellent introduction to chaos in dynamical systems. It is accessible and contains a lot of interesting material. It is a valuable book to both students and others who are interested in chaos.

According to MathSciNet, "The text is clearly written, contains numerous exercises, and has ample suggestions for further reading. The goals are explicitly stated in the introduction ('the basic ideas of the field should be accessible to junior and senior mathematics majors as well as to graduate students and scientists in other disciplines') and in the reviewer's opinion they are attained. The reader is also cautioned that results come only with much more difficulty when one attempts to treat more general dynamical systems. Indeed, computations with specific nonlinear ordinary differential equations are next to impossible."

L. Perko, *Differential Equations and Dynamical Systems*, Third edition, Texts in Applied Mathematics, 7, Springer-Verlag, New York, 2001. xiv+553 pp.

This is a popular textbook for advanced undergraduate and beginning graduate courses. It gives a systematic account of the qualitative and geometric theory of nonlinear differential equations and dynamical systems.

According to MathSciNet, "This excellent book covers those topics necessary for a clear understanding of the qualitative theory of ordinary differential equations and of the concept of a dynamical system... This book is to be highly recommended for advanced undergraduates and beginning graduate students. Researchers from other fields who use the qualitative theory of ordinary differential equations will also find this book valuable."

W. de Melo, S. van Strien, *One-dimensional Dynamics*, Ergebnisse der Mathematik und ihrer Grenzgebiete (3), 25, Springer-Verlag, Berlin, 1993. xiv+605 pp.

One-dimensional dynamics can provide guides and motivations for the general higher dimensional dynamical systems, in particular with respect to non-hyperbolic dynamical systems. This book gives a complete and coherent description of one-dimensional dynamics. This accessible book is well-organized and contains many valuable historical remarks and a good selection of references, and hence is a good introduction to beginners and a valuable reference for experts.

From the book description, "This monograph gives an account of the state of the art in one-dimensional dynamical systems... Moreover, the exciting new developments on universality and renormalization due to D. Sullivan, are presented here in full detail for the first time."

According to MathSciNet, "This is a lengthy book, a survey and monograph on one-dimensional dynamics. It unifies the existing theory, and contains several simplifications and precise expositions of very deep recent results. It proceeds from the beginning to the frontiers of research..."

P. Collet, J.P. Eckmann, *Iterated Maps on the Interval as Dynamical Systems*, Progress in Physics, 1, Birkhäuser, Boston, Mass., 1980. vii+248 pp.

This book gives a good and reliable introduction and survey of some earlier results on dynamics of iteration of smooth self-maps of the interval. This is a valuable reference for people both in mathematics and in physics working on related topics.

From the book description, "Iterations of continuous maps of an interval to itself serve as the simplest examples of models for dynamical systems. These models present an interesting mathematical structure going far beyond the simple equilibrium solutions one might expect. If, in addition, the dynamical system depends on an experimentally controllable parameter, there is a corresponding mathematical structure revealing a great deal about interrelations between the behavior for different parameter values. This work

10.3. DYNAMICAL SYSTEMS

explains some of the early results of this theory to mathematicians and theoretical physicists, with the additional hope of stimulating experimentalists to look for more of these general phenomena of beautiful regularity, which oftentimes seem to appear near the much less understood chaotic systems. Although continuous maps of an interval to itself seem to have been first introduced to model biological systems, they can be found as models in most natural sciences as well as economics."

According to MathSciNet, "This book is a thorough and readable introduction to some aspects of the theory of one-dimensional dynamical systems... This is an important and beautiful exposition, both as an orientation for the reader unfamiliar with this theory and as a prelude to studying in greater depth some of the hard papers on the subject."

D. Lind, B. Marcus, *An Introduction to Symbolic Dynamics and Coding*, Cambridge University Press, Cambridge, 1995. xvi+495 pp.

This is the first general introduction to symbolic dynamics and its applications to coding for both engineers and mathematicians. It is an excellent and successful book. It was written as a textbook and has also become a valuable reference.

Symbolic dynamics originated as a method to study general dynamical systems. Indeed, it models a topological or smooth dynamical system by a discrete space consisting of infinite sequences of abstract symbols, each of which corresponds to a state of the system, and the dynamics is given by the shift operator. It is a rapidly growing area of dynamical systems and has found significant uses in coding for data storage and transmission as well as in linear algebra.

According to MathSciNet, "It is concerned mostly with the sofic subshifts and subshifts of finite type and presents both classical results and the recent developments in the field... The book is well organized, and lucidly written. It is completely rigorous and yet easy to read. The new concepts are introduced only when needed (thus the concept of subshift is explained without any reference to topology; later, when general dynamical systems are introduced, compactness appears as a very natural concept). There is a wealth of examples illustrating the definitions and theorems. They are very instructive in helping to understand the principles involved in the proofs. There are also many exercises ranging from simple verifications to hard problems."

L. Arnold, *Random Dynamical Systems*, Springer Monographs in Mathematics, Springer-Verlag, Berlin, 1998. xvi+586 pp.

This book gives the first systematic, self-contained account of the theory of measure-preserving random dynamical systems. It contains numerous analytical and numerical examples to illustrate the theories, and provides a very comprehensive foundation l for further study and applications. It is a very valuable introduction for beginners and a reference for others.

From the book description, "this book includes products of random mappings as well as random and stochastic differential equations. The basic multiplicative ergodic theorem is presented, providing a random substitute for linear algebra. On its basis, many applications are detailed. Numerous instructive examples are treated analytically or numerically."

N.P. Bhatia, G.P. Szegö, *Stability Theory of Dynamical Systems*, Die Grundlehren der mathematischen Wissenschaften, Band 161, Springer-Verlag, New York-Berlin, 1970. xi+225 pp.

This is a classical introduction to stability concepts and Lyapunov functions in the theory of continuous flows in a metric space. This is both a good textbook and a valuable reference due to excellent comments at the end of each chapter.

According to MathSciNet, "This is an introductory book intended for beginning graduate students or, perhaps advanced undergraduates... The book has many good points: clear organization, historical notes and references at the end of every chapter, and an excellent bibliography. The text is well written, at a level appropriate for the intended audience, and it represents a very good introduction to the basic theory of dynamical systems."

According to Bulletin de la Societ Mathematique de Belgique, "The exposition is remarkably clear, definitions are separated explicitly, theorems are often provided together with the motivation for changing one or other hypothesis, as well as the relevance of certain generalisations... This study is an excellent review of the current situation for problems of stability of the solution of differential equations. It is addressed to all interested in non-linear differential problems, as much from the theoretical as from the applications angle."

J. Guckenheimer, P. Holmes, *Nonlinear Oscillations, Dynamical Systems, and Bifurcations of Vector Fields*, Applied Mathematical Sciences, 42, Springer-Verlag, New York, 1983. xvi+453 pp.

This book gives an attractive account of chaotic behavior in nonlinear oscillations. It is a valuable survey and a reference for both graduate students and experts.

According to MathSciNet, "The book contains essentially a synopsis of results obtained before 1983 in the theory of abstract dynamic systems, and reduced in scope to third order ODEs... The presentation is intended for readers working in applied fields, and who are familiar with abstract terminology."

From the book description, "Taking their cue from Poincaré, the authors stress the geometrical and topological properties of solutions of differential equations and iterated maps. Numerous exercises, some of which require nontrivial algebraic manipulations and computer work, convey the important analytical underpinnings of problems in dynamical systems and help readers develop an intuitive feel for the properties involved."

10.3. DYNAMICAL SYSTEMS

M. Bohner, A. Peterson, *Dynamic Equations on Time Scales. An Introduction with Applications*, Birkhäuser Boston, Inc., Boston, MA, 2001. x+358 pp.

This book gives a self-contained and accessible account of dynamic equations on time scales, which tries to unify continuous and discrete dynamic equations. This book was written as an textbook for both undergraduate and graduate students and also as a reference for researchers. It has turned out to be very successful.

According to MathSciNet, "The monograph under review comes at an excellent time in the rapid development of dynamic equations on time scales. Both authors are authorities in this field of study and they have produced an excellent introduction to it."

Y. Kuang, *Delay Differential Equations with Applications in Population Dynamics*, Mathematics in Science and Engineering, 191, Academic Press, Inc., Boston, MA, 1993. xii+398 pp.

This is a successful book on the qualitative theory of delay differential equations and applications to global qualitative properties of population dynamics models. It is clearly written and accessible. It is a valuable book for people at all levels who are interested in this and related subject.

According to MathSciNet, "The book is written in two parts, The theory of delay differential equations and Applications in population biology... The material contained in this book gives a marvelous insight to many problems and questions in mathematical ecology together with techniques that can be utilized in solving them. The reviewer believes that the book will be a good source of techniques, results, and questions of interest to researchers and students for years to come."

H.L. Smith, *Monotone Dynamical Systems. An Introduction to the Theory of Competitive and Cooperative Systems*, Mathematical Surveys and Monographs, 41, American Mathematical Society, Providence, RI, 1995. x+174 pp.

This book was written when ideas using monotonicity (or order preservence) have been systematically developed and integrated with the theory of dynamical systems, and have become a subject called the theory of monotone dynamical systems. This book gives a systematic account of this relatively new subject. It is clearly written and can serve both as a good introduction to and reference on monotone dynamical systems.

According to MathSciNet, "The book under review, written by one of the leading experts in the field, provides a comprehensive and lucid introduction to the theory of monotone dynamical systems with continuous time. It is particularly appreciated by the reviewer that each chapter is complemented by examples illustrating the application of the theory to biological problems. The book is to be highly recommended as a graduate text as well as a reference for researchers working both in the theory and in applications of monotone dynamical systems."

J.D. Murray, *Mathematical Biology*, Second edition, Biomathematics, 19, Springer-Verlag, Berlin, 1993. xiv+767 pp.

This is an excellent book on mathematics biology, on applying mathematics to modelling biological phenomena. It covers a wide range of topics from mathematical biology, and can serve both as an introduction and as a reference for all people who are interested in the subject. It contains many good practical examples and will complement well standard books on ordinary differential equations and nonlinear partial differential equations. This book covers a diversity of topics and is suitable for a beginning graduate class.

According to MathSciNet, "The readers of this book are intended to be biologists as well as mathematicians. No previous knowledge of biology is assumed, but a brief description of the biological background of each topic is given... The author considers mathematical biology to be the most exciting modern application of mathematics. He believes that mathematical biology research, to be useful and interesting, must be relevant biologically. Thus the emphasis throughout the book is on the practical application of mathematical models in helping to unravel the underlying mechanisms involved in the biological processes. The main purpose of the book is to present some of the basic and, to a large extent, generally accepted theoretical frameworks for a variety of biological models."

10.3.3 Infinite-dimensional dynamical systems

R. Temam, *Infinite-dimensional Dynamical Systems in Mechanics and Physics*, Second edition, Applied Mathematical Sciences, 68, Springer-Verlag, New York, 1997. xxii+648 pp.

This book, written by a leading expert, is a systematic account of the onset of turbulence in infinite-dimensional dynamical systems or flows. It is a valuable reference for researchers on this important and active field.

According to MathSciNet, "This book is a balanced and authoritative treatment of attractors in infinite- (and finite-) dimensional dynamical systems with specific applications to a variety of equations of mechanics, such as those of Navier-Stokes, MHD, Cahn-Hilliard, Kuramoto-Sivashinsky, etc... the reviewer especially liked the concrete nature of this book. For instance, in the standard Lorenz attractor, the Hausdorff dimension upper estimate 2.538 is derived analytically. The author is to be congratulated on putting together this fine book."

J. Hale, *Asymptotic Behavior of Dissipative Systems*, Mathematical Surveys and Monographs, 25, American Mathematical Society, Providence, RI, 1988. x+198 pp.

This book is a detailed introduction to basic results in dissipative systems, where a dissipative system in a Banach space V is defined by a nonlinear evolution equation with

10.3. DYNAMICAL SYSTEMS

the following property: there is a bounded set of V where every orbit eventually enters and remains. It is a very valuable introduction for students and a reference for theirs. It has been very successful in attracting people to this subject by discussing advanced and recent applications, and important open problems.

According to MathSciNet, "Most of the issues in the modern theory of finite-dimensional dynamical systems have been obtained by marrying analysis and geometry (Poincaré, Birkhoff, Lyapunov, Andronov, Smale, Arnold, etc. On the other hand e.g. physical problems lead one to the consideration of infinite-dimensional dynamical systems, mainly systems generated by partial differential equations (P.D.E.s) and also delay-differential equations (D.D.E.s)...

The goal (achieved) of this monograph is to show how many of the concepts of dynamical systems on locally compact spaces can be adapted to infinite-dimensional dynamical systems that are dissipative...

Although advanced, this book is a very good introduction to the subject, and the reading of the abstract part, which is elegant, is pleasant. Applications cover a wide range and give to the reader a quite good idea of the power of the abstract theory."

A.V. Babin, M.I. Vishik, *Attractors of Evolution Equations*, Translated and revised from the 1989 Russian original by Babin, Studies in Mathematics and its Applications, 25, North-Holland Publishing Co., Amsterdam, 1992. x+532 pp.

This book gives a systematic account of the asymptotic behaviour of solutions of evolution equations, which is an important part of the modern theory of infinite-dimensional dynamical systems. Indeed, evolution equations include the Navier-Stokes system, magneto-hydrodynamics equations, reaction-diffusion equations, and damped semilinear wave equations. This book is a valuable introduction and reference.

According to MathSciNet, "The subject of the book is global attractors for autonomous dissipative nonlinear partial differential equations. There is some overlap with similar books by J. K. Hale, *Asymptotic behavior of dissipative systems*, 1988, and R. Temam, *Infinite-dimensional dynamical systems in mechanics and physics*, 1988, but the book is distinguished by a more detailed discussion of the extensive work of the present authors in this subject. Many examples are discussed, and there is very little overlap with the examples treated in the other two books. Although recent topics such as inertial manifolds and approximate inertial manifolds are not treated and non-autonomous equations are not mentioned, the book is an excellent introduction to a difficult subject."

R. Curtain, H. Zwart, *An Introduction to Infinite-dimensional Linear Systems Theory*, Texts in Applied Mathematics, 21, Springer-Verlag, New York, 1995. xviii+698 pp.

The book is an introductory text on state-space and frequency domain aspects of infinite-dimensional control systems. Both aspects are studied in an integrated fashion. It contains a very extensive appendix on background material, which makes it a self-contained book for almost all readers. As the authors explained, "An important consideration was that it should be accessible to mathematicians and to graduate engineers with a minimal background in functional analysis. Moreover, for the majority of the students this would be their only acquaintance with infinite-dimensional systems." It is a good and time tested textbook.

According to MathSciNet, "The book under review, written at the level of a graduate textbook, presents an introduction to infinite-dimensional system theory. Particular emphasis is given to classical topics in linear control theory such as controllability, stabilizability and the linear-quadratic regulator problem... In summary, the book is a welcome addition to infinite-dimensional system theory. It should prove a valuable tool for those wishing to use the text for a course as well as for researchers in other areas who wish to enter the field."

V. Komornik, *Exact Controllability and Stabilization, The Multiplier Method*, Research in Applied Mathematics, Masson, Paris; John Wiley & Sons, Ltd., Chichester, 1994. viii+156 pp.

This book gives an excellent summary of the status of exact controllability and stabilizability up to the time when the book was written.

According to MathSciNet, "This text provides an introduction to the field of controllability and stabilization of linear evolutionary systems of partial differential equations of conservative type. As the title of the book says, the methods developed are essentially based on the 'multiplier technique'. Multipliers are a way of obtaining norm inequalities for the solutions of PDEs which are not obvious a priori and that do not hold as a consequence of classical techniques for existence and uniqueness... The text is very carefully written and the author has made an important effort to simplify the proofs of the already known results. This makes the book very useful not only for post-graduate students or for those who want to know about this field for the first time, but also to the experts in the field."

T. Kailath, *Linear Systems*, Prentice-Hall Information and System Sciences Series, Prentice-Hall, Inc., Englewood Cliffs, N.J., 1980. xxi+682 pp.

This book gives a self-contained, comprehensive account of basic methods for analysis and application of linear systems that arise in signal processing problems in communications, control, system identification and digital filtering. It may not be easy, but it is a very valuable introduction and reference.

10.3. DYNAMICAL SYSTEMS

10.3.4 R. Thom, Structural Stability and Morphogenesis. An Outline of a General Theory of Models, Translated from the French by D. H. Fowler, With a foreword by C. H. Waddington, Advanced Book Classics, Addison-Wesley Publishing Company, Advanced Book Program, Redwood City, CA, 1989. xxxvi+348 pp.

This is a classic. It is the first systematic account of the author's attempt to identify the language of biology and more general growth phenomenon in the sense that mathematics is the language of nature and analysis is the language of physics. It is not a standard mathematics book and will be valuable to readers who are more interested in philosophical aspects of mathematics.

According to an Amazon review, "Rene Thom was the primary force behind the creation of Catastrophe Theory, and this is the book that brought his ideas forth.

Thom uses topology as his primary mathematical tool for describing and investigating catastrophes and stabilities. As such, this is primarily a qualitative, not a quantitative, approach. The advantage of this is that it allows Thom to spend a great deal of time discussing general characteristics without getting bogged down in formulae or proofs. Indeed, the book is largely text, interspersed with graphs and only the occasional polynomial or differential equations.

And Thom is very insightful when discussing those general characteristics. Along the way he comments on everything from embryology to societal structures as viewed from the perspective of catastrophe theory. Much of the value of the book is buried not in the technical results as it is in his understated comments and speculations regarding the myriad of applications he investigates. A lot of careful and deep thought is present, and there is a significant amount of pregnant wisdom inside..."

According to a review of Bulletin of AMS, "René Thom has written a provocative book. It contains much of interest to mathematicians and has already had a significant impact upon mathematics, but *Stabilite Structurelle et Morphogenese* is not a work of mathematics. Because Thom is a mathematician, it is tempting to apply mathematical standards to the work. This is certainly a mistake since Thom has made no pretense of having tried to meet these standards. He even ends the book with a plea for the freedom to write vaguely and intuitively without being ostracized by the mathematical community for doing so. Instead of insisting that Thorn's style conform to prevailing norms, we should applaud him for sharing his wonderful imagination with us.

The book touches upon an enormous spectrum of material from developmental biology to optics to linguistics as well as mathematics...

Underlying the entire theory is a fundamental hypothesis of structural stability. This stability manifests itself in a local constancy of qualitative structures. Thus two members of the same species can be quite far from being metrically congruent but qualitatively we sense their similarity. How can one express this kind of similarity? Thom suggests that ordinary language does it well. Indeed he asserts that structurally stability is inherent in language: it is impossible to describe structurally unstable processes in simple language. So Thom accepts structural stability as an a priori requirement for his models and examines the mathematical possibilities consistent with this hypothesis..."

According to MathSciNet, "Debate over the significance of the book as a theory of mathematical models in biology and other sciences continues and will continue for years to come; its significance to mathematics in sowing the seeds of many new developments in the theory of singularities of smooth mappings is beyond doubt. It will be a very long time before anything like a definitive review can be given."

10.3.5 Functional-differential equations

J. Hale, *Theory of Functional Differential Equations*, Second edition, Applied Mathematical Sciences, Vol. 3, Springer-Verlag, New York-Heidelberg, 1977. x+365 pp.

This book gives an extensive account of the motivations of and results on functional differential equations, which are of the form $x'(t) = f(t, x_t)$, where $x_t(\theta) = \varphi(t + \theta)$. A basic principle in science is the principle of causality: the future state of a system is independent of the past states and is determined solely by the present state. This leads to the usual differential equations. The importance of functional differential equations comes from the realization or fact that the principle of causality is only an approximation to the true situation, and past states will have impact on the future. There are subtle differences between functional differential equations and ordinary differential equations, and their difference is well-illustrated by examples. This is a clearly written and accessible book. It is an extension of the author's earlier book *Functional differential equations* and can serve a good introduction and reference.

J. Hale, S.Verduyn Lunel, *Introduction to Functional-differential Equations*, Applied Mathematical Sciences, 99, Springer-Verlag, New York, 1993. x+447 pp.

This book is a continuation and revision of the previous book. Both editions of this classical book has been successful. From the preface: "The present book builds upon an earlier work of Hale, *Theory of functional-differential equations*. We try to maintain the spirit of that book and retain approximately one third of the material intact."

10.3.6 Complex dynamics

Complex dynamics is the study of dynamical systems defined by iteration of functions on complex number spaces, in particular, by iteration of polynomials and rational functions on domains of the complex place $\mathbb{C}$ and rational functions on Riemann surfaces. It is

10.3. DYNAMICAL SYSTEMS

both a classical and new subject. It was created by Fatou and Julia in 1918-20 and then laid dormant for a long time until it experienced a renaissance in the early 1980s. Due to its special features, it has been an area of very active research in dynamics system since then.

J. Milnor, *Dynamics in One Complex Variable. Introductory Lectures*, Third edition, Annals of Mathematics Studies, 160, Princeton University Press, Princeton, NJ, 2006. viii+304 pp.

This is a book by a master expositor that gives an excellent introduction into this exciting subject.

According to a review in Bulletin of AMS, "Since about 1980, the subject has undergone explosive growth. Why did the subject go to sleep? One reason is surely that during the period 1920-1975 differential equations and all associated dynamical questions lost the central position they had formerly occupied in mathematics (except in the Soviet Union). But there is another reason, which explains much of why the subject came back to life: computer graphics."

According to MathSciNet, "This book is very clearly written; it should be accessible to readers with a background of basic courses in complex variables and topology, and some differential geometry. I know of several student-run seminars that have worked through the book on their own. This is an active and beautiful subject, and I highly recommend Milnor's book to anyone interested in the field (no need to recommend it to the experts: they have it on their bookshelves already)."

L. Carleson, T. Gamelin, *Complex Dynamics*, Universitext: Tracts in Mathematics, Springer-Verlag, New York, 1993. x+175 pp.

This book gives a comprehensive account to complex dynamics by starting with the basic theory of Julia and Fatou, and ending up with most results up the time the book was written. It may not be an easy introduction, but it is a very valuable reference.

According to MathSciNet, "This is an excellent book on iteration of one-dimensional complex functions (mainly rational functions). The book is written at an advanced graduate level, and having a strong background in complex analysis is recommended for reading. The authors present a more analytic point of view than other expositions ... and give many results not found in them"

A. Beardon, *Iteration of Rational Functions. Complex Analytic Dynamical Systems*, Graduate Texts in Mathematics, 132, Springer-Verlag, New York, 1991. xvi+280 pp.

This is first textbook that is entirely devoted to iteration of rational functions. It concentrates on the mathematical aspects of the subject and presents the classical theory of Fatou and Julia as well as more recent results. It is well-written and accessible, and is a very good first introduction to the subject of complex dynamics.

According to the preface: "This is not a book for experts, nor is it written by one; it is a modest attempt to lay down the basic foundations of the theory of iteration of rational maps in a clear, precise, complete and rigorous way." The time shows that he has achieved his goal well.

C. McMullen, *Complex Dynamics and Renormalization*, Annals of Mathematics Studies, 135, Princeton University Press, Princeton, NJ, 1994. x+214 pp.

This is a research monograph and is a valuable reference written by a leading expert on complex dynamics and renormalization. It is clearly written and readable. One of the main conjectures in complex dynamics is that hyperbolic rational functions are dense in the space of rational functions of a fixed degree, which is equivalent to the conjecture that the Julia set of a quadratic polynomial carries no invariant line field. It was known that this is true for quadratic polynomials which are not infinitely renormalizable, and one purpose of this book is to show that this is also true for a certain class of infinitely renormalizable polynomials.

From the book description, "this book presents a study of renormalization of quadratic polynomials and a rapid introduction to techniques in complex dynamics. Its central concern is the structure of an infinitely renormalizable quadratic polynomial $f(z) = z^2 + c$. As discovered by Feigenbaum, such a mapping exhibits a repetition of form at infinitely many scales. Drawing on universal estimates in hyperbolic geometry, this work gives an analysis of the limiting forms that can occur and develops a rigidity criterion for the polynomial f. This criterion supports general conjectures about the behavior of rational maps and the structure of the Mandelbrot set."

C. McMullen, *Renormalization and 3-manifolds Which Fiber Over the Circle*, Annals of Mathematics Studies, 142, Princeton University Press, Princeton, NJ, 1996. x+253 pp.

This book is a sequel of the book above, *Complex Dynamics and Renormalization*, and gives a unified approach to the construction of fixed-points of renormalization and to the construction of hyperbolic 3-manifolds fibering over the circle. It reenforces the analogy between complex dynamics and hyperbolic geometry. This is an important and valuable reference for experts in both subjects.

According to MathSciNet, "In this monograph, the author presents a comprehensive study of a theory which brings into parallel two recent and very deep theorems, involving geometry and dynamics. These are Thurston's theorem on the existence of hyperbolic metrics on three-manifolds which fiber over the circle with pseudo-Anosov monodromy, and Sullivan's theorem on the convergence of the renormalization map for real quadratic mappings..."

10.4 Ergodic theory

Gibbs was probably the first great scientist from USA. He made important theoretical contributions to physics, chemistry, and mathematics. For example, his work on thermodynamics was crucial to transform physical chemistry into a rigorous science, and together with Maxwell and Boltzmann, he created statistical mechanics and also coined the name. In mathematics, he invented the vector calculus, independently of Heaviside, who did similar work during the same period. There is also the Gibbs phenomenon on the limiting behavior of partial sums of the Fourier series at a discontinuous point.

Figure 10.4: Gibbs' book.

Ergodic theory studies dynamical systems which admit invariant measures. It was motivated by problems of statistical physics. A central problem of ergodic theory is concerned with the long time behavior of a dynamical system. The first important result is the Poincaré recurrence theorem, which states that almost all points in any subset of the phase space eventually revisit the subset. More precise information is provided by various ergodic theorems which roughly state that the time average of a function along the trajectories exists almost everywhere and is equal to the space average. Two of the most important ergodic theorems are due to Birkhoff and von Neumann. There are also stronger properties, such as mixing and equidistribution, and various results on them have been obtained.

P. Walters, *An Introduction to Ergodic Theory*, Graduate Texts in Mathematics, 79, Springer-Verlag, New York-Berlin, 1982. ix+250 pp.

This is a standard introduction to ergodic theory. Besides basic topics in ergodic theory, it also includes an overview of many applications to differential dynamics, differential geometry, number theory, probability, von Neumann algebras, and statistical mechanics. It is an excellent introduction and a valuable reference for people at different levels.

According to MathSciNet, "This handsomely produced hardcover book is a completely revised and expanded version of the author's earlier book [Ergodic theory—introductory lectures, Lecture Notes in Math., 458, 1975]. The almost equal division of the earlier book between the measure-theoretic ergodic theory and the topological dynamics is carried over to the new book, although each part is expanded in different directions...

The earlier book has been very well received by students in ergodic theory, as it had been very carefully and clearly written and as it contained many examples. This new book should be even more popular. In fact, first, both the style and the mathematical exposition are much more polished and the book now seems to be even easier to read. Secondly, the author makes an admirable job of collecting almost all the prerequisites from measure theory, topology and functional analysis as separate theorems, with references for their proofs or, sometimes, with their proofs. Thirdly the author now includes more interesting recent material, especially in the topological dynamics section. Finally there are now many very well organized comments on the further developments, with references, that will provide an excellent guide for the reader to recent literature in ergodic theory."

I.P. Cornfeld, S.V. Fomin, Y.G. Sinai, *Ergodic Theory*, Translated from the Russian by A. B. Sosinskii, Grundlehren der Mathematischen Wissenschaften, 245. Springer-Verlag, New York, 1982. x+486 pp.

This is the first modern book on ergodic theory that studies in detail many important examples of measure-preserving dynamical systems. In fact, studying examples is the guiding principle of this important book. It fills in a gap in the literature on ergodic theory. This book contains a lot of interesting material and explains ideas well. It is a readable introduction and a valuable reference.

According to MathSciNet, "Ergodic theory for the purposes of this book means the study of measure preserving transformations (dynamical systems) which may be in the form of individual transformations, endomorphisms or automorphisms, or in the form of semiflows or flows indexed by the positive reals or the reals, respectively. In many examples the dynamical system is related to some space with additional structure and there is some preliminary, not necessarily easy, work to do to establish the existence of suitable invariant measures... This is a substantial work and the authors have made a genuine attempt to keep the chapters as independent units to facilitate their use for reference purposes."

K. Petersen, *Ergodic Theory*, Corrected reprint of the 1983 original, Cambridge Studies in Advanced Mathematics, 2, Cambridge University Press, Cambridge, 1989. xii+329 pp.

This is another good introduction to ergodic theory by emphasizing measure theoretic and functional analytic aspects. It also includes some discussion of topological dynamics.

According to MathSciNet, "It is a well-written treatment of basic ergodic theory with further development in several areas which are presently actively researched... Good use is made of clarifying diagrams and at the beginning of long and complex proofs or constructions an overview of what is to follow is usually given... This book is a real contribution and welcome addition to the literature of ergodic theory."

10.4. ERGODIC THEORY

R. Bowen, *Equilibrium States and the Ergodic Theory of Anosov Diffeomorphisms*, With a preface by David Ruelle. Edited by Jean-René Chazottes, Second revised edition, Lecture Notes in Mathematics, 470, Springer-Verlag, Berlin, 2008. viii+75 pp.

This book gives an efficient and elegant account of the work of Bowen, Sinai, Ruelle and others on the ergodic theory of Anosov diffeomorphisms, which provide important examples of structurally stable transformations. It is a valuable reference.

According to MathSciNet, "This book is devoted to studying the properties of continuous and smooth dynamical systems some of whose trajectories are asymptotically exponentially unstable. There are two fields of applications of ergodic theory for the exploration of continuous and smooth systems. First, it is used for the study of properties of almost all (relative to a certain 'natural' invariant measure, e.g., the measure induced by a phase volume for classical mechanical systems) trajectories of the system. The second field of applications concerns the investigation of properties of the set of all the invariant measures for a given dynamical system and the finding of some remarkable measures in this set. The subject of this book belongs to the second direction... The book is clearly and precisely written throughout and makes for comfortable reading. It is a good source of information about recent work in this field."

R. Mane, *Ergodic Theory and Differentiable Dynamics*, Translated from the Portuguese by Silvio Levy, Ergebnisse der Mathematik und ihrer Grenzgebiete (3), 8, Springer-Verlag, Berlin, 1987. xii+317 pp.

This book is another excellent introduction to ergodic theory, with emphasis on connection between the ergodic theory and the theory of differentiable dynamical systems. It is a valuable introduction and reference.

According to MathSciNet, This book presents the theory of smooth ergodic theory, i.e. the part of modern ergodic theory which is connected with differential dynamical systems. P. Walters book, *An Introduction to Ergodic Theory*, 1982, is the most similar, but it only covers about two-thirds of the material. Much of this recent material is available in numerous research articles, but not a systematic textbook like the one under review... The book is well written with problems at the end of each section. It is self-contained and includes a rapid review of the theorems needed from measure theory. The statements of the theorems are clear and easily found. This book provides an accessible introduction to an important and active area of current research."

According to a review in Bulletin of AMS, "It presents the right material in a coherent framework with just the right balance of general theory and concrete examples. The topics covered are well chosen and there is a wealth of material to be found here... It must be said that some of the topics covered are technical and difficult. This is probably a difficulty with any book presenting a young and active field, however. The author in his introduction acknowledges that some proofs are arid and demanding' and encourages the reader to concentrate on their complement in the text. This is good advice, and I found it quite feasible to do so and still learn a great deal."

H. Furstenberg, *Recurrence in Ergodic Theory and Combinatorial Number Theory*, M. B. Porter Lectures, Princeton University Press, Princeton, N.J., 1981. xi+203 pp.

This is an elegantly written classic and has had a huge impact on several subjects, in particular on combinatorial number theory in the past few years. It is an original book written by a very original mathematician.

According to MathSciNet, "This very readable book discusses some recent applications, due principally to the author, of dynamical systems and ergodic theory to combinatorics and number theory. It is divided into three parts. In Part I, entitled 'Recurrence and uniform recurrence in compact spaces', the author gives an introduction to recurrence in topological dynamical systems, and then proves the multiple Birkhoff recurrence theorem... Part II carries the title 'Recurrence in measure preserving systems'. After a short introduction to the relevant part of measure-theoretic ergodic theory, this section is devoted to a proof of the multiple recurrence theorem. Part III, called 'Dynamics and large sets of integers', investigates the connections between recurrence in topological dynamics and combinatorial results concerning finite partitions of the integers (e.g., Hindman's theorem, Rado's theorem). Here the notion of proximality plays a central role. In reading this book, the reviewer found that the first part tickled his imagination and made him want to continue, the second part provided a good deal of work and tested his technical ability, while the last part led him to imagine the future possibilities for research. An excellent work!"

F. Greenleaf, *Invariant Means on Topological Groups and Their Applications*, Van Nostrand Mathematical Studies, No. 16, Van Nostrand Reinhold Co., New York-Toronto, Ont.-London, 1969. ix+113 pp.

This is a standard introduction and reference on invariant means on topological groups and their applications.

According to MathSciNet, "The author unifies and clarifies the study of the various algebras $L^\infty(G)$, $CG(G)$, $UCB_r(G)$ and $UCB(G)$ of essentially bounded Borel measurable, bounded continuous, bounded right uniformly continuous and bounded two-sided uniformly continuous complex-valued functions, respectively, on the locally compact group G in which amenability questions have been considered. Some of the proofs and some of the results themselves have not hitherto appeared in the literature... Applications to the representation theory of locally compact groups and to ergodic theory, and a brief sketch of the theory of weakly almost periodic semigroups of operators and weakly almost periodic (WAP) functions on topological semigroups are presented..."

XI. A display room of the Clements Library of the University of Michigan

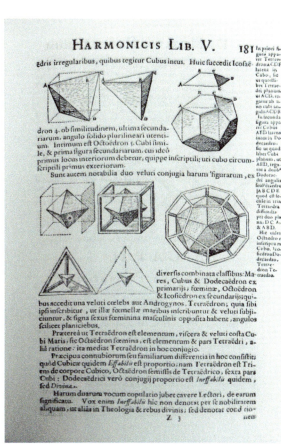

One important and famous problem in discrete mathematics is the Kepler conjecture, which says that no arrangement of equally sized spheres in the three-dimensional Euclidean space has a greater average density than that of the cubic close packing (face-centered cubic) and hexagonal close packing arrangements. It was solved by Hales in 1998 with a computer aided proof. The refereeing of the paper took several years, and the referees were basically convinced the correctness of the proof.

Combinatorics is an ancient subject and was studied as far back as the sixth century BC. Though the arithmetical triangle was presented by mathematicians in books several centuries before Pascal, it is usually called Pascal's triangle. One reason might be that in the Middle Ages, combinatorics was studied largely outside of the European civilization, and during the Renaissance combinatorics enjoyed a rebirth. Pascal's book gave a systematical exposition of known results on the triangle, and he used them to solve problems in probability theory, therefore giving it a prominent place in mathematics.

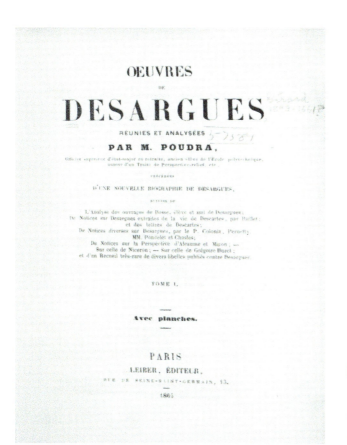

Desargues was a mathematician and engineer. He made important contributions to projective geometry and finite geometry such as Desargues' theorem, the Desargues graph and Desarguesian planes. He is considered one of the founders of projective geometry. His work on perspective and geometrical projections can be viewed as a culmination of centuries of scientific study from the classical time in optics to Kepler. Discrete geometry is often connected with the continuous geometry. One recent important example is the Tits building, which has been applied to a wide range of subjects and is related to the Desarguesian planes.

Collected Papers of
SRINIVASA RAMANUJAN

Edited by
G. H. HARDY
P. V. SESHU AIYAR
and
B. M. WILSON

CHELSEA PUBLISHING COMPANY
NEW YORK

With almost no formal training in mathematics, Ramanujan made highly original contributions to analysis, number theory, infinite series, and other subjects. Living in India with no access to the larger mathematical community, Ramanujan developed his own mathematical research in isolation. (This might not be the only example since the Japanese mathematics also reached its height during the isolation caused by the wars). He had a unique ability and intuition to conceive and derive various identifies. He also made highly original conjectures. For example, the Ramanujan conjecture is still one of the most important open problems in the theory of automorphic forms.

Chapter 11

Discrete mathematics and combinatorics

A basic point of calculus is to understand continuous change of functions and related mathematical objects and structures. On the other hand, discrete mathematics studies mathematical structures that are fundamentally discrete rather than continuous. For example, any object consisting of finitely many points should be considered as discrete. But the distinction is not so clear. For example, all possible configurations of finitely many points in space and their motions are not discrete anymore.

Combinatorics studies finite or countable discrete structures and is an integral part of discrete mathematics. Another important part concerns graph theory. In this chapter, we broadly classify books in discrete mathematics according to combinatorics, polytopes, lattices of the Euclidean spaces, and graph theory.

11.1 Combinatorics

Combinatorics studies finite or countable discrete structures. It includes (1) enumerative combinatorics: counting the structures of a given kind and size, (2) combinatorial designs and matroid theory: deciding when certain criteria can be met, and constructing and analyzing objects meeting the criteria, (3) extremal combinatorics and combinatorial optimization: finding "largest", "smallest", or "optimal" objects, (4) algebraic combinatorics: studying combinatorial structures arising in an algebraic context, or applying algebraic techniques to combinatorial problems.

Combinatorial problems arise in many areas of mathematics such as algebra, probability theory, topology, and geometry, and combinatorics also has many applications in optimization, computer science, ergodic theory and statistical physics. Many combinatorial questions have historically been considered in isolation, giving an ad hoc solution to a problem arising in some mathematical context. In the later twentieth century, however, powerful and general theoretical methods were developed, making combinatorics into an independent branch of mathematics in its own right. One of the oldest and most accessible parts of combinatorics is graph theory, which also has numerous natural connections to other areas. Combinatorics is used frequently in computer science to obtain formulas and estimates in the analysis of algorithms.

Cayley is one of the most famous English mathematicians, though he worked as a lawyer for 14 years before he became a full time mathematician. In fact, he wrote almost 300 papers when he was a lawyer. While he was preparing for the bar examination, he went to Hamilton's lectures in Dublin on quaternions. His friend Sylvester, another famous mathematician, was then an actuary. They used to walk together round discussing the theory of invariants and covariants. He is probably best known for the Cayley number and the Cayley graphs. But he also contributed to many other subjects such as combinatorics and permutation groups. He was very productive. For example, his Collected Mathematical papers number 13 quarto volumes, and contain 967 papers.

Figure 11.1: Cayley's book.

11.1.1 R. Stanley, Enumerative Combinatorics, Vol. I., Vol. 2., With a foreword by Gian-Carlo Rota, The Wadsworth & Brooks/Cole Mathematics Series, 1986. xiv+306 pp., 1999. xii+581 pp.

This two volume set is a standard introduction to and a reference on enumerative combinatorics written by a leading expert. It is both accessible and comprehensive.

According to the MathSciNet review of the first volume, "In a little over 300 pages of careful exposition, the author has packed a tremendous amount of information into this book. This has been possible partly due to the fact that over one third of the book is devoted to 180 graded exercises (many with multiple parts) and their solutions (or references to solutions). The book is very up to date, is very detailed, and quite general without seeking the greatest generality possible. The book demands of the reader considerable mathematical maturity as well as considerable familiarity with undergraduate mathematics."

For the second volume, "The text gives an excellent account of the basic topics of enumerative combinatorics not covered in Volume 1, and the exercises cover an enormous amount of additional material... As in Volume 1, there are historical notes at the end

11.1. COMBINATORICS

of each chapter, and these demonstrate Stanley's uniquely broad knowledge of the field. The numerous exercises go far beyond the text in introducing a large number of related topics. Many exercises contain material appearing here for the first time, and each exercise has either a solution or a reference to the literature... Stanley's book is a valuable contribution to enumerative combinatorics."

L. Comtet, *Advanced Combinatorics. The Art of Finite and Infinite Expansions*, Revised and enlarged edition, D. Reidel Publishing Co., Dordrecht, 1974. xi+343 pp.

This is a classical book on combinatorics concentrating on graphs and configurations.

According to MathSciNet, "This is a welcome addition to the growing library of books on its subject. In a remarkable feat of compression, the author gives in two pocket-size paperback volumes at least a taste (and often much more) of the great variety of combinatorial results, both enumerative and existential, of traditional and current interest."

E. Bannai, T. Ito, *Algebraic Combinatorics. I. Association Schemes*, The Benjamin/Cummings Publishing Co., Inc., Menlo Park, CA, 1984. xxiv+425 pp.

This book is not a general account of algebraic combinatorics, but rather focuses a particular topic and gives a detailed treatment: the abstract theory of general association schemes, its relations to group theory, and the state of the classification problem for (P and Q)-polynomial association schemes of large diameter. It is a valuable reference on this topic.

11.1.2 Polytopes, convex polytopes and geometric arrangements

A polytope is a region of $\mathbb{R}^n$ with flat sides. For example, a polygon is a polytope in two dimensions, a polyhedron in three dimensions. An n-dimensional polytope in $\mathbb{R}^n$ is bounded by a number of $(n-1)$-dimensional facets, and these facets are themselves polytopes of dimension $n-1$. A polytope may or may not be convex. The convex polytopes are the simplest kind of polytopes, and form the basis for different generalizations of the concept of polytopes. A convex polytope is sometimes defined as the intersection of a set of half-spaces. An abstract polytope is a partially ordered set of elements or members (which represent facets and higher codimension sides of geometric polytopes), which obeys certain rules. Besides being basic objects in combinatorics, polytopes have important applications in optimization and linear programming.

G. Ziegler, *Lectures on Polytopes*, Graduate Texts in Mathematics, 152, Springer-Verlag, New York, 1995. x+370 pp.

This book is an excellent introduction to basic methods and modern tools of the theory of polytopes. It is also valuable to researchers.

According to the MathSciNet, "Ziegler has given us a well-written and engaging text with plenty of good problems (some open). Although the book is not encyclopedic, extensive notes and a 500-entry bibliography with pointers to the text make it an essential reference for the most recent results and methods in polytope theory."

B. Grünbaum, *Convex Polytopes* , Second edition, Prepared and with a preface by Volker Kaibel, Victor Klee and Günter M. Ziegler, Graduate Texts in Mathematics, 221. Springer-Verlag, New York, 2003. xvi+468 pp.

This is a classic on convex polytopes. It was the first systematic modern account of theory of convex polytopes.

According to MathSciNet, "The author has produced an excellent and much-needed book about this recently resurgent area of geometry. Stimulated by the development of linear programming, interest in the field of convex polytopes has been renewed and a great many results have been discovered or re-discovered in recent years. However, prior to this book, the theory was scattered about in various research papers or out-of-print volumes. The author has done an extraordinarily thorough job of collecting these numerous results and organizing them into readable form."

According to MathSciNet of the second edition, "Every chapter of the book is supplied with a section entitled 'Additional notes and comments', prepared by Volker Kaibel, Victor Klee, and Günter Ziegler; these notes summarize the most important developments with respect to the topics treated by Grunbaum. Together with additional bibliographical references, this material represents 74 pages of the book. The reader is thus provided both with the source text and with very valuable information on the subsequent development generated by the book.

Each of the 19 chapters of the monograph contains, besides the main material, historical notes and an unusually large number of exercises, which may also serve as an additional source of interesting results... The new edition of Grunbaum's book is an excellent gift for all geometry lovers."

J. Matousek, *Lectures on Discrete Geometry*, Graduate Texts in Mathematics, 212, Springer-Verlag, New York, 2002. xvi+481 pp.

This is a lively and stimulating introduction to some aspects of discrete geometry such as combinatorial properties of configurations of geometric objects, arrangements of points, or hyperplanes and convex sets.

According to MathSciNet, "This book, aimed at early graduate students, is an excellent introduction to discrete geometry understood as the study of combinatorial properties of arrangements of points, lines, hyperplanes or convex sets in the Euclidean d-space. Relations to discrete transformation groups and to the classical area of packing and covering are neglected. The book can also be used as a valuable resource to researchers in various areas (for example, graph theory, computational geometry and combinatorial optimization)."

11.1.3 Gröbner bases

Given an ideal of a polynomial ring, a Gröbner basis is a special basis, i.e., a generating subset which enjoys many desirable properties. It can be viewed as a multivariate, non-linear generalization of the Euclidean algorithm for computation of univariate greatest common divisors and the Gaussian elimination for linear systems. Gröbner bases appear naturally in computational algebraic geometry, computational commutative algebra, and computer algebra.

B. Sturmfels, *Gröbner Bases and Convex Polytopes*, University Lecture Series, 8, American Mathematical Society, Providence, RI, 1996. xii+162 pp.

This book gives a comprehensive introduction to the interplay of Gröbner bases with convex polytopes and toric varieties, and applications of Gröbner bases to many other topics such as combinatorics, convexity, algebraic topology, probability, and integer programming. It is a valuable introduction to this beautiful rapidly developing subject and is also a useful reference.

According to MathSciNet, "During the last ten years Gröbner basis theory has become an important tool for several algorithms in algebraic geometry and commutative algebra. There are introductory texts on the subject... The aim of the present monograph—grown out of a series of the author's lectures—consists in new and specific applications of the theory in the domains of integer programming and computational statistics. The textbook underlines the interdisciplinary nature of Gröbner bases as a tool for the interplay of computational algebra and the theory of convex polytopes."

D. Cox, J. Little, D. O'Shea, *Using Algebraic Geometry*, Second edition, Graduate Texts in Mathematics, 185. Springer, New York, 2005. xii+572 pp.

The book gives an attractive introduction to computational aspects of several classical and modern topics of commutative algebra and algebraic geometry. The first edition has been successful in establishing a bridge connecting modern computer science and classical algebraic geometry. This book is a good introduction and a useful reference.

According to MathSciNet, "By virtue of its foundations algebraic geometry is a rather sophisticated subject based on the construction of schemes and cohomological aspects as initiated by A. Grothendieck. Recently there have been particular results in this field, in particular related to arithmetic questions. The book under review follows another recent trend. The discovery of new algorithms for dealing with polynomials, coupled with their implementation on inexpensive yet fast computers, has sparked a minor revolution in the study and practice of algebraic geometry. A milestone in this direction has been the

Buchberger algorithm for computing Gröbner bases and its implementation in several computer algebra systems. These algorithmic methods have also led to exciting new applications of algebraic geometry in several fields outside of algebraic geometry and related to various computational and practical problems.

The main goal of the present book is to illustrate the many uses of algebraic geometry and highlight the more recent applications of Gröbner bases and resultants."

T. Becker, V. Weispfenning, *Gröbner Bases. A Computational Approach to Commutative Algebra*, In cooperation with Heinz Kredel, Graduate Texts in Mathematics, 141, Springer-Verlag, New York, 1993. xxii+574 pp.

This book gives a self-contained introduction to the theory and applications of Gröbner bases. This book is carefully written with detailed explanations.

According to MathSciNet, "The authors intend to bridge the gap between the conventional (theoretical) commutative algebra and the (practical) computer algebra approaches. In their own words, the purpose of the book is 'to give a self-contained, mathematically sound introduction to the theory of Gröbner bases and to some of its applications, stressing both theoretical and computational aspects'. They also want to show that on the other hand Gröbner bases are not only a powerful tool for actual computations, but also a cornerstone of commutative algebra... The authors succeed well in giving a 'mathematically sound' foundation for the theory of Gröbner bases and their applications. The emphasis is on proving that things can be done algorithmically rather than on exhibiting more efficient but less easy to grasp 'state of the art' algorithms... The thoroughness of the book should make it an excellent reference guide to Gröbner bases and their application, and it also is very suitable for an introduction to Gröbner bases for researchers or students with some prior commutative algebra knowledge. It may also be used by computer algebra users who wish to investigate the correctness of various algorithms and approaches."

11.1.4 Matroid theory

J. Oxley, *Matroid Theory*, Second edition, Oxford Graduate Texts in Mathematics, 21, Oxford University Press, Oxford, 2011. xiv+684 pp.

This is a standard book on matroid theory written by an expert. It is both a good introduction and a valuable reference.

From the book description, " (1) What is the essence of the similarity between linearly independent sets of columns of a matrix and forests in a graph? (2) Why does the greedy algorithm produce a spanning tree of minimum weight in a connected graph? (3) Can we test in polynomial time whether a matrix is totally unimodular?

11.1. COMBINATORICS

Matroid theory examines and answers questions like these. Seventy-five years of study of matroids has seen the development of a rich theory with links to graphs, lattices, codes, transversals, and projective geometries. Matroids are of fundamental importance in combinatorial optimization and their applications extend into electrical and structural engineering."

According to Bulletin of AMS, "Matroids were invented by Hassler Whitney in the mid-1950s, followed immediately by Garrett Birkhoff's work on their lattice-theoretic equivalent, geometric lattices. Some of the deepest work in the field was done in the 1950s by W. T. Tutte, who gave an excluded minor characterization of several important classes of matroids, including the regular or unimodular matroids. These matroids are defined as those which may be represented as a vector matroid over every field. Matroids enjoyed an explosion of interest in the 1970s and 1980s. This interest now has somewhat lessened and been replaced partially by the current interest in oriented matroids, a variation in which each (ordered) basis is given a sign indicating its orientation. A prototypical example would be any subset of 3-dimensional real affine space, in which the bases are the affinely independent subsets of cardinality four, and the orientation of such a basis depends on whether it forms a right- or left-handed tetrahedron. Oriented matroids have a wide variety of applications from convex polytopes to Pontryagin classes...

[This book] includes more background, such as finite fields and finite projective and affine geometries, and the level of the exercises is well suited to graduate students. The book is well written and includes a couple of nice touches... this is a very useful book. I recommend it highly both as an introduction to matroid theory and as a reference work for those already seriously interested in the subject, whether for its own sake or for its applications to other fields."

According to MathSciNet, "This book is an excellent graduate textbook and reference work on matroid theory. The care that went into the writing of the book is evident by the quality of its exposition. The first half of the book provides an overview of matroid theory suitable for a one- or two-semester graduate course. The book contains numerous examples and exercises. The second half of the book contains recent research results and efficient proofs of all the major theorems of matroid theory... It is an excellent first book on the subject due to its comprehensive nature. There is a wealth of material to mine for graduate students, graph theorists, and researchers in the area."

D.J. Welsh, *Matroid Theory*, L. M. S. Monographs, No. 8, Academic Press [Harcourt Brace Jovanovich, Publishers], London-New York, 1976. xi+433 pp.

This is a classical book on matroid theory and is still a very useful reference now.

The difference between this and the previous book is explained by Oxley in his introduction: "[Welsh's book] appeared during my second year at Oxford and it has been my constant companion ever since. When I contemplated writing this book, the first question I had to answer was how should it differ from Welsh's book. This book attempts to blend Welsh's very graph-theoretic approach to matroids with the geometric approach of Rota's school that I learnt from Brylawski. Unfortunately, I cannot emulate Welsh's feat of providing, in a single volume, a complete survey of the current state of knowledge in matroid theory; the subject has grown too much. Therefore I have had to be selective. While the basic topics virtually select themselves, the more advanced topics covered here reflect my own research interests."

While there is some overlap between these two books, Welsh's book contains a lot of material which is not available elsewhere.

M. Aigner, *Combinatorial Theory*, Grundlehren der Mathematischen Wissenschaften, 234, Reprint of the 1979 original, Classics in Mathematics, Springer-Verlag, Berlin, 1997. viii+483 pp.

This is a classical introduction to combinatorics, and it covers most aspects of enumeration and order theory. For example, it contains two substantial chapters on matroids. It is one of the first books that try to map out the scope of combinatorics and put them into a coherent theory.

11.1.5 L. Lovász, Combinatorial Problems and Exercises, Corrected reprint of the 1993 second edition, AMS Chelsea Publishing, Providence, RI, 2007. 642 pp.

This is a very good book consisting of problems in combinatorics, in particular graph theory, written by a leading expert. It is a valuable source for both students and teachers. Indeed, according to the book description, "The main purpose of this book is to provide help in learning existing techniques in combinatorics. The most effective way of learning such techniques is to solve exercises and problems."

According to MathSciNet, "This book is a classic. The author masterfully analyzes the proof techniques utilized in 607 different results partitioned into the following fifteen sections...

The plan of the book is to present first 80 pages containing only the statements of the 607 problems. Then there are 48 pages of brief hints, followed by the bulk of the book: 387 pages of concisely presented solutions to these problems, quite a number of which are due to the author and are very elegant indeed. The book concludes with a 13-page 'dictionary' giving the necessary definitions, together with indexes of notations, subjects, and authors. It is entirely appropriate to quote from the very eloquent preface, which expresses many sentiments with which the reviewer concurs."

11.2 Discrete mathematics

Euler is a master of everything, and he is certainly a master of both the continuous mathematics and discrete mathematics. His contributions to the graph theory can be considered as some of the most beautiful results. The subject of discrete mathematics is both old and new. The emergence of the computer science sheds a new light on discrete mathematics. For example, one of the most important and highly priced problems is the $P = NP$ problem.

Figure 11.2: Euler's book.

Discrete mathematics studies structures that are fundamentally discrete rather than continuous. For example, basic objects include integers, graphs, lattices. Discrete objects can often be enumerated by integers. It is often said that discrete mathematics deals with countable sets.

11.2.1 J. Conway, N. Sloane, Sphere Packings, Lattices and Groups, Third edition, With additional contributions by E. Bannai, R. E. Borcherds, J. Leech, S. P. Norton, A. M. Odlyzko, R. A. Parker, L. Queen and B. B. Venkov, Grundlehren der Mathematischen Wissenschaften, 290. Springer-Verlag, New York, 1999. lxxiv+703 pp.

This is the standard book on sphere packings and lattices in the Euclidean space, or rather lattice sphere packings. It is very comprehensive and detailed. It is considered the "Bible" of the subject by many people.

According to MathSciNet, "The book is a landmark in the literature on sphere packings. It is mainly concerned with the problem of packing spheres in d-dimensional Euclidean space E^d. The authors also study closely related problems as, e.g., the kissing number problem (how many spheres touch a given central sphere?); the problem of covering E^d in the least dense way by equal spheres; and the classification of lattices and quadratic forms. The book also deals with applications of these geometric problems to other areas of mathematics (mainly number theory) and to areas outside mathematics, mainly the channel coding problem, but also crystals and quasicrystals.

Two mathematical objects play a central role in this book, namely the famous sphere packings in the E_8-lattice and in the Leech lattice E_{24}. As the authors formulate, one could say "that the book is devoted to studying these two lattices and their properties". Also remarkable are the informative tables and graphs which are helpful for the reader.

The book contains a lot of new material, mainly from papers by the two authors, but also remarkable contributions by others (e.g. by Bannai, Leech, Norton, Odlyzko, Parker, Queen, and Venkov). In fact the book was originally planned as a collection of some important papers by these authors which can be seen from its structure and the different chapters. However, in its final form it has become much more than such a collection—it is a successful synthesis of these papers."

11.2.2 Graph theory

Graph theory is a classical subject and is probably more accessible than some other subjects. Some open problems are solved by people who have not gone through the usual training. There are many good books on different aspects of it.

B. Bollobás, *Modern Graph Theory*, Graduate Texts in Mathematics, 184, Springer-Verlag, New York, 1998. xiv+394 pp.

This is a standard modern introduction to graph theory as the title suggests. It was written by a major figure of the subject. Its comprehensive coverage also makes it a valuable reference.

According to MathSciNet, "The back cover and the preface of this excellent book state well its goals and format: "The time has now come when graph theory should be part of the education of every serious student of mathematics and computer science, both for its own sake and to enhance the appreciation of mathematics as a whole. This book is an in-depth account of graph theory, written with such a student in mind; it reflects the current state of the subject and emphasizes connections with other branches of mathematics, like optimization theory, group theory, matrix algebra, probability theory, logic, and knot theory. The volume grew out of the author's earlier book *Graph theory*, 1979, but its length is well over twice that of its predecessor, allowing it to reveal many exciting new developments in the subject."

B. Bollobás, *Graph Theory. An Introductory Course*, Graduate Texts in Mathematics, 63, Springer-Verlag, New York-Berlin, 1979. x+180 pp.

It was written as an accessible introduction to students and has achieved its goal well.

From the preface, "This book is intended for the young student who is interested in graph theory and wishes to study it as part of his mathematical education. Experience at Cambridge shows that none of the currently available texts meet this need. Either they are too specialized for their audience or they lack the depth and development needed to reveal the nature of the subject. We start from the premise that graph theory is one of several courses which compete for the student's attention and should contribute to his appreciation of mathematics as a whole. Therefore, the book does not consist merely of a catalogue of results but also contains extensive descriptive passages designed to convey the flavor of the subject and to arouse the student's interest. Those theorems which are vital to the development are stated clearly, together with full and detailed proofs. The book thereby offers a leisurely introduction to graph theory which culminates in a thorough grounding in most aspects of the subject."

According to MathSciNet, "The student who fully masters the contents of this introductory book will have a background in graph theory which is sufficient for reading specialized books on the subject as well as current research papers. This monograph is worthy of serious study."

R. Diestel, *Graph Theory*, Fourth edition, Graduate Texts in Mathematics, 173, Springer, Heidelberg, 2010. xviii+437 pp.

This is another excellent modern introduction to graph theory.

According to MathSciNet, "This book covers all of the central topics of modern graph theory. It can serve as study material for introductory courses of graph theory but it is strongly recommended as a textbook for graduate students in mathematics. It would be also welcomed by more advanced readers who wish to quickly obtain deeper knowledge in a specific area of graph theory...

The book is written very carefully and in clear style. Chapters begin with a brief introduction of the presented area, continue with a presentation of classic results followed by some deeper and more complicated theorems. At the end of each chapter, there is a section with a set of challenging problems and some historical and bibliographical notes that bring benefit to readers with deeper interest.

The book contains some material that normally cannot be found in textbooks. Many of the proofs are new and simplified in comparison to the classical ones. Some results are proved in a few different ways. The presentation is rather original. Each theorem contains marginal notes with a list of the previous theorems it uses and the upcoming theorems that will depend on it. This is a feature that helps with the preparation of courses and individual study."

D. West, *Introduction to Graph Theory*, Prentice Hall, Inc., Upper Saddle River, NJ, 1996. xvi+512 pp.

This is an accessible and thorough introduction to graph theory suitable for undergraduate and graduate students, or anyone who wants to understand the basics of graph theory, or just is curious about graph theory. It is also a valuable and popular reference.

From the preface: "Graph theory is a delightful playground for the exploration of proof techniques in discrete mathematics, and its results have applications in many areas of the computing, social, and natural sciences. The design of this book permits its use in a one-semester introduction at the undergraduate or beginning graduate level, or in a patient two-semester introduction. No previous knowledge of graph theory is assumed. Many algorithms and applications are included, but the focus is on understanding the structure of graphs and the techniques used to analyze problems in graph theory."

F. Harary, *Graph Theory*, Addison-Wesley Publishing Co., Reading, Mass.-Menlo Park, Calif.-London, 1969. ix+274 pp.

This book is a classical introduction to graph theory, and the beginning part is especially accessible. This is considered an essential book for graph theorists.

According to Zentralblatt MATH, "It is difficult to imagine that a graph theorist exists whose personal library does not include *Graph Theory* by Frank Farary, for there is no other book on graph theory which contains the wealth of material that this does. Furthermore, the writing style of the author makes the reading easy and, on the whole, enjoyable."

According to MathSciNet, "Some graph theorists conceive of their field as deeply imbedded in combinatorial mathematics, set theory, algebra, or even topology. W. T. Tutte and the late Oystein Ore represent this point of view. Others regard graph theory as standing apart from, although 'intimately related to many branches of mathematics, including group theory, matrix theory,...and combinatorics'. The quote is from the Preface of the author's book which places him as perhaps the staunchest proponent of the latter school. Whereas to members of the former group, graphs are essentially intuitively appealing representations or motivation for and special cases of more abstract systems, to the author they are the means and the end in themselves. The book under review fully reflects this philosophy..."

N. Biggs, *Algebraic Graph Theory*, Second edition, Cambridge Mathematical Library, Cambridge University Press, Cambridge, 1993. viii+205 pp.

This is another classical and popular introduction to graph theory.

According to MathSciNet, "With the continuing exponential growth of research activity in graph theory, there is a current trend to the preparation of monographs that cover special topics in depth and present an up-to-date exposition. This is such a book on algebraic graph theory and it is very welcome indeed."

According to the book description, "Dr. Biggs aims to express properties of graphs in algebraic terms, then to deduce theorems about them. In the first section, he tackles the applications of linear algebra and matrix theory to the study of graphs; algebraic constructions such as adjacency matrix and the incidence matrix and their applications are discussed in depth. There follows an extensive account of the theory of chromatic polynomials, a subject that has strong links with the 'interaction models' studied in theoretical physics, and the theory of knots. The last part deals with symmetry and regularity properties. Here there are important connections with other branches of algebraic combinatorics and group theory."

C. Berge, *Graphs*, Second revised edition of part 1 of the 1973 English version, North-Holland Mathematical Library, 6-1, North-Holland Publishing Co., Amsterdam, 1985. xiii+413 pp.

This is another classical book on graph theory. It is one of the first books which presented a coherent theory of graphs and contributed substantially to the development of graph theory.

According to MathSciNet, "The book is a classic in its area and has been instrumental in introducing mathematicians to graphs and taking them to the frontiers of current research... the revisions succeed admirably in bringing the reader up to date with the current situation in graph theory and then launching him or her on many interesting research problems."

J. Bondy, U. Murty, *Graph Theory with Applications*, American Elsevier Publishing Co., Inc., New York, 1976. x + 264 pp.

This is a classical introductory book on graph theory which emphasizes applications. It is still a very valuable introduction and reference.

From the preface: "The applications appearing at the end of each chapter actually make use of theory developed earlier in the same chapter. We also stress the importance of efficient methods of solving problems. Several good algorithms are included and their efficiencies are analyzed. We do not, however, go into the computer implementation of these algorithms."

According to MathSciNet, "This is the best book dealing with various applications of graph theory that has appeared yet...

The authors make no claim to present a comprehensive overview of graph theory and have chosen to omit almost entirely such intriguing topics as the interaction between and mutual applicability of graphs and groups, and of graphs and matrices. However, they have included proofs of almost every theorem stated in the book, and several of these are new and elegant—some being due to the work of Lászlo Lovász... All things considered, it is the high quality and importance of the applications which make this a most welcome book. Every chapter includes at least one application, and the list is impressive: the shortest path problem; Sperner's lemma; the connector problem; construction of reliable communication networks; the Chinese postman problem; the traveling salesman problem; the personnel assignment problem; the optimal assignment problem; the timetabling problem; a geometry problem; a storage problem; a planarity algorithm; ranking the participants in a tournament; feasible flows; and perfect squares."

C. Godsil, G. Royle, *Algebraic Graph Theory*, Graduate Texts in Mathematics, 207, Springer-Verlag, New York, 2001. xx+439 pp.

This book on graph theory explains how many clever and tricky arguments in graph theory are just unexpected and useful applications of algebraic methods. It is a well-written book containing a wealth of useful material. It is a very valuable introduction (especially the first half) for students and an essential reference for graph theorists.

According to MathSciNet, "There are a number of standard introductions to algebraic graph theory... This new text is a welcome addition to the literature, being beautifully written and wide-ranging in its coverage."

A.E. Brouwer, A.M. Cohen, A. Neumaier, *Distance-regular Graphs*, Ergebnisse der Mathematik und ihrer Grenzgebiete (3) [Results in Mathematics and Related Areas (3)], 18, Springer-Verlag, Berlin, 1989. xviii+495 pp.

This book is the only one devoted entirely to distance-regular graphs. It is well-written and is valuable both as an introduction and as a reference.

According to MathSciNet, "Distance-regular graphs arise naturally in a number of contexts. There are connections to coding theory, design theory, geometry and group theory, as well as to other areas of graph theory. As a result, this has been an active area of research for the last 15–20 years."

From the preface: "The main emphasis of this book is on describing the known distance-regular graphs, on classifying and, if possible, characterizing them. The structure of these graphs is touched upon insofar as necessary to describe the graphs, determine their intersection arrays, or characterize them."

T. Haynes, S. Hedetniemi, P. Slater, *Fundamentals of Domination in Graphs*, Monographs and Textbooks in Pure and Applied Mathematics, 208, Marcel Dekker, Inc., New York, 1998. xii+446 pp.

This is the only comprehensive account of domination in graphs, which, started from the recreational chessboard problems of W. Rouse-Ball in the 19th century, has been steadily developing and has many applications in several branches of mathematics such as coding theory and in practical situations.

According to MathSciNet, "This long-awaited book provides the first comprehensive treatment of domination in graphs. It is an essential work in which a vast amount of recent work on domination has been extracted from the literature and re-organized into one comprehensive volume. To make the book as self-contained as possible, it begins with a very brief overview of the graph-theoretic terminology and results required as background. This is followed with an introduction to domination via well-known problems such as the queens' domination problem, school bus routing problems, computer networks, and so on... The book is suitable as a textbook for a second course in graph theory at the graduate level and essential as a reference work for researchers in the area."

11.2. DISCRETE MATHEMATICS

M.C. Golumbic, *Algorithmic Graph Theory and Perfect Graphs*, With a foreword by Claude Berge, Second edition, Annals of Discrete Mathematics, 57, Elsevier Science B.V., Amsterdam, 2004. xxvi+314 pp.

This is a very valuable book on perfect graphs, their structural properties, construction of efficient algorithms for recognizing these graphs and for solving coloring, clique, stable set, and clique-cover problems. Though perfect graphs are special graphs, they have been actively studied since 1960. This book is an important source and reference on the subject.

According to MathSciNet, "For the second edition, this volume is, as was its predecessor, an excellent and motivating introduction to the world of perfect graphs; it is accessible to young researchers who will appreciate that most of the misprints have been removed..."

11.2.3 Graphs and their spectra

D. Cvetković, M. Doob, H. Sachs, *Spectra of Graphs, Theory and Application*, Third edition, Johann Ambrosius Barth, Heidelberg, 1995. ii+447 pp.

This is the first book on spectra of graphs. Though the authors did not claim that this was a systematic treatment on graph spectra, it is comprehensive and has been a standard introduction to and reference on this important topic. It contains practically all results achieved in this field up to the time of publication.

F. Chung, *Spectral Graph Theory*, CBMS Regional Conference Series in Mathematics, 92, Published for the Conference Board of the Mathematical Sciences, Washington, DC; by the American Mathematical Society, Providence, RI, 1997. xii+207 pp.

This is an introduction to a new direction of spectra of graphs. It includes results which are motivated by spectral theory of compact Riemannian manifolds. It does not overlap much with the previous book.

According to MathSciNet, "The analogy between the spectrum on graphs and the spectrum on Riemannian manifolds has had a major influence on spectral graph theory (and, the reviewer would like to add, on spectral geometry as well). The goal of this book is to capture this analogy in a broad, unified perspective, and to put it to work in understanding graphs... Even though the point of view of the book is quite geometric, the methods and exposition are purely graph-theoretic. As a result, the book is quite accessible to a reader who does not have any background in geometry."

11.2.4 Random graphs

A random graph is a graph that is generated by some random process. The theory of random graphs uses methods from graph theory and probability theory to study properties of typical random graphs. Random graphs were first defined in 1959.

B. Bollobás, *Random Graphs*, Second edition, Cambridge Studies in Advanced Mathematics, 73, Cambridge University Press, Cambridge, 2001. xviii+498 pp.

This book is written by an expert and gives a comprehensive account of random graphs. It is a classic and has been the standard reference on this subject since its publication. It has played an important role in the development of the theory of random graphs.

From the book description: "It is self-contained, and with numerous exercises in each chapter, is ideal for advanced courses of self-study."

According to Math Review of the first edition, "This book is the first systematic and extensive account of a substantial body of methods and results from the theory of random graphs. It is largely self-contained, with wide coverage. It deals with probabilistic rather than enumerative approaches, and thus, for example, omits the extensive theory of random trees. It is very well written and the material is covered in depth, though the presentation is at times quite terse."

S. Janson, T. Luczak, A. Rucinski, *Random Graphs*, Wiley-Interscience Series in Discrete Mathematics and Optimization, Wiley-Interscience, New York, 2000. xii+333 pp.

This book can be considered as a up-to-date sequel to the previous classic by Bollobás. It is a well-written book by major contributors to the subject and contains the most recent results in the asymptotic theory of random graphs. It is a valuable reference for this active subject.

According to MathSciNet, "The success of Bollobás's monograph spurred further activity in the area, ensuring, ironically, that it would lose its comprehensive quality quite rapidly. Nevertheless, it remained the definitive treatment (certainly in the English-speaking world) until the arrival of the book under review. The new book is a very worthy companion to the old. I say 'companion' because it is not designed to displace the earlier text; indeed, the authors do not include a topic unless there is significant progress to report. The authors have all been active at the highest level in this subject, and they write with authority. The development is everywhere clear, and great attention is paid to conveying the key ideas... It is fifteen years since Bollobás's monograph appeared, and this new definitive work should take us through the next fifteen. Such is the importance and appeal of this book that it should find its way onto the shelves not only of those working directly in the area of random graphs, but also anyone with a more general interest in combinatorics, probability theory, or certain aspects of computer science."

XII. The New Underground Law Library of the University of Michigan

Cardano was one of the founders of probability. One reason was that his gambling and his need for money motivated him to understand the mathematics behind the gambling better. His book about games of chance, *Liber de ludo aleae* (Book on Games of Chance) was written around 1564, but not published until 1663. It contains the first systematic treatment of probability, together with a section on effective cheating methods.

Though the book by Cardano contained some discussions about probability, the theory of probability dated to the correspondence between Fermat and Pascal in 1654. Because of this, Fermat and Pascal are now regarded as joint founders of probability theory. Fermat is credited with carrying out the first rigorous calculation in probability, in order to answer a question posed by a professional gambler. He made deep contributions to several subjects in mathematics including calculus, analytic geometry, number theory and optics. To most people, Fermat is best known for Fermat's Last Theorem.

One of the most basic and important result in probability is the law of large numbers. Such a phenomenon was observed by Cardano, i.e., the accuracies of empirical statistics tend to improve with the number of trials. The first version of the law of large numbers was proved by Jacob Bernoulli. It took him over 20 years to develop a sufficiently rigorous proof, which was published in his book *Ars Conjectandi* (*The Art of Conjecturing*) in 1713. He named this his "Golden Theorem" but it became generally known as "Bernoulli's Theorem". In 1837, Poisson further named it "la loi des grands nombres" (The law of large numbers).

Huygens gave the earliest known scientific treatment of probability as a subject and wrote the first book on probability theory titled *De ratiociniis in ludo aleae* (On Reasoning in Games of Chance). It was published in 1657 by another mathematician. In this picture, Huygens wrote a preface for Bernouli's book *Ars Conjectandi*. Huygens is better known for his contributions to other subjects such as for explaining the nature of the rings of Saturn and for the discovery of its moon Titan.

Chapter 12

Probability and applications

Probability is a relatively young branch of mathematics, though ideas related to probability were used by people right at the beginning of the civilization. For example, the design of casting sticks in the ancient Chinese book *I Ching* (also called *Book of Changes*) from the third to the second millennium BC might involve sophisticated probability in order to ensure a fair and hence mystical outcome.

Besides some elementary work by Cardano in the sixteen century, probability as a subject of mathematics started with correspondences of Pierre de Fermat and Blaise Pascal in the seventeenth century on the mathematical foundation of gambling.

Since probability and its application have experienced many changes in the past few decades, we broadly classify books in this chapter into (1) more classical books on probability, (2) more modern books, (3) more specialized books n probability, and (4) applications of probability.

12.1 Probability

Chebyshev is often considered as a founding father of Russian mathematics and his students include Dmitry Grave, Aleksandr Korkin, Aleksandr Lyapunov, and Andrei Markov. He is known for his important work in the field of probability, statistics, mechanics, and number theory. For example, the Chebyshev inequality can be used to prove the weak law of large numbers, and the Chebyshev polynomials in the theory of orthogonal polynomials are also named after him.

Figure 12.1: Chebyshev.

Probability theory is concerned with the analysis of random phenomena, which can be quantitively described in terms of probability. The basic concepts of probability theory are random variables and stochastic processes. As a mathematical foundation for statistics, probability theory is essential to all problems and subjects that involve quantitative analysis of large sets of data. Methods of probability theory also apply to descriptions of complex systems provided with only partial knowledge of their state, as in statistical mechanics.

The mathematical theory of probability has its roots in attempts to analyze games of chance by Gerolamo Cardano in the sixteenth century, and by Pierre de Fermat and Blaise Pascal in the seventeenth century, and slightly later by Christiaan Huygens. Initially, probability theory mainly considered discrete events, and its methods were mainly combinatorial. Eventually, analytical considerations compelled the incorporation of continuous variables into the theory. This culminated in modern probability theory, on foundations laid by A. N. Kolmogorov.

12.1.1 A.N. Kolmogorov, Foundations of the Theory of Probability, Translation edited by Nathan Morrison, with an added bibliography by A. T. Bharucha-Reid, Chelsea Publishing Co., New York, 1956. viii+84 pp. Translation of Grundbegriffe der Wahrscheinlichkeits-rechnung, Springer, Berlin, 1933.

This book is historically very important. It gave the foundation of modern probability theory and built up probability theory in a rigorous way similar to what Euclid did with geometry.

He formalized the intuitive notion of probability in terms of measure theory. This measure-theoretic formalization of probability has served and still serves as the standard foundation of probability theory. Indeed, almost all current mathematical work on probability uses this measure-theoretic approach.

According to a review in Bulletin of AMS in 1934, "It is the purpose of this monograph to develop probability theory from a postulational standpoint. For this purpose a probability field is defined as an assemblage with a definite ordering of numbers that satisfy the system of axioms. A brief exposition is given of the construction of such fields and of the manner in which the framework of the postulational system can be related to the applications to phenomena. The addition and multiplication theorems follow at once. Moreover, the theorem of Bayes, concerning whose validity there have been many controversies, is also an almost immediate consequence of the system of postulates, but the reviewer does not think this derivation of the theorem of Bayes settles the old contention relative to the validity of inferring the characteristics of a statistical population from a sample by means of the theorem of Bayes.

Use is made of the Lebesgue theories of measure and of integration. Indeed, it is held that it seemed almost hopeless to deal with the logical foundations of probability without these theories.

The development includes infinite probability fields by means of an additional axiom, distribution functions in space of many dimensions, differentiation and integration of mathematical expectations, and the law of large numbers. This little book seems to the reviewer to be an important contribution directed towards securing a more logical development of probability theory."

12.1.2 W. Feller, An Introduction to Probability Theory and its Applications, Vol. I, Vol. II. Third edition, John Wiley & Sons, Inc., New York-London-Sydney, 1968. xviii+509 pp., 1971, xxiv+669 pp.

This two volume set is a classic written by a major contributor during the earlier development of the subject. This is the first systematic and accessible introduction to probability theory. A lot of results appeared for the first time in book form. Many examples and exercises are also high original. Its impact on the development of probability has been huge, and it was called "the most successful treatise on probability ever written" by Gian-Carlo Rota.

According to a review in Bulletin of AMS in 1951, "This is the first volume of a projected two-volume work. In order to avoid questions of measurability and analytic difficulties, this volume is restricted to consideration of discrete sample spaces. This does not prevent the inclusion of an enormous amount of material, all of it interesting, much of it not available in any existing books, and some of it original. The effect is to make the book highly readable even for that part of the mathematical public which has no prior knowledge of probability. Thus the book amply justifies the first part of its title in that it takes a reader with some mathematical maturity and no prior knowledge of probability, and gives him a considerable knowledge of probability with the necessary background for going further. The proofs are in the spirit of probability theory and should help give the student a feeling for the subject.

Probability theory is now a rigorous and flourishing branch of analysis, distinguished from, say, measure theory, by the character and interest of its problems. It is true that probability theory, like geometry, had its origin in certain practical problems. However, like geometry, the theory now concerns itself with problems of interest per se, many of which are very idealized, and have only a remote connection or no presently visible connection, with practical problems. At the same time the development of the science is continually stimulated by challenging problems arising in the various fields of application. This book contains a huge number of examples illustrating almost every aspect of the theory developed. These examples are very interesting and not at all of the ad hoc variety. It is no mean feat to present so many interesting examples between two covers. They enhance the interest of the theory even for the pure mathematician, except perhaps for the extreme diehard of the 'God save mathematics from its applications' school."

12.1.3 More classical books on probability

M. Loéve, *Probability Theory. I., II.*, Fourth edition, Graduate Texts in Mathematics, Vol. 45, 46. Springer-Verlag, New York-Heidelberg, 1977. xvii+425 pp., 1978. xvi+413 pp.

This is another classic written by a major probabilist.

According to MathSciNet, "This popular book, on which many of today's probabilists cut their eye teeth, appears in its fourth edition. The first edition appeared in 1955 and each successive revision has included new material. The size has consequently increased so as to necessitate the present edition appearing in two volumes... The current edition treats several topics not in earlier editions. Chief among these are Brownian motion, functional central limit theorems and the invariance principle, random walk and fluctuation theory.

The author has attempted to retain the encyclopedic character of his treatise, starting off with the elements of measure and integration theory and proceeding to the end. But his task has become more and more difficult as time progresses because of the continuing great expansion of his subject. Thus there are a number of topics of intense current interest not mentioned... The author's compromise of topics covered seems to be about right."

K.L. Chung, *A Course in Probability Theory*, Third edition, Academic Press, Inc., San Diego, CA, 2001. xviii+419 pp.

This is another classic by a real expert in probability. It has educated generations of probabilists and is still a very good introduction.

According to MathSciNet, "In spite of the increasing flow of books on various topics in probability, there are almost no course textbooks like this one which, assuming a moderate background of measure theory but no knowledge of probability, cover the standard fundamental topics... This book is definitely a course textbook, not a compendium of theorems. The subject has been pruned to the content of a two semester course. Although it starts from the beginning, the book is severely mathematical. Readers interested in applications or in permutation-coin-die-urn problems will have to look elsewhere. The book was intended to be and is a solid and clearly written mathematical textbook."

12.1. PROBABILITY

L. Breiman, *Probability*, Corrected reprint of the 1968 original, Classics in Applied Mathematics, 7, Society for Industrial and Applied Mathematics (SIAM), Philadelphia, PA, 1992. xiv+421 pp.

This is a classical textbook for graduate students written by a famous statistician. One feature of this book is to present most important results in different directions instead of details.

From the book description, "Well known for the clear, inductive nature of its exposition, this reprint volume is an excellent introduction to mathematical probability theory. It may be used as a graduate-level text in one- or two-semester courses in probability for students who are familiar with basic measure theory, or as a supplement in courses in stochastic processes or mathematical statistics. Designed around the needs of the student, this book achieves readability and clarity by giving the most important results in each area while not dwelling on any one subject. Each new idea or concept is introduced from an intuitive, common-sense point of view. Students are helped to understand why things work, instead of being given a dry theorem-proof regime."

According to MathSciNet, "In writing such a text, an author necessarily has the awkward choice of treating a restricted number of subjects in reasonable detail or of attempting a wider coverage more superficially. In opting for the second alternative the author has succeeded quite well, and the student is introduced, albeit often sketchily, to a number of modern developments not normally found in a beginning text. The compression has often been achieved by giving only an idea or a sketch of a proof, or by simply referring the student to the original paper for a proof... The style of writing is very informal and chatty. With a few exceptions this goes over quite well. Though one might reproach the author with an undue haste in keeping the reader's attention, and in skimming the cream of many subjects, the overall effect is good and may to some extent alleviate the discouragement of many non-specialists who want a reasonably modern account not overly encumbered with heavy analytic and set-theoretic preliminaries."

M. Kac, *Statistical Independence in Probability, Analysis and Number Theory*, The Carus Mathematical Monographs, No. 12, Published by the Mathematical Association of America, Distributed by John Wiley and Sons, Inc., New York, 1959. xiv+93 pp.

This is not a textbook or introduction to probability. Instead it is a wonderful expository book to show the usefulness of the concept of statistical independence through a great variety of applications in different fields. It was written by a great expert.

According to MathSciNet, "The author states in the preface that his aim is to show that '(a) extremely simple observations are often the starting point of rich and fruitful theories and (b) many seemingly unrelated developments are in reality variations on the same simple theme.' This aim is achieved in the most instructive way...

Many problems throughout the book provide other applications to analysis and number theory. Bibliographical indications allow the reader to pursue further the study of the ideas and methods."

12.1.4 More modern books on probability

O. Kallenberg, *Foundations of Modern Probability*, Second edition, Probability and its Applications (New York), Springer-Verlag, New York, 2002. xx+638 pp.

This is a comprehensive modern introduction to probability. It is a great introduction to and a comprehensive reference on probability for advanced graduate students and researchers. It has been a popular book since its publication.

According to MathSciNet, "To sum it up, one can perhaps see a distinction among advanced probability books into those which are original and path-breaking in content, such as Loévy's and Doob's well-known examples, and those which aim primarily to assimilate known material, such as Loéve's and more recently Rogers and Williams'. Seen in this light, Kallenberg's present book would have to qualify as the assimilation of probability par excellence. It is a great edifice of material, clearly and ingeniously presented, without any non-mathematical distractions. Readers wishing to venture into it may do so with confidence that they are in very capable hands".

R. Durrett, *Probability: Theory and Examples*, Fourth edition, Cambridge Series in Statistical and Probabilistic Mathematics, Cambridge University Press, Cambridge, 2010. x+428 pp.

This book was written as a textbook for graduate courses. It contains a wealth of information and is a good book for more advanced or experienced students. It has been a popular advanced textbook.

From the book description, "This book is an introduction to probability theory covering laws of large numbers, central limit theorems, random walks, martingales, Markov chains, ergodic theorems, and Brownian motion. It is a comprehensive treatment concentrating on the results that are the most useful for applications. Its philosophy is that the best way to learn probability is to see it in action, so there are 200 examples and 450 problems."

A.N. Shiryaev, *Probability*, Translated from the Russian by R. P. Boas, Second edition, Graduate Texts in Mathematics, 95, Springer-Verlag, New York, 1996. xvi+623 pp.

This is indeed a good graduate text book on probability as the title of the book series says. It is also a good reference.

According to Zentralblatt MATH, "Generally speaking, it is a delicate problem which material to include into any course, and how to teach it. Both these problems are solved very successfully in the book under review. The author has chosen suitable material from probability theory, mathematical statistics and stochastic processes, and, which is essential, he has found probably the best form of presentation... It is clear that this book contains important and interesting results obtained through a long time period... Let us note especially that the great number of ideas, notions and statements in the book are well-motivated, explained in detail and illustrated by suitably chosen examples and a large number of exercises. Thus, the present book is a synthesis of all significant classical ideas and results, and many of the major achievements of modern probability theory."

S. Ross, *Introduction to Probability Models*, Seventh edition, Harcourt/Academic Press, Burlington, MA, 2000. xvi+693 pp.

This book is a classical introduction to probability for undergraduate students. It has been a standard textbook and has been revised seven times.

From the book description, "It provides an introduction to elementary probability theory and stochastic processes, and shows how probability theory can be applied to the study of phenomena in fields such as engineering, computer science, management science, the physical and social sciences, and operations research."

12.1.5 Probability and analysis

D. Stroock, *Probability Theory, an Analytic View*, Second edition, Cambridge University Press, Cambridge, 2011. xxii+527 pp.

This is a very good introduction to probability by emphasizing the analytic foundation. Its merit and comparison with other books are well explained by the review in MathSciNet, "anyone who can read it without undue effort will profit. Besides, it is uniformly well written and well spiced with comments to aid the intuition, so the readership should include a wide range, both of students and of professional probabilists. The treatment is relatively abstract and sophisticated, however, and it builds up to a point where the number of back references begins to threaten the 'undue effort' requirement.

In an ideal world, the present book would be a good answer to what every probabilist should learn. It is solidly mainstream probability from an analytical standpoint, and the methods are modern (at the present date). It stops barely short of stochastic analysis (no Ito formula, etc.), but this gives it an honesty concerning sets of probability 0 which, it seems to the reviewer, is perhaps preferable to a more stochastic analytic approach. While it is perhaps a bit less original than Doob's (first) book, and a bit less useful as a reference than Loéve's, it is more enjoyable to read than either. We can expect it to take its place alongside the classics of probability theory."

R. Dudley, *Real Analysis and Probability*, Revised reprint of the 1989 original, Cambridge Studies in Advanced Mathematics, 74, Cambridge University Press, Cambridge, 2002. x+555 pp.

This book gives a good introduction to the main parts of real analysis which may become important at some stage for probabilists together with some important results in probability. It is a very suitable textbook for courses on measure theoretic probability.

According to MathSciNet, "This is a remarkable textbook on real analysis and probability. It contains the core of the standard real analysis and probability first-year graduate courses (one semester of each) at most of the leading American universities, and more. What makes the book special, in the reviewer's view, is the care and scholarship with which the material is treated, and the choice of additional topics (that is, topics not usually covered in first year graduate courses)."

P. Billingsley, *Probability and Measure*, Third edition. Wiley Series in Probability and Mathematical Statistics, A Wiley-Interscience Publication, John Wiley & Sons, Inc., New York, 1995. xiv+593 pp.

This is a well written self-contained textbook on measure and probability theory and is very suitable for advanced graduate courses. It "interweaves material on probability and measure, so that probability problems generate an interest in measure theory and measure theory is then developed and applied to probability."

According to MathSciNet, "Starting from a knowledge of elementary probability theory, the author sets out, in the compass of 500 pages or so, to develop both measure theory and measure-theoretic probability in tandem; in his words, 'probability motivating measure theory and measure theory generating further probability'; again, 'Halmos and Saks have been the strongest measure-theoretic and Doob and Feller the strongest probabilistic influences on this book, and the spirit of Kac's small volume [on statistical independence] has been very important'."

12.1.6 More specialized books in probability

P. Billingsley, *Convergence of Probability Measures*, John Wiley & Sons, Inc., New York-London-Sydney, 1968. xii+253 pp.

This is a classic which describes weak-convergence methods in metric spaces and their applications. It is well-organized and clearly written.

From the book description, the author "incorporates many examples and applications that illustrate the power and utility of this theory in a range of disciplines-from analysis and number theory to statistics, engineering, economics, and population biology. With an emphasis on the simplicity of the mathematics and smooth transitions between topics, the Second Edition boasts major revisions of the sections on dependent random variables

12.1. PROBABILITY

as well as new sections on relative measure, on lacunary trigonometric series, and on the Poisson-Dirichlet distribution as a description of the long cycles in permutations and the large divisors of integers."

According to the preface, "This book is about weak-convergence methods in metric spaces, with applications sufficient to show their power and utility...

Although standard measure-theoretic probability and metric-space topology are assumed, no general (non-metric) topology is used, and the few results required from functional analysis are proved in the text or in an appendix.

Mastering the impulse to hoard the examples and applications till the last, thereby obliging the reader to persevere to the end, I have instead spread them evenly through the book to illustrate the theory as it emerges in stages."

M. Fukushima, Y. Oshima, M. Takeda, *Dirichlet Forms and Symmetric Markov Processes*, Second revised and extended edition, de Gruyter Studies in Mathematics, 19, Walter de Gruyter & Co., Berlin, 2011. x+489 pp.

This book has been a standard reference on the theory of Dirichlet forms since its publication.

According to MathSciNet, "This book ... constitutes a comprehensive reference text of the theory of symmetric Dirichlet forms, its connections with potential theory, symmetric Hunt processes, and stochastic analysis by means of additive functionals. The state space is assumed to be locally compact, but this condition can be relaxed by making use of regular representations that are developed in the appendix...

The theory is constantly illustrated by a variety of interesting examples which follow chapter after chapter. The book appears to be the best reference on this important subject at the interface between functional analysis and probability theory."

L. Kuipers, H. Niederreiter, *Uniform Distribution of Sequences*, Pure and Applied Mathematics, Wiley-Interscience [John Wiley & Sons], New York-London-Sydney, 1974. xiv+390 pp.

This book gives the first comprehensive account of uniform distribution. It is a clearly written and is an accessible book.

According to MathSciNet, "This is the first book on uniform distribution... The present authors have collected most papers on the subject (to be found in the enormous bibliography, with more than 1500 references) and have written a very readable and nice book. This book will not only serve as a useful introduction to uniform distribution but also as a reference source for researchers in the field. The choice of the material presented here is thoroughly thought out and from time to time the most general result is not presented but rather the authors describe the underlying principles and ideas; other results and related material are often described in the notes at the end of each section. These notes play an important role in this book as they supply many additional references and ideas relating to the results in the section. There are also many exercises (sometimes hard) which are applications of and supplements to the text... This book contains not only new theorems but often provides new and simpler proofs and a unified treatment of many results."

M. Ledoux, M. Talagrand, *Probability in Banach Spaces. Isoperimetry and Processes*, Ergebnisse der Mathematik und ihrer Grenzgebiete (3), 23, Reprint of the 1991 edition, Classics in Mathematics, Springer-Verlag, Berlin, 2011. xii+480 pp.

This is a classic on probability in Banach spaces, which is a branch of modern mathematics that emphasizes the geometric and functional analytic aspects of probability theory.

According to MathSciNet, "This book gives an excellent, almost complete account of the whole subject of probability in Banach spaces, a branch of probability theory that has undergone vigorous development during the last thirty years. There is no doubt in the reviewer's mind that this book will become a classic...

The notes and references at the end of every chapter provide exceptionally complete and accurate accounts of the history of the subject... Although this is mainly a reference book for researchers and teachers of probability and linear analysis, it can also be used as an advanced graduate text."

12.1.7 Random walks

A random walk is a mathematical model of a trajectory that results from of taking successive random steps. The name random walk was first introduced by Karl Pearson in 1905. Random walks have been used in many fields: ecology, economics, psychology, computer science, physics, chemistry, and biology, and explain the observed behaviors of processes in these fields. In mathematics, random walks on finitely generated groups have been extensively studied and have played an important role in geometric group theory.

F. Spitzer, *Principles of Random Walks*, Second edition, Graduate Texts in Mathematics, Vol. 34, Springer-Verlag, New York-Heidelberg, 1976. xiii+408 pp.

This is the first complete treatment of random walk on the lattice points of the Euclidean space. It has become a standard reference on this interesting topic.

According to MathSciNet, "This book presents a complete, interesting and quite self-contained treatment of random walk... The author explains why this high degree of specialization is worthwhile, as follows: 'The theory of such random walks is far more complete than that of any larger class of Markov chains. Random walk occupies such a privileged position primarily because of a delicate interplay between methods from harmonic analysis on one hand, and from potential theory on the other hand.'... Probabilistically speaking, "potential theory is basically concerned with the probability laws governing the time and position of a Markov process when it first visits a specified subset of its state space... This book certainly covers almost all major topics in the theory of random walk. It will be invaluable to both pure and applied probabilists, as well as to many people in analysis."

W. Woess, *Random Walks on Infinite Graphs and Groups*, Cambridge Tracts in Mathematics, 138, Cambridge University Press, Cambridge, 2000. xii+334 pp.

This is the first book on random walks on general infinite graphs and (non-abelian) groups, which is a relatively new subject. It was written by a major contributor.

From the book description, "Wolfgang Woess considers Markov chains whose state space is equipped with the structure of an infinite, locally-finite graph, or of a finitely generated group. He assumes the transition probabilities are adapted to the underlying structure in some way that must be specified precisely in each case. He also explores the impact the particular type of structure has on various aspects of the behavior of the random walk. In addition, the author shows how random walks are useful tools for classifying, or at least describing, the structure of graphs and groups."

According to a review in Bulletin of AMS, "the standpoint of this book is to start with a graph, groups, etc. and investigate the interplay between the behaviour of random walks on those objects on one hand and properties of the underlying structure itself on the other... As pointed out by the author in the preface, this book is not self-contained, so that it is intended for graduate students and researchers working in stochastic processes. I think, however, that from the nature of the materials and their presentations, the book is also accessible to undergraduate students and a motivated reader with some basic knowledge of probability and functional analysis."

P. Doyle, J.L. Snell, *Random Walks and Electric Networks*, Carus Mathematical Monographs, 22, Mathematical Association of America, Washington, DC, 1984. xiv+159 pp.

This is an enjoyable book which uses relatively elementary methods to study an important problem in the theory of random walks. It is both fun and informative to read, and provides "instructive example of the interplay between physical and mathematical concepts; it demonstrates how ideas from one scientific area can throw light on another", according to Zentralblatt MATH.

According to MathSciNet, "For many probabilists the relationship between electricity and probability is exemplified by the deep connections between Brownian motion and Newtonian potential theory. It does, however, require a good bit of sophistication to appreciate these connections... In contrast, the present authors set out to study similar connections between Markov chains and electrical networks which require only a little knowledge of probability theory."

G. Lawler, *Intersections of Random Walks: Probability and its Applications*, Birkhäuser Boston, Inc., Boston, MA, 1991. 219 pp.

This is a systematic exposition of an important special topic in random walks: intersections of random walks, which have important applications in statistical physics and statistical chemistry. This is an accessible book written by a major contributor. It can serve as an introduction for students and a reference for researchers.

From the book description, "Focusing on a number of problems related to the intersection of random walks and the self-avoiding walk... With the inclusion of a self-contained introduction to the properties of simple random walks, and an emphasis on rigorous results, this text should be of use to researchers in probability and statistical physics, and to graduate students interested in basic properties of random walk."

According to MathSciNet, "Some of the hardest and most intriguing problems in probability come from analyzing the self-avoiding random walk... The book has a good chance of becoming one of the standard references in the field for a long time as rapid progress in the area seems unlikely without the emergence of significantly different techniques."

12.2 Stochastic analysis

The notions of Borel algebra (or Borel σ-algebra) and Borel measure have played a basic role in measure theory and probability. They are named after Emile Borel and totally unrelated to Borel subalgebras and Borel subgroups in the theory of Lie algebras and linear algebraic groups, which are named after Armand Borel. The importance of Borel's work can also be seen in other results in probability named after him such as Borel-Cantelli Lemma and Borel's law of large numbers. He was also special in the sense that he was also a successful politician.

Figure 12.2: Borel.

After Kolmogorov provided the first axiomatic foundation for the theory of probability, Doob was one of the first people who realized that probability theory became measure theory with its own problems and terminology, that this would make it possible to give rigorous proofs for known results in probability, and that the tools of measure theory would also lead to new probability results.

12.2. STOCHASTIC ANALYSIS

12.2.1 J.L. Doob, Stochastic Processes, Reprint of the 1953 original, Wiley Classics Library, A Wiley-Interscience Publication, John Wiley & Sons, Inc., New York, 1990. viii+654 pp.

Doob wrote a series of papers on the foundations of probability and stochastic processes such as martingales, Markov processes, and stationary processes. After that, he realized that there was a real need for a book showing what is known about the various types of stochastic processes, and decided to write the famous book "Stochastic Processes" below. Right after its publication, its importance was recognized, and it has become one of the most influential books in the modern probability theory.

From the book description, "The theory of stochastic processes has developed so much in the last twenty years that the need for a systematic account of the subject has been felt, particularly by students and instructors of probability. This book fills that need. While even elementary definitions and theorems are stated in detail, this is not recommended as a first text in probability and there has been no compromise with the mathematics of probability."

According to MathSciNet, "In this valuable book the author defines a stochastic process as 'any process running along in time and controlled by probabilistic laws', or, more precisely, as 'any family of random variables $\{x_t : t \in T\}$' where 'a random variable is... simply a measurable function'. He observes that 'probability theory is simply a branch of measure theory, with its own special emphasis and field of application', and adheres uncompromisingly to this point of view throughout the book. It follows that one really is working all the time with genuinely probabilistic arguments, instead of performing analytical tricks with distribution functions, but a thorough familiarity with measure theory has to be assumed... Very few readers will work steadily through the whole of this difficult book, but it cannot fail to exercise a decisive influence on the development of its subject."

J.L. Doob, *Classical Potential Theory and Its Probabilistic Counterpart*, Grundlehren der Mathematischen Wissenschaften, 262, Reprint of the 1984 edition, Classics in Mathematics, Springer-Verlag, Berlin, 2001. xxvi+846 pp.

This is a much longer book written by Doob after his retirement. He shows in this book that his two favorite subjects: martingales and potential theory can be studied by the same mathematical tools. This is not as original or influential as the earlier book, but one has to admire the author.

According to MathSciNet, "It is divided into two quite distinct parts, each occupying about half the volume. The first half concerns the potential theory of the Laplace operator Δ (i.e. classical potential theory) and of the heat operator $(\sigma^2/2)\Delta - d/ds$ and its adjoint $(\sigma^2/2)\Delta + d/ds$ (i.e. parabolic potential theory), while the second half treats the probabilistic counterparts (interpreted liberally) to the objects in the first half."

W. Fleming, H.M. Soner, *Controlled Markov Processes and Viscosity Solutions*, Applications of Mathematics (New York), 25, Second edition, Stochastic Modelling and Applied Probability, 25, Springer, New York, 2006. xviii+429 pp.

This book is an important reference and gives a reasonably self-contained introduction to optimal stochastic control for continuous time Markov processes from the point of view of dynamic programming, and to the theory of viscosity solutions.

According to MathSciNet, "This book is a comprehensive introduction to the theory of optimal stochastic control for continuous Markov processes. By means of the dynamic programming technique, a nonlinear evolution equation should be satisfied by the value function... This book is highly recommended to anyone who wishes to learn the dynamic principle applied to optimal stochastic control for diffusion processes. Without any doubt, this book most likely is going to become a classic in the area, enlarging the previous version written by Fleming and R. W. Rishel [Deterministic and stochastic optimal control, 1975]."

S. Meyn, R. Tweedie, *Markov Chains and Stochastic Stability*, Second edition, With a prologue by Peter W. Glynn, Cambridge University Press, Cambridge, 2009. xxviii+594 pp.

This book provides a good introduction to stochastic dynamical systems. From the book description, "This publication deals with the action of Markov chains on general state spaces. It discusses the theories and the use to be gained, concentrating on the areas of engineering, operations research and control theory. Throughout, the theme of stochastic stability and the search for practical methods of verifying such stability, provide a new and powerful technique. This does not only affect applications but also the development of the theory itself. The impact of the theory on specific models is discussed in detail, in order to provide examples as well as to demonstrate the importance of these models. This book can be used as a textbook, ... the book can serve as a research resource and active tool for practitioners."

According to MathSciNet, "The authors undertake two goals. They give an up-to-date and comprehensible treatment of the theory of general Markov chains as it has developed during the last decades since the pioneering works by W. Doeblin, J. L. Doob, K. L. Chung and S. Orey. The second goal is for them still more important. Their ambition is to demonstrate the wide and often surprisingly accessible applicability of this theory. In order to achieve this goal the authors place great emphasis on the style of presentation. The necessary theoretical concepts and results are introduced first in simple contexts like the familiar countable Markov chains or deterministic systems."

12.2. STOCHASTIC ANALYSIS

12.2.2 Brownian motions and stochastic processes

Brownian motion is the mathematical model which describes random movements such as the random drifting of particles suspended in a liquid. Brownian motion is among the simplest of the continuous-time stochastic processes, and is related to the simpler random walk. It is described by the Wiener process: a continuous-time stochastic process named in honor of Norbert Wiener.

K. Ito, H. McKean, *Diffusion Processes and Their Sample Paths*, Classics in Mathematics, Die Grundlehren der mathematischen Wissenschaften, Band 125, Springer-Verlag, Berlin-New York, 1974. xv+321 pp.

This is a classic written by two top experts. It has a profound impact on the development on diffusion processes.

According to the book description, "Since its first publication in 1965 this book has had a profound and enduring influence on research into the stochastic processes associated with diffusion phenomena. Generations of mathematicians have appreciated the clarity of the descriptions given of one- or more- dimensional diffusion processes and the mathematical insight provided into Brownian motion."

According to MathSciNet, "This book certainly is a most successful work on current probability theory. Although the details of proofs are often omitted and left to the 'industrious reader', the leading idea is well-explained with simple models. The authors pay special attention to 'problems for solution' in choice and arrangement. They often play a role as a guide to other subjects which are not discussed in the book systematically. Therefore, if the reader is a graduate student, it may be recommended that he try to solve a series of problems of similar character, which are taken up here and there in the book, by different methods."

H. McKean, *Stochastic Integrals*, Reprint of the 1969 edition, with errata, AMS Chelsea Publishing, Providence, RI, 2005. xvi+141 pp.

This is a classical introduction to stochastic integrals and has educated many people.

According to MathSciNet, "This little book is a brilliant introduction to an important boundary field between the theory of probability and that of differential equations. Its subject may be best described by quoting from the preface:

'This book deals with a special topic in the field of diffusion processes: differential and integral calculus based upon Brownian motion. Roughly speaking, it is the same as the customary calculus of smooth functions, except that in taking the differential of a smooth function f of the 1-dimensional Brownian path $t \to b(t)$, it is necessary to keep two terms in the power series expansion and to replace $(db)^2$ by $dt \ldots$'

The same subject was treated in the recent book by I. I. Gihman and A. V. Skorohod, *Stochastic Differential Equations*, 1968. The author's book is smaller, contains more examples and applications, and is therefore much better suited to beginners."

D. Revuz, M. Yor, *Continuous Martingales and Brownian Motion*, Third edition, Grundlehren der Mathematischen Wissenschaften, 293, Springer-Verlag, Berlin, 1999. xiv+602 pp.

This is not an easy book to learn the subject from, but it is a substantial book which contains a wealth of information and is a valuable reference on more advanced readers and researchers.

According to MathSciNet, "Here the authors have combed the literature for results which are of real interest and yet can be worked out fairly easily, and the exercises comprise a veritable treasure of such material. Inevitably, the indexing system is not always adequate to locate all of these results by name, so the reader must be ready to do some digging... The real unifying principle of the book, if there is one, is the predictable stochastic integral and Ito's formula, which has 'revolutionized the study of Brownian motion' (p. 139). This gives the book an almost magical quality, in which Ito's formula acts as a kind of hidden spring leading to all sorts of novel and unexpected results. Problems which, in the Markovian case, would formerly have been reduced to solving ordinary differential equations, are treated instead by stochastic differential equations, for which the existence and uniqueness theory is carefully developed...

The outcome is a unique blend of theory and practice, which truly exhibits mathematics as the 'art of the possible'... The authors have produced a text which can be a delight to read, where even the most complicated results are presented so efficiently as not to necessitate use of pencil and paper. However, it is not a text for everyone, and prospective readers will probably discover quite soon whether or not they find this style congenial. In any case it will be invaluable as an access work for a large modern literature on continuous stochastic integration..."

J. Jacod, A. Shiryaev, *Limit Theorems for Stochastic Processes*, Grundlehren der Mathematischen Wissenschaften, 288, Second edition, Grundlehren der Mathematischen Wissenschaften, 288, Springer-Verlag, Berlin, 2003. xx+661 pp.

This book gives an excellent self-contained account of the theory of convergence in law of stochastic processes, in particular semimartingales. It has been an essential reference to people in the field.

According to MathSciNet, "This book is devoted to the convergence in law of stochastic processes from the point of view of semimartingale theory: this class of processes is broad enough to accommodate most common processes, and the stochastic calculus is nowadays a powerful tool for studying them. ... The authors describe very general situ-

12.2. STOCHASTIC ANALYSIS

ations to cover many various cases. That often makes statements and proofs somewhat technical, but they have done a great service by giving numerous examples and explanations in order to make the reading easier. This is an interesting up-to-date book, and it has been written by authors whose recent contribution to the question is essential."

P. Hall, C.C. Heyde, *Martingale Limit Theory and its Application*, Probability and Mathematical Statistics, Academic Press, Inc., New York-London, 1980. xii+308 pp.

This is a standard reference on classical limit theorems for discrete parameter martingales. It contains a substantial number of applications, many of which appear in book form for the first time.

According to MathSciNet, "The book starts with a brisk treatment of the a.s. convergence theory, laws of large numbers and square function inequalities... It gives a good view of the 'state of the art' in the theory, and a wide sample of applications; it should succeed in becoming the standard reference work in this area."

J. Bertoin, *Lévy Processes*, Cambridge Tracts in Mathematics, 121, Cambridge University Press, Cambridge, 1996. x+265 pp.

This book gives a self-contained account of Lévy processes, which are stochastic processes with independent and stationary increments in the Euclidean framework and can be thought of as random walks in continuous time, and they have many applications in areas such as queues, dams, mathematical finance and risk estimation. It is a standard reference on the subject.

According to MathSciNet, "There has long been a need for a book-length treatment of the theory of Levy processes ($\mathbb{R}^d$-valued processes with stationary independent increments)... In just over 250 pages [the author] presents a clear vision of this important area of probability theory. The development is reasonably self-contained, and it is clear that the author has taken pains to simplify proofs wherever possible... This book would serve wonderfully as a follow-up to an introductory graduate probability course— many of the topics served up in such a course (characteristic functions, independence, Poisson processes, and martingales) play important roles in the theory developed here, and the book illustrates the application of these techniques to an important 'case study'. The book should also prove valuable to probability researchers, as a concise reference work on Levy processes."

K. Sato, *Lévy Processes and Infinitely Divisible Distributions*, Translated from the 1990 Japanese original, Revised by the author, Cambridge Studies in Advanced Mathematics, 68, Cambridge University Press, Cambridge, 1999. xii+486 pp.

This book gives a detailed, accessible account of the essential results on $\mathbb{R}^d$-valued Lévy processes, the infinitely divisible distributions and related topics. It starts from basics, and all notions used are introduced, and all results are proven. It is a standard reference on the subject.

According to MathSciNet, "This book is a work of scrupulous scholarship, and an impressively detailed compendium of knowledge. One could question, e.g., the decision to prefer analytic to pathwise methods in places (such as the derivation of the Levy-Ito decomposition), but here and generally the book happily complements that of J. Bertoin, *Lévy processes*, 1996. I recommend that the books be used in tandem, and regard them as essential reading."

G. Samorodnitsky, M. Taqqu, *Stable non-Gaussian Random Processes. Stochastic Models with Infinite Variance. Stochastic Modeling*, Chapman & Hall, New York, 1994. xxii+632 pp.

This is a comprehensive and standard reference on stable non-Gaussian random processes.

According to MathSciNet, "The authors convince us that the class of stable processes with infinite variance is as rich and interesting as the class of Gaussian processes. They treat many different types of stable processes, such as fractional stable processes, stable moving averages, stable harmonisable processes, the stable Ornstein-Uhlenbeck process, and stationary stable processes. The reader is introduced into a new world of processes with large jumps and wild oscillations. Because of this erratic behaviour, stable processes with infinite variance seem appropriate models for real phenomena with heavy-tailed distributions, such as crashes and catastrophes. Therefore they have attracted the attention of many applied workers in meteorology, economics, finance, insurance, physics, etc. There is a vast amount of mathematical work on stable processes. The book under review is the first comprehensive treatment of the topic. It collects a wealth of results and applications, including the substantial contributions from the research of the authors... In summary, this is an excellent book on stable processes. It will become the standard reference on the topic."

D. Stroock, S.R. Varadhan, *Multidimensional Diffusion Processes*, Grundlehren der Mathematischen Wissenschaften, 233. Reprint of the 1997 edition, Classics in Mathematics, Springer-Verlag, Berlin, 2006. xii+338 pp.

This book gives a very complete account of diffusion processes. It is a useful reference to advanced readers and experts, especially those interested in Markov processes from a more theoretical point of view.

According to Bulletin of AMS, "Both the Markov-process approach and the Ito approach have been immensely successful in diffusion theory. The Stroock-Varadhan book, developed from the historic 1969 papers by its authors, presents the martingale-problem approach as a more powerful and, in certain regards, more intrinsic-means of studying the foundations of the subject... the authors make the uncompromising decision not 'to proselytise by intimidating the reader with myriad examples demonstrating the full scope

12.2. STOCHASTIC ANALYSIS

of the techniques, but rather to persuade the reader 'with a careful treatment of just one problem to which they apply. Most of the main tools of stochastic-processes theory are used, ... but it is the formidable combination of probability theory with analysis which is the core of the work... For immediate confirmation of the subject's sparkle, virtuosity, and depth, see McKean's 1969 book."

S. Ethier, T. Kurtz, *Markov Processes. Characterization and Convergence*, Wiley Series in Probability and Mathematical Statistics: Probability and Mathematical Statistics, John Wiley & Sons, Inc., New York, 1986. x+534 pp.

This book gives an accessible, comprehensive account of Markov process theory. It has been an essential reference for people working on this subject.

According to Zentralblatt MATH, "The book is written as a self-contained reference text for those already acquainted with the elements of probability, measure and functional analysis. As such, it succeeds admirably, providing a handy source of useful and detailed information on the basic theory required for each of the three techniques, in addition to a careful and exhaustive treatment of the subject matter proper... There is no question but that space should immediately be reserved for the book on the library shelf. Those who aspire to mastery of the contents should also reserve a large number of long winter evenings."

According to MathSciNet, "Diffusion processes are used in many situations to approximate various other stochastic processes which are sometimes Markov and often not. This book deals with various methods that are used to establish convergence theorems for stochastic processes. Any limit theorem is dependent on a suitable characterization of the limiting process. Typically, for the given sequence of stochastic processes one establishes some weak compactness results, proves some properties for an arbitrary limit point and then uses these properties to characterize the limit point uniquely. The book is motivated to a very large extent by this point of view."

A. van der Vaart, J. Wellner, *Weak Convergence and Empirical Processes. With Applications to Statistics*, Springer Series in Statistics, Springer-Verlag, New York, 1996. xvi+508 pp.

This book is carefully written and well-thought out and gives a complete and detailed exposition of general empirical process theory and its applications. It is a very valuable reference for mathematical statisticians and probabilists, and is also appropriate as a graduate textbook.

According to MathSciNet, "This book is a very good text on empirical processes. There are some previous monographs in this subject ... The book under review is a more complete and detailed exposition on empirical processes and their applications. It also covers the most recent developments in the area."

12.2.3 Stochastic calculus and equations

N. Ikeda, S. Watanabe, *Stochastic Differential Equations and Diffusion Processes*, Second edition, North-Holland Mathematical Library, 24, North-Holland Publishing Co., Amsterdam; Kodansha, Ltd., Tokyo, 1989. xvi+555 pp.

This book gives a modern, systematic and comprehensive account of the theory of stochastic differential equations and diffusion processes. It is a valuable reference for people working on stochastic differential equations and applications.

According to MathSciNet, "The purpose of this book is to present a systematic, rigorous and modern theory of stochastic integrals, stochastic differential equations and (associated) diffusion processes... The theory is developed along the lines of Ito's original approach, but within the modern (semi-)martingale framework (the Stroock-Varadhan martingale method can also be seen throughout). This book comprises most of the major results on the topic and the presentation is very lucid. This is not just another book added to the list of great books such as those of Dynkin, Ito and McKean, Stroock and Varadhan, and Friedman. In our opinion, it is the best book, to date, that introduces the reader to the theory of stochastic differential equations and diffusion processes; and we feel that every student who plans to work in this area should read this book (and own a copy of it). We are convinced that every specialist in the area already has a copy of this book."

G. Da Prato, J. Zabczyk, *Stochastic Equations in Infinite Dimensions*, Encyclopedia of Mathematics and its Applications, 44, Cambridge University Press, Cambridge, 1992. xviii+454 pp.

This book gives a systematic and self-contained account of stochastic equations in Hilbert and Banach spaces where the equations are defined by the semigroup concept. It has been an essential reference on stochastic evolution equations and related topics.

According to MathSciNet, "Altogether, this is an excellent book which covers a large part of stochastic evolution equations with clear proofs and a very interesting analysis of their properties. The exposition is self-contained, dealing with a wide range of infinite-dimensional stochastic and deterministic analysis. In my opinion this book will become an indispensable tool for everybody working on stochastic evolution equations and related areas."

I. Karatzas, S. Shreve, *Brownian Motion and Stochastic Calculus*, Second edition, Graduate Texts in Mathematics, 113, Springer-Verlag, New York, 1991. xxiv+470 pp.

This book gives a comprehensive and updated account of all the essential results of the Brownian motion, stochastic calculus and their applications. It is clearly written and well-organized and is a very useful introduction and a valuable reference, i.e., a good book for people at all levels.

12.2. STOCHASTIC ANALYSIS

According to MathSciNet, "The title may say it all. This book is much more than a book on Brownian motion. Although it is always a good idea to start reading any book from its preface, here it is particularly so. The preface will tell the reader clearly what this book is about and what to find in it. ... A number of good problems in the heart of the subject matter are given in each chapter. Solutions to some problems are supplied. The notes at the end of each chapter are very important. These notes are more than comments on what the authors have done. They indicate some of the topics which have been omitted. This book is a good text for a course covering the topics reviewed above—a valuable book for every graduate student studying stochastic processes, and for those who are interested in pure and applied probability."

H. Kunita, *Stochastic Flows and Stochastic Differential Equations*, Cambridge Studies in Advanced Mathematics, 24, Cambridge University Press, Cambridge, 1990. xiv+346 pp.

This is the first comprehensive book devoted to stochastic differential equations and stochastic flow of diffeomorphisms and properties of stochastic flows. It is written by an expert. It can serve as an advanced introduction and a useful reference.

According to MathSciNet, "This is the first book to expound, in a detailed way, the relation between stochastic flows and stochastic differential calculus. The new points, compared with the many existing treatises on stochastic calculus, are that it emphasizes the study of random flows of diffeomorphisms rather than semimartingales or diffusion processes. One should note that it was only recently observed that the most natural examples of stochastic flows (namely the isotropic ones) could not be described by stochastic differential equations (SDEs) involving a finite number of real-valued Brownian motions. For example, Brownian flows, i.e., flows with independent increments, are described by stochastic differential equations driven by vector-field-valued Brownian motions."

I. Karatzas, S. Shreve, *Methods of Mathematical Finance*, Applications of Mathematics (New York), 39, Springer-Verlag, New York, 1998. xvi+407 pp.

This book gives a systematic account of some applications of stochastic analysis and optimal control theory to finance models. It is useful both as an advanced introduction and valuable reference.

According to MathSciNet, "In contrast to several other books on mathematical finance which appeared in recent years, this book deals not only with the so-called partial equilibrium approach (i.e., the arbitrage pricing of European and American contingent claims) but also with the general equilibrium approach (i.e., with the equilibrium specification of prices of primary assets). A major part of the book is devoted to solving valuation and portfolio optimization problems under market imperfections, such as market incompleteness and portfolio constraints... Undoubtedly, the book constitutes a valuable research-level text which should be consulted by anyone interested in the area. Unlike other currently available monographs, it provides an exhaustive and up-to-date treatment of portfolio optimization and valuation problems under constraints. It is also quite suitable as a textbook for an advanced course on mathematical finance."

12.2.4 Large deviations

The theory of large deviations studies the asymptotic behavior of remote tails of sequences of probability distributions, in particular with the exponential decay of the probability measures of certain kinds of extreme or tail events, as the number of observations grows arbitrarily large.

J.D. Deuschel, D. Stroock, *Large Deviations*, Pure and Applied Mathematics, 137, Academic Press, Inc., Boston, MA, 1989. xiv+307 pp.

This is a valuable reference for experts, but probably not suitable as an introduction.

According to Zentralblatt MATH, "This is not a book that will make you fall in love with Large Deviation Theory (LDT), but for those already acquainted with the theory, it could help the relationship mature. The book provides a sound base for LDT and answers questions and clears up technical problems found in articles previously written on the subject."

A. Dembo, O. Zeitouni, *Large Deviations Techniques and Applications*, Corrected reprint of the second (1998) edition, Stochastic Modelling and Applied Probability, 38, Springer-Verlag, Berlin, 2010. xvi+396 pp.

This is a successful advanced introduction to the theory of large deviations and applications. It is written by two experts and can also serve as a reference.

According to the summary on the back-cover, "Originally developed in the context of statistical mechanics and of (random) dynamical systems, [the theory of large deviations] proved to be a powerful tool in the analysis of systems where the combined effects of random perturbations lead to a behavior significantly different from the noiseless case. The volume complements the central elements of this theory with selected applications in communication and control systems, bio-molecular sequence analysis, hypothesis testing problems in statistics, and the Gibbs conditioning principle in statistical mechanics."

According to MathSciNet, "This book covers some advanced topics and gives several new applications. Its main feature however is a clear and well-balanced presentation of the various aspects of large deviations. It may also be used in teaching an undergraduate (Chapters 1 to 3) or graduate course."

R. Ellis, *Entropy, Large Deviations, and Statistical Mechanics*, Grundlehren der Mathematischen Wissenschaften [Fundamental Principles of Mathematical Sciences], 271, Springer-Verlag, New York, 1985. xiv+364 pp.

12.2. STOCHASTIC ANALYSIS

This is a valuable reference on large deviations and statistical mechanics. It is carefully written with lots of detail and is self-contained.

According to MathSciNet, "Besides the fact that the author's treatment of large deviations is a nice contribution to the literature on the subject, his book has the virtue that it provides a beautifully unified and mathematically appealing account of certain aspects of statistical mechanics. In particular, he carries out the program suggested by O. E. Lanford, III and substantiates, once again, that good mathematics often originates from good physics. Furthermore, he does not make the mistake of assuming that his mathematical audience will be familiar with the physics and has done an admirable job of explaining the necessary physical background. Finally, it is clear that the author's book is the product of many painstaking hours of work; and the reviewer is confident that its readers will benefit from his efforts."

12.2.5 Malliavin calculus

P. Malliavin, *Stochastic Analysis*, Grundlehren der Mathematischen Wissenschaften, 313, Springer-Verlag, Berlin, 1997, xii+343 pp.

This is a book written by the founder of the subject, *Malliavin Calculus*. It contains a nonstandard approach to some well-known subjects and clear description of other subjects.

The Malliavin calculus extends the calculus of variations from functions to stochastic processes. It is also called the stochastic calculus of variations and allows the computation of derivatives of random variables. Malliavin invented his calculus in order to give a stochastic proof of the result that Hörmander's condition implies the existence of a density for the solution of a stochastic differential equation; Hörmander's original proof was based on the theory of partial differential equations. Malliavin calculus allowed him to prove regularity bounds for the solution's density and has been applied to stochastic partial differential equations. It also allows integration by parts with random variables, which is used in mathematical finance to compute the sensitivities of financial derivatives and has other applications as well.

D. Nualart, *The Malliavin Calculus and Related Topics*, Second edition, Probability and its Applications (New York), Springer-Verlag, Berlin, 2006. xiv+382 pp.

This book gives a very readable account of the Malliavin calculus and some of its applications. It is an advanced, yet accessible introduction, and it demonstrates well the power of the Malliavin calculus.

According to MathSciNet, "Malliavin calculus is a special differential calculus for functionals on a Gaussian probability space, but the distinguishing features of the subject are its applications, notably the study of probability densities of functionals of Gaussian processes, hypoellipticity of partial differential operators, and the theory of non-adapted stochastic integrals... Nualart's text is a very welcome addition to the Malliavin calculus literature. It brings the older monographs of S. Watanabe and D. Bell up to date and is unique among expositions in the breadth and accessibility of its coverage."

12.3 Applications of probability

Wiener was a famous child prodigy and an early researcher in stochastic and noise processes. For example, he did foundational work on the mathematics of the Brownian motion by proving many results now widely known such as the non-differentiability of the paths. Because of these results, the one-dimensional version of Brownian motion is called the Wiener process, which is the best known of stochastic processes. Besides his other contributions in mathematics such as Paley-Wiener theorem, he also worked on electronic engineering, electronic communication, and control systems.

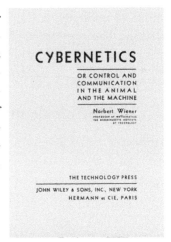

Figure 12.3: Wiener's book.

12.3.1 Probabilistic methods and applications

N. Alon, J. Spencer, *The Probabilistic Method*, With an appendix by Paul Erdős, Third edition, Wiley-Interscience Series in Discrete Mathematics and Optimization, John Wiley & Sons, Inc., Hoboken, NJ, 2008. xviii+352 pp.

This is an essential reference on probabilistic methods in combinatorics. It is clearly written and contains many illustrating examples and practical exercises, and it is a very valuable resource to many users in different subjects.

According to MathSciNet, "The focus of this well-written book consists of probabilistic techniques useful in graph theory, number theory, geometry, and combinatorics. Some objects of study are themselves random, such as a graph on finitely many points, each pair of vertices being connected by an edge with some fixed probability. On the other hand, Ramsey numbers and many other objects having no intrinsic randomness are treated with probabilistic tools.

The book can be used very constructively in a study group of graduate students and faculty. It is organized so that one could master one technique at a time, see how it has

12.3. APPLICATIONS OF PROBABILITY

been used in the book in a variety of settings, and then pose other questions susceptible to the same technique."

J. Kahane, *Some Random Series of Functions*, Second edition, Cambridge Studies in Advanced Mathematics, 5, Cambridge University Press, Cambridge, 1985. xiv+305 pp.

It is an important book on random series. The first edition of this book has stimulated a lot of work both in Fourier series and in random processes. The second edition is substantially longer and provides a very readable account of the interplay between classical real function theory and stochastic processes.

From the book description, "The subject matter of this book is important and has wide application in mathematics, statistics, engineering, and physics. Professor Kahane's presentation is suitable even for beginning graduate students in probability and analysis (exercises are provided throughout), as well as nonspecialists in the other disciplines to which this subject has application."

According to MathSciNet, "The monograph was very influential, no doubt due to its clarity of exposition and the many interesting problems that it posed..."

Percolation studies the movement and filtering of fluids through porous materials. An extensive mathematical model of percolation, called percolation theory, was introduced by Broadbent and Hammersley in 1957. Percolation theory has been an active area of research for mathematicians and physicists, it has brought new understanding and techniques to a broad range of topics in physics, materials science, complex networks, epidemiology as well as in geography.

G. Grimmett, *Percolation*, Second edition, Grundlehren der Mathematischen Wissenschaften, 321, Springer-Verlag, Berlin, 1999. xiv+444 pp.

This book gives a self-contained account of percolation theory including many recent results. It is clearly written and includes physical background. It is a useful introduction and reference for mathematicians and physicists.

According to MathSciNet, "Remarkable progress has been made recently, especially during the last five years. This book presents a self-contained account of percolation theory, readily accessible to students and nonspecialists, which contains many of the exciting new developments in the field... The book gives an excellent overview of percolation theory, collecting (and often simplifying) the most important results and techniques in one place. There are numerous references to the literature, with detailed comments on the development and history of the subject. It is an important contribution to the field."

B. Hughes, *Random Walks and Random Environments. Vol. 1. Random Walks, Vol. 2. Random Environments*, Oxford Science Publications, The Clarendon Press, Oxford University Press, New York, 1995. xxii+631 pp., 1996. xxiv+526 pp.

These two books give a largely self-contained account of some of the most useful, important, and beautiful results for structurally disordered systems. They are linked by notions such as scaling, universality, and dimensionality. Volume 1 emphasizes random walks, and volume 2 percolation theory. It is a comprehensive reference for people at all levels.

According to MathSciNet, "This well-written two-volume book is an advanced work on the applications of probability theory and methods to physical problems of structurally disordered systems, especially on the topics of random walk, self-avoiding walk, percolation, fractals, etc... This whole volume is mathematically solid; the content of it is sufficiently rigorous and diverse to satisfy the needs of applied mathematicians, physicists, engineers and graduate students in the related disciplines. Due to the attractive topics as well as the informal and vivid style often used in the presentation and description of the challenging mathematical and physical problems, they will find this book a very readable one and will benefit a lot from it."

"The origin of the problems and the physical background and applications of the models are always described in detail. The book also provides an up-to-date view of the most challenging problems in the subject. Due to the elegant and beautiful presentation, this book can be heartily recommended to graduate students and physicists in related fields, and even experienced mathematicians will certainly learn something new from this book."

D. Voiculescu, K. Dykema, A. Nica, *Free Random Variables. A Noncommutative Probability Approach to Free Products with Applications to Random Matrices, Operator Algebras and Harmonic Analysis on Free Groups*, CRM Monograph Series, 1, American Mathematical Society, Providence, RI, 1992. vi+70 pp.

This book gives an introduction to the free probability theory, a relatively recent branch of non-commutative probability theory, together with some applications to random matrices and a new construction of type II_1 von Neumann factors. This is a valuable reference on this specialized topic.

According to MathSciNet, "The book consists mainly of a collection of all relevant results of Voiculescu and his coauthors on this subject (which covers almost all of the development in this field), put together in a unified way and enriched with some introductory and motivating remarks and examples. Although the original papers arose out of an operator algebraic context, here the main emphasis is put on the algebraic level, where the essential ideas become most transparent. Thus the treatment is also accessible to persons without any background in operator algebras. The fundamental theorems are usually given with proof, whereas additional results are provided with a short idea or sketch of the proof... This book can be recommended to everyone who wants to get an impression of the beauty and fruitfulness of the concept of freeness."

12.3. APPLICATIONS OF PROBABILITY

T. Liggett, *Interacting Particle Systems*, Grundlehren der Mathematischen Wissenschaften, 276, Reprint of the 1985 original, Classics in Mathematics, Springer-Verlag, Berlin, 2005. xvi+496 pp.

This book is the first book that gives a complete and accessible account of an important class of random processes, interacting particle systems, which can serve as models of many different things such as ferromagnetism, spread of infection or turbulence in liquids. It has been a standard introduction and reference on this relatively new subject.

According to MathSciNet, "This book presents a complete treatment of a new class of random processes, which have been studied intensively during the last fifteen years. None of this material has ever appeared in book form before. The class of processes treated goes under such names as 'interacting particle systems', 'systems with locally interacting components', or 'interacting Markov processes'. Thus, the basic feature is the presence of some kind of interaction between the elements of an infinite particle system, or between a countable collection of sites which may be in one of several states (e.g., vacant or occupied); alternatively, the interaction may be superimposed upon an infinite system of Markov processes by changing the transition rates of each process in a way depending on the instantaneous location of the other processes... This example is typical of open questions in this field which are easy to state but hard to answer. Over sixty open problems are listed at the end of the appropriate chapters. The high quality of this work, on a technically difficult subject, makes a fascinating subject and its open problems as accessible as possible."

T. Liggett, *Stochastic Interacting Systems: Contact, Voter and Exclusion Processes*, Grundlehren der Mathematischen Wissenschaften, 324, Springer-Verlag, Berlin, 1999. xii+332 pp.

This is a natural continuation of the previous book and gives a good overview of the developments after that book. It is a very valuable and updated reference.

V. Jikov, S. Kozlov, O. Oleinik, *Homogenization of Differential Operators and Integral Functionals*, Translated from the Russian by G. A. Yosifian, Springer-Verlag, Berlin, 1994. xii+570 pp.

This book gives a comprehensive account of the well developed theory of homogenization. It provides the mathematical foundations of the theory and explains various relations with other theories of mathematics, physics and mechanics such as the ergodic theory, the theory of composite materials. It is a very valuable advanced introduction and valuable reference.

According to MathSciNet, "... The authors consider new types of homogenization problems concerning differential operators with random coefficients, diffusion in random media, G-convergence, estimates of Hashin and Strikman type, stratified structures, spectral problems, Γ-convergence, variational problems, plasticity, etc. The stochastic point of view on composite structures is present in most chapters..."

12.3.2 Random matrices

A random matrix is a matrix-valued random variable. In physics, random matrices have been applied to nuclear physics, solid-state physics, quantum chaos and quantum gravity. For example, in 1955, Eugene Wigner used random matrices to model effectively the spectra of heavy atoms by proposing that the spacings between the lines in the spectrum of a heavy atom should resemble the spacings between the eigenvalues of a random matrix, and should depend only on the symmetry class of the underlying evolution. In mathematics, random matrices have been applied to understand the nontrivial zeros of the Riemann zeta function such as spacing and correlations between them.

M. Mehta, *Random Matrices*, Third edition, Pure and Applied Mathematics (Amsterdam), 142, Elsevier/Academic Press, Amsterdam, 2004. xviii+688 pp.

This is a classic and has played an important role in the development and applications of random matrices, which has becoming a really active subject in the past few years. It has been a standard reference and introduction.

From the book description, "This book gives a coherent and detailed description of analytical methods devised to study random matrices. These methods are critical to the understanding of various fields in in mathematics and mathematical physics, such as nuclear excitations, ultrasonic resonances of structural materials, chaotic systems, the zeros of the Riemann and other zeta functions. More generally they apply to the characteristic energies of any sufficiently complicated system and which have found, since the publication of the second edition, many new applications in active research areas such as quantum gravity, traffic and communications networks or stock movement in the financial markets."

From the Math Review of the first edition, "The random-matrix model, although deceptively simple in its formulation, has not one but many consequences, the precise derivation of which is by no means a straightforward matter. Nevertheless, numerous experimentally verifiable results have been obtained by an exceptionally skillful exercise of classical analysis, and these developments are authoritatively and systematically presented in this book by one of the chief contributors... The book under review is a scholarly work which provides an admirably true picture of what has been accomplished in the field by analytical means."

G. Anderson, A. Guionnet, O. Zeitouni, *An Introduction to Random Matrices*, Cambridge Studies in Advanced Mathematics, 118, Cambridge University Press, Cambridge, 2010. xiv+492 pp.

This book gives a rigorous introduction to the basic theory of random matrices and some modern developments. It is also a very valuable reference.

According to MathSciNet, "The field of random matrices is a sprawling one, which originated in statistics and nuclear physics, but which nowadays has many deep and interesting connections with combinatorics, complex analysis, high-dimensional geometry, concentration of measure, integrable systems, Lie groups, and number theory. As such, it is nearly impossible to write a text on the subject that covers all aspects of random matrix theory, especially given that many parts of the subject are still evolving and not yet at a mature state of understanding. Nevertheless, the authors here have done an admirable job in presenting in a coherent and self-contained fashion a significant number of 'core' topics of random matrix theory...

Great efforts have been made in the text to make it self-contained, with numerous appendices on linear algebra (including quaternionic linear algebra), Riemannian geometry, operator algebras, probability theory, and point-set topology. Nevertheless, each chapter is also supported by an extensive list of references. As such, this is a very valuable new reference for the subject, incorporating many modern results and perspectives that are not present in earlier texts on this topic. While many topics are not covered in full detail in the text,... this book would serve as an excellent foundation with which to begin studying other aspects of random matrix theory."

P. Deift, *Orthogonal Polynomials and Random Matrices: a Riemann-Hilbert Approach*, Courant Lecture Notes in Mathematics, 3, New York University, Courant Institute of Mathematical Sciences, New York; American Mathematical Society, Providence, RI, 1999. viii+273 pp.

This book gives an accessible introduction to some results on universality for a variety of statistical quantities arising in the theory of random matrix models. It is also a useful reference to answer the central question: Why do very general ensembles of random $n \times n$ matrices exhibit universal behavior as $n \to \infty$? It is a valuable reference on such asymptotic results.

According to MathSciNet, this book prepares "the reader to understand more advanced expositions and modern research literature on applications of asymptotic methods for the matrix Riemann-Hilbert factorization problems to a variety of different problems in analysis, the field where the author together with his students and collaborators has made many important contributions. The presentation is clear and pedagogically very well done."

N. Katz, P. Sarnak, *Random Matrices, Frobenius Eigenvalues, and Monodromy*, American Mathematical Society Colloquium Publications, 45, American Mathematical Society, Providence, RI, 1999. xii+419 pp.

This book gives a very complete account on the deep relation connecting the spacings between zeros of the Riemann zeta and L-functions with spacings between eigenvalues of random elements of large compact classical groups. It is a very useful reference on and an introduction to this beautiful subject, written by two well-known experts.

According to MathSciNet, "This book is fascinating in many aspects: First, its rigorous, systematic and accessible exposition of the subject makes it a bright landmark at the crossroads of arithmetic and mathematical physics; no doubt it will become a basic reference in random matrix theory. Second, it offers its reader a bouquet of beautiful new results but also leaves the door open to many challenging conjectures."

XIII. The East Hall of the University of Michigan, the home of the Mathematics Department

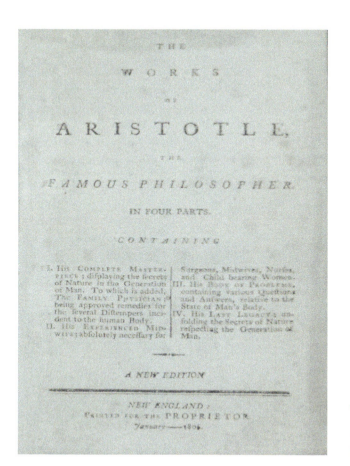

Aristotle was a Greek philosopher and polymath, a student of Plato and teacher of Alexander the Great. He wrote extensively and contributed to almost all subjects including physics, metaphysics, logic etc. Logic was established as a formal discipline by Aristotle and was given a fundamental place in philosophy. Aristotelian logic became widely accepted in science and mathematics and used in the west until the early nineteenth century. It was responsible for several basic methods and terminologies such as predicable, syllogisms and propositions in logic.

Besides his famous 23 problems at the ICM in Paris, 1900, Hilbert is also famous for his failed attempt to formalize mathematics, i.e., a formalization of all of mathematics in axiomatic form, together with a proof that this axiomatization of mathematics is consistent. He was motivated by the foundational crisis of mathematics at that time. Unfortunately, Gödel's incompleteness theorems in 1931 showed that Hilbert's program was not possible. It was said that Hilbert was a bit angry when he first heard of Gödel's result. It was a blow to Hilbert, since shortly before that he confidently announced to the world "Wir müssen wissen. Wir werden wissen." (We must know. We will know.) These words are carved on his tomestone.

Godöl is probably the most important logician in the twentieth century, but his life towards the end was a sad one. He basically starved himself to death after his wife died, who was six years older than he was. Gödel is best known for his two incompleteness theorems, published in 1931 when he was only 25 years old, one year after obtaining Ph.D. at the University of Vienna. These results might be made over popular by many popular books and other media forms. Another major contribution of Gödel is his theorem that neither the axiom of choice nor the continuum hypothesis can be disproved from the standard axioms of set theory, assuming these axioms are consistent. It might be interesting to note that he had some difficulty in getting a permanent job at the Institute for Advanced Study, Princeton.

Neumann made major contributions to many subjects both inside and outside mathematics such as foundations of mathematics, functional analysis, ergodic theory, geometry, topology, numerical analysis, quantum mechanics, hydrodynamics, fluid dynamics, game theory, computer science, and statistics. Together with Morgenstern, he wrote the classic *Theory of Games and Economic Behavior*. In mathematics, his most important contribution lies in operator algebras, in particular von Neumann algebras. He was also one of the few mathematicians who had some real political influence in the history of USA.

Chapter 13

Foundations of math, computer science, numerical math

Though mathematical logic is an essential part of the foundation of mathematics, it is probably not so familiar to many mathematicians. Indeed, foundations of mathematics refer to fields of mathematics such as mathematical logic, axiomatic set theory, proof theory, model theory, type theory and recursion theory; and mathematical logic usually consists of similar fields such as set theory, model theory, recursion theory, and proof theory. The search for foundations of mathematics is also a central question of the philosophy of mathematics.

Theoretical computer science is a branch of computer science that focuses on more abstract or mathematical aspects of computing. Among many mathematical subjects, numerical mathematics is one that has probably the closest connection with computer science.

In this chapter, we discuss some books that are related to foundation of mathematics, computer science and numerical mathematics.

13.1 Mathematical logic

Mathematical logic studies logic and application of formal logic to other areas of mathematics, and has close connection to foundations of mathematics, theoretical computer science and philosophical logic. It is often divided into set theory, model theory, recursion theory, and proof theory.

Frege was a mathematician, logician and philosopher. He made major contributions to the foundations of mathematics and is regarded as one of the founders of modern logic. Though he was ignored first by the logic community, his work was introduced to later generations by Peano, Russell and Ludwig Wittgenstein. He is also regarded as one of the founders of analytic philosophy.

Figure 13.1: Frege's book.

13.1.1 Mathematical logic

A. Church, *Introduction to Mathematical Logic*, Reprint of the second (1956) edition, Princeton Landmarks in Mathematics, Princeton Paperbacks, Princeton University Press, Princeton, NJ, 1996. x+378 pp.

This is a classic written by one of the founders of mathematical logic.

According to MathSciNet, "It is intended to be used as a textbook by students with substantial mathematical background and also, within limitations, as a reference work. In both roles it seems to succeed admirably."

From the book description, "Alonzo Church was a pioneer in the field of mathematical logic, whose contributions to number theory and the theories of algorithms and computability laid the theoretical foundations of computer science. His first Princeton book, *The Calculi of Lambda-Conversion*, 1941, established an invaluable tool that computer scientists still use today.

Even beyond the accomplishment of that book, however, his second Princeton book, *Introduction to Mathematical Logic*, defined its subject for a generation... Although new results in mathematical logic have been developed and other textbooks have been published, it remains, sixty years later, a basic source for understanding formal logic."

J. Shoenfield, *Mathematical Logic*, Reprint of the 1973 second printing, Association for Symbolic Logic, Urbana, IL; A K Peters, Ltd., Natick, MA, 2001. viii+344 pp.

This book introduces the central topics of mathematical logic and is a classical introduction for students and others.

From the book description, "This classic introduction to the main areas of mathematical logic provides the basis for a first graduate course in the subject. It embodies the viewpoint that mathematical logic is not a collection of vaguely related results, but a coherent method of attacking some of the most interesting problems, which face the mathematician. The author presents the basic concepts in an unusually clear and accessible fashion, concentrating on what he views as the central topics of mathematical logic: proof theory, model theory, recursion theory, axiomatic number theory, and set theory... This book has played a role in the education of many mature and accomplished researchers."

According to MathSciNet, "In the reviewer's opinion this is the only text in existence which is suitable for a comprehensive graduate course in mathematical logic. Like other graduate texts in mathematics, it presents non-trivial theorems and proofs in each main topic, rather than a particularly coherent point of view..."

J. van Heijenoort, *From Frege to Gödel. A Source Book in Mathematical Logic, 1879–1931*, Reprint of the third printing of the 1967 original, Edited by Jean van Heijenoort, Harvard University Press, Cambridge, MA, 2002. xii+665 pp.

This book is a very valuable source book on mathematical logic of the classical period.

According to Library Journal, "Collected here in one volume are some thirty-six high quality translations into English of the most important foreign-language works in mathematical logic, as well as articles and letters by Whitehead, Russell, Norbert Weiner and Post. This book is, in effect, the record of an important chapter in the history of thought. No serious student of logic or foundations of mathematics will want to be without it."

According to MathSciNet, "Originally published in 1967, *From Frege to Gödel* has been lauded ever since as an exemplar of what a source book should be: a compilation of seminal papers illustrating the development of a discipline, drawn from a wide variety of sources (including many not readily accessible), translated with meticulous care, and prefaced by detailed introductory notes that place the papers in historical context and provide detailed commentary on their contents. In addition, *From Frege to Gödel* introduced a concise and unobtrusive format for reference citations that, in the reviewer's opinion, is the best that has ever been devised... From Frege to Gödel made logicians aware of their heritage. That it has remained in print for thirty-five years is a testament to its lasting value to students and professionals alike."

13.1.2 D. Hofstadter, Gödel, Escher, Bach: an Eternal Golden Braid, Basic Books, Inc., Publishers, New York, 1979. xxi+777 pp.

This is hugely popular book for readers with different backgrounds. It is a book on the philosophy of mind written by a researcher in artificial intelligence. It touches many fascinating topics. It popularity was beyond the original expectation of the author.

According to MathSciNet, "The title of this large book only hints at its contents, which include such research topics as DNA and protein synthesis, Zen Buddhism, Ramanujan and the Church-Turing thesis, the behavior of ants and the sphex wasp, the development of the pianoforte, non-standard models and 'supernatural' numbers, problems of translating Dostoyevsky from the Russian, recursive and r.e. sets. What brings Gödel, Escher, and Bach together is the author's interest in 'strange loops', which transport one 'upwards (or downwards) through the levels of some hierarchical system [until] we unexpectedly find ourselves right back where we started' (p. 10). Such loops are visually realized by M. C. Escher in such disturbing lithographs as 'Drawing Hands', 'Ascending and Descending', and 'Reptiles' (reproduced, along with many others, in this volume). Musical examples can be found in the canons and fugues of J. S. Bach, perhaps most strikingly in the canon at the unison in retrograde motion and the modulating canon at the fifth from 'The Musical Offering'. And in logic there are those puzzling paradoxes of self-reference such as 'This sentence is false', which the genius of K. Gödel developed into a proof of the incompleteness of formal arithmetic and stronger systems (a result whose proof and interpretation occupies a central place in this work)..."

13.1.3 B. Russell, Introduction to Mathematical Philosophy, Reprint of the 1920 second edition, Dover Publications, Inc., New York, 1993. viii+208 pp.

This is a classic written by a master.

According to MathSciNet, "This is the Dover edition of a true classic: that which Russell devoted to giving a semipopular exposition of his logicist programme, a few years after Principia mathematica was published. The content of the book is especially useful for understanding the sense of Russell's expression 'mathematical philosophy', that is, the actual construction of mathematics from logic in such a way that the philosophical aspects of the logicist programme were much more conspicuous. So the book, originally published in 1919, was intended to be the philosophical companion to the Principia. This has, however, to be correctly understood; the book is not properly devoted to philosophical problems, with the partial exception of some paragraphs in the final chapters, but rather to showing the many interesting philosophical implications of the actual logicist construction..."

13.1.4 Set theory

Set theory is the foundation of mathematics, often in the form of Zermelo-Fraenkel set theory with the axiom of choice, and the language of set theory is used in the definitions of almost all mathematical objects. The modern study of set theory was initiated by Georg

Cantor and Richard Dedekind in the 1870s. After the discovery of paradoxes in naive set theory, numerous axiom systems were proposed in the early twentieth century, of which the Zermelo-Fraenkel axioms, with the axiom of choice, is the best-known. Beyond its foundational role, set theory is a branch of mathematics in itself.

P. Halmos, *Naive Set Theory*, Reprint of the 1960 edition, Undergraduate Texts in Mathematics, Springer-Verlag, New York-Heidelberg, 1974. vii+104 pp.

This is a classic introduction to set theory written by a known expositor.

According to MathSciNet, "In an almost conversational though not uncritical style, the author quickly proceeds from basic facts to more involved ones, by means of a great variety of instructive examples. If one disregards problems of the foundations and needs only 'some' set theory (as a basic standard tool only), this seems to be the best mode. The specific manner of exposition of the standard material may be seen from the contents... Concluding, the reviewer wants to say that the book is to be recommended as one of the best modern introductory text books of set theory."

According to an Amazon review, "This book is an excellent primer on the basics of set theory that all graduate students need, but are not necessarily obtained in the general undergraduate curriculum. Halmos writes in an abbreviated, yet effective style that imparts the necessary details without an excess of words. Theorems and exercises are very few, so it really cannot be used as a textbook. If you need a great deal of explanations, then it is not for you. However, if your need is for a book that distills the essence of set theory down to the shortest possible size, then this book should be yours, either in your college or personal library."

F.R. Drake, D. Singh, *Intermediate Set Theory*, John Wiley & Sons, Ltd., Chichester, 1996. x+234 pp.

This is a good, intermediate level textbook on set theory.

According to MathSciNet, "Every mathematician needs to speak a little set theory and those of us working in topology or set theory itself need to speak a great deal more. As stated in the preface, the authors' aim was to 'write a text which was not a first introduction to set theory (this seemed to be catered for by many undergraduate texts which introduced the elements of set theory), but would take students from that point to a point where they would be able to read the many graduate texts which set out the latest researches (or really the researches of the last thirty years or so)'... The authors have more than succeeded in achieving this aim. The book is well written and would be a valuable addition to every graduate student library."

A. Kechris, *Classical Descriptive Set Theory*, Graduate Texts in Mathematics, 156, Springer-Verlag, New York, 1995. xviii+402 pp.

This book is an excellent self-contained introduction to classical descriptive set theory and provides ideas on connections with other areas of mathematics. It is also a valuable reference.

According to MathSciNet, "The author presents a modern image of classical descriptive set theory in Polish spaces... In a very concise style the author takes the reader from the definition of a topological space through a broad spectrum of topics... A brief review preceding every chapter makes the text easier to read. Another thing making it a very good graduate textbook are the 400 exercises, many with hints at the end of the book. On the other hand a very broad scope of material, numerous applications, and the Epilogue indicating directions for further research make it an excellent reference for researchers in descriptive set theory."

Y. Moschovakis, *Descriptive Set Theory*, Second edition, Mathematical Surveys and Monographs, 155, American Mathematical Society, Providence, RI, 2009. xiv+502 pp.

This is a comprehensive book on descriptive set theory. It is an important reference.

According to MathSciNet, "This is an excellent example of the right book written at the right time by the right person. Descriptive set theory was founded around the turn of the century and enjoyed a very active life into the 1930's as a mixture of point-set topology, real analysis, and set theory. Then for over thirty years there were only sporadic results because, as we know now, the most interesting remaining problems have turned out to be unsolvable on the basis of commonly accepted mathematical principles as formulated in any standard axiomatic set theory. Logicians entered the field in the role of 'spoilers', as the author puts it, by proving that these standard techniques were insufficient, but soon found, again in the author's words, a 'friendlier role' by finding solutions to many of the old problems using new set-theoretic hypotheses. Logicians transformed the theory in another way by adding a new dimension of effectiveness using notions arising from the study of recursive functions and hierarchies. These developments enormously revitalized descriptive set theory and it again became a very active area in the late 1960s and throughout the 1970s. The book under review is a report on the status of the theory in 1980..."

T. Jech, *Set Theory*, The third millennium edition, revised and expanded, Springer Monographs in Mathematics, Springer-Verlag, Berlin, 2003. xiv+769 pp.

This is a very good introduction to set theory for experienced readers and a very valuable reference for specialists. It contains a wealth of information.

From the book description, "Set Theory has experienced a rapid development in recent years, with major advances in forcing, inner models, large cardinals and descriptive set theory. The present book covers each of these areas, giving the reader an understanding of the ideas involved."

According to an Amazon review, "The author tried to cover everything a set theorist should master plus a representative selection of topics of current interest. That makes for a lot of ground to cover, but Jech did a great job. The writing is very well organized and clear. Every short chapter has many exercises, often with hints. There are extensive sections on applications of forcing. The indexes are really good...

Because of the brevity, it is a bit hard to learn from, but it makes a great secondary reference. For example, its explanations are often clearer and more direct than in Kunen and with more detailed proofs. It you are going to have any more exposure to set theory than an introductory course, you will probably want to buy a copy."

P. Cohen, *Set Theory and the Continuum Hypothesis*, W. A. Benjamin, Inc., New York-Amsterdam, 1966 vi+154 pp.

This book gives a complete proof of the author's famous theorem on the independence of continuum hypothesis. It is a unique book written by the founder of the subject and can serve as a valuable introduction for students and reference for experts.

From the book description, "it employs intuitive explanations as well as detailed mathematical proofs in a self-contained treatment. This unique text and reference is suitable for students and professionals."

According to an Amazon review, "Paul Cohen's *Set Theory and the Continuum Hypothesis* is not only the best technical treatment of his solution to the most notorious unsolved problem in mathematics, it is the best introduction to mathematical logic (though Manin's *A Course in Mathematical Logic* is also remarkably excellent and is the first book to read after this one)."

According to MathSciNet, "The book is intended for the experienced mathematician, the novice should read it only with a good guide. This book presents a fresh and intuitive approach and it gives some glimpses into the mental process that led the author to his discoveries. The reader will find in this book just the right amount of philosophical remarks for a mathematical monograph. In some places where routine details have to be checked the author passes the buck to the reader, who would rather prefer to have the author do this part..."

K. Gödel, *The Consistency of the Continuum Hypothesis*, Annals of Mathematics Studies, no. 3, Princeton University Press, Princeton, N. J., 1940. 66 pp.

This is a classic written by the absolute authority of the modern mathematical logic. Gödel explained his solution to the famous problem on *the consistency of the continuum hypothesis* in this book. It states that there is no set whose cardinality is strictly between that of the integers and that of the real numbers Gödel showed that the continuum hypothesis cannot be disproved from the standard Zermelo-Fraenkel set theory, even if the axiom of choice is adopted (ZFC). Later Paul Cohen showed that the continuum hypothesis cannot be proven from those same axioms either. Hence, the continuum hypothesis is independent of ZFC.

S. Kleene, *Mathematical Logic*, Reprint of the 1967 original [Wiley, New York], Dover Publications, Inc., Mineola, NY, 2002. xiv+398 pp.

According to Zentralblatt MATH, "Despite the large number of textbooks and monographs devoted to this subject in the last 35 years, Kleene's book keeps some remarkable advantages with respect to simplicity and clarity making it fresh and attractive. Moreover, paying equal attention to the classical parts of mathematical logic (propositional calculus and predicate calculus) as well as to the problems of computability and decidability, so important in computer science (beside their intrinsic interest), this book can be recommended equally to students in mathematics and to those in computer science. In contrast to many books of this profile, Kleene's book is not written almost exclusively in symbols and formulas, it pays much attention to motivations, explanations and historical and philosophical comments, so it can be recommended to students in philosophy too. Obviously, some parts should have been supplemented with those important results that were obtained after the year 1966."

Y. Manin, *A Course in Mathematical Logic*, Second edition, Chapters I–III translated from the Russian by Neal Koblitz, With new chapters by Boris Zilber and the author, Graduate Texts in Mathematics, 53, Springer, New York, 2010. xviii+384 pp.

This is a nonstandard book on mathematical logic written by a mathematician. It is suitable for mathematicians since mathematical logic is presented as a part of mathematics. It is both a useful and enjoyable book.

According to MathSciNet, "As one might expect from a graduate text on logic by a very distinguished algebraic geometer, this book assumes no previous acquaintance with logic, but proceeds at a high level of mathematical sophistication."

K. Kunen, *Set Theory. An Introduction to Independence Proofs*, Reprint of the 1980 original, Studies in Logic and the Foundations of Mathematics, 102, North-Holland Publishing Co., Amsterdam, 1983. xvi+313 pp.

This book was written as an advanced introduction to set theory, in particular, includes many independence proofs after the work of P. Cohen. It is a standard textbook and is also a very useful reference.

From the book description, "Many branches of abstract mathematics have been affected by the modern independence proofs in set theory. This book provides an intro-

13.1. MATHEMATICAL LOGIC

duction to relative consistency proofs in axiomatic set theory, and is intended to be used as a text in beginning graduate courses in that subject."

According to an Amazon review, "This is the most widely used textbook for graduate-level set theory, and with good reason. Kunen manages to cover all the essentials of set theory in a quick 300 pages–and he does so with exceptional clarity and depth... The main topics in the book are constructibility (developed on the basis of an understanding of ordinal definability) and forcing, with a final chapter on iterated forcing. Loads of material can be found in the vast number of exercises which, especially in the later chapters, provide a quick survey of important results in the literature.

Kunen's style is both entertaining and precise... One can easily go back to this book year after year and expect new layers of insight to unfold. Kunen demonstrates both his mastery of set theory and mastery of the language in this superb text."

S. Shelah, *Proper and Improper Forcing*, Second edition, Perspectives in Mathematical Logic, Springer-Verlag, Berlin, 1998.

This book is a reliable reference written by a top expert.

According to MathSciNet, "Saharon Shelah's proper forcing has become a staple part of the methods of modern set theory, with its applications powerful and wide-ranging and the development of its theory a fount of continuing research. Shelah came to proper forcing in the late 1970's, and in a timely tract (the first edition of the present text) *Proper forcing*, 1982, communicated the subject to an excited set theory community. The tract in fact went far beyond its title to bring together much of Shelah's work in set theory to that time, and the wealth of information and techniques therein greatly stimulated set-theoretic research in the ensuing years."

From the book description, "The main aim of the book is thus to enable a researcher interested in an independence result of the appropriate kind, to have much of the work done for him, thus allowing him to quote general results."

13.1.5 Model theory, non-standard analysis, recursive functions

The model theory studies classes of mathematical structures such as groups, fields, graphs, universes of set theory using tools from mathematical logic. A recursive function is a function that is defined by recursion. One important application of the model theory is to the theory of non-standard analysis, which gives a solid foundation to an approach to analysis (or calculus) via a notion of an infinitesimal number, used vaguely in the initial work of Leibniz and also Newton.

C.C. Chang, H.J. Keisler, *Model Theory*, Third edition, Studies in Logic and the Foundations of Mathematics, 73, North-Holland Publishing Co., Amsterdam, 1990. xvi+650 pp.

This has been one of the most successful textbooks on model theory. It is probably out-dated.

According to an Amazon review, "this book is that it was first written in 1973, and it shows. Although this is the third edition of the book, its original structure is still largely the same. The field of model theory has changed a lot since the early 70's. For instance, in 1978 Shelah wrote a famous book that simultaneously answered many open questions in model theory and changed the direction of the whole subject, and the 1990's have seen many new applications to algebra and other areas of pure math. However, these important developments aren't reflected much in this book. The new sections added to this edition aren't exactly on the cutting edge..."

W. Hodges, *Model Theory*, Encyclopedia of Mathematics and its Applications, 42, Cambridge University Press, Cambridge, 1993. xiv+772 pp.

This is a very comprehensive book on model theory. It emphasizes constructions rather than classification, and many results and methods appeared for the first time in a book. It is an important reference.

According to MathSciNet, "As a subject, model theory is too vast to be captured in a single tome. The author well understands this. He views model theory as 'the study of the construction and classification of structures within specified classes of structures'. Any class of structures of interest to a mathematician is fair game (not necessarily just those described by first-order means). The emphasis in this book is on construction and definability. In twelve chapters and an appendix the author sets forth the basic tools that the fledgling model theorist should master. For those at the expert level he gives a shape to what has already come, some lines of current research, and intimations of what lies in the future... The text itself is rich with examples (mostly algebraic in flavor) and with the author's views as expressed in quotations from the gurus of model theory and in his own trenchant remarks."

M.D. Marker, *Model theory. An Introduction*, Graduate Texts in Mathematics, 217. Springer-Verlag, New York, 2002. viii+342 pp.

This book is an advanced introduction to model theory. It presents the basic material and illustrates how the two traditional themes of model theory interact. It is a successful textbook.

According to MathSciNet, "This is an extremely fine graduate level textbook on model theory. There is a careful selection of topics, with a route leading to a substantial treatment of Hrushovski's proof of the Mordell-Lang conjecture for function fields... There is a strong focus on the meaning of model-theoretic concepts in mathematically interesting examples."

A. Robinson, *Non-standard Analysis*, Reprint of the second (1974) edition, With a foreword by Wilhelmus A. J. Luxemburg, Princeton Landmarks in Mathematics, Princeton University Press, Princeton, NJ, 1996. xx+293 pp.

This is a classic on non-standard analysis written by its creator.

When Newton and in particular Leinniz developed calculus, they often formulated their theories using expressions such as infinitesimal number and vanishing quantity. They were intuitive and convenient. But these formulations were not rigorous and were widely criticized by many people. It was a challenging problem to develop a solid foundation to infinitesimals and such an approach to analysis. Robinson was the first person who did this in a satisfactory way. The theory of non-standard analysis has also been applied to other subjects in mathematics, for example, number theory and group theories etc. Non-standard analysis often gives more streamlined proofs, but there are some theorems which have only known proofs by non-standard analysis.

According to the book description, "Non-standard analysis grew out of Robinson's attempt to resolve the contradictions posed by infinitesimals within calculus. He introduced this new subject in a seminar at Princeton in 1960, and it remains as controversial today as it was then. This paperback reprint of the 1974 revised edition is indispensable reading for anyone interested in non-standard analysis. It treats in rich detail many areas of application..."

According to Zentralblatt MATH, "Robinson had considered his new theory of infinitesimals as an application of model theory, eventually vindicating Leibnizian differential calculus. Though ultrapowers and axiomatic approaches have made nonstandard analysis both more easily accessible and more generally applicable, the text is still unsurpassed in the originality of its ideas, in the surprising range of applications, and in its philosophy of relevant mathematics. Also, Chapter 10, concerning the history of the calculus, has influenced historiography by initiating a debate on Cauchy and his infinitesimals. Though Gödel's belief, that nonstandard analysis would be 'the analysis of the future', has not yet become true the topic is of lasting interest, and the use of the method in applications has steadily grown."

R. Soare, *Recursively Enumerable Sets and Degrees. A Study of Computable Functions and Computably Generated Sets*, Perspectives in Mathematical Logic, Springer-Verlag, Berlin, 1987. xviii+437 pp.

This book gives a coherent introduction to the current theory for the recursively enumerable sets and degrees, which is a complex and difficult area of logic. It is an accessible introduction and useful reference.

According to MathSciNet, "The book under review is an excellent and thorough treatment of that part of the subject which lies closest to its origins in the work of Gödel, Church, Kleene, and Turing in the 1930s. The focus is almost exclusively on functions and sets of natural numbers and more particularly on those sets which are either recursively enumerable or whose nature is somehow determined by recursive enumerability."

H. Rogers, *Theory of Recursive Functions and Effective Computability*, Second edition, MIT Press, Cambridge, MA, 1987. xxii+482 pp.

This has been a standard textbook on recursion theory.

According to the MathSciNet review of the first edition, "This important book deals, in a semi-formal way, with the analysis of recursively enumerable sets initiated by E. L. Post in 1944... The author gives an up to date report on solved and unsolved problems on recursively enumerable sets... In 1944 Post raised the question whether any two non-recursive, recursively enumerable sets must be of the same degree, a question which was answered in the negative independently by Friedberg and Muchnik, who showed in 1956 that there are more than two recursively enumerable degrees. The author makes an extensive study of recent work on degrees, including new concepts introduced by C. E. M. Yates...

A valuable feature of the book is the very substantial set of exercises—these are of four kinds, unmarked exercises which are straightforward tests of the readers understanding of the text, and three categories of marked examples which range from harder exercises on the text through exercises (with hints) which extend the text to more difficult problems which are sometimes discussed later in the book."

13.2 Computer science

Turing is most famous for the Turing Machine, which gives a rigorous formalization of the familiar but vague concepts of "algorithm" and "computation" and has been crucial in the development of computer science. Because of this, Turing is sometimes called the father of computer science and artificial intelligence. But his personal life was a tragedy. One story says that when he visited USA, he watched the Disney movie "Snow White" several dozen times. When Turing tried to commit suicide, he put poison into an apple and took a bit of the apple, imitating the death of Snow White. The story continues that, because of this, the Apple Computer uses a bitten apple as their famous logo.

Figure 13.2: Turing's book.

13.2. COMPUTER SCIENCE 595

13.2.1 D. Knuth, The Art of Computer Programming. Vol. 1-IV, Second printing, Addison-Wesley Publishing Co., 1969. xxi+634 pp.

This is the "Bible" on computer programming written by the leading expert. It contains the most important results in computer science and explains them with mathematical rigor. It is a valuable book for everyone who is interested in algorithms and programming.

Knuth has been called the "father" of the analysis of algorithms and contributed to the development of the rigorous analysis of the computational complexity of algorithms and systematized formal mathematical techniques for it. In addition to his fundamental contributions to theoretical computer science, Knuth is the creator of the TeX computer typesetting system, the related METAFONT font definition language and rendering system, and the Computer Modern family of typefaces.

Publisher's description: "This multivolume work on the analysis of algorithms has long been recognized as the definitive description of classical computer science. The three complete volumes published to date already comprise a unique and invaluable resource in programming theory and practice. Countless readers have spoken about the profound personal influence of Knuth's writings. Scientists have marveled at the beauty and elegance of his analysis, while practicing programmers have successfully applied his 'cookbook' solutions to their day-to-day problems. All have admired Knuth for the breadth, clarity, accuracy, and good humor found in his books.

13.2.2 R. Graham, D. Knuth, O. Patashnik, Concrete Mathematics. A Foundation for Computer Science, Second edition, Addison-Wesley Publishing Company, Reading, MA, 1994. xiv+657 pp.

This book is not a traditional one on discrete mathematics, rather a compendium of some of the basic techniques for solving the very specific problems that arise in all branches of the mathematical sciences. It is a very valuable reference.

From the book description, "The primary aim of its well-known authors is to provide a solid and relevant base of mathematical skills - the skills needed to solve complex problems, to evaluate horrendous sums, and to discover subtle patterns in data. It is an indispensable text and reference not only for computer scientists - the authors themselves rely heavily on it! - but for serious users of mathematics in virtually every discipline... The subject matter is primarily an expansion of the Mathematical Preliminaries section in Knuth's classic *Art of Computer Programming*, but the style of presentation is more leisurely, and individual topics are covered more deeply. Several new topics have been added, and the most significant ideas have been traced to their historical roots. The book includes more than 500 exercises, divided into six categories. Complete answers are provided for all exercises, except research problems, making the book particularly valuable for self-study... Readers will appreciate the informal style of Concrete Mathematics. Particularly enjoyable are the marginal graffiti contributed by students who have taken courses based on this material. The authors want to convey not only the importance of the techniques presented, but some of the fun in learning and using them."

According to MathSciNet, "this book contains precisely the kind of (elementary) mathematics that a student will need before successfully delving into the subject of 'analysis of algorithms', as exercised in Knuth's *The art of computer programming*."

13.2.3 T. Cover, J. Thomas, Elements of Information Theory, Second edition, Wiley-Interscience [John Wiley & Sons], Hoboken, NJ, 2006. xxiv+748 pp.

This book was written as "a simple and accessible book on information theory". This goal has been achieved, probably the original expectation. It has also turned out to be a very valuable reference.

From the book description, "The Second Edition of this fundamental textbook maintains the book's tradition of clear, thought-provoking instruction. Readers are provided once again with an instructive mix of mathematics, physics, statistics, and information theory.

All the essential topics in information theory are covered in detail... The authors provide readers with a solid understanding of the underlying theory and applications. Problem sets and a telegraphic summary at the end of each chapter further assist readers. The historical notes that follow each chapter recap the main points."

According to MathSciNet, "This book was a pleasure to read. It is written as a textbook for a senior undergraduate and first-year graduate course in information theory. Thus most of the material has already appeared in book form or in the research literature. However, the presentation is frequently novel, with new or improved versions of proofs and with new connections made between different topics. There are also new results at several points in the book.

The intended audience includes students of communication theory, computer science and statistics... The expository style is suitable for the broad audience at which the book is directed. There is frequently a heuristic development or a special case which is designed to develop intuition, followed by careful statements and proofs of theorems. The coverage of recent results, especially in the developing area of network information theory, is quite good, with descriptions of results published as late as 1989."

I. Csiszár, J. Körner, *Information Theory. Coding Theorems for Discrete Memoryless Systems*, Second edition, Cambridge University Press, Cambridge, 2011. xxii+499 pp.

13.2. COMPUTER SCIENCE 597

This book gives a comprehensive, rigorous and unified account of information theory. It has been a very valuable book for both graduate students and experts, especially for those people who are seriously interested in Shannon theory. It has some impact on the development of mathematical information theory.

According to MathSciNet, "This is an excellent mathematical book on information theory, extremely well-written, covering a relatively restricted area (coding theorems for discrete memoryless systems) but going very deeply, including the most important developments of the last decade, and giving rigorous proofs. The authors analyse only discrete memoryless communication systems; this allows them to use a unified basically combinatorial approach which leads to stronger results and simpler proofs."

From the book description, "It includes in-depth coverage of the mathematics of reliable information transmission, both in two-terminal and multi-terminal network scenarios... It is an ideal resource for graduate students and researchers in electrical and electronic engineering, computer science and applied mathematics."

M. Nielsen, I. Chuang, *Quantum Computation and Quantum Information*, Cambridge University Press, Cambridge, 2000. xxvi+676 pp.

This is first comprehensive introduction to quantum computation and quantum information. It has been a standard textbook on the subject.

From the book description, "This comprehensive textbook describes such remarkable effects as fast quantum algorithms, quantum teleportation, quantum cryptography and quantum error-correction. Quantum mechanics and computer science are introduced before moving on to describe what a quantum computer is, how it can be used to solve problems faster than 'classical' computers and its real-world implementation. It concludes with an in-depth treatment of quantum information. Containing a wealth of figures and exercises, this well-known textbook is ideal for courses on the subject, and will interest beginning graduate students and researchers in physics, computer science, mathematics, and electrical engineering."

13.2.4 N. Wiener, Cybernetics, or Control and Communication in the Animal and the Machine, Actualités Sci. Ind., no. 1053, Hermann et Cie., Paris; The Technology Press, Cambridge, Mass.; John Wiley & Sons, Inc., New York, 1948. 194 pp.

This is a classic written by a master. It has had a huge impact on the development of many branches of science.

According to The Saturday Review of Literature, "It appears impossible for anyone seriously interested in our civilization to ignore this book. It is a 'must' book for those in every branch of science... in addition, economists, politicians, statesmen, and businessmen cannot afford to overlook cybernetics and its tremendous, even terrifying implications. It is a beautifully written book, lucid, direct, and despite its complexity, as readable by the layman as the trained scientist."

According to MathSciNet of the first edition, "The author has undertaken the ambitious program of making a survey of the field of control and communication (the word cybernetics is taken from a Greek word meaning steersman) from the point of view of modern man and society."

For the second edition, "The Second World War focused the attention of many scientists and technologists upon problems of the design of communication systems and servomechanisms. That situation prompted the author of the book under review to make mathematico-philosophical investigations of broadness and depth into the subject of communication and control. The outcome of that inquiry became the first edition of the book... Since then there have been many changes in cybernetics..."

N. Wiener, *Extrapolation, Interpolation, and Smoothing of Stationary Time Series. With Engineering Applications*, The Technology Press of the Massachusetts Institute of Technology, Cambridge, Mass; John Wiley & Sons, Inc., New York, N. Y.; Chapman & Hall, Ltd., London, 1949. ix+163 pp.

This is a famous classic written by Wiener. It gives the first attempt to combine the theory of statistical time series with communicating engineering, and it has succeeded well. As explained in the introduction, "This book represents an attempt to unite the theory and practice of two fields of work which are of vital importance in the present emergency, and which have a complete natural methodological unity, but which have up to the present drawn their inspiration from two entirely distinct traditions, and which are widely different in their vocabulary and the training of their personnel. These two fields are those of time series in statistics and of communication engineering."

The book description gives a good summary, "It has been the opinion of many that Wiener will be remembered for his Extrapolation long after Cybernetics is forgotten. Indeed few computer-science students would know today what cybernetics is all about, while every communication student knows what Wiener's filter is. The work was circulated as a classified memorandum in 1942, as it was connected with sensitive war-time efforts to improve radar communication. This book became the basis for modern communication theory, by a scientist considered one of the founders of the field of artificial intelligence. Combining ideas from statistics and time-series analysis, Wiener used Gauss's method of shaping the characteristic of a detector to allow for the maximal recognition of signals in the presence of noise. This method came to be known as the 'Wiener filter'."

C. Shannon, W. Weaver, *The Mathematical Theory of Communication*, The University of Illinois Press, Urbana, Ill., 1949. vi+117 pp.

13.2. COMPUTER SCIENCE

This is a classic written by a master and has had a huge impact on the theory of communication and information.

From the book description, "Scientific knowledge grows at a phenomenal pace-but few books have had as lasting an impact or played as important a role in our modern world as 'The Mathematical Theory of Communication', published originally as a paper on communication theory in the 'Bell System Technical Journal' more than fifty years ago. Republished in book form shortly thereafter, it has since gone through four hardcover and sixteen paperback printings. It is a revolutionary work, astounding in its foresight and contemporaneity..."

According to Scientific American, "Before this there was no universal way of measuring the complexities of messages or the capabilities of circuits to transmit them. Shannon gave us a mathematical way... invaluable ... to scientists and engineers the world over."

According to an Amazon review, "While being referenced in many courses and textbooks, few have read it unfortunately. This is not the kind of book that will change your life but it is amongst the ones that are part of the CULTURE of anyone far or less involved in communication theory.

The content is certainly very conceptual but it provides a different view of what information is. In this world where content is king, it will refresh your notion of syntax and semantics, and the difference between just words and the information that lies within them.

Even if it is quite small, it's not the book you'll read from the beginning to the end without a stop. It is very deep and has profound implications on everyday's computer scientist's life."

13.2.5 M. Petkovsek, H. Wilf, D. Zeilberger, $A = B$, With a foreword by Donald E. Knuth, With a separately available computer disk, A K Peters, Ltd., Wellesley, MA, 1996. xii+212 pp.

This is a very useful source and reference book. According to Zentralblatt MATH, "This book is an essential resource for anyone who ever encounters binomial coefficient identities, for anyone who is interested in how computers are being used to discover and prove mathematical identities, and for anyone who simply enjoys a well-written book that presents interesting, cutting edge mathematics in an accessible style."

From the book description, "This book is of interest to mathematicians and computer scientists working in finite mathematics and combinatorics. It presents a breakthrough method for analyzing complex summations. Beautifully written, the book contains practical applications as well as conceptual developments that will have applications in other areas of mathematics."

According to MathSciNet, "Summarizing, the authors not only skillfully explained their methods to a computer, they also did a truly outstanding job in explaining these recent developments to a broad audience ranging from students to researchers. In particular, this book is a must for all those who at least once have struggled with a binomial sum."

According to an Amazon review, "What the authors have done is to discover a way of using computer algebra and some mathematical ideas they and others had earlier to make it possible to find the sum of most of the hypergeometric and basic hypergeometric series which can be summed. This method also leads to recurrence relations when polynomial series cannot be summed. It does not work for all series, but most of the time it does work. This book and the ideas in it should be part of the working knowledge of anyone who uses special functions of hypergeometric or basic hypergeometric type."

13.2.6 I. MacWilliams, N. Sloane, The Theory of Error-correcting Codes, I, II, North-Holland Mathematical Library, Vol. 16. North-Holland Publishing Co., Amsterdam-New York-Oxford, 1977. pp. i–xv and 1–369, pp. i–ix and 370–762.

This is a very successful textbook for students and a reference for researchers on the theory of error-correcting codes.

According to a review in Bulletin of AMS, "The first few sentences of the preface are as follows: 'Coding theory began in the late 1940s with the work of Golay, Hamming and Shannon. Although it has its origins in an engineering problem, the subject has developed by using more and more sophisticated mathematical techniques. It is our goal to present the theory of error-correcting codes in a simple, easily understandable manner, and yet also to cover all the important aspects of the subject.' The authors have been eminently successful in attaining their goal. For this reason these volumes are excellent as a text. They are also excellent as a reference for people working in coding as well as other mathematicians who are interested in applications of algebra or combinatorics or just interested in this new, fascinating subject. Since Shannon first demonstrated, using probabilistic methods, that one could communicate as reliably as desired by using long enough error-correcting codes, much work has gone into this subject by both mathematicians and electrical engineers. This has resulted in the construction and analysis of various codes and families of codes and the devising of practical decoding algorithms. It has also resulted in a growing mathematical theory of error-correcting codes which uses techniques from a variety of different areas as well as its own techniques. Due to the great activity in coding in the last thirty years and the exciting nature of the developments, there is now a great body of material on this subject. Even though there are other excellent books on coding, much appears here in book-form for the first time... These volumes are very well-written with many nice exercises and open research problems."

13.2.7 More books on coding theory

Codes are used for data compression, cryptography, error-correction and network coding, and the coding theory studies properties of codes and their fitness for specific applications. There are essentially two aspects of the coding theory: (1) data compression, (2) error correction.

J. van Lint, *Introduction to Coding Theory*, Third edition, Graduate Texts in Mathematics, 86, Springer-Verlag, Berlin, 1999. xiv+227 pp.

This is a very well-written introduction to coding theory for mathematics students and mathematicians.

According to MathSciNet, "This is a concise, self-contained and neat introduction to the subject of coding theory suitable for students of mathematics... The highly concise style and the excellent organization of the material make this book a useful and handy quick reference for the expert as well as a textbook suitable for an intermediate to advanced teacher-assisted course on coding theory."

H. Stichtenoth, *Algebraic Function Fields and Codes*, Second edition, Graduate Texts in Mathematics, 254, Springer-Verlag, Berlin, 2009. xiv+355 pp.

This book gives an excellent self-contained introduction to the theory of algebraic curves from an algebraic point of view with applications to coding theory. It is carefully written and contains many illustrative examples.

According to MathSciNet, "This book presents self-contained introductions to algebraic function fields in one variable and to algebraic-geometric (Goppa) codes... The language used is that of places, adéles, and Weil differentials, although divisors here are written additively..."

T. Richardson, R. Urbanke, *Modern Coding Theory*, Cambridge University Press, Cambridge, 2008. xvi+572 pp.

This book gives an up-to-dated account of iterative coding, and is a very useful reference for researchers and practitioners in coding theory. It is also a good advanced introduction for graduate students.

According to MathSciNet, "This book is devoted to the study of iterative channel decoding. Covering the most recent advances, this book is an important resource for researchers in iterative codes (including LDPC codes) and turbo codes at universities and communications companies."

Error correction coding allows the detection and correction of errors occurring during the transmission of data in digital communication systems. Iterative techniques have revolutionized the theory and practice of error correction coding and have been adopted in the majority of next-generation communications standards. Originally, iterative channel decoding was conceived by Gallager in 1960, and was rediscovered in the mid-1990s by several groups of researchers."

According to an Amazon review, "This book is your 'one stop shop' to learn all about low-density parity-check (LDPC) codes and iterative decoding. This seems to be the primary intention of the authors since they do not talk about anything that isn't related to LDPC codes or iterative decoding. This book is not meant as an introductory course in coding theory. It is very good for an advanced course in iterative decoding and as a good reference material for research."

13.2.8 Algorithm and automata

An algorithm is an effective method expressed as a finite list of well-defined instructions for calculating a function. Algorithms are used for calculation, data processing, and automated reasoning.

Automata theory is the study of abstract machines, or rather mathematical machines or systems, and the computational problems that can be solved using these machines. These abstract machines are called automata.

M. Garey, D. Johnson, *Computers and Intractability*, A guide to the theory of NP-completeness, A Series of Books in the Mathematical Sciences, W. H. Freeman and Co., San Francisco, Calif., 1979. x+338 pp.

This book provides an excellent guide to the theory of NP-completeness with some emphasis on applications to practical problems.

According to MathSciNet, "There are a variety of computational problems, which are strongly suspected to be intractable, for which there is no proof that they are intractable, but which are known to be equivalent from the point of view of complexity (i.e., if one of them is tractable, then any of them is). (The famous $P = NP$ problem is, essentially, the question of whether one (= each) of these problems has an efficient algorithm.) These problems are said to be NP-complete. It is often quite simple to prove that a problem is NP-complete; even in case a more sophisticated construction is needed, almost no knowledge of the theory of computational complexity is required...

The book consists of two basic parts: a theoretical section, and a list of more than 300 NP-hard problems (i.e., problems at least as hard as NP-complete)... The theoretical part is written as a textbook for algorithm designers who once in a while come across an intractable problem (which usually turns out to be NP-complete)... Summing up, the book is extremely interesting both for people 'employed in the halls of industry', and for people proving NP-completeness or only musing about whether $P = NP$."

According to an Amazon review, "It is amazing how, after all these years, this book remains a fundamental one to be introduced on what can be effectively and efficiently solved by computers and above all on what it seems not efficiently solvable, independently of the advances of technology. Other texts have been published after that one... Nevertheless, the Garey-Johnson book remains the fundamental book for a clear introduction to this central problem of what is tractable by computers."

T. Cormen, C. Leiserson, R. Rivest, *Introduction to Algorithms*, Third edition, MIT Press, Cambridge, MA, 2009. xx+1292 pp.

The book gives an accessible and in-depth account of a broad range of algorithms. It is a standard reference for experts and a textbook for students in the field.

According to MathSciNet, "This is the most readable, accurate, comprehensive, and voluminous textbook available today on the subject of algorithm design and analysis... And while the book's presentation style is quite traditional, its breadth of coverage puts it in a class by itself. Almost all of the most useful algorithms can be found in the book, making it a valuable reference for undergraduate and graduate students, as well as practitioners."

S.A. Eilenberg, *Automata, Languages, and Machines. Vol. A., Vol. B.*, Pure and Applied Mathematics, Vol. 58, 59, Academic Press, New York, 1974. xvi+451 pp., 1976. xiii+387 pp.

This book gives a systematic account of automata and formal language theory by emphasizing the mathematical aspect. It was written for both pure mathematicians and computer scientists. It is an important reference.

According to MathSciNet, "The work includes a unifying mathematical presentation of almost all major topics of automata and formal language theory. Volume A contains a study of rational structure, that is of objects (sets, functions, relations) that can be recognized by finite state devices. The second volume is dedicated to a profound algebraic study of objects which occur in the realm of rational structures."

J. Hopcroft, J. Ullman, *Introduction to Automata Theory, Languages, and Computation*, Addison-Wesley Series in Computer Science, Addison-Wesley Publishing Co., Reading, Mass., 1979. x+418 pp.

This book is a classical reference and a textbook on theoretical computer science with the main emphasis on automata theory, formal language theory and the theory of computation.

According to the author, the selection criteria for the material in this book is: "Rather than attempt to be encyclopedic, we have been brutal in our editing of the material, selecting only topics central to the theoretical development of the field or with importance to engineering applications."

C. Papadimitriou, *Computational Complexity*, Addison-Wesley Publishing Company, Reading, MA, 1994. xvi+523 pp.

This is a successful advanced introduction to computational complexity for beginners. It is also a valuable reference for experts.

According to MathSciNet, "The area of computational complexity deals with the study of the existence of efficient algorithms... The current book, although it contains a few algorithms, is mostly about the rest of computational complexity... In other words, it treats topics such as models of computation and relations between different complexity measures (such as time, space and nondeterminism) and also considers different complexity classes... The number of books treating computational complexity is very small and a new book is very welcome.

The book contains a very wide range of subjects and treats many to a depth that takes one almost to current research questions. In addition to the text of the book, each section contains further results (often as exercises), references, and pointers to the literature. Thus anybody mastering the entire book will obtain a very good grasp of contemporary complexity theory... as an introductory book I think the book is only suitable for motivated and mathematically mature readers. The average computer science graduate students who just want an introduction to the theory of computation might have a hard time.

I think, however, that the book is excellent for a second course which goes more in depth. In such a case the second half of the book would be used. Almost everything one would like to discuss in such a course is included... It is an excellent and much needed book for an advanced course in computational complexity, while it might be a bit tough for an introductory course. It will also be of great value as a reference for researchers."

13.3 Game theory and optimization

Game theory is a mathematical method for analyzing calculated circumstances, such as in games, where a person's success is based upon the choices of others. More formally, it studies mathematical models of conflict and cooperation between intelligent rational decision-makers. Game theory is mainly used in economics, political science, and psychology, and other, more prescribed sciences, like logic or biology.

Optimization, or mathematical programming, studies the problem on how to select the best elements from some set of available alternatives under the given criterion and conditions. In the simplest case, an optimization problem consists of maximizing or minimizing a real function, and this can be solved by systematically choosing input

13.3. GAME THEORY AND OPTIMIZATION

values from within an allowed set and computing the value of the function. Optimization theory and its generalizations form an important subject in applied mathematics and have broad applications in sciences.

Fermat made important contributions to the development of calculus and was one of the first people who studied the problem of maximum and minimum values of a function. In Methodus ad disquirendam maximam et minima *and in* De tangentibus linearum curvarum, *he developed a method for determining maxima, minima, and tangents to various curves, which was equivalent to determining critical points by differentiation. Fermat's principle of least time, which was used by him to derive Snell's law of refraction in 1657, was the first variational principle in physics since a principle of least distance was stated in the first century CE by Hero of Alexandria, who was ancient Greek mathematician and engineer and is considered the greatest experimenter of antiquity and his work represented the Hellenistic scientific tradition.*

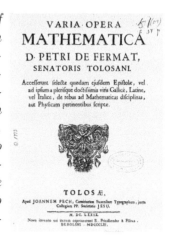

Figure 13.3: Fermat's book.

13.3.1 J. von Neumann, O. Morgenstern, Theory of Games and Economic Behavior, Fourth printing of the 2004 sixtieth-anniversary edition, With an introduction by Harold W. Kuhn and an afterword by Ariel Rubinstein, Princeton University Press, Princeton, NJ, 2007. xxxii+739 pp.

This book is a classic written by founders of the subjects. It has had a huge impact on the game theory.

From the book description, "This is the classic work upon which modern-day game theory is based. What began more than sixty years ago as a modest proposal that a mathematician and an economist write a short paper together blossomed, in 1944, when Princeton University Press published *Theory of Games and Economic Behavior.* In it, John von Neumann and Oskar Morgenstern conceived a groundbreaking mathematical theory of economic and social organization, based on a theory of games of strategy. Not only would this revolutionize economics, but the entirely new field of scientific inquiry it yielded–game theory–has since been widely used to analyze a host of real-world phenomena from arms races to optimal policy choices of presidential candidates, from vaccination policy to major league baseball salary negotiations. And it is today established throughout both the social sciences and a wide range of other sciences."

According to MathSciNet, "Although Cournot's analysis is more general than Waldegrave's, game theory did not really exist as a unique field until John von Neumann published a series of papers in 1928. While the French mathematician Emile Borel did some earlier work on games, Von Neumann can rightfully be credited as the inventor of game theory... Von Neumann's work in game theory culminated in the 1944 book *Theory of Games and Economic Behavior*. This profound work contains the method for finding mutually consistent solutions for two-person zero-sum games. During this time period, work on game theory was primarily focused on cooperative game theory, which analyzes optimal strategies for groups of individuals, presuming that they can enforce agreements between them about proper strategies."

According to a review of Bulletin of AMS, "Posterity may regard this book as one of the major scientific achievements of the first half of the twentieth century. This will undoubtedly be the case if the authors have succeeded in establishing a new exact science—the science of economics. The foundation which they have laid is extremely promising. Since both mathematicians and economists will be needed for the further development of the theory it is in order to comment on the background necessary for reading the book. The mathematics required beyond algebra and analytic geometry is developed in the book. On the other hand the non-mathematically trained reader will be called upon to exercise a high degree of patience if he is to comprehend the theory. The mathematically trained reader will find the reasoning stimulating and challenging. As to economics, a limited background is sufficient."

13.3.2 More books on game theory and optimization

A. Schrijver, *Theory of Linear and Integer Programming*, Wiley-Interscience Series in Discrete Mathematics, Wiley-Interscience Publication, John Wiley & Sons, Ltd., Chichester, 1986. xii+471 pp.

This book gives a comprehensive account of theory of linear and integer programming, emphasizing the more theoretic aspects. It is a very solid and reliable introduction and reference.

According to MathSciNet, "Many important books exist on linear and integer linear programming. Consequently a prospective reader would like to know what is special about the book under review. The author says: 'The emphasis of this book is on the more theoretical aspects, and it aims at complementing the more practically oriented books.' This is a suitable characterization, but the reviewer thinks we have at hand a more comprehensive publication. The author has written the book of linear and integer programming. A reference work has now appeared. If we have a question in the field of linear and integer programming, this book will give an answer or will show us the way to an answer. The author describes new connections between several mathematical

13.3. GAME THEORY AND OPTIMIZATION

questions, proves many interesting results, mentions lots of surprising historical facts, and leads us through the theory and algorithms of linear and integer linear programming... This encyclopedic book is distinguished by its high scientific level and its concentrated manner."

J. Hiriart-Urruty, C. Lemaréchal, *Convex Analysis and Minimization Algorithms. I. Fundamentals., II. Advanced theory and bundle methods*, Grundlehren der Mathematischen Wissenschaften, 305, 306, Springer-Verlag, Berlin, 1993. xviii+417 pp., xviii+346 pp.

This book gives a very accessible account of the fundamentals of convex analysis in finite-dimensional vector spaces. It is carefully written and beautifully illustrated. It is a very valuable introduction and reference.

According to MathSciNet, "The first volume is an excellent exposition of finite-dimensional convex analysis and convex optimization. The book is very well written, nicely illustrated, and clearly understandable even for senior undergraduate students of mathematics as well as for engineers, economists, applied scientists, etc. On the other hand, this presentation is undoubtedly interesting and useful for experts in convex analysis, optimization, and related fields. Throughout the book, the authors carefully follow the recommendation by A. Einstein: 'Everything should be made as simple as possible, but not simpler.' The two major topics in the title are distinct but highly interrelated, as is demonstrated by this presentation. The authors mainly consider variational aspects of convex analysis and its relationship with continuous optimization... [The second volume] contains an excellent exposition of advanced numerical algorithms in convex optimization based on classical and new developments in convex analysis."

J.H. Conway, *On Numbers and Games*, Second edition, A K Peters, Ltd., Natick, MA, 2001. xii+242 pp.

This is an unusual book written by a very original, or unusual, mathematician.

From the book description, it "is one of those rare publications that sprang to life in a moment of creative energy and has remained influential for over a quarter of a century. Originally written to define the relation between the theories of transfinite numbers and mathematical games, the resulting work is a mathematically sophisticated but eminently enjoyable guide to game theory. By defining numbers as the strengths of positions in certain games, the author arrives at a new class, the surreal numbers, that includes both real numbers and ordinal numbers. These surreal numbers are applied in the author's mathematical analysis of game strategies. The additions to the Second Edition present recent developments in the area of mathematical game theory, with a concentration on surreal numbers and the additive theory of partizan games."

According to MathSciNet, "Overall, this book is a momentous addition to the mathematical literature: a new, exciting, and highly original theory is expounded by its creator in a style that is at once concise, literate, and delightfully whimsical. The new theory of real numbers, in particular, is a profound and revolutionary contribution to the foundations of analysis."

S. Boyd, L. Vandenberghe, *Convex Optimization*, Cambridge University Press, Cambridge, 2004. xiv+716 pp.

This book is an accessible introduction to theory of convex optimization and could be the first book one starts with.

According to MathSciNet, "During the last decade we were blessed with many books on convex optimization. Some of them present the main results and methods of convex optimization from interesting, new viewpoints. Others focus on specific topics and go deep into the theory and applications of various classes of methods for solving convex programming problems. Most of them are well written treatises addressed to readers able to handle advanced mathematical tools. Few of those books are really intended to be text books for the uninitiated in the subtleties of the theory and applications of mathematical programming and even fewer are written with the economy of mathematical pre-requisites which will make them accessible to practitioners without advanced mathematical training. The book by Boyd and Vandenberghe reviewed here is one of those few and, in this class, the best I have ever seen. It complements two other outstanding text books of convex optimization: that of D. P. Bertsekas, *Nonlinear programming*, 1999, which is accessible to readers with undergraduate level of mathematical training in analysis and algebra, and that of A. Ben-Tal and A. Nemirovskii, *Lectures on modern convex optimization*, 2001, which is addressed to postgraduate students and researchers 'in possession of the basic elements of mathematical culture'. These three books represent the three stairs the student should climb in order to enter the world of the modern theory of optimization and learn the basics of the art of modelling and solving practical optimization problems. The book under review should be the first the student will climb. It is a gentle, but rigorous, introduction to the basic concepts and methods of the field..."

E. Berlekamp, J. Conway, R. Guy, *Winning Ways for Your Mathematical Plays. Vol. 1. Games in General, Vol. 2. Games in Particular, Vol. 3, Vol. 4*, Academic Press, Inc., London-New York, 1982. xxxi+426+xi pp., pp. i–xxxiii and 429–850 and i–xix.; A K Peters, Ltd., Natick, MA, 2003. pp. i–xxii and 461–801, 2004. pp. i–xvi and 803–1004.

These volumes contain a wealth of material on mathematical games, and are valuable introduction and reference on a mathematical approach to playing games. This four volume set is an expansion of the classic two-volume book on combinatorial games which was published in 1982.

According to MathSciNet, they "are crammed to the brim with information, colored illustrations and examples...

Each chapter ends with a section called 'Extras', where underlying principles or additional details are given. Instead of formal proofs, short convincing arguments or examples are provided. This tends to increase considerably the amount of material packed in the 850 pages of the book. Together with the wit, humour and originality of approach, it also increases the readability or apparent readability. To really understand and prove everything in the book, not to mention to attempt solutions of the many questions inspired on every page of the book, will engage many people for many years.

As the authors state, the book is not an encyclopedia, since there are many games, theories and puzzles not included in it. In fact, there are certain directions not pursued in the book, such as transpolynomiality of games or questions of undecidability or computability of strategies. The thrust of the book lies in the direction of formulating exact or suboptimal polynomial strategies for very broad classes of combinatorial games."

J. Nocedal, S. Wright, *Numerical Optimization*, Second edition, Springer Series in Operations Research and Financial Engineering, Springer, New York, 2006. xxii+664 pp.

This book gives a comprehensive description of the effective methods in continuous optimization. It was written in response to the growing interest in optimization in engineering, science and business by focusing on methods that are best suited for practical applications.

According to MathSciNet, "I find the book under review to be a well-written treatment of continuous nonlinear optimization and I would recommend its use for upper level undergraduate or graduate level courses in nonlinear optimization. The book is sufficiently detailed to be useful to researchers, but its real merit is as an educational resource..."

M. Grötschel, L. Lovász, A. Schrijver, *Geometric Algorithms and Combinatorial Optimization. Algorithms and Combinatorics: Study and Research Texts, 2*, Springer-Verlag, Berlin, 1988. xii+362 pp.

This book gives a detailed account of two recent powerful geometric techniques in combinatorial optimization, ellipsoid method and basis reduction. It has been an essential introduction and reference for people interested in fundamental complexity issues of mathematical programming.

From the book description, "This book develops geometric techniques for proving the polynomial time solvability of problems in convexity theory, geometry, and - in particular - combinatorial optimization. It offers a unifying approach based on two fundamental geometric algorithms: - the ellipsoid method for finding a point in a convex set and - the basis reduction method for point lattices. The ellipsoid method was used by Khachiyan to show the polynomial time solvability of linear programming. The basis reduction method yields a polynomial time procedure for certain diophantine approximation problems. A combination of these techniques makes it possible to show the polynomial time solvability of many questions concerning polyhedra - for instance, of linear programming problems having possibly exponentially many inequalities. Utilizing results from polyhedral combinatorics, it provides short proofs of the polynomial time solvability of many combinatorial optimization problems. For a number of these problems, the geometric algorithms discussed in this book are the only techniques known to derive polynomial time solvability. This book is a continuation and extension of previous research of the authors for which they received the Fulkerson Prize, awarded by the Mathematical Programming Society and the American Mathematical Society."

L. Cesari, *Optimization—Theory and Applications. Problems with Ordinary Differential Equations*, Applications of Mathematics (New York), 17, Springer-Verlag, New York, 1983. xiv+542 pp.

This book gives a systematic and modern account of the calculus of variations for problems governed by ordinary differential equations and some applications to engineering and economics. It is a valuable introduction and reference.

According to MathSciNet, "This book is an encyclopedic treatment dedicated to the exposition of both the classical and modern theories... It is devoted to problems in one independent variable, and in particular to nonparametric problems ... Among the books available on the current subject, this work is unique both in scope and scholarly approach. The vast bibliography and the chapter-by-chapter bibliographical comments are invaluable. Equally useful is the large number of illustrative examples, used quickly in one place to set off a detail of proof, or more extensively elsewhere to apply the theory."

13.4 Numerical analysis and matrix computation

Numerical analysis studies algorithms that produce numerical approximation to the problems in mathematical analysis when exact solutions are difficult or impossible to be obtained by symbolic manipulations. One of the earliest mathematical writings, the Babylonian tablet BC 7289, gives a numerical approximation of $\sqrt{2}$, the length of the diagonal in a unit square. Numerical analysis continues this long tradition of practical mathematical calculations, and is concerned with obtaining approximate solutions while maintaining reasonable bounds on errors. Numerical analysis naturally finds applications in all fields of engineering and the physical sciences, the life sciences and social sciences.

13.4. NUMERICAL ANALYSIS AND MATRIX COMPUTATION

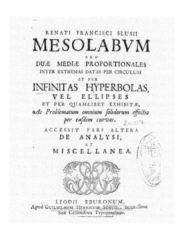

Figures 13.4: Newton's book and Euler's book.

To most students, the first numerical method is Newton's method to find approximate values of roots of a real valued function. It was first described by Isaac Newton in De analysi per aequationes numero terminorum infinitas, *which was written in 1669, published in 1711 by William Jones, and in* De metodis fluxionum et serierum infinitarum, *which was written in 1671, translated and published as Method of Fluxions in 1736 by John Colson. The other basic numerical method is Euler's method for solutions of ordinary differential equations, which was first introduced in his book* Institutionum calculi integralis (Foundations of differential calculus).

13.4.1 Numerical analysis

F. Acton, *Numerical Methods that Work*, Corrected reprint of the 1970 edition, Mathematical Association of America, Washington, DC, 1990. xx+549 pp.

This is an accessible and fun-to-read introduction to numerical analysis.

From the introduction: "This book discusses efficient numerical methods for the solution of equations – algebraic, transcendental, and differential – assuming that an electronic computer is available to perform the bulk of the arithmetic. I wrote it for upper-class students in engineering and the physical sciences. More importantly, I wrote it for students whose motivations lie in the physical world, who would get answers to 'real' problems. This intended audience shapes both the content of the book and its expository method... Finally, pedagogical expedience strongly suggests that the teaching of programming of digital computers be clearly separated from the teaching of numerical methods for solving problems, at least at the elementary levels."

From the book description, "Acton deals with a commonsense approach to numerical algorithms for the solution of equations: algebraic, transcendental, and differential. He assumes that a computer is available for performing the bulk of the arithmetic. The book is written with clarity and precision, intended for practical rather than theoretical use. This book will interest mathematicians, both pure and applied, as well as any scientist or engineer working with numerical problems."

According to Korner, "At the highest level, numerical analysis is a mixture of science, art and bar-room brawl. The most that a general mathematical course can hope to do for students in this direction is to make them 'educated consumers', able to make sensible use of package programs and able, if the need arises, to cobble together something that will work as it is supposed to (even if rather more slowly than an expert's version). Acton's book is an excellent, relaxed and good-humored introduction to the problems and tools of the numerical analyst." "

A. Björck, G. Dahlquist, *Numerical Methods*, Translated from the Swedish by Ned Anderson, Reprint of the 1974 English translation. Dover Publications, Inc., Mineola, NY, 2003. xviii+573 pp.

This is a valuable and comprehensive textbook. It is well-balanced between theoretical treatment and practical applications.

According to MathSciNet, "This is a substantial, detailed and rigorous textbook of numerical analysis, in which an excellent balance is struck between the theory, on the one hand, and the needs of practitioners (i.e., the selection of the best methods—for both large-scale and small-scale computing) on the other. The prerequisites are slight (calculus and linear algebra and preferably some acquaintance with computer programming) so that some of the finer theoretical points (those at which numerical analysis becomes applied functional analysis, for example) are outside the scope of the book. However, the class of readers for whom the book is intended are admirably served. "

J.H. Wilkinson, *The Algebraic Eigenvalue Problem*, Monographs on Numerical Analysis, Oxford Science Publications, The Clarendon Press, Oxford University Press, New York, 1988. xviii+662 pp.

This book gives a good summary of a basic and mature subject and is an important and valuable reference written by a top expert.

According to MathSciNet, "The author has been engaged in the numerical treatment of matrices for over two decades, and was among the first to develop programs for the purpose on an electronic computer. Probably no one has had more extensive practical experience in the field. Combined with the experience is a thorough mathematical background against which the problems are viewed. This book is a product of that background and experience.

The specific objective is to describe those techniques with which the author has had direct experience, and to present for each a detailed error analysis. But in order to minimize the demands upon the reader, and not to limit his audience, he develops essentially all the special mathematics, presupposing virtually nothing other than a general acquaintance with the basic notions of linear spaces, with dependence and rank... While it can hardly be said that the book contains the answers to all the questions, it can be said that its appearance marks the end of an era in the development of the subject."

G. Golub, C. Van Loan, *Matrix Computations*, Third edition, Johns Hopkins Studies in the Mathematical Sciences, Johns Hopkins University Press, Baltimore, MD, 1996. xxx+698 pp.

This is a classical reference on matrix theory and numerical linear algebra.

According to MathSciNet, "Numerical linear algebra is a large and expanding subject which has a growing impact in mathematics and computer science. As the authors declare in the preface, they have written the present book in order to impart a sense of unity to this exciting field... Certainly the book will be of great interest not only to numerical analysts, but also to students and practicing engineers. For the first category of readers perhaps the main attraction will be the large amount of perturbation theory and error analysis and the use of the powerful singular decomposition. Students will find a large number of problems and some elementary topics of undergraduate numerical analysis. For practicing engineers there are included references to EISPACK and LINPACK subroutines and many algorithms in a computational form, which are easy to transform into a programming project."

A. Householder, *The Theory of Matrices in Numerical Analysis*, Reprint of 1964 edition, Dover Publications, Inc., New York, 1975. xi+257 pp.

This is a classical book on numerical analysis and gives a comprehensive account of those parts of matrix theory that are most useful in numerical analysis.

According to MathSciNet, "Without question, this book represents one of the real highs in scholarly attainment in numerical analysis, and as such, it belongs on the shelves of students and researchers alike in this field. The author has succeeded in bringing all the related contributions of various authors in the field of matrix theory under a single unified point of view... The reviewer did not find the book to be easy reading, basically because of an almost extreme conciseness in presentation and notation... In summary, this book contains an enormous density of results, references, and potential research items in the matrix theory of numerical analysis."

G.W. Stewart, *Introduction to Matrix Computations*, Computer Science and Applied Mathematics, Academic Press, New York-London, 1973. xiii+441 pp.

This is a valuable and accessible introduction to matrix computations for beginners.

According to MathSciNet, "Numerical linear algebra is a very broad subject and the author has wisely decided to restrict this introductory text to three fundamental problems, the solution of linear equations, linear least squares problems and in-core solution of the eigenvalue problem.

A major difficulty in writing a book on numerical linear algebra is that most readers will be familiar with the background material but usually in a much more abstract setting. In his introductory chapter the author has solved this problem in a particularly elegant fashion. The material presented here bridges the gap between the usual treatment of abstract vector spaces and the matrix theory with which a modern numerical analyst must be familiar in a way that should meet with widespread approval... Few books on numerical analysis have given me as much satisfaction as did this. It has perfect balance in an age where this quality is becoming increasingly rare. It is an admirable introduction to the advances that have taken place in numerical linear algebra since the advent of electronic computers."

Y. Saad, *Iterative Methods for Sparse Linear Systems*, Second edition, Society for Industrial and Applied Mathematics, Philadelphia, PA, 2003. xviii+528 pp.

This is a very valuable reference. From the book description, "Tremendous progress has been made in the scientific and engineering disciplines regarding the use of iterative methods for linear systems. The size and complexity of linear and nonlinear systems arising in typical applications has grown, meaning that using direct solvers for the three-dimensional models of these problems is no longer effective. At the same time, parallel computing, becoming less expensive and standardized, has penetrated these application areas. Iterative methods are easier than direct solvers to implement on parallel computers but require approaches and solution algorithms that are different from classical methods."

According to MathSciNet, "This book gives a good overview of practical iterative algorithms for solving large-scale linear sparse systems of equations. The emphasis is on Krylov subspace techniques... The book can be used as text to teach a graduate-level course on iterative methods for linear systems for mathematicians, computer scientists or engineers. The presentation is very clear. The mathematical treatment is elementary. Only knowledge of basics in analysis and linear algebra is required."

P. Davis, P. Rabinowitz, *Methods of Numerical Integration*, Computer Science and Applied Mathematics, Academic Press, New York-London, 1975. xii+459 pp.

This is a classic on numerical integration and useful to people who are working on applied mathematics and numerical analysis from both the theoretical and practical points of view.

According to MathSciNet, it is "Its outstanding features continue to be the thorough coverage of the research literature, the balanced treatment of many diverse points of view, both theoretical and practical, the excellent choice of illustrative numerical examples, the strong orientation toward computer implementation, and the inimitable delightful prose of the authors, which, although informal at times, is always informative and enlightening."

E. Hairer, S.P. Norsett, G. Wanner, *Solving Ordinary Differential Equations. I. Nonstiff Problems*, Second edition, Springer Series in Computational Mathematics, 8, Springer-Verlag, Berlin, 1993. xvi+528 pp.

This comprehensive book is a standard textbook on solving ordinary differential equations numerically.

According to MathSciNet, "This volume, on nonstiff equations, is the first of a two-volume set. It has three long chapters, one on classical mathematical theory, one on Runge-Kutta and extrapolation methods, and one on multistep and general linear methods. An appendix of Fortran codes is followed by an excellent, but admittedly incomplete, bibliography which begins with work of J. Bernoulli (1697) and continues to the present. Each chapter begins with remarks about the early work on the three topics..."

R. Temam, *Navier-Stokes Equations. Theory and Numerical Analysis*, Reprint of the 1984 edition, AMS Chelsea Publishing, Providence, RI, 2001. xiv+408 pp.

This is a classic on Navier-Stokes equations on aspects of both mathematical analysis and computational fluid dynamics.

According to Math Review of the first edition, "The emphasis of this book is on the theories of various numerical methods for solving the incompressible Navier-Stokes equations. It is an important book for it is the first systematic, book-length work on that topic. Although the treatment is not complete (for instance, high Reynolds number flow is not included), the book is a good introduction to numerical analysis of the Navier-Stokes equations. Furthermore, it includes a good discussion of finite element methods applied to the Navier-Stokes equations.

The book is written in the style of a textbook and the author has attempted to make the treatment self-contained. However, in order to understand the book, readers should already be familiar with the Navier-Stokes equations and should know a lot about Sobolev spaces. Hence, it is probably too sophisticated mathematically to be of much use as a textbook and should be primarily regarded as a reference book for researchers. This brings the reviewer to her only criticism of this otherwise excellent work. As is often true with textbooks, each theorem in this book depends on all the preceding theorems. This makes the book a bit difficult for a researcher to read."

From the book description, this book discusses "the linearized stationary case, the nonlinear stationary case, and the full nonlinear time-dependent case. The relevant mathematical tools are introduced at each stage. The new material in this book is Appendix

III ... This appendix contains a few aspects not addressed in the earlier editions, in particular a short derivation of the Navier-Stokes equations from the basic conservation principles in continuum mechanics, further historical perspectives, and indications on new developments in the area. The appendix also surveys some aspects of the related Euler equations and the compressible Navier-Stokes equations. Readers are advised to peruse this appendix before reading the core of the book."

J. Ortega, W. Rheinboldt, *Iterative Solution of Nonlinear Equations in Several Variables*, Reprint of the 1970 original, Classics in Applied Mathematics, 30, Society for Industrial and Applied Mathematics (SIAM), Philadelphia, PA, 2000. xxvi+572 pp.

This book is the first book on numerical solutions of systems of nonlinear equations and is a classic.

According to MathSciNet, "This is one of the most important books to be written in numerical mathematics...

This is the first comprehensive treatment of the numerical solution of n nonlinear equations in n unknowns. While the emphasis is on the study of iterative methods, a substantial treatment of existence theorems is also included... This book is of interest to both mathematicians and numerical analysts. It should be on the shelf of every serious student of the difficult and important problem of solving multivariate nonlinear problems".

From the book description, it "provides a survey of the theoretical results on systems of nonlinear equations in finite dimension and the major iterative methods for their computational solution. Originally published in 1970, it offers a research-level presentation of the principal results known at that time. Although the field has developed since the book originally appeared, it remains a major background reference for the literature before 1970... The results and proof techniques introduced here still represent a solid basis for this topic."

P. Kloeden, E. Platen, *Numerical Solution of Stochastic Differential Equations*, Applications of Mathematics (New York), 23, Springer-Verlag, Berlin, 1992. xxxvi+632 pp.

This book gives a good summary of the early results of the theory of the numerical integration of stochastic differential equations, including various methods of approximations, fundamental convergence theorems. It has been an important reference and is also a good introduction.

According to Zentralblatt MATH, "The book is designed to be more accessible by allowing readers, who so desire, to omit unessential theoretical discussion."

13.4.2 Finite element methods and finite difference methods

The finite element method is a relatively recent and effective numerical technique for finding approximate solutions of partial differential equations and integral equations. The idea is to either eliminate the differential equation completely (steady state problems), or rendering the PDE into an approximating system of ordinary differential equations, which are then numerically integrated using standard techniques such as the Euler method and Runge-Kutta methods. In solving partial differential equations, the primary challenge is to create an equation that approximates the equation to be solved, and is numerically stable, i.e., the errors in the input and intermediate calculations do not accumulate and cause the resulting output to be meaningless.

Besides the finite element method, there are several other methods to solve differential equations numerically, such as finite difference methods, finite volume methods, domain decomposition methods. The finite element method is a good choice for solving partial differential equations over complicated domains, when the domain changes, when the desired precision varies over the entire domain, or when the solution lacks smoothness.

Finite difference method is another powerful method to find numerical solutions of differential equations. Briefly, since the derivative of a function is the limit of the ratio of the difference of the function over the difference of the variable, the ratio for small increment of the variable gives an approximation to the derivative. Approximating derivatives by differences reduces differential equations to difference equations.

S. Brenner, L.R. Scott, *The Mathematical Theory of Finite Element Methods*, Third edition, Texts in Applied Mathematics, 15, Springer, New York, 2008. xviii+397 pp.

This book gives a detailed introduction to the mathematical theory of the finite element method. It is well-written and contains a wealth of information, and hence is a valuable reference. It is also a good but demanding introduction for beginners.

According to MathSciNet, "This book, according to the authors, is written as a basis for a graduate level course for well-prepared engineers and scientists equipped with only a course in real variables. It purports to provide a text for a first course in the mathematical theory of finite elements for such an audience. The book contains thirteen chapters, of which the first six are regarded as essential material for a course. The remaining six chapters are on special topics which reflect the specific interests and expertise of the authors. At the end of each chapter is a list of exercises that support the theoretical work presented... It is both a well-done text and a good reference."

F. Brezzi, M. Fortin, *Mixed and Hybrid Finite Element Methods*, Springer Series in Computational Mathematics, 15, Springer-Verlag, New York, 1991. x+350 pp.

This book discusses the mathematical analysis of the mixed and hybrid finite element methods and is written by two major contributors to the subject. It is a very reliable reference.

From the book description, "Research on non-standard finite element methods is evolving rapidly and in this text Brezzi and Fortin give a general framework in which the development is taking place. The presentation is built around a few classic examples: Dirichlet's problem, Stokes problem, Linear elasticity. The authors provide with this publication an analysis of the methods in order to understand their properties as thoroughly as possible."

According to MathSciNet, "This is a very rich book which will probably become a standard reference on mixed and hybrid finite element methods. The general framework has been built up through intense research during the last 15 years. The general theory for this kind of nonconforming finite element method is extensively illustrated, with deep analysis of 'basic' problems such as Dirichlet's problem (standard and nonstandard approach), the Stokes problem (extensively for incompressible materials and flow problems), problems of linear elasticity (especially for moderately thick plates), etc. Optimal error estimates are given when known. Extensive references are provided."

G. Strang, G. Fix, *An Analysis of the Finite Element Method*, Second edition, Wellesley-Cambridge Press, Wellesley, MA, 2008. x+402 pp.

This is one of the earliest books on finite element methods and is a classic. It is still a valuable reference.

According to MathSciNet, "This book is required reading for anyone interested in the mathematical analysis of the finite element method (FEM)... Chapter 1 can be a course in itself, because it is an introduction to the whole subject... Finally, and not least importantly, the book is written in a pleasant and sometimes witty style."

According to a review posted by the publisher, "Though some sections are dated, the contents of this book remain a solid foundation for understanding the behavior of finite element techniques in theory and in practice. Furthermore, while the contents are mathematically rigorous, the authors largely avoid the theorem/lemma/proof of most mathematical books, and instead describe the analysis in an almost conversational tone."

G. Forsythe, W. Wasow, *Finite-difference Methods for Partial Differential Equations*, Reprint of the 1960 original, Dover Phoenix Editions, Dover Publications, Inc., Mineola, NY, 2004. xiv+444 pp.

This is a classical book and covers standard topics in finite difference methods for partial differential equations. Though some material is outdated, it is still a standard reference for finite difference methods and for historical information on numerical methods.

According to MathSciNet, "The book is written to provide the numerical-analysis background necessary for an understanding of finite-difference methods for solving partial differential equations. It is aimed at several groups of readers including pure and applied mathematical analysts, computer programmers, engineers, physicists, and others interested in solving partial differential equations, and graduate students in these fields. The book is not intended to be complete and comprehensive. The authors have avoided, on the one hand, topics such as existence and uniqueness proofs based on finite-difference methods and, on the other hand, a detailed description of programming techniques. Only a few numerical examples are described. The authors believe that, by providing the reader with a general knowledge of the theoretical background and the known methods, they can modify and apply the techniques to particular problems."

R. Richtmyer, K. Morton, *Difference Methods for Initial-value Problems*, Second edition, Interscience Tracts in Pure and Applied Mathematics, No. 4, Reprint of the second edition, Robert E. Krieger Publishing Co., Inc., Malabar, FL, 1994. xiv+405 pp.

The book was written primarily for users of difference methods. It is a classical book and is also valuable to others who are interested in the subject.

According to MathSciNet, "For several years, the brilliant collaboration of mathematicians at New York University has been producing some elegant bridges across the gap that has separated the 'respectable' abstraction of subjects such as linear operators from the 'undignified' empiricism of applied mathematics—represented, in the present instance, by numerical computation. This work has filtered down by now to the text stratum;... and the present work is a ... example. That such an effort encounters difficulties in satisfying a wide range of readers is probably clear. In the opinion of the reviewer, the present book is a good deal more successful than most, although there will doubtless be those who find it wanting.... The entire volume makes interesting reading, and one hopes it will convince both pure mathematicians and computer experts that they have a fruitful common area of study."

V. Girault, P.A. Raviart, *Finite Element Methods for Navier-Stokes Equations*, Theory and algorithms. Springer Series in Computational Mathematics, 5. Springer-Verlag, Berlin, 1986. x+374 pp.

This is a classic and complements the more theoretical books *The finite element method for elliptic problems* by Ciarlet, and *Navier-Stokes equations* by Temam.

From the book description, "This book provides insight in the mathematics of Galerkin finite element method as applied to parabolic equations. The approach is based on first discretizing in the spatial variables by Galerkin's method, using piecewise polynomial trial functions, and then applying some single step or multistep time stepping method. The concern is stability and error analysis of approximate solutions in various norms, and under various regularity assumptions on the exact solution."

According to MathSciNet, "This is the definitive treatise on the mathematical theory of finite element methods for the approximation of solutions of the stationary Navier-Stokes equations... The book is an absolute must for anyone interested in the mathematical theory of finite element methods for the Navier–Stokes equations."

P. Ciarlet, *The Finite Element Method for Elliptic Problems*, Reprint of the 1978 original, Classics in Applied Mathematics, 40, Society for Industrial and Applied Mathematics (SIAM), Philadelphia, PA, 2002. xxviii+530 pp.

This is a classical book on the finite element method for elliptic problems.

From the book description, it "is the only book available that fully analyzes the mathematical foundations of the finite element method. It is a valuable reference and introduction to current research on the numerical analysis of the finite element method, and also a working textbook for graduate courses in numerical analysis... Although nearly 25 years have passed since this book was first published, the majority of its content remains up-to-date."

According to MathSciNet, "This beautiful book is the first text containing a complete treatment of the basic mathematical aspects of the finite element method. This book is exceptional not only for the author's mathematical precision, modernity and pedagogic mastery, but also because only actual problems and finite element methods which are most widely used in contemporary engineering applications are described and analyzed. Some parts of the book can be used by engineers for selecting and evaluating particular methods. But above all, the book is aimed at graduate students and research workers and will certainly become the basic reference text..."

According to an Amazon review, "Ciarlet's text is not the only book to analyze in depth the mathematical theory of finite element methods, but it is still one of the best. The main focus of the book is on elliptic problems, and particularly linear problems."

M. Ainsworth, J. Oden, *A Posteriori Error Estimation in Finite Element Analysis*, Pure and Applied Mathematics (New York), Wiley-Interscience [John Wiley & Sons], New York, 2000. xx+240 pp.

From the book description, it "is a lucid and convenient resource for researchers in almost any field of finite element methods, and for applied mathematicians and engineers who have an interest in error estimation and/or finite elements."

It is a good introduction to and reference on the mathematical underpinnings of many of the most effective methods for error estimation that are available today.

13.4.3 Approximation theory

Approximation theory studies how functions can best be approximated by simpler functions such as polynomials, rational functions and piecewise polynomial functions, and

13.4. NUMERICAL ANALYSIS AND MATRIX COMPUTATION

estimates the errors of the approximation. Choices of functions and what is meant by best and simpler functions will depend on the application. In order to be useful and to implement approximations, it is important that functions and procedures can be easily constructed.

R. DeVore, G. Lorentz, *Constructive Approximation*, Grundlehren der Mathematischen Wissenschaften, 303, Springer-Verlag, Berlin, 1993. x+449 pp.

This a standard reference written by leading experts. It contains a systematic and unified exposition of basic problems in approximation theory of functions of one real variable.

According to MathSciNet, "The authors are top researchers in the field who had good taste and judgement in collecting the most important results of the constructive aspects of the theory. The only prerequisite is elementary real analysis, in particular, the knowledge of Lebesgue integration and L^p spaces; other than that the book is self-contained. This monograph may well be the best book available on the subject, so it can be recommended to graduate students, mathematicians, physicists and engineers who have an interest in constructive approximation. The authors not only collect the most important results on the subject, but in many instances they also provide new proofs as well. The organization and presentation are extremely clear, and the book is suitable for graduate and even higher level undergraduate courses."

According to the book description, "Coupled with its sequel, this book gives a connected, unified exposition of Approximation Theory for functions of one real variable."

G. Lorentz, M. Golitschek, Y. Makovoz, *Constructive Approximation. Advanced Problems*, Grundlehren der Mathematischen Wissenschaften, 304, Springer-Verlag, Berlin, 1996. xii+649 pp.

From the book description, "This and the earlier book by R.A. DeVore and G.G. Lorentz (Vol. 303 of the same series), cover the whole field of approximation of functions of one real variable. The main subject of this volume is approximation by polynomials, rational functions, splines and operators. There are excursions into the related fields: interpolation, complex variable approximation, wavelets, widths, and functional analysis. Emphasis is on basic results, illustrative examples, rather than on generality or special problems."

According to MathSciNet, "Approximation theory has a history of well over 100 years, and many books on the subject have appeared (these authors refer to 90 of them). Three of the early books on the subject were written in the period 1920–1930 by Bernstein, de la Valle-Poussin, and Jackson. A spate of books appeared 25 years later in the period 1955–1970, including the classic texts by Akhiezer, Butzer-Berens, Cheney, Kolmogorov, Lorentz, Meinardus, Natanson, Nikolsky, Rice, Sard, Timan, and Walsh. During the last 25 years many more books have appeared, with the emphasis on special topics in approximation... The book ... is a companion to an earlier book..."

C. de Boor, *A Practical Guide to Splines*, Revised edition. Applied Mathematical Sciences, 27. Springer-Verlag, New York, 2001. xviii+346 pp.

This book is a classic book on splines written by a leading expert. It is both a systematic introduction for students and a valuable reference for experts.

Basically, the spline theory deals with approximations of functions by piecewise polynomials, which satisfy certain smoothness conditions at the joining points. It is an important topic in numerical analysis. This book emphasizes practical methods and their implementation on computer.

L. Schumaker, *Spline Functions: Basic Theory*, Third edition, Cambridge Mathematical Library, Cambridge University Press, Cambridge, 2007. xvi+582 pp.

This is a standard reference on splines written by an expert. It contains a wealth of information.

According to MathSciNet, "This book, written in a masterful form and style, was not intended as a textbook. Perhaps this is the reason that there are no exercises, but we should mention that the book contains many illustrative examples discussed in detail. However, the book can be read by anyone with a reasonable background in calculus and linear algebra. The author suggests that the book could be used for a one-semester introduction to splines, through a judicious choice of material. In the opinion of the reviewer, however, this book remains primarily a remarkable research monograph and reference work. The presentation is very well organized and carefully worked out. The proofs of the theorems are detailed and rigorously done."

The Nine Chapters on the Mathematical Art (Jiuzhang Suanshu) *was the first systematic Chinese treatise on mathematics. It summarizes all major results in mathematics obtained up to the Han Dynasty and also contains some original results. Since it classifies all problems into nine chapters, it is called Nine Chapters. This book contains explicit algorithms to solve solutions to various problems. In some sense, this is one of the oldest books on algorithms and numerical analysis.*

Figure 13.4: Jiuzhang Suanshu.

Appendix A: Books in Chinese version[1]

1. Adams, R.A., 索伯列夫空間, 人民教育出版社, 1983

2. Ahlfors, L.V., 複分析, 機械工業出版社, 2005

3. Aigner, M., Ziegler, G.M., 數學天書中的證明, 高等教育出版社, 2011

4. Aleksandrov (亞歷山大洛夫), A.D., Kolmogorov (柯爾莫果洛夫), A.N., Lavrent'ev (拉夫倫捷夫), M.A.等, 數學——它的內容、方法和意義(1, 2, 3), 科學出版社, 2003

5. Alexandroff (亞歷山德羅夫), P., 拓撲學基本概念(見: 直觀幾何(下冊)附), 高等教育出版社, 2013

6. Anderson, F.W., Fuller, K.R., 環與模範疇, 科學出版社, 2008

7. Apostol, T.M., 數學分析, 機械工業出版社, 2009

8. Apostol, T.M., 解析數論引論, 哈爾濱工業大學出版社, 2011

9. Arnold (阿諾爾德), V.I., 常微分方程續論: 常微分方程的幾何方法, 科學出版社, 1989

10. Arnold (阿諾爾德), V.I., 經典力學的數學方法, 高等教育出版社, 2006

11. Arnold (阿諾爾德), V.I., 常微分方程, 科學出版社, 1985

12. Artin, E., 伽羅華理論, 上海科學技術出版社, 1979; Galois 理論, 哈爾濱工業大學出版社, 2011

---

[1] The book lists in Appendices A and B may be incomplete due to the lack of information.

13. Artin, M., 代數, 機械工業出版社, 2009
14. Atiyah, M.F., MacDonald, I.G., 交換代數導引, 科學出版社, 1982
15. Axler, S., 線性代數應該這樣學, 人民郵電出版社, 2009
16. Banach, S., 線性算子理論, 科學出版社, 2011
17. Beardon, A.F., 離散群幾何, 北京理工大學出版社, 1988
18. Bell, E.T., 數學精英, 商務印書館, 1991; 數學大師, 上海科技教育出版社, 2004
19. Berlekamp, E., Conway, J., Guy, R., 穩操勝券 (上,下), 上海教育出版社, 2003
20. Birkhoff, G., Mac Lane, S., 近世代數概論(上,下), 人民教育出版社, 1979, 1980
21. Bondy, J.A., Murty, U.S.R., 圖論及其應用, 科學出版社, 1984
22. Boyer, C.B., Merzbach, U.C., 數學史 (上,下), 中央編譯出版社, 2012
23. Brezis, H., 泛函分析——理論和應用, 清華大學出版社, 2009
24. Cartan, H., 解析函數論初步, 高等教育出版社, 2008
25. Courant, R., Hilbert, D., 數學物理方法 (I, II), 科學出版社, 1958, 1977
26. Courant, R., John, F., 微積分和數學分析引論 (兩卷四分冊), 科學出版社, 2001
27. Courant, R., Robbins, H., 數學是什麼, 湖南教育出版社, 1985; 復旦大學出版社, 2012
28. Coxeter, H.S.M., Greitzer, S.L., 幾何重觀, 河南教育出版社, 1984
29. Daubechies, I., 小波十講, 國防工業出版社, 2011
30. Dauben, J.W., 康托的無窮的數學和哲學, 江蘇教育出版社, 1989; 大連理工大學出版社, 2008
31. Demidovich (吉米多維奇), B.P., 數學分析習題集, 高等教育出版社, 2010
32. Dieudonne, J., 現代分析基礎 (1, 2), 科學出版社, 1982, 1986
33. Dirac, P.A.M., 廣義相對論, 科學出版社, 1979
34. do Carmo, M.P., 曲線和曲面的微分幾何學, 上海科學技術出版社, 1988; 機械工業出版社, 2005

APPENDIX A: BOOKS IN CHINESE VERSION

35. Dubrovin (杜布洛文), B.A., Novikov (諾維可夫), S.P., Fomenko (福明柯), A.T., 現代幾何學: 方法與應用(一, 二, 三), 高等教育出版社, 2006, 2007

36. Duvaut, G., Lions, J.L., 力學和物理學中的變分不等方程, 科學出版社, 1987

37. Einstein, A., 相對論的意義, 科學出版社, 1979

38. Eves, H., 數學史概論, 哈爾濱工業大學出版社, 2009

39. Falconer, K.J., 分形幾何——數學基礎及其應用, 東北大學出版社, 1991

40. Feller, W., 概率論及其應用 (1, 2), 人民郵電出版社, 2006, 2008

41. Friedman, A., 拋物型偏微分方程, 科學出版社, 1984

42. Gantmacher (甘特馬赫爾), F.R., 矩陣論 (上, 下), 高等教育出版社, 1955

43. Gelbaum, B.R., Olmsted, J.M.H., 分析中的反例, 上海科學技術出版社, 1980

44. Gelfand (蓋爾芳特), I.M., Shilov (希洛夫), G.E., 廣義函數 (1, 2, 3), 科學出版社, 1984, 1985, 1983

45. Gelfand (盖尔芳特), I.M., Vilenkin (維列金), N.Ya., 廣義函數 (4), 科學出版社, 1984

46. Gilberg, D., Trudinger, N.S., 二階橢圓型偏微分方程, 上海科學技術出版社, 1981

47. Golub, G.H., Van Loan, C.F., 矩陣計算, 人民郵電出版社, 2011

48. Graham, R.L., Knuth, D.E., Patashnik, O., 具體數學, 西安電子科技大學出版社, 1992; 人民郵電出版社, 2013

49. Guy, R.K., 數論中未解決的問題, 科學出版社, 2003

50. Hall, M., 群論, 科學出版社, 1981

51. Halmos, P.R., 希爾伯特空間問題集, 上海科學技術出版社, 1984

52. Halmos, P.R., 我要作數學家, 江西教育出版社, 1999

53. Halmos, P.R., 測度論, 科學出版社, 1965

54. Harary, F., 圖論, 上海科學技術出版社, 1980

55. Hardy, G.H., 純數學教程, 人民郵電出版社, 2009

56. Hardy, G.H., 一個數學家的辯白, 江蘇教育出版社, 1996

57. Hardy, G.H., Littlewood, J.E., Pólya, G., 不等式, 人民郵電出版社, 2008

58. Hardy, G.H., Wright, E.M. (Heath-Brown, D.R., Silverman, J.H.), 哈代數論, 人民郵電出版社, 2010

59. Hartshorne, R., 代數幾何, 科學出版社, 2001

60. Hawking, S.W., Ellis, G.F.R., 時空的大尺度結構, 湖南科學技術出版社, 2006

61. Hecke, E., 代數數理論講義, 科學出版社, 2005

62. Hilbert, D., Cohn-Vossen, S., 直觀幾何 (上, 下), 高等教育出版社, 2013

63. Hille, E., Phillips, R.S., 泛函分析與半群 (上), 上海科學技術出版社, 1964

64. Hirsch, M.W., Smale, S., Devaney, R.L., 微分方程、動力系統與混沌導論, 人民郵電出版社, 2008

65. Hodges, A., 艾倫·圖靈傳——如謎的解謎者, 湖南科學技術出版社, 2012

66. Hoffman, P., 數字情種——埃爾德什傳, 上海科技教育出版社, 2000

67. Hofstadter (侯世達), D.R., GEB——一條永恆的金帶, 四川人民出版社, 1984; 哥德爾、艾舍爾、巴赫——集異璧之大成, 商務印書館, 1997

68. Hopf, H., 整體微分幾何, 科學出版社, 1987

69. Hörmander, L., 線性偏微分算子, 科學出版社, 1980

70. Horn, R.A., Johnson, C.R., 矩陣分析, 天津大學出版社, 1989

71. 華羅庚, 堆壘素數論, 科學出版社, 1957

72. 華羅庚, 數論導引, 科學出版社, 1975

73. Humphreys, J.E., 李代數及其表示理論導引, 上海科學技術出版社, 1981

74. Huppert, B., 有限群論 (第一卷第一、二分冊), 福建人民出版社, 1992

75. Jacobson, N., 基礎代數 (第一卷第一、二分冊), 高等教育出版社, 1987, 1988

76. Jacobson, N., 李代數, 上海科學技術出版社, 1964

77. John, F., 偏微分方程, 科學出版社, 1986

APPENDIX A: BOOKS IN CHINESE VERSION

78. Kanigel, R., 知無涯者: 拉馬努金傳, 上海科技教育出版社, 2002

79. Kelley, J.L., 一般拓撲學, 科學出版社, 1982

80. Khinchin (辛欽), A.Ya., 數論的三顆明珠, 上海科學技術出版社, 1984

81. Kinderlehrer, D., Stampacchia, G., 變分不等方程及其應用, 科學出版社, 1991

82. Klein, F., 數學在19世紀的發展(一, 二), 高等教育出版社, 2010, 2011

83. Klein, F., Klein 數學講座, 高等教育出版社, 2013

84. Kline, M., 古今數學思想 (1–4), 上海科學技術出版社, 2002

85. Knuth, D.E., 計算機程序設計藝術 (1, 2, 3), 國防工業出版社, 2002

86. Kobayashi (小林昭七), S., Nomizu (野水克己), K., 微分幾何基礎 (一), 科學出版社, 2010

87. Kolmogorov (柯爾莫哥洛夫), A.N., 概率論基本概念, 商務印書館, 1953

88. Kolmogorov (柯爾莫哥洛夫), A.N., Fomin (佛明), S., 函數論與泛函分析初步, 高等教育出版社, 2006

89. Kurosh (庫洛什), A., 群論 (上, 下), 高等教育出版社, 1987, 1982

90. Ladyzhenskaya (拉迪任斯卡婭), O.A., 粘性不可壓縮流體動力學的數學問題, 上海科學技術出版社, 1985

91. Ladyzhenskaya (拉迪任斯卡婭), O.A., Ural'tseva (烏拉利采娃), N.N., 線性和擬線性橢圓型方程, 科學出版社, 1987

92. Landau, L.D., Lifshitz, E.M., 理論物理學教程 (1–10), 高等教育出版社, 2007–2015

93. Lax, P.D., 泛函分析, 人民郵電出版社, 2010

94. Lieb, E.H., Loss, M., 分析學, 高等教育出版社, 2007

95. Littlewood, J.E., Littlewood 數學隨筆集, 高等教育出版社, 2013

96. Markushevich (馬庫雪維奇), A.I., 解析函數論, 高等教育出版社, 1957

97. Marsden, J.E., Ratiu, T.S., 力學和對稱性導論——經典力學系統初探, 清華大學出版社, 2006

98. Meyer, Y., 小波與算子 (1, 2), 世界圖書出版公司, 1992, 1995

99. Milnor, J., 莫爾斯理論, 科學出版社, 1988

100. Milnor, J., 從微分觀點看拓撲, 上海科學技術出版社, 1983; 從微分觀點看拓撲 (雙語版), 人民郵電出版社, 2008

101. Mitrinovic, D.S., 解析不等式, 科學出版社, 1987

102. Munkres, J.R., 代數拓撲基礎, 科學出版社, 2006

103. Munkres, J.R., 拓撲學基本教程, 科學出版社, 1987

104. Petrovski (彼得羅夫斯基), I.G., 常微分方程講義, 高等教育出版社, 1957

105. Pólya, G., 怎樣解題, 科學出版社, 1982

106. Pólya, G., 數學的發現, 科學出版社, 2006

107. Pólya, G., 數學與猜想 (1, 2), 科學出版社, 1984

108. Pólya, G., Szegö, G., 數學分析中的問題和定理 (1, 2), 上海科學技術出版社, 1981, 1985

109. Pontryagin (邦德列雅金), L., 連續群 (上,下), 科學出版社, 1957, 1958

110. Protter, M.H., Weinberger, H.F., 微分方程的最大值原理, 科學出版社, 1985

111. Reid, C., 庫朗: 一位數學家的雙城記, 東方出版中心, 1999

112. Reid, C., 希爾伯特: 數學世界的亞歷山大, 上海科學技術出版社, 2001

113. Reid, C., 奈曼: 來自生活的統計學家, 上海科學技術出版社, 2001

114. Riesz, F., Sz.-Nagy, B., 泛函分析講義(1, 2), 科學出版社, 1980

115. Robinson, A., 非標準分析, 科學出版社, 1980

116. Rudin, W., 泛函分析, 機械工業出版社, 2004

117. Rudin, W., 實分析與複分析, 機械工業出版社, 2006

118. Salsburg, D., 女士品茶:20世紀統計怎樣變革了科學, 中國統計出版社, 2004

119. Santaló, L., 積分幾何與幾何概率, 南開大學出版社, 1991

120. Schechter, B., 我的大腦敞開了——天才數學家保羅·愛多士傳奇, 上海譯文出版社, 2002

APPENDIX A: BOOKS IN CHINESE VERSION

121. Schoen, R. (孫理察), 丘成桐, 微分幾何講義, 高等教育出版社, 2004
122. Schoen, R. (孫理察), 丘成桐, 調和映照講義, 高等教育出版社, 2008
123. Schwartz, L., 廣義函數論, 高等教育出版社, 2010
124. Seifert, H., Threlfall, W., 拓撲學, 人民教育出版社, 1959
125. Serre, J.-P., 數論教程, 上海科學技術出版社, 1980; 高等教育出版社, 2007
126. Serre, J.-P., 有限群的線性表示, 科學出版社, 1984; 高等教育出版社, 2007
127. Shiryaev (施利亞耶夫), A.N., 概率 (一, 二), 高等教育出版社, 2007, 2008
128. Spanier, E.H., 代數拓撲學 (前三章), 上海科學技術出版社, 1987
129. Spivak, M., 流形上的微積分, 科學出版社, 1981; 流形上的微積分 (雙語版), 人民郵電出版社, 2006
130. Stanley, R.P., 計數組合學 (一), 高等教育出版社, 2009
131. Stein, E.M., Weiss, G., 歐氏空間上的Fourier分析引論, 上海科學技術出版社, 1987
132. Stewart, I., 對稱的歷史, 上海人民出版社, 2011
133. Stillwell, J., 數學及其歷史, 高等教育出版社, 2011
134. Struik, D.J., 數學簡史, 科學出版社, 1956
135. Tenenbaum, G., 解析與概率數論導引, 高等教育出版社, 2011
136. Titchmarsh, E.C., 函數論, 科學出版社, 1962
137. (Ulam, S.M.) Mauldin, R.D., 數學探索——蘇格蘭咖啡館數學問題集, 四川教育出版社, 1987
138. Ulam, S.M., 一位數學家的經歷, 上海科學技術出版社, 1989
139. van der Waerden, B.L., 代數學(I, II), 科學出版社, 2009
140. van Lint, J.H., 編碼理論導引, 科學出版社, 1988
141. von Neumann, J., 計算機與人腦, 商務印書館, 1965; 北京大學出版社, 2010
142. Warner, F.W., 微分流形和李群基礎, 科學出版社, 2008

143. Weil, A., 數論——從漢穆拉比到勒讓德的歷史導引, 高等教育出版社, 2010
144. West, D.B., 圖論導引, 機械工業出版社, 2006
145. Weyl, H., 對稱, 商務印書館, 1986; 上海科技教育出版社, 2005
146. Wiener, N., 我是一個數學家, 上海科學技術出版社, 1987
147. Wiener, N., 控制論 (或關於在動物和機器中控制和通訊的科學), 科學出版社, 2009
148. Wiener, N., 昔日神童, 上海科學技術出版社, 1982
149. Wilkinson, J.H., 代數特徵值問題, 科學出版社, 2006
150. Yosida (吉田耕作), K., 泛函分析, 人民教育出版社, 1981

Appendix B: Books reprinted in the mainland of China

1. Adams, R.A., John J.F. Fournier, *Sobolev Spaces*, 世界圖書出版公司, 2009

2. Ahlfors, L.V., *Complex Analysis*, 機械工業出版社, 2004

3. Aigner, M., Ziegler, G.M., *Proofs from THE BOOK*, 世界圖書出版公司, 2006

4. Anderson, F.W., Fuller, K.R., *Rings and Categories of Modules*, 世界圖書出版公司, 2004

5. Apostol, T.M., *Calculus*, 機械工業出版社, 2004

6. Apostol, T.M., *Introduction to Analytic Number Theory*, 世界圖書出版公司, 2012

7. Arnold, V.I., *Geometrical Methods in the Theory of Ordinary Differential Equations*, 世界圖書出版公司, 1992

8. Arnold, V.I., *Mathematical Methods of Classical Mechanics*, 世界圖書出版公司, 1999

9. Arnold, V.I., *Ordinary Differential Equations*, 世界圖書出版公司, 2003

10. Artin, M., *Algebra*, 機械工業出版社, 2004

11. Atiyah, M.F., *K-theory* (見: Michael Atiyah Collected Works, Vol. 2), 世界圖書出版公司, 1988

12. Axler, S., *Linear Algebra Done Right*, 世界圖書出版公司, 2008

13. Barth, W.P., Hulek, K., Peters, C.A.M., Van de Ven, A., *Compact Complex Surfaces*, 世界圖書出版公司, 2011

14. Beardon, A.F., *The Geometry of Discrete Groups*, 世界圖書出版公司, 2011

15. Benedetti, R., Petronio, C., *Lectures on Hyperbolic Geometry*, 世界圖書出版公司, 2012

16. Berline, N., Getzler, E., Vergne, M., *Heat Kernels and Dirac Operators*, 世界圖書出版公司, 2009

17. Besse, A.L., *Einstein Manifolds*, 世界圖書出版公司, 1989

18. Bhatia, R., *Matrix Analysis*, 世界圖書出版公司, 2011

19. Billingsley, P., *Probability and Measure*, 世界圖書出版公司, 2007

20. Bix, R., *Conics and Cubics*, 世界圖書出版公司, 2011

21. Bollobás, B., *Graph Theory: An Introductory Course*, 世界圖書出版公司, 1999

22. Bollobás, B., *Modern Graph Theory*, 科學出版社, 2001; 世界圖書出版公司, 2003

23. Bollobás, B., *Random Graphs*, 世界圖書出版公司, 2003

24. Borel, A., *Linear Algebraic Groups*, 世界圖書出版公司, 2009

25. Borwein, P., Erdélyi, T., *Polynomials and Polynomial Inequalities*, 世界圖書出版公司, 1997

26. Bott, R., Tu, L.W., *Differential Forms in Algebraic Topology*, 世界圖書出版公司, 2009

27. Bredon, G. E., *Topology and Geometry*, 世界圖書出版公司, 2008

28. Brocker, T., Dieck, T.t., *Representations of Compact Lie Groups*, 世界圖書出版公司, 1999

29. Brown, K.S., *Cohomology of Groups*, 世界圖書出版公司, 2009

30. Burris, S., Sankappanavar, H.P., *A Course in Universal Algebra*, 世界圖書出版公司, 1988

31. Cartan, H., Eilenberg, S., *Homological Algebra*, 世界圖書出版公司, 2011

32. Chavel, I., *Riemannian Geometry: A Modern Introduction*, 世界圖書出版公司, 2000

APPENDIX B: BOOKS REPRINTED IN THE MAINLAND OF CHINA

33. Chern, S.S. (陳省身), *Complex Manifolds Without Potential Theory*, 世界圖書出版公司, 2008

34. Chevalley, C., *Theory of Lie Groups*, 世界圖書出版公司, 2013

35. Chriss, N., Ginzburg, V., *Representation Theory and Complex Geometry*, 世界圖書出版公司, 2012

36. Chung, K.L. (鍾開萊), *A Course in Probability Theory*, 機械工業出版社, 2010

37. Cohn, D.L., *Measure Theory*, 世界圖書出版公司, 2012

38. Connes, A., *Noncommutative Geometry*, 世界圖書出版公司, 2008

39. Conway, J.B., *A Course in Functional Analysis*, 世界圖書出版公司, 2003

40. Conway, J.H., Sloane, N.J.A., *Sphere Packings, Lattices and Groups*, 世界圖書出版公司, 2008

41. Courant, R., John, F., *Introduction to Calculus and Analysis*, 世界圖書出版公司, 2008

42. Crowell, R.H., Fox, R.H., *Introduction to Knot Theory*, 世界圖書出版公司, 2000

43. Davenport, H., *Multiplicative Number Theory*, 世界圖書出版公司, 1988

44. Deimling, K., *Nonlinear Functional Analysis*, 世界圖書出版公司, 1988

45. Diestel, R., *Graph Theory*, 世界圖書出版公司, 2003

46. Dirac, P.A.M., *General Theory of Relativity*, 世界圖書出版公司, 2011

47. Dixon, J.D., Mortimer, B., *Permutation Groups*, 世界圖書出版公司, 1997

48. Do Carmo, M.P., *Differential Geometry of Curves and Surfaces*, 機械工業出版社, 2004

49. Do Carmo, M.P., *Riemannian Geometry*, 世界圖書出版公司, 2008

50. Doob, J.L., *Classical Potential Theory and Its Probabilistic Counterpart*, 世界圖書出版公司, 2013

51. Dubrovin, B.A., Fomenko, A.T., Novikov, S.P., *Mondern Geometry—Methods and Applications*, 世界圖書出版公司, 1999

52. Durrett, R., *Probability: Theory and Examples*, 世界圖書出版公司, 2007

53. Eisenbud, D., *Commutative Algebra: with a View Toward Algebraic Geometry*, 世界圖書出版公司, 2008

54. Eisenhart, L.P., *Riemannian Geometry*, 世界圖書出版公司, 2011

55. Faddeev, L.D., Takhtajan, L.A., *Hamiltonian Methods in the Theory of Solitons*, 世界圖書出版公司, 2013

56. Farkas, H.M., Kra, I., *Riemann Surfaces*, 世界圖書出版公司, 2003

57. Federer, H., *Geometric Measure Theory*, 世界圖書出版公司, 2004

58. Folland, G.B., *Real Analysis: Modern Techniques and Their Applications*, 世界圖書出版公司, 2007

59. Forster, O., *Lectures on Riemann Surfaces*, 世界圖書出版公司, 1988

60. Fraleigh, J.B., *A First Course in Abstract Algebra*, 世界圖書出版公司, 2008

61. Frankel, T., *The Geometry of Physics: An Introduction*, 清華大學出版社, 2005

62. Fried, M.D., Jarden, M., *Field Arithmetic*, 世界圖書出版公司, 1989

63. Fulton, W., Harris, J., *Representation Theory: A First Course*, 世界圖書出版公司, 2005

64. Gallot, S., Hulin, D., Lafontaine, J., *Riemannian Geometry*, 世界圖書出版公司, 2008

65. Garnett, J.B., *Bounded Analytic Functions*, 世界圖書出版公司, 2010

66. Gilberg, D.T., Trudinger, N.S., *Elliptic Partial Differential Equations of Second Order*, 世界圖書出版公司, 2003

67. Gillman, L., Jerison, M., *Rings of Continuous Functions*, 世界圖書出版公司

68. Godsil, C., Royle, G., *Algebraic Graph Theory*, 世界圖書出版公司, 2004

69. Golub, G.H., Van Loan, C.F., *Matrix Computations*, 人民郵電出版社, 2009

70. Golubitsky, M., Guillemin, V., *Stable Mappings and Their Singularities*, 世界圖書出版公司, 1988

71. Gowers, T., Barrow-Green, J., Leader, I., *The Princeton Companion to Mathematics*, 世界圖書出版公司, 2013

72. Graham, R.L., Knuth, D.E., Patashnik, O., *Concrete Mathematics: A Foundation for Computer Science*, 機械工業出版社, 2012

73. Grauert, H., Remmert, R., *Coherent Analytic Sheaves*, 世界圖書出版公司, 1989

74. Griffiths, P., Harris, J., *Principles of Algebraic Geometry*, 世界圖書出版公司, 2007

75. Gromov, M., *Partial Differential Relations*, 世界圖書出版公司, 1989

76. Guy, R.K., *Unsolved Problems in Number Theory*, 科學出版社, 2007

77. Halmos, P.R., *A Hilbert Space Problem Book*, 世界圖書出版公司, 2009

78. Halmos, P.R., *Finite-Dimensional Vector Spaces*, 世界圖書出版公司, 2007

79. Halmos, P.R., *Measure Theory*, 世界圖書出版公司, 2007

80. Halmos, P.R., *Naive Set Theory*, 世界圖書出版公司, 2008

81. Hardy, G.H., *A Course of Pure Mathematics*, 機械工業出版社, 2005

82. Hardy, G.H., Littlewood, J.E., Pólya, G., *Inequalities*, 世界圖書出版公司, 2004

83. Hardy, G.H., Wright, E.M., *An Introduction to the Theory of Numbers*, 人民郵電出版社, 2007

84. Harris, J., *Algebraic Geometry: A First Course*, 世界圖書出版公司, 2000

85. Harris, J., Morrison, I., *Moduli of Curves*, 世界圖書出版公司, 2011

86. Hartshorne, R., *Algebraic Geometry*, 世界圖書出版公司, 2011

87. Hasse, H., *Number Theory*, 世界圖書出版公司, 2010

88. Hatcher, A., *Algebraic Topology*, 清華大學出版社, 2005

89. Hawking, S.W., Ellis, G.F.R., *The Large Scale Structure of Space-time*, 湖南科學技術出版社, 2006

90. Hecke, E., *Lectures on the Theory of Algebraic Numbers*, 世界圖書出版公司, 2000

91. Hirsch, M.W., *Differential Topology*, 世界圖書出版公司, 2000

92. Hirzebruch, F., *Topological Methods in Algebraic Geometry*, 世界圖書出版公司, 2004

93. Hörmander, L., *An Introduction to Complex Analysis in Several Variables*, 人民郵電出版社, 2007

94. Hörmander, L., *Lectures on Nonlinear Hyperbolic Differential Equations*, 世界圖書出版公司, 2003

95. Hörmander, L., *The Analysis of linear Partial Differential Operators (I–IV)*, 世界圖書出版公司, 2005

96. Horn, R.A., Johnson, C.R., *Matrix Analysis*, 人民郵電出版社, 2005

97. Horn, R.A., Johnson, C.R., *Topics in Matrix Analysis*, 人民郵電出版社, 2005

98. Humphreys, J.E., *Introduction to Lie Algebras and Representation Theory*, 世界圖書出版公司, 2006

99. Humphreys, J.E., *Linear Algebraic Groups*, 世界圖書出版公司, 2009

100. Huybrechts, D., *Complex Geometry: An Introduction*, 世界圖書出版公司, 2010

101. Ireland, K., Rosen, M., *A Classical Introduction to Modern Number Theory*, 世界圖書出版公司, 2003

102. Jost, J., *Riemannian Geometry and Geometric Analysis*, 世界圖書出版公司, 2005

103. Kac, V.G., *Infinite Dimensional Lie Algebras*, 世界圖書出版公司, 2006

104. Karatzas, I., Shreve, S.E., *Brownian Motion and Stochastic Calculus*, 世界圖書出版公司, 2006

105. Kassel, C., *Quantum Groups*, 世界圖書出版公司, 2000

106. Katz, V., *A History of Mathematics: An Introduction*, 機械工業出版社, 2012

107. Kelley, J.L., *General Topology*, 世界圖書出版公司, 2000

108. Knapp, A.W., *Representations Theory of Semisimple Groups: An Overview Based on Examples (1, 2)*, 世界圖書出版公司, 2011

109. Knuth, D.E., *The Art of Computer Programming (1, 2, 3)*, 清華大學出版社, 2002

110. Kobayashi, S. (小林昭七), *Differential Geometry of Complex Vector Bundles*, 世界圖書出版公司, 1987

111. Kobayashi, S. (小林昭七), *Transformation Groups in Differential Geometry*, 世界圖書出版公司, 2010

112. Koblitz, N., *p-adic Numbers, p-adic Analysis and Zeta-functions*, 世界圖書出版公司, 2009

113. Kodaira, K., *Complex Manifolds and Deformation of Complex Structures*, 世界圖書出版公司, 2008

114. Lang, S., *Algebra*, 世界圖書出版公司, 2004

115. Lang, S., *Algebraic Number Theory*, 世界圖書出版公司, 2003

116. Lang, S., *Cyclotomic Fields I and II*, 世界圖書出版公司, 2009

117. Lang, S., *Elliptic Functions*, 世界圖書出版公司, 2011

118. Lawson, H.B., Michelsohn, M.-L., *Spin Geometry*, 世界圖書出版公司, 2011

119. Lax, P.D., *Functional Analysis*, 高等教育出版社, 2007

120. Lee, J.M., *Riemannian Manifolds: An Introduction to Curvature*, 世界圖書出版公司, 2003

121. Lindenstrauss, J., Tzafriri, L., *Classical Banach Spaces (I, II)*, 世界圖書出版公司, 2010

122. Loéve, M., *Probability Theory (1, 2)*, 世界圖書出版公司, 2000

123. Mac Lane, S., *Categries for the Working Mathematician*, 世界圖書出版公司, 1988

124. Malliavin, P., *Stochastic Analysis*, 世界圖書出版公司, 2003

125. Mac Lane, S., *Homology*, 世界圖書出版公司, 2009

126. Manin, Y.I., *A Course in Mathematical Logic*, 世界圖書出版公司, 1988

127. Marsden, J.E., Ratiu, T.S., *Introduction to Mechanics and Symmetry: A Basic Exposition of Classical Mechanical Systems*, 世界圖書出版公司, 2008

128. Matoušek, J., *Lectures on Discrete Geometry*, 世界圖書出版公司, 2011

129. Milnor, J., *Dynamics in One Complex Variable*, 世界圖書出版公司, 2013

130. Milnor, J., *Morse Theory*, 世界圖書出版公司, 2011

131. Milnor, J., *Topology from the Differentiable Viewpoint* (雙語版), 人民郵電出版社, 2008

132. Milnor, J., Stasheff, J.D., *Characteristic Classes*, 世界圖書出版公司, 2009

133. Morgan, F., *Geometric Measure Theory: A Beginner's Guide*, 世界圖書出版公司, 2009

134. Morrey, C.B. Jr., *Multiple Integrals in the Calculus of Variations*, 世界圖書出版公司, 2013

135. Mumford, D., *Algebraic Geometry I: Complex Projective Varieties*, 世界圖書出版公司, 2008

136. Mumford, D., Fogarty, J., Kirwan, F., *Geometric Invariant Theory*, 世界圖書出版公司, 2012

137. Munkres, J.R., *Topology*, 機械工業出版社, 2004

138. Nathanson, M.B., *Additive Number Theory: Inverse Problems and the Geometry of Sumsets*, 世界圖書出版公司, 2012

139. Olver, P.J., *Applications of Lie Groups to Differential Equations*, 世界圖書出版公司, 1999

140. Pazy, A., *Semigroups of Linear Operators and Applications to Partial Differential Equations*, 世界圖書出版公司, 2006

141. Petersen, P., *Riemannian Geometry*, 科學出版社, 2007

142. Pierce, R.S., *Associative Algebras*, 世界圖書出版公司, 1988

143. Pólya, G., Szegö, G., *Problems and Theorems in Analysis (I, II)*, 世界圖書出版公司, 2004

144. Robinson, D.J.S., *A Course in the Theory of Groups*, 世界圖書出版公司, 2000

145. Rockafellar, T., *Convex Analysis*, 世界圖書出版公司, 2011

146. Rosen, M., *Number Theory in Function Fields*, 世界圖書出版公司, 2011

147. Rosenberg, J., *Algebraic K-Theory and Its Applications*, 世界圖書出版公司, 2010

148. Rudin, W., *Function Theory in the Unit Ball of* $\mathbb{C}^n$, 世界圖書出版公司, 2013

149. Rudin, W., *Functional Analysis*, 機械工業出版社, 2012

150. Rudin, W., *Real and Complex Analysis*, 世界圖書出版公司, 1990

151. Sachs, R.K., Wu, H.-H., *General Relativity for Mathematicians*, 世界圖書出版公司, 2000

152. Sagan, B.E., *The Symmetric Group: Representations, Combinatorial Algorithms, and Symmetric Functions*, 世界圖書出版公司, 2012

153. Sakai, S., C^*-*Algebras and* W^*-*Algebras*, 世界圖書出版公司, 2009

154. Santaló, L., *Integrals Geometry and Geometric Probability*, 世界圖書出版公司, 2009

155. Schaefer, H.H., *Topological Vector Spaces*, 世界圖書出版公司, 2009

156. Serre, J.-P., *A Course in Arithmetic*, 世界圖書出版公司, 2009

157. Serre, J.-P., *Linear Representations of Finite Groups*, 世界圖書出版公司, 2008

158. Serre, J.-P., *Local Fields*, 世界圖書出版公司, 2008

159. Shafarevich, I.R., *Basic Algebraic Geometry (1, 2)*, 世界圖書出版公司, 1998

160. Shafarevich, I.R., *Basic Notions of Algebra*, 科學出版社, 2006

161. Sharpe, R.W., *Differential Geometry: Cartan's Generalization of Klein's Erlangen Program*, 世界圖書出版公司, 2011

162. Shiryaev, A.N., *Probability*, 世界圖書出版公司, 1988

163. Silverman, J.H., *Advanced Topics in the Arithmetic of Elliptic Curves*, 世界圖書出版公司, 2010

164. Silverman, J.H., *The Arithmetic of Elliptic Curves*, 世界圖書出版公司, 2011

165. Spanier, E.H., *Algebraic Topology*, 世界圖書出版公司, 2008

166. Spitzer, F., *Principles of Random Walk*, 世界圖書出版公司, 1988

167. Spivak, M., *Calculus on Manifolds* (雙語版), 人民郵電出版社, 2006

168. Springer, T.A., *Linear Algebraic Groups*, 世界圖書出版公司, 2013

169. Stanley, R.P., *Enumerative Combinatorics (1, 2)*, 機械工業出版社, 2004

170. Steenrod, N., *The Topology of Fibre Bundles*, 世界圖書出版公司, 2011

171. Stein, E.M., *Harmonic Analysis*, 世界圖書出版公司, 2006

172. Stein, E.M., *Singular Integrals and Differentiability Properties of Functions*, 世界圖書出版公司, 2011

173. Stein, E.M., Weiss, G., *Introduction to Fourier Analysis on Euclidean Spaces*, 世界圖書出版公司, 2009

174. Sternberg, S., *Group Theory and Physics*, 世界圖書出版公司, 2000

175. Struwe, M., *Variational Methods: Applications to Nonlinear Partial Differential Equations and Hamiltonian Systems*, 世界圖書出版公司, 2012

176. Switzer, R.M., *Algebraic Topology: Homotopy and Homology*, 世界圖書出版公司, 2004

177. Taylor, M.E., *Partial Differential Equations (I, II, III)*, 世界圖書出版公司, 1999

178. Thirring, W., *Classical Mathematical Physics: Dynamical Systems and Field Theories*, 世界圖書出版公司, 2008

179. van Lint, J.H., *Introduction to Coding Theory*, 世界圖書出版公司, 2003

180. Varadarajan, V.S., *Lie Groups, Lie Algebras and Their Representations*, 世界圖書出版公司, 2008

181. Vaughan, R.C., *The Hardy–Littlewood Method*, 世界圖書出版公司, 1998

182. Walters, P., *An Introduction to Ergodic Theory*, 世界圖書出版公司, 2000

183. Warner, F.W., *Foundations of Differentiable Manifolds and Lie Groups*, 世界圖書出版公司, 2004

184. Washington, L.C., *Introduction to Cyclotomic Fields*, 世界圖書出版公司, 1997

185. Weil, A., *Basic Number Theory*, 世界圖書出版公司, 2010

186. Weil, A., *Elliptic Functions According to Eisenstein and Kronecker*, 世界圖書出版公司, 2009

187. Wells, R.O.Jr., *Differential Analysis on Complex Manifolds*, 世界圖書出版公司, 2004

APPENDIX B: BOOKS REPRINTED IN THE MAINLAND OF CHINA

188. Weyl, H., *The Classical Groups: Their Invariants and Representations*, 世界圖書出版公司, 2011

189. Whitehead, G.W., *Elements of Homotopy Theory*, 世界圖書出版公司, 1988

190. Whittaker, E.T., Watson, G.N., *A Course of Modern Analysis*, 世界圖書出版公司, 2008

191. Yosida, K., *Functional Analysis*, 世界圖書出版公司, 1999

192. Zariski, O., *Algebraic Surfaces*, 世界圖書出版公司, 2010

193. Zariski, O., Samuel, P., *Commutative Algebra (1, 2)*, 世界圖書出版公司, 2000

194. Zeidler, E., *Nonlinear Functional Analysis and Its Applications*, 世界圖書出版公司, 2009

195. Ziemer, W.P., *Weakly Differentiable Functions: Sobolev Spaces and Functions of Bounded Variation*, 世界圖書出版公司, 1999

196. Zygmund, A., *Trigonometric Series*, 機械工業出版社, 2005

Index

A

Abikoff
    The Real Analytic Theory of Teichmüller Space, 136
Abraham, Marsden
    Foundations of Mechanics, 512
Abramenko, Brown
    Buildings. Theory and Applications, 455
Acton
    Numerical Methods that Work, 611
Adams, J.
    Stable Homotopy and Generalised Homology, 332
Adams, R., Fournier
    Sobolev Spaces, 427
Adem, Milgram
    Cohomology of Finite Groups, 336
Agmon
    Lectures on Elliptic Boundary Value Problems, 422
Ahlfors
    Complex Analysis, 115
    Conformal Invariants: Topics in Geometric Function Theory, 133
    Lectures on Quasiconformal Mappings, 132
Aigner
    Combinatorial Theory, 544
Aigner, Ziegler
    Proofs from The Book, 107
Ainsworth, Oden
    A Posteriori Error Estimation in Finite Element Analysis, 620
Akhiezer, Glazman
    Theory of Linear Operators in Hilbert Space, 170
Akivis, Rosenfeld
    Élie Cartan (1869–1951), 75

Aleksandrov, Kolmogorov, Lavrent'ev
    Mathematics: Its Content, Methods, and Meaning, 12
Alexandroff
    Elementary Concepts of Topology, 38
Alexandroff, Hopf
    Topologie I, 316
Alexandrov (or Alexandroff)
    Combinatorial Topology, 317
Alfsen
    Compact Convex Sets and Boundary Integrals, 312
Alligood, Sauer, Yorke
    Chaos. An Introduction to Dynamical Systems, 521
Alon, Spencer
    The Probabilistic Method, 576
Ambrosio, Fusco, Pallara
    Functions of Bounded Variation and Free Discontinuity Problems, 432
Anderson, Fuller
    Rings and Categories of Modules, 216
Anderson, Guionnet, Zeitouni
    An Introduction to Random Matrices, 581
Andrews
    The Theory of Partitions, 406
Apostol
    Calculus, 84
    Introduction to Analytic Number Theory, 377
Arbarello, Cornalba, Griffiths
    Geometry of Algebraic Curves, II, 298
Arbarello, Cornalba, Griffiths, Harris
    Geometry of Algebraic Curves, I, 297
Arnold, L.
    Random Dynamical Systems, 523
Arnold, V.
    Geometrical Methods in the Theory of Ordinary Differential Equations, 410

Mathematical Methods of Classical Mechanics, 512
Ordinary Differential Equations, 410
Artin, E.
Galois Theory, 364
Geometric Algebra, 452
Artin, M.
Algebra, 199
Ash, Gross
Fearless Symmetry. Exposing the Hidden Patterns of Numbers, 33
Astala, Iwaniec, Martin
Elliptic Partial Differential Equations and Quasiconformal Mappings in the Plane, 134
Atiyah
K-theory, 333
Atiyah, MacDonald
Introduction to Commutative Algebra, 219
Attouch
Variational Convergence for Functions and Operators, 269
Aubin
Nonlinear Analysis on Manifolds. Monge Ampére Equations, 260
Some Nonlinear Problems in Riemannian Geometry, 260
Aubin, Cellina
Differential Inclusions, 311
Aubin, Ekeland
Applied Nonlinear Analysis, 102
Aubin, Frankowska
Set-valued Analysis, 102
Auslander, Reiten, Smalo
Representation Theory of Artin Algebras, 492
Axler
Linear Algebra Done Right, 190

B

Babin, Vishik
Attractors of Evolution Equations, 527
Baker
Principles of Geometry, 283
Transcendental Number Theory, 382
Ballmann, Gromov, Schroeder
Manifolds of Nonpositive Curvature, 253

Banach
Theory of Linear Operations, 163
Bannai, Ito
Algebraic Combinatorics. I. Association Schemes, 539
Bardi, Capuzzo-Dolcetta
Optimal Control and Viscosity Solutions of Hamilton-Jacobi-Bellman Equations, 434
Barth, Hulek, Peters, Ven
Compact Complex Surfaces, 299
Bass
Algebraic K-theory, 335
Baxter
Exactly Solved Models in Statistical Mechanics, 511
Beardon
Iteration of Rational Functions, 531
The Geometry of Discrete Groups, 468
Beauville
Surfaces Algebriques Complexes, 300
Beckenbach, Bellman
Inequalities, 111
Becker, Weispfenning
Gröbner Bases. A Computational Approach to Commutative Algebra, 542
Bell
Men of Mathematics, 42
Benedetti, Petronio
Lectures on Hyperbolic Geometry, 348
Bennett, Sharpley
Interpolation of Operators, 157
Benson
Representations and Cohomology, 336
Bensoussan, Lions, Papanicolaou
Asymptotic Analysis for Periodic Structures, 417
Berge
Graphs, 549
Bergh, Löfström
Interpolation Spaces. An Introduction, 156
Berlekamp, Conway, Guy
Winning Ways for Your Mathematical Plays, 608
Berline, Getzler, Vergne
Heat Kernels and Dirac Operators, 262
Berman, Plemmons
Nonnegative Matrices in the Mathematical

Sciences, 193
Bertoin
    Lévy Processes, 569
Besicovitch
    Almost Periodic Functions, 143
Besse
    Einstein Manifolds, 248
    Manifolds all of Whose Geodesics are Closed, 248
Bhatia
    Matrix Analysis, 193
Bhatia, Szegö
    Stability Theory of Dynamical Systems, 524
Biggs
    Algebraic Graph Theory, 548
Billingsley
    Convergence of Probability Measures, 560
    Probability and measure, 560
Bingham, Goldie, Teugels
    Regular variation, 94
Birkhoff, Ga.
    Lattice Theory, 200
Birkhoff, Ge.
    Dynamical Systems, 519
Birkhoff, Mac Lane
    A Survey of Modern Algebra, 196
Birman
    Braids, Links, and Mapping Class Groups, 351
Bishop, Crittenden
    Geometry of Manifolds, 236
Bix
    Conics and Cubics, 289
Bjorck, Dahlquist
    Numerical Methods, 612
Blair
    Riemannian Geometry of Contact and Symplectic Manifolds, 292
Bochnak, Coste, Roy
    Real Algebraic Geometry, 290
Bohner, Peterson
    Dynamic Equations on Time Scales, 525
Bohr
    Almost Periodic Functions, 144
Bollobás
    Graph Theory. An Introductory Course, 546
    Modern Graph Theory, 546
    Random Graphs, 552
Bolza
    Lectures on the Calculus of Variations, 270
Bombieri, Gubler
    Heights in Diophantine Geometry, 390
Bondy, Murty
    Graph Theory with Applications, 549
Borel
    Essays in the History of Lie Groups and Algebraic Groups, 74
    Introduction aux groupes arithmétiques, 463
    Linear Algebraic Groups, 445
Borel, Casselman
    Automorphic Forms, Representations and L-functions, 403
Borel, Wallach
    Continuous Cohomology, Discrete Subgroups, and Representations of Reductive Groups, 404
Borevich, Shafarevich
    Number Theory, 362
Borwein, Erdélyi
    Polynomials and Polynomial Inequalities, 105
Bosch, Güntzer, Remmert
    Non-Archimedean Analysis. A Systematic Approach to Rigid Analytic Geometry, 393
Bosch, Lütkebohmert, Raynaud
    Néron Models, 393
Bott, Tu
    Differential Forms in Algebraic Topology, 341
Bourbaki
    Commutative Algebra, 220
    Elements of the History of Mathematics, 63
    Lie Groups and Lie Algebras, 442
Bowen
    Equilibrium States and the Ergodic Theory of Anosov Diffeomorphisms, 535
Boyd, El Ghaoui, Feron, Balakrishnan
    Linear Matrix Inequalities in System and Control Theory, 113
Boyd, Vandenberghe
    Convex Optimization, 608
Boyer
    A History of Mathematics, 61
Bredon
    Introduction to Compact Transformation

Groups, 475
Topology and Geometry, 327
Breiman
Probability, 557
Brenner, Scott
The Mathematical Theory of Finite Element Methods, 617
Brezis
Analyse Fonctionnelle. Théorie et Applications, 184
Functional Analysis, Sobolev Spaces and Partial Differential Equations, 184
Operateurs Maximaux Monotones et Semi-groupes de Contractions dans les espaces de Hilbert, 185
Brezzi, Fortin
Mixed and Hybrid Finite Element Methods, 617
Bridson, Haefliger
Metric Spaces of Non-positive Curvature, 255
Brieskorn, Knörrer
Plane Algebraic Curves, 298
Brocker, tom Dieck
Representations of Compact Lie Groups, 492
Brouwer, Cohen, Neumaier
Distance-regular Graphs, 550
Browder
Surgery on Simply-connected Manifolds, 352
Brown
Buildings, 454
Cohomology of Groups, 335
Bruns, Herzog
Cohen-Macaulay Rings, 223
Buhler
Gauss. A Biographical Study, 53
Bump
Automorphic Forms and Representations, 404
Burago, D., Burago, Y., Ivanov
A Course in Metric Geometry, 254
Burago, Zalgaller
Geometric Inequalities, 112
Burde, Zieschang
Knots, 329
Burnside
Theory of Groups of Finite Order, 207
Burris, Sankappanavar
A Course in Universal Algebra, 203

Busemann
The Geometry of Geodesics, 240
Buser
Geometry and Spectra of Compact Riemann Surfaces, 139
Byron, Fuller
Mathematics of Classical and Quantum Physics, 498

C

Caffarelli, Cabré
Fully Nonlinear Elliptic Equations, 429
Canary, Epstein, Green
Notes on notes of Thurston, 346
Canuto, Hussaini, Quarteroni, Zang
Spectral Methods in Fluid Dynamics, 432
Carleson, Gamelin
Complex Dynamics, 531
Carslaw, Jaeger
Conduction of Heat in Solids, 426
Cartan
Elementary Theory of Analytic Functions of One or Several Complex Variables, 115
Cartan, Eilenberg
Homological Algebra, 226
Carter
Finite Groups of Lie Type. Conjugacy Classes and Complex Characters, 206
Simple Groups of Lie Type, 205
Cassels
An Introduction to Diophantine Approximation, 383
An Introduction to the Geometry of Numbers, 373
Local Fields, 371
Rational Quadratic Forms, 393
Cassels, Fröhlich
Algebraic Number Theory, 367
Casson, Bleiler
Automorphisms of Surfaces after Nielsen and Thurston, 347
Castaing, Valadier
Convex Analysis and Measurable Multifunctions, 310
Cercignani
The Boltzmann Equation and Its Applications, 510

INDEX

Cercignani, Illner, Pulvirenti
    The Mathematical Theory of Dilute Gases, 510
Cesari
    Optimization—Theory and Applications. Problems with Ordinary Differential Equations, 610
    symptotic Behavior and Stability Problems in Ordinary Differential Equations, 409
Chang
    Infinite-dimensional Morse Theory and Multiple Solution Problems, 420
Chang, Keisler
    Model Theory, 591
Chavel
    Eigenvalues in Riemannian Geometry, 261
    Riemannian Geometry–A Modern Introduction, 237
Cheeger, Ebin
    Comparison Theorems in Riemannian Geometry, 241
Chen, Shaw
    Partial Differential Equations in Several Complex Variables, 125
Chern
    Complex Manifolds Without Potential Theory, 273
Chevalley
    Theory of Lie Groups. I. , 440
Chihara
    An Introduction to Orthogonal Polynomials, 106
Chow, Hale
    Methods of Bifurcation Theory, 411
Chriss, Ginzburg
    Representation Theory and Complex Geometry, 491
Chung
    A Course in Probability Theory, 556
    Spectral Graph Theory, 551
Church
    Introduction to Mathematical Logic, 584
Ciarlet
    The Finite Element Method for Elliptic Problems, 620
Clarke
    Optimization and Nonsmooth Analysis, 103
Clemens
    A Scrapbook of Complex Curve Theory, 298
Clifford, Preston
    The Algebraic Theory of Semigroups, 203
Coddington, Levinson
    Theory of Ordinary Differential Equations, 408
Cohen, H.
    A Course in Computational Algebraic Number Theory, 370
    Advanced Topics in Computational Number Theory, 370
Cohen, P.
    Set Theory and the Continuum Hypothesis, 589
Cohn
    Measure Theory, 98
Collet, Eckmann
    Iterated Maps on the Interval as Dynamical Systems, 522
Colton, Kress
    Integral Equation Methods in Scattering Theory, 517
    Inverse Acoustic and Electromagnetic Scattering Theory, 419
Comtet
    Advanced Combinatorics. The Art of Finite and Infinite Expansions, 539
Conley
    Isolated Invariant Sets and the Morse Index, 412
Connes
    Noncommutative Geometry, 178
Constantin, Foias
    Navier-Stokes Equations, 434
Conway, J.B.
    A Course in Functional Analysis, 167
Conway, J.H.
    On Numbers and Games, 607
Conway, J.H., Burgiel, Goodman-Strauss
    The Symmetries of Things, 30
Conway, J.H., Curtis, Norton, Parker, Wilson
    Atlas of Finite Groups. Maximal Subgroups and Ordinary Characters for Simple Groups, 212
Conway, Sloane
    Sphere Packings, Lattices and Groups, 545
Cormen, Leiserson, Rivest
    Introduction to Algorithms, 603

Cornfeld, Fomin, Sinai
  Ergodic Theory, 534
Corwin, Greenleaf
  Representations of Nilpotent Lie Groups and Their Applications. Part I. Basic Theory and Examples, 492
Courant, Friedrichs
  Supersonic Flow and Shock Waves, 433
Courant, Hilbert
  Methods of Mathematical Physics. Vol. I., Vol. II., 496
Courant, John
  Introduction to Calculus and Analysis, 83
Courant, Robbins
  What Is Mathematics?, 6
Cover, Thomas
  Elements of Information Theory, 596
Cowen, MacCluer
  Composition Operators on Spaces of Analytic Functions, 155
Cox
  Primes of the Form $x^2 + ny^2$. Fermat, Class Field Theory and Complex Multiplication, 58
Cox, Little, O'Shea
  Ideals, Varieties, and Algorithms, 224
  Using Algebraic Geometry, 541
Coxeter
  Introduction to Geometry, 39
  Projective Geometry, 40
  Regular Polytopes, 41
Coxeter, Greitzer
  Geometry Revisited, 40
Coxeter, Moser
  Generators and Relations for Discrete Groups, 470
Crilly
  Arthur Cayley, 54
Croft, Falconer, Guy
  Unsolved Problems in Geometry, 110
Crowell, Fox
  Introduction to Knot Theory, 330
Csiszár, Körner
  Information Theory. Coding Theorems for Discrete Memoryless Systems, 597
Curtain, Zwart
  An Introduction to Infinite-dimensional Linear Systems Theory, 528
Curtis
  Pioneers of Representation Theory: Frobenius, Burnside, Schur, and Brauer, 68
Curtis, Reiner
  Methods of Representation Theory, 215
Curtis, Reiner
  Representation Theory of Finite Groups and Associative Algebras, 214
Cvetković, Doob, Sachs
  Spectra of Graphs, 551

D

Da Prato, Zabczyk
  Stochastic Equations in Infinite Dimensions, 572
Dacorogna
  Direct Methods in the Calculus of Variations, 268
Dafermos
  Hyperbolic Conservation Laws in Continuum Physics, 436
Dal Maso
  An Introduction to Γ-convergence, 269
Daubechies
  Ten Lectures on Wavelets, 160
Dauben
  Georg Cantor, 50
Davenport
  Multiplicative Number Theory, 381
  The Higher Arithmetic. An Introduction to the Theory of Numbers, 361
David, Semmes
  Analysis of and on Uniformly Rectifiable Sets, 101
Davies
  Heat Kernels and Spectral Theory, 262
Davis
  Circulant Matrices, 194
  The Geometry and Topology of Coxeter Groups, 475
Davis, Rabinowitz
  Methods of Numerical Integration, 614
de Azcárraga, Izquierdo
  Lie Groups, Lie Algebras, Cohomology and Some Applications in Physics, 483
de Boor

A Practical Guide to Splines, 622
de la Harpe
    Topics in Geometric Group Theory, 470
de Melo, van Strien
    One-dimensional Dynamics, 522
Deift
    Orthogonal Polynomials and Random Matrices: a Riemann-Hilbert Approach, 581
Deimling
    Nonlinear Functional Analysis, 185
Deligne
    Cohomologie étale, 391
Demazure, Gabriel
    Groupes algébriques. Tome I: Géométrie algébrique, généralités, groupes commutatifs, 447
    Introduction to Algebraic Geometry and Algebraic Groups, 447
Dembo, Zeitouni
    Large Deviations Techniques and Applications, 574
Dembowski
    Finite Geometries, 457
Demidovich
    Problems in Mathematical Analysis, 108
Dereziński, Gérard
    Scattering Theory of Classical and Quantum N-particle Systems, 515
Derksen, Kemper
    Computational Invariant Theory, 451
Deuschel, Stroock
    Large Deviations, 574
Devaney
    An Introduction to Chaotic Dynamical Systems, 521
DeVore, Lorentz
    Constructive Approximation, 621
Dickson
    History of the Theory of Numbers. I, II, III, 70
Diestel
    Graph Theory, 547
Diestel, Jarchow, Tonge
    Absolutely Summing Operators, 168
Diestel, Uhl
    Vector Measures, 100
Dieudonné
    A History of Algebraic and Differential Topology 1900–1960, 72
    Foundations of Modern Analysis, 91
Dirac
    General Theory of Relativity, 504
Dixmier
    C^*-algebras, 181
    von Neumann Algebras, 180
Dixon, Mortimer
    Permutation Groups, 210
do Carmo
    Differential Geometry of Curves and Surfaces, 237
    Riemannian Geometry, 237
Dolgachev
    Lectures on Invariant Theory, 294
Donaldson, Kronheimer
    The Geometry of Four-manifolds, 350
Doob
    Classical Potential Theory and Its Probabilistic Counterpart, 565
    Stochastic Processes, 565
Doyle, Snell
    Random Walks and Electric Networks, 563
Drake, Singh
    Intermediate Set Theory, 587
Du Sautoy
    Symmetry: A Journey into the Patterns of Nature, 28
Dubrovin, Fomenko, Novikov
    Modern Geometry—Methods and Applications, 238
Dudley
    Real Analysis and Probability, 560
Dugundji
    Topology, 320
Dunford, Schwartz
    Linear Operators, 175
Dunham
    Euler: the Master of Us All, 54
Duoandikoetxea
    Fourier Analysis, 142
Duren
    Theory of H^p Spaces, 153
Durrett
    Probability: Theory and Examples, 558
Duvaut, Lions

Inequalities in Mechanics and Physics, 114
Dym, McKean
    Fourier Series and Integrals, 143

E

Edwards
    Riemann's Zeta Function, 375
Eichler, Zagier
    The Theory of Jacobi Forms, 405
Eilenberg
    Automata, Languages, and Machines, 603
Eilenberg, Steenrod
    Foundations of Algebraic Topology, 325
Einstein
    The Meaning of Relativity, 503
Eisenbud
    Commutative Algebra. With a View Toward Algebraic Geometry, 221
Eisenhart
    Riemannian Geometry, 239
Ekeland, Témam
    Convex Analysis and Variational Problems, 310
Ellis
    Entropy, Large Deviations, and Statistical Mechanics, 574
Engel, Nagel
    One-parameter Semigroups for Linear Evolution Equations, 176
Engelking
    General Topology, 320
Engl, Hanke, Neubauer
    Regularization of Inverse Problems, 419
Enriques
    Le Superficie Algebriche, 301
Epstein, Cannon, Holt, Levy, Paterson, Thurston
    Word Processing in Groups, 471
Erdélyi, Magnus, Oberhettinger, Tricomi
    Higher Transcendental Functions, 86
Ethier, Kurtz
    Markov Processes, 571
Evans
    Partial Differential Equations, 413
Evans, Gariepy
    Measure Theory and Fine Properties of Functions, 99
Eves
    An Introduction to the History of Mathematics, 62

F

Faddeev, Takhtajan
    Hamiltonian Methods in the Theory of Solitons, 517
Falconer
    Fractal Geometry. Mathematical Foundations and Applications, 270
    The Geometry of Fractal Sets, 271
Faltings, Chai
    Degeneration of Abelian Varieties, 387
Faraut, Korányi
    Analysis on Symmetric Cones, 257
Farkas, Kra
    Riemann Surfaces, 131
Federer
    Geometric Measure Theory, 265
Feit
    The Representation Theory of Finite Groups, 213
Feller
    An Introduction to Probability Theory and its Applications, 555
Fenchel, Nielsen
    Discontinuous Groups of Isometries in the Hyperbolic Plane, 468
Feynman, Leighton, Sands
    The Feynman Lectures on Physics, 502
Field, Golubitsky
    Symmetry in Chaos. A Search for Pattern in Mathematics, Art, and Nature, 31
Fine
    Basic Hypergeometric Series and Applications, 406
Fleming, Soner
    Controlled Markov Processes and Viscosity Solutions, 566
Fletcher, Markovic
    Quasiconformal Maps and Teichmüller Theory, 138
Folland
    Harmonic Analysis in Phase Space, 160
    Real Analysis. Modern Techniques and Their Applications, 92
Forster

INDEX

Lectures on Riemann Surfaces, 131
Forsythe, Wasow
    Finite-difference Methods for Partial Differential Equations, 618
Fraleigh
    A First Course in Abstract Algebra, 200
Frankel
    The Geometry of Physics, 239
Frenkel, Lepowsky, Meurman
    Vertex Operator Algebras and the Monster, 481
Fricke, Klein
    Vorlesungen über die Theorie der elliptischen Modulfunctionen, 397
    Vorlesungen uber die Theorie der automorphen Funktionen, 396
Fried, Jarden
    Field Arithmetic, 394
Friedman
    Partial Differential Equations of Parabolic Type, 426
Fuchs
    Infinite Abelian Groups, 473
Fukushima, Oshima, Takeda
    Dirichlet Forms and Symmetric Markov Processes, 561
Fulton
    Algebraic Curves. An Introduction to Algebraic Geometry, 288
    Intersection Theory, 302
    Introduction to Intersection Theory in Algebraic Geometry, 304
    Introduction to Toric Varieties, 307
    Young Tableaux, 486
Fulton, Harris
    Representation Theory. A first Course, 489
Furstenberg
    Recurrence in Ergodic Theory and Combinatorial Number Theory, 536

G

Galdi
    An Introduction to the Mathematical Theory of the Navier-Stokes Equations, 434
Gallot, Hulin, Lafontaine
    Riemannian Geometry, 238
Gantmacher
    The Theory of Matrices, 192
Garcia-Cuerva, Rubio de Francia
    Weighted Norm Inequalities and Related Topics, 113
Gardiner
    Teichmüller Theory and Quadratic Differentials, 137
Gardiner, Lakic
    Quasiconformal Teichmüller Theory, 138
Gardner
    Geometric Tomography, 312
Garey, Johnson
    Computers and Intractability, 602
Garnett
    Bounded Analytic Functions, 152
Garnett, Marshall
    Harmonic Measure, 153
Garrett
    Buildings and Classical Groups, 455
Gasper, Rahman
    Basic Hypergeometric Series, 405
Geck
    An Introduction to Algebraic Geometry and Algebraic Groups, 449
Gelbart
    Automorphic Forms on Adéle Groups, 401
Gelbaum, Olmsted
    Counterexamples in Analysis, 96
Gelfand
    Lectures on Linear Algebra, 188
Gelfand, Fomin
    Calculus of Variations, 268
Gelfand, Graev, Pyatetskii-Shapiro
    Representation Theory and Automorphic Functions, 172, 399
Gelfand, Kapranov, Zelevinsky
    Discriminants, Resultants, and Multidimensional Determinants, 308
Gelfand, Manin
    Methods of Homological Algebra, 229
Gelfand, Shilov, Vilenkin, Graev
    Generalized Functions, 172
Gelfond
    Transcendental and Algebraic Numbers, 383
George, Askey, Roy
    Special Functions, 87
Georgii

Gibbs Measures and Phase Transitions, 510
Giaquinta
  Multiple Integrals in the Calculus of Variations and Nonlinear Elliptic Systems, 269
Gilbarg, Trudinger
  Elliptic Partial Differential Equations of Second Order, 421
Gilkey
  Invariance Theory, the Heat Equation, and the Atiyah-Singer Index Theorem, 263
Gillman, Jerison
  Rings of Continuous Functions, 183
Girault, Raviart
  Finite Element Methods for Navier-Stokes Equations, 619
Giusti
  Minimal Surfaces and Functions of Bounded Variation, 252
Glimm, Jaffe
  Quantum Physics. A Functional Integral Point of View, 511
Godel
  The Consistency of the Continuum Hypothesis, 589
Godement
  Topologie algébrique et théorie des faisceaux, 127
Godsil, Royle
  Algebraic Graph Theory, 550
Goebel, Kirk
  Topics in Metric Fixed Point Theory, 321
Gohberg, Krein
  Introduction to the Theory of Linear Nonselfadjoint Operators, 170
Goldman
  Complex Hyperbolic Geometry, 468
Goldstein, Poole, Safko
  Classical Mechanics, 513
Golub, Van Loan
  Matrix Computations, 613
Golubitsky, Guillemin
  Stable Mappings and Their Singularities, 341
Golumbic
  Algorithmic Graph Theory and Perfect Graphs, 551
Goluzin
  Geometric Theory of Functions of a Complex Variable, 121
Gompf, Stipsicz
  4-manifolds and Kirby Calculus, 350
Goodman, Wallach
  Representations and Invariants of the Classical Groups, 488
  Symmetry, Representations, and Invariants, 489
Gorenstein
  Finite Groups, 208
Gorenstein, Lyons, Solomon
  The Classification of the Finite Simple Groups, 211
Goresky, MacPherson
  Stratified Morse theory, 343
Goursat
  A Course in Mathematical Analysis, 81
Gowers, Barrow-Green, Leader
  The Princeton Companion to Mathematics, 14
Graham, Knuth, Patashnik
  Concrete Mathematics. A Foundation for Computer Science, 595
Granas, Dugundji
  Fixed Point Theory, 321
Gratzer
  General Lattice Theory, 201
Grauert, Remmert
  Coherent Analytic Sheaves, 126
Greenleaf
  Invariant Means on Topological Groups and Their Applications, 536
Griffiths, Harris
  Principles of Algebraic Geometry, 286
Grimmett
  Percolation, 577
Grisvard
  Elliptic Problems in Nonsmooth Domains, 422
Grochenig
  Foundations of Time-frequency Analysis, 144
Gromov
  Metric Structures for Riemannian and Non-Riemannian Spaces, 253
  Asymptotic Invariants of Infinite Groups. Geometric Group Theory, 471
  Partial Differential Relations, 264
Grothendieck
  The Éléments de géométrie algébrique, 279
Grotschel, Lovász, Schrijver

Geometric Algorithms and Combinatorial Optimization, 609

Gruber
  Convex and Discrete Geometry, 311
Gruber, Lekkerkerker
  Geometry of Numbers, 373
Grünbaum
  Convex Polytopes, 540
Grunbaum, Shephard
  Tilings and Patterns, 31
  Tilings and Patterns. An Introduction, 32
Guckenheimer, Holmes
  Nonlinear Oscillations, Dynamical Systems, and Bifurcations of Vector Fields, 524
Guillemin, Pollack
  Differential Topology, 341
Guillemin, Sternberg
  Geometric Asymptotics, 416
  Symplectic Techniques in Physics, 290
Gunning
  Introduction to Holomorphic Functions of Several Variables, 124
Gunning, Rossi
  Analytic Functions of Several Complex Variables, 123
Guy
  Unsolved Problems in Number Theory, 109

H

Hairer, Norsett, Wanner
  Solving Ordinary Differential Equations, 615
Halberstam, Richert
  Sieve Methods, 379
Hale
  Asymptotic Behavior of Dissipative Systems, 526
  Theory of Functional Differential Equations, 530
Hale, Verduyn Lunel
  Introduction to Functional-differential Equations, 530
Hall
  The Theory of Groups, 209
Hall, Heyde
  Martingale Limit Theory and its Application, 569
Halmos
  A Hilbert Space Problem Book, 169
  Finite-dimensional Vector Spaces, 189
  I Want to Be a Mathematician. An Automathography in Three Parts, 22
  Measure Theory, 97
  Naive Set Theory, 587
Harary
  Graph Theory, 548
Hardy
  A Course of Pure Mathematics, 80
  A Mathematician's Apology, 19
  Divergent Series, 94
Hardy, Littlewood, Pólya
  Inequalities, 110
Hardy, Wright
  An Introduction to the Theory of Numbers, 356
Hargittai, I., Hargittai, M.
  Symmetry Through the Eyes of a Chemist, 35
  Symmetry: A Unifying Concept, 36
Harris
  Algebraic Geometry. A First Course, 287
Harris, Morrison
  Moduli of Curves, 296
Hartman
  Ordinary Differential Equations, 409
Hartshorne
  Algebraic Geometry, 285
Hasse
  Number Theory, 362
Hatcher
  Algebraic Topology, 326
Hawking, Ellis
  The Large Scale Structure of Space-time, 506
Hawkins
  Emergence of the Theory of Lie Groups. An Essay in the History of Mathematics 1869–1926, 73
Hayman
  Meromorphic Functions, 122
Haynes, Hedetniemi, Slater
  Fundamentals of Domination in Graphs, 550
Hecke
  Lectures on the Theory of Algebraic Numbers, 366
Heinonen
  Lectures on Analysis on Metric Spaces, 100

Heinonen, Kilpeläinen, Martio
  Nonlinear Potential Theory of Degenerate Elliptic Equations, 151
Helgason
  Differential Geometry and Symmetric Spaces, 256
  Differential Geometry, Lie Groups, and Symmetric Spaces, 256
  Geometric Analysis on Symmetric Spaces, 257
Hempel
  3-manifolds, 349
Henrici
  Applied and Computational Complex Analysis, 119
Henry
  Geometric Theory of Semilinear Parabolic Equations, 425
Herstein
  Topics in Algebra, 200
Hewitt, Ross
  Abstract Harmonic Analysis, 145
Hilbert
  The Theory of Algebraic Number Fields, 365
  Theory of Algebraic Invariants, 450
Hilbert, Cohn-Vossen
  Geometry and the Imagination, 37
Hille, Phillips
  Functional Analysis and Semi-groups, 177
Hiller
  Geometry of Coxeter Groups, 474
Hindry, Silverman
  Diophantine Geometry. An Introduction, 389
Hiriart-Urruty, Lemaréchal
  Convex Analysis and Minimization Algorithms, 607
Hirsch
  Differential Topology, 340
Hirsch, Smale
  Differential Equations, Dynamical Systems, and Linear Algebra, 411
Hirschfeld
  Projective Geometries Over Finite Fields, 456
Hirschfeld, Thas
  General Galois Geometries, 456
Hirzebruch
  Topological Methods in Algebraic Geometry, 289
Hodge
  The Theory and Applications of Harmonic Integrals, 342
Hodge, Pedoe
  Methods of Algebraic Geometry, 282
Hodges A.
  Alan Turing: the Enigma, 47
Hodges W.
  Model Theory, 592
Hofer, Zehnder
  Symplectic Invariants and Hamiltonian Dynamics, 292
Hoffman P.
  The Man Who Loved Only Numbers, 49
Hoffman, K.
  Banach Spaces of Analytic Functions, 154
Hofstadter
  Gödel, Escher, Bach: an Eternal Golden Braid, 585
Hopcroft, Ullman
  Introduction to Automata Theory, Languages, and Computation, 603
Hopf
  Differential Geometry in the Large, 234
Hörmander
  An Introduction to Complex Analysis in Several Variables, 272
  Lectures on Nonlinear Hyperbolic Differential Equations, 415
  Linear Partial Differential Operators, 414
  Notions of Convexity, 415
  The Analysis of Linear Partial Differential Operators. I-IV, 414
Horn, Johnson
  Matrix Analysis, 191
  Topics in Matrix Analysis, 192
Householder
  The Theory of Matrices in Numerical Analysis, 613
Hovey
  Model Categories, 337
Howie
  Fundamentals of Semigroup Theory, 204
Hsiang
  Cohomology Theory of Topological Transformation Groups, 476
Hua

INDEX

Additive Theory of Prime Numbers, 380
Introduction to Number Theory, 358
Hubbard
    Teichmüller Theory and Applications to Geometry, Topology, and Dynamics, 137
Hughes
    Random Walks and Random Environments, 577
Humphreys
    Introduction to Lie Algebras and Representation Theory, 443
    Linear Algebraic Groups, 446
    Reflection Groups and Coxeter Groups, 474
Huppert
    Endliche Gruppen. I., 210
Huppert, Blackburn
    Finite groups. II, III, 210
Hurewicz, Wallman
    Dimension Theory, 322
Huybrechts
    Complex Geometry. An Introduction, 274
Huybrechts, Lehn
    The Geometry of Moduli Spaces of Sheaves, 295

I

Ikeda, Watanabe
    Stochastic Differential Equations and Diffusion Processes, 572
Imayoshi, Taniguchi
    An Introduction to Teichmüller Spaces, 136
Ince
    Ordinary Differential Equations, 410
Ireland, Rosen
    A Classical Introduction to Modern Number Theory, 359
Isaacs
    Character Theory of Finite Groups, 214
Isakov
    Inverse Problems for Partial Differential Equations, 418
Ito, McKean
    Diffusion Processes and their Sample Paths, 567
Ivic
    The Riemann Zeta-function, 375
Iwaniec
    Topics in Classical Automorphic Forms, 402
Iwaniec, Kowalski
    Analytic Number Theory, 378

J

Jaco
    Lectures on Three-manifold Topology, 349
Jacobson
    Basic Algebra, 198
    Lie Algebras, 445
Jacod, Shiryaev
    Limit Theorems for Stochastic Processes, 568
Jacquet, Langlands
    Automorphic Forms on GL(2), 400
James, Kerber
    The Representation Theory of the Symmetric Group, 486
Janson, Svante, Luczak, Rucinski
    Random Graphs, 552
Jantzen
    Representations of Algebraic Groups, 452
Jech
    Set Theory, 588
Jikov, Kozlov, Oleinik
    Homogenization of Differential Operators and Integral Functionals, 579
John
    Partial Differential Equations, 416
Jost
    Riemannian Geometry and Geometric Analysis, 260
Joyce
    Compact Manifolds with Special Holonomy, 249
Jurdjevic
    Geometric Control Theory, 431

K

Kac, M.
    Enigmas of Chance, 25
    Statistical Independence in Probability, Analysis and Number Theory, 557
Kac, M., Ulam
    Mathematics and Logic, 10
Kac, V.
    Infinite-dimensional Lie Algebras, 479
Kadison, Ringrose

Fundamentals of the Theory of Operator
Algebras, 182
Kahane
Some Random Series of Functions, 577
Kailath
Linear Systems, 528
Kallenberg
Foundations of Modern Probability, 558
Kanigel
The Man Who Knew Infinity. A Life of the Genius Ramanujan, 49
Kapovich
Hyperbolic Manifolds and Discrete Groups, 348
Karatsuba, Voronin
The Riemann Zeta-function, 376
Karatzas, Shreve
Brownian Motion and Stochastic Calculus, 572
Methods of Mathematical Finance, 573
Karoubi
K-theory. An Introduction, 334
Kashiwara, Schapira
Sheaves on Manifolds, 127
Kassel
Quantum Groups, 480
Kato
Perturbation Theory for Linear Operators, 428
Katok, Hasselblatt
Introduction to the Modern Theory of Dynamical Systems, 520
Katz
A History of Mathematics. An Introduction, 60
Katz, Sarnak
Random Matrices, Frobenius Eigenvalues, and Monodromy, 581
Katznelson
An Introduction to Harmonic Analysis, 141
Kechris
Classical Descriptive Set Theory, 588
Kelley
General Topology, 319
Kempf, Knudsen, Mumford, Saint-Donat
Toroidal Embeddings. I, 307
Khinchin
Continued Fractions, 364
Three Pearls of Number Theory, 363
Kinderlehrer, Stampacchia
An Introduction to Variational Inequalities and Their Applications, 417
Kirby, Siebenmann
Foundational Essays on Topological Manifolds, Smoothings, and Triangulations, 353
Klee, Wagon
Old and New Unsolved Problems in Plane Geometry and Number Theory, 18
Kleene
Mathematical Logic, 590
Klein
Development of Mathematics in the 19th Century, 76
Lectures on Mathematics, 77
Lectures on the Icosahedron and the Solution of Equations of the Fifth Degree, 398
Klimek
Pluripotential Theory, 150
Kline
Mathematical Thought from Ancient to Modern Times, 65
Klingenberg
Riemannian Geometry, 242
Kloeden, Platen
Numerical Solution of Stochastic Differential Equations, 616
Knapp
Lie Groups Beyond an Introduction, 444
Representation Theory of Semisimple Groups. An Overview Based on Examples, 490
Knopp
Theory of Functions, 118
Knuth
The Art of Computer Programming, 595
Kobayashi
Differential Geometry of Complex Vector Bundles, 249
Hyperbolic Complex Spaces, 275
Hyperbolic Manifolds and Holomorphic Mappings, 275
Transformation Groups in Differential Geometry, 476
Kobayashi, Nomizu
Foundations of Differential Geometry, 245
Koblitz
p-adic Numbers, p-adic Analysis, and Zeta-functions, 360

Kodaira
    Complex Manifolds and Deformation of Complex Structures, 273
Kolchin
    Differential Algebra and Algebraic Groups, 460
Kollár
    Rational Curves on Algebraic Varieties, 296
Kollár, Mori
    Birational Geometry of Algebraic Varieties, 297
Kolmogorov
    Foundations of the Theory of Probability, 554
Kolmogorov, Fomin
    Elements of the Theory of Functions and Functional Analysis, 164
Komornik
    Exact Controllability and Stabilization, 528
Korner
    Fourier Analysis, 11
    The Pleasures of Counting, 10
Kosmann-Schwarzbach
    The Noether Theorems. Invariance and Conservation Laws in the Twentieth Century, 35
Krantz
    Function Theory of Several Complex Variables, 125
Kuang
    Delay Differential Equations with Applications in Population Dynamics, 525
Kuipers, Niederreiter
    Uniform Distribution of Sequences, 561
Kunen
    Set Theory. An Introduction to Independence Proofs, 590
Kunita
    Stochastic Flows and Stochastic Differential Equations, 573
Kuratowski
    Topology, 318
Kurosh
    The Theory of Groups, 209

L

Ladyenskaja, Solonnikov, Uralceva
    Linear and Quasilinear Equations of Parabolic Type, 426
Ladyzhenskaya
    The Mathematical Theory of Viscous Incompressible Flow, 435
Ladyzhenskaya, Uraltseva
    Linear and Quasilinear Elliptic Equations, 423
Lakshmikantham, Bainov, Simeonov
    Theory of Impulsive Differential Equations, 417
Lam
    Introduction to Quadratic Forms Over Fields, 391
Lamb
    Hydrodynamics, 437
Lancaster, Tismenetsky
    The Theory of Matrices, 193
Landau, Lifshitz
    Course of Theoretical Physics, 499
Landkof
    Foundations of Modern Potential Theory, 149
Lang
    Algebra, 197
    Algebraic Number Theory, 368
    Cyclotomic Fields, 369
    Elliptic Functions, 386
    Fundamentals of Diophantine Geometry, 389
    Introduction to Modular Forms, 401
Lange, Birkenhake
    Complex Abelian Varieties, 388
Langlands
    Base Change for GL(2), 403
    On the Functional Equations Satisfied by Eisenstein Series, 402
Lawler
    Intersections of Random Walks, 564
Lawson, Michelsohn
    Spin Geometry, 247
Lax
    Functional analysis, 165
    Hyperbolic Systems of Conservation Laws and the Mathematical Theory of Shock Waves, 433
Lazarsfeld
    Positivity in Algebraic Geometry, 305
Lebedev
    Special Functions and Their Applications, 89
Lederman, Hill
    Symmetry and the Beautiful Universe, 34
Ledoux, Talagrand
    Probability in Banach Spaces. Isoperimetry

Lee
- and Processes, 562
- Riemannian Manifolds. An Introduction to Curvature, 236

Lefschetz
- Algebraic Topology, 319

Lehner
- Discontinuous Groups and Automorphic Functions, 398

Lehto
- Univalent Functions and Teichmüller Spaces, 135

Lehto, Virtanen
- Quasiconformal Mappings in the Plane, 133

Lickorish
- An Introduction to Knot Theory, 330

Lidl, Niederreiter
- Finite Fields, 359

Lieb, Loss
- Analysis, 93

Lifshitz, Pitaevskii
- Course of Theoretical Physics, 501

Liggett
- Interacting Particle Systems, 579
- Stochastic Interacting Systems: Contact, Voter and Exclusion Processes, 579

Lind, Marcus
- An Introduction to Symbolic Dynamics and Coding, 523

Lindenstrauss, Tzafriri
- Classical Banach Spaces, 167

Lions
- Mathematical Topics in Fluid Mechanics, 435
- Quelques méthodes de résolution des problèmes aux limites non linéaires, 430

Littlewood
- Littlewood's Miscellany, 21

Loéve
- Probability Theory, 556

Loday
- Cyclic Homology, 231

Lojasiewicz
- Introduction to Complex Analytic Geometry, 126

Lorentz, Golitschek, Makovoz
- Constructive Approximation. Advanced Problems, 621

Lovász
- Combinatorial Problems and Exercises, 544

Lubotzky
- Discrete Groups, Expanding Graphs and Invariant Measures, 467

Lueck
- L^2-invariants: Theory and Applications to Geometry and K-theory, 478

Lunardi
- Analytic Semigroups and Optimal Regularity in Parabolic Problems, 178

Lusztig
- Introduction to Quantum Groups, 480

Lyndon, Schupp
- Combinatorial Group Theory, 470

M

Mac Lane
- Categories for the Working Mathematician, 230
- Homology, 227

Macdonald
- Symmetric Functions and Hall Polynomials, 484

Mackey
- Unitary Group Representations in Physics, Probability, and Number Theory, 497

MacLane, Birkhoff
- Algebra, 197

MacWilliams, Sloane
- The Theory of Error-correcting Codes, 600

Magnus
- Noneuclidean Tesselations and Their Groups, 397

Majid
- Foundations of Quantum Group Theory, 480

Malliavin
- Stochastic Analysis, 575

Mandelbrot
- The Fractal Geometry of Nature, 271

Mane
- Ergodic Theory and Differentiable Dynamics, 535

Manin
- A Course in Mathematical Logic, 590
- Cubic Forms. Algebra, Geometry, Arithmetic, 394

Marchenko

INDEX

Sturm Liouville Operators and Applications, 516

Marcus
  Number Fields, 367
Marden
  Outer Circles. An Introduction to Hyperbolic 3-manifolds, 348
Margulis
  Discrete Subgroups of Semisimple Lie Groups, 466
Marker
  Model theory. An Introduction, 592
Markushevich
  Theory of Functions of a Complex Variable, 120
Marsden, Ratiu
  Introduction to Mechanics and Symmetry. A Basic Exposition of Classical Mechanical Systems, 501
Marshall, Olkin
  Inequalities: Theory of Majorization and Its Applications, 112
Maskit
  Kleinian Groups, 469
Maslov, Fedoriuk
  Semiclassical Approximation in Quantum Mechanics, 416
Matousek
  Lectures on Discrete Geometry, 540
Matsumura
  Commutative Algebra, 221
  Commutative Ring Theory, 222
Mattila
  Geometry of Sets and Measures in Euclidean Spaces. Fractals and Rectifiability, 266
Mawhin, Willem
  Critical Point Theory and Hamiltonian Systems, 420
May
  Simplicial Objects in Algebraic Topology, 328
Maz'ja
  Sobolev Spaces, 427
McCleary
  User's Guide to Spectral Sequences, 230
McConnell, Robson
  Noncommutative Noetherian Rings, 224
McDuff, Salamon
  J-holomorphic Curves and Symplectic Topology, 291
  Introduction to Symplectic Topology, 291
McKean
  Stochastic Integrals, 567
McMullen
  Complex Dynamics and Renormalization, 532
  Renormalization and 3-manifolds Which Fiber Over the Circle, 532
McWeeny
  Symmetry: An Introduction to Group Theory and Its Applications, 30
Mehta
  Random Matrices, 580
Melrose
  The Atiyah-Patodi-Singer Index Theorem, 424
Meyer, Coifman
  Wavelets and Operators, 161
Meyn, Tweedie
  Markov Chains and Stochastic Stability, 566
Milman, Schechtman
  Asymptotic Theory of Finite-dimensional Normed Spaces, 313
Milne
  Étale Cohomology, 390
Milnor
  Dynamics in One Complex Variable. Introductory Lectures, 531
  Introduction to Algebraic K-theory, 334
  Lectures on the h-cobordism Theorem, 339
  Morse Theory, 339
  Singular Points of Complex Hypersurfaces, 289
  Topology from the Differentiable Viewpoint, 340
Milnor, Stasheff
  Characteristic Classes, 324
Misner, Thorne, Wheeler
  Gavitation, 507
Mitrinović
  Analytic Inequalities, 111
Miyake
  Modular Forms, 401
Moeglin, Waldspurger
  Spectral Decomposition and Eisenstein Series. Une paraphrase de l'Écriture, 403
Monastyrsky
  Riemann, Topology, and Physics, 52

Montgomery, D., Zippin
    Topological Transformation Groups, 477
Montgomery, H.
    Ten Lectures on the Interface Between Analytic Number Theory and Harmonic Analysis, 378
Montgomery, H., Vaughan
    Multiplicative Number Theory. I. Classical Theory, 381
Montgomery, R.
    A Tour of Subriemannian Geometries, Their Geodesics and Applications, 247
Montgomery, S.
    Hopf Algebras and Their Actions on Rings, 481
Moore
    Foundations of Point Set Theory, 319
Morgan
    Geometric Measure Theory, A Beginner's Guide, 267
Morrey
    Multiple Integrals in the Calculus of Variations, 267
Morrow, Kodaira
    Complex Manifolds, 274
Morse
    The Calculus of Variations in the Large, 339
Moschovakis
    Descriptive Set Theory, 588
Moser
    Stable and Random Motions in Dynamical Systems, 519
Mostow
    Strong Rigidity of Locally Symmetric Spaces, 464
Motohashi
    Spectral Theory of the Riemann Zeta-function, 376
Mukai
    An Introduction to Invariants and Moduli, 295
Mumford
    Abelian Varieties, 387
    Algebraic Geometry. I, 284
    Lectures on Curves on an Algebraic Surface, 302
    Tata Lectures on Theta, 387
    The Red book of Varieties and Schemes, 283
Mumford, Fogarty, Kirwan
    Geometric Invariant Theory, 293
Mumford, Series, Wright
    Indra's Pearls, 36
Munkres
    Elements of Algebraic Topology, 327
    Topology: A First Course, 327
Murray
    Mathematical Biology, 526

N

Nadler
    Continuum Theory. An Introduction, 320
Nag
    The Complex Analytic Theory of Teichmüller Spaces, 135
Nagata
    Local Rings, 223
    Modern Dimension Theory, 322
Nakajima
    Lectures on Hilbert Schemes of Points on Surfaces, 294
Nathanson
    Additive Number Theory. Inverse Problems and the Geometry of Sumsets, 380
Necas
    Les méthodes directes en théorie des équations elliptiques, 421
Nehari
    Conformal Mapping, 121
Neugebauer
    A History of Ancient Mathematical Astronomy, 70
Neukirch
    Algebraic Number Theory, 369
Neukirch, Schmidt, Wingberg
    Cohomology of Number Fields, 372
Newman
    The World of Mathematics. Vols. I-IV, 66
Newstead
    Introduction to Moduli Problems and Orbit Spaces, 293
Nielsen, Chuang
    Quantum Computation and Quantum Information, 597
Nocedal, Wright
    Numerical Optimization, 609

INDEX

Novikov, Manakov, Pitaevskii, Zakharov
: Theory of Solitons. The Inverse Scattering Method, 515

Nualart
: The Malliavin Calculus and Related Topics, 575

O

O'Meara
: Introduction to Quadratic Forms, 392

O'Neill
: Semi-Riemannian Geometry, 246

Oda
: Convex Bodies and Algebraic Geometry, 306

Olver, F.
: Asymptotics and Special Functions, 88

Olver, P.
: Applications of Lie Groups to Differential Equations, 461
: Classical Invariant Theory, 451
: Equivalence, Invariants, and Symmetry, 462

Onishchik, Vinberg
: Lie Groups and Algebraic Groups, 443

Orlik, Terao
: Arrangements of Hyperplanes, 344

Ortega, Rheinboldt
: Iterative Solution of Nonlinear Equations in Several Variables, 616

Oxley
: Matroid Theory, 542

P

Paley, Wiener
: Fourier Transforms in the Complex Domain, 143

Papadimitriou
: Computational Complexity, 604

Parshall, Rowe
: The Emergence of the American Mathematical Research Community, 1876–1900, 71

Passman
: The Algebraic Structure of Group Rings, 217

Payne, Thas
: Finite Generalized Quadrangles, 456

Pazy
: Semigroups of Linear Operators and Applications to Partial Differential Equations, 176

Pedersen
: C^*-algebras and Their Automorphism Groups, 182

Penrose
: The Road to Reality. A complete Guide to the Laws of the Universe, 508

Perko
: Differential equations and dynamical systems, 522

Pesin
: Dimension Theory in Dynamical Systems, 323

Petersen, K.
: Ergodic Theory, 534

Petersen, P.
: Riemannian Geometry, 241

Petkovsek, Wilf, Zeilberger
: $A = B$, 599

Petrovski
: Ordinary Differential Equations, 409

Phillips
: Scattering Theory for Automorphic Functions, 402

Pierce
: Associative Algebras, 215

Pisier
: The Volume of Convex Bodies and Banach Space Geometry, 313

Platonov, Rapinchuk
: Algebraic Groups and Number Theory, 448

Podlubny
: Fractional Differential Equations, 95

Pólya
: How to Solve it. A New Aspect of Mathematical Method, 15
: Mathematical Discovery. On Understanding, Learning, and Teaching Problem Solving, 17
: Mathematics and Plausible Reasoning, 16

Pólya, Szegö
: Isoperimetric Inequalities in Mathematical Physics, 263
: Problems and Theorems in Analysis, 104

Pommerenke
: Boundary Behaviour of Conformal Maps, 123

Pontryagin

Topological Groups, 145
Pressley, Segal
    Loop Groups, 479
Protter, Weinberger
    Maximum Principles in Differential Equations, 422

R

Rabinowitz
    Minimax Methods in Critical Point Theory with Applications to Differential Equations, 419
Rademacher, Toeplitz
    The Enjoyment of Mathematics; Selections from Mathematics for the Amateur, 8
Raghunathan
    Discrete Subgroups of Lie Groups, 464
Ransford
    Potential Theory in the Complex Plane, 150
Ratcliffe
    Foundations of Hyperbolic Manifolds, 348
Ravenel
    Complex Cobordism and Stable Homotopy Groups of Spheres, 332
Reed, Simon
    Methods of Modern Mathematical Physics, 514
Reid
    Julia. A Life in Mathematics, 45
    Courant in Göttingen and New York, 45
    Hilbert, 44
    Neyman–From Life, 45
    The Search for E.T. Bell, 45
Reiner
    Maximal Orders, 217
Remmert
    Classical Topics in Complex Function Theory, 117
    Theory of Complex Functions, 116
Revuz, Yor
    Continuous Martingales and Brownian Motion, 568
Richardson, Urbanke
    Modern Coding Theory, 601
Richtmyer, Morton
    Difference Methods for Initial-value Problems, 619
Riesz, Sz.-Nagy
    Functional Analysis, 164
Ritt
    Differential Algebra, 460
Roberts, Varberg
    Convex Functions, 311
Robinson, A.
    Non-standard Analysis, 593
Robinson, D.
    A Course in the Theory of Groups, 210
Rockafellar
    Convex Analysis, 309
Rockafellar, Wets
    Variational analysis, 101
Rogers
    Hausdorff Measures, 99
    Packing and Covering, 373
    Theory of Recursive Functions and Effective Computability, 594
Rolfsen
    Knots and Links, 328
Ronan
    Lectures on Buildings, 454
    Symmetry and the Monster. One of the Greatest Quests of Mathematics, 32
Rosen, J.
    Symmetry Discovered: Concepts and Applications in Nature and Science, 29
    Symmetry in Science. An Introduction to the General Theory, 30
    Symmetry Rules. How Science and Nature Are Founded on Symmetry, 30
Rosen, M.
    Number Theory in Function Fields, 359
Rosenberg
    Algebraic K-theory and Its Applications, 334
Ross
    Introduction to Probability Models, 559
Rotman
    An Introduction to Homological Algebra, 228
Rourke, Sanderson
    Introduction to Piecewise-linear Topology, 351
Rudin
    Function Theory in the Unit Ball of $\mathbb{C}^n$, 90
    Functional Analysis, 166
    Real and Complex Analysis, 90
    The Way I Remember It, 26
Ruelle
    Statistical Mechanics: Rigorous Results, 509

Thermodynamic Formalism. The Mathematical Structures of Classical Equilibrium Statistical Mechanics, 510

Russell
  Introduction to Mathematical Philosophy, 586

S

Saad
  Iterative Methods for Sparse Linear Systems, 614
Sachs, Wu
  General Relativity for Mathematicians, 505
Sagan
  The Symmetric Group. Representations, Combinatorial Algorithms, and Symmetric Functions, 487
Sakai
  C^*-algebras and W^*-algebras, 181
Salsburg
  The Lady Tasting Tea. How Statistics Revolutionized Science in the Twentieth Century, 45
Samko, Kilbas, Marichev
  Fractional Integrals and Derivatives. Theory and Applications, 95
Samorodnitsky, Taqqu
  Stable non-Gaussian Random Processes, 570
Santaló
  Integral Geometry and Geometric Probability, 251
Sarnak
  Some Applications of Modular Forms, 467
Sato
  Lévy Processes and Infinitely Divisible Distributions, 569
Schaefer
  Banach Lattices and Positive Operators, 202
  Topological Vector Spaces, 168
Scharlau
  Quadratic and Hermitian Forms, 392
Scharlau, Opolka
  From Fermat to Minkowski. Lectures on the Theory of Numbers and its Historical Development, 59
Schechter

My Brain is Open. The Mathematical Journeys of Paul Erdős, 48
Schneider
  Convex Bodies: the Brunn-Minkowski Theory, 312
Schoen, Yau
  Lectures on Differential Geometry, 259
  Lectures on Harmonic Maps, 259
Schrijver
  Theory of Linear and Integer Programming, 606
Schroeder
  Number Theory in Science and Communication, 360
Schumaker
  Spline Functions: Basic Theory, 622
Schwartz
  Théorie des Distributions, 171
Seifert, Threlfall
  Seifert and Threlfall: A Textbook of Topology, 318
Serre
  A Course in Arithmetic, 385
  Complex Semisimple Lie Algebras, 441
  Galois Cohomology, 372
  Linear Representations of Finite Groups, 484
  Local Algebra, 220
  Local Fields, 371
  Trees, 466
Shafarevich
  Basic Algebraic Geometry, 281
  Basic Notions of Algebra, 204
Shannon, Weaver
  The Mathematical Theory of Communication, 599
Sharpe
  Differential Geometry, 250
Shelah
  Proper and Improper Forcing, 591
Shilov
  Linear Algebra, 191
Shimura
  Introduction to the Arithmetic Theory of Automorphic Functions, 384
  The Map of My Life, 24
Shiryaev
  Probability, 558

Shoenfield
    Mathematical Logic, 584
Shohat, Tamarkin
    The Problem of Moments, 96
Shubin
    Pseudodifferential Operators and Spectral
        Theory, 424
Siegel
    Topics in Complex Function Theory, 276
Siegel, Moser
    Lectures on Celestial Mechanics, 518
Silverman
    Advanced Topics in the Arithmetic of
        Elliptic Curves, 386
    The Arithmetic of Elliptic Curves, 385
Smith
    Monotone Dynamical Systems, 525
Smith, Kahanpää, Kekïáinen, Traves
    An Invitation to Algebraic Geometry, 288
Smoller
    Shock Waves and Reaction-diffusion
        Equations, 433
Soare
    Recursively Enumerable Sets and Degrees.
        A Study of Computable Functions and
        Computably Generated Sets, 593
Spanier
    Algebraic Topology, 326
Spicci
    Beyond the Limit: The Dream of Sofya
        Kovalevskaya, 47
Spitzer
    Principles of Random Walks, 562
Spivak
    A Comprehensive Introduction to Differential
        Geometry, 243
    Calculus, 82
    Calculus on Manifolds. A Modern Approach
        to Classical Theorems of Advanced
        Calculus, 82
Springer, G.
    Introduction to Riemann Surfaces, 130
Springer, T.
    Linear Algebraic Groups, 447
Stanley
    Enumerative Combinatorics, 538
Steenrod
    The Topology of Fibre Bundles, 325
Stein
    Harmonic Analysis: Real-variable Methods,
        Orthogonality, and Oscillatory
        Integrals, 159
    Singular Integrals and Differentiability
        Properties of Functions, 157
Stein, Weiss
    Introduction to Fourier Analysis on Euclidean
        Spaces, 158
Steinberg
    Lectures on Chevalley Groups, 448
Sternberg
    Group Theory and Physics, 482
Stewart
    Introduction to Matrix Computations, 613
    Why Beauty is Truth, 28
Stichtenoth
    Algebraic Function Fields and Codes, 601
Stillwell
    Mathematics and its History, 63
Strang, Fix
    An Analysis of the Finite Element Method,
        618
Strebel
    Quadratic Differentials, 139
Stroock
    Probability Theory, an Analytic View, 559
Stroock, Varadhan
    Multidimensional Diffusion Processes, 570
Struik
    A Concise History of Mathematics, 60
Struwe
    Variational Methods. Applications to Non-
        linear Partial Differential Equations and
        Hamiltonian Systems, 431
Stubhaug
    Niels Henrik Abel and His Times, 46
Sturmfels
    Gröbner Bases and Convex Polytopes, 541
Sweedler
    Hopf Algebras, 481
Switzer
    Algebraic Topology—Homotopy and
        Homology, 327
Sz.-Nagy, Foias
    Harmonic Analysis of Operators on Hilbert Space,

146
Szegö
    Orthogonal Polynomials, 106

T

Takesaki
    Theory of operator algebras, 180
Tauvel, Yu
    Lie Algebras and Algebraic Groups, 449
Taylor
    Partial Differential Equations, 423
    Pseudodifferential Operators, 424
    Pseudodifferential Operators and Nonlinear PDE, 423
Temam
    Infinite-dimensional Dynamical Systems in Mechanics and Physics, 526
    Navier-Stokes Equations. Theory and Numerical Analysis, 615
Tenenbaum
    Introduction to Analytic and Probabilistic Number Theory, 377
Thirring
    A Course in Mathematical Physics. Vol. I. Classical Dynamical Systems, 513
Thom
    Structural Stability and Morphogenesis. An Outline of a General Theory of Models, 529
Thurston
    The Geometry and Topology of Three Manifolds, 345
    Three-dimensional Geometry and Topology, Vol. 1, 346
Titchmarsh
    The Theory of Functions, 117
    The Theory of the Riemann Zeta-Function, 375
Tits
    Buildings of Spherical Type and Finite BN-pairs, 453
tom Dieck
    Transformation Groups, 476
Totik
    Logarithmic Potentials with External Fields, 151
Triebel
    Interpolation Theory, Function spaces, Differential Operators, 155
    Theory of Function Spaces, 154
Tsuji
    Potential Theory in Modern Function Theory, 149
Turaev
    Quantum Invariants of Knots and 3-manifolds, 331

U

Ulam
    A Collection of Mathematical Problems, 108
    Adventures of a Mathematician, 23

V

van der Put, Singer
    Galois Theory of Linear Differential Equations, 461
van der Vaart, Wellner
    Weak Convergence and Empirical Processes. With Applications to Statistics, 571
van der Waerden
    Algebra, 194
van Heijenoort
    From Frege to Gödel. A Source Book in Mathematical Logic, 1879–1931, 585
van Lint
    Introduction to Coding Theory, 601
Varadarajan
    Lie Groups, Lie Algebras, and Their Representations, 444
Varopoulos, Saloff-Coste, Coulhon
    Analysis and Geometry on Groups, 147
Vaughan
    The Hardy-Littlewood Method, 379
Vignéras
    Arithmétique des algèbres de quaternions, 384
Villani
    Optimal Transport, Old and New, 430
    Topics in Optimal Transportation, 430
Voiculescu, Dykema, Nica
    Free Random Variables, 578
von Neumann
    Mathematical Foundations of Quantum Mechanics, 498

The Computer and the Brain, 12
von Neumann, Morgenstern
    Theory of Games and Economic Behavior, 605

W

Waldschmidt
    Diophantine Approximation on Linear Algebraic Groups. Transcendence Properties of the Exponential Function in Several Variables, 383

Walker
    Algebraic Curves, 288

Wall
    Surgery on Compact Manifolds, 351

Wallach
    Real Reductive Groups, 491

Walters
    An Introduction to Ergodic Theory, 533

Warner
    Foundations of Differentiable Manifolds and Lie Groups, 235

Washington
    Introduction to Cyclotomic Fields, 369

Watson
    A Treatise on the Theory of Bessel Functions, 86

Wehrfritz
    Infinite Linear Groups. An Account of the Group-Theoretic Properties of Infinite Groups of Matrices, 472

Weibel
    An Introduction to Homological Algebra, 228

Weil
    Basic Number Theory, 368
    Elliptic Functions According to Eisenstein and Kronecker, 57
    Number theory. An Approach through History. From Hammurapi to Legendre, 55
    The Apprenticeship of a Mathematician, 22

Weinberger
    The Topological Classification of Stratified Spaces, 344

Wells
    Differential Analysis on Complex Manifolds, 275

Welsh
    Matroid Theory, 543

West
    Introduction to Graph Theory, 547

Weyl
    Space, Time, Matter, 505
    Symmetry, 27
    The Classical Groups. Their Invariants and Representations, 487
    The Concept of a Riemann Surface, 128
    The Theory of Groups and Quantum Mechanics, 497

Wheeden, Zygmund
    Measure and Integral, 98

Whitehead
    Elements of Homotopy Theory, 337

Whitham
    Linear and Nonlinear Waves, 436

Whittaker, Watson
    A Course of Modern Analysis, 85

Whyburn
    Analytic Topology, 478

Widder
    The Laplace Transform, 145

Wiener
    I am a Mathematician. The later Life of a Prodigy, 23
    Cybernetics, or Control and Communication in the Animal and the Machine, 597
    Ex-prodigy. My Childhood and Youth, 23
    Extrapolation, Interpolation, and Smoothing of Stationary Time Series, 598

Wilkinson
    The Algebraic Eigenvalue Problem, 612

Willem
    Minimax Theorems, 420

Woess
    Random Walks on Infinite Graphs and Groups, 563

Wolf
    Spaces of Constant Curvature, 245

Y

Yafaev
    Mathematical Scattering Theory, 516

Yaglom
    Felix Klein and Sophus Lie, 52

Yosida
    Functional analysis, 174

INDEX

Z

Zariski
    Algebraic Surfaces, 301
Zariski, Samuel
    Commutative Algebra, 219
Zassenhaus
    The Theory of Groups, 207
Zee
    Fearful Symmetry. The Search for Beauty in
        Modern Physics, 34
Zeidler
    Nonlinear Functional Analysis and Its
        Applications, 186
Zhu
    Operator Theory in Function Spaces, 155
Ziegler
    Lectures on Polytopes, 539
Ziemer
    Weakly Differentiable Functions. Sobolev Spaces
        and Functions of Bounded Variation, 428
Zimmer
    Ergodic Theory and Semisimple Groups, 465
    Essential Results of Functional Analysis, 166
Zygmund
    Trigonometric Series, 147